D1429335

KIRK-OTHMER

ENCYCLOPEDIA OF CHEMICAL TECHNOLOGY

FOURTH EDITION

SUPPLEMENT VOLUME

AEROGELS
TO
XYLYLENE POLYMERS

EXECUTIVE EDITOR
Jacqueline I. Kroschwitz

EDITOR
Mary Howe-Grant

KIRK-OTHMER

ENCYCLOPEDIA OF CHEMICAL TECHNOLOGY

FOURTH EDITION

SUPPLEMENT VOLUME

AEROGELS
TO
XYLYLENE POLYMERS

A Wiley-Interscience Publication
JOHN WILEY & SONS
New York • Chichester • Weinheim • Brisbane • Singapore • Toronto

This book is printed on acid-free paper. ⊗

Copyright © 1998 by John Wiley & Sons, Inc. All rights reserved.

Published simultaneously in Canada.

Library of Congress Cataloging-in-Publication Data

Encyclopedia of chemical technology/executive editor, Jacqueline
  I. Kroschwitz; editor, Mary Howe-Grant.—4th ed.
    p. cm.
  At head of title: Kirk-Othmer.
  "A Wiley-Interscience publication."
  Contents: Supplement, Aerogels to xylylene polymers
  ISBN 0471-52696-7 (Supplement)
  1. Chemistry, Technical—Encyclopedias.   I. Kirk, Raymond E.
  (Raymond Eller), 1890–1957.   II. Othmer, Donald F. (Donald
  Frederick), 1904–1995.   III. Kroschwitz, Jacqueline I., 1942– .
  IV. Howe-Grant, Mary, 1943– .   V. Title: Kirk-Othmer encyclopedia
  of chemical technology.
  TP9.E685 1992   91-16789
  660′.03—dc20

Printed in the United States of America

10 9 8 7 6 5 4 3 2 1

CONTENTS

EDITORIAL STAFF
FOR SUPPLEMENT VOLUME

Executive Editor: **Jacqueline I. Kroschwitz**
Editor: **Mary Howe-Grant**
Associate Managing Editor: **Lindy Humphreys**
Copy Editors: **Lawrence Altieri**
Michalina Bickford
Assistant Managing Editor: **Brendan A. Vilardo**

CONTRIBUTORS
TO SUPPLEMENT VOLUME

R. D. Allen, *IBM Research Division, San Jose, California*, Lithographic resists

Hazel Aranha-Creado, *Pall Corporation, Port Washington, New York*, Microbial and viral filtration

Mitchell A. Avery, *University of Mississippi, University*, Hormones, adrenal-cortical

William F. Beach, *Alpha Metals, Bridgewater, New Jersey*, Xylylene polymers

John K. Borchardt, *Shell Chemical Company, Houston, Texas*, Recycling, plastics

Sebastian Borrello, *Texas Instruments, Dallas, Texas,* Thermography

Randy J. Bowers, *Edison Welding Institute, Columbus, Ohio,* Welding

Christian Butzke, *University of California, Davis,* Wine

William J. Cannella, *Chevron Research & Technology Company, Richmond, California,* Xylenes and ethylbenzene

G. R. "Buck" Coleman, *Consultant, Houston, Texas,* Urea

Stephen C. DeVito, *U.S. Environmental Protection Agency, Washington, D.C.,* Nitriles

Edward F. Ezell, *The BOC Group, Inc., Murray Hill, New Jersey,* High purity gases

Otto Frank, *Air Products & Chemicals, Inc., Allentown, Pennsylvania,* Solvent recovery, condensation

J. Edward Glass, *North Dakota State University, Fargo,* Water-soluble polymers

Karl F. Graff, *Edison Welding Institute, Columbus, Ohio,* Welding

James A. Harvey, *Hewlett-Packard Company; Oregon Graduate Institute of Science & Technology, Corvallis,* Smart materials

William D. Hinsberg, *IBM Research Division, San Jose, California,* Lithographic resists

Shuen-Cheng Hwang, *The BOC Group, Inc., Murray Hill, New Jersey,* High purity gases

Edmond I. Ko, *Carnegie Mellon University, Pittsburgh, Pennsylvania,* Aerogels

Roger H. Kottke, *Great Lakes Chemical Corporation, West Lafayette, Indiana,* Furan derivatives

Michael R. Ladisch, *Purdue University, West Lafayette, Indiana,* Bioseparations

Tom Leahy, *Mettler-Toledo, Inc., Inman, South Carolina,* Weighing and proportioning

William F. Lehmann, *Consultant, Federal Way, Washington,* Wood-based composites and laminates

Gabriel P. López, *University of New Mexico, Albuquerque,* Nanotechnology

José A. López, *Baylor College of Medicine, Houston, Texas,* Nanotechnology

Ivo Mavrovic, *Consultant, New York, New York,* Urea

Ronald J. McKinney, *E. I. du Pont de Nemours & Co., Inc., Wilmington, Delaware,* Nitriles

K. Oshima, *Pall Corporation, Port Washington, New York,* Microbial and viral filtration

Jeanne E. Pemberton, *University of Arizona, Tucson,* Surface and interface analysis

Roger C. Prince, *Exxon Research & Engineering Company, Annandale, New Jersey,* Bioremediation

Salvatore Profeta, Jr., *Monsanto, St. Louis, Missouri,* Molecular modeling

A. Ray Shirley, Jr., *Applied Chemical Technology, Muscle Shoals, Alabama,* Urea

Scott Sibley, *Goucher College, Baltimore, Maryland,* Semiconductors, organic

Vernon L. Singleton, *University of California, Davis,* Wine

Duane H. Smith, *Technical Solutions, Morgantown, West Virginia,* Microemulsions

Robert R. Stickney, *Texas A&M University, Bryan, Texas,* Aquaculture

Apryll M. Stalcup, *University of Cincinnati, Ohio,* Chiral separations

Kenneth S. Suslick, *University of Illinois, Urbana*, Sonochemistry

Rajesh Vaidya, *University of New Mexico, Albuquerque*, Nanotechnology

G. M. Wallraff, *IBM Research Division, San Jose, California*, Lithographic resists

Edwin Weber, *Technische Universität Bergakademie Freiberg, Institut Fürorganische Chemie, Germany*, Molecular recognition

Walter H. Whitlock, *The BOC Group, Inc., Murray Hill, New Jersey*, High purity gases

Paul Wight, *Zeneca Specialties, Manchester, England*, Xanthene dyes

John R. Woolfrey, *University of Mississippi, University*, Hormones, adrenal-cortical

NOTE ON CHEMICAL ABSTRACTS SERVICE REGISTRY NUMBERS AND NOMENCLATURE

Chemical Abstracts Service (CAS) Registry Numbers are unique numerical identifiers assigned to substances recorded in the CAS Registry System. They appear in brackets in the *Chemical Abstracts* (CA) substance and formula indexes following the names of compounds. A single compound may have synonyms in the chemical literature. A simple compound like phenethylamine can be named β-phenylethylamine or, as in *Chemical Abstracts*, benzeneethanamine. The usefulness of the *Encyclopedia* depends on accessibility through the most common correct name of a substance. Because of this diversity in nomenclature careful attention has been given to the problem in order to assist the reader as much as possible, especially in locating the systematic CA index name by means of the Registry Number. For this purpose, the reader may refer to the CAS Registry Handbook—Number Section which lists in numerical order the Registry Number with the *Chemical Abstracts* index name and the molecular formula; eg, **458-88-8**, Piperidine, 2-propyl-, (*S*)-, $C_8H_{17}N$; in the *Encyclopedia* this compound would be found under its common name, coniine [*458-88-8*]. Alternatively, this information can be retrieved electronically from CAS Online. In many cases molecular formulas have also been provided in the *Encyclopedia* text to facilitate electronic searching. The Registry Number is a valuable link for the reader in retrieving additional published information on substances and also as a point of access for on-line data bases.

In all cases, the CAS Registry Numbers have been given for title compounds in articles and for all compounds in the index. All specific substances indexed in *Chemical Abstracts* since 1965 are included in the CAS Registry System as are a large number of substances derived from a variety of reference works. The CAS Registry System identifies a substance on the basis of an unambiguous computer-language description of its molecular structure including stereochemical detail. The Registry Number is a machine-checkable number (like a Social Security number) assigned in sequential order to each substance as it enters the registry system. The value of the number lies in the fact that it is a concise and unique means of substance identification, which is independent of, and therefore bridges, many systems of chemical nomenclature. For polymers, one Registry Number may be used for the entire family; eg, polyoxyethylene (20) sorbitan monolaurate has the same number as all of its polyoxyethylene homologues.

Cross-references are inserted in the index for many common names and for some systematic names. Trademark names appear in the index. Names that are incorrect, misleading, or ambiguous are avoided. Formulas are given very frequently in the text to help in identifying compounds. The spelling and form used, even for industrial names, follow American chemical usage, but not always the usage of *Chemical Abstracts* (eg, *coniine* is used instead of *(S)-2-propylpiperidine*, *aniline* instead of *benzenamine*, and *acrylic acid* instead of *2-propenoic acid*).

There are variations in representation of rings in different disciplines. The dye industry does not designate aromaticity or double bonds in rings. All double bonds and aromaticity are shown in the *Encyclopedia* as a matter of course. For example, tetralin has an aromatic ring and a saturated ring and its structure

appears in the *Encyclopedia* with its common name, Registry Number enclosed in brackets, and parenthetical CA index name, ie, tetralin [*119-64-2*] (1,2,3,4-tetrahydronaphthalene). With names and structural formulas, and especially with CAS Registry Numbers, the aim is to help the reader have a concise means of substance identification.

CONVERSION FACTORS, ABBREVIATIONS, AND UNIT SYMBOLS

SI Units (Adopted 1960)

The International System of Units (abbreviated SI), is being implemented throughout the world. This measurement system is a modernized version of the MKSA (meter, kilogram, second, ampere) system, and its details are published and controlled by an international treaty organization (The International Bureau of Weights and Measures) (1).

SI units are divided into three classes:

BASE UNITS

length	meter[†] (m)
mass	kilogram (kg)
time	second (s)
electric current	ampere (A)
thermodynamic temperature[‡]	kelvin (K)
amount of substance	mole (mol)
luminous intensity	candela (cd)

SUPPLEMENTARY UNITS

plane angle	radian (rad)
solid angle	steradian (sr)

[†]The spellings "metre" and "litre" are preferred by ASTM; however, "-er" is used in the *Encyclopedia*.

[‡]Wide use is made of Celsius temperature (t) defined by

$$t = T - T_0$$

where T is the thermodynamic temperature, expressed in kelvin, and $T_0 = 273.15$ K by definition. A temperature interval may be expressed in degrees Celsius as well as in kelvin.

DERIVED UNITS AND OTHER ACCEPTABLE UNITS

These units are formed by combining base units, supplementary units, and other derived units (2–4). Those derived units having special names and symbols are marked with an asterisk in the list below.

Quantity	Unit	Symbol	Acceptable equivalent
*absorbed dose	gray	Gy	J/kg
acceleration	meter per second squared	m/s^2	
*activity (of a radionuclide)	becquerel	Bq	1/s
area	square kilometer	km^2	
	square hectometer	hm^2	ha (hectare)
	square meter	m^2	
concentration (of amount of substance)	mole per cubic meter	mol/m^3	
current density	ampere per square meter	$A//m^2$	
density, mass density	kilogram per cubic meter	kg/m^3	g/L; mg/cm^3
dipole moment (quantity)	coulomb meter	C·m	
*dose equivalent	sievert	Sv	J/kg
*electric capacitance	farad	F	C/V
*electric charge, quantity of electricity	coulomb	C	A·s
electric charge density	coulomb per cubic meter	C/m^3	
*electric conductance	siemens	S	A/V
electric field strength	volt per meter	V/m	
electric flux density	coulomb per square meter	C/m^2	
*electric potential, potential difference, electromotive force	volt	V	W/A
*electric resistance	ohm	Ω	V/A
*energy, work, quantity of heat	megajoule	MJ	
	kilojoule	kJ	
	joule	J	N·m
	electronvolt[†]	eV[†]	
	kilowatt-hour[†]	kW·h[†]	
energy density	joule per cubic meter	J/m^3	
*force	kilonewton	kN	
	newton	N	$kg·m/s^2$

[†]This non-SI unit is recognized by the CIPM as having to be retained because of practical importance or use in specialized fields (1).

Quantity	Unit	Symbol	Acceptable equivalent
*frequency	megahertz	MHz	
	hertz	Hz	1/s
heat capacity, entropy	joule per kelvin	J/K	
heat capacity (specific), specific entropy	joule per kilogram kelvin	J/(kg·K)	
heat-transfer coefficient	watt per square meter kelvin	W/(m^2·K)	
*illuminance	lux	lx	lm/m^2
*inductance	henry	H	Wb/A
linear density	kilogram per meter	kg/m	
luminance	candela per square meter	cd/m^2	
*luminous flux	lumen	lm	cd·sr
magnetic field strength	ampere per meter	A/m	
*magnetic flux	weber	Wb	V·s
*magnetic flux density	tesla	T	Wb/m^2
molar energy	joule per mole	J/mol	
molar entropy, molar heat capacity	joule per mole kelvin	J/(mol·K)	
moment of force, torque	newton meter	N·m	
momentum	kilogram meter per second	kg·m/s	
permeability	henry per meter	H/m	
permittivity	farad per meter	F/m	
*power, heat flow rate, radiant flux	kilowatt	kW	
	watt	W	J/s
power density, heat flux density, irradiance	watt per square meter	W/m^2	
*pressure, stress	megapascal	MPa	
	kilopascal	kPa	
	pascal	Pa	N/m^2
sound level	decibel	dB	
specific energy	joule per kilogram	J/kg	
specific volume	cubic meter per kilogram	m^3/kg	
surface tension	newton per meter	N/m	
thermal conductivity	watt per meter kelvin	W/(m·K)	
velocity	meter per second	m/s	
	kilometer per hour	km/h	
viscosity, dynamic	pascal second	Pa·s	
	millipascal second	mPa·s	
viscosity, kinematic	square meter per second	m^2/s	
	square millimeter per second	mm^2/s	

Quantity	Unit	Symbol	Acceptable equivalent
volume	cubic meter	m^3	
	cubic diameter	dm^3	L (liter) (5)
	cubic centimeter	cm^3	mL
wave number	1 per meter	m^{-1}	
	1 per centimeter	cm^{-1}	

In addition, there are 16 prefixes used to indicate order of magnitude, as follows:

Multiplication factor	Prefix	Symbol	Note
10^{18}	exa	E	
10^{15}	peta	P	
10^{12}	tera	T	
10^{9}	giga	G	
10^{6}	mega	M	
10^{3}	kilo	k	
10^{2}	hecto	h[a]	[a]Although hecto, deka, deci, and centi
10	deka	da[a]	are SI prefixes, their use should be
10^{-1}	deci	d[a]	avoided except for SI unit-multiples
10^{-2}	centi	c[a]	for area and volume and nontech-
10^{-3}	milli	m	nical use of centimeter, as for body
10^{-6}	micro	μ	and clothing measurement.
10^{-9}	nano	n	
10^{-12}	pico	p	
10^{-15}	femto	f	
10^{-18}	atto	a	

For a complete description of SI and its use the reader is referred to ASTM E380 (4) and the article UNITS AND CONVERSION FACTORS which appears in Vol. 24.

A representative list of conversion factors from non-SI to SI units is presented herewith. Factors are given to four significant figures. Exact relationships are followed by a dagger. A more complete list is given in the latest editions of ASTM E380 (4) and ANSI Z210.1 (6).

Conversion Factors to SI Units

To convert from	To	Multiply by
acre	square meter (m^2)	4.047×10^3
angstrom	meter (m)	1.0×10^{-10}†
are	square meter (m^2)	1.0×10^{2}†

†Exact.

To convert from	To	Multiply by
astronomical unit	meter (m)	1.496×10^{11}
atmosphere, standard	pascal (Pa)	1.013×10^{5}
bar	pascal (Pa)	$1.0 \times 10^{5\dagger}$
barn	square meter (m²)	$1.0 \times 10^{-28\dagger}$
barrel (42 U.S. liquid gallons)	cubic meter (m³)	0.1590
Bohr magneton (μ_B)	J/T	9.274×10^{-24}
Btu (International Table)	joule (J)	1.055×10^{3}
Btu (mean)	joule (J)	1.056×10^{3}
Btu (thermochemical)	joule (J)	1.054×10^{3}
bushel	cubic meter (m³)	3.524×10^{-2}
calorie (International Table)	joule (J)	4.187
calorie (mean)	joule (J)	4.190
calorie (thermochemical)	joule (J)	$4.184^{\dagger}$
centipoise	pascal second (Pa·s)	$1.0 \times 10^{-3\dagger}$
centistokes	square millimeter per second (mm²/s)	$1.0^{\dagger}$
cfm (cubic foot per minute)	cubic meter per second (m³/s)	4.72×10^{-4}
cubic inch	cubic meter (m³)	1.639×10^{-5}
cubic foot	cubic meter (m³)	2.832×10^{-2}
cubic yard	cubic meter (m³)	0.7646
curie	becquerel (Bq)	$3.70 \times 10^{10\dagger}$
debye	coulomb meter (C·m)	3.336×10^{-30}
degree (angle)	radian (rad)	1.745×10^{-2}
denier (international)	kilogram per meter (kg/m)	1.111×10^{-7}
	tex‡	0.1111
dram (apothecaries')	kilogram (kg)	3.888×10^{-3}
dram (avoirdupois)	kilogram (kg)	1.772×10^{-3}
dram (U.S. fluid)	cubic meter (m³)	3.697×10^{-6}
dyne	newton (N)	$1.0 \times 10^{-5\dagger}$
dyne/cm	newton per meter (N/m)	$1.0 \times 10^{-3\dagger}$
electronvolt	joule (J)	1.602×10^{-19}
erg	joule (J)	$1.0 \times 10^{-7\dagger}$
fathom	meter (m)	1.829
fluid ounce (U.S.)	cubic meter (m³)	2.957×10^{-5}
foot	meter (m)	$0.3048^{\dagger}$
footcandle	lux (lx)	10.76
furlong	meter (m)	2.012×10^{-2}
gal	meter per second squared (m/s²)	$1.0 \times 10^{-2\dagger}$
gallon (U.S. dry)	cubic meter (m³)	4.405×10^{-3}
gallon (U.S. liquid)	cubic meter (m³)	3.785×10^{-3}
gallon per minute (gpm)	cubic meter per second (m³/s)	6.309×10^{-5}
	cubic meter per hour (m³/h)	0.2271

†Exact.
‡See footnote on p. xiii.

To convert from	To	Multiply by
gauss	tesla (T)	1.0×10^{-4}
gilbert	ampere (A)	0.7958
gill (U.S.)	cubic meter (m^3)	1.183×10^{-4}
grade	radian	1.571×10^{-2}
grain	kilogram (kg)	6.480×10^{-5}
gram force per denier	newton per tex (N/tex)	8.826×10^{-2}
hectare	square meter (m^2)	$1.0 \times 10^{4\dagger}$
horsepower (550 ft·lbf/s)	watt (W)	7.457×10^2
horsepower (boiler)	watt (W)	9.810×10^3
horsepower (electric)	watt (W)	$7.46 \times 10^{2\dagger}$
hundredweight (long)	kilogram (kg)	50.80
hundredweight (short)	kilogram (kg)	45.36
inch	meter (m)	$2.54 \times 10^{-2\dagger}$
inch of mercury (32°F)	pascal (Pa)	3.386×10^3
inch of water (39.2°F)	pascal (Pa)	2.491×10^2
kilogram-force	newton (N)	9.807
kilowatt hour	megajoule (MJ)	$3.6^\dagger$
kip	newton (N)	4.448×10^3
knot (international)	meter per second (m/S)	0.5144
lambert	candela per square meter (cd/m^3)	3.183×10^3
league (British nautical)	meter (m)	5.559×10^3
league (statute)	meter (m)	4.828×10^3
light year	meter (m)	9.461×10^{15}
liter (for fluids only)	cubic meter (m^3)	$1.0 \times 10^{-3\dagger}$
maxwell	weber (Wb)	$1.0 \times 10^{-8\dagger}$
micron	meter (m)	$1.0 \times 10^{-6\dagger}$
mil	meter (m)	$2.54 \times 10^{-5\dagger}$
mile (statute)	meter (m)	1.609×10^3
mile (U.S. nautical)	meter (m)	$1.852 \times 10^{3\dagger}$
mile per hour	meter per second (m/s)	0.4470
millibar	pascal (Pa)	1.0×10^2
millimeter of mercury (0°C)	pascal (Pa)	$1.333 \times 10^{2\dagger}$
minute (angular)	radian	2.909×10^{-4}
myriagram	kilogram (kg)	10
myriameter	kilometer (km)	10
oersted	ampere per meter (A/m)	79.58
ounce (avoirdupois)	kilogram (kg)	2.835×10^{-2}
ounce (troy)	kilogram (kg)	3.110×10^{-2}
ounce (U.S. fluid)	cubic meter (m^3)	2.957×10^{-5}
ounce-force	newton (N)	0.2780
peck (U.S.)	cubic meter (m^3)	8.810×10^{-3}
pennyweight	kilogram (kg)	1.555×10^{-3}
pint (U.S. dry)	cubic meter (m^3)	5.506×10^{-4}
pint (U.S. liquid)	cubic meter (m^3)	4.732×10^{-4}

†Exact.

To convert from	To	Multiply by
poise (absolute viscosity)	pascal second (Pa·s)	$0.10^\dagger$
pound (avoirdupois)	kilogram (kg)	0.4536
pound (troy)	kilogram (kg)	0.3732
poundal	newton (N)	0.1383
pound-force	newton (N)	4.448
pound force per square inch (psi)	pascal (Pa)	6.895×10^3
quart (U.S. dry)	cubic meter (m^3)	1.101×10^{-3}
quart (U.S. liquid)	cubic meter (m^3)	9.464×10^{-4}
quintal	kilogram (kg)	$1.0 \times 10^{2\dagger}$
rad	gray (Gy)	$1.0 \times 10^{-2\dagger}$
rod	meter (m)	5.029
roentgen	coulomb per kilogram (C/kg)	2.58×10^{-4}
second (angle)	radian (rad)	$4.848 \times 10^{-6\dagger}$
section	square meter (m^2)	2.590×10^6
slug	kilogram (kg)	14.59
spherical candle power	lumen (lm)	12.57
square inch	square meter (m^2)	6.452×10^{-4}
square foot	square meter (m^2)	9.290×10^{-2}
square mile	square meter (m^2)	2.590×10^6
square yard	square meter (m^2)	0.8361
stere	cubic meter (m^3)	$1.0^\dagger$
stokes (kinematic viscosity)	square meter per second (m^2/s)	$1.0 \times 10^{-4\dagger}$
tex	kilogram per meter (kg/m)	$1.0 \times 10^{-6\dagger}$
ton (long, 2240 pounds)	kilogram (kg)	1.016×10^3
ton (metric) (tonne)	kilogram (kg)	$1.0 \times 10^{3\dagger}$
ton (short, 2000 pounds)	kilogram (kg)	9.072×10^2
torr	pascal (Pa)	1.333×10^2
unit pole	weber (Wb)	1.257×10^{-7}
yard	meter (m)	$0.9144^\dagger$

†Exact.

Abbreviations and Unit Symbols

Following is a list of common abbreviations and unit symbols used in the *Encyclopedia*. In general they agree with those listed in *American National Standard Abbreviations for Use on Drawings and in Text* (*ANSI Y1.1*) (6) and *American National Standard Letter Symbols for Units in Science and Technology* (*ANSI Y10*) (6). Also included is a list of acronyms for a number of private and government organizations as well as common industrial solvents, polymers, and other chemicals.

Rules for Writing Unit Symbols (4):

1. Unit symbols are printed in upright letters (roman) regardless of the type style used in the surrounding text.
2. Unit symbols are unaltered in the plural.
3. Unit symbols are not followed by a period except when used at the end of a sentence.
4. Letter unit symbols are generally printed lower-case (for example, cd for candela) unless the unit name has been derived from a proper name, in which case the first letter of the symbol is capitalized (W, Pa). Prefixes and unit symbols retain their prescribed form regardless of the surrounding typography.
5. In the complete expression for a quantity, a space should be left between the numerical value and the unit symbol. For example, write 2.37 lm, *not* 2.37lm, and 35 mm, *not* 35mm. When the quantity is used in an adjectival sense, a hyphen is often used, for example, 35-mm film. *Exception:* No space is left between the numerical value and the symbols of degree, minute, and second of plane angle, degree Celsius, and the percent sign.
6. No space is used between the prefix and unit symbol (for example, kg).
7. Symbols, not abbreviations, should be used for units. For example, use "A," not "amp," for ampere.
8. When multiplying unit symbols, use a raised dot:

$$N \cdot m \quad \text{for} \quad \text{newton meter}$$

In the case of W·h, the dot may be omitted, thus:

$$Wh$$

An exception to this practice is made for computer printouts, automatic typewriter work, etc, where the raised dot is not possible, and a dot on the line may be used.

9. When dividing unit symbols, use one of the following forms:

$$m/s \quad or \quad m \cdot s^{-1} \quad or \quad \frac{m}{s}$$

In no case should more than one slash be used in the same expression unless parentheses are inserted to avoid ambiguity. For example, write:

$$J/(mol \cdot K) \quad or \quad J \cdot mol^{-1} \cdot K^{-1} \quad or \quad (J/mol)/K$$

but *not*

$$J/mol/K$$

10. Do not mix symbols and unit names in the same expression. Write:

$$\text{joules per kilogram} \quad or \quad \text{J/kg} \quad or \quad \text{J·kg}^{-1}$$

but *not*

$$\text{joules/kilogram} \quad nor \quad \text{joules/kg} \quad nor \quad \text{joules·kg}^{-1}$$

ABBREVIATIONS AND UNITS

A	ampere		AOAC	Association of Official
A	anion (eg, HA)			Analytical Chemists
A	mass number		AOCS	American Oil Chemists'
a	atto (prefix for 10^{-18})			Society
AATCC	American Association of		APHA	American Public Health
	Textile Chemists and			Association
	Colorists		API	American Petroleum
ABS	acrylonitrile–butadiene–			Institute
	styrene		aq	aqueous
abs	absolute		Ar	aryl
ac	alternating current, *n.*		*ar-*	aromatic
a-c	alternating current, *adj.*		*as-*	asymmetric(al)
ac-	alicyclic		ASHRAE	American Society of
acac	acetylacetonate			Heating, Refrigerating,
ACGIH	American Conference of			and Air Conditioning
	Governmental Industrial			Engineers
	Hygienists		ASM	American Society for
ACS	American Chemical			Metals
	Society		ASME	American Society of
AGA	American Gas Association			Mechanical Engineers
Ah	ampere hour		ASTM	American Society for
AIChE	American Institute of			Testing and Materials
	Chemical Engineers		at no.	atomic number
AIME	American Institute of		at wt	atomic weight
	Mining, Metallurgical,		av(g)	average
	and Petroleum		AWS	American Welding Society
	Engineers		*b*	bonding orbital
AIP	American Institute of		bbl	barrel
	Physics		bcc	body-centered cubic
AISI	American Iron and Steel		BCT	body-centered tetragonal
	Institute		Bé	Baumé
alc	alcohol(ic)		BET	Brunauer-Emmett-Teller
Alk	alkyl			(adsorption equation)
alk	alkaline (not alkali)		bid	twice daily
amt	amount		Boc	*t*-butyloxycarbonyl
amu	atomic mass unit		BOD	biochemical (biological)
ANSI	American National			oxygen demand
	Standards Institute		bp	boiling point
AO	atomic orbital		Bq	becquerel

C	coulomb	DIN	Deutsche Industrie
°C	degree Celsius		Normen
C-	denoting attachment to	*dl*-; DL-	racemic
	carbon	DMA	dimethylacetamide
c	centi (prefix for 10^{-2})	DMF	dimethylformamide
c	critical	DMG	dimethyl glyoxime
ca	circa (approximately)	DMSO	dimethyl sulfoxide
cd	candela; current density;	DOD	Department of Defense
	circular dichroism	DOE	Department of Energy
CFR	Code of Federal	DOT	Department of
	Regulations		Transportation
cgs	centimeter-gram-second	DP	degree of polymerization
CI	Color Index	dp	dew point
cis-	isomer in which	DPH	diamond pyramid
	substituted groups are		hardness
	on same side of double	dstl(d)	distill(ed)
	bond between C atoms	dta	differential thermal
cl	carload		analysis
cm	centimeter	(*E*)-	entgegen; opposed
cmil	circular mil	ϵ	dielectric constant
cmpd	compound		(unitless number)
CNS	central nervous system	*e*	electron
CoA	coenzyme A	ECU	electrochemical unit
COD	chemical oxygen demand	ed.	edited, edition, editor
coml	commercial(ly)	ED	effective dose
cp	chemically pure	EDTA	ethylenediaminetetra-
cph	close-packed hexagonal		acetic acid
CPSC	Consumer Product Safety	emf	electromotive force
	Commission	emu	electromagnetic unit
cryst	crystalline	en	ethylene diamine
cub	cubic	eng	engineering
D	debye	EPA	Environmental Protection
D-	denoting configurational		Agency
	relationship	epr	electron paramagnetic
d	differential operator		resonance
d	day; deci (prefix for 10^{-1})	eq.	equation
d	density	esca	electron spectroscopy for
d-	*dextro*-, dextrorotatory		chemical analysis
da	deka (prefix for 10^1)	esp	especially
dB	decibel	esr	electron-spin resonance
dc	direct current, *n*.	est(d)	estimate(d)
d-c	direct current, *adj*.	estn	estimation
dec	decompose	esu	electrostatic unit
detd	determined	exp	experiment, experimental
detn	determination	ext(d)	extract(ed)
Di	didymium, a mixture of all	F	farad (capacitance)
	lanthanons	*F*	faraday (96,487 C)
dia	diameter	f	femto (prefix for 10^{-15})
dil	dilute		

FAO	Food and Agriculture Organization (United Nations)	hyd	hydrated, hydrous
		hyg	hygroscopic
		Hz	hertz
fcc	face-centered cubic	i (eg, Pr^i)	iso (eg, isopropyl)
FDA	Food and Drug Administration	i-	inactive (eg, i-methionine)
		IACS	International Annealed Copper Standard
FEA	Federal Energy Administration	ibp	initial boiling point
FHSA	Federal Hazardous Substances Act	IC	integrated circuit
		ICC	Interstate Commerce Commission
fob	free on board		
fp	freezing point	ICT	International Critical Table
FPC	Federal Power Commission		
		ID	inside diameter; infective dose
FRB	Federal Reserve Board		
frz	freezing	ip	intraperitoneal
G	giga (prefix for 10^9)	IPS	iron pipe size
G	gravitational constant = 6.67×10^{11} N·m^2/kg^2	ir	infrared
		IRLG	Interagency Regulatory Liaison Group
g	gram		
(g)	gas, only as in H_2O(g)	ISO	International Organization Standardization
g	gravitational acceleration		
gc	gas chromatography	ITS-90	International Temperature Scale (NIST)
gem-	geminal		
glc	gas−liquid chromatography	IU	International Unit
		IUPAC	International Union of Pure and Applied Chemistry
g-mol wt; gmw	gram-molecular weight		
GNP	gross national product	IV	iodine value
gpc	gel-permeation chromatography	iv	intravenous
		J	joule
GRAS	Generally Recognized as Safe	K	kelvin
		k	kilo (prefix for 10^3)
grd	ground	kg	kilogram
Gy	gray	L	denoting configurational relationship
H	henry		
h	hour; hecto (prefix for 10^2)	L	liter (for fluids only) (5)
ha	hectare	l-	$levo$-, levorotatory
HB	Brinell hardness number	(l)	liquid, only as in NH_3(l)
Hb	hemoglobin	LC_{50}	conc lethal to 50% of the animals tested
hcp	hexagonal close-packed		
hex	hexagonal	LCAO	linear combination of atomic orbitals
HK	Knoop hardness number		
hplc	high performance liquid chromatography	lc	liquid chromatography
		LCD	liquid crystal display
HRC	Rockwell hardness (C scale)	lcl	less than carload lots
		LD_{50}	dose lethal to 50% of the animals tested
HV	Vickers hardness number		

LED	light-emitting diode	N-	denoting attachment to nitrogen
liq	liquid		
lm	lumen	n (as n_D^{20})	index of refraction (for 20°C and sodium light)
ln	logarithm (natural)		
LNG	liquefied natural gas	^{n} (as Bu^n),	
log	logarithm (common)	n-	normal (straight-chain structure)
LOI	limiting oxygen index		
LPG	liquefied petroleum gas	n	neutron
ltl	less than truckload lots	n	nano (prefix for 10^9)
lx	lux	na	not available
M	mega (prefix for 10^6); metal (as in MA)	NAS	National Academy of Sciences
M	molar; actual mass	NASA	National Aeronautics and Space Administration
$\overline{M}_w$	weight-average mol wt		
$\overline{M}_n$	number-average mol wt	nat	natural
m	meter; milli (prefix for 10^{-3})	ndt	nondestructive testing
		neg	negative
m	molal	NF	*National Formulary*
m-	meta	NIH	National Institutes of Health
max	maximum		
MCA	Chemical Manufacturers' Association (was Manufacturing Chemists Association)	NIOSH	National Institute of Occupational Safety and Health
		NIST	National Institute of Standards and Technology (formerly National Bureau of Standards)
MEK	methyl ethyl ketone		
meq	milliequivalent		
mfd	manufactured		
mfg	manufacturing		
mfr	manufacturer	nmr	nuclear magnetic resonance
MIBC	methyl isobutyl carbinol		
MIBK	methyl isobutyl ketone	NND	New and Nonofficial Drugs (AMA)
MIC	minimum inhibiting concentration		
		no.	number
min	minute; minimum	NOI-(BN)	not otherwise indexed (by name)
mL	milliliter		
MLD	minimum lethal dose	NOS	not otherwise specified
MO	molecular orbital	nqr	nuclear quadruple resonance
mo	month		
mol	mole	NRC	Nuclear Regulatory Commission; National Research Council
mol wt	molecular weight		
mp	melting point		
MR	molar refraction	NRI	New Ring Index
ms	mass spectrometry	NSF	National Science Foundation
MSDS	material safety data sheet		
mxt	mixture	NTA	nitrilotriacetic acid
μ	micro (prefix for 10^{-6})	NTP	normal temperature and pressure (25°C and 101.3 kPa or 1 atm)
N	newton (force)		
N	normal (concentration); neutron number		

NTSB	National Transportation Safety Board	qv	quod vide (which see)
O-	denoting attachment to oxygen	R	univalent hydrocarbon radical
o-	ortho	(R)-	rectus (clockwise configuration)
OD	outside diameter	r	precision of data
OPEC	Organization of Petroleum Exporting Countries	rad	radian; radius
o-phen	o-phenanthridine	RCRA	Resource Conservation and Recovery Act
OSHA	Occupational Safety and Health Administration	rds	rate-determining step
owf	on weight of fiber	ref.	reference
Ω	ohm	rf	radio frequency, n.
P	peta (prefix for 10^{15})	r-f	radio frequency, adj.
p	pico (prefix for 10^{-12})	rh	relative humidity
p-	para	RI	Ring Index
p	proton	rms	root-mean square
p.	page	rpm	rotations per minute
Pa	pascal (pressure)	rps	revolutions per second
PEL	personal exposure limit based on an 8-h exposure	RT	room temperature
		RTECS	Registry of Toxic Effects of Chemical Substances
pd	potential difference	s (eg, Bu^s); sec-	secondary (eg, secondary butyl)
pH	negative logarithm of the effective hydrogen ion concentration	S	siemens
		(S)-	sinister (counterclockwise configuration)
phr	parts per hundred of resin (rubber)	S-	denoting attachment to sulfur
p-i-n	positive-intrinsic-negative	s-	symmetric(al)
pmr	proton magnetic resonance	s	second
p-n	positive-negative	(s)	solid, only as in $H_2O(s)$
po	per os (oral)	SAE	Society of Automotive Engineers
POP	polyoxypropylene		
pos	positive	SAN	styrene-acrylonitrile
pp.	pages	sat(d)	saturate(d)
ppb	parts per billion (10^9)	satn	saturation
ppm	parts per million (10^6)	SBS	styrene–butadiene–styrene
ppmv	parts per million by volume		
ppmwt	parts per million by weight	sc	subcutaneous
PPO	poly(phenyl oxide)	SCF	self-consistent field; standard cubic feet
ppt(d)	precipitate(d)		
pptn	precipitation	Sch	Schultz number
Pr (no.)	foreign prototype (number)	sem	scanning electron microscope(y)
pt	point; part		
PVC	poly(vinyl chloride)	SFs	Saybolt Furol seconds
pwd	powder	sl sol	slightly soluble
py	pyridine	sol	soluble

soln	solution	*trans-*	isomer in which substituted groups are on opposite sides of double bond between C atoms
soly	solubility		
sp	specific; species		
sp gr	specific gravity		
sr	steradian		
std	standard	TSCA	Toxic Substances Control Act
STP	standard temperature and pressure (0°C and 101.3 kPa)		
		TWA	time-weighted average
		Twad	Twaddell
sub	sublime(s)	UL	Underwriters' Laboratory
SUs	Saybolt Universal seconds	USDA	United States Department of Agriculture
syn	synthetic		
^{t} (eg, Bu^t), *t-, tert-*	tertiary (eg, tertiary butyl)	USP	*United States Pharmacopeia*
		uv	ultraviolet
T	tera (prefix for 10^{12}); tesla (magnetic flux density)	V	volt (emf)
		var	variable
t	metric ton (tonne)	*vic-*	vicinal
t	temperature	vol	volume (not volatile)
TAPPI	Technical Association of the Pulp and Paper Industry	vs	versus
		v sol	very soluble
		W	watt
TCC	Tagliabue closed cup	Wb	weber
tex	tex (linear density)	Wh	watt hour
T_g	glass-transition temperature	WHO	World Health Organization (United Nations)
tga	thermogravimetric analysis		
		wk	week
THF	tetrahydrofuran	yr	year
tlc	thin layer chromatography	(Z)-	zusammen; together; atomic number
TLV	threshold limit value		

Non-SI (Unacceptable and Obsolete) Units		Use
Å	angstrom	nm
at	atmosphere, technical	Pa
atm	atmosphere, standard	Pa
b	barn	cm^2
bar†	bar	Pa
bbl	barrel	m^3
bhp	brake horsepower	W
Btu	British thermal unit	J
bu	bushel	m^3; L
cal	calorie	J
cfm	cubic foot per minute	m^3/s
Ci	curie	Bq
cSt	centistokes	mm^2/s
c/s	cycle per second	Hz

†Do not use bar (10^5 Pa) or millibar (10^2 Pa) because they are not SI units, and are accepted internationally only for a limited time in special fields because of existing usage.

Non-SI (Unacceptable and Obsolete) Units		Use
cu	cubic	exponential form
D	debye	$C \cdot m$
den	denier	tex
dr	dram	kg
dyn	dyne	N
dyn/cm	dyne per centimeter	mN/m
erg	erg	J
eu	entropy unit	J/K
°F	degree Fahrenheit	°C; K
fc	footcandle	lx
fl	footlambert	lx
fl oz	fluid ounce	m^3; L
ft	foot	m
ft·lbf	foot pound-force	J
gf den	gram-force per denier	N/tex
G	gauss	T
Gal	gal	m/s^2
gal	gallon	m^3; L
Gb	gilbert	A
gpm	gallon per minute	(m^3/s); (m^3/h)
gr	grain	kg
hp	horsepower	W
ihp	indicated horsepower	W
in.	inch	m
in. Hg	inch of mercury	Pa
in. H_2O	inch of water	Pa
in.-lbf	inch pound-force	J
kcal	kilo-calorie	J
kgf	kilogram-force	N
kilo	for kilogram	kg
L	lambert	lx
lb	pound	kg
lbf	pound-force	N
mho	mho	S
mi	mile	m
MM	million	M
mm Hg	millimeter of mercury	Pa
$m\mu$	millimicron	nm
mph	miles per hour	km/h
μ	micron	μm
Oe	oersted	A/m
oz	ounce	kg
ozf	ounce-force	N
η	poise	$Pa \cdot s$
P	poise	$Pa \cdot s$
ph	phot	lx
psi	pounds-force per square inch	Pa
psia	pounds-force per square inch absolute	Pa
psig	pounds-force per square inch gage	Pa
qt	quart	m^3; L
°R	degree Rankine	K
rd	rad	Gy
sb	stilb	lx
SCF	standard cubic foot	m^3
sq	square	exponential form
thm	therm	J
yd	yard	m

BIBLIOGRAPHY

1. The International Bureau of Weights and Measures, BIPM (Parc de Saint-Cloud, France) is described in Appendix X2 of Ref. 4. This bureau operates under the exclusive supervision of the International Committee for Weights and Measures (CIPM).
2. *Metric Editorial Guide (ANMC-78-1)*, latest ed., American National Metric Council, 5410 Grosvenor Lane, Bethesda, Md. 20814, 1981.
3. *SI Units and Recommendations for the Use of Their Multiples and of Certain Other Units (ISO 1000-1981)*, American National Standards Institute, 1430 Broadway, New York, 10018, 1981.
4. Based on *ASTM E380-89a (Standard Practice for Use of the International System of Units (SI))*, American Society for Testing and Materials, 1916 Race Street, Philadelphia, Pa. 19103, 1989.
5. *Fed. Reg.*, Dec. 10, 1976 (41 FR 36414).
6. For ANSI address, see Ref. 3.

R. P. LUKENS
ASTM Committee E-43 on SI Practice

AEROGELS

Aerogels, solid materials that are so porous that they contain mostly air, were first prepared in 1931. Kistler used a technique known as supercritical drying to remove the solvent from a gel (a solid network that encapsulates its solvent) such that, as stated in his paper in *Nature*, "no evaporation of liquid can occur and consequently no contraction of the gel can be brought about by capillary forces at its surface (1)." In the six decades since then, there have been significant advances made both in the use of new precursors to form gels and in the removal of solvent from them. These advances have greatly simplified the preparation of aerogels and, in turn, improved their economic viability for commercial applications. Increasingly an aerogel is defined in terms of its properties and not the way in which it is prepared.

Almost all applications of aerogels are based on the unique properties associated with a highly porous network. Envision an aerogel as a sponge consisting of many interconnecting particles. The particles are so small and so loosely connected that the void space in the sponge, the pores, can make up for over 90% of the sponge volume. As an example, a silica aerogel contains particles that are of the order of 10 nm and each particle is connected to two or three other particles on average. Such a material has a typical density of about 100 kg/m^3 and accessible surface area of about 1000 m^2/g.

The ability to prepare materials of such low density, and perhaps more importantly, to vary the density in a controlled manner, is indeed what make aerogels attractive in many applications such as thermal insulation, detection of high energy particles, and catalysis. Thus, this article begins with a discussion of sol–gel chemistry used to form the wet gel. The intention is not to discuss in detail the fundamental chemistry involved, but to provide the basic principles that explain the formation of a porous network and the effect of preparative variables on its microstructure (often referred to as nanostructure because the relevant length scale is on the order of nanometer), ie, factors that impact on density. The general applicability of these principles are illustrated by examples of inorganic, organic, and inorganic–organic gels.

The section on preparation and manufacturing continues to focus on the microstructure of a gel, in particular its evolution during the removal of solvent. In addition to the original supercritical drying used by Kistler, there are now safer and more cost-effective methods that do not involve supercritical drying. More importantly, these methods ensure that the products possess the defining characteristics of an aerogel (ultrafine pores, small interconnected particles, and high porosity).

This article also aims at establishing the structure–property–application relationships of aerogels. Selected examples are given to show what some desirable properties are and how they can be delivered by design based on an understanding of the preparation and preservation of a gel's microstructure.

Sol–Gel Chemistry

Inorganic Materials.   Sol–gel chemistry involves first the formation of a sol, which is a suspension of solid particles in a liquid, then of a gel, which is a diphasic material with a solid encapsulating a solvent. A detailed description of the fundamental chemistry is available in the literature (2–4). The chemistry involving the most commonly used precursors, the alkoxides ($M(OR)_m$), can be described in terms of two classes of reactions:

Hydrolysis

$$-M\text{-}OR + H_2O \longrightarrow -M\text{-}OH + ROH$$

Condensation

$$-M\text{-}OH + XO\text{-}M\text{-} \longrightarrow -M\text{-}O\text{-}M + XOH$$

where X can either be H or R, an alkyl group

The important feature is that a three-dimensional gel network comes from the condensation of partially hydrolyzed species. Thus, the microstructure of a gel is governed by the rate of particle (cluster) growth and their extent of crosslinking or, more specifically, by the *relative* rates of hydrolysis and condensation (3).

The gelation of silica from tetraethylorthosilicate (TEOS) serves as an example of the above principle. Under acidic conditions, hydrolysis occurs at a faster rate than condensation and the resulting gel is weakly branched. Under basic conditions, the reverse is true and the resulting gel is highly branched and contains colloidal aggregates (5,6). Furthermore, acid-catalyzed gels contain higher concentrations of adsorbed water, silanol groups, and unreacted alkoxy groups than base-catalyzed ones (7). These differences in microstructure and surface functionality, shown schematically in Figure 1, result in different responses to heat treatment. Acid- and base-catalyzed gels yield micro- (pore width less than 2 nm) and meso-porous (2–50 nm) materials, respectively, upon heating (8). Clearly an acid-catalyzed gel which is weakly branched and contains surface functionalities that promote further condensation collapses to give micropores. This example highlights a crucial point: *the initial microstructure and surface functionality of a gel dictates the properties of the heat-treated product.*

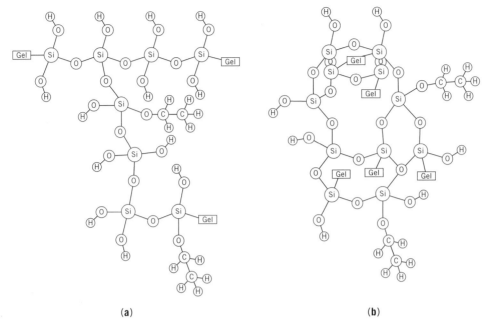

Fig. 1. Schematics of (**a**) acid-catalyzed and (**b**) base-catalyzed silica gels showing the differences in microstructure and surface functional groups. Reproduced from ref. 7. Courtesy of the American Ceramic Society.

Besides pH, other preparative variables that can affect the microstructure of a gel, and consequently, the properties of the dried and heat-treated product include water content, solvent, precursor type and concentration, and temperature (9). Of these, water content has been studied most extensively because of its large effect on gelation and its relative ease of use as a preparative variable. In general, too little water (less than one mole per mole of metal alkoxide) prevents gelation and too much (more than the stoichiometric amount) leads to precipitation (3,9). Other than the amount of water used, the rate at which it is added offers another level of control over gel characteristics.

The principles discussed so far are valid for silicates as well as nonsilicates, although there are more data available for the former. The alkoxides of transition metals tend to be more reactive than silicon alkoxides because both hydrolysis and condensation are nucleophilic substitution reactions and their cations have a more positive partial charge than silicon (3). This difference in reactivity presents both a challenge and an opportunity in the preparation of two-component systems. In a two-component system, the minor component can either be a network modifier or a network former. In the latter case, the distribution of the two components, or mixing, at a molecular level is governed by the *relative* precursor reactivity. Qualitatively good mixing is achieved when two precursors have similar reactivities. When two precursors have dissimilar reactivities, the sol−gel technique offers several strategies to prepare well-mixed two-component gels. Two such strategies are prehydrolysis (10), which involves prereacting a less reactive precursor with water to give it a head start, and chemical modification (11), which involves slowing down a more reactive precursor by substituting some of its alkoxy groups with bulkier, less reactive groups such as acetate. The ability to control microstructure *and* component mixing is what sets sol–gel apart from other methods in preparing multicomponent solids.

Organic Materials.   The first organic aerogel was prepared by the aqueous polycondensation of resorcinol with formaldehyde using sodium carbonate as a base catalyst (12). Figure 2**a** shows that the chemistry is similar to the sol–gel chemistry of inorganic materials. Subsequent to the reaction between resorcinol and formaldehyde, the functionalized resorcinol rings (ie, those possessing -CH_2OH groups) condense with each other to form nanometer-sized clusters, which then crosslink to form a gel. The process is influenced by typical sol–gel parameters such as pH, reactant concentration, and temperature. The resorcinol–catalyst molar ratio turns out to be the dominant factor that affects the microstructure of the gel (13).

Resorcinol–formaldehyde gels are dark red in color and do not transmit light. A colorless and transparent organic gel can be prepared by reacting melamine with formaldehyde using sodium hydroxide as a base catalyst (14). The chemistry, shown in Figure 2**b**, involves first the formation of -CH_2OH (hydroxymethyl) groups and then the formation of -$NHCH_2NH$- (diamino methylene) and -$NHCH_2OCH_2NH$- (diamino methylene ether) bridges through crosslinking reactions. The solution pH is again a key parameter, it is necessary to add hydrochloric acid in the second part of the process to promote condensation and gel formation.

Both the preparations of resorcinol–formaldehyde and melamine–formaldehyde gels are aqueous-based. Since water is deleterious to a gel's structure at high temperatures and immiscible with carbon dioxide (a commonly used supercritical drying agent), these gels cannot be supercritically dried without a tedious solvent-exchange step. In order to circumvent this problem, an alternative synthetic route of organic gels that is based upon a phenolic–furfural reaction using an acid catalyst has been developed (15). The solvent-exchange step is eliminated by using alcohol as a solvent. The phenolic–furfural gels are dark brown in color.

Carbon aerogels can be prepared from the organic gels mentioned above by supercritical drying with carbon dioxide and a subsequent heat-treating step in an inert atmosphere. For example, the following heating schedule has been used in flowing nitrogen (13,15): 22 to 250°C in 2 h, held at 250°C for 4 h, 250 to 1050°C in 9.5 h, held at 1050°C for 4 h, and cooled to room temperature in 16 h. This treatment results in a volume shrinkage of the sample of 65–75% and a mass of 45–50%. Despite these changes, the carbon aerogels are similar in morphology to their organic precursors, underscoring again the importance of structural control in the gelation step. Furthermore, changing the sol–gel conditions can lead to aerogels that have a wide range of physical properties. As a specific example, surface area for the phenolic–furfural aerogels is about 385 ± 16 m^2/g over a density range of 100–250 kg/m^3, whereas the corresponding carbon aerogels have surface areas of 512 ± 40 m^2/g over a density range of 300–450 kg/m^3.

Inorganic–Organic Hybrids.   One of the fastest growing areas in sol–gel processing is the preparation of materials containing both inorganic and organic components. The reason is that many applications demand special properties

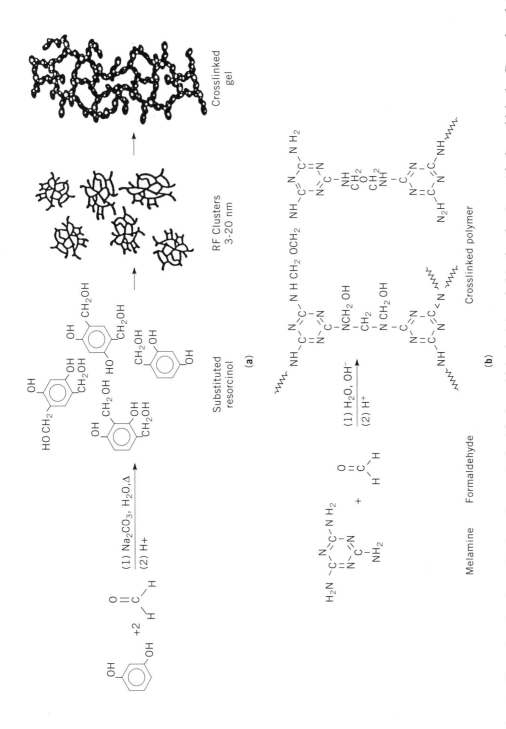

Fig. 2. The sol–gel polymerization of resorcinol with formaldehyde (**a**) and (**b**) of melamine with formaldehyde. Reproduced from refs. 13 and 14, respectively. Courtesy of the Materials Research Society.

5

that pure materials can seldom provide. The combination of inorganic and organic materials is, thus, an attractive way to deliver materials that have desirable physical, chemical, and structural characteristics. In this regard, sol–gel chemistry offers a real advantage because its mild preparation conditions do not degrade organic polymers, as would the high temperatures that are associated with conventional ceramic processing techniques. The voluminous literature on the sol–gel preparation of inorganic–organic hybrids can be found in several recent reviews (16–20) and the references therein; only a qualitative sketch is given in this section.

There are several ways to classify inorganic–organic materials, all of which depend on the strength of interaction between the two components. The interaction can range from nonexistent or weak, such as organic species embedded (or entrapped) in an inorganic network, to very strong, such as systems involving covalent bonds. For weakly interacting systems, the general observation is that most molecules can be entrapped in sol–gel matrices and, once entrapped, these molecules retain most of their characteristic physical and chemical properties (18). The entrapment of bioorganic materials such as enzymes, whole cells, antibodies, and other proteins (21) falls into this category.

For strongly interacting systems, chemical bonding can be induced by using functionalized precursors. There are three basic types of precursors: inorganically functionalized preformed organic polymers, organically functionalized oxides, and precursors containing both inorganic and organic functional groups. Examples of commonly used precursors are organofunctional metal alkoxides $(RO)_n$-E-X-A (19) and bridged polysilsesquioxanes (20). In $(RO)_n$-E-X-A, A is a functional organic group and X is a hydrolytically stable spacer linking A and the metal alkoxide which provides the inorganic function. The use of such precursors has allowed the control of very small domain sizes, often in the nanometer range, in the preparation of inorganic–organic materials (16,19,20). The challenge is to achieve a high degree of mixing of the two phases, thus, enabling the manipulation of interfacial properties at a molecular level. Another promising strategy to provide better homogeneity between the two phases is to form the inorganic and organic phases simultaneously, leading to what is known as a simultaneous interpenetrating network (16).

All the methods described so far involve introducing the organic phase prior to the formation of a solid phase. There is an interesting alternative to prepare a nanocomposite by a sequential approach (22). In this approach a silica aerogel was first prepared, then carbon was deposited in it by the decomposition by hydrocarbons (eg, methane, acetylene) at a temperature range of 500–850°C. This example demonstrates the feasibility of preparing hybrid materials by depositing an inorganic or organic phase onto an organic or inorganic substrate, respectively.

Preparation and Manufacturing

Supercritical Drying.   The development of aerogel technology from the original work of Kistler to about late 1980s has been reviewed (23). Over this period, supercritical drying was the dominant method in preparing aerogels and, for this reason, aerogels are synonymous with supercritically dried materials.

As noted earlier, supercritical drying could be an insufficient definition of an aerogel because it might not lead to the defining characteristics. Kistler used inorganic salts and a large amount of water in his work (1,24), making the subsequent salt removal and solvent-exchange steps time-consuming (water has to be removed because it would dissolve the gel structure at high temperatures). A significant time savings came when alkoxides were used as precursors in organic solvents, thereby requiring a minimum amount of water and eliminating the tedious solvent-exchange step (25,26). The introduction of carbon dioxide, which has a lower critical temperature and pressure than alcohols, as a supercritical drying agent allowed the drying step to be done under milder conditions and improved its safety (27). The development of a semicontinuous drying process further facilitated the preparation of aerogels (28). Together these advances, summarized in Table 1, have made possible the relatively safe supercritical drying of aerogels in a matter of hours. In recent years, the challenge has been to produce aerogel-like materials without using supercritical drying at all in an attempt to deliver economically competitive products. This topic will be discussed in more detail later.

As stated earlier, the main idea behind supercritical drying is to eliminate the liquid–vapor interface inside a pore, thereby removing the accompanying capillary pressure which acts to collapse a gel network. The value of this approach is demonstrated by the fact that aerogels do have higher porosities, higher specific surfaces areas, and lower apparent densities than xerogels, materials that are prepared by evaporative drying. However, it is incorrect to think that a gel remains static during supercritical drying. Rather, supercritical drying should be considered as part of the aging process, during which events such as condensation, dissolution, and reprecipitation can occur. The extent to which a gel undergoes aging during supercritical drying depends on the structure of the initial gel network. For example, it has been shown that a higher drying temperature changes the particle structure of base-catalyzed silica aerogels but not that of acid-catalyzed ones (29). It is also known that gels that have uniform-sized pores can withstand the capillary forces during drying better because of a more uniform stress distribution. Such gels can be prepared by a careful manipulation of sol–gel parameters such as pH and solvent or by the use of so-called drying control chemical additives (DCCA) (30). Clearly, an understanding of the

Table 1. Important Developments in the Preparation of Aerogels

Decade	Developments
1930	Using inorganic salts as precursors, alcohol as the supercritical drying agent, and a batch process; a solvent-exchange step was necessary to remove water from the gel.
1960	Using alkoxides as precursors, alcohol as the supercritical drying agent, and a batch process; the solvent exchange step was eliminated.
1980	Using alkoxides as precursors, carbon dioxide as the drying agent, and a semicontinuous process; the drying procedure became safer and faster. Introduction of organic aerogels.
1990	Producing aerogel-like materials without supercritical drying at all; preparation of inorganic–organic hybrid materials.

interrelationship between preparative and drying parameters is important in controlling the properties of aerogels.

The most widely studied supercritical drying variable is temperature simply because different solvents have different supercritical temperatures. Specifically, since alcohols have higher supercritical temperatures than carbon dioxide, there have been many recent reports on the effect of drying agent on the textural properties and crystallization behavior of aerogels (31–33). These results demonstrate nicely the accelerated aging that a gel undergoes at high temperatures and pressures. For this reason carbon dioxide is the drying agent of choice if the goal is to stabilize kinetically constrained structure, and materials prepared by this low-temperature route are referred to by some people as *carbogels*. In general, carbogels are also different from aerogels in surface functionality, in particular hydrophilicity, which impacts on the moisture sensitivity of these materials and their subsequent transformations under heat treatment.

It is less well known, but certainly no less important, that even with carbon dioxide as a drying agent, the supercritical drying conditions can also affect the properties of a product. For example, in the preparation of titania aerogels, temperature, pressure, the use of either liquid or supercritical CO_2, and the drying duration have all been shown to affect the surface area, pore volume, and pore size distributions of both the as-dried and calcined materials (34,35). The specific effect of using either liquid or supercritical CO_2 is shown in Figure 3 as an illustration (36).

Other important drying variables include the path to the critical point, composition of the drying medium, and depressurization (37). The rates of heating and depressurization are especially important in the preparation of monoliths because if the pore liquid does not have sufficient time to flow out of the gel network, it could lead to excessive stresses that cause cracking (38,39). The container of a gel is by itself a source of stresses by preventing the radial expansion and flow of liquid. Quantitative models describing these phenomena are available and have been tested against experimental data for silica gels (38–40).

For some applications it is desirable to prepare aerogels as thin films that are either self-supporting or supported on another substrate. All common coating methods such as dip coating, spin coating, and spray coating can be used to prepare gel films. However, for highly porous films (ie, porosity > 75%), special care is necessary to minimize the rate of solvent evaporation both during and after gel formation. One way to do so is to perform the coating processes within an enclosure that is filled with the saturated vapor of the working solvent and a partial pressure of ammonium hydroxide that catalyzes the gelation of the films (41). The subsequent supercritical drying step can be done in either alcohol or carbon dioxide. The choice depends on the desired properties of the aerogels and, in the case of supported films, the thermal stability of the substrate materials.

In all the processes discussed above, the gelation and supercritical drying steps are done sequentially. Recently a process that involves the direct injection of the precursor into a strong mold body followed by rapid heating for gelation and supercritical drying to take place was reported (42). By eliminating the need of forming a gel first, this entire process can be done in less than three hours per cycle. Besides the saving in time, gel containment minimizes some stresses and makes it possible to produce near net-shape aerogels and precision surfaces. The

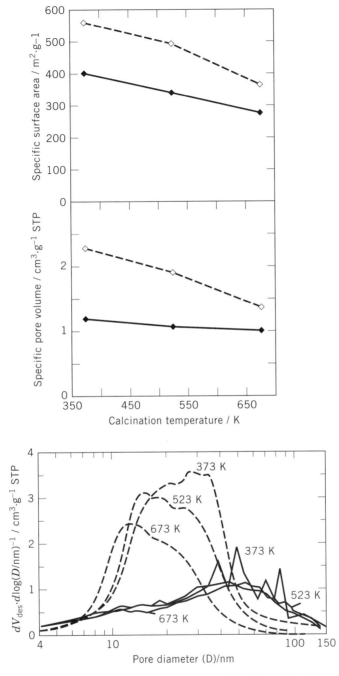

Fig. 3. Effect of using either liquid or supercritical carbon dioxide on the textural properties of titania aerogels calcined at the temperatures shown. (—), dried with liquid carbon dioxide at 6 MPa and 283 K; (- - -), dried with supercritical carbon dioxide at 30 MPa and 323 K. Reproduced from ref. 36. Courtesy of Marcel Dekker, Inc.

optical and thermal properties of silica aerogels thus prepared are comparable to those prepared with conventional methods (42).

Ambient Preparations. Supercritical drying with alcohols incurs high capital and operating costs because the process is run at high temperatures and pressures and needs to remove a large amount of pore liquid. Carbon dioxide allows drying to be done under milder conditions but its use is limited to miscible solvents. Thus, economic and safety considerations have provided a strong motivation for the development of techniques that can produce aerogel-like materials at ambient conditions, ie, without supercritical drying. The strategy is to minimize the deleterious effect of capillary pressure which is given by:

$$P = 2\sigma \cos(\theta)/r$$

where P is capillary pressure, σ is surface tension, θ is the contact angle between liquid and solid, and r is pore radius.

The equation above suggests that one approach would be to use a pore liquid that has a low surface tension. Indeed, two-step acid–base or acid–acid catalyzed silica gels have been made, aged in ethanol or water, washed with various aprotic solvents, and finally evaporatively dried at 323 K for 48 hours and then at 383 K for 48 hours (43). The aprotic solvents used and their corresponding surface tension in N/m at room temperature (shown in parentheses) are: tetrahydrofuran (23.1×10^{-3}); acetone (23.7×10^{-3}); cyclohexane (25.3×10^{-3}); acetonitrile (29.3×10^{-3}); nitromethane (32.7×10^{-3}); and 1:4 dioxane (33.6×10^{-3}). For both the acid-catalyzed and base-catalyzed gels, it was found that increasing surface tension causes a decrease in surface area and total pore volume. However, for acid-catalyzed gels with an intermediate water wash, the micropore volume increases with increasing surface tension. It was suggested that the water wash leads to a hydroxylated surface which, upon further condensation, gives a stiffer network and consequently a larger fraction of micropore volume. The important point is that with a pore liquid that has a sufficiently small surface tension, ambient pressure aerogels can have comparable pore volume and bulk density to those prepared with supercritical drying (see Fig. 4) (44).

For base-catalyzed silica gels, it has been shown that modifying the surface functionality is an effective way to minimize drying shrinkage (45,46). In particular, surface hydroxyl groups, the condensation of which leads to pore collapse, can be "capped off" via reactions with organic groups such as tetraethoxysilane and trimethylchlorosilane. This surface modification approach (also referred to as surface derivatization), initially developed for bulk specimens, has recently been applied to the preparation of thin films (47,48). The process involves the following steps: (*1*) prepare a base-catalyzed gel, (*2*) wash the gel with hexane, (*3*) replace the surface hydroxyls with organosilicon groups, (*4*) reliquify the surface-modified gel with ultrasound, (*5*) dip-coat the redispersed sols onto a silicon substrate, and (*6*) heat treat the film in air at 450°C. During drying these materials exhibit reversible shrinkage in a gradual dilation, or "spring-back," of the film. The extent of spring-back is a function of processing conditions. Films with porosity in the range of 30–99% can be prepared via a proper control of the washing, surface modification, dip-coating, and heat treatment conditions (48).

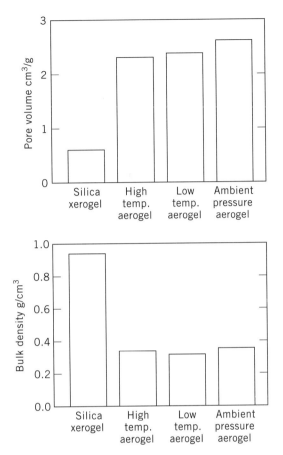

Fig. 4. Comparison of physical properties of silica xerogels and aerogels. Note the similar properties of the aerogels prepared with and without supercritical drying. Reproduced from ref. 44. Courtesy of the Materials Research Society.

In changing surface hydroxyls into organosilicon groups, surface modification has an additional advantage of producing hydrophobic gels. This feature, namely the immiscibility of surface-modified gel with water, has led to the development of a rapid extractive drying process shown in Figure 5 (49). The basic idea is to submerge a wet gel into a pool of hot water that is above the boiling point of the pore fluid. After the pore fluid boils out of the gel, the gel floats to the surface because it is not wetted by water and can be easily recovered. Other working fluids, such as ethylene glycol and glycerol, can be used instead of water as long as they have a high boiling point and do not wet the gel. This ambient pressure process offers improved heat transfer rates and, in turn, greater energy efficiency without compromising desirable aerogel properties.

Another approach to produce aerogels without supercritical drying is freeze drying, in which the liquid–vapor interface is eliminated by freezing a wet gel into a solid and then subliming the solvent to form what is known as a *cryogel*. Some potential problems are that cracks may develop in the frozen gel and sublimation, if done too fast, can melt the solvent. Cryogels of silica and nickel

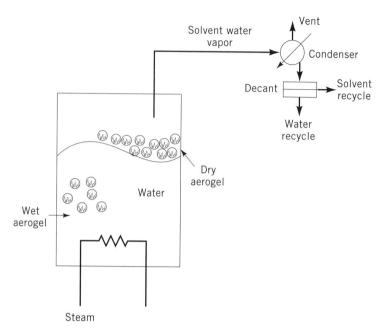

Fig. 5. Schematic diagram of an extractive drying process that produces aerogels at ambient pressure. Reproduced from ref. 49. Courtesy of the Materials Research Society.

oxide–alumina have similar, but not identical, properties to aerogels of the same materials (50). In general, there has not been a lot of study on freeze drying and the limited data available suggest that it might not be as attractive as the above ambient approaches in producing aerogels on a commercial scale.

Properties

A detailed discussion of the properties of aerogels can be found in several recent review articles (51–55) and the references therein. This section provides a physical basis for these properties by focusing on the microstructure of an aerogel. The intent is to provide a bridge between the two previous sections, which discuss the preparative and drying parameters that affect microstructure, and the next one, which outlines the potential applications made possible by unique structural features. The emphasis is on silica aerogels because they have been the most extensively characterized.

All properties of aerogels are related to the extremely high porosity of these materials. Consider an aerogel as consisting of strings of pearls; each pearl is a structural unit and connected to only two to three other units on average. The size of each structural unit depends on the gelation and drying conditions and varies between 1 to 100 nm in diameter. Correspondingly, pore sizes in an aerogel range from 1 to 100 nm. For silica aerogels skeletal (ie, the solid phase) densities, ρ_s, of 1700–2100 kg/m^3 have been reported, whereas the overall densities, ρ, are in the range of 3–500 kg/m^3 (54). The fact that the overall density is much lower than the skeletal density suggests that an aerogel is full of open space, or pores.

Quantitatively the porosity, P, is given by

$$P = 1 - (\rho/\rho_s)$$

Thus, the porosity of an aerogel is in excess of 90% and can be as high as 99.9%. As a consequence of such a high porosity, aerogels have large internal surface area and pore volume.

Since the pores in an aerogel are comparable to, or smaller than, the mean free path of molecules at ambient conditions (about 70 nm), gaseous conduction of heat within them is inefficient. Coupled with the fact that solid conduction is suppressed due to the low density, a silica aerogel has a typical thermal conductivity of 0.015 W/(m·K) without evacuation. This value is at least an order of magnitude lower than that of ordinary glass and considerably lower than that of CFC(chlorofluorocarbon)-blown polyurethane foams (54).

The low density and high porosity of aerogels lead to several other unique properties. Sound travels in silica aerogels at a longitudinal velocity of 100–300 m/s (the corresponding value in ordinary glass is about 500 m/s) (56). The low sound velocity and low density combine to give a low acoustic impedance, which is the product of the two quantities. Silica aerogels also have a refractive index that is very close to unity, meaning light can enter and leave a piece of aerogel without appreciable reflective losses and refractive effects. And for porosity larger than 0.7, silica aerogels have a very low dielectric constant (less than 2) (57). Finally, compared to silica glass, silica aerogels have an extremely small Young's modulus (10^6–10^7 N/m^2, several orders of magnitude less than nonporous materials), and thus, can be compressed easily (51).

From a practical viewpoint it is not only the low density of aerogels but also the variability of density over a wide range that offers interesting possibilities. As discussed in "Inorganic Materials," a silica gel can be formed under either acidic or basic conditions. Densities that can be obtained with these one-step processes are in the range of 25–500 kg/m^3. To produce ultralow-density silica aerogels, a two-step process has been developed to extend the lower limit to 3 kg/m^3 (58). In the first step, either tetramethyl- or tetraethyl-orthosilicate is reacted with a substoichiometric amount of water to form a partially hydrolyzed, partially condensed silica precursor. After distilling off the alcohol solvent, this precursor is stored in a nonalcoholic solvent. The second step involves the further hydrolysis of the precursor under basic conditions to form a gel. The ability to vary density over two orders of magnitude is significant because almost all the properties discussed in this section vary with density. Specifically (52,54),

Property	Value
refractive index	$n = 1 + 2.1 \times 10^{-4}\,\rho$; ρ in kg/m^3
sound velocity	$v_s \propto \rho^{\beta}$
Young's modulus	$Y \propto \rho^{\alpha}$
where for silica	$\beta = (\alpha - 1)/2$
	$\alpha = 3.6$; $\beta = 1.3$

Note that scaling exponents depend on preparative conditions.

Table 2 summarizes the key physical properties of silica aerogels. A range of values is given for each property because the exact value is dependent on the preparative conditions and, in particular, on density.

Table 2. Typical Values of Physical Properties of Silica Aerogels

Property	Values
density, kg/m^3	3–500
surface area, m^2/g	800–1000
pore sizes, nm	1–100
pore volume, cm^3/g	3–9
porosity, %	75–99.9
thermal conductivity, $W/(m \cdot K)$	0.01–0.02
longitudinal sound velocity, m/s	100–300
acoustic impedance, $kg/(m^2 \cdot s)$	10^3–10^6
dielectric constant	1–2
Young's modulus, N/m^2	10^6–10^7

Applications

Thermal Insulation. In addition to their low thermal conductivity, as discussed in the section above, silica aerogels can be prepared to be highly transparent in the visible spectrum region. Thus, they are promising materials as superinsulating window-spacer. To take further advantage of its high solar transmission, a silica aerogel layer sandwiched between two glass panes can be used to collect solar energy passively. Figure 6 shows a specific arrangement in which an insulating layer of transparent silica aerogel is placed in front of a brick wall, the surface of which is painted black. Solar radiation collected at the black surface is mostly transferred as heat into the house because heat loss through the aerogel layer is minimal. A shade is put into place to prevent overheating if necessary (53). The same principle applies when silica aerogels are used as insulating covers for solar panels.

The thermal conductivity of silica aerogels can be further reduced by minimizing the radiation leakage with an opacifier such as carbon black (54). The introduction of an opacifier makes the material opaque and unsuitable for window insulation. On the other hand, opaque silica aerogels can be used as insulating materials in appliances such as refrigerators and freezers. Compared with CFC-blown insulating foams, which could release chlorine into the atmosphere, silica aerogels pose no such environmental hazard and are nonflammable. Moreover, the thermal conductivity of opaque silica aerogels is a weak function of temperature (52), making them useful as insulating materials for heat-storage systems. Another promising application of silica aerogels is as a filler in vacuum panel because they do not require a high vacuum for good thermal performance. The high surface area of these materials further allow them to act as a "getter" by adsorbing gases in the panel.

The commercial viability of silica aerogels as thermal insulators depends on the ability to produce them at a competitive price. After all, in the 1950s, the production of Monsanto's Santocel stopped after a lower-cost process to manufacture fumed silica was developed (59). Recently an initial economic analysis

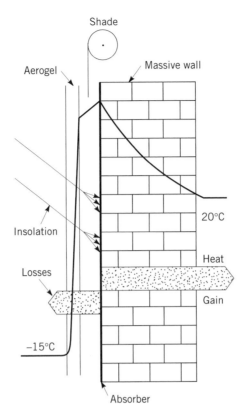

Fig. 6. Schematic diagram showing the use of transparent silica aerogel in passive solar collection. Reproduced from ref. 53. Courtesy of Elsevier Science-NL.

that consists of six factors in the manufacturing of aerogels: starting material, solvent, energy, wage, equipment, and facility was published (60). The results show that the dominant cost is the cost of the starting material and that aerogels could be competitive with commercial insulating materials on a cost per R value basis. Indeed, BASF has developed a silica aerogel which has the registered trademark Basogel (61). Since supercritical drying, even with carbon dioxide at a lower temperature, is an energy-intensive process, NanoPore, Inc. is developing an ambient approach to make silica aerogels (49,59). Technological progresses in the next several years will be critical in determining whether aerogels can capture a significant share of the commercial insulation market, which is probably their largest potential area of application. At least two U.S. companies are currently developing aerogels as insulating materials. Aspen Systems manufactures silica aerogels in the forms of powders, monoliths, and blankets. Their present (1996) price range is from $100 to $2,000 per cubic foot, depending on the size of the order (62). Aerojet Corp. has collaborated with different end-users in evaluating the market potential of organic aerogels (59), which have even lower thermal conductivities than their silica counterparts.

Catalysis. Kistler explored the catalytic applications of aerogels in the 1930s because of the unique pore characteristics of aerogels (24), but this area of research stayed dormant for about three decades until less tedious procedures to produce the materials were introduced (25,26). Three recent review articles summarize the flurry of research activities since then (63–65). Table 3 is a much abbreviated list of what has been cited in these three articles to demonstrate simply the wide range of catalytic materials and reactions that have been studied.

Most of the studies on catalytic aerogels reported in the open literature involve testing powder samples in experimental reactors. For the potential scale-up to commercial operations, aerogels have several limitations which, ironically, arise from their unique properties. First, even though aerogels have high specific surface area, they also have low densities. The product of the two quantities, which is surface area per unit volume, does not offer a significant advantage over other materials that have been used as adsorbents or catalysts. Second, the low thermal conductivity of aerogels means that it would be difficult to transfer heat in or out of a catalytic packed bed. Third, aerogels are fragile and do not withstand mechanical stress well. Attempts have been made to overcome the last two limitations by supporting or combining aerogels with materials that are either more thermally conducting or more rigid (63). Finally, as in the case of thermal insulation, catalytic aerogels need to be cost-competitive, even though the economic pressure is not as severe because the cost of a catalyst is usually a small fraction of the value of its derived products.

Table 3. Examples of Materials and Reactions Involving Catalytic Aerogels

Materials	Reactions	Examples
Type I, simple oxides		silica, alumina, titania, zirconia
Type II, mixed oxides		nickel oxide–alumina, titania–silica, zirconia–silica, chromia–alumina
Type III, ternary oxides		nickel–oxide–silica–alumina, magnesia–alumina–silica, titania–silica–vanadia, alumina–chromia–thoria
Type IV, supported metals		palladium–alumina, ruthenium–silica, platinum–titania, palladium–silica–alumina
Type V, doped oxides (dopant not an oxide)		zirconia–sulfate, zirconia–phosphate, niobia–phosphate, $TiCl_4$–alumina
Types I, II, III, IV	partial oxidation	isobutylene to methyacrolein, acetaldehyde to acetic acid
Type IV	hydrogenation	cyclopendiene to cyclopentene, benzene to cyclohexane
Types I, II, V	isomerization	1-butene to *cis*- and *trans*-2-butene, *n*-butane to isobutane
Type II	epoxidatiion	1-hexene and cyclohexene to the corresponding epoxides
Type IV	hydrotreating	hydrodenitrogenation and hydrodesulfurization
Type V	polymerization	ethylene to polyethylene
Types II and III	nitric oxide reduction	reduction of NO by ammonia to nitrogen

It is commonly believed that catalytic aerogels are interesting because their composition, morphology, and structure can be controlled at a microscopic level. But this feature is generic to any sol–gel-derived materials, so perhaps what set aerogels apart is that during solvent removal, there is another level control over the physical and chemical properties of the products. For example, supercritically dried materials are usually mesoporous and, as such, should be good catalysts for liquid-phase reactions for which there could be diffusional limitations. And for multicomponent materials, the distribution (or mixing state) of the various components could be better preserved without compromising the integrity of the porous network. A nice illustration of these effects is the recent results on epoxidation (66–68).

For the epoxidation of olefins over catalysts containing titania and silica, two factors are considered to be crucial: the accessibility of large pores for the bulky olefins (containing six to ten carbons) and the presence of Ti–O–Si linkages. It has been shown (66–68) that both of these desirable properties can be obtained only by supercritically drying an optimized gel at low temperatures with carbon dioxide. High temperature supercritical drying led to the segregation of titania and destruction of Ti–O–Si linkages. Conventional evaporative drying maintained the density of Ti–O–Si linkages but resulted in microporous xerogels that were inactive. Figure 7 (69) illustrates these observations for the epoxidation of cyclohexene. Besides reporting similar results for other olefins such as cyclododecene, norbonene, and limonene, these researchers were able to establish a semiquantitative correlation between activity and Ti–O–Si connectivity.

The Ti–O–Si linkages in titania–silica also have a large impact on the acidic properties of this mixed oxide. It was recently demonstrated that, over the entire composition range, the extent of mixing as controlled by prehydrolysis changes the acid site density, acid site type (Lewis versus Brønsted), and 1-butene isomerization activity of titania–silica aerogels (70). In fact, these observations appear to be general for other mixed oxide pairs in that sol–gel chemistry affords a higher degree of control over the intimacy of molecular-scale mixing that is not available by other preparative methods (71). Intimate mixing is actually undesirable in some instances, as shown for the selective catalytic reduction of NO with NH_3 (72). For this system, titania crystallites are believed to be good for stabilizing a two-dimensional overlayer of vanadia. Still, for this reaction the role of oxide–oxide interactions in affecting surface acidic properties remains critical (73). The fact that multicomponent materials of specific pore characteristics can be prepared to be either well-mixed or poorly-mixed, depending on the application, represents a unique advantage of catalytic aerogels.

Scientific Research. There are some applications that require such specific and unique properties that an aerogel is the only alternative. In these situations price is no longer a factor and aerogels have been and will continue to be produced for scientific purposes. One example is the use of silica aerogel monoliths as radiators in Chernekov counters, or detectors, in high energy physics and nuclear astrophysics experiments (23). In order to measure precisely the momentum of elementary particles, which are produced by particle interactions in high energy accelerators, over different momentum ranges, it is necessary to vary the refractive index of radiators. The reason is that a charged particle produces light in passing through a medium only if its velocity is higher than the velocity of

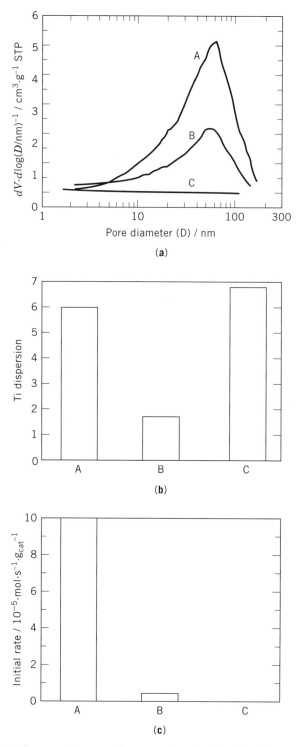

Fig. 7. The effect of preparation on the pore size distribution (**a**), titanium dispersion (**b**), and the activity for epoxidation of cyclohexene (**c**) of titania–silica containing 10 wt % titania and calcined in air at 673 K. Sample A, low-temperature aerogel; Sample B, high-temperature aerogel; Sample C, xerogel. Reproduced from ref. 69. Courtesy of Marcel Dekker, Inc.

light in that medium (52). As pointed out in the section, "Properties", the refractive index of silica aerogels is related to density. Thus, by varying the density from about 50 to 300 kg/m^3, which is easily attainable with the control of sol–gel chemistry, materials of indices of refraction between 1.01 and 1.06 can be delivered to measure low values of momentum. Indeed, two large blocks of silica aerogels have been prepared as particle detectors for use in the CERN Intersecting Storage Rings and the Deutschen Electronen-Synchrotron (DESY) (23).

Aerogels are also among the few materials that can capture cosmic particles intact (74). When a spacecraft approaches a source (eg, the corona of a comet), the cosmic particles have relative speeds of the order of tens of km/s. These hypervelocity particles tend to melt or vaporize upon collision with a solid. Mesoporous materials that are of very low density can capture these particles relatively intact, but a silica aerogel is the only one that allows the captured particles to be located easily because of its transparency. In fact, as a particle enters a silica aerogel at hypervelocity, it creates a carrot-shaped tract that points to where the particle stops. The high surface area of an aerogel provides the added advantage of adsorbing any volatiles that are either in space or generated during capture.

Panels of silica aerogels have already been flown on several Space Shuttle missions (74). Currently a STARDUST mission has been planned by NASA to use aerogels to capture cometary samples (>1000 particles of >15 micron diameter) and interstellar dust particles (>100 particles of >0.1 micron diameter). The mission involves the launch of a spacecraft in February 1999 to encounter the comet Wild 2 in December 2003 and to return the captured samples to earth in January 2006. Silica aerogels of ultralow density, high uniformity, and high purity need to be prepared to meet the primary objectives of the mission. Furthermore, there are challenges to produce nonsilicate aerogels of the same high quality and aerogels that can perform chemical as well as physical capture.

Besides being used as a tool for scientific research, silica aerogels can be the cause for new scientific phenomena. For example, the long-range correlations of the disorder in silica aerogels are believed to be responsible for the intriguing observations of the superfluid transitions in ^{4}He and ^{3}He and on the ordering of ^{3}He–^{4}He mixtures (75).

Other Applications. There are several other applications that take advantage of the unique properties of silica aerogels shown in Table 2. In a piezoceramic ultrasound transducer, the signal is reduced as sound waves cross the interface of two materials that have very different acoustic impedances (eg, the acoustic impedances of air and piezoelectric ceramics differ by several orders of magnitude). With its tunable acoustic impedance, a layer of silica aerogel sandwiched between a piezoceramic and air minimizes the mismatch and enhances the signal (54). The low dielectric constant of silica aerogels suggests their usefulness as layer materials in integrated circuits. For silica aerogels to replace the currently used silica glasses, several issues such as the control of porosity, mechanical strength, thermal stability, and process integration problems need to be addressed (57). Finally, silica aerogels are effective dehydrating agents because of their high capacity for absorbing moisture and chemical inertness. For household insect control, silica aerogels work by absorbing waxes from the cuticle of insects, which then die from dehydration; their potency can be enhanced with the doping of insecticides (76).

The introduction of organic aerogels has led to interesting areas of applications. Recent data (77,78) showing the thermal properties of these materials and, more importantly, how these properties can be varied with preparative conditions, should help their further development as thermal insulators. Carbon aerogels are also promising electrode materials in electrochemical double-layer capacitors. Large specific capacities and capacitance densities have been demonstrated in a device consisting of two carbon aerogel wafers; this performance is attributed to the high specific surface area and contiguous structure of the carbon electrode (79). These results are encouraging for the use of carbon aerogels in building low-cost, high-power, and high-energy density capacitors (often referred to as "supercapacitors"). Another potential application of carbon-aerogel electrodes is the capacitive deionization of water. In a recently developed process, water with various anions and cations is pumped through an electrochemical cell consisting of a stack of carbon aerogel electrodes (80). The large specific surface area of the electrodes allows the electrostatic capture of ions. After water is purified, the cell can be regenerated by a simple electrical discharge.

Summary

As of 1996 the commercial market for aerogels remains very small. However, aerogels have the potential of being marketable both as a commodity chemical (eg, in thermal insulation) and as a specialty chemical (eg, in electronic applications) because of their unique and tailorable properties. The next few years will be critical in assessing whether aerogels can penetrate and grow in either end of the market, as the field is changing rapidly with the development of cost-competitive technologies and novel applications.

BIBLIOGRAPHY

1. S. S. Kistler, *Nature*, **127**, 741 (1931).
2. C. J. Brinker and G. W. Scherer, *Sol-Gel Science: The Physics and Chemistry of Sol-Gel Processing*, Academic Press, New York, 1990.
3. J. Livage, M. Henry, and C. Sanchez, *Prog. Solid State Chem.* **18**, 259 (1988).
4. R. C. Mehrota, *J. Non-Cryst. Solids* **145**, 1 (1992).
5. H. D. Gesser and P. C. Goswami, *Chem. Rev.* **89**, 765 (1989).
6. R. K. Iler, *The Chemistry of Silica*, John Wiley & Sons, New York, 1979, Chapt. 3.
7. J. Y. Ying, J. B. Benziger, and A. Navrotsky, *J. Am. Ceram. Soc.* **76**, 2571 (1993).
8. B. Handy, K. L. Walther, A. Wokaun, and A. Baiker, in P. A. Jacobs, P. Grange, and B. Delmon, eds., *Preparation of Catalysts V*, Elsevier, Netherlands, 1991, pp. 239–246.
9. E. I. Ko, in G. Ertl, H. Knözinger, and J. Weitkamp, eds., *Handbook of Heterogeneous Catalysts*, VCH, Germany, 1997, Vol. 1, Sec. 2.1.4.
10. B. E. Yoldas, *J. Mater. Sci.* **14**, 1843 (1979).
11. J. Livage and C. Sanchez, *J. Non-Cryst. Solids* **145**, 11 (1992).
12. U.S. Pat. 4,873,218 (Oct. 10, 1989); U.S. Pat. 4,997,804 (Mar. 5, 1991), R. W. Pekala.
13. R. W. Pekala and C. T. Alviso, *Mat. Res. Soc. Symp. Proc.* **270**, 3 (1992).
14. R. W. Pekala and C. T. Alviso, *Mat. Res. Soc. Symp. Proc.* **180**, 791 (1990).
15. R. W. Pekala, C. T. Alviso, X. Lu, J. Gross, and J. Fricke, *J. Non-Cryst. Solids* **188**, 34 (1995).

16. B. M. Novak, *Adv. Mater.* **5**, 422 (1993).

17. C. Sanchez and F. Ribot, *New J. Chem.* **18**, 1007 (1994).

18. D. Avnir, *Acc. Chem. Res.* **28**, 328 (1995).

19. U. Schubert, N. Hüsing, and A. Lorenz, *Chem. Mater.* **7**, 2010 (1995).

20. D. A. Loy and K. J. Shea, *Chem. Rev.* **95**, 1431 (1995).

21. D. Avnir, S. Braun, O. Lev, and M. Ottolenghi, *Chem. Mater.* **6**, 1605 (1994).

22. W. Cao, X. Y. Song, and A. J. Hunt, *Mat. Res. Soc. Symp. Proc.* **349**, 87 (1994).

23. A. J. Ayen and P. A. Iacobucci, *Rev. Chem. Eng.* **5**, 157 (1988).

24. S. S. Kistler, S. Swan, Jr., and E. G. Appel, *Ind. Eng. Chem.* **26**, 388 (1934).

25. S. J. Teichner, *CHEMTECH*, **21**, 372 (1991).

26. S. J. Teichner, G. A. Nicolaon, M. A. Vicarini, and G. E. E. Gardes, *Adv. Colloid Interface Sci.* **5**, 245 (1976).

27. U.S. Pat. 4,610,863 (Sept. 9, 1986), P. H. Tewari and A. J. Hunt.

28. U.S. Pat. 4,619,908 (Oct. 28, 1986), C. P. Cheng, P. A. Iacobucci, and E. N. Walsh.

29. P. Wang, A. Emmerling, W. Tappert, O. Spormann, J. Fricke, and H. G. Haubold, *J. Appl. Cryst.* **24**, 777 (1991).

30. D. R. Ulrich, *J. Non-Cryst. Solids* **100**, 174 (1988).

31. D. M. Smith, R. Desphande, and C. J. Brinker, *Mat. Res. Symp. Proc.* **271**, 553 (1992).

32. G. Cogliati, M. Guglielmi, T. M. Che, and T. J. Clark, *Mat. Res. Symp. Proc.* **180**, 329 (1990).

33. M. Beghi, P. Chiurlo, L. Costa, M. Palladina, and M. F. Pirini, *J. Non-Cryst. Solids* **145**, 175 (1992).

34. C. J. Brodsky and E. I. Ko, *J. Mater. Chem.* **4**, 651 (1994).

35. D. C. M. Dutoit, M. Schneider, and A. Baiker, *J. Porous Mater.* **1**, 165 (1995).

36. M. Schneider and A. Baiker, *Catal. Rev.-Sci. Eng.* **37**(4), 529 (1995).

37. G. W. Scherer, in A. S. Mujumdar, ed., *Drying*, Elsevier, Netherlands, 1992, pp. 92–113.

38. G. W. Scherer, *J. Non-Cryst. Solids*, **145**, 33 (1992).

39. G. W. Scherer, *J. Sol-Gel Sci. Tech.* **3**, 127 (1994).

40. T. Woignier, G. W. Scherer, and A. Alaoue, *J. Sol-Gel Sci. Tech.*, **3**, 141 (1994).

41. L. W. Hrubesh and J. F. Poco, *J. Non-Cryst. Solids*, **188**, 46 (1995).

42. J. F. Poco, P. R. Coronado, R. W. Pekala, and L. W. Hrubesh, *Mat. Res. Soc. Symp. Proc.* **431**, 297 (1996).

43. R. Deshpande, D. M. Smith, and C. J. Brinker, *Mat. Res. Soc. Symp. Proc.* **271**, 553 (1992).

44. C. J. Brinker, R. Sehgal, N. K. Raman, S. S. Prakash, and L. Delattre, *Mat. Res. Soc. Symp. Proc.* **368**, 329 (1995).

45. D. M. Smith, R. Deshpande, and C. J. Brinker, *Mat. Res. Soc. Symp. Proc.* **271**, 567 (1992).

46. D. M. Smith, R. Deshpande, and C. J. Brinker, *Ceram. Trans.* **31**, 71 (1993).

47. S. S. Prakash, C. J. Brinker, A. J. Hurd, and S. M. Rao, *Nature* **374**, 439 (1995).

48. S. S. Prakash, C. J. Brinker, and A. J. Hurd, *J. Non-Cryst. Solids* **190**, 264 (1995).

49. D. M. Smith, W. C. Akerman, R. Roth, A. Zimmerman, and F. Schwertfeger, *Mat. Res. Soc. Symp. Proc.* **431**, 291 (1996).

50. D. Klvana, J. Chaouki, M. Repellin-Lacroix, and G. M. Pajonk, *J. Phys. Colloq.*, **C4**, 29 (1989).

51. H. D. Gesser and P. C. Goswami, *Chem. Rev.* **89**, 765 (1989).

52. J. Fricke and A. Emmerling, *J. Am. Ceram. Soc.* **75**(8), 2027 (1992).

53. J. Fricke, *J. Non-Cryst. Solids* **147/148**, 356 (1992).

54. J. Fricke and J. Gross, *Mater. Eng.* **8**, 311 (1994).

55. T. Heinrich, U. Klett, and J. Fricke, *J. Porous Mater.* **1**, 7 (1995).

56. J. Fricke, *Sci. Amer.* **256**(5), 92 (1988).

57. D. M. Smith, J. Anderson, C. C. Cho, G. P. Johnston, and S. P. Jeng, *Mat. Res. Soc. Symp. Proc.* **381**, 261 (1995).

58. T. M. Tillotson and L. W. Hrubesh, *J. Non-Cryst. Solids* **145**, 45 (1992).

59. C. M. Caruana, *Chem. Engr. Prog.* (June 11, 1995).

60. G. Carlson, D. Lewis, K. McKinley, J. Richardson, and T. Tillotson, *J. Non-Cryst. Solids* **186**, 372 (1995).

61. G. Herrmann, R. Iden, M. Mielke, F. Teich, and B. Ziegler, *J. Non-Cryst. Solids* **186**, 380 (1995).

62. http://www.aspensystems.com/aerogel.html

63. G. M. Pajonk, *Appl. Catal.* **72**, 217 (1991).

64. G. M. Pajonk, *Catal. Today*, **35**, 319 (1997).

65. M. Schneider and A. Baiker, *Catal. Rev.-Sci. Eng.* **37**(4), 515 (1995).

66. D. Dutoit, M. Schneider, and A. Baiker, *J. Catal.* **153**, 165 (1995).

67. R. Hutter, T. Mallat, and A. Baiker, *J. Catal.* **153**, 177 (1995).

68. R. Hutter, T. Mallat, and A. Baiker, *J. Catal.* **157**, 665 (1995).

69. From ref. 65, p. 544.

70. J. B. Miller, S. T. Johnston, and E. I. Ko, *J. Catal.* **150**, 311 (1994).

71. J. B. Miller and E. I. Ko, *Catal. Today*, **35**, 269 (1997).

72. B. E. Handy, A. Baiker, M. Schramal-Marth, and A. Wokaun, *J. Catal.* **133**, 1 (1992).

73. R. J. Willey, C. T. Wang, and J. B. Peri, *J. Non-Cryst. Solids* **186**, 408 (1995).

74. P. Tsou, *J. Non-Cryst. Solids* **186**, 415 (1995).

75. M. Chan, N. Mulders, and J. Reppy, *Physics Today* (Aug. 30, 1996).

76. http://www.ent.agri.umn.edu/academics/classes/ipm/chapters/ware.htm

77. X. Lu, R. Caps, J. Fricke, C. T. Alviso, and R. W. Pekala, *J. Non-Cryst. Solids* **188**, 226 (1995).

78. V. Bock, O. Nilsson, J. Blumm, J. Fricke, *J. Non-Cryst. Solids* **185**, 233 (1995).

79. S. T. Mayer, R. W. Pekala, and J. K. Kaschmitter, *J. Electrochem. Soc.* **140**(2), 446 (1993).

80. J. C. Farmer, D. V. Fix, G. V. Mack, R. W. Pekala, and J. F. Poco, *J. Electrochem. Soc.* **143**(1), 159 (1996).

EDMOND I. KO
Carnegie Mellon University

AQUACULTURE

One definition of aquaculture is the rearing of aquatic organisms under controlled or semicontrolled conditions (1). Another, promulgated by the Food and Agriculture Organization of the United Nations, is that aquaculture is, "the farming of aquatic organisms, including fish, molluscs, crustaceans, and aquatic plants" (2). Included within those broad definitions are activities in fresh, brackish, marine, and even hypersaline waters. The term mariculture is often used in conjunction with aquaculture in the marine environment.

Public sector aquaculture involves production of aquatic animals to augment or establish recreational and commercial fisheries. Public sector aquaculture is widely practiced in North America and to a lesser extent in other parts of the world. The FAO definition of aquaculture also indicates that farming implies ownership of the organisms being cultured, which would seem to exclude public sector aquaculture.

In recent years, aquaculture has been increasingly used as a means of aiding in the recovery of threatened and endangered species. Those efforts are currently public sector activities, although there is interest in the private sector to become involved. As global awareness of endangered species issues grows, recovery programs for aquatic threatened and endangered species may arise in many more countries. Going hand in hand with attempts to recover endangered species are enhancement stocking programs aimed at releasing juvenile animals to build back stocks of aquatic animals that have been reduced due to overfishing. Examples of enhancement programs currently in existence include the stocking of cod in Norway, flounders in Japan, and red drum in the United States.

The bulk of global production from aquaculture is utilized directly as human food, with public aquaculture playing a minor role in many nations or being absent. Private aquaculture is not only about human food production, however. There is, in some regions, well-developed private sector aquaculture involved in the production of bait and ornamental fishes and invertebrates.

Aquatic plants are cultured in many regions of the world. In fact, aquatic plants, primarily seaweeds, account for nearly 25% of the world's aquaculture production (3). Most of the information available in the literature relates to the production of such aquatic animals as molluscs, crustaceans, and finfish.

The origins of aquaculture are rooted in China and may go back some 4000 years. Asia dominates the world in aquaculture production, with China producing over 10 million metric tons and Japan and India each producing well over 1 million metric tons (4). In North America, the culture of fish began in the mid-nineteenth century and grew rapidly in the public sector after the establishment of the U.S. Fish and Fisheries Commission in 1871 (5). Private aquaculture existed as a minor industry for many decades, coming into prominence beginning in the 1960s. Since then the United States has become one of the leaders in aquaculture research and development, although production, while significant at over 400,000 metric tons by 1992 (4), amounted to only about 2% of the world's total of nearly 19 million metric tons. The United States commercial aquaculture industry is dominated by channel catfish, trout, salmon, minnows, oysters, mussels, clams, and crawfish. A number of other fishes and invertebrates are also being reared. Included are tilapia, striped bass and hybrid striped bass, red drum, goldfish, tropical fishes, and shrimp. In the public sector, hatcheries produce large numbers of such species as salmon, trout, largemouth and smallmouth bass, sunfish, crappie, northern pike, muskellunge, walleye, and catfish for stocking or growout.

Aquaculture production continues to grow annually, but increasing competition for suitable land and water, problems associated with wastewater from aquaculture facilities, disease outbreaks, and potential shortages of animal protein for aquatic animal feeds are having, or may have, negative effects on future growth. New technology, including the application of molecular genetic approaches to improving performance and disease resistance in aquaculture species, along with the development of water reuse (recirculating) systems and the establishment of offshore facilities, may provide the impetus for a resurgence in growth of the industry.

Economics

The production of aquatic animals for recreation, in nations where that type of aquaculture exists, is typically funded through user fees such as fishing licenses that support hatcheries and the personnel to run them. In order for most private aquaculture companies to get started, outside funding is required. Funding may come through banks and other commercial lending sources or from venture capitalists. The high risks associated with aquaculture have made it difficult for many firms to obtain bank loans, although that situation is changing as bankers become more knowledgeable and comfortable with underwriting aquaculture ventures.

A key factor in obtaining funding support for aquaculture is development of a sound business plan. The plan needs to demonstrate that the prospective culturist has identified all costs associated with establishment of the facility and its day-to-day operation. One or more suitable sites should have been identified and the species to be cultured selected before the business plan is submitted. Cost estimates should be verifiable. Having actual bids for a specific task at a specific location; eg, pond construction, well drilling, building construction, and vehicle costs helps strengthen the business plan.

Land costs vary enormously both between and within countries. Compare the cost of coastal land in south Florida where it might be possible to consider rearing shrimp with that of Mississippi farmland suitable for catfish farming. The former might be thousands of dollars for every meter of ocean front, while the latter may be obtained for one or two thousand dollars per hectare.

The amount of land required varies as well, not only as a function of the amount of production that is anticipated, but also on the type of culture system that is used. It may take several hectares of static culture ponds to produce the same biomass of animals as one modest size raceway through which large volumes of water are constantly flowed. Construction costs vary from one location to another. Local labor and fuel costs must be factored into the equation. The experience of contractors in building aquaculture facilities is another factor to be considered.

The need for redundancy in the culture system needs to be assessed. Failure of a well pump that brings up water to supply a static pond system may not be a serious problem in countries where new pumps can be purchased in a nearby town. However it can be disastrous in developing countries where new pumps and pump parts are often not available, but must be ordered from another country. Several weeks or months may pass before the situation can be remedied unless the culturist maintains a selection of spares.

The business plan needs to provide projections of annual production. Based on those estimates and assumed food conversion rates (food conversion is calculated by determining the amount of feed consumed by the animals for each kilogram of weight gain), an estimate of feed costs can be made. For many aquaculture ventures, between 40 and 50% of the variable costs involved in aquaculture can be attributed to feed.

Aquaculturists may elect to purchase animals for stocking or maintain their own broodstock and hatchery. The decision may rest on such factors as the availability and cost of fry fish, postlarval fish, oyster spat, or other early life history stages in the location selected for the aquaculture venture.

Land purchases and many of the costs associated with facility development can be accomplished with long-term loans of 15 to 30 years. Equipment such as pumps and trucks are usually depreciated over a few years and are funded with shorter-term loans. Operating expenses for such items as feed, chemicals, fuel, utilities, salaries, taxes, and insurance may require periodic short term loans to keep the business solvent. The projected income should be based on a realistic estimate of farmgate value of the product and an accurate assessment of anticipated production. Each business plan should project income and expenses projected over the term of all loans in order to demonstrate to the lending agency or venture capitalist that there is a high probability the investment will be repaid.

Regulation

The extent to which governments regulate aquaculture varies greatly from one nation to another. In some parts of the world, particularly in developing nations, there has historically been little or no regulation. Inexpensive land and labor, low taxes, excellent climates, and a lack of government interference have drawn many aquaculturists to underdeveloped countries, most of which are in the tropics. Unregulated expansion of aquaculture in some countries has led to pollution problems, destruction of valuable habitats such as mangrove swamps, and has enhanced the spread of disease from one farm to another. The need for imposing regulations is now becoming evident around the world. Response to that need varies considerably from one nation to another.

In developed countries there may or may not be a standardized set of national regulations. The United States is an example of a mixture of local, state, and federal regulations. Permits from a country, state, or federal agency may be required for drilling wells, pumping water, releasing water, use of exotic species, constructing facilities, etc. In the United States, most permits can be obtained at the local or state level. In some instances the federal government has deferred permitting authority to the states when state regulations are as rigorous or more so than national regulations. Federal agencies become involved when aquaculture projects are conducted in navigable waters (U.S. Army Corps of Engineers) or might impact threatened or endangered species (National Biological Service).

In general, it is easier to establish an aquaculture facility on private land than in public waters such as a lake or coastal embayment. Prospective aquaculturists who want to establish facilities in public waters may be confronted at public hearings by outraged citizens who do not want to see an aquaculture facility in what they consider to be their water. The issue is highly contentious in some nations (eg, the United States). In other countries, aquaculture in public waters is seen as not only a good use of natural resources, but can be considered an amenity (eg, Japan).

Obtaining permits is often not simple. Few states have one office that can accommodate the prospective aquaculturist. In most cases it is necessary to contact a number of state agencies to apply for permits. Public hearings may be required before permits are approved. The process can take months or even years to complete. The costs involved in going through the process may be prodigious. After the expenditure of considerable amounts of time and money, there is no guarantee that the permits will ultimately be granted.

Most states now have an aquaculture coordinator, usually housed in the state department of agriculture, who can assist prospective aquaculturists in finding a path through the permitting process. Anyone considering development of an aquaculture facility should become educated on the permitting process of the state or nation in which the facility will be developed. In cases where the process is involved, it should be initiated well in advance of the anticipated time of actual facility construction.

Species Under Cultivation

This article emphasizes aquatic animal production, but many hundreds of thousands of people are involved, worldwide, in aquatic plant production. The quantity of brown seaweeds, red seaweeds, green seaweeds, and other algae produced in 1992 was estimated at over 4.8 million metric tons (Table 1). Miscellaneous aquatic plants such as watercress and water chestnuts contributed an additional 600,000 metric tons. Microscopic algae and cyanobacteria are sometimes marketed as food or as a nutritional supplement (eg, *Spirulina* sp.). In addition, an undocumented quantity of algae (mostly of the single-celled variety) is produced for use as food for filter-feeding aquatic animals (primarily molluscs and zooplankton). Planktonic organisms such as rotifers are reared on algae

Table 1. World Aquaculture Production in 1992 for Selected Groups of Aquaculture Species[a]

Species group	Production, 10^3 t
carp and other cyprinids	6,652
tilapia and other cichlids	473
salmon and trout	628
flatfish	120
freshwater crustaceans	624
shrimp	884
oysters	954
mussels	109
scallops	549
clams, cockles, arkshells	765
brown seaweeds	3,640
red seaweeds	1,133
green seaweeds and other algae	17
miscellaneous aquatic plants	600
Total[b]	*19,311*

[a] Ref. 4.
[b] Includes species categories not listed.

and then used to feed the young of crustaceans and fishes that do not accept prepared feeds.

Animal aquaculture is concentrated on finfish, molluscs, and crustaceans. Sponges, echinoderms, tunicates, turtles, frogs, and alligators are being cultured, but production is insignificant in comparison with the three principal groups. Common and scientific names of many of the species of the finfish, molluscs, and crustaceans currently under culture are presented in Table 2. Included are examples of bait, recreational, and food animals.

Various species of carp and other members of the family *Cyprinidae* lead the world in terms of quantity of animals produced. In 1992 the total was over 6.6 million metric tons (Table 1). China is the leading carp producing nation, and is the world's leading aquaculture nation over all (Table 3). Significant amounts of carp are also produced in India and parts of Europe.

Fishes in the family *Salmonidae* (trout and salmon) are in high demand, with the interest in salmon being greatest in developed nations. Salmon, mostly Atlantic salmon, are produced in Canada, Chile, Norway, New Zealand, Scotland, and the United States. Fishes in the family *Cichlidae*, which includes several cultured species of tilapia, are reared primarily in the tropics, but have been widely introduced throughout both the developed and developing world.

Catfish are not a major contributor to aquaculture production globally, but the channel catfish industry dominates United States aquaculture. United States catfish production, primarily channel catfish, was 209,090 metric tons in 1992 (4).

Among the invertebrates, most of the world's production is associated with mussels, oysters, shrimp, scallops, and clams. Crawfish culture is of considerable importance in the United States, but amounted to only 24,211 metric tons in 1992 (4), insignificant compared to some other invertebrate species.

Small amounts of crabs, lobsters, and abalone are being cultured in various nations. All three bring good prices in the marketplace but have drawbacks associated with their culture. Crabs and lobsters are highly cannibalistic. Rearing them separately to keep them from consuming one another during molting has precluded economic culture in nearly every instance. Abalone eat seaweeds and can only be reared in conjunction with a concurrent seaweed culture facility or in regions where large supplies of suitable seaweeds are available from nature. In some instances the value of the seaweed for direct human consumption may make the highest and best use of the plants.

In all, there are perhaps 100 species of aquatic animals under culture. Many researchers have turned their attention to species for which there is demand by consumers, but for which the technology required for commercial production is not available. Examples are dolphin, also known as mahimahi (*Coryphaena hippurus*), Pacific halibut (*Hippoglossus stenolepis*), southern flounder (*Paralichthys lethostigma*), winter flounder (*Pseudopleuronectes americanus*), American lobster (*Homarus americanus*), and blue crab (*Callinectes sapidus*). Each of the species mentioned is marine and has small eggs and larvae. Providing the first feeding stages with acceptable food has been a common problem, as has the fragility of the early life stages of many species, and the problem of cannibalism.

Fishes with large eggs, such as trout, salmon, catfish, and tilapia were among the first to be economically successful in modern times. Small eggs do

Table 2. Examples of Common and Scientific Names of Selected Aquaculture Species

Type of organism	Common name	Scientific name
finfish	African catfish	*Clarias gariepinus*
	Atlantic halibut	*Hippoglossus hippoglossus*
	Atlantic salmon	*Salmo salar*
	Bighead carp	*Aristichthys nobilis*
	Bigmouth buffalo	*Ictiobus bubalus*
	Black crappie	*Pomoxis nigromaculatus*
	Blue catfish	*Ictalurus furcatus*
	Blue tilapia	*Tilapia aurea*
	Bluegill	*Lepomis macrochirus*
	Brook trout	*Salvelinus fontinalis*
	Brown trout	*Salmo trutta*
	Catla	*Catla catla*
	Channel catfish	*Ictalurus punctatus*
	Chinook salmon	*Oncorhynchus tshawytscha*
	Chum salmon	*Oncorhynchus keta*
	Coho salmon	*Oncorhynchus kisutch*
	Common carp	*Cyprinus carpio*
	Fathead minnow	*Pimephales promelus*
	Gilthead sea bream	*Sparus aurata*
	Goldfish	*Carassius auratus*
	Grass carp	*Ctenopharyngodon idella*
	Largemouth bass	*Micropterus salmoides*
	Milkfish	*Chanos chanos*
	Mossambique tilapia	*Tilapia mossambica*
	Mrigal	*Cirrhinus mrigala*
	Mud carp	*Cirrhina molitorella*
	Muskellunge	*Esox masquinongy*
	Nile tilapia	*Tilapia nilotica*
	Northern pike	*Esox lucius*
	Pacu	*Colossoma metrei*
	Pink salmon	*Oncorhynchus gorbuscha*
	Plaice	*Pleuronectes platessa*
	Rabbitfish	*Siganus* spp.
	Rainbow trout	*Oncorhynchus mykiss*
	Red drum	*Sciaenops ocellatus*
	Rohu	*Labeo rohita*
	Sea bass	*Dicentrarchus labrax*
	Shiners	*Notropis* spp.
	Silver carp	*Hypophthalmichthys molitrix*
	Smallmouth bass	*Micropterus dolomieui*
	Sole	*Solea solea*
	Steelhead	*Oncorhynchus mykiss*
	Striped bass	*Morone saxatilis*
	Walking catfish	*Clarias batrachus*
	Walleye	*Stizostedion vitreum vitreum*
	White crappie	*Pomoxis annularis*
	Yellow perch	*Perca flavescens*
	Yellowtail	*Seriola quinqueradiata*
molluscs	American oyster	*Crassostrea virginica*
	Bay scallop	*Aequipecten irradians*
	Blue mussel	*Mytilus edulis*

Table 2. (*Continued*)

Type of organism	Common name	Scientific name
	Northern quahog	*Mercenaria mercenaria*
	Pacific oyster	*Crassostrea gigas*
	Southern quahog	*Mercenaria campechiensis*
crustaceans	Freshwater shrimp	*Macrobrachium rosenbergii*
	Blue shrimp	*Penaeus stylirostris*
	Kuruma shrimp	*Penaeus japonicus*
	Pacific white shrimp	*Penaeus vannamei*
	Red swamp crawfish	*Procambarus clarkii*
	Tiger shrimp	*Penaeus monodon*
	White river crawfish	*Procambarus acutus acutus*
	White shrimp	*Penaeus setiferus*
algae (seaweeds)	California giant kelp	*Macrocystis pyrifera*
	Eucheuma	*Eucheuma cottoni*
	False Irish moss	*Gigartina stellata*
	Gracilaria	*Gracilaria* sp.
	Irish moss	*Chondrus crispus*
	Laminaria	*Laminaria* spp.
	Nori or laver	*Porphyra* spp.
	Wakame	*Undaria* spp.

Table 3. Top Aquaculture Producing Nations in 1992 and Amount of Production in Each[a]

Nation	Production[b], 10^3 t
China	10,409
Japan	1,397
India	1,374
Republic of Korea	955
Philippines	737
Indonesia	690
United States	413
Thailand	359

[a] Ref. 4.
[b] Includes aquatic plants.

not necessarily mean that sophisticated research is required to develop the technology required for successful culture. Carp, which have been cultured in China for millenia, have extremely small eggs. At the time the methodology for carp culture was developed, there were no research scientists, although there must have been dedicated farmers who used their common sense and trial-and-error methods to establish aquaculture.

Culture Systems

At one extreme aquaculture can be conducted with a small amount of intervention from humans and the employment of little technology. At the other is total environmental control and the use of computers, molecular genetics, and complex modern technology. Many aquaculturists operate between the extremes. The range of culture approaches can be described as running from extensive to intensive, or even hyper-intensive, with extensive systems being relatively simple and intensive systems being complex to very complex. In general, as the level of culture intensity increases, stocking density, and as a consequence, production per unit area of culture system or volume of water, increases.

The most extensive types of aquaculture involve minimal human intervention to promote increases in natural productivity. Good examples can be found relative to oyster and pond fish culture. With respect to oysters, one of the most extensive forms of culture involves placing oyster, clam, or other types of shell (cultch) on the bottom in intertidal areas that are known to have good oyster reproduction, but limited natural cultch material. The additional substrate may subsequently be colonized by oyster larvae (spat), thereby potentially increasing productivity. The next level of intensity might involve placing bags of cultch out in nature to collect spat in a productive area that already has sufficient quantities of natural cultch. After spat settlement, the bags of shells would be moved to an area where limited natural substrate availability has led to low productivity. The growout area may be held as a common resource, or it may be made commercially available on a leasehold basis.

The next step in increasing oyster culture intensity might involve hatchery production and settling of spat on cultch. Once again, the cultch would be later distributed over a bed leased or owned by the oyster culturist (Fig. 1). Control of predators such as starfish and oyster drills could easily be a part of culture at all levels.

The highest level of intensity with respect to oysters involves hatchery production of spat and the rearing of them suspended from rafts, long-lines, or as

Fig. 1. Bags of oyster shell used as cultch for the settling of oyster spat. The spat-laden shell is ultimately distributed on leased oyster beds.

cultchless oysters in trays. In the raft and longline techniques cultch material to which oyster spat or larval mussels are attached is strung on ropes (longlines) suspended from floating rafts or from other ropes held parallel to the water surface with buoys. The lines suspended from the ropes (called strings) of growing oysters are of such a length that the young shellfish are kept within the photic zone and not allowed to touch the bottom where starfish, oyster drills, and other predators can attack. Scallops, which do not attach to cultch, are sometimes grown in bags suspended from long lines or rafts. Similarly, young mussels can be held in proximity to strings with fine mesh materials that retains the animals in place until they attach (by means of what is called a byssus thread) to the string.

The stocking of ponds, lakes, and reservoirs to increase the production of desirable fishes that depend on natural productivity for their food supply and are ultimately captured by recreational fishermen or for subsistence is another example of extensive aquaculture. Some would consider such practices as lying outside of the realm of aquaculture, but since the practice involves human intervention and often employs fishes produced in hatcheries, recreational or subsistence level stocking is associated with, if not a part of aquaculture. Similarly, stocking new ponds or water bodies which have been drained or poisoned to eliminate undesirable species prior to restocking, can lead to increased production of desirable species.

Most of the aquaculture practiced around the world is conducted in ponds (Fig. 2). Ponds range in size but production units are generally 0.1 to 10 ha in area. The intensity of aquaculture in ponds can range from a few kg/ha to thousands of kg/ha of annual production.

Fertilization of ponds to increase productivity is the next level of intensity with respect to fish culture, followed by provision of supplemental feeds.

Fig. 2.   Aquaculture ponds are often rectangular in shape. They should be equipped with plumbing for both inflow and drainage of water.

Supplemental feeds are those that provide some additional nutrition but cannot be depended upon to supply all the required nutrients. Provision of complete feeds, those that do provide all of the nutrients required by the fish, translates to another increase in intensity. Associated with one or more of the stages described might be the application of techniques that lead to the maintenance of good water quality. Examples are continuous water exchange, mechanical aeration, and the use of various chemicals used to adjust such factors as pH, alkalinity, and hardness.

With the application of increased technology and control over the culture system, intensity continues to increase. Utilization of specific pathogen-free animals, provision of nutritionally complete feeds, careful monitoring and control of water quality, and the use of animals bred for good performance, can lead to impressive production levels. The United States channel catfish industry is a good example. During the early 1960s pond production levels were typically about 1500 kg/ha/yr. As better feeds and management practices were developed, production increased to an average of about 3000 kg/ha/yr by the next decade. In the 1980s, nutritionally complete feeds were perfected, better prediction and amelioration of water quality problems had been developed, and diseases were being better avoided or controlled. Catfish farmers typically produced 4000 kg/ha/yr and some, who used aeration and exchanged water during part of the growing season, were able to produce 10,000 kg/ha/yr or more.

Where water is plentiful and inexpensive, raceway culture is an attractive option and one which allows for production levels well in excess of what is possible in ponds. Trout are frequently reared in linear raceways from hatching to market size. Linear raceways are longer than they are wide, and are usually no deeper than 1–2 m (Fig. 3). High density raceways used in production facilities

Fig. 3. A commercial trout facility in Idaho, U.S. Linear raceways are commonly used for the production of trout from fry to either release or market and for salmon from fry to smolt size.

are commonly constructed of poured concrete. Small raceways of the type used in hatcheries and research facilities may be constructed of fiberglass or other resilient materials. Water is introduced at one end and flows by gravity through the raceway to exit the other end. Circular raceways, called tanks (Fig. 4), are also used by aquaculturists. Tanks are usually no more than 2 m deep and may be from less than 1 m to as much as 10 m in diameter. Concrete tanks can be found, but most are constructed of fiberglass, metal, or wood that is sealed and covered with epoxy or some other waterproof material. Plastic liners are commonly used in metal or wood tanks to prevent leakage, and in the case of metal, to avoid exposing the aquaculture animals to trace element toxicity.

Linear raceways are commonly used by trout and salmon culturists both for commercial production and for hatchery programs conducted by government agencies. Large numbers of state and federal salmon hatcheries in Washington and Oregon, along with governmental and private hatcheries in Alaska collect and fertilize eggs, hatch them, and rear the young fish to the smolt stage at which time they become physiologically adapted to enter seawater.

Commercial salmon culturists can rear their fish to market size in freshwater raceways although most salmon are grown from smolt to market size or adulthood in the marine environment, either as free roaming fish or in confinement. Since salmon instinctively return to spawn in the waters where they were hatched, it is possible to establish hatcheries and smolt-rearing facilities that take advantage of the homing instinct. The technique is known as ocean ranching. When the fish that had been released as smolts return to spawn, sufficient numbers of adults are collected for use as broodfish to continue the cycle. The remainder may be harvested by the aquaculturist or by commercial fishermen after which the fish are processed and marketed.

Fig. 4. Circular tanks are a raceway option. They can be placed outdoors or used in conjunction with indoor water systems. Circular tanks are commonly used in recirculating systems.

Salmon, steelhead trout, and a variety of marine fishes are currently being reared in net-pens (Fig. 5). The typical salmon net-pen is several meters on each side and may be as much as 10 m deep (1). Smaller units, called cages, are sometimes used by freshwater culturists. Cages tend to have volumes of no more than a few cubic meters.

Net-pen technology was developed in the 1960s, but has only been widely employed commercially for salmon production since the 1980s when the Norwegian salmon farming industry was developed. The Japanese began producing large numbers of sea bream and yellowtail in net-pens during the 1960s. Other nations have employed the technology as well. Most net-pens are located in protected waters since they are easily damaged or destroyed by storms.

Competition by various user groups for space in protected coastal waters in much of the world has led to strict controls and in some cases prohibitions against the establishment of inshore net-pen facilities. As a result, there is growing interest in developing the technology to move offshore. Various designs for offshore net-pens have been developed and a few have been tested (Fig. 6). A number of different designs, including systems that are semi- or totally submersible, have been able to withstand storm waves of at least 6 m, but the costs of those systems are very high compared with inshore net-pens, so commercial viability has yet to be demonstrated.

The highest levels of intensity that can be found in aquaculture systems are associated with totally closed systems, often called recirculating systems. In these systems, all water passing through the chambers in which the finfish or shellfish are held is continuously treated and reused. Once filled initially, closed systems can theoretically be operated for long periods of time without water replacement. It is necessary to add some water to such systems to make up for that lost to evaporation, splashout, and in conjunction with solids removal.

Fig. 5. Marine net-pens such as the ones shown here in Puget Sound, Washington U.S. are used for the rearing of salmon by commercial fish farmers.

Fig. 6. A salmon net-pen in Scotland designed for use offshore, and in this case, exposed coastal waters.

Many of the recirculating systems in use today are operated in a mode between entirely closed and completely open. In most there is a significant percentage of replacement water added either continuously or intermittently on a daily basis. Such partial recirculating systems may exchange from a few percent to several hundred percent of system volume each day.

The heart of a recirculating water system is the biofilter, a device that contains a medium on which bacteria that help purify the water become established (Fig. 7). Fish and aquatic invertebrates produce ammonia as a primary metabolite. If not removed or converted to a less toxic chemical, ammonia can quickly reach lethal levels. Two genera of bacteria are responsible for ammonia removal in biofilters. The first, *Nitrosomonas*, converts ammonia (NH_3) to nitrite (NO_2^-). The second, *Nitrobacter*, converts nitrite to nitrate (NO_3^-). Nitrite is highly toxic to aquatic animals, although nitrate can be allowed to accumulate to relatively high levels. If both genera of bacteria are active, the conversion from ammonia through nitrite to nitrate is so rapid that nitrite levels remain within the safe range.

Other than the biofilter and culture chambers, recirculating systems typically also employ one or more settling chambers or mechanical filters to remove solids such as unconsumed feed, feces, and mats of bacteria that slough from the biofilter into the water. Each recirculating system requires a mechanical means of moving water from component to component. That usually means mechanical pumping, though air-lifts can also be used.

Control of circulating bacteria and oxidation of organic matter can be obtained through ozonation of the water. Ozone (O_3) is highly toxic to aquatic organisms. Ozone must be allowed to dissipate prior to exposing the water to the aquaculture animals. With time, and with the assistance of aeration, ozone can be driven off or converted to molecular oxygen. Various commercial firms

Fig. 7. A bead filter, one of many types of biological filters, shown in association with a laboratory-scale recirculating water system. Small plastic beads inside the fiberglass chamber provide surface area for colonization by bacteria that convert ammonia to nitrate.

market ozone generators and can assist aquaculturists in selecting the proper equipment to meet system needs.

Ultraviolet (uv) light has also been used to sterilize the water in aquaculture systems. The effectiveness of uv decreases with the thickness of the water column being treated, so the water is usually flowed past uv lights as a thin film (alternatively, the water may flow through a tube a few cm in diameter that is surrounded by uv lights). Uv systems require more routine maintenance than ozone systems. Uv bulbs lose their power with time and need to be changed periodically. In addition, organic material exposed to uv light foul the surface of the transparent quartz (sometimes plastic) tubes past or through which the water flows, thereby causing a film between the water and the uv source that reduces the effectiveness of the uv light.

Recirculating systems often feature other types of apparatus, such as foam strippers and supplemental aeration. The technology for denitrifying nitrate to nitrogen gas has developed to the point that it may find a place in commercial culture systems in the near future. Computerized water-quality monitoring systems that will sound alarms and call emergency telephone numbers to report systems failures to the culturists are finding increased use among culturists using recirculating systems.

The technology involved makes recirculating systems expensive to construct and operate. Redundancy in the system, ie, providing backups for all critical components, and automation are important considerations. When a pump fails, for example, the failure must be instantly communicated to the culturist and the culturist must have the ability to keep the system operating while the problem is being addressed. Loss of a critical component for even a few minutes can result in the loss of all animals within the system.

Recirculating systems can make aquaculture feasible in locations where conditions would otherwise not be conducive to successful operations. Such systems can also be used to reduce transportation costs by making it possible to grow animals near markets. In areas where there are concerns about pollution or the use of exotic species, closed systems provide an alternative approach to more extensive types of operations.

Water Sources and Quality

Sources of water for aquaculture include municipal supplies, wells, springs, streams, lakes, reservoirs, estuaries, and the ocean. The water may be used directly from the source or it may be treated in some fashion prior to use (see WATER).

Many municipal water sources are chlorinated and contain sufficiently high levels of chlorine so as to be toxic to aquatic life. Chlorine can be removed by passing the water through activated charcoal filters or through the use of sodium thiosulfate metered into the incoming water. Municipal water is usually not used in aquaculture operations that utilize large quantities of water, either continuously or periodically, because of the initial high cost of the water and the cost of pretreatment to remove chlorine.

If polled, most aquaculturists would probably indicate a preference for well water. Both freshwater and saline wells are common sources of water for aquaculture. The most commonly used pretreatments of well water include temperature alteration (either heating or cooling); aeration to add oxygen or to remove or oxidize such substances as carbon dioxide, hydrogen sulfide, and iron and increasing salinity (in mariculture systems). Pretreatment may also include adjusting pH, hardness, and alkalinity through the application of appropriate chemicals.

To heat or cool water requires large amounts of energy. A major consideration in locating an aquaculture facility is to have not only a sufficient supply of water, but to have water at or near the optimum temperature for growing the species that has been selected. The vast supply of spring water of almost perfect temperature in the Hagerman Valley of Idaho supports the majority of the trout production in the United States. Where geothermal water is available, tropical species can be grown in locations where ambient winter temperatures would otherwise not allow them to survive.

Another large cost associated with incoming water is associated with its movement. Many aquaculture facilities that utilize surface waters and those that obtain their water from wells other than artesian wells are required to pump the water into their facilities. Pumping costs can be a major expense, particularly when the facility requires continuous inflow.

Surface water can sometimes be obtained through gravity flow by locating aquaculture facilities at elevations below those of adjacent springs, streams, lakes, or reservoirs. Coastal facilities may be able to obtain water through tidal flow.

The most common treatment of incoming surface water is removal of particulate matter. This can be effected through the use of settling basins or filtration. Particle removal may involve the reduction or elimination of suspended inorganic material such as clay, silt, and sand. It may also involve removal of organic material, including living organisms. Organisms that enter aquaculture facilities if not filtered from the incoming water include phytoplankton and zooplankton, plants and plant parts, macroinvertebrates, and fishes. Some of the organisms, if not removed, can survive and grow to become predators on, or competitors with, the target aquaculture species. Very small organisms, such as bacteria, can be removed mechanically. However, other forms of sterilization, such as ozonation and the use of uv radiation, are more efficient and effective.

For many freshwater species that can be characterized as warmwater (such as channel catfish and tilapia) or coldwater (such as trout), the conditions outlined in Table 4 should provide an acceptable environment. So-called midrange species are those with an optimum temperature for growth of about 25°C (examples are walleye, northern pike, muskellunge, and yellow perch). Typically they do well under the conditions, other than temperature, specified in Table 4 for coldwater species. Some species have higher or lower tolerances than others. For example, tilapia can tolerate temperatures in excess of 34°C, but have poor tolerance for low temperature. Most tilapia species die when the temperature

Table 4. General Water Quality Requirements for Trout and Warmwater Aquatic Animals in Fresh Water[a]

Variable	Acceptable level or range	
	Cold water	Warm water
temperature, °C	<20	26–30
alkalinity, mg/L	10–400	50–400
dissolved oxygen, mg/L	>5	≥5
hardness, mg/L	10–400	50–400
pH	6.5–8.5	6.5–8.5
total ammonia, mg/L	<0.1	<1.0
ferrous iron, mg/L	0	0
ferric iron, mg/L	0.5	0–0.5
carbon dioxide, mg/L	0–10	0–15
hydrogen sulfide, mg/L	0	0
cadmium, μg/L	<10	<10
chromium, μg/L	<100	<100
copper, μg/L	<25	<25
lead, μg/L	<100	<100
mercury, μg/L	<0.1	<0.1
zinc, μg/L	<100	<100

[a]Refs. 1, 6, 7.

falls below about 12°C. Tilapia have a remarkably high tolerance for ammonia compared with such species as trout and salmon, which have a high tolerance for cold water, but cannot tolerate water temperatures much above 20°C. Marine fish may be able to tolerate a wide range of salinity (such euryhaline species include flounders, red drum, salmon, and some species of shrimp), or they may have a narrow tolerance range (they are called stenohaline species, examples of which are dolphin, halibut, and lobsters). Recommended water quality conditions for marine fish production systems are presented in Table 3.

The water quality criteria for each species should be determined from the literature or through experimentation when literature information is unavailable. Synergistic effects that occur among water quality variables can have an influence on the tolerance a species has under any given set of circumstances. Ammonia is a good example. Ionized ammonia (NH_4^+) is not particularly lethal to aquatic animals, but unionized ammonia (NH_3) can be toxic even when present at a fraction of a part per million (depending on species). The percentage of unionized ammonia in the water at any given total ammonia concentration changes in relation to such factors as temperature and pH. As either temperature or pH increase, so does the percentage of unionized ammonia relative to the level of total ammonia.

Another example is dissolved oxygen (DO). The amount of DO water can hold at saturation is affected by both temperature and salinity. The warmer and/or saline the water, the lower the saturation DO level. Oxygen saturation is also affected by atmospheric pressure. The saturation oxygen level decreases as elevation increases.

Biocides should not be present in water used for aquaculture. Sources of herbicides and pesticides are runoff from agricultural land, contamination of the water table, and spray drift from crop-dusting activity. Excessive levels of phosphorus and nitrogen may occur where runoff from fertilized land enters an aquaculture facility either from surface runoff or groundwater contamination. Trace metal levels should be low as indicated in Tables 4 and 5.

Most aquaculture facilities release water constantly or periodically into the environment without passing it through a municipal sewage treatment plant. The effects of those effluents on natural systems have become a subject of intense scrutiny in recent years and have, in some instances, resulted in opposition to further development of aquaculture facilities in some locales. There have even been demands that some existing operations should be shut down.

Regulation of aquaculture varies greatly both between and within nations. Some governmental agencies with jurisdiction over aquaculture have placed severe restrictions on the levels of such nutrients as phosphorus and nitrogen that can be released into receiving waters. Regulations on suspended solids levels in effluent water are also common. The installation of settling ponds, exposure of the water to filter feeding animals that will remove solids, and mechanical filtration have been used to treat effluents. Reduction or removal of dissolved nutrients through tertiary treatment is possible, but is generally not economically feasible at present. Research is currently underway to develop feeds containing reduced levels of nutrients or to provide nutrients in forms that can better be utilized by the culture animals. The goal in both approaches is to reduce losses of nutrients to the environment through excretion.

Table 5. Suggested Water Quality Conditions for Marine Fish Production Facilities[a]

Variable	Acceptable level or range
temperature, °C	1–40 (depends on species)
salinity, g/kg	1–40 (depends on species)
dissolved oxygen, mg/L	>6
pH	<7.9–8.2
total ammonia, μg/L as NH_3	<10
iron, μg/L	100
carbon dioxide, mg/L	<10
hydrogen sulfide, μg/L	<1
cadmium, μg/L	<3
chromium, μg/L	<25
copper, μg/L	<3
mercury, μg/L	<0.1
nickel, μg/L	<5
lead, μg/L	<4
zinc, μg/L	<25

[a]Ref. 8.

Nutrition and Feeding

Problems associated with excessive levels of nutrients and unwanted nuisance species have already been mentioned. There are cases in which intentional fertilization is used by aquaculturists in order to produce desirable types of natural food for the species under culture. Examples of this approach include inorganic fertilizer applications in ponds to promote phytoplankton and zooplankton blooms that provide food for young fish such as channel catfish, the development of algal mats through fertilization of milkfish ponds, and the use of organic fertilizers (from livestock and human excrement) in Chinese carp ponds to encourage the growth of phytoplankton, macrophytes, and benthic invertebrates. In the latter instance, various species of carp with different food habits are stocked to ensure that all of the types of natural foods produced as a result of fertilization are consumed.

Provision of live foods is currently necessary for the early stages of many aquaculture species because acceptable prepared feeds have yet to be developed. Algae is routinely cultured for the early stages of molluscs produced in hatcheries. Once the molluscs are placed in growout areas, natural productivity is depended upon to provide the algae upon which the shellfish feed.

In cases where zooplankton are reared as a food for predatory larvae or fry, it may be necessary to maintain three cultures. Though wild zooplankton have been used successfully in some instances (eg, in Norway wild zooplankton have been collected and fed to larval Pacific halibut), the normal process involves culturing algae to feed to zooplankton that are fed to a young shrimp or fish.

The most popular live foods for first feeding animals such as shrimp and marine fishes that have small eggs and larvae are rotifers and brine shrimp nauplii. After periods ranging from several days to several weeks, depending on the species being reared, the aquaculture animals will become sufficiently large to accept pelleted feeds and can be weaned onto prepared diets. Problems associated with utilizing prepared feeds from first feeding include difficulty in providing very small particles that contain all the required nutrients, loss of soluble nutrients into the water from small particles before the animals consume the feed, and in some cases, the fact that prepared feeds do not behave the same as live foods when placed in the water. For species that are sight feeders, behavior of the food is an important factor.

Some of the most popular aquaculture species accept prepared feeds from first feeding. Included are catfish, tilapia, salmon, and trout. All of the fishes listed have relatively large eggs (several mm diameter) that develop into fry that have large yolk sacs. The nutrients in the yolk sac lead to production of first-feeding fry with well-developed digestive tracts that produce the enzymes required to efficiently digest diets that contain the same types of ingredients used for larger animals.

Fish nutritionists have, over the last few decades, successfully determined the nutritional requirements of many aquaculture species and have developed practical feed formulations based on those requirements. For species such as Atlantic salmon, various species of Pacific salmon and trout, common carp, channel catfish and tilapia sufficient information exists to design diets precisely suited to each species. There is always interest among aquaculturists to develop new species. In each instance, the nutritional requirements of the new species must be investigated. Although there are many similarities among aquatic animals, diets that produce the best growth at the least cost vary significantly and can only be formulated when the nutritional requirements are known. Determination of those requirements may require several years of research, although suitable diets based on existing formulations can be employed while the research is being conducted.

Requirements for energy, protein, carbohydrates, lipids, vitamins and minerals have been determined for the species commonly cultured (9). As a rule of thumb, trout and salmon diets will, if consumed, support growth and survival in virtually any aquaculture species. Such diets often serve as the control against which experimental diets are compared.

Since feeds contain other substances than those required by the animals of interest, studies have also been conducted on antinutritional factors in feedstuffs and on the use of additives. Certain feed ingredients contain chemicals that retard growth or may actually be toxic. Examples are gossypol in cottonseed meal and trypsin inhibitor in soybean meal. Restriction on the amount of the feedstuffs used is one way to avoid problems. In some cases, as is true of trypsin inhibitor, proper processing can destroy the antinutritional factor. In this case, heating of soybean meal is effective.

Animals that do not readily accept pelleted feeds may be enticed to do so if the feed carries an odor that induces ingestion. Color development is an important consideration in aquarium species and some animals produced for human food. External coloration is desired in aquarium species. Pink flesh in cultured salmon is desired by much of the consuming public. Coloration, whether external or of the flesh, can be achieved by incorporating ingredients that contain pigments or by adding extracts or synthetic compounds. One class of additives that impart color is the carotenoids.

Prepared feeds are marketed in various forms from very fine particles through crumbles, flakes, and pellets. Pelleted rations may be hard, semimoist, or moist. Hard pellets typically contain less than 10% water and can be stored under cool, dry conditions for at least 90 days without deterioration of quality. Semimoist pellets are chemically stabilized to protect them from degradation and mold if they are properly stored, while moist pellets must be frozen if they are not used immediately after manufacture. Moist feeds are produced in machines similar to sausage grinders.

Hard pellets are the type preferred if the species under culture will accept them. Semimoist feeds are most commonly used in conjunction with feeding young fishes and species that find hard pellets unpalatable. Moist feeds, which contain high percentages of fresh fish, are usually available only in the vicinity of fish-processing plants.

The most widely used types of prepared feeds are produced by pressure pelleting or extrusion. Pressure pelleting involves pushing the ground and mixed feed ingredients through holes in a die that is a few centimeters thick to produce spaghettilike strands of the desired diameter. The strands are cut to length as they exit the die. Steam is often injected into the pellet mill in a location that exposes the feed mixture to moist heat just before the mix enters the die. Exposure to steam improves binding and extends pellet water stability.

Extruded pellets are produced by exposing the ground and mixed ingredients to much higher heat and pressure and for a longer time than is the case with pressure pellets. In the extrusion process the ingredients undergo some cooking that can be beneficial in reducing the levels of certain antinutritional factors, such as trypsin inhibitor. There may be concomitant losses of heat labile nutrients such as vitamin C, so overfortification to obtain the desired level in the final product may be required.

Crumbles are formed by grinding pellets to the desired sizes. Specialty feeds such as flakes can be made by running newly manufactured pellets through a press or through use of a double drum dryer. The latter type of flakes begin as a slurry of feed ingredients and water. When the slurry is pressed between the hot rollers of the double drum dryer, wafer thin sheets of dry feed are produced that are then broken into small pieces. The different colors observed in some tropical fish foods represent a mixture of flakes, each of which contains one or more different additives that impart color.

Pressure pellets sink when placed in water, whereas under the proper conditions, floating pellets can be produced through the extrusion process. That is accomplished when the feed mixture contains high levels of starch that expands and traps air as the cooked pellets leave the barrel of the extruder. This gives the pellets a density of less than 1.0. Floating pellets are desirable for species that come to the surface to feed since the aquaculturist can visually determine that the fish are actively feeding and can control daily feeding rates based on observed consumption.

Sinking extruded pellets are used for shrimp and other species that will not surface to obtain food. Shrimp consume very small particles, so they will nibble pieces from a pellet over an extended period of time. For that reason, both pressure and extruded pellets need to have high water stability. Extruded feeds, whether sinking or floating, may remain intact for up to 24 hours after being placed in the water. Pressure pellets begin to disintegrate after a few minutes, unless supplemental binders are incorporated into the feed mixture. As previously indicated, the use of steam in conjunction with pressure pelleting also enhances pellet stability.

Nearly all aquaculture feeds contain at least some animal protein since the amino acid levels in plant proteins cannot meet the requirements of most aquatic animals. Fish meal is the most commonly used source of animal protein in aquaculture feeds, though blood meal, poultry by-product meal, and meat and bone meal have also been successfully used. Commonly used plant proteins include corn meal, cottonseed meal, peanut meal, rice, soybean meal, and wheat. A number of other ingredients have also been used, many of which are only locally available. Most formulations contain a few percent of added fat from such sources as fish oil, tallow, or more commonly, oilseed oils such as corn oil and soybean oil. Complete rations contain added vitamins and minerals. Purified amino acids, binders, carotenoids, and antioxidants are other components found in many feeds. Growth hormone and antibiotics are sometimes used. Regulations on the incorporation of hormones along with other chemicals and drugs into aquatic animal feeds are in place in the United States and some other countries (Table 6). Few such regulations have been promulgated in developing nations.

Feeding practices vary from species to species. It is important not to over-feed since waste feed not only means wasted money, it can also lead to degradation of water quality. Most species require only three to four percent of body weight in dry feed daily for optimum growth. Very young animals are an exception. They are fed at a higher rate because they are growing rapidly and consume a greater percentage of body weight daily than older animals. It is important to have food readily available to them. Food should be spread evenly over the culture chamber area so the young animals do not have to expend a great deal of energy searching for a meal. Feeding rates as high as 50% of body weight daily are not uncommon for young animals. Since total biomass is small, even in intensively stocked units such as raceways, the economic cost is not high. Water quality in raceways can be maintained by siphoning out waste feed periodically. In ponds, any unconsumed feed acts as fertilizer and the quantities used are not high enough to affect water quality adversely.

Young animals may be fed several times daily. Examples include the standard practices of feeding fry channel catfish every three hours and young northern pike as frequently as every few minutes. Keeping carnivorous species such as northern pike satiated helps reduce the incidence of cannibalism.

Reproduction and Genetics

Species such as carp, salmon, trout, channel catfish, and tilapia have been bred for many generations in captivity though they usually differ little in appearance or genetically from their wild counterparts. A few exceptions exist, such as the leather carp, a common carp strain selectively bred to produce only one row of scales, and the Donaldson trout, a strain of rainbow trout developed over numerous generations to grow more rapidly to larger size and with a stouter body than its wild cousins.

Selective breeding has been long-practiced as a mean of improving aquaculture stocks. In some instances it has not been possible or it is quite difficult and expensive to produce broodstock and spawn them in captivity, so culturists continue to rear animals obtained from nature. Most of the species that are being reared in significant quantities around the world are produced in hatcheries using either captured or cultured broodstock. Milkfish is a notable exception. The species has been spawned in captivity, but most of the fish reared in confinement are collected as juveniles in seines and sold to fish culturists. Wild shrimp postlarvae continue to be used to stock ponds in some parts of the world though hatcheries may also be available in the event sufficient numbers of wild postlarvae are unavailable in a given year.

Spawning techniques vary widely from one species to another. Tilapia and catfish are typically allowed to spawn in ponds. Fertilized eggs can be collected from the mouths of female tilapia, but it is common practice to collect schools of fry after they are released from the mother's mouth to forage on their own. Catfish lay eggs in adhesive masses. Spawning chambers such as milk cans and grease cans are placed in ponds and may be examined every few days for the presence of egg masses. Some catfish farmers allow the eggs to hatch in the pond, though most collect eggs and incubate them in a hatchery. Adult Pacific salmon die after spawning. Females are usually sacrificed by cutting open the abdomen to release the eggs. Milt is obtained by squeezing the belly of males. Trout and Atlantic salmon can be reconditioned to spawn annually. Eggs are usually obtained from those species in the same fashion as from male Pacific salmon (1).

Unlike catfish, tilapia, trout and salmon, that produce several hundred to several thousand eggs per female, many marine species produce large numbers of very small eggs. Hundreds of thousands to millions of eggs are produced by such species as halibut, flounders, red drum, striped bass, and shrimp. Catfish, salmon, and trout spawn once a year, while tilapia and some marine species spawn repeatedly if the proper environmental conditions are maintained (1). Red drum, for example, spawn every few days for periods of several months when light and temperature and properly controlled (10).

Fish breeders have worked with varying degrees of success to improve growth and disease resistance in a number of species. As genetic engineering techniques are adapted to aquatic animals, dramatic and rapid changes in the genetic makeup of aquaculture species may be possible. However, since it is virtually impossible to prevent escapement of aquacultured animals into the natural environment, potential negative impacts of such organisms on wild populations cannot be ignored.

For some species, one sex may grow more rapidly than the other. A prime example is tilapia, which mature at an early age (often within six months of hatching). At maturity, submarketable females divert large amounts of food energy to egg production. Also, since they are mouth brooders (holding the eggs and fry within their mouths for about two weeks) and repeat spawners (spawning about once a month if the water temperature is suitable), the females grow very slowly once they mature. All-male, or predominantly male, populations of tilapia can be produced by feeding androgens to fry, which are undifferentiated sexually. Various forms of testosterone have been used effectively in sex reversing tilapia and other fishes.

In species such as flatfishes, females may grow more rapidly than males and ultimately reach much larger sizes. For them, producing all-female populations for growout might be beneficial.

Diseases and Their Control

Aquatic animals are susceptible to a variety of diseases including those caused by viruses, bacteria, fungi, and parasites. A range of chemicals and vaccines has been developed for treating the known diseases, although some conditions have resisted all control attempts to date and severe restrictions on the use of therapeutants in some nations has impaired that ability of aquaculturists to control disease outbreaks. The United States is a good example of a nation in which the variety of treatment chemicals is limited (Table 6).

Maintenance of conditions in the culture environment that keep stress to a minimum is one of the best methods of avoiding diseases. Vaccines have been developed against several diseases and more are under development. Selective breeding of animals with disease resistance has met with only limited success. Good sanitation and disinfection of contaminated facilities are important avoidance and control measure. Some disinfectants are listed in Table 6. Pond soils can be sterilized with burnt lime (CaO), hydrated lime [Ca(OH)$_2$], or chlorine compounds (12).

When treatment chemicals have to be employed, they may be incorporated in the food, used in dips, flushes and baths, or allowed to remain in the water for extended periods. Since one of the first responses of aquatic animals to disease is reduction or cessation of feeding, treatments with medicated feeds must be initiated as soon as development of an outbreak is suspected. Antibiotics, such as terramycin, can be dissolved in the water, but may be less effective than when given orally.

Table 6. Therapeutants and Disinfecting Agents Approved for Use in United States Aquaculture[a]

Name of compound	Use of compound
Therapeutants	
copper	antibacterial for shrimp
formalin	parasiticide for various species
furanace (Nifurpyrinol)	antibiotic for aquarium fishes
oxytetracycline (Terramycin)	antibiotic for fishes and lobsters
sodium chloride	osmoregulatory enhancer for fishes
sulfadimethoxine (Romet)	antibacterial for salmonids and catfish
trichlorofon (Masoten)	parasiticide for baitfish and goldfish
Disinfectants	
calcium hypochlorite (HTH)	used in raceways and on equipment
didecyl dimethyl ammonium chloride (Sanaqua)	used in aquaria and fish-holding equipment
povidone–iodine compounds (Argentyne, Betadine, Wescodyne)	disinfection of fish eggs

[a]Ref. 11.

Vaccines can be administered through injections, orally, or by immersion. Injection is the most effective means of vaccinating aquatic animals but it is stressful, time-consuming, and expensive. The time and expense may be acceptable for use in conjunction with broodfish and other valuable animals. Oral administration of vaccines may be ineffective as many vaccines are deactivated in the digestive tract of the animals the vaccines are intended to protect. Dip treatment by which the vaccines enter the animals through diffusion from the water are not generally as effective as injection but can be used to vaccinate large numbers of animals in short periods of time.

Harvesting, Processing, and Marketing

Harvesting techniques vary depending on the type of culture system involved. Seines are often used to capture fish from ponds, or the majority of the animals can be collected by draining the pond through netting. Fish pumps are available that can physically remove aquatic animals directly onto hauling trucks from ponds, raceways, cages and net-pens without causing damage to the animals.

Aquaculturists may harvest, and even process their own crops, although custom harvesting and hauling companies are often available in areas where the aquaculture industry is sufficiently developed to support them. Some processing plants also provide harvesting and live-hauling services.

Some species, with channel catfish being a good example, can develop off-flavors. A characteristic off-flavor in catfish is often described as an earthy-musty, or muddy flavor. The problem is associated with the chemical geosmin and related compounds that are produced by certain types of algae (1). Processors often require that a sample fish from each pond scheduled for harvest be brought to the plant about two weeks prior to harvest for a taste test. Subsequent samples are taken to the processor three days before and during the day of processing. A portion of the sample fish is cooked and tasted. If off-flavor is detected, the fish will be rejected. Once the source of geosmin is no longer present, the fish will metabolize the compound. The process may involve moving the fish into clean well water or by merely waiting until the algae bloom dissipates, after which the geosmin will be rapidly metabolized. Within a few days the fish can be retested and if no off-flavor is detected, they can be harvested and processed.

Centralized processing plants specifically designed to handle regional aquaculture crops are established in areas where production is sufficiently high. In coastal regions, aquacultured animals are often processed in plants that also service capture fisheries.

Marketing can be done by aquaculturists who operate their own processing facilities. Most aquaculture operations depend on a regional processing plant to market the final product. In all cases aquaculturists should remember that their job is not complete until the product reaches the consumer in prime condition.

BIBLIOGRAPHY

"Aquaculture" in *ECT* 3rd ed., Vol. 3, pp. 194–213, by Howard P. Clemens and Michael Conway, University of Oklahoma.

1. R. R. Stickney, *Principles of Aquaculture*, John Wiley & Sons, Inc., New York, 1994.
2. FAO, *Agriculture Production Statistics, 1974–1993*, Food and Agriculture Organization of the United Nations, 1995.
3. FAO, *FAO Fisheries Circular 815, revision 2*, Food and Agriculture Organization of the United Nations, 1991.
4. *Aquaculture Buyer's Guide '95 and Industry Directory*, Vol. 8, 1995 Aquaculture Magazine, Asheville, NC.
5. R. R. Stickney, *Aquaculture in the United States: A Historical Review*, John Wiley & Sons, Inc., New York, 1996.
6. R. G. Piper, I. B. McElwain, L. E. Orme, J. P. McCraren, L. G. Fowler, and J. R. Leonard, *Fish Hatchery Management*, U.S. Fish and Wildlife Service, Washington, D.C., 1982.
7. C. E. Boyd, *Water Quality in Ponds for Aquaculture*, Alabama Agricultural Experiment State, Auburn University, 1990.
8. J. Huegenin and J. Colt, *Design and Operating Guide for Aquaculture Seawater Systems*, Elsevier, New York, 1989.
9. National Research Council, *Nutrient Requirements of Fish*, National Academy Press, Washington, D.C., 1993.
10. A. Henderson-Arzapalo, *Rev. Aquat. Sci.* **6**, 479 (1992).
11. F. P. Meyer and R. A. Schnick, *Rev. Aquatic Sci.* **1**, 693 (1989); a review of chemicals used for the control of fish diseases.
12. C. E. Boyd, *Bottom Soils, Sediment, and Pond Aquaculture*, Chapman and Hall, New York, 1995.

ROBERT STICKNEY
Texas A&M University

BIOREMEDIATION

Bioremediation is the process of judiciously exploiting biological processes to minimize an unwanted environmental impact; usually it is the removal of a contaminant from the biosphere. Like most definitions in biology, that of bioremediation is the subject of some debate. A narrow definition might focus on the conversion of contaminating organic molecules to carbon dioxide, water and inorganic ions, and the oxidation or reduction of contaminating inorganic ions. A broader definition would include biological processes for ameliorating extremes of pH, concentrating contaminants so that they can be more easily removed by physical techniques, converting toxic species to less toxic or less bioavailable forms that pose less of a threat to the environment, and restoring functional ecosystems to contaminated or disturbed sites when the contaminants or disturbance cannot be removed.

The concept of "judiciously exploiting biological approaches" is also a subject of debate:

Some would restrict it to providing a nutrient that is otherwise limiting the most effective growth of organisms catalyzing the desired reaction, whether it is the degradation of an organic compound, the reduction or oxidation of an inorganic ion, or the accumulation of a contaminant. This simple approach has been successful with a range of contaminants. The nutrient might be a fertilizer providing nitrogen, phosphorus and other essential minerals, or an electron acceptor such as oxygen.

Others would extend the fertilizer concept to the simultaneous addition of readily biodegradable substrates along with the fertilizer nutrients to stimulate the growth of contaminant-degrading organisms most rapidly, and to aid in the rapid utilization of the fertilizer nutrients before they might be leached from the contaminated area. The specific requirements for the most efficacious substrates is an area of current research.

An alternative use of added readily degradable substrates is to drive the local environment towards anaerobiosis so that reactions such as reductive dechlorinations or reductive removal of nitro-groups are promoted.

A broader view would include the addition of a substrate to stimulate the growth of organisms known to degrade the contaminant of interest only as an incidental part of their metabolism, one might almost say serendipitously. This process is sometimes called co-metabolism, and it too has had success.

Others would include the addition of materials aimed at increasing the bioavailability of the contaminant to the degrading organisms. The most studied compounds are surfactants, but cations have been reported to increase the bioavailability of some organic compounds, and sorbents and clays are also considered. The dispersion of spilled oil on water by the application of dispersants is perhaps the major commercial use of this idea.

Another important option is the addition of remediating organisms. While the addition of contaminant-degrading bacteria has not yet had much documented success with natural products such as hydrocarbons, there is reason to expect that it will be efficacious with pollutants that are more recent additions to the environment. The planting of specific plants, in the process known as phytoremediation, is also a promising approach. Not only do the plants themselves have remediating activities, such as the accumulation of certain metal ions, but they also have extensive bacterial and fungal populations associated with their root systems, and inoculation of this rhizosphere is widely practiced. It is, thus, possible that planting seeds with microbial inoculants will become an option for bioremediation. There is much talk of genetically modifying organisms so that their remediative potential is increased, and in the future this may well become an important option.

The broadest view of "aiding and abetting" includes doing nothing, but merely watching natural processes occur without further intervention. This has been termed "Intrinsic Bioremediation", and it too has met with success. From an environmental point of view, although unfortunately not always from a regulatory viewpoint, it is important that any remediation intervention yield a clear net environmental benefit. Sometimes very mild stimulation of intrinsic processes may be the most environmentally responsible option.

Bioremediation overlaps some older biotechnologies. Municipal and industrial wastewater treatment is a well-established industry, and although it can be distinguished from bioremediation in that the pollutants are under physical control during treatment, the fundamental biological processes have much in common. Similarly composting is a well-established phenomenon, currently gaining popularity in the municipal solid waste treatment industry, and the biofiltration of waste gases is becoming a useful technology. Developments of these technologies, where the contaminant is already under physical control, will undoubtedly aid the development of bioremediation as an accepted tool for dealing with similar wastes when they have escaped control. This article focuses on biological treatments for contaminants when they have escaped into the environment.

Bioremediation is already a commercially viable technology, with estimates of aggregate bioremediation revenues of $2–3 billion for the period 1994–2000 (1). There are significant opportunities to enlarge upon this success. Bioremediation has applications in the gas phase, in water, and in soils and sediments. For water and soils, the process can be carried out *in situ*, or after the contaminated medium has been moved to some sort of contained reactor (*ex situ*). The former is generally rather cheaper, but the latter may result in such a significant increase in rate that the additional cost of manipulating the contaminated material is overshadowed by the time saved. Bioremediation may explicitly exploit bacteria, fungi, algae, or higher plants. Each, in turn, may be part of a complex food-web, and optimizing the local ecosystem may be as important as focusing solely on the primary degraders or accumulators.

Bioremediation usually competes with alternative approaches to achieving an environmental goal. Bioremediation is typically among the least expensive options, but an additional important consideration is that in many cases bioremediation is a permanent solution to the contamination problem, since the contaminant is completely destroyed or collected. Some of the alternatives technologies, such as thermal desorption and destruction of organics, are also permanent solutions, but the simplest, removing the contaminant to a dump site, merely moves the problem, and may well not eliminate the potential liability. Furthermore, by its very nature bioremediation addresses the bioavailable part of any contamination, and when biodegradation or bioaccumulation ceases this probably means that the bioavailable part of the contamination has been addressed. Residual concentrations of contaminants, although perhaps detectable by today's sensitive analytical techniques, may in fact have no residual environmental impact. The same cannot necessarily be said for nonbiological technologies, which may leave bioavailable contaminants at low levels.

Bioremediation also has the advantage that is can be relatively nonintrusive, and can sometimes be used in situations where other approaches would be severely disruptive. For example, bioremediation has been used to clean up hydrocarbon spills under buildings, roads, and airport runways without interfering with the continued use of these facilities.

On the other hand, bioremediation is usually slower than most physical techniques, and may not always be able to meet some very strict clean-up standards. Nevertheless, it is becoming a widely used technology. This article addresses bioremediation in its broadest sense, focusing on the contaminants that can be treated, the underlying biological processes that can mitigate the contamination, and the technologies that have been used, or are being developed, to treat them.

General Biological Aspects

The biosphere plays an important role in the great elemental cycles of Earth (2), and bioremediation must be placed in this context if it is to be appreciated in its broadest ramifications. One of the underlying fundamental truths of biological diversity is that if there is free energy available in the metabolism of a substrate, there is probably a guild of organisms that has evolved to make use of it. This is particularly germane to the biodegradation of organic molecules. For example, crude oil seeps to both land and water have occurred for millennia, and as a consequence, aerobic oil-degrading microorganisms are ubiquitous. If biology does not yet take advantage of a source of free energy, then it can be expected that there will be a strong selection pressure in favor of any organism that develops an ability to exploit it. This has been seen with by-products of nylon manufacture, where a *Pseudomonas aeruginosa* has gained the ability to degrade the novel compound 6-aminohexanoate linear dimer, a by-product of nylon-6 manufacture, as the sole source of carbon and nitrogen (3). The successful bioremediation of xenobiotic compounds, such as pesticides and herbicides, may well represent a similar acquisition of traits.

Not all organic molecules provide a source of free energy, however. Some, such as small halogenated solvents, provide no significant source of nutrients or energy, and their aerobic destruction can only occur co-metabolically with the degradation of a more nutritious substrate (4). The white-rot fungi provide another variation on this theme. These organisms seem unique in their ability to degrade lignin, the structural polymer of higher plants. They may not gain any direct energetic benefit from lignin degradation, but it clearly allows access to cellulose which is a substrate for growth. Lignin degradation is catalyzed by a group of extracellular peroxidases that generate nonspecific oxidants, and there have been several proposals to use these systems for destroying contaminating organic compounds (5).

With successful bioremediation, organic compounds can eventually be converted to carbon dioxide, water, and biomass. Similarly, nitrogenous molecules, such as excess ammonia or nitrate in groundwater, can be mineralized to gaseous nitrogen. Alternatively they can stimulate the growth of plants, either terrestrial or marine, and the plant biomass can eventually be harvested so that the nitrogen is effectively removed from the local environment. Other nonorganic contaminants provide a different challenge for bioremediation. A few, such as mercury and selenium, are volatilized by some biological processes, but it is not clear that this is always beneficial. In some cases, such as chromium and arsenic, there is a dramatic difference in environmental toxicity depending on the redox state of the contaminant. Bioremediation has sometimes focused on this detoxification, usually by bacterial processes. A more satisfying approach would be to use a biological process to accumulate and concentrate the contaminant so that it can be removed for safe disposal. Fungi, algae, and higher plants have all been used in these efforts.

Table 1 explains a few of the biological terms that are widely used in discussing bioremediation, and which are used in the following text.

General Technological Aspects

Successful bioremediation hinges upon the effective application of the biology discussed above. Sometimes the contaminant is on the surface, so access to it is reasonably simple. Indeed the required technology may be as simple as broadcast spreaders or sprayers to apply fertilizers, or tilling the soil to allow good aeration. Of course this is not necessarily as simple as it sounds, since contaminated sites

Table 1. Some Biological Definitions Relevant to Bioremediation

Term	Explanation
aerobic	conditions with free oxygen
anaerobic	conditions scrupulously free of oxygen
anoxic	conditions with very low levels of oxygen
autotrophic	growth using atmospheric CO_2 as sole source of carbon
co-metabolic degradation	biodegradation of a contaminant only fortuitously with degradation of a true substrate
denitrification	the reduction of nitrate to gaseous nitrogen
Eukaryotes	organisms with a membrane-bound nucleus; the protozoa, fungi, plants, and animals
eutrophic	very rich nutrient conditions, especially of nitrogen compounds
heterotrophic	growth at the expense of complex organic substrates
lignolytic	growth of white-rot fungi under conditions where they synthesize lignin-degrading peroxidases
methanogenic	very anaerobic conditions, where carbon dioxide is reduced to methane
methanotrophic	aerobic growth with methane as sole source of carbon and energy
mineralization	conversion of a contaminant to its simplest forms, eg, CO_2, H_2O, CH_4, N_2, Cl^-
nitrate-reducing, denitrifying	anoxic conditions, where nitrate is reduced to nitrogen gas
nitrification	the biological oxidation of ammonia to nitrite and nitrate
oligotrophic	very low nutrient conditions
Prokaryotes	organisms lacking a membrane-bound nucleus; the bacteria and archaea
recalcitrant	very resistant to biodegradation
reductive dehalogenation, reductive dechlorination	the sequential loss of halogen substituents under anaerobic, usually methanogenic, conditions
rhizosphere	the soil around plant roots; this zone has different microbial populations from the bulk soil
sulfate-reducing	very anaerobic conditions, where sulfate is reduced, by sulfate-reducing bacteria, to sulfide
vadose zone	the part of the soil above the water table

are often very different from agricultural fields, and the technology has to be significantly stronger to "plow" the soil. Frequently the contaminant is below the surface, and applying even simple bioremediation strategies can be very involved. Table 2 lists some of the technologies in use today.

Organic Contaminants

Hydrocarbons. *Constituents.* Hydrocarbons get into the environment from biogenic and fossil sources. Methane is produced by anaerobic bacteria in enormous quantities in soils, sediments, ruminants and termites, and it is consumed by methanotrophic bacteria on a similar scale. Submarine methane seeps

Table 2. Some Technological Definitions Relevant to Bioremediation

Technology	Description
air sparging; aquifer sparging; biosparging	injection of air to stimulate aerobic degradation; may also stimulate volatilization
air stripping	injection of air to stimulate volatilization
aquifer bioremediation	*in situ* bioremediation in an aquifer, usually by adding nutrients or co-substrates
aquifer sparging	injection of air into a contaminated aquifer to stimulate aerobic degradation, may also stimulate volatilization
batch reactor	a bioreactor loaded with contaminated material, and run until the contaminant has been consumed, then emptied, and the process is repeated
bioactive barrier; bioactive zone; biowall	a zone, usually subsurface, where biodegradation of a contaminant occurs so that no contaminant passes the barrier
bioaugmentation	addition of exogenous bacteria with defined degradation potential (or rarely indigenous bacteria cultivated in a reactor and reapplied)
biofilm reactor	a reactor where bacterial communities are encouraged on a high surface area support, biofilms often have a redox gradient so that the deepest layer is anaerobic while the outside is aerobic
biofiltration	usually an air filter with degrading organisms supported on a high surface area support such as granulated activated carbon
biofluffing	augering soil to increase porosity
bioleaching	extracting metallic contaminants at acid pH
biological fluidized bed; fluidized-bed bioreactor	bioreactor where the fluid phase is moving fast enough to suspend the solid phase as a fluid-like phase
biopile; soil heaping	an engineered pile of excavated contaminated soil, with engineering to optimize air, water, and nutrient control
bioslurping	vacuum extraction of the floating contaminant, water, and vapor from the vadose zone; the air flow stimulates biodegradation
biostimulation	optimizing conditions for the indigenous biota to degrade the contaminant
biotransformation	the biological conversion of a contaminant to some other form, but not to carbon dioxide and water
biotrickling filter	a reactor where a contaminated gas stream passes up a reactor with immobilized micro-organisms on a solid support, while nutrient liquor trickles down the reactor
bioventing	vacuum extraction of contaminant vapors from the vadose zone, thereby drawing in air that stimulates the biodegradation of the remainder

53

Table 2. (*Continued*)

Technology	Description
borehole bioreactor; in-well bioreactor	the addition of nutrients and electron acceptor to stimulate biodegradation *in situ* in a contaminated aquifer
closed-loop bioremediation	groundwater recovery, a bioreactor, and low-pressure reinjection to maximize nutrient use, and maintain temperature in cold climates
composting	addition of biodegradable bulking agent to stimulate microbial activity; optimal composting generally involves self-heating to 50–60°C
constructed wetland	artificial marsh for bioremediation of contaminated water
continuous stirred tank reactor (CSTR)	a completely mixed bioreactor
digester	usually an anaerobic bioreactor for digestion of solids and sludges that generates methane
ex-situ bioremediation	usually the bioremediation of excavated contaminated soil in a biopile, compost system or bioreactor
fixed-bed bioreactor	bioreactor with immobilized cells on a packed column matrix
land-farming; land treatment	application of a biodegradable sludge as a thin layer to a soil to encourage biodegradation; the soil is typically tilled regularly
natural attenuation; intrinsic bioremediation	unassisted biodegradation of a contaminant
phytoextraction	the use of plants to remove and accumulate contaminants from soil or water to harvestable biomass
phytofiltration	the use of completely immersed plant seedlings, to remove contaminants from water
phytoremediation	the use of plants to effect bioremediation
phytostabilization	the use of plants to stabilize soil against wind and water erosion
pump and treat	pumping groundwater to the surface, treating, and reinjection or disposal
rhizofiltration	the use of roots to immobilize contaminants from a water stream
rotating biological contactor	bioreactor with rotating device that moves a biofilm through the bulk water phase and the air phase to stimulate aerobic degradation
sequencing batch reactor	periodically aerated solid phase or slurry bioreactor operated in batch mode
soil-vapor extraction	vacuum-assisted vapor extraction

support substantial oases of marine life, with a variety of invertebrates possessing symbiotic methanotrophic bacteria (6). Thus, methanotrophic bacteria are ubiquitous in aerobic environments. Plants generate large amounts of volatile hydrocarbons, including isoprene and a range of terpenes (7). These compounds provide an abundant substrate for hydrocarbon-degrading organisms.

Crude oil has been part of the biosphere for millennia, leaking from oil seeps on land and in the sea. Crude oils are very complex mixtures, primarily of hydrocarbons although some components do have heteroatoms such as nitrogen (eg, carbazole) or sulfur (eg, dibenzothiophene). Chemically, the principal components of crude oils and refined products can be classified as aliphatics, aromatics, naphthenics, and asphaltic molecules. Representative examples are shown in Figure 1. The ratios of these different classes varies in different oils, but a typical crude oil might contain the four classes in a ratio of approximately 30:30:30:10. Most crude oils contain hydrocarbons ranging in size from methane to molecules with hundreds of carbons, although the lightest molecules are usually absent in oils that have been partially biodegraded in their reservoir. When crude oils reach the surface environment the lighter molecules evaporate, and are either destroyed by atmospheric photooxidation or are washed out of the

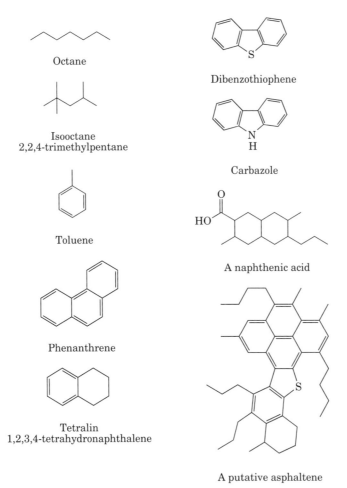

Octane

Isooctane
2,2,4-trimethylpentane

Toluene

Phenanthrene

Tetralin
1,2,3,4-tetrahydronaphthalene

Dibenzothiophene

Carbazole

A naphthenic acid

A putative asphaltene

Fig. 1.   Some representative hydrocarbons found in crude oil.

atmosphere in rain, and are biodegraded. Some molecules, such as the smaller aromatics (benzene, toluene, etc) have significant solubilities, and can be washed out of floating slicks, whether these are at sea, or on terrestrial water-tables. Fortunately the majority of molecules in crude oils, and refined products made from them, are biodegradable, at least under aerobic conditions.

Biodegradation. Methane and the volatile plant terpenes are fully biodegradable by aerobic organisms, and most refined petroleum products are essentially completely biodegradable under aerobic conditions. Estimates for crude oil biodegradability range up to 90% (8), and the least biodegradable material, principally polar molecules and asphaltenes, lacks the "oily" feel and properties that are associated with oil. These are essentially impossible to distinguish from more recent organic material in soils and sediments, such as the humic and fulvic acids, and appear to be biologically inert.

Numerous bacterial and fungal genera have species able to degrade hydrocarbons aerobically and the pathways of degradation of representative aliphatic, naphthenic and aromatic molecules have been well characterized in at least some species (8). Other organisms, such as algae and plants, do not seem to play a very important role in the biodegradation of hydrocarbons. It is a truism that the hallmark of an oil-degrading organism is its ability to insert oxygen atoms into the hydrocarbon, and there are many ways in which this is achieved. Figures 2 and 3 show the most well-studied. Once a hydrocarbon possesses a carboxylate or alcohol functionality it is almost invariably a readily degradable compound. A simple example at the human level is the difference between oleic acid, a high calorie food, and octadecane, present in mineral oil, which is so inert that it serves as an intestinal lubricant!

For many years it was assumed that oil biodegradation was an exclusively aerobic process, since any degradation must involve oxidation. Indeed the

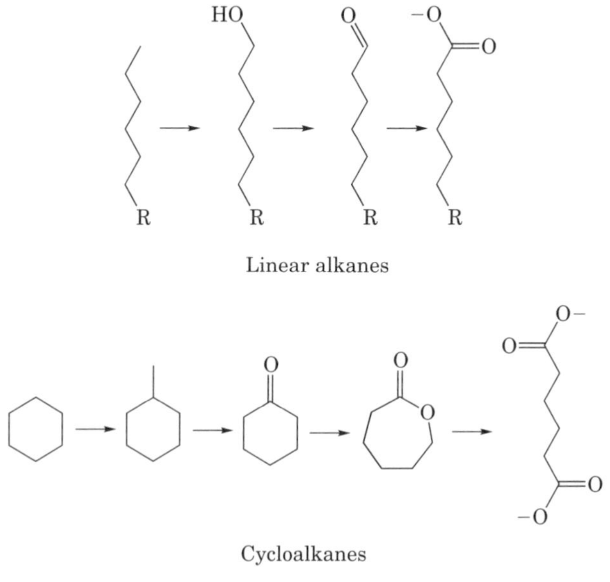

Linear alkanes

Cycloalkanes

Fig. 2. Initial steps in the biodegradation of linear and cyclic alkanes.

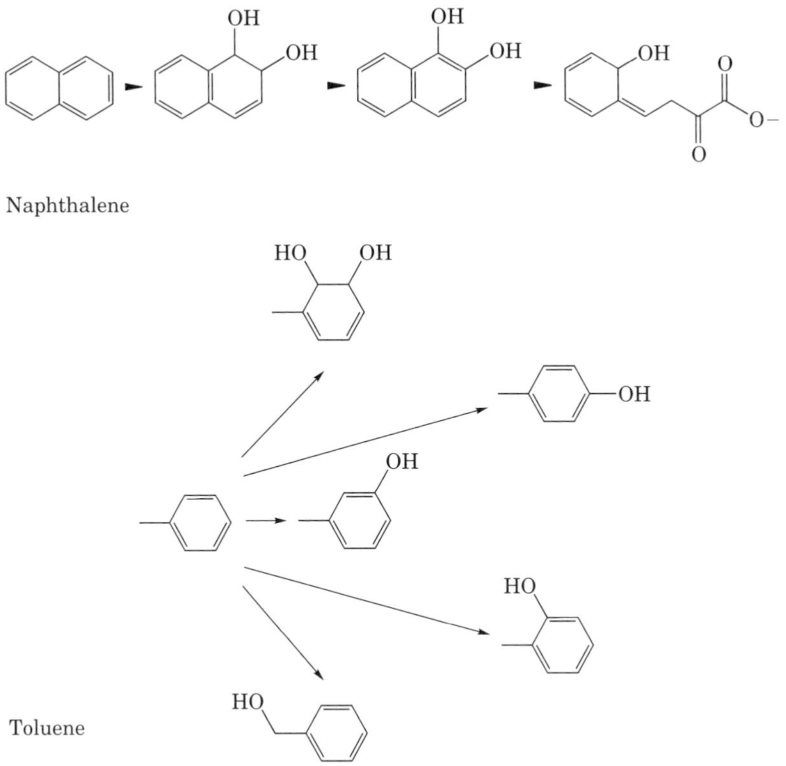

Fig. 3. Initial steps in the aerobic degradation of naphthalene, as a representative multiringed aromatic, and toluene. The different initial steps of toluene degradation are examples of the diversity found in different organisms.

very existence of oil reservoirs indicates that anaerobic degradative processes in such environments must be very slow. Nevertheless, in recent years it has become clear that at least some hydrocarbons are oxidized by bacteria under completely anaerobic conditions, where the oxygen is probably coming from water. Limited hydrocarbon biodegradation has now been shown under sulfate-, nitrate-, carbon dioxide- and ferric iron-reducing conditions (Table 3). The phenomenon is still poorly understood, however, and at present the largest molecules demonstrated to undergo biodegradation under these conditions are hexadecane, heptadecene, and phenanthrene. The pathways of degradation are only beginning to be addressed. Figure 4 shows the intermediates identified in anaerobic toluene degradation in different organisms. It is noteworthy that while organisms capable of aerobic oil biodegradation seem to be ubiquitous, organisms capable of the anaerobic degradation of hydrocarbon have to date only been found in a few places.

Although the majority of molecules in crude oils and refined products are hydrocarbons, the U.S. Clean Air Act amendment of 1990 mandated the addition of oxygenated compounds to gasoline in many parts of the United States. The requirement is usually that 2% (w/w) of the fuel be oxygen, which requires that

Table 3. Hydrocarbons That Have Been Shown to be Biodegraded Under Anaerobic Conditions

Electron acceptor	Substrate
nitrate (to nitrogen)	heptadecene
	toluene, ethylbenzene, xylene
	naphthalene
	terpenes
iron(III) (to iron(II))	toluene
manganese(IV) (to Mn(II))	toluene
sulfate (to sulfide)	hexadecane, alkylbenzenes
	benzene
	naphthalene, phenanthrene
CO_2 (to methane)	toluene, xylene

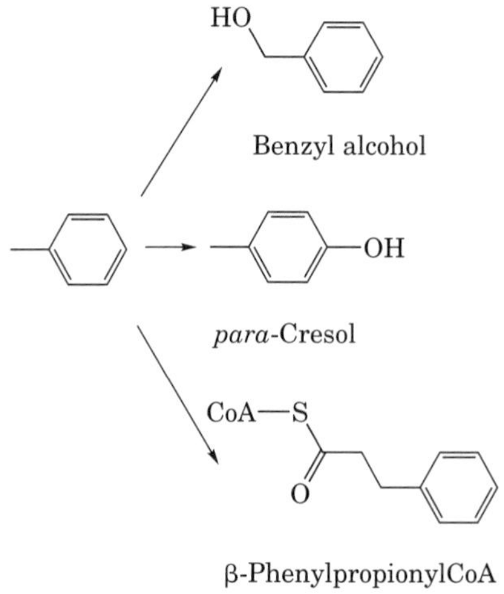

Fig. 4. Proposed initial steps in the anaerobic biodegradation of toluene in different organisms.

5–15% (v/v) of the gasoline be an oxygenated additive (eg, methanol, ethanol, methyl *tert*-butyl ether (MTBE), etc). Although methanol and ethanol are readily degraded under aerobic conditions, the degradability of MTBE remains something of an open question. The compound was previously very rare in the environment, but now it is one of the major chemicals in commerce. At first it seemed that the compound was completely resistant to biodegradation, but complete mineralization has now been reported (9). Whether biodegradation can be optimized for effective bioremediation remains to be seen.

Bioremediation. Crude oil and refined products are readily biodegradable under aerobic conditions, but they are only incomplete foods since they lack any significant nitrogen, phosphorus, and essential trace elements. Bioremediation strategies for removing large quantities of hydrocarbon must therefore include the addition of fertilizers to provide these elements in a bioavailable form.

Air. Hydrocarbon vapors in air are readily treated with biofilters. These are typically rather large devices with a very large surface area provided by bulky material such as a bark or straw compost. The contaminated air, perhaps from a soil vapor-extraction treatment, or from a factory using hydrocarbon solvents, is blown through the filter, and organisms, usually indigenous to the filter material or provided by a soil or commercial inoculum, grow and consume the hydrocarbons. Adequate moisture must be maintained for effective operation. Alternatively, trickling biofilters with recycled water are also in use. Both bacteria and fungi readily colonize such filters, and they can be very effective. Nevertheless, biofilters are usually equipped with a small granulated activated carbon "backup" filter to handle any sudden pulse loads that might overwhelm the biological capacity of the filter. Biofilters compete with granulated activated carbon filters, and are often cheaper because they minimize the cost of the granulated activated carbon, and the energy required to destroy the contaminant and the granulated activated carbon when the latter is saturated (10). Potential problems include plugging and uneven air or water flow, but successful designs work for many years with minimal maintenance except the occasional addition of nutrients and stirring of the bed.

Sea. Crude oil spills at sea are perhaps the most widely covered environmental incidents in the national and international media. Despite their notoriety, catastrophic tanker spills and well blow-outs are fortunately rather rare, and their total input into the world's oceans is approximately equivalent to that from natural seeps; significantly more oil reaches the world's oceans from municipal sewers (11). Physical collection of the spilled oil is the preferred remediation option, but if skimming is unable to collect the oil, biodegradation and perhaps combustion or photooxidation are the only routes for elimination of the spill. One approach to stimulating biodegradation is to disperse the oil with chemical dispersants. Early dispersants had undesirable toxicity, but modern dispersants and application protocols can stimulate biodegradation by increasing the surface area of the oil available for microbial attachment, and perhaps providing nutrients to stimulate microbial growth (12). Patents have been issued for dispersant formulations that specifically include nitrogen and phosphorus nutrients (13), but the products are not currently commercially available.

Bioremediation by the addition of oil-degrading microbes is often promoted as a treatment option for floating spills, but this approach has not yet met with any documented success (13).

Shorelines. The successful bioremediation of shorelines affected by the spill from the *Exxon Valdez* in Prince William Sound, Alaska, was perhaps the largest bioremediation project to date (14,15). More than 73 miles of shoreline were treated in 1989 and similar amounts of fertilizer were used in 1990. Oil had typically penetrated into the surface gravel on these shorelines, occasionally getting as deep as 30 cm into the sediment. Since the gravel was typically very permeable, oxygen availability was unlikely to be the limiting factor for biodegradation, and indeed this was subsequently shown to be correct. Bioremediation thus focused on the addition of nitrogen and phosphorus fertilizers to partially remove the nutrient-limitation on oil degradation. Of course the addition of fertilizers was complicated by the fact that oiled shorelines were washed by tides twice a day. These tides would have rapidly removed any soluble fertilizer, so a strategy was sought that would provide nutrients for a significant length of time. Various approaches to applying fertilizers were tried, including both standard and slow release nutrients, oleophilic nutrients and solutions of liquid fertilizers. Two fertilizers were used in the full-scale applications; one, an oleophilic product known as Inipol EAP22 (trademark of CECA, Paris, France), was a microemulsion of a concentrated solution of urea in an oil phase of oleic acid and trilaurethphosphate, with butoxyethanol as a cosolvent. This product was designed to adhere to oil, and to release its nutrients to bacteria growing at the oil-water interface. The other fertilizer was a slow-release formulation of inorganic nutrients, primarily ammonium nitrate and ammonium phosphate, in a polymerized vegetable oil skin. This product, known as Customblen (trademark of Grace-Sierra, Milpitas, California), released nutrients with every tide, and these were distributed throughout the oiled zone as the tide fell. Fertilizer application rates were carefully monitored so that the nutrients would cause no harm, and the rate of oil biodegradation was stimulated between two- and five-fold (14,15).

A wide range of fertilizers, including agricultural and horticultural fertilizers, and bone and fish meals have been tried at the pilot scale, usually with at least modest success (13). Some current work is aimed at addressing whether providing a readily degradable substrate with the fertilizer nutrients helps immobilize the nutrients in biomass at the oiled site. Of course nutrient-supplementation is only likely to markedly stimulate the rate of biodegradation where nutrient levels are naturally low. It is unlikely that fertilizers will have a dramatic effect in situations where agricultural or municipal run-off maintains elevated levels of nutrients, such as happens in some estuaries and bays. Here aeration is likely to be most effective.

Bioremediation by the addition of oil-degrading microbes has been promoted as a treatment option for oiled shorelines, but this approach has not yet met with any documented success (13).

Areas where there are currently few remediation options include oiled marshes, mangroves, and coral reefs. These environments are generally easily damaged by human intrusion and physical cleaning options may not provide any net environmental benefit. Bioremediation may provide some attractive options, and some success has been claimed, on a small scale, with fertilizer applications (13). Marshes and mangroves offer the additional complication that they are typically anoxic. Perhaps the anaerobic degradation of oil could be stimulated by inoculation with anaerobic hydrocarbon degrading microbes, or perhaps gentle aeration or the addition of slow release oxygen compounds, such as some inorganic peroxides, might stimulate aerobic degradation without significantly changing the redox balance of these environments. This is an area where research is very much in its infancy and there are no well-documented success stories to date.

Bioremediation also offers options for dealing with oiled material, such as seaweed, that gets stranded on shorelines; composting has been shown to be effective.

Groundwater.   Spills of refined petroleum product on land, and leaking underground storage tanks, sometimes contaminate groundwater. Bioremediation is becoming an increasingly popular treatment for such situations.

Hydrocarbons typically have a specific gravity of less than 1, and refined products usually float on the water table if they penetrate soil that deeply. In the parlance of the remediation industry, such floating spills are often called NAPLs (nonaqueous phase liquids). Indeed they are sometimes known as LNAPLs for light nonaqueous phase liquids, to distinguish them from more dense materials, such as halogenated compounds, which are more likely to sink in ground water. Stand-alone bioremediation is an option for these situations, but "pump and treat" is the more usual treatment. Contaminated water is brought to the surface, free product is removed by flotation, and the cleaned water re-injected into the aquifer or discarded. Adding a bioremediation component to the treatment, typically by adding oxygen and low levels of nutrients, is an appealing and cost-effective way of stimulating the degradation of the residual hydrocarbon not extracted by the pumping. This approach is becoming widely used.

Hydrocarbons are not very soluble in water, but the most soluble components leach out of a spill if there is continual flushing. Typically only small aromatic molecules, the infamous BTEX (benzene, toluene, ethylbenzene, and xylenes, Fig. 5), are soluble enough to contaminate groundwater. Although with the advent of oxygenated gasolines, it is expected that these oxygenates (ethanol, methanol, MTBE (methyl-*tert*-butyl ether) etc) will also be found in groundwater. In the past, remediation of such situations has usually used pump-and-treat methodologies. These methods are slow and may leave reservoirs of contaminants in pockets that are poorly connected to the main water body. Of course the contaminant is biodegradable, and some biodegradation is probably already occurring when the contamination is discovered. The cheapest approach to remediation is, thus, to allow this intrinsic process to continue. Evidence that it is indeed occurring can be found in the selective disappearance of the most biodegradable compounds in the contaminant mixture, and the concomitant disappearance of electron acceptors from the groundwater. Thus oxygen is depleted as the preferential terminal electron acceptor for metabolism, followed by nitrate, ferric iron, sulfate, and finally CO_2 for methanogenesis (16).

Intrinsic bioremediation is becoming an acceptable option in locations where the contaminated groundwater poses little threat to environmental health. Nevertheless, although intrinsic bioremediation is appealingly simple, it may not be

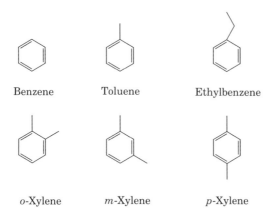

Fig. 5. The components of BTEX.

the lowest cost option if there are extensive monitoring and documentation costs involved for several years. In such cases it may well be more cost effective to optimize conditions for biodegradation.

One approach is to optimize the levels of electron acceptors. Oxygen can be pumped in as the pure gas, or as air, although this is relatively energy intensive since oxygen is so poorly soluble. Hydrogen peroxide has been used in some situations, but there have been problems with biomass plugging near the injection wells. Slow release formulations of inorganic peroxides, such as magnesium peroxide, have recently been used with success (17). Nitrate may be added, although there are sometimes regulatory limitations on the amount of this material that may be added to groundwater (18). Ferric iron availability may be manipulated by adding ligands (19).

If there are significant amounts of both volatile and nonvolatile contaminants, remediation may be achieved by a combination of liquid and vapor extraction of the former, and bioremediation of the latter. This combination has been termed "bioslurping", where the act of pumping out the liquid contaminant phase draws in air at other wells to stimulate aerobic degradation (20). Such bioremediation requires that there be enough nutrients to allow microbial growth, and fertilizer nutrients are frequently added at the air injection wells. Bioslurping has had a number of well-documented successes.

The majority of remediation operations include stopping the source of the contamination, but in some cases this is impossible, either because of the location of the spill, or because it is over a large area, and not a point source. In these situations it may be possible to intercept the flow of contaminated groundwater off-site, and ensure that no contamination passes. The simplest intervention is a line of wells for pump and treat, but including a biological component may be more cost effective. This can range from the installation of a sparge line for aerating the contaminated plume, to installing some form of semicontained bioreactor where nutrients can be applied with some modicum of control. Often these designs are combined with barriers to ensure that all the contaminated plume passes through the reactive zone. These designs have a variety of names, including biowall, trench biosparge, funnel and gate, bubble curtain, sparge curtain and engineered trenches and gates. Both aerobic and anaerobic designs have been successfully installed.

Where there are large volumes of contaminated water under a small site, it is sometimes most convenient to treat the contaminant in a biological reactor at the surface. Considerable research has gone into reactor optimization for different situations and a variety of stirred reactors, fluidized bed reactors, and trickling filters have been developed. Such reactors are usually much more efficient than *in situ* treatments, although correspondingly more expensive.

Of course the presence of a liquid phase of hydrocarbon in a soil gives rise to vapor contamination in the vadose zone above the water table. This can be treated by vacuum extraction, and the passage of the exhaust gases through a biofilter (see above) can be a cheap and effective way of destroying the contaminant permanently.

Soil. Hydrocarbon contamination of soils runs the gamut from crude oils at production well and pipeline spills, to the full slate of refined products at refineries, distribution centers, service stations and accident sites. Significant hydrocarbon contamination is also often found at manufactured gas plants, now mainly abandoned, wood treatment facilities, railroad rights of way and terminals, and various military bases. Sometimes the contamination is the result of leaking underground storage tanks and pipelines, leading to subsurface contamination, but surface spills also occur. Physical removal of gross contamination is an obvious first step at all locations, and bioremediation is an appealing option for remediating residual contamination in many of these sites.

Spills from production facilities and pipelines often involve both oil and brine, since most oil reservoirs float on top of concentrated brines, and both are produced in later stages of production. The brine is typically separated from the oil and re-injected into the reservoir, but some is retained in many production pipelines. The environmental impact of spilled brine can be quite deleterious. Not only is salt toxic to most plants, and can inhibit many soil bacteria, but it also can have a major effect on the soil structure by altering the physical properties of clays. Successful bioremediation strategies must therefore include remediating the brine. In wet regions the salt is eventually diluted by rainfall, but in arid regions, and to speed the process in wetter regions, gypsum is often added to restore soil porosity.

Many hydrocarbons bind quite tightly to soil components, and are thereby less available to microbial degradation. The kinetics of binding seem to be complex, and the process of "aging" is only poorly understood. Nevertheless, it seems clear that hydrocarbons that have been in contact with soil for a long time are not as available for biodegradation as fresh spills. Several groups of researchers have suggested the addition of surfactants to overcome this limitation, but this approach is not yet widely used. A significant potential concern is that the surfactant will be degraded in preference to the contaminant of concern.

Intrinsic biodegradation occurs, but it usually only removes the lightest refined products, such as gasoline, diesel and jet fuel. Active intervention is typically required. Usually the least expensive approach is *in situ* remediation, typically with the addition of nutrients, and the attempted optimization of moisture and oxygen by tilling. Various approaches to applying fertilizers have been tried, including both standard and slow-release nutrients, oleophilic nutrients and solutions of liquid fertilizers. Oxygen is a likely limiting nutrient in many cases, and soil tilling is widely practiced. This *in situ* bioremediation of hydrocarbon-contaminated soils is akin to the old practice of "land-farming", wherein sludges and other refinery wastes were deliberately spread onto soil and tilled and fertilized to stimulate biodegradation. Although this practice is now discontinued in the United States, it was quite widely used.

Deeper contamination may be remedied with bioventing, where air is injected through some wells, and extracted through others to both strip volatiles and provide oxygen to indigenous organisms. Fertilizer nutrients may also be added. This is usually only a viable option with lighter refined products.

A recent suggestion has been to use plants to stimulate the microbial degradation of the hydrocarbon (hydrocarbon phytoremediation). This has yet to receive clear experimental verification, but the plants are proposed to help deliver air to the soil microbes, and to stimulate microbial growth in the rhizosphere by the release of nutrients from the roots. The esthetic appeal of an active phytoremediation project can be very great.

When soil contamination extends to some depth it may be preferable to excavate the contaminated soil and put it into "biopiles" where oxygen, nutrient and moisture levels are more easily controlled. Biopiles can also be kept warm during winter months, increasing the amount of time available for biodegradation in colder climates. Since the soil is well mixed during the construction of the pile, there is an opportunity to add selected microbial and fungal strains in an additional attempt to maximize biodegradation.

Composting by the addition of readily degradable bulking agents is also a useful option for relatively small volumes of excavated contaminated soil. Since efficient composting invariably involves self-heating as biodegradation proceeds, this also offers an option for extending the bioremediation season into the winter months in cold climates. A potential drawback of composting is that it usually increases the volume of contaminated material, but if fully successful the finished compost can be returned to the site as a positive contribution to soil quality.

Slurry bioreactors offer the most aggressive approach to maximizing contact between the contaminated soil and the degrading organisms. Both lagoons and reactor vessels have been used, but the former are often not optimally designed for all the soil to be partially suspended by the mixing impellers. Contained reactor designs include mixing tank, airlift, and fluidized-bed aeration. A major advantage of contained slurry bioreactors is the potential ability to optimize nutrients, aeration and degradative inocula as fresh soil is added, and the control of waste materials, including gases continually. Slurry bioreactors are usually the most expensive bioremediation option because of the large power requirements, but under some conditions this cost is offset by the rapid biodegradation that can occur.

In all these cases it is important to bear in mind that although the majority of hydrocarbons are readily biodegraded, some, such as the steranes and hopanes, are very resistant to microbial attack. Estimates of oil biodegradation range from 60–95% for different crude oils, so fresh spills of crude oils are readily treated by bioremediation (21). Refined products, such as gasoline, diesel, jet fuels, and heating oils are usually more biodegradable than typical whole oils, but the various heavy fractions of crude oils, such as the asphalts, are far less biodegradable, and are not such attractive targets for bioremediation. Some crude oils have already been extensively biodegraded in their reservoirs, and these are also poor targets for bioremediation. An example is Orimulsion, a heavy oil in water emulsion (70% bitumen) stabilized by low levels of surfactants, used as a fuel for electricity generation. Similarly, old spills may have already undergone significant biodegradation and the residue may be relatively biologically inert. It is thus important to run laboratory studies to ensure that the contaminant is sufficiently biodegradable that clean-up targets can be met.

Halogenated Organic Solvents. *Constituents.* Halogenated organic solvents are widely used in metal processing, electronics, dry cleaning and paint, paper and textile manufacturing, and some representative examples are shown in Figure 6. These solvents have been used for more than fifty years, and unfortunately they are fairly widespread contaminants. Unlike the hydrocarbons, which usually float on water, the halogenated solvents typically have specific gravities greater than 1, and they generally sink to the bottom of any ground-

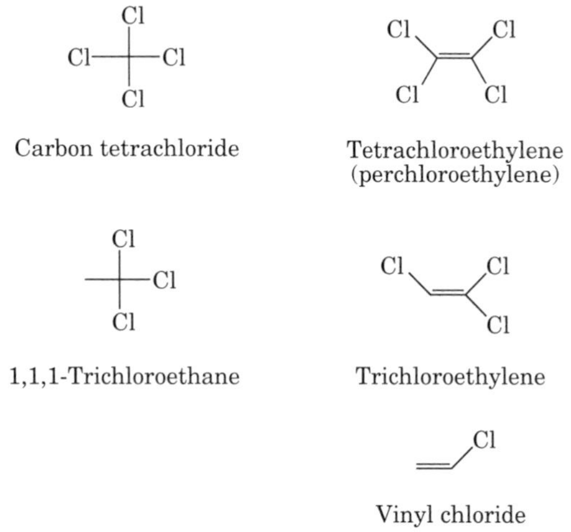

Cl—|—Cl with Cl above and Cl below (Carbon tetrachloride)

Carbon tetrachloride

Tetrachloroethylene
(perchloroethylene)

1,1,1-Trichloroethane

Trichloroethylene

Vinyl chloride

Fig. 6. Some representative halogenated solvents.

water, and float on the bedrock. For this reason they are sometimes known as DNAPLs for dense nonaqueous phase liquids.

Biodegradation. Halogenated solvents are degraded under aerobic and anaerobic conditions. The anaerobic process is typically a reductive dechlorination that progressively removes one halide at a time (Fig. 7). For example, under methanogenic conditions, carbon tetrachloride is sequentially dechlorinated to chloroform, dichloromethane, methyl chloride and methane, while trichloroethylene is sequentially reduced to ethylene. Some of these compounds are also dehalogenated under sulfate-reducing conditions, and under denitrifying conditions there are reports that the final product can be CO_2 (22). Chloromethane and dichloromethane have been shown to be the sole carbon source for several anaerobic organisms (23), and it seems there is much to be learned about the microbial diversity of anaerobic microorganisms capable of dechlorinating solvents.

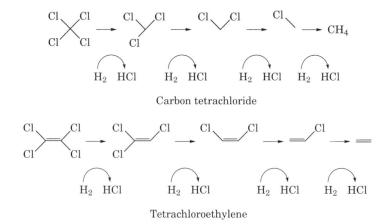

Fig. 7. Reductive dechlorination of carbon tetrachloride and tetrachloroethylene.

The simplest chlorinated alkanes, alkenes, and alcohols (eg, chloromethane, dichloromethane, chloroethane, 1,2-dichloroethane, vinyl chloride, and 2-chloro-ethanol) serve as substrates for aerobic growth for some bacteria, but the majority of halogenated solvents cannot support growth (24). Nevertheless these compounds are mineralized under aerobic conditions, albeit with no apparent benefit to the degrading organism. Indeed, the oxidation appears to be fortuitous and it occurs during the metabolism of a growth substrate. The phenomenon is therefore known as co-metabolism or co-oxidation. Numerous bacteria are able to catalyze the oxidation of trichloroethylene; some use monooxygenases (for example methane and ammonia oxidizing species); others contain dioxygenases (eg, some toluene oxidizing species). The difference between these two classes of enzymes is the fate of the two atoms of molecular oxygen. Monooxygenases insert one oxygen atom into their substrate, and reduce the other to water. Dioxygenases insert both atoms into their substrate. The effect of these two types of enzyme is illustrated in Figure 8. In either case, biodegradation proceeds to complete mineralization (25).

The biodegradation of trichloroethylene is the most studied since this is probably the most widespread halogenated solvent contaminant. Several substrates drive trichlorethylene co-oxidation, including methane, propane, propylene, toluene, isopropylbenzene, and ammonia (25). The enzymes that metabolize these substrates have subtly different selectivities with regard to the halogenated solvents, and to date none are capable of co-oxidizing carbon tetrachloride or tetrachloroethylene. Complete mineralization of these compounds can, however, be achieved by sequential anaerobic and aerobic processes.

Bioremediation. Air.   Biofilters are an effective way of dealing with air from industrial processes that use halogenated solvents such chloromethane, dichloromethane, chloroethane, 1,2-dichloroethane and vinyl chloride, that support aerobic growth (26). Both compost-based dry systems and trickling filter wet systems are in use. Similar filters could be incorporated into pump-and-treat operations.

Groundwater and Soil.   Halogenated solvents have contaminated soils and groundwater throughout the industrialized world, and remediation has a high priority. Since the solvents are so dense, they are typically found on the bedrock underlying aquifers. Pumping out the liquid phase is an obvious first step if the contaminant is likely to be mobile, but *in situ* bioremediation is a promising

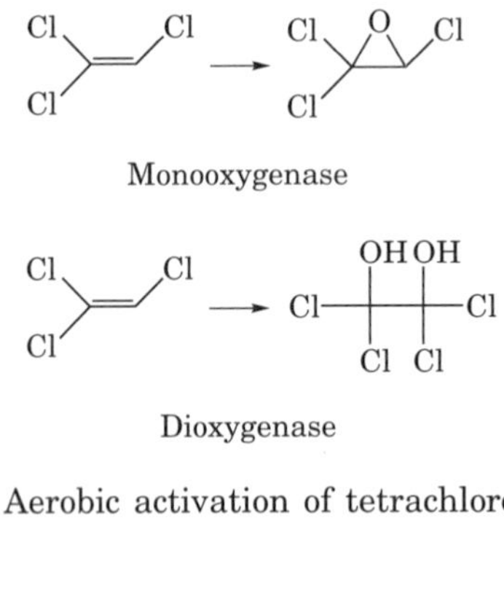

Monooxygenase

Dioxygenase

Fig. 8.   Aerobic activation of tetrachloroethylene.

option. Thus, the U.S. Department of Energy is investigating the use of anaerobic *in situ* degradation of carbon tetrachloride in an aquifer some 76 m below the Hanford, Washington site, with nitrate as electron acceptor, and acetate as electron donor (22).

Trichloroethylene is the most frequent target of remediation, and as discussed above, this is only metabolized cometabolically. Remediation operations thus incorporate the addition of the cometabolized substrate. Methane was used successfully at the U.S. Department of Energy site at Savannah River, near Aiken, South Carolina (27) which had both an air-stripping and a biological component. Horizontal wells were used to pump methane and air below the contaminant, while an upper horizontal well in the vadose zone was used to withdraw these gases through the contaminated zone. Optimum biodegradation performance seemed to come from alternate injection of air and methane in air, and the inclusion of nitrous oxide and triethylphosphate, both gases, to give a C:N:P ratio of 100:10:1.

Plants may have a role to play in enhancing microbial biodegradation of halogenated solvents, for it has recently been shown that mineralization of radiolabelled trichlorothylene is substantially greater in vegetated rather than unvegetated soils (28), indicating that the rhizosphere provides a favorable environment for microbial degradation of organic compounds.

Methane has also been used in aerobic bioreactors that are part of a pump-and-treat operation, and toluene and phenol have also been used as co-substrates at the pilot scale (29). Anaerobic reactors have also been developed for treating trichloroethylene. For example, Wu and co-workers (30) have developed a successful upflow anaerobic methanogenic bioreactor that converts trichloroethylene and several other halogenated compounds to ethylene.

Groundwater contaminated with other halogenated solvents can also be treated in aboveground reactors. Aerobic reactors are useful for those compounds that can support growth. For example, a membrane reactor has been designed for treating 1,2-dichloroethane (31), and bubble columns and packed bed reactors have been developed for the aerobic degradation of 2-chloroethanol (32). As mentioned above, sequential anaerobic and aerobic reactors are capable of mineralizing tetrachloroethylene (33).

Halogenated Organic Compounds. *Constituents.* Complex halogenated organic compounds have been widely used in commerce in the last fifty years. A few representative examples are shown in Figure 9; pentachlorophenol has been widely used as a wood preservative, and also for termite control. (2,4-Dichlorophenoxy)acetate (2,4-D) is widely used as a broad-leaf herbicide, DDT was widely used as an insecticide, and hexachlorophene has been widely used as a germicide. Polychlorinated biphenyls (PCBs) were sold with varying levels of chlorination for a range of purposes. They ranged from light oily fluids (with two, three, or four chlorines) to viscous oils (five chlorines) to greases and waxes (six or more chlorines), and their names indicated the level of chlorination. Thus Aroclor 1242 (trademark of Monsanto, U.S.), Clophen A30 (trademark of Farbenfabriken Bayer AG, Germany) and Kanechlor 300 (trademark of Kanegafuchi Chemical Industries, Japan) all contained 42% chlorine by weight, and an average of three chlorines per biphenyl. An important property shared by all these compounds is their relative resistance to biodegradation, so at first glance

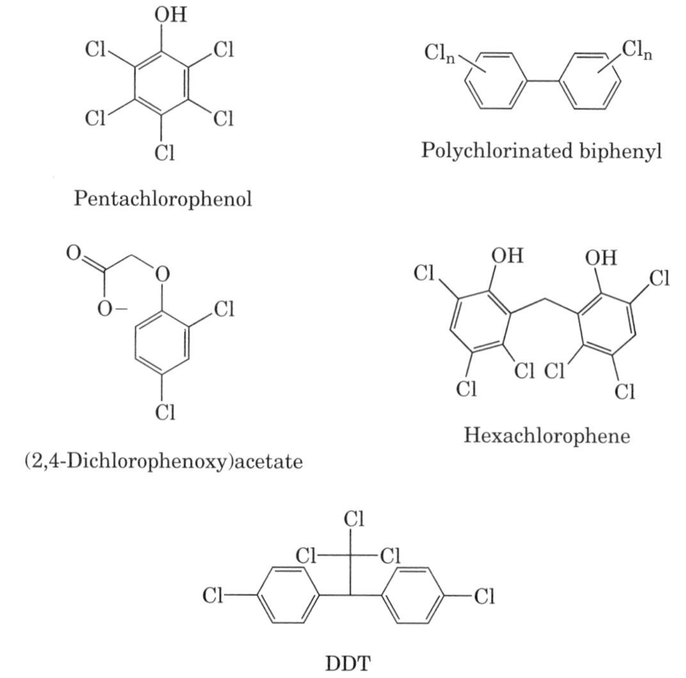

Fig. 9. Representative halogenated organic contaminants.

they may not seem a good target for bioremediation. Indeed complex halogenated organic compounds were widely thought to be almost exclusively anthropogenic in origin, so that there would have been little time for biodegradation pathways to evolve. This view is being corrected, for in fact a variety of organisms, particularly marine algae and some fungi, produce significant quantities of these compounds (34). There is, thus, good reason to expect that halogenated-organic degrading organisms will be found in the biosphere, and this has been borne out in practice.

Biodegradation. An important characteristic of degradation is the cleavage of carbon–chlorine bonds, and the enzymes that catalyze these reactions, the dehalogenases, are being characterized (35). The reductive dechlorination seen with carbon tetrachloride and tetrachloroethylene (Fig. 7) seems to be a general phenomenon, and even compounds as persistent as DDT and the polychlorinated biphenyls are reductively dechlorinated under some conditions, particularly under methanogenic conditions. Some compounds, such as pentachlorophenol, can be completely mineralized under anaerobic conditions, but the more recalcitrant ones require aerobic degradation after reductive dehalogenation.

Pentachlorophenol can be mineralized aerobically and anaerobically, and both processes have been exploited for bioremediation. Under methanogenic conditions a reductive dehalogenation, analogous to that seen with halogenated solvents, eventually generates phenol, although some of the intermediate congeners are quite recalcitrant (36). The phenol is further mineralized to methane and carbon dioxide, and sulfate-reducing bacteria may be involved (37).

Several bacterial isolates are able to grow aerobically using pentachlorophenol as sole source of carbon and energy, but many grow rather better when supplemented with a more nutritions substrate. Unfortunately, in many cases it seems that the more vigorous growth with these substrates does not enhance the biodegradation of pentachlorophenol (38). The initial step in pentachlorophenol biodegradation in one *Flavobacterium* species is an NADPH-dependent oxygenolytic dechlorination, where the *para*-chloro group is replaced by a hydroxyl to generate tetrachlorohydroquinone. Other species seem to produce the same metabolite by a hydrolytic process (38). Fungal degradation of pentachlorophenol, apparently using the lignolytic apparatus, has also been reported (39).

(2,4-Dichlorophenoxy)acetate (2,4-D) has been one of the world's most popular herbicides. Although it is somewhat resistant to biodegradation, it is biodegraded by several bacterial isolates. It is a general truism that the more halogens on a molecule, the slower its biodegradation, and this is borne out with the related herbicide 2,4,5-T ((2,4,5-trichlorophenoxy)acetate). Nevertheless, bacterial degradation has been seen under both aerobic and anaerobic conditions, the latter involving reductive dechlorination via 2,4-D. Aerobic degradation removes acetate from (2,4-dichlorophenoxy)acetate to yield 2,4-dichlorophenol, which is subsequently hydroxylated to 3,5-dichlorocatechol, followed by ring cleavage and complete mineralization (40). Genetic engineering has been used to construct strains that are particularly adept at consuming these compounds, but whether these will overcome the regulatory hurdles to allow their use outside the laboratory remains to be seen.

DDT (1,1,1-trichloro-2,2-bis(*p*-chlorophenyl)ethane) is a remarkably resistant molecule, which explains both its efficacy as an insecticide, and its accumulation at the top of the food-chain. Nevertheless, there are indications that it can be biodegraded, both anaerobically with initial dechlorination and aerobically with initial ring hydroxylation (40). DDT and its partially degraded congeners are very hydrophobic, and biodegradation seems to be stimulated by adding surfactants. White-rot fungi also degrade DDT under lignolytic conditions, although there is little mineralization to CO_2 (5).

The polychlorinated biphenyls are quite recalcitrant. Some lightly chlorinated biphenyls are readily mineralized under aerobic conditions (41), and indeed the structure of an enzyme that catalyzes the key ring-cleavage oxygenation has recently been determined by x-ray crystallography (42). More chlorinated congeners are resistant to aerobic degradation, but they are reductively dechlorinated under anaerobic conditions. Complete degradation of the commercial mixtures, thus, generally requires an anaerobic process followed by an aerobic one. A major issue for the oxidative process is that biphenyl seems to be required for significant expression of the biodegradative system (41); the chlorinated compounds do not induce the enzymes that would degrade them. The biochemistry of the biodegradative process is only beginning to be unraveled, but already there are suggestions that there is considerable diversity in the enzyme systems able to degrade these compounds. There is a lot of effort aimed at engineering organisms to degrade polychlorinated biphenyls, and in finding ways to make these very hydrophobic compounds more bioavailable by the use of chemical oxidants, such as Fenton's Reagent (43), or surfactants.

Bioremediation.   Soil.   Pentachlorophenol has been the target of bioremediation at a number of wood-treatment facilities, and good success has been achieved in several applications. It is rarely the sole contaminant, and is often present with polynuclear hydrocarbons from coal creosote. *In situ* degradation has been stimulated by bioventing, where air is injected through some wells, and extracted through others to both strip volatiles and provide oxygen to indigenous organisms (44). Just as with the halogenated solvents, it seems that plants stimulate microbial degradation of pentachlorophenol in the rhizosphere (45).

The kinetics of such *in situ* degradation are rather slow, however, and more active bioremediation is usually attempted. For example, contaminated soil at the Champion Superfund site in Libby, Montana, was placed into 1-acre land treatment units in 6-in. layers, and irrigated, tilled, and fertilized. Under these conditions, the half-lives of pentachlorophenol, pyrene, and several other polynuclear aromatic hydrocarbons, initially present at around 100–200 ppm, were on the order of 40 days (46). This success relied on the indigenous microbial populations in the soil, but many groups are focusing on the addition of organisms (eg, 47), perhaps immobilized on some sort of carrier. Composting, and bioremediation focusing on the use of white-rot fungi, has also met with success at the pilot scale. Others have used fed-batch or fluidized-bed bioreactors to stimulate the biodegradation of pentachlorophenol. This allows significant optimization of the process and increases in rates of degradation by tenfold (48).

A major concern when remediating wood-treatment sites is that pentachlorophenol was often used in combination with metal salts, and these compounds, such as chromated-copper-arsenate, are potent inhibitors of at least some pentachlorophenol degrading organisms (49). Sites with significant levels of such inorganics may not be suitable candidates for bioremediation.

The phenoxy-herbicide, 2,4-D, has been successfully bioremediated in a soil contaminated with such a high level of the compound (710 ppm) that it was toxic to microorganisms (50). There were essentially no indigenous bacteria in the soil. Success relied on washing a significant fraction of the contaminant off the soil and adding bacteria enriched from a less contaminated site. Success was achieved in remediating both soil washwater and soil in a bioslurry reactor (50). 2,4-D is also effectively degraded in composting, with about half being completely mineralized, and the other half becoming incorporated in a nonextractable form in the residual soil organic matter (51).

The bioremediation of polychlorinated biphenyls in soils in receiving significant attention because these compounds are quite widely distributed in the environment, either from leaking electrical transformers or sometimes because they were applied as part of road maintenance. In the latter case, the contamination usually includes petroleum hydrocarbons, and unfortunately it seems that the two contaminants inhibit the degradation of each other. Nevertheless, cultures are being found that can degrade both polychlorinated biphenyls and petroleum hydrocarbons. There is also interest in the role of rhizosphere organisms in polychlorinated biphenyl degradation, particularly since some plants exude phenolic compounds into the rhizosphere that can stimulate the aerobic degradation of the less chlorinated biphenyls.

Groundwater.   A successful groundwater bioremediation of pentachlorophenol is being carried out at the Libby Superfund site described above. A shallow aquifer is present at 5.5 to 21 m below the surface, and a contaminant plume is nearly 1.6 km in length. Nutrients and hydrogen peroxide were added at the source area and approximately half way along the plume, and pentachlorophenol concentrations decreased from 420 ppm to 3 ppm where oxygen concentrations were successfully raised. A membrane oxygen dissolution system has now been installed to replace the hydrogen peroxide additions, and costs have been substantially lowered without an apparent decrease in remediation performance (52).

Pentachlorophenol is readily degraded in biofilm reactors (53), so bioremediation is a promising option for the treatment of contaminated groundwater brought to the surface as part of a pump-and-treat operation.

River and Pond Sediments.   Much of the work on polychlorinated biphenyls has focused on the remediation of aquatic sediments, particularly from rivers, estuaries, and ponds. As noted above, a few of the most lightly chlorinated compounds are mineralized under aerobic conditions, but the more chlorinated species seem completely resistant to aerobic degradation, even by white rot fungi (54). On the other hand, there is extensive dechlorination of highly chlorinated forms under anaerobic conditions, particularly methanogenic conditions. Bioremediation thus requires anaerobic and aerobic regimes. Intrinsic biodegradation of polychlorinated biphenyls can be recognized by the changing "fingerprint" of the individual isomers as biodegradation proceeds (41,55). The anaerobic dechlorination of the most recalcitrant congeners can apparently be primed by adding a readily dehalogenated congener, such as 2,5,3′,4′-tetrachlorobiphenyl (56), but whether this is a realistic approach for *in situ* bioremediation remains to be seen. Harkness and co-workers (57) have successfully stimulated aerobic biodegradation in large caissons in the Hudson River by adding inorganic nutrients, biphenyl, and hydrogen peroxide, but found that repeated addition of a polychlorinated-biphenyl degrading bacterium (*Alcaligenes eutrophus* H850) had no beneficial effect. Essentially no biodegradation occurred in the stirred control caissons, but losses on the order of 40% were seen in the caissons that received nutrients and peroxide, regardless of whether the stirring was aggressive or rather gentle. Whether this approach can be scaled-up for large-scale use, with a net environmental benefit, remains to be seen.

Nonchlorinated Pesticides and Herbicides. *Constituents.* A vast number of compounds are used as herbicides, pesticides, fungicides, etc, and a few are shown in Figure 10. In order to be effective they must have their effect before they are degraded in the environment, but on the other hand they must not be so resistant to degradation that they accumulate where they are used, or accumulate in food chains. It is unusual for these compounds to become contaminants where they are applied correctly, but manufacturing facilities, storage depots and rural airfields where crop-dusters are based have had spills that can lead to long lasting contamination. Bioremediation is a promising technology to remediate such sites. There are also some locations were groundwater has become contaminated by these chemicals, and again bioremediation may be a cost-effective remediation strategy.

Biodegradation. The vast majority of pesticides, herbicides, fungicides, and insecticides in use today are biodegradable, although the intrinsic biodegradability of individual compounds is one of the variables used in deciding which

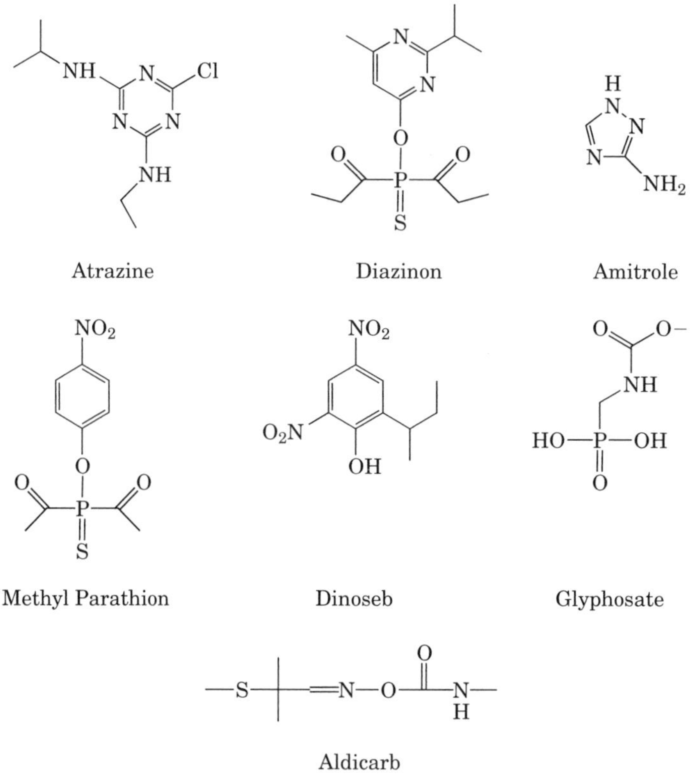

Fig. 10. Examples of herbicides and pesticides amenable to bioremediation.

compound to use for which task. Some herbicides are acutely toxic to plants that absorb them, but are so readily degraded in soil that seeds can be planted at the same time as the herbicide application. Other herbicides are known to be effective at preventing plant growth for many months, and are used in situations where this is the desired goal. Very few degradation pathways of these compounds have been worked out in detail, but some generalizations can be made.

Compounds with organophosphate moieties, such as Diazinon, Methyl Parathion, Coumaphos and Glyphosate are usually hydrolyzed at the phosphorus atom (40,58). Indeed several *Flavobacterium* isolates are able to grow using parathion and diazinon as sole sources of carbon.

Triazines pose rather more of a problem, probably because the carbons are in an effectively oxidized state so that no metabolic energy is obtained by their metabolism. Very few pure cultures of microorganisms are able to degrade triazines such as Atrazine, although some *Pseudomonads* are able to use the compound as sole source of nitrogen in the presence of citrate or other simple carbon substrates. The initial reactions seem to be the removal of the ethyl or isopropyl substituents on the ring (41), followed by complete mineralization of the triazine ring.

Nitroaromatic compounds, such as Dinoseb, are degraded under aerobic and anaerobic conditions (59). The nitro group may be cleaved from the molecule as nitrite, or reduced to an amino group under either aerobic or anaerobic conditions. Alternatively, the ring may be the subject of reductive attack. Thus, while these molecules are sometimes quite long-lived in the environment, they can be completely mineralized under appropriate conditions (59). Recent work has isolated a *Clostridium bifermentans* able to anaerobically degrade dinoseb cometabolically in the presence of a fermentable substrate. The dinoseb was degraded to below detectable levels, although only a small fraction was actually mineralized to CO_2 (60).

Carbamates such as Aldicarb undergo degradation under both aerobic and anaerobic conditions. Indeed the oxidation of the sulfur moiety to the sulfoxide and sulfone is part of the activation of the compound to its most potent form. Subsequent aerobic metabolism can completely mineralize the compound, although this process is usually relatively slow so that it is an effective insecticide, acaricide and nematocide. Anaerobically these compounds are hydrolyzed, and then mineralized by methanogens (61).

Bioremediation. Groundwater. Atrazine dominated the world herbicide market in the 1980s, and contamination of groundwater has been reported in several locations in the U.S., Europe, and South Africa. There are several reports that once in groundwater it is very recalcitrant, suggesting that atrazine-degrading organisms are not widespread. Nevertheless, successful biodegradation has been achieved with indigenous organisms in laboratory mesocosms after a lag phase, and once activity was found, it remained (62). Interestingly the degradation was somewhat slowed by the addition of low concentrations of readily assimilate carbon, such as lactate, and it is not clear how biodegradation might be stimulated in the field. Nevertheless, it is clear that intrinsic remediation is likely to lead to the disappearance of atrazine from groundwaters once atrazine-utilizers have become abundant, and perhaps inoculation with atrazine-metabolizing species will be effective.

If more active treatment is required, such as pump and treat, it is possible that biological reactors will be a cost-effective replacement for activated carbon filters (63).

Marsh and Pond Sediments. Herbicides and pesticides are detectable in marsh and pond sediments, but intrinsic biodegradation is usually found to be occurring. Little work has yet been presented where the biodegradation of these compounds has been successfully stimulated by a bioremediation approach.

Soil. Herbicides and pesticides are of course metabolized in the soil to which they are applied, and there are many reports of isolating degrading organisms from such sites. Degradation activities are typically much higher at sites that have seen product application, indicating that natural enrichment processes occur. Much current effort is aimed at assessing the diversity of degradative pathways, and in many cases it seems that several different natural metabolic pathways can degrade individual pollutants. Little work has yet been presented where the biodegradation of these compounds has been successfully stimulated by a bioremediation approach, but inoculation with active organisms may be a promising approach (64,65).

It is a general observation that herbicide degradation occurs more readily in cultivated than fallow soil, suggesting that rhizosphere organisms are effective herbicide degraders. Whether this can be effectively exploited in a phytoremediation strategy remains to be seen.

Military Chemicals. *Constituents.* The military use a range of chemicals as explosives and propellants, which are sometimes termed "energetic molecules". Generally speaking, modern explosives are cyclic, often heterocyclic, composed of carbon, nitrogen and oxygen. Perhaps the most well known is 2,4,6-trinitrotoluene (Fig. 11), but RDX (Royal Demolition eXplosive; hexahydro-1,3,5-trinitro-1,2,3-triazine) and HMX (High Melting eXplosive; octahydro-1,3,5,7-tetranitro-1,3,5,7-tetrazocine), are even more powerful. *N*,*N*-Dimethylhydrazine is used as a solid rocket fuel. These compounds are sometimes present at quite high levels in soils and groundwater on military bases and production sites. One quite infamous problem at the latter is "pink water", a relatively undefined mixture of photodegradation products of TNT. Bioremediation is a promising new technology for treating sites contaminated with such compounds.

Bioremediation may also be an appropriate tool for dealing with chemical agents such as the mustards and organophosphate neurotoxins (66), but little work on actual bioremediation has been published.

Biodegradation. Natural nitrosubstituted organic compounds are quite unusual, and it was once thought that their degradation was principally by abiotic processes. As shown above in the case of Dinoseb, however, nitrosubstituted compounds are subject to a variety of degradative processes. The biodegradation of TNT is well established (67). Under anaerobic conditions it is readily reduced to the corresponding aromatic amines and subsequently deaminated to toluene. As shown in the section on hydrocarbons, the latter can be mineralized under anaerobic conditions, leading to the potentially complete mineralization of TNT in the absence of oxygen.

Under aerobic conditions TNT can be mineralized by a range of bacteria and fungi, often co-metabolically with the degradation of a more degradable substrate. There is even evidence that some plants are able to deaminate TNT reductively. However, there is also ample evidence that under some conditions,

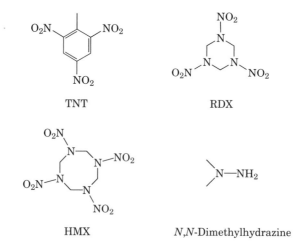

Fig. 11. Military explosives and a rocket propellant.

the TNT is converted to insoluble large molecular weight compounds. This probably occurs by addition reactions to soil components, such as humic and fulvic acids, or cellular material in the case of plants. Lignolytic fungi are also yielding promising results for the degradation of both TNT and "pink water", particularly if the latter is pretreated with uv irradiation.

RDX and HMX are rather more recalcitrant, especially under aerobic conditions, but there are promising indications that biodegradation can occur under some conditions, especially composting (67). Several strains of bacteria able to use RDX (and Triazine) as a sole source of nitrogen for growth have recently been isolated, and this is an area where rapid progress is being made.

Little work has been reported on the biodegradation of dimethylhydrazine, but it may become an important target for remediation at some sites (68).

Bioremediation. Groundwater. Nitrotoluenes have been detected in groundwater in some areas, and intrinsic remediation may be occurring at some sites by anaerobic degradation. Research into whether this can be stimulated with a net environmental benefit is in its very earliest stages, and no clear evidence for success has been presented.

A commercial technology (69), the SABRE process, treats contaminated water and soil in a two-stage process by adding a readily degradable carbon and an inoculum of anaerobic bacteria able to degrade the contaminant. An initial aerobic fermentation removes oxygen so that the subsequent reduction of the contaminant is not accompanied by oxidative polymerization.

Soil. Composting of soils contaminated by high explosives is being carried out at the Umatilla Army Depot near Hermiston, Oregon (70). Soil from munitions washout lagoons is being treated indoors in compost rows of 2,000 m^3, and the estimated cost is less than one-third the estimated cost of incineration. If this is successful, there are 30 similar sites on the National Priority List that could be treated in a similar way.

Other Organic Compounds. The majority of organic compounds in commerce are biodegradable, so bioremediation is a potential option for cleaning up after industrial and transportation accidents. For example, Ref. 71 reports the successful bioremediation of 23,000 liters of vinyl acetate spilled from a railroad tank car in Albany, New York, at a cost approximately half that of excavation and disposal of the 1,100 m^3 of contaminated soil. They also report that *in situ* biological treatment has been used to remediate spills of other organics, including acrylonitrile, styrene, 2-butoxyethanol, and ethacrylate, at other sites where there had been railroad accidents. Bioremediation is thus already an important tool in remediating accidental spills of organic compounds.

Bioremediation is also an option when spills of such compounds contaminate groundwater. For example, bioremediation seems a feasible treatment for aquifers contaminated with alkylpyridines (72) and phenol (73).

Inorganic Contaminants

Nitrogen Compounds. *Constituents.* Nitrogen-containing compounds are of concern for several reasons. Nitrate levels are regulated in groundwater because of concerns for human and animal health. Ammonia is regulated in streams and effluents as a potential fish toxicant, and any nitrogenous

be immobilized as an insoluble polymetallic sulfide by sulfate reducing bacteria, presumably adventitiously due to the production of hydrogen sulfide (82). Indeed many contaminant metal and metalloid ions can be immobilized as metal sulfides by sulfate reducing bacteria.

A rather more specific mechanism of microbial immobilization of metal ions is represented by the accumulation of uranium as an extracellular precipitate of hydrogen uranyl phosphate by a *Citrobacter* species (83). Staggering amounts of uranium can be precipitated; more than 900% of the bacterial dry weight! Recent work has shown that even elements that do not readily form insoluble phosphates, such as nickel and neptunium, may be incorporated into the uranyl phosphate crystallites (84). The precipitation is driven by the production of phosphate ions at the cell surface by an external phosphatase. Although the process requires the addition of a phosphate donor, such as glycerol-2-phosphate, it may be a valuable tool for cleaning water contaminated with radionuclides. An alternative mode of uranium precipitation is driven by sulfate-reducing bacteria such as *Desulfovibrio desulfuricans*, which reduce U(VI) to insoluble U(IV). When combined with bicarbonate extraction of contaminated soil, this may provide an effective treatment for removing uranium from contaminated soil (85).

Microbial processes can also detoxify mercury ions and organic compounds by reducing the mercury to the elemental form, which is volatile (86). This certainly reduces the environmental impact of compounds such as methylmercury, however, such a bioprocess would have to include a mercury capture system before it could be exploited on a large scale with public support.

Rhizofiltration.   Rhizofiltration is the use of plant root systems to remove contaminating metal ions from water (87). The plants must have substantial root systems, and success has been reported with both hydroponically grown terrestrial plants, such as sunflowers (*Helianthus annuus*), and floating aquatic plants such as water hyacinth (*Eichhornia crassipes*), and water milfoil (*Myriophyllum spicatum*). A filamentous cyanobacterium, *Phormidium* sp., has also proven quite effective at removing metals from water. Understanding rhizofiltration at the molecular level is at an early stage, but it is already clear that several processes are involved. Adsorption of the contaminant to the root surface must occur first, and presumably happens in all plants, probably to polygalacturonic acids for cations, and positively charged polypeptides for anions. Some plants actively translocate the metal ions into their cells and some transport the metal to the leafy shoots. This can be exploited in phytoextraction of metal ions from soils (see below). This phenomenon may actually be undesirable in rhizofiltration, because if the metals are retained in the roots there will be a smaller volume of contaminated material to treat at the end of the water treatment. Perhaps of more importance for rhizofiltration is the process of root assisted precipitation, where metal ions are precipitated on the roots as insoluble inorganic complexes. For example, lead is precipitated as lead phosphates on roots of Indian Mustard (*Brassica juncea*) to a remarkable level; up to 45% of dry weight (87). This phenomenon seems to require active exudation of precipitants from the roots, for it is dramatically reduced when dead roots are used.

Phytoextraction.   The fact that certain plant species are able to survive on soils with such high levels of metal ions that the growth of most plants is inhibited has been known for centuries. However, it was not until this century that it became clear that some of these plants actually accumulated the potentially toxic metals and somehow resisted their toxic effects. These plants, known as hyperaccumulators, are exemplified by Alpine Pennycress, *Thlaspi caerulescens*, which is able to accumulate a wide range of metals, including cadmium, chromium, copper, lead, nickel and zinc, more than several thousand fold in its roots, and somewhat less in its leaves (88). Unfortunately for a use in bioremediation, most hyperaccumulators are small, slow growing plants. Researchers have, therefore, turned to more rapidly growing plants which, while they might not be quite as effective as the hyperaccumulators on a weight basis, might be able to extract more metal in a growing season (88). Indian mustard, *Brassica juncea*, was found to be almost as effective as *T. caerulescens* and to grow much faster. This plant is now being used in field trials of phytoextraction.

Phytoextraction of metals from soils requires plants that have substantial root systems to maximize contact with the contaminated soil, and effective transport mechanisms to get the metal from the root to harvestable biomass. Merely accumulating the metal in or on the roots would be less desirable, because harvesting would likely increase the environmental impact of the contaminant by creating dust. It has been known for some time that metals such as cadmium are often stored in plant tissues as phytochelatin-complexes. Phytochelatins are small peptides rich in cysteines that chelate metals via the cysteinyl sulfur (89). Other elements are stored as metalphytate precipitates (88), and in both cases seem to be concentrated in the vacuolar sap. But how the metals are transported in the xylem sap from the root to the shoot is only beginning to be explored. Cadmium seems to be transported as a soluble salt of organic acids (90), at least in *B. juncea*, while nickel is thought to be transported as a soluble histidine complex in *Alyssum* species (91). Optimizing this process will be an important consideration for the commercial success of phytoremediation.

It will also be important to understand the rhizosphere ecology around the roots of metal accumulating plants fully. Maximizing the bioavailability of the contaminant metals in this zone may require the optimization of the microbial communities, or perhaps the addition of soil amendments. There are early indications that such intervention may be beneficial (88), but research in this area is at a very early stage.

Phytoremediation is also being developed for dealing with soils contaminated with high levels of selenium in California; again *B. juncea* seems to be particularly effective in accumulating the contaminant from soil, and all plants tested were more effective at removing selenate than selenite (92). This is an interesting contrast to bacterial systems, where selenite reduction is more commonly found than selenate reduction.

There have also been reports that some plants, including *B. juncea*, stimulate volatilization of dimethylselenide, although it seems that this is an indirect effect where the plant roots stimulate bacterial evolution of the gas (93). Presumably the selenium is eventually oxidized in the atmosphere and returned to the soil as rain. Since much of the world is marginally selenium deficient, such a process might have no deleterious environmental effects. A similar volatilization approach may eventually be used for mercury contamination. Ref. 94 describes the effective transfer of bacterial mercury reduction genes into a plant. In the future, this approach might offer an option for cleaning mercury-contaminated soils, albeit with some form of mercury capture technology.

Phytostabilization.   In some cases it may be important to stabilize a contaminated site to minimize harmful environmental impacts. Barren soils are more prone to erosion, leaching, and dust formation than vegetated soils, so establishing plants as a ground cover may be a very beneficial treatment, even if the contaminant remains in place. In some cases, planting metal tolerant species is enough to establish a plant community, but in other situations, such as mine waste piles, it may be essential to restore fertility to the soil by adjusting the pH and adding organic composts. Where phytoextraction requires plants that actively transport metals to the above-ground biomass, this would be an undesirable phenomenon in phytostabilization programs because it would promote the movement of the contaminating metals into the biosphere.

It may also be possible to stimulate bioreductions in soils, such as Cr(VI) to Cr(III), more effectively by growing plants than by adding bacteria or nutrients to stimulate the microbial process.

Bioremediation. Water.   Groundwater can be treated in anaerobic bioreactors that encourage the growth of sulfate reducing bacteria, where the metals are reduced to insoluble sulfides, and concentrated in the sludge. For example, such a system is in use to decontaminate a zinc smelter site in the Netherlands (95).

Bacterial remediation of selenium oxyanions in San Joaquin, California, drainage water is under active investigation (96,97), but has not yet been commercialized. Agricultural drainage rich in selenium is also typically rich in nitrates, so bioremediation must also include conditions that stimulate denitrification (98). Phytoextraction of selenium is also being tested, but is not yet being used on a large scale.

Phytoremediation is also not yet being used commercially, but results at several field trial suggest that commercialization is not far away. Perhaps the biggest success to date is the successful rhizofiltration of radionuclides from a Department of Energy site at Ashtabula, Ohio, where uranium concentrations of 350 ppb were reduced to less than 5 ppb, well below groundwater standards, by Sunflower roots (99). Small field tests also suggest that rhizofiltration will be a remediation option for removing radionuclides, such as ^{137}Cs and ^{90}Sr, from contaminated lakes and ponds near the Chernobyl nuclear reactor in Ukraine.

Mine Drainage.   Natural drainage water that come into contact with active and abandoned metal and coal mines can become seriously contaminated with a range of heavy metal ions, and/or often become quite acidic, with a pH near 2. While underground, the water is typically anoxic, and any iron is present as soluble ferrous species. When this mixes with aerobic surface water, the iron precipitates as bright orange ferric hydroxide, and this can have a serious environmental impact. In recent years it has become clear that the environmental impact of acid mine drainage can be minimized by the construction of artificial wetlands that combine geochemistry and biological treatments. These systems are being designed for a range of wastewaters, most of which fall outside the scope of this article.

The precipitation of ferric hydroxide is typically biologically mediated by iron-oxidizing bacteria at acid pH, but is usually rather slow. Abiological oxidation becomes more important at pH values above 5, and this is usually much faster than the biological process. Most constructed wetlands for treating acid mine waters thus start with a zone designed to raise the pH. A bed of crushed limestone often suffices to raise the pH significantly, and it is important that this be kept anoxic to prevent rust precipitation on the limestone, which would prevent further production of alkalinity. Once the pH is near neutral, the water is discharged into an aerobic wetland to encourage the precipitation of iron and aluminum oxides, and the co-precipitation of arsenic, if this is present (100,101).

If heavy metals are present in the mine water, the iron-free water can be made to flow into an anaerobic part of the constructed wetland, where organic material, such as compost, manure or sawdust, provides reductants to sulfate-reducing bacteria that become established. These bacteria reduce sulfate to sulfide, which precipitates the heavy metal ions as insoluble sulfides. It is important that this part of the constructed wetland be kept anaerobic, to prevent oxidation and remobilization of the precipitated metals. For this reason, this part of the wetland is typically kept flooded and free of aquatic plants that might introduce oxygen through their roots (100). Finally, an aerobic algal mat can act as a polishing step to complete the removal of contaminants, particularly manganese (101,102). Mine drainage that is not acidic may not need such complicated systems and individual parts of the treatment train described above may suffice in some situations.

Soil. The first reported field trial of the use of hyperaccumulating plants to remove metals from a soil contaminated by sludge applications has been reported (103). The results were positive, but the rates of metal uptake suggest a time scale of decades for complete cleanup. Trials with higher biomass plants, such as *B. juncea*, are underway at several chromium and lead contaminated sites (88), but data are not yet available.

The bacterial reduction of Cr(VI) to Cr(III) discussed above is also being used to reduce the hazards of chromium in soils and water (104).

Conclusions

Bioremediation has been successfully used to treat a wide range of contaminants, including crude oils and refined petroleum products, halogenated solvents, pesticides, herbicides, military chemicals, and mine waters. Much of this success has come by small adjustments of the local environment to encourage the growth of remediating organisms. Fertilizer addition has been a successful treatment for terrestrial and marine oil spills, and the addition of co-substrates, particularly methane, has been a successful treatment for remediating halogenated solvents, such as trichloroethylene. Composting is proving to be a successful treatment for a range of contaminants, and constructed wetlands are successfully treating a range of wastewaters, including those emanating from mines.

One area where there is considerable disagreement between academic scientists and engineering practitioners of bioremediation is the area of bioaugmentation, the addition of selected microorganisms to a site to encourage biodegradation. Dozens of companies sell bacteria for this purpose and claim success in the field, but efforts to demonstrate effectiveness rigorously have met with little success (105,57). Perhaps the most startling test was performed by the

U.S. EPA when testing potential inoculants for stimulating oil biodegradation in Alaska following the *Exxon Valdez* oil spill (106). Eight products were tested in small laboratory reactors that allowed substantial degradation of a test oil by the indigenous organisms of Prince William Sound. All eight microbial inocula had a greater stimulatory effect on alkane degradation, at least for the first 11 days, when they were sterilized by autoclaving prior to addition! This suggests that the indigenous organisms readily out-competed the added products, but that autoclaving the products released some trace nutrient that was able to stimulate the growth of the endogenous organisms.

Of course, bioaugmentation may prove more effective with contaminants that have only recently entered the biosphere, such as methyl-*tert*-butyl ether, or with organisms that have a dramatic selective advantage over indigenous organisms. For example, modern molecular biology may offer opportunities for moving effective degradation pathways into organisms native to a contaminated site, improving biodegradation pathways by broadening the substrate range of degradative enzymes, or removing regulatory constraints to maximize degradative activities. There are, however, many technical and regulatory hurdles to be surmounted before this potential can be tested.

Another area where there is controversy is in the role of surfactants. Many bacteria, particularly those that degrade hydrocarbons, produce surfactants as they grow. Some release them into the medium, others incorporate them into their cell exterior, and there have been elegant experiments to show that inhibiting the production of these compounds inhibits the ability of the bacteria to degrade oil. There have thus been many suggestions to add surfactants, either bacterial or synthetic, to stimulate biodegradation. This seems to be beneficial in the case of some oil dispersants at sea (12), but there have not yet been any clear demonstrations of efficacy on a large scale in soils. The role of surfactants in bioremediation is an area that requires further study.

Bioremediation has many advantages over other technologies, both in cost and in effectively destroying or extracting the pollutant. An important issue is thus when to consider it, and a series of questions may lead to the appropriate answer (see Table 4).

If the answers to the questions in Table 4 lead to the selection of bioremediation, it then becomes important to assess the success of the bioremediation strategy in achieving the clean-up criteria. A major disadvantage of bioremediation is that it is typically rather slower than competing technologies such as thermal treatments. How can regulators and responsible parties gain confidence during this time that success will indeed be achieved? The National Research Council (107) has recently addressed this issue, and suggested a three-fold strategy for "proving" bioremediation: (*1*) a documented loss of contaminants from the site; (*2*) laboratory tests showing the potential of endogenous microbes to catalyze the reactions of interest; and (*3*) some evidence that this potential is achieved in the field.

Although laboratory tests to demonstrate the potential of endogenous organisms are relatively straightforward, the other two are often not as simple as they seem, especially in soils and sediments. Documenting the loss of contaminant is often difficult because of the heterogeneous distribution of the contaminant in the field; large numbers of samples may be needed in order to be able

Table 4. Will Bioremediation be a Suitable Treatment for a Site Contaminated with Organic, Nitrogenous, or Organic Contaminants?

Organic

Is the contaminant biodegradable? If the contaminant is a complex mixture of components, are the individual chemical species biodegradable? If the contaminant has been at the site for some time, biodegradation of the most readily degradable components may have already occurred. Is the residual contamination biodegradable?

Are degrading organisms present at the site?

What is limiting their growth and activity? Can this be added effectively?

Are the levels of contaminant amenable to bioremediation? Are they toxic to microorganisms? Are they so abundant that even substantial microbial activity will take too long to clean the site?

Are the clean-up standards reasonable? Are biological processes known to degrade substrates down to the levels required?

Nitrogenous

Are appropriate microorganisms present at the site?

What is limiting their growth and activity? Can this be added effectively?

Are the levels of contaminant amenable to bioremediation? Are they toxic to microorganisms? Are they so abundant that even substantial microbial activity will take too long to clean the site?

Can the nitrogenous compound be used by plants?

Are the clean-up standards reasonable? Are biological processes known to degrade substrates down to the levels required?

Inorganic

Can the contaminant be made less hazardous by changing its redox state?

Can the contaminant be brought to a reactor or constructed wetland where biological systems, microbial or plant, can extract and immobilize the contaminant?

Can plants extract the contaminant from the soil matrix?

Are the clean-up standards reasonable? Are biological processes known to accumulate contaminants down to the levels required?

to detect statistically significant decreases in absolute concentration. Providing evidence that biodegradation has indeed been stimulated is also a challenge, but if the contaminant is a complex mixture, such as a crude oil, refined petroleum, or a mixture of polychlorinated biphenyls, the least degradable compounds in the mixture can be used as conserved internal markers for quantifying the degradation of the more degradable components.

For example, hopane (Fig. 12) is a conserved marker in crude oils in at least the early stages of biodegradation (up to 80% degradation) (108); its concentration increases in the residual oil as biodegradation proceeds. Basing estimates of biodegradation on hopane allowed us to quantify the effect of the successful bioremediation strategy following the *Exxon Valdez* oil-spill (14,15). Even in refined products, the least degradable detectable analyte can serve this role. Thus, the trimethyl-phenanthrenes can be used to estimate qualitatively the degradation of diesel oils in the environment, even though these molecules are themselves biodegradable (109). Of course, since these molecules are themselves biodegradable, estimates of the rates of biodegradation of more readily biodegradable compounds are systematically underestimated using this approach, but it can

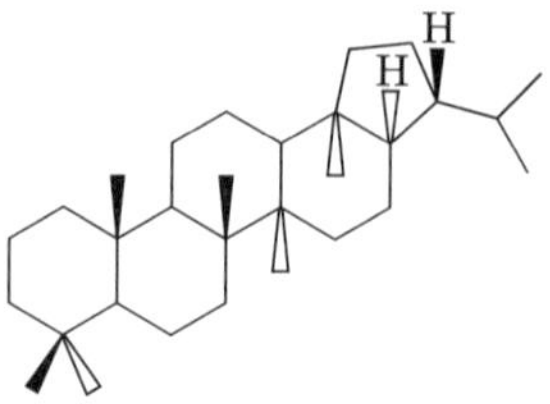

Fig. 12. $17\alpha(H),21\beta(H)$-hopane.

still provide very valuable information. The more halogenated polychlorinated biphenyls can serve a similar role in assessing the environmental fate of these products (55), and benzene is typically the last compound in BTEX plumes to be degraded, particularly under anaerobic conditions.

In situations where conserved internal markers cannot be used, such as in spills of essentially pure compounds, the evidence for enhanced biodegradation may have to be more indirect. Oxygen consumption, increases in microbial activity or population, and carbon dioxide evolution have all been used with success.

Finally a caveat. Despite its documented success in many situations, bioremediation may not always be able to meet current clean up criteria for a particular site. Some standards are so tight that they are essentially "detection limit" standards, and it is not clear that biological processes will be able to remove contaminants to such low levels. For example, the level of contaminant may be so low that it does not induce the microorganisms to produce the enzymes necessary for biodegradation. Or perhaps the contaminant is bound to soil or sediment particles in such a way that it is not available for biodegradation, although it is still extractable with aggressive solvents in analytical procedures. These are areas that require further research, but bioremediation will be more likely to fulfill its promise as an important tool in contaminated site remediation if there is progress towards standards based on bioavailability and net environmental benefit from the clean up, rather than on arbitrary absolute standards.

BIBLIOGRAPHY

1. K. Devine, in R. E. Hinchee, J. A. Kittel, and H. J. Reisinger, eds., *Applied Bioremediation of Hydrocarbons*, Battelle Press, Columbus, Ohio, 1995, pp. 53–59.
2. S. S. Butcher, R. J. Charlson, G. H. Orians, and G. V. Wolfe, *Global Biogeochemical Cycles*, Academic Press, New York, 1992.
3. I. D. Prijambada, S. Negoro, T. Yomo, and I. Urabe, *Appl. Environ. Microbiol.* **61**, 2020–2022 (1995).
4. P. Adriaens and T. M. Vogel, in L. Y. Young and C. E. Cerniglia, eds., *Microbial Transformation and Degradation of Toxic Organics*, Wiley-Liss, New York, 1995, pp. 435–486.
5. A. Paszczynski and R. L. Crawford, *Biotechnol. Progr.* **11**, 368–379 (1995).
6. J. J. Childress, *Amer. Zool.* **35**, 83–90 (1995).
7. T. D. Sharkey and E. L. Singsaas, *Nature* **374**, 769 (1995).
8. R. C. Prince, Bioremediation of Crude Oil, in R. A. Meyers, ed., *Encyclopedia of Environmental Analysis and Remediation*, John Wiley & Sons, Inc. (in press) (1998).
9. J. P. Salanitro, *Curr. Opini. Biotechnol.* **6**, 337–340 (1995).

10. J. M. Yudelson and P. D. Tinari, in R. E. Hinchee, G. D. Sayles and R. S. Skeen, eds., *Biological Unit Processes for Hazardous Waste Treatment*, Battelle Press, Columbus, Ohio, 1995, pp. 205–209.

11. National Research Council *Oil in the Sea: Inputs, Fates and Effects*, National Academy Press, Washington, D.C., 1985.

12. R. Varadaraj, M. L. Robbins, J. Bock, S. Pace, and D. MacDonald, in *Proceedings of the 1995 International Oil Spill Conference*, American Petroleum Institute, Washington, D.C., 1995, pp. 101–106.

13. R. C. Prince, *Critical Reviews Microbiology* **19**, 217–242 (1993).

14. R. C. Prince, J. R. Clark, J. E. Lindstrom, E. L. Butler, E. J. Brown, G. Winter, M. J. Grossman, R. R. Parrish, R. E. Bare, J. F. Braddock, W. G. Steinhauer, G. S. Douglas, J. M. Kennedy, P. Barter, J. R. Bragg, E. J. Harner, and R. M. Atlas, in R. E. Hinchee, B. C. Alleman, R. E. Hoeppel and R. N. Miller, eds., *Hydrocarbon Remediation*, Lewis Publishers, Boca Raton, Fla., 1994, pp. 107–124.

15. J. R. Bragg, R. C. Prince, E. J. Harner, and R. M. Atlas, *Nature* **368**, 413–418 (1994).

16. R. C. Borden, C. A. Gomez, and M. T. Becker, *Ground Water* **33**, 180–189 (1995).

17. R. D. Norris, in Ref. 1, pp. 483–487 (1995).

18. G. Battermann and M. Meier-Löhr, in Ref. 1, pp. 155–164.

19. D. R. Lovley, J. C. Woodward, and F. H. Chapelle, *Nature* **370**, 128–131 (1994).

20. B. A. Keet, in Ref. 1, pp. 329–334.

21. S. J. McMillen, N. R. Gray, J. M. Kerr, A. G. Requejo, T. J. McDonald, and G. S. Douglas, in R. E. Hinchee, G. S. Douglas, and S. K. Ong, eds., *Monitoring and Verification of Bioremediation*, Battelle Press, Columbus, Ohio, 1995, pp. 1–9.

22. B. M. Peyton, M. J. Truex, R. S. Skeen, and B. S. Hooker, in R. E. Hinchee, A. Leeson, and L. Semprini, eds., *Bioremediation of Chlorinated Solvents*, Battelle Press, Columbus, Ohio, 1995, pp. 111–116.

23. D. Kohler-Staub, S. Frank, and T. Leisinger, *Biodegradation* **6**, 229–236 (1995).

24. D. B. Janssen, F. Pries, and J. R. van der Ploeg, *Ann. Rev. Microbiol.* **48**, 163–191 (1994).

25. D. J. Arp, *Curr. Opin. Biotechnol.* **6**, 352–358 (1995).

26. S. J. Ergas, K. Kinney, M. E. Fuller, and K. M. Scow, *Biotechnol. Bioengineer.* **44**, 1048–1054 (1994).

27. K. H. Lombard, J. W. Borten, and T. C. Hazen, in R. E. Hinchee, ed., *Air Sparging for Site Remediation*, Lewis Publishers, Boca Raton, Fla., 1994, pp. 81–96.

28. T. A. Anderson and B. T. Walton, *Environ. Toxicol. Chem.* **14**, 2041–2047 (1995).

29. J. A. Oleskiewicz and M. Elektorowicz, *J. Soil Contam.* **2**, 205–208 (1993).

30. W. M. Wu, J. Nye, R. F. Hickey, M. K. Jain and J. G. Zeikus, in R. E. Hinchee, A. Leeson and L. Semprini, eds., *Bioremediation of Chlorinated Solvents*, Battelle Press, Columbus, Ohio, 1995, pp. 45–52.

31. L. M. F. Dossantos and A. G. Livingston, *Appl. Microbiol. Biotechnol.* **42**, 421–431 (1994).

32. M. Knippschild and H. J. Rehm, *Appl. Microbiol. Biotechnol.* **44**, 253–258 (1995).

33. J. Gerritse, V. Renard, J. Visser, and J. C. Gottschal, *Appl. Microbiol. Biotechnol.* **43**, 920–928 (1995).

34. G. W. Gribble, *J. Chem. Ed.* **71**, 907–911 (1994).

35. S. Fetzner and F. Lingens, *Microbiol. Rev.* **58**, 641–673 (1994).

36. P. Juteau, R. Beaudet, G. Mcsween, F. Lepine, and J. G. Bisaillon, *Can. J. Microbiol.* **41**, 862–868 (1995).

37. C. Kennes, W. M. Wu, L. Bhatnagar and J. G. Zeikus, *Appl. Microbiol. Biotechnol.* **44**, 801–806 (1996).

38. K. A. McAllister, H. Lee, and J. T. Trevors, *Biodegradation* **7**, 1–40 (1996).

39. B. C. Okeke, A. Paterson, J. E. Smith, and I. A. Watson-Craik, *Letts. Appl. Microbiol.* **19**, 284–287 (1994).

40. J. Aislabie and G. Lloyd-Jones, *Aust. J. Soil. Res.* **33**, 925–942 (1995).
41. D. A. Abramowicz, *Crit. Rev. Biotechnol.* **10**, 241–251 (1990).
42. S. Han, L. D. Eltis, K. N. Timmis, S. W. Muchmore, and J. T. Bolin, *Science* **270**, 976–980 (1995).
43. B. N. Aronstein and L. E. Rice, *J. Chem. Technol. Biotechnol.* **63**, 321–328 (1995).
44. J. L. Gentry and T. J. Simpkin, in R. E. Hinchee, R. M. Miller, and P. C. Johnson, eds., *In situ Aeration: Air Sparging, Bioventing and Related Remediation Processes*, Battelle Press, Columbus, Ohio, pp. 283–289 (1995).
45. A. M. Ferro, R. C. Sims, and B. Bugbee, *J. Environ. Qual.* **23**, 272–279 (1994).
46. S. G. Huling, D. F. Pope, J. E. Matthews, J. L. Sims, R. C. Sims, and D. L. Sorenson, in R. E. Hinchee, R. E. Hoeppel and D. B. Anderson, eds., *Bioremediation of Recalcitrant Organics*, Battelle Press, Columbus, Ohio, 1995, pp. 101–109.
47. G. M. Colores, P. M. Radehaus, and S. K. Schmidt, *Appl. Biochem. Biotechnol.* **54**, 271–275 (1995).
48. M. P. Otte, J. Gagnon, Y. Comeau, N. Maatte, C. W. Greer, and R. Samson, *Appl. Microbiol. Biotechnol.* **40**, 926–932 (1994).
49. A. J. Wall and G. W. Stratton, *Water Air Soil Poll.* **82**, 723–737 (1995).
50. F. Baud-Grasset and T. M. Vogel, in R. E. Hinchee, J. E. Fredrickson and B. C. Alleman, eds., *Bioaugmentation for Site Remediation*, Battelle Press, Columbus, Ohio, 1995, pp. 39–48.
51. F. C. Michel, C. A. Reddy, and L. J. Forney, *Appl. Environ. Microbiol.* **61**, 2566–2571 (1995).
52. C. J. Gantzer and D. Cosgriff, in Ref. 44, pp. 543–549.
53. R. U. Edgehill, *Water Res.* **30**, 357–363 (1996).
54. D. Dietrich, W. J. Hickey, and R. Lamar, *Appl. Environ. Microbiol.* **61**, 3904–3909 (1995).
55. D. L. Bedard and R. J. May, *Environ. Sci. Technol.* **30**, 237–245 (1996).
56. D. L. Bedard, S. C. Bunnell and L. A. Smullen, *Environ. Sci. Technol.* **30**, 687–694 (1996).
57. M. R. Harkness, J. B.. McDermott, D. A. Abramowicz, J. J. Salvo, W. P. Flanagan, M. L. Stephens, F. J. Mondello, R. J. May, H. J. Lobos, K. M. Carroll, M. J. Brennan, A. A. Bracco, K. M. Fish, G. L. Warner, P. R. Wilson, D. K. Dietrich, D. T. Lin, C. B. Morgan, C. B. and W. L. Gately, *Science* **259**, 503–507 (1993).
58. R. E. Dick and J. P. Quinn, *Appl. Microbiol. Biotechnol.* **43**, 545–550 (1995).
59. R. H. Kaake, D. L. Crawford, and R. L. Crawford, *Biodegradation* **6**, 329–337 (1995).
60. T. B. Hammill and R. L. Crawford, *Appl. Environ. Microbiol.* **62**, 1842–1846 (1996).
61. J. Kazumi and D. G. Capone, *Appl. Environ. Microbiol.* **62**, 2820–2829 (1995).
62. I. Mirgain, G. Green, and H. Monteil, *Environ. Technol.* **16**, 967–976 (1995).
63. C. J. Hapeman, J. S. Karns, and D. R. Shelton, *J. Agric. Food Chem.* **43**, 1383–1391 (1995).
64. D. R. Shelton, S. Khader, J. S. Karns, and B. M. Pogell, *Biodegradation* **7**, 129–136 (1996).
65. J. A. Entry, P. K. Donnelly, and W. H. Emmingham, *Appl. Soil Ecol.* **3**, 85–90 (1996).
66. D. L. Kaplan, *Curr. Opinion Biotechnol.* **3**, 253–260 (1992).
67. T. Gorontzy, O. Dryzga, M. W. Kahl, D. Bruns-Nagel, J. Breitung, E. von Loew and K.-H. Blotevogel, *Crit. Rev. Microbiol.* **20**, 265–284 (1994).
68. N. S.. Kasimov, V. B. Grebenyuk, T. V. Koroleva, and Y. V. Proskuryakov, *Eurasian Soil Science* **28**, 79–95 (1996).
69. U.S. Pat. 5,387,271 (1995), D. L. Crawford, T. O. Stevens and R. L. Crawford.
70. S. Eversmeyer, P. Faessler, and C. Bird, *Soil Groundwater Cleanup* (Dec. 27–29, 1995).

71. E. Flathman, B. J. Krupp, P. Zottola, J. R. Trausch, J. H. Carson, R. Yao, G. J. Laird, P. M. Woodhull, D. E. Jerger, and P. R. Lear, *Remediation* **6**, 57–79 (1996).
72. Z. Ronen, J. M. Bollag, C. H. Hsu, and J. C. Young, *Ground Water* **34**, 194–199 (1996).
73. M. H. Essa, S. Farooq, and G. F. Nakhla, *Water Air Soil Poll.* **87**, 267–281 (1996).
74. V. Cuomo, A. Perretti, I. Palomba, A. Verde, and A. Cuomo, *J. Appl. Phycol.* **7**, 479–485 (1995).
75. G. C. Schmidt and M. B. Ballew, in R. E. Hinchee, J. L. Means, and D. R. Burris, eds., *Bioremediation of Inorganics*, Battelle Press, Columbus, Ohio, 1995, pp. 109–116.
76. W. J. Hunter and R. F. Folett, in *Clean Water – Clean Environment – 21st Century*, Vol. II: *Nutrients*. American Society of Agricultural Engineers, St Joseph, Mich., 1995, pp. 79–82.
77. D. K. Ewart and M. N. Hughes, *Adv. Inorg. Chem.* **36**, 103–135 (1991).
78. B. Volesky and Z. R. Holan, *Biotechnol. Prog.* **11**, 235–250 (1995).
79. C. E. Turick, W. A. Apel, and N. S. Carmiol, *Appl. Microbiol. Biotechnol.* **44**, 683–688 (1996).
80. W. T. Frankenberger and U. Karlson, *Geomicrobiol. J.* **12**, 265–278 (1994).
81. D. Ahmann, A. L. Roberts, L. R. Krumholz, and F. M. M. Morel, *Nature* **371**, 750 (1994).
82. K. A. Rittle, J. I. Drever, and P. J. S. Colberg, *Geomicrobiol. J.* **13**, 1–11 (1995).
83. L. E. Macaskie, R. M. Empson, F. Lin, and M. R. Tolley, *J. Chem. Technol. Biotechnol.* **63**, 1–16 (1995).
84. K. M. Bonthrone, G. Basnakova, F. Lin, and L. E. Macaskie, *Nature Biotechnology* **14**, 635–638 (1996).
85. E. J. P. Phillips, E. R. Landa, and D. R. Lovley, *J. Ind. Microbiol.* **14**, 203–207 (1995).
86. E. Saouter, M. Gillman, and T. Barkay, *J. Ind. Microbiol.* **14**, 343–348 (1995).
87. V. Dushenkov, P. B. A. N. Kumar, H. Motto, and I. Raskin, *Environ. Sci. Technol.* **29**, 1239–1245 (1995).
88. D. E. Salt, M. Blaylock, N. P. B. A. Kumar, V. Dushenkov, B. D. Ensley, I. Chet, and I. Raskin, *Bio-Technology* **13**, 468–474 (1995).
89. W. E. Rauser, *Ann. Rev. Biochem.* **59**, 61–86 (1990).
90. D. E. Salt, R. C. Prince, I. J. Pickering, and I. Raskin, *Plant Physiol.* **109**, 1427–1433 (1995).
91. U. Kramer, J. D. Cotterhowells, J. M. Charnock, A. J. M. Baker, and J. A. C. Smith, *Nature* **379**, 635–638 (1996).
92. G. S. Banuelos and D. W. Meek, *J. Environ. Qual.* **19**, 772–777 (1990).
93. A. M. Zayed and N. Terry, *J. Plant. Physiol.* **143**, 8–14 (1994).
94. C. L. Rugh, H. D. Wilde, N. M. Stack, D. M. Thomson, A. O. Summers, and R. B. Meagher, *Proc. Natl. Acad. Sci. USA* **93**, 3182–3187 (1996).
95. L. J. Barnes, P. J. M. Scheeren, and C. J. N. Buisman, in J. L. Means and R. E. Hinchee, eds., *Emerging Technology for Bioremediation of Metals*, Lewis Publishers, Boca Raton, Fla., 1994, pp. 38–49.
96. J. M. Macy, S. Lawson, and H. DeMott-Decker, *Appl. Microbiol. Biotechnol.* **40**, 588–594 (1993).
97. L. P. Owens, K. C. Kovac, J. A. L. Kipps, and D. W. J. Hayes, in Ref. 75, pp. 89–94.
98. J. L. Kipps, in J. L. Means and R. E. Hinchee, eds., *Emerging Technology for Bioremediation of Metals*, Lewis Publishers, Boca Raton, Fla., 1994, pp. 105–109.
99. C. M. Cooney, *Environ. Sci. Technol.* **30**, A194 (1996).
100. G. Robb and J. Robinson, *Mining Environ. Manage.*, 19–21 (Sept. 1995).
101. M. E. Dodds-Smith, C. A. Payne, and J. J. Gusek, *Mining Environ. Manage.*, 22–24 (Sept. 1995).
102. J. Bender and P. Phillips, *Mining Environ. Manage.*, 25–27 (Sept. 1995).

of process liquid chromatographic separations starting from a minimum amount of laboratory data are given.

Economic Aspects

The development of biotechnology processes in the biopharmaceutical and bio-product industries is driven by the precept of being first to market while achieving a defined product purity, and developing a reliable process to meet validation requirements. The economics of bioseparations are important, but are likely to be secondary to the goal of being first to market. The cost of a lost opportunity in a tightly focused market where there is room for only a few manufacturers can be devastating for products which take 5 to 10 years and $100 to $200 $\times$ 10^6 to develop. After process and product are validated, the cost of change in any portion of the procedure can also be great, if only to satisfy regulatory constraints. Hence, once the manufacturing process is in place, changes are likely to be considered only if significant improvements result.

The three main sources of competitive advantage in the manufacture of high value protein products are first to market, high product quality, and low cost (3). The first company to market a new protein biopharmaceutical, and the first to gain patent protection, enjoys a substantial advantage. The second company to enter the market may find itself enjoying only one-tenth of the sales. In the absence of patent protection, product differentiation becomes very important. Differentiation reflects a product that is purer, more active, or has a greater lot-to-lot consistency.

Biopharmaceuticals and Protein Products.   Purification of proteins is a critical and expensive part of the production process, often accounting for ≥50% of total production costs (2). Hence, bioseparation processes have a significant impact on manufacturing costs. For small-volume, very high value biotherapeutics (Table 1), however, these costs may be considered secondary to the first to market principle unless a lower cost competitor surfaces. Annual 1995 sales were $700 million for human insulin (5), $300 million for tissue plasminogen activator, and $220 million for human growth hormone (6). The most successful bioproduct in biotechnology history, recombinant erythropoetin (EPO), had worldwide sales estimated at $1.6 to $2.6 billion in 1995 (5,7,8). Epogen is a genetically

Table 1. Unit Values and Relative Production Quantities for Selected Approved Biopharmaceuticals, 1990–1991[a]

Product	Year approved	Selling price, $/g[b]	Quantity for $200 $\times$ 10^6 in sales, kg
human insulin	1982	375	530.0
tissue plasminogen activator	1987	23,000	8.7
human growth hormone	1985	35,000	5.7
erythropoetin (Epogen)	1989	840,000	0.24
GM CSF	1991	384,000	0.52
G-CSF	1991	450,000	0.44

[a]Adapted from Ref. 2 with additional data from Ref. 4.
[b]Values are approximate and are likely to decrease.

engineered version of erythropoetin [*11096-26-7*], which is produced by the kidneys and stimulates blood stem cells to mature into red blood cells. Epogen can reverse the severe anemia often caused by kidney disease. Amgen's sales of this product, together with Neupogen (a recombinant protein that directs blood stem cells to become bacteria-fighting neutrophils), was about $1.8 billion in 1995 (9).

Bioproduct Separations

The task of quickly specifying, designing, and scaling-up a bioproduct separation is not simple. These separations are carried out in a liquid phase using macromolecules which are labile, and where conformation and heterogeneous chemical structure undergoing even subtle change during purification may result in an unacceptable product. A typical purification scheme for biopharmaceutical proteins involves the harvesting of protein-containing material or cells, concentration of protein using ultrafiltration (qv), initial chromatographic steps, viral clearance steps, additional chromatographic steps, again concentration of protein using ultrafiltration, and finally formulation (10).

Biosynthetic Human Insulin from *E. coli*.   Insulin [*9004-10-8*], a polypeptide hormone, stimulates anabolic reactions for carbohydrates, proteins, and fats thereby producing a lowered blood glucose level. Porcine insulin [*12584-58-6*] and bovine insulin [*11070-73-8*] were used to treat diabetes prior to the availability of human insulin [*11061-68-0*]. All three insulins are similar in amino acid sequence. Eli Lilly's human insulin was approved for testing in humans in 1980 by the U.S. FDA and was placed on the market by 1982 (11,12).

Human insulin was the first animal protein to be made in bacteria in a sequence identical to the human pancreatic peptide. Expression of separate insulin A and B chains were achieved in *Escherichia coli* K-12 using genes for the insulin A and B chains synthesized and cloned in frame with the β-galactosidase gene of plasmid pBR322 (13,14). Insulin's small size, 21 amino acids for the A-chain, mol wt = 2300; and 30 for the B-chain, mol wt = 3400, together with the absence of methionine (Met) and tryptophan (Trp) residues, were critical elements both in the decision to undertake cloning of this peptide hormone and in the rapid development of the manufacturing process. The Met and Trp residues, produced as a consequence of engineering and expression in *E. coli*, are hydrolyzed by reagents used during the recovery process. The presence of these amino acids in insulin would have resulted in the hydrolysis and destruction of the product (12).

Recovery and Purification.   The production of Eli Lilly's human insulin requires 31 principal processing steps of which 27 are associated with product recovery and purification (13). The production process for human insulin, based on a fermentation which yields proinsulin, provides an instructive case study on the range of unit operations which must be considered in the recovery and purification of a recombinant product from a bacterial fermentation. Whereas the exact sequence has not been published, the principle steps in the purification scheme are outlined in Figure 1**a**.

The fermentation product is a fusion protein where a portion of the Trp protein is connected to proinsulin through a Met residue (Fig. 1**b**). The *E. coli*

Yield Losses.   The numerous steps incur a built-in yield loss. For example, if only 2% yield loss were to be associated with each step, the overall yield for a purification sequence of 10 steps would be as in equation 1:

$$\eta = 100(1 - L/100)^n = 100(1 - 0.02)^{10} = 81.7\% \tag{1}$$

where η denotes yield, L the percent yield loss, and n the number of steps. If the yield loss at each step were 5%, the overall yield would only be 60%. Maximizing recovery at each step is important.

The purification of human insulin involves five separate alterations in the molecular structure, and hence, changes in physicochemical properties during its recovery and purification. The various forms are fusion protein, denatured aggregate, denatured monomers, properly folded proinsulin, and finally insulin. Whereas various purification procedures are used repeatedly, thus introducing more steps in the process, the change of removing contaminants is maximized because the contaminants are not as likely to change chemically in the same way as the insulin molecule. The final purification steps rely on multiple properties of the insulin, such as size, hydrophobicity, ionic charge, and crystallizability (13). The final purity level is reported to be >99.99% (15).

Tissue Plasminogen Activator from Mammalian Cell Culture.   Tissue-type plasminogen activator or tissue plasminogen activator [105857-23-6] (t-PA) was originally identified in tissue extracts in the late 1940s (15). Other known plasminogen activators include streptokinase from bacteria, urokinase from urine, and prourokinase from plasma (16). In 1981 the Bowes melanoma cell line was found to secrete t-PA (known as mt-PA) at 100 × higher concentrations, making possible the isolation and purification of this enzyme in sufficient quantities that antibodies could be generated and assays developed to lead to cloning of the gene for this enzyme and subsequent expression of the enzyme in both *E. coli* and a chinese hamster ovary (CHO) cell line (15,17).

Comparison of the melanoma and recombinant forms of the enzymes showed the same activity toward dissolution of blood clots. Comparison to urokinase, another thrombolytic agent, served as the basis for introducing recombinant t-PA into clinical trials in 1984 (17). Two pilot studies demonstrated that mt-PA resulted in thrombolysis without significant fibrinogenolysis. Fibrinogen, the precursor to fibrin, is important to the clotting of blood. Because mt-PA was available in limited quantities, recombinant t-PA (rt-PA) was used to carry out the first significant clinical trial. Doses of 0.375–0.75 mg rt-PA/kg body weight was found to be effective in humans for achieving 70% recannalization. Another pilot study confirmed that a dose of 80 mg over three hours gave the same results (17). The comparison of rt-PA (injected intravenously) to streptokinase IV (injected intracoronary) produced sufficiently favorable results to end the trial early, and make the results public in 1985, resulting in the use of t-PA for heart attacks (15). A trial completed in 1996 showed that t-PA administered within three hours of a stroke caused by a clot in the brain facilitated full recovery of 31% of stroke patients. Hence, another use of rt-PA is likely to develop (18).

Approximately 500,000 Americans suffer strokes each year. Many of the 80% that survive suffer paralysis and impaired vision and speech, often needing rehabilitation and/or long-term care. Hence, whereas treatment using rt-PA is likely to be expensive (costs are $2200/dose for treating heat attacks), the benefits of rt-PA could outweigh costs. In the case of heart attacks, the 10 times less expensive microbially derived streptokinase can be used. There is currently no competing pharmaceutical for treatment of strokes (18,19). Consequently, the cost of manufacture of rt-PA may not be as dominant an issue as would be the case of other types of bioproducts.

Characteristics of t-PA. Tissue plasminogen activator, a proteolytic, hydrophobic enzyme, has a molecular weight of 66,000, 12 disulfide bonds, 4 possible glycosylation sites, and a bridge of 6 amino acids connecting the principal protein structures (17,20,21). Only three of the sites (Asn-117–118, -448) are actually glycosylated (16). When administered to heart attack victims it dissolves clots consisting of platelets in a fibrin protein matrix and acts by clipping plasminogen, an active precursor protein found in the blood, to form plasmin, a potent protease that degrades fibrin (17,21). Whereas plasminogen activator is found in blood and tissues, concentrations are low (17).

The concentration of t-PA in human blood is 2–5 ng/mL, ie, 2–5 ppb. Plasminogen activation is accelerated in the presence of a clot, but the rate is slow. The dissolution of a clot requires a week or more during normal repair of vascular damage (17). Prevention of irreversible tissue damage during a heart attack requires that a clot, formed by rupture of an atherosclerotic plaque, be dissolved in a matter of hours. This rapid thrombolysis (dissolution of the clot) must be achieved without significant fibrinogenolysis elsewhere in the patient.

rt-PA is derived from a biological source, transformed CHO cells, and by definition is a biologic, not a drug. It is generally not possible to define biologics as discrete chemical entities or demonstrate a unique composition. Other biologics include blood fractionation products such as albumin and Factor VIII, and both live and killed viral vaccines.

The process used to make a biologic is closely monitored and regulated by regulatory agencies, because a significant change in the process may result in a product which is different from that previously reviewed and regulated, and hence may require a new license. Process changes made during the investigational new drug (IND) development stage, and before the license is approved, are more easily incorporated into a new product (from a regulatory point of view) than after the license is generated (15).

t-PA Production. Recombinant technology provides the only practical means of rt-PA production. The amount of t-PA required per dose is on the order of 100 mg. Cell lines of transformed CHO cells, selected for high levels of rt-PA expression using methotrexate, are grown in large fermenters (21). The purification steps for rt-PA must therefore separate out cells, virus, and DNA. The literature on the industrial practice of recovery and purification of rt-PA generated by suspension culture of chinese hamster ovary cells is limited (15). Recovering a protein derived from mammalian cells involves a number of steps (15). One possible scheme is shown in Figure 2. The culture medium is separated from the cell by sterile filtration (see MICROBIAL AND VIRAL FILTRATION). This is followed by additional removal by cross-flow filtration, ultrafiltration (qv), and chromatography to remove DNA and remaining viruses. The product protein then undergoes purification by chromatography.

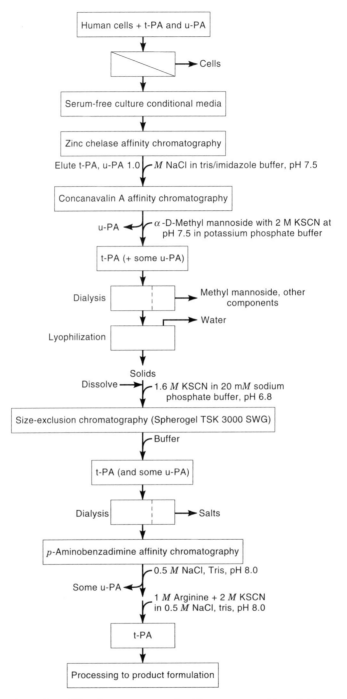

Fig. 3. Overview of purification sequence for the nonrecombinant tissue plasminogen activator (t-PA) which also contains urokinase plasminogen activator (u-PA). Serum-free culture conditional media is from normal human cell line. The temperature for all steps, except for size-exclusion chromatography (22°C), was 4°C. Adapted from Ref. 16.

for fractionating the two proteins. The yield is only 20 mg from 1400 L, illustrating the critical role of a recombinant cell line in obtaining both high yields and higher selectivity in producing a specific type of protein.

Because the culture media contained both t-PA and u-PA, this separation required several extra affinity chromatography steps, as well as dialysis/buffer exchange between the different chromatography columns (see Fig. 3). The salts and buffers added during the purification sequence must also be removed from the product at various points, adding significant complexity to the purification sequence. Desalting the buffer exchange constitute significant separation steps in the production of almost all biotechnology protein products.

Adsorption of t-PA to process equipment surfaces consisting of either stainless steel or glass was minimized by adding the detergent polyoxyethylene sorbitan monooleate (Tween 80) to the serum-free culture conditioned media at 0.01% (vol/vol). The equipment was also rinsed, before use, with phosphate buffered saline (PBS) containing 0.01% Tween 80. Hydrophilic, plastic equipment was used whenever possible. All buffers were sterile filtered. Sterile filtration of liquids and gases is usually carried out using 0.2 or 0.45 μm filters.

Manufacture of Biologics and Government Regulation

The difference between biologics and drugs is not only a matter of definition, it is also a process design issue. To compensate for the incomplete analytical capability to define biologics, regulatory agencies include parameters of the process used to make biologics in the control and monitoring. Changes in the process may yield a different product from that previously reviewed and approved. A different product requires a new license (15). Thus substantial barriers exist in terms of effort, money, and time to making significant changes in processes used to produce licensed biologics. Process changes are to be expected during the investigational new drugs (IND) phase and before the license is approved, but significant changes are rarely made after licensing. The time which can elapse between conception of an idea for a process change and granting of a new license can be as much as two years and cost several million dollars.

The definition of biologics versus drugs continues to evolve. Assignment is made on a case by case basis (25). Section 351 of the Public Health Service Act defines a biologic product as "any virus, therapeutic serum, toxin, antitoxin, vaccine, blood, blood component or derivative, allergenic product, or analogous product . . . applicable to the prevention, treatment, or cure of diseases or injuries in man." Biologics are subject to licensing provisions that require that both the manufacturing facility and the product be approved. All licensed products are subject to specific requirements for lot release by the FDA. In comparison, drugs are approved under section 505 of the FD & C act (21 USC 301-392), where there is not lot release by the FDA except for insulin products. Insulin, growth hormone, and many other hormones have been treated as drugs, whereas erythropoietin (EPO), which also fulfills the criteria of a hormone, was reviewed in the biologic division of the FDA. Insulin is derived from a bacterial fermentation; EPO is obtained from mammalian cell culture. Hormones, for the most part, are expected to be reviewed as drugs.

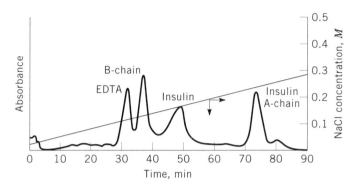

Fig. 5. Anion-exchange separation of insulin, and insulin A- and B-chains, over diethy-laminoethyl (DEAE) in a 10.9×200 mm column having a volume of 18.7 mL. Sample volume is 0.5 mL and protein concentration in 16.7 mM Tris buffer at pH 7.3 is 1 mg/mL for each component in the presence of EDTA. Eluent (also 16.7 mM Tris buffer, pH 7.3) flow rate is 1.27 mL/min, and protein detection is by uv absorbance at 280 nm. The straight line depicts the salt gradient. Courtesy of the American Chemical Society (48).

Ion-Exchange Chromatography. Ion-exchange chromatography is initiated by eluting an injected sample through a column using a buffer but no NaCl or other displacing salt. The protein, which has charged sites spread over its surface, displaces anions or cations previously equilibrated on the stationary phase, ie, the protein sites exchange with the salt counterions associated with the ion-exchange stationary phase. A protein having a greater number and/or density of charged sites displaces or exchanges more ions and hence binds more strongly than a protein having a lower charge number or charge density.

Proteins deform and change shape in response to the environment. Hence, a protein left on the surface of an ion-exchange resin for a day or longer may slowly start to unfold exposing an increasing number of charged sites to bind with the ion-exchange resin. It is possible that this process can continue until the protein binds so strongly that it is impossible to desorb the protein without dissolving it, in NaOH, for example, and destroying it. To prevent such a situation, ion-exchange chromatography must be completed in a matter of hours or less.

After the column is loaded, proteins of similar size and shape are separated by differential desorption from the ion exchanger by using an increasing salt gradient of the mobile phase. The more weakly bound macromolecules elute first; the most tightly bound elute last, at the highest salt concentration. Figure 5 is an example of an anion-exchange separation (48). Prior to injection of the sample, the column was equilibrated with the 16.7 mM Tris buffer; the EDTA stabilized the solubility of the insulin sample injected onto the column. All of the proteins are initially retained on the anion-exchange stationary phase (DEAE 650 M) during loading of the sample onto the column. Subsequent application of the NaCl gradient, formed by the controlled mixing of buffers from two reservoirs of mobile phase, elutes the proteins. One reservoir contains only the 16.7 mM Tris buffer; the second contains 0.5 M NaCl in the same buffer. Following the elution of the last peak, the column may be flushed using a buffer at a high (2.5 M NaCl) salt concentration to verify that all proteins are desorbed. In some cases, a cleaning procedure is performed by passing methanolic NaOH through the column. The column is then re-equilibrated using the salt-free buffer, by pumping approximately 10 column volumes of the buffer through the stationary phase, or until the pH of the effluent and influent are the same to prepare the column for another injection.

Amphoteric Properties Determine Conditions for Ion-Exchange Chromatography. Proteins, amphoteric polymers of acidic, basic, and neutral amino acid residues, carry both negatively and positively charged groups on the surface, the ratio depending on pH (30). The isoelectric point (pI) is the pH at which a protein has an equal number of positive and negative charges. Proteins in solutions at a pH > pI have a net negative charge. Below the pI, proteins have a net positive charge. Many proteins have a pI < 7 and are processed using buffers having a pH of 7 to 8. Thus anion exchangers (positively charged stationary phases) are popular for protein chromatography. Ion-exchange matrices derivatized having negatively charged groups are cation exchangers. These bind positively charged proteins, ie, cations, when the mobile phase pH is < pI.

The selection of the pH of the buffer, as well as the type of ion-exchange (anion or cation) stationary phase is a function of the amphoteric nature of the protein and protein stability as a function of pH. For example, for a protein stable at pH > 6.5, having pI = 5.5, an anion exchanger is appropriate when the separation is run in a buffer of pH > 6.5. If this protein were stable at pH = 5, and the pI = 5.5, a cation exchanger and buffer of pH < 5.0 would be appropriate.

The ion-exchange (qv) groups used to derivatize stationary phases for the purification of proteins are summarized in Table 2. Corresponding buffers are given in Table 3. Strong anion and cation exchangers are almost fully ionized at pH = 3–11 and coincide with the pH range of protein purification. Weak anion and cation exchangers have a narrower pH range over which they are ionized. Anion exchangers are preferred because desorption of the protein is more readily accomplished at lower salt concentrations.

Table 2. Ion-Exchange Groups Used in Protein Purification[a]

Name	Abbreviation	Formula
	Weak anion	
aminoethyl	AE	$-C_2H_4NH_3^+$
diethylaminoethyl	DEAE	$-C_2H_4NH(C_2H_5)_2^+$
	Weak cation	
carboxy	C	$-COO^-$
carboxymethyl	CM	$-CH_2COO^-$
	Strong anion	
trimethylaminoethyl	TAM	$-CH_2N(CH_3)_3^+$
triethylaminoethyl	TEAE	$-C_2H_4N(C_2H_5)_3^+$
diethyl-2-hydroxypropylaminoethyl	QAE	$-C_2H_4N^+(C_2H_5)_2CH_2CH(OH)CH_3$
	Strong cation	
sulfo	S	$-SO_3^-$
sulfomethyl	SM	$-CH_2SO_3^-$
sulfopropyl	SP	$-C_3H_6SO_3^-$

[a]Courtesy of IRL Press (30).

Table 3. Buffers for Ion-Exchange Chromatography[a]

Buffer	pK	.	Buffering range
Anion exchange			
L-histadine	6.15		5.5–6.0
imidazole	7.00		6.6–7.1
triethanolamine	7.77		7.3–7.7
Tris	8.16		7.5–8.0
diethanolamine	8.80		8.4–8.8
Cation exchange			
acetic acid	4.76		4.8–5.2
citric acid	4.76		4.2–5.2
Mes	6.15		5.5–6.7
phosphate	7.20		6.7–7.6
hepes	7.55		7.6–8.2

[a]Courtesy of IRL Press (30).

Size Exclusion (Gel-Permeation) Chromatography.   Size-exclusion chromatography is often referred to as gel-permeation chromatography because the stationary phases are usually made up of soft spherical particles which resemble gels. Separation occurs by a molecular sieving effect (see MOLECULAR SIEVES; SIZE SEPARATION). The larger molecules, which explore less of the intraparticle void fraction (ie, pores) than smaller molecules, elute first because the former spend less time inside the stationary phase than the latter. Separation can be achieved if the porosity of the stationary phase is properly selected and there is a significant difference in the sizes of the molecules to be separated. This size difference is measured in terms of hydrodynamic ratio. To select the stationary phase having the appropriate pore size requires that the size of the proteins be known.

The apparatus utilized to carry out size-exclusion (gel-permeation) chromatography is analogous to that used for isocratic operating conditions (see Fig. 4). The column is packed with a gel-filtration stationary phase, selected according to the molecular weight of the protein of interest (31). A variety of commercially available gel-filtration matrices facilitates separations ranging from molecular weights of 50 to 10^8 (Fig. 6). However, a single gel having a porosity which is capable of sieving molecules over the entire separation range does not exist.

An example of a size-exclusion chromatogram is given in Figure 7 for both a bench-scale (23.5 mL column) separation and a large-scale (86,000 mL column) run. The stationary phase is Sepharose CL-6B, a cross-linked agarose with a nominal molecular weight range of $\sim$5000–2×10^6 (see Fig. 6) (31).

Buffer Exchange and Desalting.   A primary use of size-exclusion chromatography (sec) is for removal of salt or buffer from the protein, ie, desalting and buffer exchange (32). The difference in molecular weights is large; salts generally have a mol wt <200, whereas mol wts of proteins are between 10,000 and 60,000.

Alternative methods of desalting and buffer exchange include continuous diafiltration, countercurrent dialysis (ccd), a membrane separation technique, and cross-flow filtration, which uses membranes (see MICROBIAL AND VIRAL FILTRATION). Both of those methods rely on filtration at a molecular scale, using

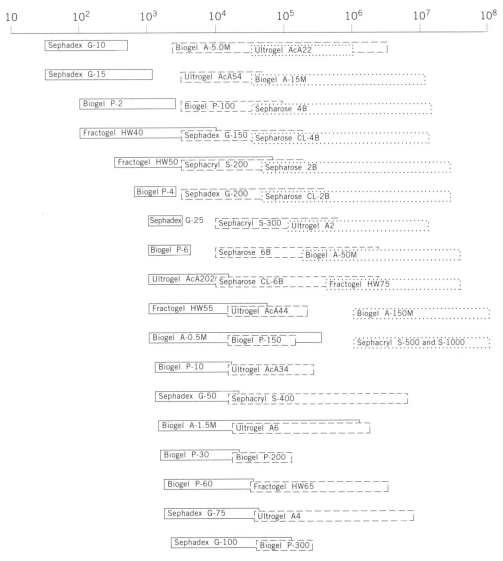

Fig. 6. Fractionation ranges of commercially available gel-filtration matrices: (□) small, (⊏⊐) medium, and (⋯) large (31).

membranes having porosities which reject proteins but allow passage of salts. Membrane methods are often preferred for an unspecified protein because these procedures are less costly and enable higher throughput than size-exclusion chromatography. Buffer exchange, used to remove denaturing agents in order to induce refolding of proteins, to remove buffers between purification steps, or to remove buffers and other reagents from the final product, is usually carried out at later steps in a recovery sequence (see Figs. 2 and 3). Equations for calculating separation efficiencies and recovery yields for all three methods for a specific case study using a recombinant protein of 160,000 mol wt are available (32). Size-exclusion chromatography using Sephadex G-25M gel-filtration media in

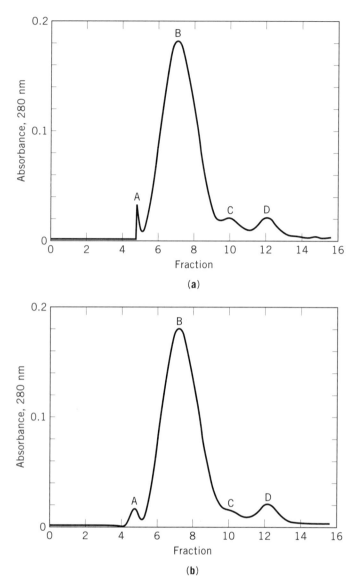

Fig. 7. Chromatograms of size-exclusion separation of IgM (mol wt = 800,000) from albumin (69,000) where A–D correspond to IgM aggregates, IgM, monomer units, and albumin, respectively, using (**a**) FPLC Superose 6 in a 1 × 30-cm long column, and (**b**) Sepharose CL-6B in a 37-cm column. Courtesy of American Chemical Society (24).

this case had disadvantages compared to the membrane filtration techniques giving 100 × less complete ion removal, 130–1200% greater dilution, 30% higher cost, 66% higher (eluting) buffer requirement, 50% higher space on the plant floor, 50% higher operating time, and half the throughput. However, cross- or tangential-flow filtration (tff) required up to 90 passes through a pump whereas sec and ccd were single pass. Other disadvantages of tff include frequent change-out of membranes and relatively large volumes of protein feed being required for laboratory-scale tests, a particular disadvantage for recombinant products. The more recent testing of a novel size-exclusion stationary phase, however, which facilitates rapid preparative (process-scale) separation of salts from protein in less than seven minutes, shows that process size-exclusion chromatography is capable of high throughput while reducing the volume needed to obtain proper refolding of the protein. Salts causing the protein to be denatured were rapidly removed and reduced to a level where protein refolding occurred (33). The use of sec is likely to continue to be a widely practiced technology in industry. Rapid size-exclusion columns for the purpose of buffer exchange have been developed which enable desalting to be achieved at linear velocities of 500–600 cm/h (33), significantly increasing throughput and reducing operating time and plant floor space. Further, sec using gel-filtration media on cellulosic-based material has a special niche for the partial and controlled separation of denaturing salts from recombinant proteins for purposes of refolding. The development of such rapid desalting techniques is important because of the larger volumes of proteins needing to be processed in industry.

Column Size. Size-exclusion chromatography columns are generally the largest column on a process scale. Separation is based strictly on diffusion rates of the molecules inside the gel particles. No proteins or other solutes are adsorbed or otherwise retained owing to adsorption, thus, significant dilution of the sample of volume can occur, particularly for small sample volumes. The volumetric capacity of this type of chromatography is determined by the concentration of the proteins for a given volume of the feed placed on the column.

The volume of the solvent between the point of injection and the peak maximum of the eluting protein is defined as the elution volume, V_e (Fig. 8). The fluid volume between the particles of the stationary phase is the extraparticulate void volume or exclusion volume, V_o. The porosity of the stationary phase, determined by the extent of cross-linking of the polymers which make up the particles of a gel-permeation matrix, determines the extent to which a protein or other solute can explore the intraparticulate void volume. The higher the cross-linking, the smaller the effective pore size and the lower the molecular weight (or size) of the molecule which is excluded from the gel. Hence, the apparent porosities of a gel-permeation column are a function of the molecular probes used to measure it. For a large molecule that is completely excluded, the void volume is equivalent to the extraparticulate void volume, V_o. Molecules that are small enough to penetrate the gel have an elution volume $> V_o$. A small molecule, such as a salt, can potentially explore almost all of the bed volume and has the following elution volume:

$$V_e = V_o + K_d V_s \qquad (2)$$

where K_d represents the fraction of the volume of the mobile phase inside the particle, V_s, which can be explored by the molecular probe. For a probe small enough to explore all of the intraparticle void volume, K_d is 1 and the elution volume is $V_o + V_s$. Because the combined volume of the fluid between the particles, V_o, and inside the particles, V_s, cannot exceed the total volumes of the column, V_t, V_e must be less than column volume V_t. All components injected into a size-exclusion column thus elute in one column volume.

imbibe significant quantities of water and have bed volumes ranging from 2 to 3 mL/g dry weight of stationary phase for Sephadex G-10 (nominal molecular weight cutoff of 10,000) to 20–25 mL/g dry weight of stationary-phase Sephadex G-200 (nominal molecular weight cutoff of 200,000). Their structures resemble a cross-linked spider web, where the extent of cross-linking or association between hydrated polymer chains, rather than specific pore sizes, determine the apparent pore size distribution. The hydrophilic character of the polymers which make up these gels require cross-linking to prevent dissolution. The hydrophilic character is compatible with the majority of industrially relevant proteins, most of which can be denatured by hydrophobic surfaces but preserved in active confirmation at hydrophilic conditions. This property can be offset by poor flow properties, however, particularly for lightly cross-linked gels, because these gels are soft and have a tendency to compress when flow rates exceed a threshold which decreases with decreasing extents of cross-linking. Hence, Sephadex G-10 has the highest cross-linking and flow stability, and the lowest specific bed volume, but also has the lowest effective pore size or porosity, limiting its sieving capabilities. Sephadex G-200 has the lowest cross-linking and flow stability and the highest specific bed volume and effective pore size. Sephadex 200 is useful for separating high molecular weight proteins, but at relatively low flow rates (Table 5).

The distribution coefficient, K_d, represents the fractional volume of a specific stationary phase explored by a given solute, represented by equation 3:

$$K_d = \frac{V_e - V_o}{V_s} \qquad (3)$$

where V_o is the void volume, V_s is the volume of the solvent (usually acqueous buffer) inside the gel which is available to very small molecules, and V_e is the elution volume of a small volume of injected molecular probe. The measurement

Table 5. Comparison of Gel-Permeation Stationary Phase[a]

Sephadex G-X[b]	Specific volume water mL/g dry gel	Permeability, K_o	Operating pressure[c], kPa[d]	Flow rate[c] water, mL/(cm^2·h)
10	2–3	19	f	f
15	2.5–3.5	18	f	f
25	4–6	9–290[e]	f	f
50	9–11	13.5–400[e]	f	f
75	12–15		160	77
100	15–20		96	50
150	20–30		36	23
200	30–40		16	12

[a] Adapted from Ref. 34.
[b] Corresponds to the nominal cutoff value for wt $\times 10^3$, eg, G-10 has a mol wt cutoff value of ~10,000.
[c] Value is maximum unless otherwise noted.
[d] To convert kPa to cm H_2O, multiply by 10.2.
[e] Depends on particle size (dp); as dp increases, K_o increases.
[f] May be calculated using Darcy's law: $U = K_o (\Delta P/L)$, where U is linear flow as mL/(cm^2·h), L is bed length in cm, ΔP is pressure drop over gel bed in kPa[d], and the maximum pressure is 30.4 kPa[d] (310 cm H_2O).

of V_s is difficult, requiring use of an ion or small molecule which freely diffuses into all of the fluid volume inside the gel particles and then is readily detected at the outlet of the column. D_2O and radioactive ^{23}Na have been used. The latter is detected by a refractive index detector. An indirect measurement of V_s is more convenient and adequate. The column void volume (Fig. 9) may be measured using a soluble, high molecular weight target molecule which, because it does not explore any of the internal fluid volume of the stationary phase, it only distributed in the mobile phase. Blue dextran, a water-soluble, sulfonated, blue-colored dextran having mol wt > 669,000, manufactured by Pharmacia, and DNA (Type III from salmon tests) have been employed (26). The total column volume, V_t, can be calculated from the dimensions of the bed, although the direct measurement of column volume using water displacement before packing is more accurate. (It should be noted that total column volume is also represented by V_c in the recent literature.) The difference, $V_t - V_o$, is then taken as an approximation of V_s. On this basis, K_{av}, the fraction of stationary phase volume available for a given solute species, is defined as in equation 4:

$$K_{av} = \frac{V_e - V_o}{V_t - V_o} \tag{4}$$

The constant K_{av}, is not a true partition coefficient because of difference, $V_t - V_o$, includes the solids and the fluid associated with the gel or stationary phase. By definition, V_s represents only the fluid inside the stationary-phase particles and does not include the volume occupied by the solids which make up the gel. Thus K_{av} is a property of the gel, and like K_d it defines solute behavior independently of the bed dimensions. The ratio of K_{av} to K_d should be a constant for a given gel packed in a specific column (34).

Selectivity curves result from measured values of K_{av} plotted vs log (mol wt) enabling molecular weight determination of globular proteins having similar asymmetry factors. A sphere has an asymmetry factor of 1; an ellipsoid has

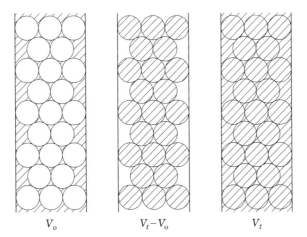

Fig. 9. Diagrammatic representation of V_t and V_o. $V_t - V_o$ includes the volume of the solid material forming the matrix of each bead. Courtesy of Pharmacia (34).

stationary phases having 1–5 μm particle sizes and for sample volumes which typically range from 1 to 10 μL.

One purpose in monitoring a protein product is to detect the presence of a change in which as little as one amino acid has been chemically or biologically altered or replaced during the manufacturing process. Variant amino acid(s) in a protein may not affect protein retention during reversed-phase chromatography if the three-dimensional structure of the polypeptide shields the variant residue from the surface of the reversed-phase support (20). Reversed-phase chromatography discriminates between different molecules on the basis of hydrophobicity. Because large proteins may contain only small patches of hydrophobic residues, these patches may not correlate to the molecular modifications which a reversed-phase analytical method seeks to detect. The reversed-phase method must therefore be completely validated, and preferably combined with controlled chemical and/or proteolytic hydrolysis followed by chromatography or electrophoresis (see ELECTROSEPARATIONS) of the cleared protein to give a map of the resulting peptide fragments (20,43).

A peptide map is generated by cleaving a previously purified protein using chemicals or enzymes. Hydrolytic agents having known specificity are used to perform limited proteolysis followed by resolution and identification of all the peptide fragments formed. Identification of changes, and reconstruction of the protein's primary structure, is then possible. Reagents and enzymes which cleave specific bonds are discussed in the literature (44).

An example of a peptide map generated by trypsin hydrolysis of recombinant tissue plasminogen activator (rt-PA) is shown in Figure 12. The chromatogram shows the resolving power or reversed-phase high performance liquid chromatography in separating peptides obtained from t-PA in which the disulfide bonds had been reduced and alkylated prior to enzyme hydrolysis. The small peptides formed have little or no three-dimensional structure. Hence, measurable shifts in elution profiles occur when there are variant amino acids because a single amino acid change in a peptide has a larger effect on its solubility and retention than the same change has in a protein. The replacement of arginine at position 275 in a normal t-PA molecule with glutamic acid results in a significant peak shift (see Fig. 12) (43), showing how tryptic mapping can be a suitable method for monitoring lot-to-lot consistency of this particular recombinant product (20).

Reversed-phase high performance liquid chromatography has come into use for estimating the purity of proteins and peptides as well. However, before employed, a high performance liquid chromatographic (hplc) profile of a given protein must be completely validated (43).

Insulin Purification.   An example of the purification of recombinant product by reversed-phase chromatography is recombinant insulin, a polypeptide hormone. Insulin consists of 51 amino acid residues in two chains and is relatively small. Reversed-phase chromatography is used after most of the other impurities have been removed by a prior ion-exchange step (see Fig. 1) (12). The method utilizes a process-grade C_8 reversed-phase support (Zorbex) having a particle size of 10 μm (14). Partially purified insulin crystals, dissolved in a water-rich mobile phase, are applied to the column and then eluted in a linear gradient generated by mixing 0.25 M aqueous acetic acid to 60% acetonitrile. The acidic mobile

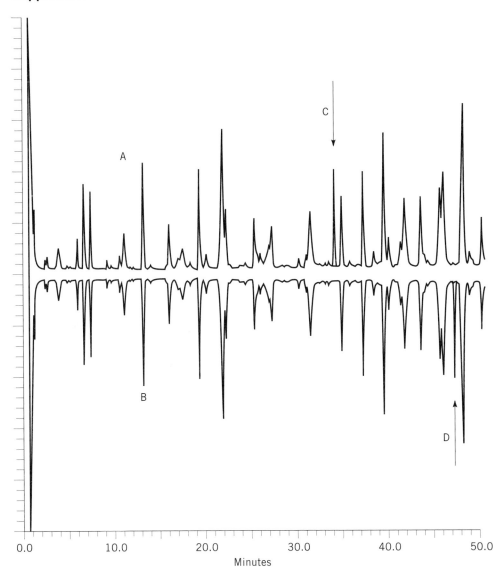

Fig. 12. Tryptic map of rt-PA (mol wt = 66,000) showing peptides formed from hydrolysis of reduced, alkylated rt-PA. Separation by reversed-phase octadecyl (C$_{18}$) column using aqueous acetonitrile with an added acidic agent to the mobile phase. Arrows show the difference between A, normal, and B, mutant rt-PA where the glutamic acid residue, D, has replaced the normal arginine residue, C, at position 275. Courtesy of Marcel Dekker (43).

phase gives excellent resolution of insulin from structurally similar insulin-like components. The ideal pH is from 3.0 to 4.0, below insulin's isoelectric point, pI = 5.4. Under mildly acidic conditions insulin may deamidate to monodesamido insulin, but if the reversed-phase separation is done within a matter of hours, the deamidation can be minimized.

This reversed-phase chromatography method was successfully used in a production-scale system to purify recombinant insulin. The insulin purified by reversed-phase chromatography has a biological potency equal to that obtained from a conventional system employing ion-exchange and size-exclusion chromatographies (14). The reversed-phase separation was, however, followed by a size-exclusion step to remove the acetonitrile eluent from the final product (12,14).

Whereas recombinant proteins produced as inclusion bodies in bacterial fermentations may be amenable to reversed-phase chromatography (42), the use of reversed-phase process chromatography does not appear to be widespread for higher molecular weight proteins.

Reversed-Phase Process Chromatography. Polypeptides, peptides, antibiotics, alkaloids, and other low molecular weight compounds are amenable to process chromatography by reversed-phase methods. There are numerous examples of bioproducts which have been purified using reversed-phase chromatography. The manufacture of salmon calcitonin, a 32-residue peptide used for treatment of post-menopausal osteoporosis, hypercalcemia, and Paget's disease of the bone, includes reversed-phase chromatography. This peptide, commercially prepared on a kilogram scale by a solid-phase synthesis, is then purified by a multimodal purification train. Reversed-phase chromatography is the dominant technique used by Rhône-Poulenc Rorer (45).

Another example is the purification of a β-lactam antibiotic, where process-scale reversed-phase separations began to be used around 1983 when suitable, high pressure process-scale equipment became available. A reversed-phase microparticulate (55–105 μm particle size) C_{18} silica column, with a mobile phase of aqueous methanol having 0.1 M ammonium phosphate at pH 5.3, was able to fractionate out impurities not readily removed by liquid–liquid extraction (37). Optimization of the separation resulted in recovery of product at 93% purity and 95% yield. This type of separation differs markedly from protein purification in feed concentration ($\approx$50–200 g/L for cefonicid vs 1 to 10 g/L for protein), molecular weight of impurities ($<$5000 compared to 10,000–100,000 for proteins), and throughputs ($\approx$1–2 mg/(g stationary phase·min) compared to 0.01–0.1 mg/(g·min) for proteins).

Reversed-phase separation was also found to purify diastereomer precursors used in the chemical synthesis of the insect sex phermone of *Limantria dispar*, a pest which attacks oak trees. The liquid chromatography columns tested had dimensions of up to 15 cm id by 130 cm long, and were able to purify up to 708 g of starting material in 4.1 L sample using a column having 23 L of stationary phase. The throughput is estimated to have been on the order of 0.2–0.4 mg/(g·min), where separation was obtained using a gradient of hexane and diethyl ether.

Small Particle Silica Columns. Process-scale reversed-phase supports can have particle sizes as small as 5–25 μm. Unlike polymeric reversed-phase sorbents, these small-particle silica-based reversed-phase supports require high pressure equipment to be properly packed and operated. The introduction of axial compression columns has helped promote the use of high performance silica supports on a process scale. Resolution approaching that of an analytical-scale separation can be achieved using these columns that can also be quickly packed.

These columns consist of a plunger fitted into a stainless steel column. The particles are placed into the column in a slurry. The plunger then squeezes or compacts the bed in an axial direction to give a stable, tightly packed bed. This type of column must be operated at pressures of up to 10 MPa (100 bar), but also gives excellent resolution in run times of an hour or less (36).

Hydrophobic Interaction Chromatography. Hydrophobic interactions of solutes with a stationary phase result in their adsorption on neutral or mildly hydrophobic stationary phases. The solutes are adsorbed at a high salt concentration, and then desorbed in order of increasing surface hydrophobicity, in a decreasing kosmotrope gradient. This characteristic follows the order of the lyotropic series for the anions: phosphates > sulfates > acetates > chlorides > nitrates > thiocyanates. Anions which precipitate proteins less effectively than chloride (nitrates and thiocyanates) are chaotropes or water structure breakers, and have a randomizing effect on water's structure; the anions preceding chlorides, ie, phosphates, sulfates, and acetates, are polar kosmotropes or water structure makers. These promote precipitation of proteins. Kosmotropes also promote adsorption of proteins and other solutes onto a hydrophobic stationary phase (46). These kosmotropes have other beneficial characteristics which include increasing the thermal stability of enzymes, decreasing enzyme inactivation, protecting against proteolysis, increasing the association of protein subunits, and increasing the refolding rate of denatured proteins. Hence, utilization of hydrophobic interaction chromotography is attractive for purification of proteins where recovery of a purified protein in an active and stable conformation is desired (46,47).

Salt Effects. The definition of a capacity factor k' in hydrophobic interaction chromatography is analogous to the distribution coefficient, K_{av}, in gel permeation chromatography:

$$k' = \frac{V_e - V_o}{V_o} \tag{5}$$

However, because protein retention owing to adsorption can occur, the value of k' can be greater than one, ie, elution of the most retained peak need not occur after one column volume of mobile phase has passed through the column. The retention behavior of lysozyme on a polymeric hydrophobic interaction support follows the preferential interaction parameter of the lyotropic series of anions (Fig. 13). The preferential interaction parameter is a measure of the net salt inclusion or exclusion of the hydration layer. The higher the value, the larger the disrupting effect of the salt. This analysis led to derivation and experimental validation of the capacity factor for lysozyme with respect the lyotropic number of the anion for a hydrophilic vinyl polymer support having an average particle diameter of 30 μm, and average pore size of 100 nm. This capacity factor has the following form:

$$k' = a[C]^d[N_x - b] + h \tag{6}$$

where a, b, d, and h are protein specific parameters, N_x is the lyotropic number, and C is salt concentration in M. Hydrophobic interaction parameters can then

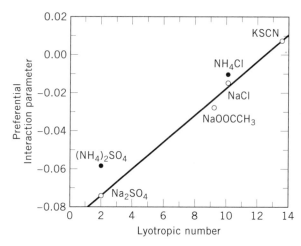

Fig. 13. Preferential interaction parameter vs lyotropic number for lysozyme on (o) bovine serum albumin and (•) Toyopearl. Courtesy of American Chemical Society (47).

be estimated for experimental peak retention data as changes in retention time upon a change in salt or salt concentration. An example of how salt type and concentration affect retention of lysozyme is illustrated in Figure 14. A similar functional relation was found for myoglobin with respect to a hydrophilic vinyl polymer derivatized using butyl groups (47).

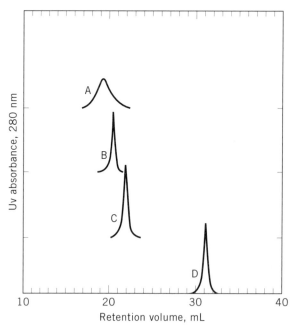

Fig. 14. Chromatographic retention of 20 μL of a 3 mg/mL solution of lysozyme on Toyopearl HW-65S using a 50 cm $\times$ 8 min ID column in 1.3 M ammonium salt, 20 mM Tris mobile phase at 1 mL/min for A, NH$_4$I; B, NH$_4$Cl; C, NH$_4$OOCCH$_3$; and D, (NH$_4$)$_2$-SO$_4$. Courtesy of American Chemical Society (47).

Various types of proteins have been purified using hydrophobic interaction chromatography including alkaline phophatase, estrogen receptors, isolectins, strepavidin, calmodulin, epoxide hydrolase, proteoglycans, hemoglobins, and snake venom toxins (46). In the case of cobra venom toxins, the order of elution of the six cardiotoxins supports the hypothesis that the mechanism of action is related to hydrophobic interactions with the phospholipids in the membrane.

The recovery of recombinant chymosin from a yeast fermentation broth showed that large-scale hydrophobic interaction chromatography could produce an acceptable product in one step. Chymosin, which used to be obtained from the lining of the stomachs of calves, is used in cheesemaking, and its cost is an issue. Because the capacity of the hydrophobic interaction stationary-phase is limited, an alternative method has been developed in which the enzyme is extracted into a two-phase polyethylene glycol (PEG) salt system. The partition for the chymosin into PEG coefficient is 100, and hence enables efficient recovery of this in one step. Together with a subsequent ion-exchange step, this method gives a suitably purified chymosin. The use of hydrophobic interaction chromatography may have helped to indicate that two-phase extraction is a viable approach (10).

Affinity Chromatography.   The concept of affinity chromatography, credited to the discovery of biospecific adsorption in 1910, was reintroduced as a means to purify enzymes in 1968 (49). Substrates and substrate inhibitors diffuse into the active sites of enzymes irreversibly or reversibly binding there. Conversely, if the substrate or substrate inhibitor is immobilized through a covalent bond to a solid particle of stationary phase having large pores, the enzyme should be able to diffuse into the stationary phase and bind with the substrate or inhibitor. Because the substrate is small (mol wt <500) and the enzyme large ($>15,000$), the diffusion of the enzyme to its binding partner at a solid surface can be sterically hindered. The placement of the substrate at the end of an alkyl or glycol chain tethered to the stationary phase's surface reduces hindrance and forms the basis of affinity chromatography. This concept has also been applied to ion-exchange chromatography under the names of tentacle or fimbriated stationary phases.

The realization that enzymes could be selectively retained in a chromatography column packed with particles of immobilized substrates or substrate analogues led to experiments with other pairs of binding partners. Numerous applications of affinity chromatography developed, given the specific and reversible yet strong affinity of biological macromolecules for numerous specific ligands or effectors. These interactions have been exploited for purposes of highly selective, but often expensive protein purifications, recovery of messenger ribonucleic acid (mRNA) in some recombinant DNA applications, and study of mechanisms of protein binding with effector molecules (49).

Minimization of Nonspecific Binding.   The purpose of affinity chromatography is the highly selective adsorption and subsequent recovery of the target biomolecule. Loss of specificity occurs when macromolecules, other than the targeted materials, adsorb onto the stationary phase owing to hydrophobic or ionic interactions. For example, a spacer arm, which allows the binding ligand on the column to be located away from the matrix surface, can improve accessibility and reduce steric hindrance to the immobilized ligand, often decreasing selectivity. Hexamethylenediamine, a common spacer arm used initially in affinity

separation is reduced to separating two different species. For example, in the case of enzyme-catalyzed esterification, the originally difficult enantiomeric separation is reduced to the separation of the ester of one optical isomer from the alcohol or acid of the other optical isomer of the original starting material, and may be accomplished using a variety of conventional separation methodologies (7). One disadvantage of enzymatically based methods is that only one enantiomer is obtained and there is usually no analogous method for producing the opposite enantiomer.

An alternative method of creating a chiral environmental is to derivatize a chiral analyte with an optically pure reagent, thus, producing diastereomers. The resultant diastereomers, containing more than one chiral center, have slightly different melting and boiling points and can often be separated using conventional methods. A number of chiral derivatizing agents, as well as the types of compounds for which they are useful, have been developed and are listed in Table 1. Limitations of this approach include lack of suitable functionality in the analyte that can be derivatized with an appropriate enantiomerically pure derivatizing agent, unavailability of a suitable derivatizing agent of sufficiently high or at least known optical purity, difficulty of removing the derivatizing group after the desired separation has been accomplished, enantiodiscrimination during derivatization, potential racemization either during derivatization or removal or the chiral derivatizing group (which is not always possible), and the additional validation required to confirm that the enantiomeric ratio of the final product corresponds to the original enantiomeric ratio.

Use of Chiral Additives

Another method for creating a chiral environment is to add an optically pure chiral selector to a bulk liquid phase. Historically, many of the chiral selectors currently available as chiral stationary phases for high performance liquid chromatography originated as chiral mobile-phase additives, particularly in thin-layer chromatography (tlc). Chiral additives have several advantages over chiral stationary phases and continue to be the predominant mode for chiral separations by tlc (8) and capillary electrophoresis (ce) (9). First of all, the chiral selector added to a bulk liquid phase can be readily changed. The use of chiral additives allows chiral separations to be done using less expensive, conventional stationary phases. A wider variety of chiral selectors are available to be used as chiral additives than are available as chiral stationary phases, thus, providing the analyst with considerable flexibility. Finally, the use of chiral additives may provide valuable insight into the chromatographic conditions and/or likelihood of success with a potential chiral stationary-phase chiral selector. This is particularly important for the development of new chiral stationary phases because of the difficulty and cost involved.

Chiral additives, however, do pose some unique problems. Many chiral agents are expensive or are not commercially available, and therefore, must be synthesized. The presence of the chiral additive in the bulk liquid phase may also interfere with detection or recovery of the analytes. Finally, the presence of enantiomeric impurity in the chiral additive may add analytical complications (10).

Table 1. Analyte Functional Groups and Chiral Derivatizing Reagents

Analyte functional group	Derivatizing agent	Product	Examples of derivatizing agents
carboxylic acid (acid or base catalyzed)	alcohol	ester	(−)-menthol
amine (1°)	amine	amide	1-phenylethylamine 1-(1-naphthyl)ethylamine
amine (1° and 2°)	aldehyde	isoindole	o-phthaldialdehyde–2-mercaptoethanol
	anhydrides	amide	γ-butyloxycarbonyl-L-leucine anhydride O,O-dibenzoyltartaric anhydride
	acyl halides	amide	(R)-(−)-methylmandelic acid chloride α-methoxy-α-trifluoromethylphenylacetyl chloride
	isocyanates	urea	α-methylbenzyl isocyanate 1-(1-naphthyl)ethyl isocyanate
	isothiocyanate	thiourea	2,3,4,6-tetra-O-acetyl-β-D-glucopyranosyl isothiocyanate α-methylbenzyl isothiocyanate
(1°, 2°; can N-dealkylate 3°)	chloroformates	carbamate	(−)-menthyl chloroformate (+)-1-(9-fluorenyl)ethylchloroformate
alcohols	acyl halides	ester	(−)-menthoxy acid chloride (S)-O-propionylmandelyl chloride
	anhydrides	ester	(S,S)-tartaric anhydride
	chloroformate	carbonate	(−)-menthyl chloroformate
	isocyanate	carbamate	α-methylbenzyl isocyanate

Thin-Layer Chromatography.   Thin-layer chromatography (tlc) offers several advantages for chiral separations and in the development of new chiral stationary phases. Besides being inexpensive, tlc can be used to screen mobile-phase conditions rapidly (ie, organic modifier content, pH, etc), chiral selectors, and analytes. Several different analytes may be run simultaneously on the same plate. Usually, no preequilibration of the mobile phase and stationary phase is required. In addition, only small amounts of mobile phase, and therefore, chiral mobile-phase additive, are required. Another significant advantage is that the analyte can always be unambiguously found on the tlc plate.

Two mechanisms for chiral separations using chiral mobile-phase additives, analogous to models developed for ion-pair chromatography, have been proposed to explain the chiral selectivity obtained using chiral mobile-phase additives. In one model, the chiral mobile-phase additive and the analyte enantiomers form "diastereomeric complexes" in solution. As noted previously, diastereomers may have slightly different physical properties such as mobile phase solubilities or slightly different affinities for the stationary phase. Thus, the chiral separation can be achieved with conventional columns.

An alternative model has been proposed in which the chiral mobile-phase additive is thought to modify the conventional, achiral stationary phase *in situ*, thus, dynamically generating a chiral stationary phase. In this case, the enantioseparation is governed by the differences in the association between the enantiomers and the chiral selector in the stationary phase.

Several different types of chiral additives have been used including (1R)-(−)-ammonium-10-camphorsulfonic acid (11), cyclodextrins (12,13), proteins, and various amino acid derivatives such as N-benzoxycarbonyl-glycyl-L-proline as well as macrocyclic antibiotics (14). Chiral counter-ions such as (1R)-(−)-ammonium-10-camphorsulfonic acid and N-benzoxycarbonyl-glycyl-L-proline have been used under normal phase conditions (eg, ca 2.5 mM with 1 mM triethylamine in methylene chloride on a diol tlc plate), promoting ion-pair associations (15). In contrast, the cyclodextrins (16,17), proteins, and amino acid derivatives have been used exclusively under aqueous mobile phase conditions. In the case of the cyclodextrins, the limited aqueous solubility of β-cyclodextrin (~0.17 M at room temperature), the most commonly used cyclodextrin, can be enhanced by using a saturated urea solution. In addition, it is recommended that 0.6 M NaCl can be used to stabilize the binder by which the stationary phase is attached to the glass support (18).

Chiral separation validation in tlc may be accomplished by recovering the individual analyte spots from the plate and subjecting them to some type of chiroptical spectroscopy such as circular dichroism or optical rotary dispersion. Alternatively, the plates may be analyzed using a scanning densitometer. Scanning densitometers irradiate the surface of the plate at a specified wavelength (in the ultraviolet or visible regions) and can measure the intensity of the reflected beam. A trace of the reflected beam vs distance has the general appearance of a chromatogram and an example is shown in Figure 2 (19). The relative peak heights or areas of the two enantiomers obtained at two or more different wavelengths should remain constant because the extinction coefficients of the enantiomers are identical at every wavelength.

Capillary Electrophoresis.   Capillary electrophoresis (ce) or capillary zone electrophoresis (cze), a relatively recent addition to the arsenal of analytical

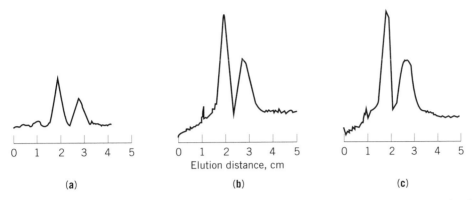

Fig. 2. Tlc densitometer scans showing the resolution of isoproterenol on an hptlc silica-gel plate obtained using a mobile phase consisting of 6.8 mM (1R)-(−)-ammonium-10-camphorsulfonic acid in 75:25 (v/v) methylene chloride:methanol. (**a**) 254 nm, (**b**) 275 m, (**c**) 300 nm.

techniques (20,21), has also been demonstrated as a powerful chiral separation method. Its high resolution capability and lower sample loading relative to hplc makes it ideal for the separation of minute amounts of components in complex biological mixtures (22,23).

In a ce experiment, a thin capillary is filled with a run buffer and a voltage is applied across the capillary. Although a complete treatment of the fundamental principles of ce is beyond the scope of this article, it can be said that the underlying impetus for separations in ce is, in general, derived from the fact that charged species migrate in response to an applied electric field proportionately to their charge and inversely proportionately to their size. Thus, given equivalent charges, lighter analytes have higher electrophoretic mobilities than heavier analytes and, given equivalent sizes, more highly charged species have higher mobilities than lesser charged or neutral species. In fact, neutral species have no intrinsic electrophoretic mobility. Species having opposite charges have electrophoretic mobilities in opposing directions.

Chiral separations by ce have been performed almost exclusively using chiral additives to the run buffer. The advantages of this approach are identical to the advantages mentioned previously with regard to using chiral mobile-phase additives in tlc. Many of the chiral selectors used successfully as mobile-phase additives in tlc and as immobilized ligands in hplc have been used successfully in ce including proteins (24), native (25) and functionalized cyclodextrins (26,27), various carbohydrates (28,29), assorted functionalized amino acids (30), chiral-ion pairing agents (31), and macrocyclic antibiotics (32). Other ce chiral selectors which have not been used as immobilized chiral selectors in hplc include bile salts (33), chiral surfactants (34), and dextran sulfate (35).

Although chiral ce is most commonly performed using aqueous buffers, there has been some work using organic solvents such as methanol, formamide, N-methylformamide or N,N-dimethylformamide with chiral additives such as quinine (36) or cyclodextrins (37,38). Nonaqueous ce requires that the background electrolyte be prepared using organic acids (eg, citric acid or acetic acid) and organic bases (eg, tetraalkylammonium halides or tris(hydroxymethyl)-aminomethane).

Theoretical models (39,40), as expressed in equation 1, where μ represents the mobility

$$\mu_1 - \mu_2 = \frac{[\mu_{1,c} - \mu_{1,f}][K_1 - K_2][CA]}{[1 + K_1[CA]][1 + K_2[CA]]} \tag{1}$$

of the analyte in the free and complexed states, K represents the binding constants of enantiomers 1 and 2 and $[CA]$ is the molar concentration of the additive, reveal that, in general, two conditions must be met to achieve chiral separations by ce. First of all, there must be differences in the binding constants of the two enantiomers with the chiral selector. Second, because the intrinsic electrophoretic mobilities of the enantiomers are identical in the free state, there must be a significant difference in the mobilities of the analyte in the complexed and free state. Chiral selectors have generally been shown to be the most effective when the intrinsic electrophoretic mobility of the additive is in the opposite direction of the intrinsic electrophoretic mobility of the analyte.

Chiral separations by ce is a rapidly growing field and offers the analyst tremendous flexibility with regard to chiral selector choice. In addition, because the additive is typically in the run buffer, there is virtually no column pre-equilibration. However, ce instruments tend to cost more than most chromatographic systems, sample capacity is much smaller for ce than for an analogous hplc method, sample recovery is not a trivial problem in ce and run-to-run reproducibility for ce tends to be much worse than for most chromatographic methods. Nevertheless, the flexibility, minimal sample and/or chiral selector required and the extremely high resolving power of ce ensure that this technique will continue to play an important role in chiral separations in the future.

Chiral Stationary Phases

Most chiral chromatographic separations are accomplished using chromatographic stationary phases that incorporate a chiral selector. The chiral separation mechanisms are generally thought to involve the formation of transient diastereomeric complexes between the enantiomers and the stationary phase chiral ligand. Differences in the stabilities of these complexes account for the differences in the retention observed for the two enantiomers. Often, the use of a chiral stationary phase allows for the direct separation of the enantiomers without the need for derivatization. One advantage offered by the use of chiral stationary phases is that the chiral selector need not be enantiomerically pure, only enriched. In addition, for chiral stationary phases having a well understood chiral recognition mechanism, assignment of configuration (eg, R or S) may be possible even in the absence of optically pure standards. However, chiral stationary phases have some limitations. The specificity required for chiral discrimination limits the broad applicability of most chiral stationary phases; thus there is no "universal" chiral stationary phase. The cost of most chiral columns are typically much higher ($\sim$3x) than for conventional columns. In contrast to conventional chromatographic columns, chiral stationary phases are generally not as robust, require more careful handling than conventional columns and usually, once column performance has begun to deteriorate, cannot be returned to their original performance levels. In many cases, chromatographic column choice or mobile phase optimization for chiral stationary phases is not as straightforward as with conventional stationary phases. In conventional chromatography, there is usually a well-behaved relationship between retention and mobile phase composition or column temperature. However, in many of the chiral stationary phases, the stationary phase present a multitude of different types of sites with not necessarily equivalent populations for interaction with the analytes. In the case of some liquid chromatographic stationary phases, the different types of sites may result in normal phase type behavior under very nonpolar mobile phase conditions and reversed-phase type behavior under highly polar mobile phase conditions. The multiplicity of types and numbers of sites also confounds thermodynamic considerations (41). Often, there is a narrow window of mobile-phase conditions under which enantioselectivity is observed and these conditions may be unique for a particular chiral analyte. Thus, for many of the chiral stationary phases, adequate chiral recognition models, used to guide selection of the appropriate column for a given separation, have yet to be developed. Column selection, therefore, is often reduced to identifying structurally similar analytes for which chiral resolution methods have been reported in the scientific literature or chromatographic supply catalogues and adapting a reported method for the chiral pair to be resolved.

An additional complication, sometimes arising with the use of chiral stationary phases, may occur when the analytes either exist as *conformers* or can undergo inversion during the chromatographic analysis. Figure 3 illustrates a typical chromatogram obtained for oxazepam, one of the chiral benzodiazepines that can undergo ring opening and inversion at the chiral center (42). As can be seen from Figure 3, the peaks appear to have a plateau between them and are sometimes referred to as "Batman peaks". This effect can sometimes be suppressed by lowering the column temperature. Although the appearance of Batman peaks is not unique to chiral separations, the specificity of chiral analyte–chiral selector interactions may increase the frequency of their occurrence.

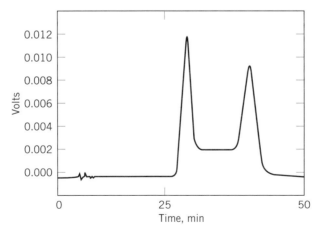

Fig. 3. The chiral separation obtained for oxazepam on a sulfated cyclodextrin hplc column (4.6 mm ID × 25 cm) using a 10% acetonitrile/buffer (25 mM ammonium acetate, pH 7).

Thin-Layer Chromatography. Chiral stationary phases have been used less extensively in tlc as in high performance liquid chromatography (hplc). This may, in large part, be due to lack of availability. The cost of many chiral selectors, as well as the accessibility and success of chiral additives, may have inhibited wide-spread commercialization. Usually, nondestructive visualization of the sample spots in tlc is accomplished using iodine vapor, uv or fluorescence. However, the presence of the chiral selector in the stationary phase can mask the analyte and interfere with detection (43).

Chiral stationary phases in tlc have been primarily limited to phases based on normal or microcrystalline cellulose (44,45), triacetylcellulose sorbents or silica-based sorbents that have been chemically modified (46) or physically coated to incorporate chiral selectors such as amino acids (47,48) or macrocyclic antibiotics (49) into the stationary phase.

Of the silica-based materials, only the ligand-exchange phases are commercially available (Chiralplate, tlc plates are available through Alltech Associates, Inc.) Supelco, Inc., the Aldrich Chemical Company, and Bodman Industries are all based on ligand exchange. Typically in the case of the ligand-exchange type tlc plates, the ligand-exchange selector is comprised of an amino acid residue to which a long hydrocarbon chain has been attached (eg, $(2S,4R,2'RS)$4-hydroxy-1-(2-hydroxydodecyl)proline) (50). The hydrocarbon chain of the functionalized amino acid is either chemically bonded to the substrate or intercalates in between the chains of a reversed phase-stationary phase thus immobilizing the chiral selector. The bidentate amino acid chiral selector is thought to reside close to the surface of the stationary phase and participates as a ligand in the formation of a bi-ligand complex with a divalent metal ion (eg, Cu^{2+}) and the chiral bidentate analyte (Fig. 4). Analytes enantioresolvable using ligand exchange are usu-

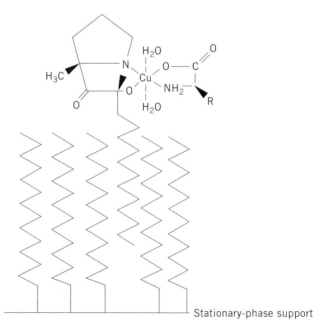

Fig. 4. A ligand-exchange chiral selector complexed with a chiral analyte.

ally restricted to 1,2-diols, α-amino acids, α-amino alcohols, and α-hydroxyacids (51,52). Again, differences in the stabilities of the diastereomeric complexes thus formed give rise to the chiral separation.

High Performance Liquid Chromatography. Although chiral mobile phase additives have been used in high performance liquid chromatography (hplc), the large amounts of solvent, thus chiral mobile phase additive, required to pre-equilibrate the stationary phase renders this approach much less attractive than for tlc and is not discussed here.

The last decade has seen the commercialization of a large number of different types of chiral stationary phases including the cyclodextrin phases (53), the chirobiotic phases (54), the π-π interaction phases (55,56), the protein phases (57–61), as well as the cellulosic and amylosic phases (62,63) and chiral crown ether phases (64,65). Currently, there are over 50 different chiral columns that are commercially available for hplc. Table 2 briefly summarizes the types of columns available as well as typical applications and mobile-phase conditions. Each of these chiral stationary phases are very successful at separating large numbers of enantiomers, which in many cases, are unresolvable using any of the other chiral stationary phases. Unfortunately, despite the large number and variety of chiral stationary phases currently available, there remains a large number of enantiomeric compounds that are unresolvable by any of the existing chiral stationary phases. In addition, incomplete understanding of the chiral recognition mechanisms of many of these chiral stationary phases limits the realization

Table 2. Classes of Hplc Chiral Stationary Phases

Column chiral selector	Typical mobile phase conditions	Typical analyte features required
pirkle	nonpolar organic; 2-propanol–hexane	π-acid or π-basic moieties for charge transfer complex; hydrogen-bonding or dipole stacking capability near chiral center
protein	phosphate buffers	aromatic near chiral center; organic acids or bases; cationic drugs
cyclodextrin	aqueous buffers; polar organic	good "fit" between chiral cavity or chiral mouth of cyclodextrin and hydrophobic moiety; hydrogen-bonding capability near chiral center
ligand exchange	aqueous buffers	α-hydroxy or α-amino acids near chiral center; can do nonaromatic
chiral crown ether	0.01 N perchloric acid	primary amines near chiral center; can do nonaromatic
macrocyclic antibiotics	aqueous buffers, nonpolar and polar organic	amines, amides, acids, esters; aromatic; hydrophobic moiety
cellulosic and amylosic	nonpolar organic	aromatic

of the full potential of the existing chiral stationary phases and hampers development of new chiral stationary phases.

Ligand-Exchange Phases

Among the earliest reports of chiral separations by liquid chromatography were based on work done by Davankov using ligand exchange (66). These types of columns are available from Phenomenex, J. T. Baker, and Regis Technologies, Inc. As noted previously in the discussion regarding ligand exchange in tlc, chiral separations by ligand exchange in hplc is accomplished using bidentate amino acid ligands, immobilized on a chromatographic substrate, and a divalent metal cation which participates in the formation of a diastereomeric complex with a bidentate chiral analyte and the ligand. Although almost any amino acid can form the basis for the chiral selector, proline and hydroxyproline exhibit the most widespread utility. Also, although other metals can be used, copper(II) is usually the metal of choice and is added to the aqueous buffer mobile phase.

The dependence of chiral recognition on the formation of the diastereomeric complex imposes constraints on the proximity of the metal binding sites, usually either an hydroxy or an amine α to a carboxylic acid, in the analyte. Principal advantages of this technique include the ability to: assign configuration in the absence of standards; enantioresolve nonaromatic analytes; use aqueous mobile phases; acquire a stationary phase with the opposite enantioselectivity; and predict the likelihood of successful chiral resolution for a given analyte based on a well-understood chiral recognition mechanism.

Pirkle Phases

The first commercially available chiral column for liquid chromatography was introduced in 1980. This was the first generation of the "Pirkle phases", named after their originator, and was based on N-(3,5-dinitrobenzoyl)phenylglycine which was immobilized on a silica support (67). Of all of the commercially available chiral stationary phases for liquid chromatography, the chiral recognition mechanism for the "Pirkle" phases are among the best understood. Chiral recognition on Pirkle phases is thought to depend upon complimentary interactions between the analyte and the selector. These interactions may be π-π, steric, hydrogen-bonding, or dipole–dipole interactions and contribute to the overall stability of the diastereomeric association complexes that form between the individual enantiomers and the chiral selector in the stationary phase. The π-π interactions arise through the association of aromatic systems with complementary electron withdrawing (eg, nitro) and electron donating (eg, alkyl) substituents. The electron deficient aromatic system is often referred to as π-acidic; the electron rich system is usually referred to as π-basic. Three unique interactions emanating from the chiral centers of the analyte and their chiral ligand in the stationary phase, seem to be required for successful chiral recognition. A model invoking three unique points of interaction is sometimes referred to as the 3-point interaction model first proposed by Dalgliesh (68). To promote analyte–selector interactions, functional groups are often introduced into the analyte through achiral derivatization. For example, amines may be derivatized with 3,5-dinitrobenzoyl chloride to introduce a π-acid aromatic group to promote diastereomeric complexation with a π-basic (R)-N-(2-naphthyl)-alanine chiral selector in the stationary phase. Derivatization often has the additional benefit of enhancing solute solubility.

Nonpolar organic mobile phases, such as hexane with ethanol or 2-propanol as typical polar modifiers, are most commonly used with these types of phases. Under these conditions, retention seems to follow normal phase-type behavior (eg, increased mobile phase polarity produces decreased retention). The normal mobile phase components only weakly interact with the stationary phase and are easily displaced by the chiral analytes thereby promoting enantiospecific interactions. Some of the Pirkle-types of phases have also been used, to a lesser extent, in the reversed phase mode.

Reciprocity, an important concept introduced by Pirkle (69,70), exploited the notion that analytes that were well resolved using a particular chiral selector would likely be good candidates for chiral selectors to enantioresolve analytes similar to the original chiral selector. For instance, the first generation Pirkle phase incorporating N-(3,5-dinitrobenzoyl)phenylglycine was very successful at enantioresolving compounds containing naphthyl moieties near the stereogenic center. This insight spawned a second generation of Pirkle phases based on N-(2-naphthyl)-α-amino acids (71). These phases were very successful at enantioresolving analytes containing a 3,5-dinitrobenzoyl group, such as 3,5-dinitrophenyl carbamates, and ureas of chiral alcohols and amines (72). These columns are available through a variety of sources including Phenomenex, Regis Technologies, Inc., J. T. Baker, Inc., and Supelco, Inc.

The structure of the Whelk-O-1 phase, the most recent addition to this type of chiral stationary phase, is illustrated in Figure 5. This selector has a wedge-like chiral surface with one edge offering the π-basic tetrahydrophenanthrene ring system; the other edge is comprised of a 3,5-dinitrobenzoyl π-acidic moiety. The amide linkage between the two ring systems presents dipole stacking and hydrogen-bonding interaction sites. The presence of both π-acid and π-base features, as well as the inherent rigidity of the chiral selector, confers greater versatility than any of the previous Pirkle-type phases, imposing fewer constraints on both analyte structural features required for successful enantioresolution and

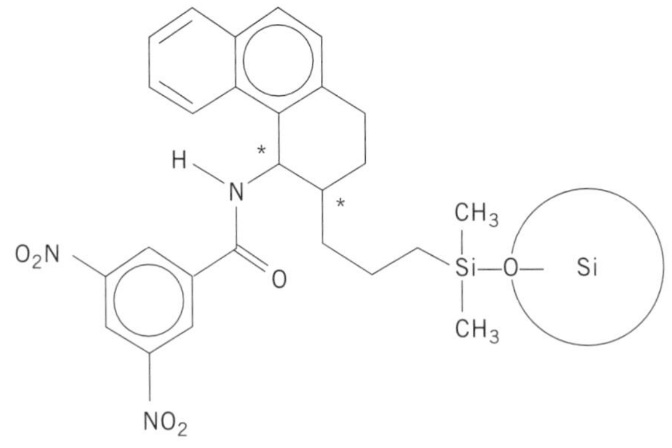

Fig. 5. The structure of the chiral selector in the Whelk-O-1 chiral stationary phase.

mobile phase conditions. Indeed, this chiral stationary phase has demonstrated considerable chiral selectivity for naproxen, warfarin, and its p-chloro analogue under nonaqueous reversed-phase conditions (73) and reversed-phase conditions (74,75). An additional advantageous feature of this phase is its availability with either the (R,R) or (S,S) configuration, thus, permitting the enantiomeric elution order to be readily changed. The small size of the chiral selector also promotes fairly high bonded ligand densities in the stationary phase, which coupled with the high enantioselectivities often achieved with these phases, facilitates their use for preparative-scale separations (76).

Cyclodextrin Phases

Cyclodextrins are macrocyclic compounds comprised of D-glucose bonded through 1,4-α-linkages and produced enzymatically from starch. The greek letter which proceeds the name indicates the number of glucose units incorporated in the CD (eg, α = 6, β = 7, γ = 8, etc). Cyclodextrins are toroidal shaped molecules with a relatively hydrophobic internal cavity (Fig. 6). The exterior is relatively hydrophilic because of the presence of the primary and secondary hydroxyls.

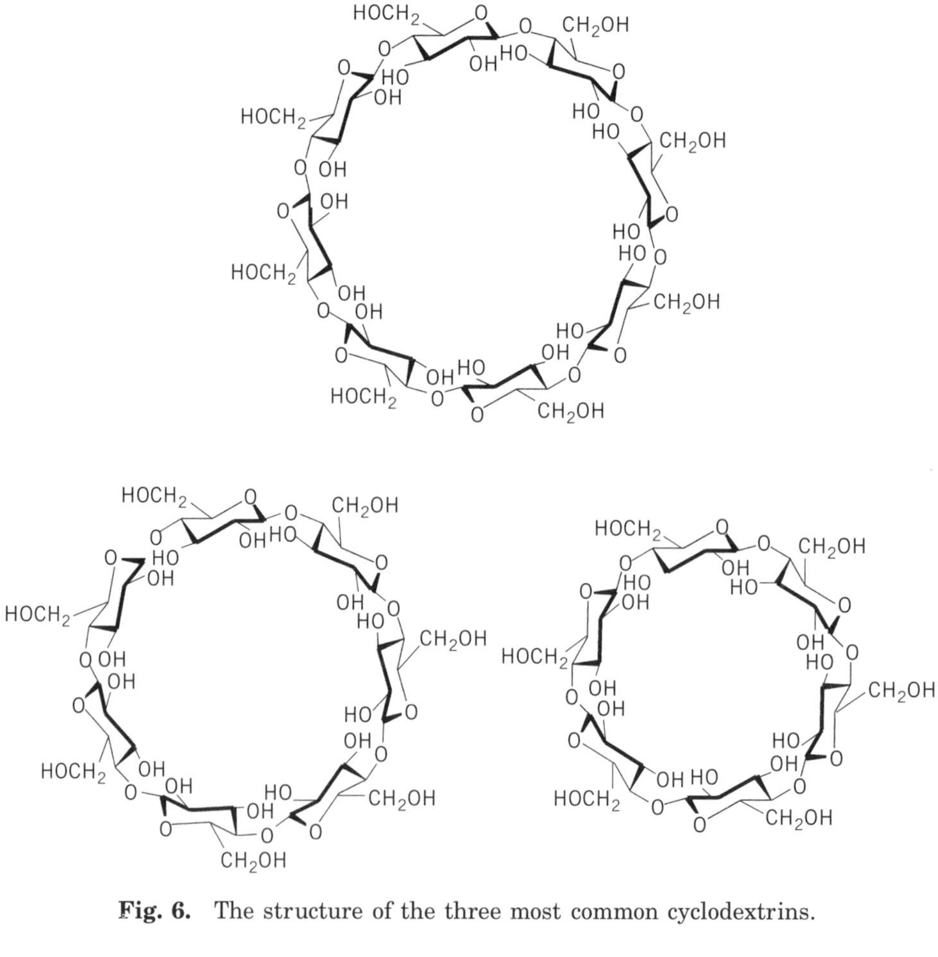

Fig. 6. The structure of the three most common cyclodextrins.

The primary C-6 hydroxyls are free to rotate and can partially block the CD cavity from one end. The mouth of the opposite end of the CD cavity is encircled by the C-2 and C-3 secondary hydroxyls. The restricted conformational freedom and orientation of these secondary hydroxyls is thought to be responsible for the chiral recognition inherent in these molecules (77).

Among the most successful of the liquid chromatographic reversed-phase chiral stationary phases have been the cyclodextrin-based phases, introduced by Armstrong (78,79) and commercially available through Advanced Separation Technologies, Inc. or Alltech Associates. The most commonly used cyclodextrin in hplc is the β-cyclodextrin. In the bonded phases, the cyclodextrins are thought to be tethered to the silica substrate through one or two spacer ligands (Fig. 7). The mechanism thought to be responsible for the chiral selectivity observed with these phases is based upon the formation of an inclusion complex between the hydrophobic moiety of the chiral analyte and the hydrophobic interior of the cyclodextrin cavity (Fig. 8). Preferential complexation between one optical isomer and the cyclodextrin through stereospecific interactions with the secondary

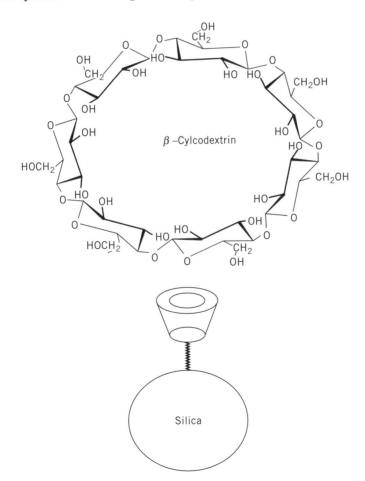

Fig. 7. A tethered cyclodextrin and the structure of β-cyclodextrin, the most common cyclodextrin used as a bonded ligand in liquid chromatography.

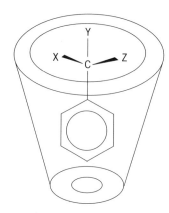

Fig. 8. A hydrophobic inclusion complex between a chiral analyte and a cyclodextrin.

hydroxyls which line the mouth of the cyclodextrin cavity results in the enantiomeric separation. Unlike the Pirkle-type phases, enantiospecific interactions between the analyte and the cyclodextrin are not the result of a single, well-defined association, but more of a statistical averaging of all the potential interactions with each interaction weighted by its energy or strength of interaction (80).

Vast amounts of empirical data suggest that chiral recognition on cyclodextrin phases in the reversed phase mode require the presence of an aromatic moiety that can fit into the cyclodextrin cavity, that there be hydrogen bonding groups in the molecule, and that the hydrophobic and hydrogen-bonding moieties should be in close proximity to the stereogenic center. Chiral recognition seems to be enhanced if the stereogenic center is positioned between two π-systems or incorporated in a ring.

Most of the chiral separations reported to date using the native cyclodextrin-based phases have been accomplished in the reversed-phase mode using aqueous buffers containing small amounts of organic modifiers. However, polar organic mobile phases have gained in popularity recently because of their ease of removal from the sample and reduced tendency to accelerate column degradation relative to the hydroorganic mobile phases (81). In these cases, because the more nonpolar component of the mobile phase is thought to occupy the cyclodextrin cavity, the analyte is thought to sit atop the mouth of the cyclodextrin much like a "lid".

Limitations with the chiral selectivity of the native cyclodextrins fostered the development of various functionalized cyclodextrin-based chiral stationary phases, including acetylated (82,83), sulfated (84), 2-hydroxypropyl (85), 3,5-dimethylphenylcarbamoylated (86) and 1-naphthylethylcaarbamoylated (87) cyclodextrin. Each of the glucose residues contribute three hydroxyl groups to which a substituent may be appended; thus, each cyclodextrin contributes multiple sites for derivatization. Typical degrees of substitution per β-cyclodextrin (with 21 hydroxyls) range from three to ten. Hence, there are many residual hydroxyls on each cyclodextrin.

The substituents of these functionalized cyclodextrins seem to play a variety of roles in enhancing chiral recognition. In some cases, the substituent may only serve to enlarge the chiral cavity or may provide alternative interaction sites.

For instance, in the case of the naphthylethylcaarbamoylated cyclodextrin, the naphthyl ring provides a π-basic site and the carbamate linkage provides additional hydrogen bonding and dipole interaction sites not available with the native cyclodextrin. On the sulfated cyclodextrin phase, the sulfate group presents the potential for ion-pair formation unavailable with the native cyclodextrin. The introduction of 2-hydroxypropyl- and 1-naphthylethylcarbamoyl substituents incorporates additional stereogenic centers onto the cyclodextrin. In some cases, the configuration of the substituent dominated the enantiomeric elution order. However, in other cases, the enantiomeric elution order was independent of the configuration of the substituent. In addition, in some cases, the chiral selectivity of the cyclodextrin seemed to synergistically augment the chiral selectivity of one configuration of the substituent while antagonizing the chiral selectivity of the oppositely configured substituent. A particularly attractive feature of these functionalized cyclodextrins is that many of them exhibit enantioselectivity under hydroorganic reversed phase, as well as normal phase and polar organic mobile-phase conditions, and that each set of conditions can provide chiral separations for analytes which are not resolved under any of the other type of mobile phase conditions. Further, the chromatographic mode (eg, reversed phase to normal phase) can be readily changed with no deleterious impact on chiral recognition as long as routine care is taken to avoid problems with solvent immiscibility. The naphthylethyl-carbamoylated cyclodextrin phase was considered to be one of the first "multi-modal" chiral stationary phases (88).

Cellulosic and Amylosic Phases

Cellulose and amylose are comprised of the same glucose subunits as the cyclodextrins. In the case of cellulose, the glucose units are attached through 1,4-β-linkages resulting in a linear polymer. In the case of amylose, the 1,4-α-linkages, as are found in the cyclodextrins, are thought to confer helicity to the polymeric chain.

As mentioned previously, cellulosic phases as well as amylosic phases have also been used extensively for enantiomeric separations more recently (89,90). Most of the work in this area has been with various derivatives of the native carbohydrate. The enantioresolving abilities of the derivatized cellulosic and amylosic phases are reported to be very dependent upon the types of substituents on the aromatic moieties that are appended onto the native carbohydrate (91). Table 3 lists some of the cellulosic and amylosic derivatives that have

Table 3. Carbohydrate Derivatives Used as Hplc Chiral Stationary Phases

Cellulosic	Amylosic
triacetate	
tribenzoate	
tribenzylether	
tricinnamate	
triphenylcarbamate	triphenylcarbamate
tris-3,5-dichlorophenylcarbamate	
tris-3,5-dimethylphenylcarbamate	tris-3,5-dimethylphenylcarbamate
tris-1-phenylethylcarbamate	tris-1-phenylethylcarbamate

been used. These columns are available through Chiral Technologies, Inc. and J. T. Baker, Inc.

With the exception of the microcrystalline cellulose I triacetate (92) and tribenzoate materials, which are sufficiently robust to be used directly as packing material, most of the commercially available cellulosic and amylosic phases are comprised of mixtures of exhaustively derivatized polymers which are coated onto large pore γ-aminopropyl silica. These coated polymeric phases exhibit admirable enantioselectivities, but as is the case with all commercially available chiral stationary phases, they also have some potential disadvantages. The large polymer size requires the use of fragile, large pore silica. The fact that the chiral selector for these phases is coated onto the silica sometimes restricts the types of mobile phases that can be used. In addition, the secondary structure of the polymer, which seems to be important in the chiral recognition mechanism, may be altered irreversibly by storing the columns in polar solvents leading to disastrous consequences for chiral separations. The polymeric nature of the chiral selectors for these phases and the importance of the secondary structure also hamper the development of models for the chiral recognition mechanism for these phases (93). Despite these factors, the cellulosic and amylosic phases have enjoyed tremendous success at enantioresolving structurally diverse compounds (94,95) and have some of the highest capacities of all the chiral commercially available chiral stationary phases, thus, rendering them among the most suitable for preparative chromatography (96).

The chiral recognition sites on these polymeric carbohydrate phases are thought to be channels or grooves in the polymer matrix and that analytes are included into these channels. Evidence for inclusion is provided by the enhanced chiral recognition observed for many analytes as the steric bulk of the alcohol mobile phase modifier increases. Chiral recognition seems to require the presence of an aromatic ring, for π-π interactions, and polar sites of unsaturation or hydrogen bonding functionalities. As in the case of the Pirkle-type phases, these chiral stationary phases are usually used in the normal phase mode and mobile phases typically consist of hexane and 2-propanol although there have been some reports of these phases being used in the reversed phase mode (97).

Protein-Based Phases

Proteins, amino acids bonded through peptide linkages to form macromolecular biopolymers, used as chiral stationary phases for hplc include bovine and human serum albumin, α_1-acid glycoprotein, ovomucoid, avidin, and cellobiohydrolase. The bovine serum albumin column is marketed under the name Resolvosil and can be obtained from Phenomenex. The human serum albumin column can be obtained from Alltech Associates, Advanced Separation Technologies, Inc., and J. T. Baker. The α_1-acid glycoprotein and cellobiohydrolase can be obtained from Advanced Separation Technologies, Inc. or J. T. Baker, Inc.

In most cases, the protein is immobilized onto γ-aminopropyl silica and covalently attached using a cross-linking reagent such as N,N'-carbonyldiimidazole. The tertiary structure or three dimensional organization of proteins are thought to be important for their activity and chiral recognition. Therefore, mobile phase conditions that cause protein "denaturation" or loss of tertiary structure must be avoided.

Typically, the mobile phases used with the protein-based chiral stationary phases consist of aqueous phosphate buffers (98). Often small amounts of organic modifiers, such as methanol, ethanol, propanol, or acetonitrile, are added to reduce hydrophobic interactions with the analyte and to improve enantioselectivity. In some cases, dramatic changes in chiral recognition occur when small amounts of organic modifiers, such as N,N-dimethyloctylamine or octanoic acid are added to the mobile phase. It is thought that these additives may be playing an active role in enhancing chiral recognition through absorption of the organic modifier onto the protein which induces conformational changes in the overall tertiary structure of the protein. In these cases, the *allosteric* modifier-mediated conformational changes in the protein are thought to enhance chiral recognition by a variety of mechanisms including changes in the accessibility of various stereospecific sites on the protein or obstruction of nonstereospecific sites (99).

As in the case of the cyclodextrin and amylosic and cellulosic phases, the chiral recognition mechanism for these protein-based phases is not well understood. In some cases, it is thought that analytes may form inclusion complexes with hydrophobic pockets within the biopolymeric matrix. These hydrophobic interactions may couple with hydrogen bonding, electrostatic interactions, and π-π or dipole stacking to individual amino acid residues, thus, contributing to stereospecific orientational constraints within the hydrophobic pockets. Optimization of chromatographic conditions and selection of analytes that can be successfully resolved on these phases is usually done empirically. In addition, the large molecular weight of these biopolymers dictates that the amount of chiral selector that can be immobilized on the column packing material is very small. Although the protein is large, relative to the analytes, the actual region of the protein that affects the chiral separation may be very small. Thus, the capacity (amount of material resolvable during a single chromatographic run) of these columns is generally fairly small ($<{\sim}0.1$ mg) and the columns are easily overloaded.

An interesting application of the protein-based phases is various protein binding and displacement experiments which can be done fairly routinely (100). For instance, the chiral selectivity of chiral stationary phases derived from the serum albumin, one of the most abundant blood proteins which functions as a transport protein, from different animal species including rabbit, rat and human has been compared (101). This work suggests that differences in the enantioselectivity, toward a particular drug, of a column derived from human serum albumin and a column derived from some other animal serum albumin might be indicative that a particular species might not be a good animal model during drug development, thus, obviating the need for animal testing.

Chiral separations on protein-based phases may also provide useful information on drug interactions. For instance, the effect of the individual enantiomers of warfarin on the enantioselectivity of human serum albumin toward benzodiazepinones has been studied using a human serum albumin column with warfarin as a mobile phase additive (102).

Chirobiotic Phases

The chirobiotic chiral stationary phases (103,104) are based on macrocyclic antibiotics such as vancomycin (**5**) and teicoplanin (**6**).

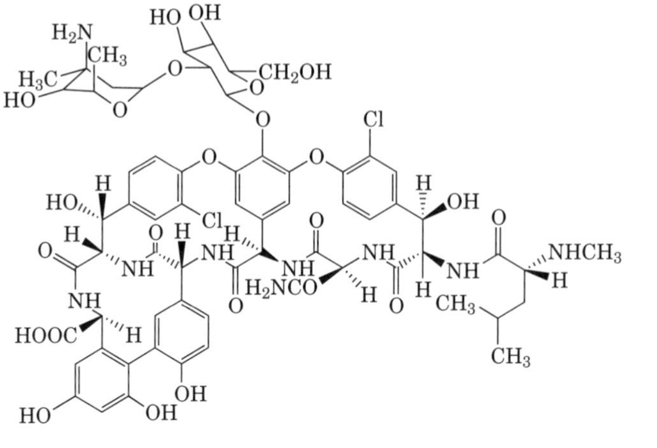

(**5**)

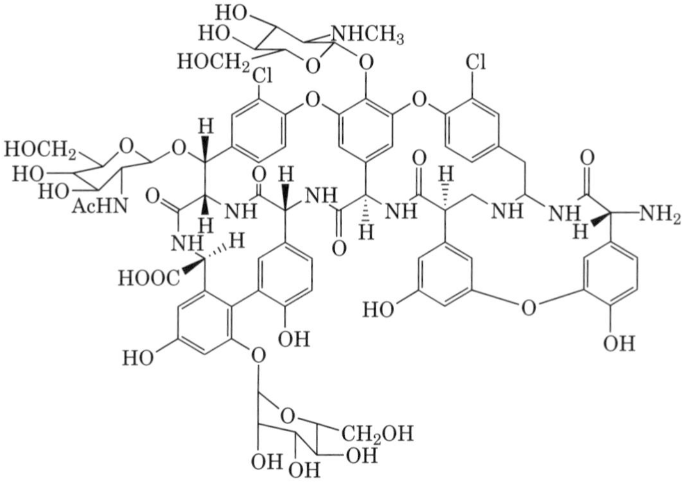

(**6**)

These chiral selectors, originally used as chiral additives in capillary zone electrophoresis, incorporate aromatic and carbohydrate, as well as peptide and ionizable moieties. The presence of aromatic groups, allowing for π-π interactions, and the macrocyclic rings, offering potential inclusion complexation, give these phases some of the advantages of the protein-based phases (eg, peptide and hydrogen bonding sites) and the carbohydrate-based phases but with greater sample capacity and greater mobile phase flexibility. Indeed, these phases seem to be truly "multimodal" in that they have demonstrated chiral selectivity in the normal, polar organic, and reversed-phase modes. In the normal and polar organic phase modes, π-π interactions, and dipole stacking are thought to play a predominant role in chiral selector–analyte interactions. In the reversed phase mode, hydrogen bonding, inclusion complexation and, for charged analytes, electrostatic interactions are thought to dominate the interactions. In addition, the use of such well-defined chiral selectors facilitate method development and optimization. These columns are commercially available through Advanced Separation Technologies, Inc. and Alltech Associates.

Chiral Crown Ether Phases

Chiral crown ethers based on 18-crown-6 (Fig. 9) can form inclusion complexes with ammonium ions and protonated primary amines. Immobilization of these chiral crown ethers on a chromatographic support provides a chiral stationary phase which can resolve most primary amino acids, amines and amino alcohols. However, the stereogenic center must be in fairly close proximity to the primary amine for successful chiral separation (105,106). Significantly, the chiral crown ether phase is unique in that it is one of the few liquid chromatographic chiral stationary phases that does not require the presence of an aromatic ring to achieve chiral separations. Although chiral recognition seems to be enhanced for analytes containing either bulky substituents or aromatic groups near the stereogenic center, only the presence of the primary amine is mandatory.

Mobile phases used with this stationary phase are typically 0.01 N perchloric acid with small amounts of methanol or acetonitrile. One significant advantage of these phases is that both configurations of the chiral stationary phase are commercially available and can be obtained from J. T. Baker Inc. and Chiral Technologies, Inc. (Crownpak CR).

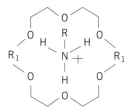

Fig. 9. An inclusion complex formed between a protonated primary amine and a chiral crown ether.

Chiral Synthetic Polymer Phases

Chiral synthetic polymer phases can be classified into three types. In one type, a polymer matrix is formed in the presence of an optically pure compound to molecularly *imprint* the polymer matrix (Fig. 10) (107,108). Subsequent to the polymerization, the chiral template is removed, leaving the polymer matrix with chiral cavities. The degree of cross-linking in the polymer matrix and degree of association between the template molecule and the monomer, is governed by the type and concentration of the monomer, the concentration of the template, the solvent and temperature or pressure under which polymerization takes place. All play a role in the chiral selectivities achieved with these phases. The selectivities achieved with these phases are generally excellent, thus, facilitating semipreparative separations. However, the applicability of these chiral stationary phases are generally limited to the analyte upon which the phase is based and a limited number of analogues. In addition, these types of phases generally exhibit

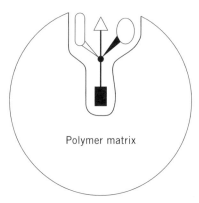

Fig. 10. The relationship between a chiral template molecule and the polymeric matrix formed in the presence of the template molecule.

poor efficiency in large part because the polymeric matrix contributes to non-sterespecific binding. Advantages of this approach include the ability to prepare reciprocal phases and the predictability of the enantiomeric elution order.

Another type of synthetic polymer-based chiral stationary phase is formed when chiral catalyst are used to initiate the polymerization. In the case of poly(methyl methacrylate) polymers, introduced by Okamoto, the chirality of the polymer arises from the helicity of the polymer and not from any inherent chirality of the individual monomeric subunits (109). Columns of this type (eg, Chiralpak OT) are available from Chiral Technologies, Inc., or J. T. Baker Inc.

A third type of synthetic polymer-based chiral stationary phase, developed by Blaschke (110), is produced when a chiral selector is either incorporated within the polymer network (111) or attached as pendant groups onto the polymer matrix. Both are analogous to methods used to produce polymeric chiral stationary phases for gc. The polymers can be either coated onto a silica substrate, comonomers bearing silane functional groups may be added for subsequent reaction with the silica, or the silica may be chemically modified to incorporate monomer-bearing silanes. More recently, L-valine-3,5-dimethylanilide has been bonded to a poly(glycidylmethacrylate-*co*-ethylenedimethacrylate polymer which formed the underlying substrate (112). Chemical bonding of the polymer to the substrate eases the mobile phase restrictions imposed on the coated chiral polymer stationary phases.

In general, the synthetic polymeric phases seem to have polarities analogous to diol-type phases and a wide range of mobile phase conditions have been used including hexane, various alcohols, acetonitrile, tetrahydrofuran, dichloromethane and their mixtures, as well as aqueous buffers.

Chiral Separation Validation for Hplc

Chiral separations present special problems for validation. Typically, in the absence of spectroscopic confirmation (eg, mass spectral or infrared data), conventional separations are validated by analyzing "pure" samples under identical chromatographic conditions. Often, two or more chromatographic stationary phases, which are known to interact with the analyte through different retention mechanisms, are used. If the pure sample and the unknown have identical retention times under each set of conditions, the identity of the unknown is assumed to be the same as the pure sample. However, often the chiral separation that is obtained with one type of column may not be achievable with any other type of chiral stationary phase. In addition, "pure" enantiomers are generally not available.

Most commonly, uv or uv–vis spectroscopy is used as the basis for detection in hplc. When using a chiral stationary phase, confirmation of a chiral separation may be obtained by either monitoring the column effluent at more than one wavelength or by running the sample more than once. The same mobile-phase conditions are used, but monitoring is done at different wavelengths. Because enantiomers have identical spectra in an achiral environment, the ratio of the peaks for the two enantiomers should be independent of wavelength. Although not absolute proof of a chiral separation, this approach does provide strong supporting evidence.

As in tlc, another method to validate a chiral separation is to collect the individual peaks and subject them to some type of optical spectroscopy, such as, circular dichroism or optical rotary dispersion. Enantiomers have mirror image spectra (eg, the negative maxima for one enantiomer corresponds to the positive maxima for the other enantiomer). One problem with this approach is that the analytes are diluted in the mobile phase. Thus, the sample must be injected several times. The individual peaks must be collected and subsequently concentrated to obtain adequate concentrations for spectral analysis.

Alternatively, a chiroptical spectroscopy can be used as the basis for detection on-line using commercially available optical rotary dispersion or circular dichroism-based detectors. Optical rotary dispersion instruments are analogous to refractive index-based detectors for conventional chromatography in that they are universal, do not require the presence of a chromophore in the analyte and have the least sensitivity of the optical detectors. Circular dichroic detection, although more sensitive than optical rotary dispersion-based detection, requires not only the presence of a uv chromophore in the analyte, but that the chromophore be not too distant from the asymmetric center of the analyte. Figure 11**a** illustrates a simulated chromatogram for an enantiomeric separation obtained using a conventional absorption detector. Figure 11**b** illustrates a simulated chromatogram for an enantiomeric separation using circular dichroic detection. Both types of chiroptical detectors produce positive and negative peaks for the two enantiomers. However, neither chiroptical detector can distinguish a fair separation (Figure 11**d**) from a poor separation (Figure 11**f**) in which there is considerable overlap of the two peaks. This is because the signals generated by the two enantiomers have opposite signs, and thus, any overlap causes cancellation of signal. Further, peak overlap results in nonlinear detector response vs concentration. Therefore, some other detection method must be used in conjunction with either of these types of detection. Nevertheless, as can be seen from Figure 11**f**, chiroptical detection can be advantageous if there is considerable overlap of the two peaks. In this case, chiroptical detection may reveal that the leading and tailing edges of the peak are enantiomerically enriched which may

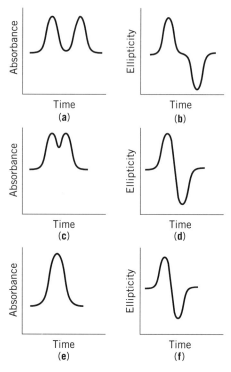

Fig. 11. Simulated chromatograms of chiral separations obtained using nonchiroptical detection (**a**, **c**, **e**) and chiroptical detection (**b**, **d**, **f**) illustrating the effect of peak overlap on the resultant chromatogram.

not be apparent from the chromatogram obtained with nonchiroptical detection (Fig. 11**e**).

Another method for validating chiral separations by lc is to couple the chromatographic system to a mass spectrometer. In mass spectrometry, high energy ions are used to bombard molecules exiting the column. The impact of the high energy ions causes the molecules to "fragment" into various ions which are then sent to a "mass discriminator". The ion fragments are detected and a fragmentation pattern or mass spectrum is reconstructed. Enantiomers have identical fragmentation patterns. Hence, identical fragmentation patterns for two peaks in the chromatogram confirms a chiral separation.

Chiral Stationary Phases for Gas Chromatography

Although chiral stationary phases for gas chromatography (gc) were introduced before liquid chromatographic chiral stationary phases, development of gc chiral stationary phases lagged behind for a variety of reasons. First of all, analysis by gc requires that the analyte be volatile and thermally stable. This condition often requires that the analyte be derivatized with an achiral reagent prior to chromatographic analysis to enhance sample volatility. In some cases, derivatization may actually enhance detector response (eg, trifluoroacetylation amplifies electron capture detection) or chiral interactions. However, it should be noted that the presence of more than one type of functionality (eg, amine and alcohol) in the analyte with differing reactivities toward the derivatizing agent may add additional complications. Typical achiral derivatizing reagents, as well as the appropriate functionality required in the analyte, are listed in Table 4.

The use of gas chromatography for chiral separations was also hampered because the high column temperatures typically used in gc tend to accelerate racemization of the stationary phase, thus, decreasing column longevity. The high column temperatures typically used in gc also tend to accelerate racemization of the analyte. In addition, the differences in the stabilities of the diastereomeric complexes formed between the enantiomers and the stationary phase tends to be overcome by the high column temperatures. Finally, preparative scale separations are generally harder to implement in gc than in hplc. However, the inability of most liquid chromatographic methods to resolve chirally small nonaromatic compounds that are frequently used as chiral synthetic building blocks, as well as improvements in gc column technology, has led to renewed interest in chiral stationary phases for gc.

Gc chiral stationary phases can be broadly classified into three categories: diamide, cyclodextrin, and metal complex.

Diamide Chiral Separations. The first chiral stationary phase for gas chromatography was reported by Gil-Av and co-workers in 1966 (113) and was based on N-trifluoroacetyl (N-TFA) L-isoleucine lauryl ester coated on an

Table 4. Analyte Functional Groups and Typical Achiral Derivatizing Reagents

Type of derivatizing agent	Analyte functional group	Examples of derivatizing agents
alkyl silyl	alcohols	N-trimethylsilylimidazole
	thiols	
	carboxylic acid	
	amines	N,O-bis(trimethylsilyl)-trifluroracetamide
acyl, haloacyl or anhydride	alcohols	acetic acid
	amines	heptafluorobutyryl chloride
	amides	trifluoroacetyl chloride
	oximes	
	thiols	
	ketones	
alcohol	carboxylic acids	methanol
alkyl halides	carboxylic acids	methyl bromide
diazoalkyl	carboxylic acids	diazomethane
	sulfonic acids	
	phenols	
isocyanate	alcohols	isopropylisocyanate
	amines	
	hydroxy acids	
phosgene	β-amino alcohols	
	β-amino thiols	
	diols	
	N-methylamino acids	
alkyl hydroxylamine	ketones	methylhydroxylamine

inert packing material. It was used to resolve the trifluoroacetylated derivatives of amino acids. Related chiral selectors used by other workers included
n-dodecanoyl-L-valine-t-butylamide and n-dodecanoyl-(S)-α-(1-naphthyl)-ethyl-
amide. The presence of the long alkyl groups lowered chiral selector volatility
thus reducing but not entirely eliminating column bleed and improving column
longevity.

The first commercially available chiral column was the Chiralsil-val
(Fig. 12), which was introduced in 1976 (114) for the separation of amino acid
type compounds by gas chromatography. It is based on a polysiloxane polymer
containing chiral side chains incorporating L-valine-t-butylamide. The polysiloxane backbone improved the thermal stability of these chiral stationary phases
relative to the original coated columns and extended the operating temperatures up to 220°C. The column is effective for the separation of perfluoroacy-
lated and esterified amino acids, amino alcohols, and some chiral sulfoxides.
Another polysiloxane-based chiral stationary phase incorporating L-valine-(R)-
α-phenylethylamide appended onto hydrolyzed XE-60 was found to be particularly successful at resolving perfluoroacetylated amino alcohol derivatives (115).
Through judicious choice of derivatizing agent, chiral separations were obtained
for a wider range of compounds, including amino alcohols, α-hydroxy acids,
diols and ketones, than had previously been obtainable using these types of
stationary phases (116).

The chiral recognition mechanism for these types of phases was attributed
primarily to hydrogen bonding and dipole–dipole interactions between the analyte and the chiral selector in the stationary phase. It was postulated that
chiral recognition involved the formation of transient five- and seven-membered
association complexes between the analyte and the chiral selector (117).

On each of these amino acid-based chiral stationary phases, the configuration of the most retained enantiomer corresponds to the configuration of the

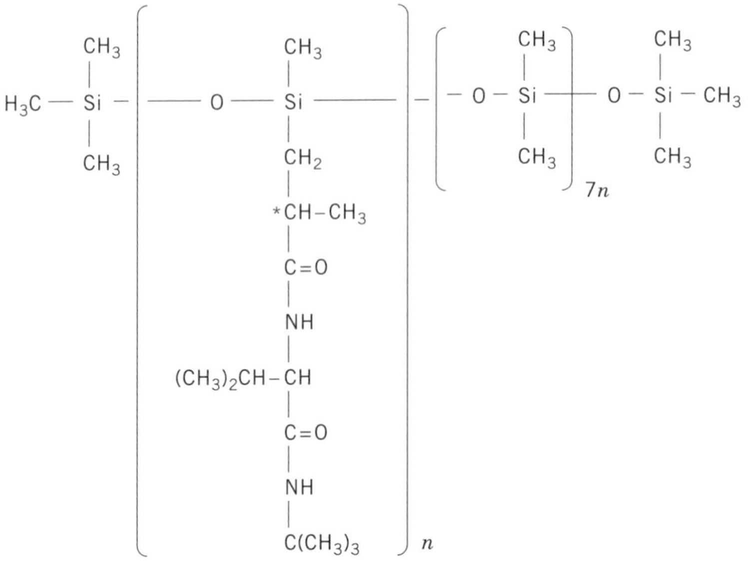

Fig. 12. Structure of Chiralsil-val.

chiral selector in the stationary phase (eg, L-amino acids were retained longer on the N-trifluoroacetyl (N-TFA) L-isoleucine lauryl ester column). Thus, configuration of the analytes can be assigned even if no optically pure material is available as long as optically pure standard materials are available for structurally related compounds. Another advantage is that stationary phases incorporating the chiral selector with either configuration may be readily prepared or are commercially available. Thus, the elution order may readily be reversed by using a column containing the chiral selector with the opposite configuration, thus, providing another tool for chiral separation validation. Also, quantitation of a very small peak in the presence of a very large peak is generally easier if the smaller peak elutes first.

Metal Complex. Complexation gas chromatography was first introduced by V. Schurig in 1980 (118) and employs transition metals (eg, nickel, cobalt, manganese or rhodium) complexed with chiral terpenoid ketoenolate ligands such as 3-trifluoroacetyl-1R-camphorate (**4**), 1R-3-pentafluoro-benzoylcamphorate or 3-heptafluorobutanoyl-(1R,2S)-pinanone-4-ate. In most cases, the chiral selector is dissolved in a polymer matrix coated on the interior walls of a capillary column. This class of chiral columns is particularly adept at enantioresolving some olefins and oxygen-containing compounds such as ketones, ethers, alcohols, spiroacetals, oxiranes, and esters. Many of these compounds lack suitable functionality for derivatization with chiral reagents, and thus, are not amenable to diastereomer formation. Unfortunately, as is the case with many of the chiral stationary phases, the chiral recognition mechanism is not sufficiently refined to allow for prediction of analyte absolute configuration on the basis of retention times except for a very limited number of cases. Nevertheless, these columns allow for the direct chiral separation of compounds that are important synthetic precursors and may be difficult to separate by any other method.

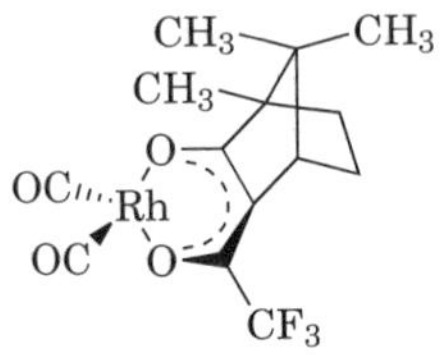

(**4**)

Cyclodextrins. As indicated previously, the native cyclodextrins, which are thermally stable, have been used extensively in liquid chromatographic chiral separations, but their utility in gc applications was hampered because their highly crystallinity and insolubility in most organic solvents made them difficult to formulate into a gc stationary phase. However, some functionalized cyclodextrins form viscous oils suitable for gc stationary-phase coatings and have been used either neat or diluted in a polysiloxane polymer as chiral stationary phases for gc (119). Some of the derivatized cyclodextrins which have been adapted to gc phases are 3-O-acetyl-2,6-di-O-pentyl, 3-O-butyryl-2,6-di-O-pentyl, 2,6-di-O-methyl-3-O-trifluoroacetyl, 2,6-dipentyl, 2-O-methyl-3,6-di-O-pentyl, permethyl, permethylhydroxypropyl, perpentyl, and propionyl. Several of these are available commercially. For instance, Advanced Separation Technologies, Inc., Alltech Associates, J. & W. Scientific, and Supelco, Inc., all carry cyclodextrin-based chiral gc columns. Although these derivatized cyclodextrins are often coated, neat, onto the capillary column inner walls, some work has been done to tether the cyclodextrins to a polysiloxane backbone to enhance the thermal stability of the resultant phase (120). Some of the separations obtained with these materials are quite remarkable and include compounds such as halogenated alkanes (Fig. 13) (121), alcohols, alkenes, bicyclic compounds, and simple alkanes.

Although the chiral recognition mechanism of these cyclodextrin-based phases is not entirely understood, thermodynamic and column capacity studies indicate that the analytes may interact with the functionalized cyclodextrins by either associating with the outside or mouth of the cyclodextrin, or by forming a more traditional inclusion complex with the cyclodextrin (122). As in the case of the metal-complex chiral stationary phase, configuration assignment is generally not possible in the absence of pure chiral standards.

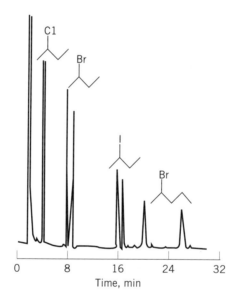

Fig. 13. Enantiomeric separations of monohalohydrocarbons on a 2,6-*O*-dipentyl-3-*O*-trifluoroacetyl-γ-cyclodextrin coated capillary column (10 m, 0.25 mm ID). Column temperature, 30°C; nitrogen carrier gas, 20.7 kPa (3 psi).

Chiral Separation Validation for Gas Chromatography

The special problems for validation presented by chiral separations can be even more burdensome for gc because most methods of detection (eg, flame ionization detection or electron capture detection) in gc destroy the sample. Even when nondestructive detection (eg, thermal conductivity) is used, individual peak collection is generally more difficult than in lc or tlc. Thus, off-line chiroptical analysis is not usually an option. Fortunately, gc can be readily coupled to a mass spectrometer and is routinely used to validate a chiral separation.

Conclusions

The field of chiral separations has grown explosively into a well-developed specialty within separation science in the last two decades. Considerable effort in the field has thus far been directed toward solving analytical separation problems and in the design and development of new chiral separation methods. However, in the future, chiral separations may have the potential to provide an invaluable tool for probing intermolecular interactions, increasing our understanding of the chemistry behind some biological processes and disease states (123,124). Also the efficacy of drug therapy can be enhanced by eliminating or minimizing untoward side effects (125) produced by distomeric ballast and allowing for better control in the application of chiral or prochiral entities to biological or environmental systems.

BIBLIOGRAPHY

1. L. Pasteur, *Comptes Rendus de l'Academie des Sciences* **26**, 535 (1848).
2. W. J. Wechter, D. G. Loughhead, R. J. Reischer, G. J. Van Giessen, and D. G. Kaiser, *Biochem. Biophys. Res. Comm.* **61**, 833 (1974).
3. R. A. Kuzel, S. K. Bhasin, H. G. Oldham, L. A. Damani, J. Murphy, P. Camilleri, and A. J. Hutt, *Chirality* **6**, 607 (1994).
4. S. A. Tobert, *Clin. Pharmacol. Ther.* **29**, 344 (1981).
5. H. Ohkawa, *Bull. Environ. Contamin. Toxicol.* **18**, 534 (1977).
6. H. G. Schmarr, A. Mosandl, and K. Grob, *Chromatographia* **29**, 125 (1990).
7. T. Aleibi-Kolbah, G. Félix, and I. W. Wainer, *Chromatographia* **35**, 264 (1993).
8. D. W. Armstrong, J. R. Faulkner, Jr., and S. M. Han, *J. Chromatogr.* **452**, 323 (1988).
9. T. J. Ward, *Anal. Chem.* **66**, 632A (1994).
10. C. Pettersson, A. Karlsson, and C. Gioeli, *J. Chromatogr.* **407**, 217 (1987).
11. J. D. Duncan, D. W. Armstrong, and A. M. Stalcup, *J. Liq. Chromatogr.* **13**, 1091 (1990).
12. D. W. Armstrong, F.-Y. He, and S. M. Han, *J. Chromatogr.* **448**, 345 (1988).
13. A. D. Cooper, T. M. Jeffries, and R. M. Gaskell, *Anal. Proc.* **29**, 258 (1992).
14. D. W. Armstrong and Y. Zhou, *J. Liq. Chromatogr.* **17**, 1695 (1994).
15. C. Pettersson and G. Schill, *J. Liq. Chromatogr.* **9**, 269 (1986).
16. M.-B. Huang, H.-K. Li, G.-L. Li, C.-T. Yan, and L.-P. Wang, *J. Chromatogr. A* **742**, 289 (1996).
17. D. W. Armstrong, J. R. Faulkner, Jr., and S. M. Han, *J. Chromatogr.* **452**, 323 (1988).
18. D. W. Armstrong, F.-Y. He, and S. M. Han, *J. Chromatogr.* **448**, 345 (1988).
19. J. D. Duncan, D. W. Armstrong, and A. M. Stalcup, *J. Liq. Chromatogr.* **13**, 1091 (1990).
20. W. G. Kuhr, *Anal. Chem.* **62**, 403R (1990).
21. B. L. Karger, *Amer. Lab.*, 23 (Oct. 1993).
22. J. W. Jorgenson, and K. D. Lukacs, *Anal. Chem.* **53**, 1298 (1981).
23. A. S. Cohen, A. Paulus, and B. L. Karger, *Chromatographia* **24**, 15 (1987).
24. P. Sun, N. Wu, G. Barker, and R. A. Hartwick, *J. Chromatogr.* **648**, 475 (1993).
25. S. Terabe, K. Otsuka, and H. Nishi, *J. Chromatogr.* **666**, 295 (1994).
26. Y. Y. Rawjee and G. Vigh, *Anal. Chem.* **66**, 619 (1994).
27. A. M. Stalcup and K. H. Gahm, *Anal. Chem.* **68**, 1360 (1996).
28. A. M. Stalcup and N. M. Agyei, *Anal. Chem.* **66**, 3054 (1994).
29. H. Soini, M. Stefansson, M.-L. Riekkola, and M. V. Novotny, *Anal. Chem.* **66**, 3477 (1994).

30. P. Gozel, E. Gassman, H. Michelson, and R. N. Zare, *Anal. Chem.* **59**, 44 (1987).
31. A. M. Stalcup and K. H. Gahm, *J. Microcolumn Sep.* **8**, 145 (1996).
32. D. W. Armstrong, K. Rundlett, and G. L. Reid, *Anal. Chem.* **66**, 1690 (1994).
33. T. O. Cole, M. J. Sepaniak, and W. L. Hinze, *J. High Resolut. Chromatogr. Chromatogr. Commun.* **13**, 570 (1990).
34. Y. Mechref and Z. El Rassi, *Chirality* **8**, 515 (1996).
35. N. M. Agyei, K. H. Gahm, and A. M. Stalcup, *Anal. Chim. Acta* **307**, 185 (1995).
36. A. M. Stalcup and K. H. Gahm, *J. Microcol. Sep.* **8**, 145 (1996).
37. R. S. Sahota and M. G. Khaledi, *Anal. Chem.* **66**, 1141 (1994).
38. F. Wang and M. G. Khaledi, *Anal. Chem.* **68**, 3460 (1996).
39. S. A. C. Wren and R. C. Rowe, *J. Chromatogr.* **603**, 235 (1992).
40. A. Guttman, A. Paulus, A. S. Cohen, N. Grinberg, and B. L. Karger, *J. Chromatogr.* **448**, 41 (1988).
41. S. Jönsson, A. Schön, R. Isaksson, C. Pettersson, and G. Pettersson, *Chirality* **5**, 505 (1992).
42. A. M. Stalcup, S. Gratz, and Y. Jin, unpublished results.
43. L. Witherow, T. D. Spurway, R. J. Ruane, I. D. Wilson, and K. Longdon, *J. Chromatogr.* **553**, 497–501 (1991).
44. H. T. K. Xuan and M. Lederer, *J. Chromatogr.* **635**, 346 (1993).
45. H. T. K. Xuan and M. Lederer, *J. Chromatogr.* **645**, 185 (1993).
46. C. A. Brunner and I. W. Wainer, *J. Chromatogr.* **472**, 277 (1989).
47. R. Bhushan and V. Parshad, *J. Chromatogr.* **721**, 369 (1996).
48. R. Bhushan and I. Ali, *Chromatographia* **35**, 679 (1993).
49. R. Bhushan and V. Parshad, *J. Chromatogr.* **736**, 235 (1996).
50. U. A. Th. Brinkman and D. Kamminga, *J. Chromatogr.* **330**, 375 (1985).
51. V. Mathur, N. Kanoongo, R. Mathur, C. K. Narang, and N. K. Mathur, *J. Chromatogr.* **685**, 360 (1994).
52. M. Remelli, R. Piazza, and F. Pulidori, *Chromatographia* **32**, 278 (1991).
53. D. W. Armstrong and W. DeMond, *J. Chromatogr. Sci.* **22**, 411 (1984).
54. D. W. Armstrong, Y. Tang, S. Chen, Y. Zhou, C. Bagwill, and J. R. Chen, *Anal. Chem.* **66**, 1473 (1994).
55. W. H. Pirkle, J. M. Finn, B. C. Hamper, J. L. Schreiner, and J. R. Pribish, *Am. Chem. Soc. Symp. Ser. No. 185*, Chapt. 18, 1982.
56. W. H. Pirkle and P. G. Murray, *J. Liq. Chromatogr.* **13**, 2123 (1990).
57. J. Hermansson and M. Eriksson, *J. Liq. Chromatogr.* **9**, 621 (1986).
58. G. Schill, I. W. Wainer, and S. A. Barkin, *J. Liq. Chromatogr.* **9**, 641 (1986).
59. S. Allenmark, *J. Liq. Chromatogr.* **9**, 425 (1986).
60. M. Okamoto and H. Nakazawa, *J. Chromatogr.* **508**, 217 (1990).
61. T. Miwa, H. Kuoda, S. Sakashita, N. Asakawa, and Y. Miyake, *J. Chromatogr.* **511**, 89 (1990).
62. R. Isaksson, P. Erlandsson, L. Hansson, A. Holmberg, and S. Berner, *J. Chromatogr.* **498**, 257 (1990).
63. Y. Okamoto, R. Aburatani, K. Hatano, and K. Hatada, *J. Liq. Chromatogr.* **11**, 2147 (1988).
64. T. Shinbo, T. Yamaguchi, K. Nishimura, and M. Sugiura, *J. Chromatogr.* **405**, 145 (1987).
65. M. Hilton and D. W. Armstrong, *J. Liq. Chromatogr.* **14**, 9 (1991).
66. V. A. Davankov, A. A. Kurganov, and A. S. Bochov, *Adv. Chromatogr.* **22**, 71 (1983).
67. W. H. Pirkle, J. M. Finn, J. L. Schreiner, and B. C. Hamper, *J. Am. Chem. Soc.* **103**, 3964 (1981).
68. C. E. Dalgliesh, *J. Chem. Soc.*, 3940 (1952).
69. W. H. Pirkle, D. W. House, and J. M. Finn, *J. Chromatogr.* **192**, 143 (1980).

70. W. H. Pirkle and T. C. Pochapsky, *J. Am. Chem. Soc.* **108**, 352 (1986).
71. W. H. Pirkle, T. C. Pochapsky, G. S. Mahler, D. E. Corey, D. S. Reno, and D. M. Alessi, *J. Org. Chem.* **51**, 4991 (1986).
72. W. H. Pirkle, G. Mahler, and M. H. Hyun, *J. Liq. Chromatogr.* **9**, 443 (1986).
73. W. H. Pirkle and C. J. Welch, *Tetrahedron Assym.* **5**, 777 (1994).
74. C. J. Welch, T. Szczerba, and S. R. Perrin, *J. Chromatogr. A* **758**, 93 (1997).
75. Regis Technologies Application Guide, pp. 30, 31.
76. C. J. Welch and S. R. Perrin, *J. Chromatogr.* **690**, 218 (1995).
77. T. J. Ward and D. W. Armstrong, "Cyclodextrin Stationary Phases" in M. Zief and L. J. Crane, eds., *Chromatographic Chiral Separations*, Marcel Dekker, Inc., New York, 1988, pp. 131.
78. S. M. Han and D. W. Armstrong, in A. M. Krstulovic, ed., *Chiral Separations by HPLC*, John Wiley & Sons, Inc., New York, 1989, pp. 208–287.
79. D. W. Armstrong, T. J. Ward, R. D. Armstrong, and T. E. Beesley, *Science* **232**, 1132 (1986).
80. R. E. Boehm, D. E. Martire, and D. W. Armstrong, *Anal. Chem.* **60**, 522 (1988).
81. D. W. Armstrong, S. Chen, C. Chang, and S. Chang, *J. Liq. Chromatogr.* **15**, 545 (1992).
82. A. M. Stalcup, J. R. Faulkner, Y. Tang, D. W. Armstrong, L. W. Levy, E. Regalado, *Biomed. Chromatogr.* **5**, 3 (1991).
83. P. Camilleri, C. A. Reid, and D. T. Manallack, *Chromatographia* **38**, 771 (1994).
84. A. M. Stalcup and K. H. Gahm, *Anal. Chem.* **68**, 1369 (1996).
85. A. M. Stalcup, S. Chang, D. W. Armstrong, and J. Pitha, *J. Chromatogr.* **513**, 181 (1990).
86. D. W. Armstrong, A. M. Stalcup, M. L. Hilton, J. D. Duncan, J. R. Faulkner, and S. C. Chang, *Anal. Chem.* **62**, 1610 (1990).
87. A. M. Stalcup, S. C. Chang, and D. W. Armstrong, *J. Chromatogr.* **540**, 113 (1991).
88. D. W. Armstrong, M. Hilton, and L. Coffin, *LC-GC* **9**, 647 (1992).
89. H. Hopf, W. Grahn, D. G. Barrett, A. Gerdes, J. Hilmer, J. Hucker, Y. Okamoto, and Y. Kaida, *Chem. Ber.* **123**, 841 (1990).
90. Y. Okamoto, Y. Kaida, R. Aburatani, and K. Hatada, *J. Chromatogr.* **477**, 367 (1989).
91. Y. Okamoto, K. Hatano, R. Aburatani, and K. Hatada, *Chem. Lett.*, 715 (1989).
92. J. M. Jansen, S. Copinga, G. Gruppen, R. Isaksson, D. T. Witte, and C. J. Grol, *Chirality* **6**, 596 (1994).
93. T. Shibata, I. Okamoto, and K. Ishii, *J. Liq. Chromatogr.* **9**, 313 (1986).
94. D. T. Witte, F. J. Bruggeman, J. P. Franke, S. Copinga, J. M. Jansen, and R. A. De Zeeuw, *Chirality* **5**, 545 (1993).
95. Y. Okamoto, T. Ohashi, Y. Kaida, and E. Yashima, *Chirality* **5**, 616 (1993).
96. J. Wagner, H.-J. Hamann, W. Döpke, A. Kunath, and E. Höft, *Chirality* **7**, 243 (1995).
97. M. Tanaka, H. Yamazaki, and H. Hakusui, *Chirality* **7**, 612 (1995).
98. J. Iredale, A.-F. Aubry, and I. W. Wainer, *Chromatographia* **31**, 329 (1991).
99. T. A. G. Noctor, I. W. Wainer, and D. S. Hage, *J. Chromatogr.* **577**, 305 (1992).
100. D. S. Hage, T. A. G. Noctor, and I. W. Wainer, *J. Chromatogr. A* **693**, 23 (1995).
101. G. Massolini, A.-F. Aubry, A. McGann, and I. Wainer, *Biochem. Pharm.* **46**, 1285 (1993).
102. E. Domenici, C. Bertucci, P. Salvadori, and I. W. Wainer, *J. Pharm. Sci.* **80**, 164 (1991).
103. D. W. Armstrong, Y. Tang, and S. Chen, *Anal. Chem.* **66**, 473 (1994).
104. D. W. Armstrong, Y. Liu, and K. H. Ekborgott, *Chirality* **7**, 474 (1995).
105. T. Shinbo, T. Yamaguchi, K. Nishimura, and M. Sugiura, *J. Chromatogr.* **405**, 145 (1987).
106. M. Hilton and D. W. Armstrong, *J. Liq. Chromatogr.* **14**, 9 (1991).

107. B. Sellergren, M. Lepistö, and K. Mosbach, *J. Am. Chem. Soc.* **110**, 5853 (1988).
108. L. Fischer, R. Müller, and B. Ekberg, *J. Am. Chem. Soc.* **113**, 9358 (1991).
109. Y. Okamoto, K. Suzuki, K. Ohta, K. Hatada, and H. Yuki, *J. Am. Chem. Soc.* **101**, 4763 (1979).
110. G. Blaschke, W. Bröker, and W. Fraenkel, *Angew. Chem.* **98**, 808 (1986).
111. S. G. Allenmark, S. Andersson, P. Möller, and D. Sanchez, *Chirality* **7**, 248 (1995).
112. Y. Liu, *Anal. Chem.* **69**, 61 (1997).
113. E. Gil-Av, B. Feibush, and R. Charles-Sigler, *Tetrahedron Lett.* **10**, 1009 (1966).
114. H. Frank, G. J. Nicholson, and E. Bayer, *J. Chromatogr. Sci.* **15**, 174 (1974).
115. W. A. König, I. Benecke, and S. Sievers, *J. Chromatogr.* **217**, 71 (1981).
116. W. A. König and E. Steinbach, and K Ernst, *J. Chromatogr.* **301**, 129 (1984).
117. B. Feibush, A. Balan, B. Altman, and Gil-Av, *J. Chem. Soc. Perkin II*, 1230 (1979).
118. V. Schurig, *Chromatographia* **13**, 263 (1980).
119. H. P. Nowotny, D. Schmalzing, D. Wistuba and V. Schurig, *J. High Resolut. Chromatogr. Chromatogr. Commun.* **12**, 383 (1989).
120. D. W. Armstrong, Y. Tang, T. Ward, and M. Nichols, *Anal. Chem.* **65**, 1114 (1993).
121. W.-Y. Li, H. L. Jin, and D. W. Armstrong, *J. Chromatogr.* **509**, 303 (1990).
122. A. Berthod, W. Li, and D. W. Armstrong, *Anal. Chem.* **64**, 873 (1992).
123. D. W. Armstrong, J. Zukowski, N. Ercal, and M. Gasper, *J. Pharm. Biomed. Anal.* **11**, 881 (1993).
124. J. Roboz, E. Nieves, and J. F. Roland, *J. Chromatogr.* **500**, 413 (1990).
125. I. W. Wainer, J. Ducharme, C. P. Granvil, H. Parenteau, and S. Abdullah, *J. Chromatogr. A* **694**, 169 (1995).

General References

C. F. Poole and S. K. Poole, *Chromatography Today*, Elsevier Science Publishers B.V., Amsterdam, The Netherlands, 1991.

I. W. Wainer, *Drug Stereochemistry: Analytical Methods and Pharmacology*, Marcel Dekker, Inc., New York, 1993.

W. A. König, *The Practice of Enantiomer Separation by Capillary Gas Chromatography*, Hüthig Verlag, Heidelberg, Germany, 1987.

P. Schreier, A. Bernreuther, and M. Huffer, *Analysis of Chiral Organic Molecules*, Walter de Gruyter & Co., Berlin, Germany, 1995.

G. Subramanian, *A Practical Approach to Chiral Separations by Liquid Chromatography*, VCH, Weinheim, Germany, 1994.

APRYLL M. STALCUP
University of Cincinnati

FURAN DERIVATIVES

Furan (**1**) is a 5-membered heterocyclic, oxygen-containing, unsaturated ring compound. From a chemical perspective it is the basic ring structure found in a whole class of industrially significant products. The furan nucleus is also found in a large number of biologically active materials. Compounds containing the furan ring (as well as the tetrahydrofuran ring) are usually referred to as furans (1). From a manufacturing standpoint, however, furfural (**2**) is the feedstock from which all of the commercial furan derivatives are derived, as it the easiest and least expensive to manufacture. In this article, the more common name furfural is used in place of the *Chemical Abstracts* name, 2-furancarboxaldehyde.

Furan is produced from furfural commercially by decarbonylation; loss of carbon monoxide from furfural gives furan directly. Tetrahydrofuran (**3**) is the saturated analogue containing no double bonds.

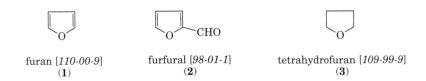

furan [*110-00-9*] furfural [*98-01-1*] tetrahydrofuran [*109-99-9*]
(**1**) (**2**) (**3**)

Furfural is derived from biomass by a process in which the hemicellulose fraction is broken down into monomeric 5-carbon sugar units which then are dehydrated to form furfural.

This article is limited primarily to simple furans in which the nucleus occurs as a free monocycle, and in general, does not include compounds in which the furan ring is fused to another ring. In recent years it has mistakenly become common practice to refer to polychlorinated isobenzofurans simply as *furans*. These isobenzofurans are comprised of chlorine-containing fused rings, are not simple furans, and are not included here. The presentation is generally limited to compounds (including resins or polymers) derived from furfural. Ring closure methods to prepare furan or tetrahydrofuran compounds are not covered.

Tetrahydrofuran (**3**) is produced commercially from furfural by decarbonylation followed by hydrogenation; it is also produced by several different methods from other raw materials. A complete discussion of tetrahydrofuran is found in Volume 9, under ETHERS. Polymers of tetrahydrofuran are covered in Volume 19 under the general topic, POLYETHERS. Several other compounds containing the tetrahydrofuran ring, which are most readily produced from furfural, are discussed here.

The furan nucleus is a cyclic, dienic ether with some aromaticity (2). It is the least aromatic of the common 5-membered heterocycles. A comparison of the aromaticity (3) of several of these compounds is shown below.

Property	Furan	Thiophene	Pyrolle	Benzene
Unified aromaticity index	53	81.5	85	100
Resonance energy, kJ/mol	113.8	180.0	169.4	191.6
kcal/mol	27.2	43.0	40.5	45.8

The balance between aromatic and aliphatic reactivity is affected by the type of substituents on the ring. Furan functions as a diene in the Diels-Alder reaction. With maleic anhydride, furan readily forms 7-oxabicyclo [2.2.1]hept-5-ene-2,3-dicarboxylic anhydride in excellent yield [*5426-09-5*] (**4**).

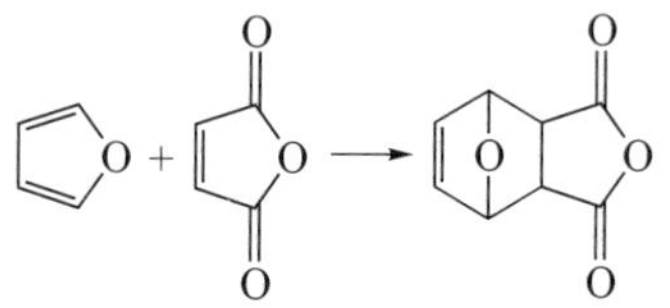

(**4**)

Alkylfurans, halofurans, alkoxyfurans, furfuryl ester and ethers, and furfural diacetate [*613-75-2*] behave similarly. Furans containing electron withdrawing constituents, for example, furfural, 2-furoic acid, and nitrofurans, fail as dienes even with very strong dienophiles.

Radicals derived from furan are named similarly to analogous radicals in the benzene series. Typical radicals are 2 (or α)-furyl (**5**), 2-furfuryl (**6**), 2-furoyl (**7**), and 2-furfurylidene (**8**):

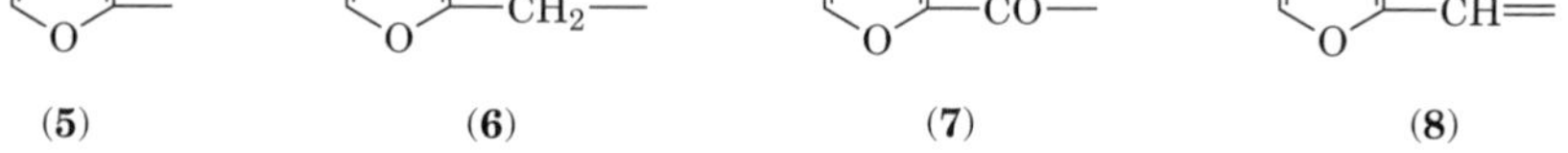

(**5**)   (**6**)   (**7**)   (**8**)

Centers of high electron density occur at the *alpha* positions of the ring which governs the behavior of furans in electrophyllic substitution reactions. Such reactions proceed readily at available alpha positions. Orientation of incoming groups is similar to that observed in benzene compounds, eg, 3-methylfuran is nitrated to yield 2-nitro-3-methylfuran (similar to *ortho*) and 3-furancarboxaldehyde gives 2-nitro-4-furancarboxaldehyde (corresponding to *meta*). Furan compounds differ from benzene analogues in the ease in which they are polymerized; resinification or polymerization is easily brought about by strong acids. Resins or polymers are the principal end products or furfural utilization and account for a high percentage of the ultimate consumption of furfural.

For reviews on furan chemistry see references 1, 4–7. The monograph by Dunlop and Peters (1), published in 1953 remains as the most comprehensive review of furan chemistry available. Unfortunately, it has never been revised, has been out of print for many years, and is not readily available. An excellent, review by Dean (7) covers advances in furan chemistry up through about 1982.

Furfural

Furfural was first isolated in the early 19th century. Dobereiner is credited with the discovery. He obtained a small amount of a yellow "oil" (too little to characterize) as a by-product in the preparation of formic acid (8). Other chemists found that the same "oil" having a characteristic aroma could be obtained by boiling finely divided vegetable materials such as oats, corn, sawdust, bran, etc, with aqueous sulfuric acid or other acids (9,10). The oil was present in the liquid resulting from condensation of the vapors produced during heating. The empirical formula $C_5H_4O_2$ was determined by Stenhouse (10). Ring structure and location of the aldehyde group were established by the efforts of Baeyer, Markwald, and Harries (11–14). For some time, furfural was a laboratory curiosity and a compounding ingredient of perfumes.

It was not until the 20th century that furfural became important commercially. The Quaker Oats Company, in the process of looking for new and better uses for oat hulls found that acid hydrolysis resulted in the formation of furfural, and was able to develop an economical process for isolation and purification. In 1922 Quaker announced the availability of several tons per month. The first large-scale application was as a solvent for the purification of wood rosin. Since then, a number of furfural plants have been built world-wide for the production of furfural and downstream products. Some plants produce as little as a few metric tons per year, the larger ones manufacture in excess of 20,000 metric tons.

Furfural can be classified as a reactive solvent. It resinifies in the presence of strong acid; the reaction is accelerated by heat. Furfural is an excellent solvent for many organic materials, especially resins and polymers. On catalyzation and curing of such a solution, a hard rigid matrix results, which does not soften on heating and is not affected by most solvents and corrosive chemicals.

Furfural is formed by a series of reactions when biomass materials containing hemicellulose are treated with acid at an elevated temperature. Hemicellulose is basically a polymer of 5-carbon sugars. Cellulose and starch are polymers of 6-carbon sugars. Most plant matter is made up of cellulose, hemicellulose, and lignin. When treated with aqueous acid, the hemicellulose is depolymerized to give primarily xylose, which under the reaction conditions loses three molecules of water and cyclizes to give furfural. The precise mechanism for the formation of furfural from C_5-sugars or their precursors has not yet been unequivocally established. Superheated steam passing through a reactor containing eg, ground oat hulls, provides heat for the conversion as well as carrying furfural away from the reaction mixture. Furfural also resinifies under the reaction conditions so the steam serves a very important function by removing the furfural as fast as it is formed. Most of the other compounds present in the reaction vessel are not volatile with steam. Condensation of the vapors gives a dilute solution of furfural in water. Furfural is isolated from the water by azeotropic distillation and phase separation, followed by dehydration and purification by distillation.

Physical Properties.   Furfural [98-01-1] (2-furancarboxaldehyde), when freshly distilled, is a colorless liquid with a pungent, aromatic odor reminiscent of almonds. It darkens appreciably on exposure to air or on extended storage. Furfural is miscible with most of the common organic solvents, but only slightly miscible with saturated aliphatic hydrocarbons. Inorganic compounds, generally, are quite insoluble in furfural.

The most important physical properties of furfural, as well as similar properties for furfuryl alcohol, tetrahydrofurfuryl alcohol and furan are given in Table 1. The tabulated properties of furfural are supplemented by a plot (Fig. 1) of the vapor–liquid compositions for the system, furfural–water (15,16).

Chemical Properties.   The chemical properties of furfural are generally characteristic of aromatic aldehydes but with some differences attributable to the furan ring. Furfural resinifies in the presence of acid and heat. Open chain compounds are formed from furfural under strong oxidizing conditions. Because furfural has a fairly low degree of aromaticity, the ring is more easily saturated than the benzene ring on catalytic hydrogenation. Depending on catalysts and conditions, products of hydrogenation can include furfuryl alcohol [98-00-0], tetrahydrofurfuryl alcohol [97-99-4], 2-methylfuran [534-22-5] and even 2-methyltetrahydrofuran [96-47-9]. Under strongly reducing conditions, the ring is opened.

Furfural is very thermally stable in the absence of oxygen. At temperatures as high as 230°C, exposure for many hours is required to produce detectable changes in the physical properties of furfural, with the exception of color (17). However, accelerating rate calorimetric data shows that a temperature above 250°C, in a closed system, furfural will spontaneously and exothermically decompose to furan and carbon monoxide with a substantial increase in pressure. The pressure may increase to 5000 psi or more, sufficient to shatter the container (18).

Furfural can be oxidized to 2-furoic acid [88-14-2], reduced to 2-furanmethanol [98-00-0], referred to herein as furfuryl alcohol, or converted to furan by decarbonylation over selected catalysts. With concentrated sodium hydroxide, furfural undergoes the Cannizzaro reaction yielding both 2-furfuryl alcohol and sodium 2-furoate [57273-36-6].

Acetals are readily formed with alcohols and cyclic acetals with 1,2 and 1,3-diols (19). Furfural reacts with poly(vinyl alcohol) under acid catalysis to effect acetalization of the hydroxyl groups (20,21). Reaction with acetic anhydride

Table 1. Physical Properties of Furan Derivatives

	Furfural [98-01-1]	Furfuryl alcohol [98-00-0]	Furan [110-00-9]	Tetrahydrofurfuryl alcohol [97-99-4]
		General properties		
molecular weight	96.09	98.10	68.08	102.13
boiling point at 101.3 kPa (1 atm), °C	161.7	170	31.36	178
freezing point, °C	−36.5		−85.6	<−80
metastable crystalline form		−29		
stable crystalline form		−14.63		
refractive index, n_D				
20°C	1.5261	1.4868	1.4214	1.4250
25°C	1.5235			1.4499
density, d_4, at 20°C	1.1598	1.1285	0.9378	1.0511
vapor pressure, 100 Pa (0.75 mm Hg)				
−15°C			130	
0°C			277	
20°C			658	
50°C			1980	
60°C	21	8.5		10
80°C	56	25		28
100°C	132	58		68
120°C	280	164		156
140°C	567	361		324
vapor density (air = 1)	3.3	3.4	2.36	3.5
critical pressure, P_c, MPa[a]	5.502		5.32	
critical temperature, T_c, °C	397		214	
solubility, wt %, in water				
20°C	8.3	∞		∞
25°C			1	
alcohol; ether	∞	∞	∞	∞
		Thermodynamic properties		
heat of vaporization, kJ/mol[b]	38.6		27.1	
specific heat (liq), J/(g·K)[b]				
20°C			1.699	
20–27°C				1.774
20–100°C	1.741			
25°C		2.100		
specific heat (vap), J/(g·K)[b]				
31.36°C			1.021	
98.99°C			1.251	
heat of combustion (liq), kJ/mol[b]	2344	2548	2092	2965

159

Table 1. (*Continued*)

	Furfural [98-01-1]	Furfuryl alcohol [98-00-0]	Furan [110-00-9]	Tetrahydrofurfuryl alcohol [97-99-4]
Fluid properties				
viscosity, mPa·s (=cP)				
20°C			0.38	6.24
25°C	1.49	4.62		
surface tension, mN/m (=dyn/cm)				
25°C		ca 38		37
29.9°C	40.7			
Electrical properties				
dielectric constant				
20°C	41.9			
23°C				13.6
Flammability properties				
explosion limits (in air), vol %	2.1–19.3	1.8–16.3	2.3–14.3	1.5–9.7
flash point, °C				
Tag closed cup	61.7	65	−35.5	
Tag open cup				83.9
ignition temperature, °C	315	391		282

[a]To convert MPa to atm, divide by 0.101.
[b]To convert J to cal, divide by 4.184.

under appropriate conditions gives the acylal, furfurylidene diacetate [613-75-2] (22,23).

Nitration and halogenation of furfural occurs under carefully controlled conditions with introduction of the substituent at the open 5-position (24,25). Nitration of furfural is usually carried out in the presence of acetic anhydride, resulting in the stable compound, 5-nitrofurfurylidene diacetate (26,27). The free aldehyde is isolated by hydrolysis and must be used immediately in a reaction because it is not very stable.

Just as most other aldehydes do, furfural condenses with compounds possessing active methylene groups such as aliphatic carboxylic esters and anhydrides, ketones, aldehydes, nitriles, and nitroparaffins.

The electron-withdrawing character of the aldehyde group makes furfural quite resistant to hydrolytic fission. Even at high temperatures, long exposure is required to effect extensive ring opening destruction of furfural by dilute acids (28). Under these conditions a competing reaction takes place: black polymer results, and the rate of its formation is dependent upon the hydrogen ion concentration and temperature.

Furfural is a resin former under the influence of strong acid. It will self-resinify as well as form copolymer resins with furfuryl alcohol, phenolic compounds, or convertible resins of these. Conditions of polymerization, whether aqueous or anhydrous, inert or oxygen atmosphere, all affect the composition of the polymer. Numerous patents have issued relating to polymerization and to

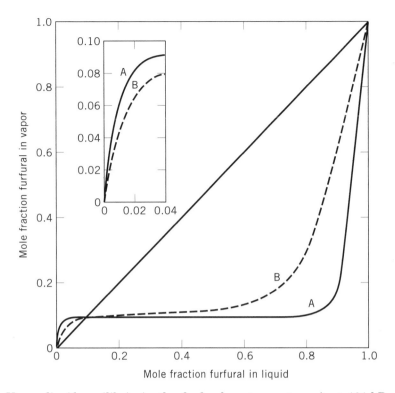

Fig. 1. Vapor–liquid equilibria in the furfural–water system. A at 101 kPa (1 atm) (15). B at 598 kPa (5.92 atm) (16).

applications. Although the resins exhibit a degree of brittleness, they have many outstanding properties; a number of applications are discussed under "Uses."

Several early interpretations of the polymerization mechanism have been proposed (1,17,29–31). Because of the complexity of this polymerization and insoluble character of the products, key intermediates have not ordinarily been isolated, nor have the products been characterized. Later work, however, on the resinification of furfural (32,33) has provided a new insight on the polymerization mechanism, particularly with respect to thermal reaction at 100–250°C in the absence of air. Based on the isolation and characterization of two intermediate products (**9**) and (**10**), structure (**11**) was proposed for the final resin. This work also explains the color produced during resinification, which always is a characteristic of the final polymer (33). The resinification chemistry is discussed in a recent review (5).

(**9**) [*57583-51-4*]   (**10**) [*72918-23-1*]

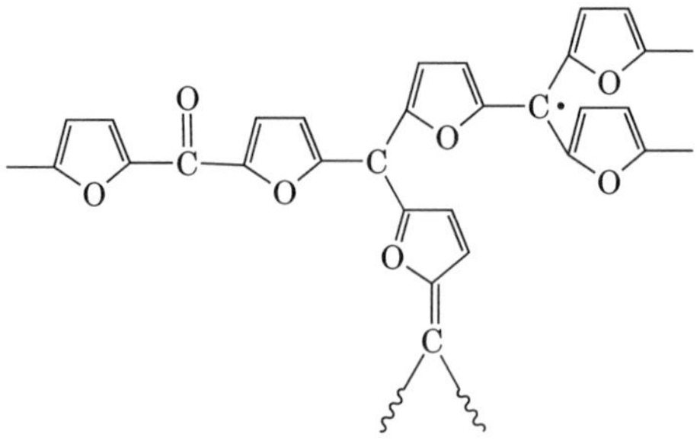

(11)

The presence of stable free radicals in the final polycondensate is supported by the observation that traces of (11) have a strong inhibiting effect on the thermal polymerization of a number of vinyl monomers. Radical polymerization was inhibited to a larger extent by a furfural resin than by typical polymerization inhibitors (34). Thermal degradative methods have been used to study the structure of furfural resinified to an insoluble and infusible state, leading to proposed structural features (35).

Manufacture.   Furfural is produced from annually renewable agricultural sources such as nonfood residues of food crops and wood wastes. The pentosan polysaccharides, xylan and arabinan, commonly known as hemicellulose, are the principal precursors of furfural and are always found together with lignin and cellulose in plant materials.

By-products such as corncobs, cottonseed hulls, oat hulls, rice hulls and cereal grasses constitute one large natural storehouse; bagasse, a by-product of sugar-cane harvesting, another; and wood and wood products yet another. The chief constituent of the pentosan fraction of these materials is xylan, a polysaccharide having backbone chains of β-D-xylopyranosyl residues (33). According to published data (34), xylan with a small amount of arabinan, a highly branched polysaccharide of α-araibinofuranosyl residues (35), accounts for 2–30% of cereal straws and grains, approximately 15–25% of deciduous woods and 5–15% of coniferous woods, based on the dry matter of these plants.

Theoretically, all pentosan-containing substances are potentially usable for the production of furfural. Only a relatively few, however, are commercially significant (1,36,37). Corncobs, oat hulls, rice hulls, and bagasse contain sufficient levels of pentosans and are often easily available in large tonnages within a limited radius of a furfural-producing plant. The simultaneous occurrence of side reactions complicates the recovery of furfural and limits its practical yield, which is far from quantitative under any commercial mode of operation. The average weight yield of furfural based on the dry weight of these raw materials is 10% or less. Thus, the economics of furfural production is highly dependent upon pentosan content, availability, and cost of raw materials, as well as the expense of collecting, shipping, and handling them.

Furfural is commercially produced in batch or continuous digesters (37) where the pentosans are first hydrolyzed to pentoses (primarily xylose (12), which are then subsequently cyclodehydrated to furfural:

CHO

H—C—OH

pentosan $\xrightarrow{\text{H}_3\text{O}^+}$ HO—C—H $\xrightarrow{-\text{H}_2\text{O}}$ [furan ring with CHO]

H—C—OH

CH$_2$OH

(12)

Strong inorganic acid can be used as a reaction catalyst, or if the temperature is raised high enough, sufficient acetic acid is produced by the heat to catalyze formation of furfural.

In all processes, raw material is charged to the digester and heated with high pressure steam. Enough excess steam is used to drive the furfural out of the reaction zone as vapor. The condensed reactor vapors are fed to a stripping column from which an enriched furfural–water distillate mixture is taken overhead and condensed. The liquid passes into a decanter where it separates into two layers. The furfural-rich lower layer containing about 6% water is processed further to obtain the furfural of commerce, and the water-rich layer containing about 8% furfural is recycled back to the stripper column as reflux. The last process steps are dehydration and distillation. The furfural isolation and recovery process is illustrated in Figure 2. Flow charts for the Escher-Wyss and Rosenlew, two slightly different processes can be found in ref. 38. In general, regardless of the process employed, the cellulose in the raw material

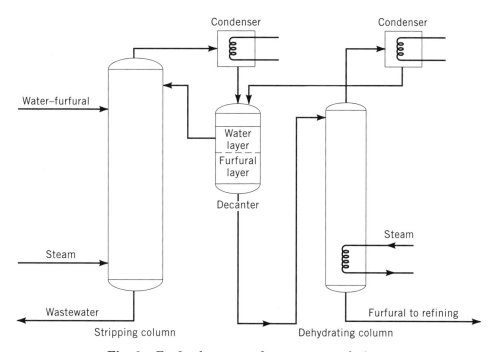

Fig. 2. Furfural recovery from aqueous solutions.

undergoes only partial degradation and remains with the lignin in the by-product residue.

Most laboratories now use gas chromatographic methods for the analysis of refined furfural. For dilute aqueous solutions, wet methods are sometimes more appropriate. A number of wet analytical methods for quantitative estimation of furfural have been reported (1), yet none have universal acceptance. For the most part, they can be divided into two general categories: those involving reactions of the aldehyde group, and those that are dependent upon the unsaturated characteristics of the furan ring. Accordingly, the determination of furfural is generally subject to interference by carbonyl compounds or other unsaturated substances. The quantitative measurement of furfural in water may be carried out by uv spectroscopy; absorbance is measured at 276 nm (39). This method is best used on dilute solutions. However, uv-absorbing impurities such as phenolic compounds interfere with the analysis. The Hughes-Acree method, based on the reaction of bromine with the double bonds in furfural was at one time the method of choice (1).

The estimation of furfural potential of various raw materials is best done by the AOAC method (1). Although liquid chromatographic methods are now available for the estimation of polymeric pentosans, results do not always correlate well with furfural formation.

Furfural is not corrosive to metal and can be shipped in mild-steel tank cars or trucks or steel drums. Storage in either aboveground or underground installations is satisfactory. For extended storage with maximum stability, cool storage conditions and nitrogen blanketing are recommended. Because furfural is an excellent solvent and penetrant, care must be taken that all joints are secure and that the pump and valve packings are in good condition. Unopened drums may be stored in cool locations for months without appreciable change in physical properties.

The flash point of furfural is 143°F by Tag Closed Cup. Because of its chemical reactivity, furfural should be kept away from strong acids, alkalies or strong oxidizing chemicals. When furfural is stored for long periods in contact with air, there is a gradual darkening of color, increase in acidity, and formation of a soluble polymer.

According to the latest available information (40), in 1992, total world-wide capacity for furfural was 240,000 metric tons (t) (530,000,000 lb). Production capability is moving away from developed countries into developing nations. The Peoples' Republic of China has become a significant factor in furfural production, with an estimated total capacity of 50,000 t, most of it being produced in small plants. New capacity has come on stream or is being planned in other Asian countries as well as in Indonesia. South African capacity is rated at 17,000 t. United States capacity was estimated at 73,000 t (30% of the world total), production at 40,000 t and consumption at 43,000. Western Europe capacity was 8,000 t (down from 27,000 in 1988). Production was 6,000 t and consumption was 36,000 t, a significant amount being imported from the western hemisphere. Japan is not presently a producer of furfural, importing approximately 9,000 t, mostly from China. At this time there is significant overcapacity worldwide; consumption is steady or declining slightly. The United States list price for furfural was $0.79/lb in 1992; it has not changed significantly since then.

Uses. Furfural is primarily a chemical feedstock for a number of monomeric compounds and resins. One route produces furan by decarbonylation. Tetrahydrofuran is derived from furan by hydrogenation. Polytetramethylene ether glycol [25190-06-1] is manufactured from tetrahydrofuran by a ring opening polymerization reaction. Another route (hydrogenation) produces furfuryl alcohol, tetrahydrofurfuryl alcohol [97-99-4], 2-methylfuran [534-22-5] and 2-methyltetrahydrofuran [96-47-9]. A variety of proprietary synthetic resins are manufactured from furfural and/or furfuryl alcohol. Other chemicals that have been derived, on a lesser scale are described in a later section. In this country more than 90% of the furfural produced is used as feedstock for other compounds. A breakdown of U.S. consumption in 1992 is shown in Table 2.

As can be seen, most of the furfural produced in this country is consumed as an intermediate for other chemicals. Hydrogenation to furfuryl alcohol is the largest use. Some of the furfuryl alcohol is further hydrogenated to produce tetrahydrofurfuryl alcohol. The next major product is furan, produced by decarbonylation. Furan is a chemical intermediate, most of it is hydrogenated to tetrahydrofuran, which in turn is polymerized to produce polytetramethylene ether glycol (PTMEG).

Furfural and furfuryl alcohol are specialty solvents. They also are reactive solvents and contribute low viscosity to resin formulations. Thermosetting resins containing furfural and/or furfuryl alcohol. "Furan resins" demonstrate specialty properties including corrosion resistance, high carbon yield, stability at elevated temperature, low fire hazard, and excellent physical strength. These properties are of industrial importance in making foundry molds and cores, fiberglass composites, mortars, cements, plastic insulation foams, refractory mixes, high carbon composites, and aggregate binders, among others.

The principal direct application of furfural is as a selective solvent. It is used for separating saturated from unsaturated compounds in petroleum refining, for the extractive distillation of butadiene and other C_4 hydrocarbons in the manufacture of synthetic rubber; and for the production of light-colored wood rosin. These applications for furfural are mature and volume is declining. Furfural is both a solvent and a processing aid for refining anthracene. It is an ingredient in furfuryl alcohol resins and in phenol–aldehyde resins. Furfural is also used as a resin solvent and wetting agent in the manufacture of abrasive wheels and brake linings. Furfural is used as a reactive solvent in cold-blending of pitch. The mixture gives resinous impregnants that interact at carbonization temperatures. These resin pitch combinations have curing and carbonization characteristics better than those of either component alone. The cold-blending

Table 2. U.S. Consumption of Furfural

Use	t	(10^6 lb)
raw material for furfuryl alcohol	24,460	(54)
raw material for tetrahydrofurfuryl alcohol	905	(2)
raw material for tetrahydrofuran	13,590	(30)
solvent applications	3,625	(8)
other uses	1,815	(4)

process significantly improves the environmental acceptance of pitch and offers industry an alternative to hot processing.

High quality carbon fibers have been produced (41) by pyrolyzing, in a nonoxidizing atmosphere, resinous fibers prepared from a viscous copolymer of furfural and pyrrole. Yield of carbon fiber having a glossy surface and only 3% voids was 66%. Compared with other graphite-type fibers, these carbon fibers had improved mechanical strength, higher oxidation resistance, and lower density.

Useful thermosetting resins are obtained by interaction of furfural with phenol. The reaction occurs under both acidic and basic catalysis. Other large uses of furfural together with phenol are in the manufacture of resin-bonded grinding wheels and coated abrasives (5).

Furfural reacts with ketones to form strong, crosslinked resins of technical interest in the former Soviet Union; the U.S. Air Force has also shown some interest (42,43). The so-called furfurylidene acetone monomer, a mixture of 2-furfurylidene methyl ketone [623-15-4] (13), bis-(2-furfurylidene) ketone [886-77-1] (14), mesityl oxide, and other oligomers, is obtained by condensation of furfural and acetone under basic conditions (44,45). Treatment of the "monomer" with an acidic catalyst leads initially to polymer of low molecular weight and ultimately to crosslinked, black, insoluble, heat-resistant resin (46).

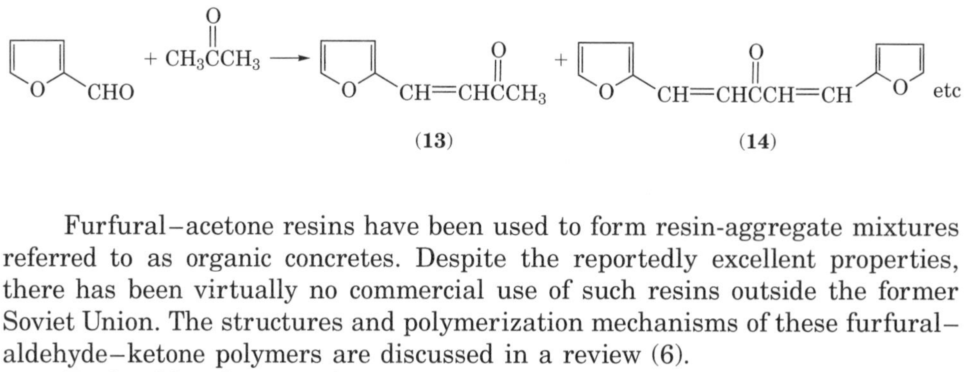

(13)   (14)

Furfural–acetone resins have been used to form resin-aggregate mixtures referred to as organic concretes. Despite the reportedly excellent properties, there has been virtually no commercial use of such resins outside the former Soviet Union. The structures and polymerization mechanisms of these furfural–aldehyde–ketone polymers are discussed in a review (6).

Furfural has been used as a component in many resin applications, most of them thermosetting. A comprehensive review of the patent literature describing these uses is beyond the scope of this review. A few, selected recent patents and journal articles have been referenced. Resins prepared from the condensation products of furfural with urea (47), formaldehyde (48), phenols (49,50), etc, modified by appropriate binders and fillers are described in the technical literature; for earlier applications, see reference 1, which contains many references in an appendix.

Furfuryl Alcohol

Physical Properties. Furfuryl alcohol (2-furanmethanol) [98-00-0] is a liquid, colorless, primary alcohol with a mild odor. On exposure to air, it gradually darkens in color. Furfuryl alcohol is completely miscible with water, alcohol, ether, acetone, and ethyl acetate, and most other organic solvents with the exception of paraffinic hydrocarbons. It is an excellent, highly polar solvent, and dissolves many resins.

The physical constants of furfuryl alcohol are listed in Table 1. When exposed to heat, acid or air the density and refractive index of furfuryl alcohol changes owing to chemical reaction (51), and the rate of change in these properties is a function of temperature and time of exposure.

Chemical Properties. The chemical properties of furfuryl alcohol are characteristic of the hydroxymethyl group; however, in the presence of acid the reactivity of the furan ring is significantly enhanced through allyl carbonium ion formation. Furfuryl alcohol undergoes the typical reactions of a primary alcohol such as oxidation, esterification, and etherification. Although stable to strong alkali, furfuryl alcohol is very sensitive to acid, thus imposing limitations on the conditions used in many of the typical alcohol reactions. For example, esterifications must be performed under neutral or basic conditions, or via base-catalyzed transesterification. Likewise, etherifications are done under basic condensation conditions to suppress polymerization of the furan ring. Yet it is this very sensitivity to acid that is the basis for most of the commercial uses of furfuryl alcohol. It is a more reactive solvent than furfural and easily resinifies under the influence of acid and heat.

Under acidic conditions, furfuryl alcohol polymerizes to black polymers, which eventually become crosslinked and insoluble in the reaction medium. The reaction can be very violent and extreme care must be taken when furfuryl alcohol is mixed with any strong Lewis acid or Brönstad acid. Copolymer resins are formed with phenolic compounds, formaldehyde and/or other aldehydes. In dilute aqueous acid, the predominant reaction is a ring opening hydrolysis to form levulinic acid [123-76-2] (52). In acidic alcoholic media, levulinic esters are formed. The mechanism for this unusual reaction in which the hydroxymethyl group of furfuryl alcohol is converted to the terminal methyl group of levulinic acid has recently been elucidated (53).

Hydrogenation of furfuryl alcohol can yield 2-tetrahydrofurfuryl alcohol [97-99-4], 2-methylfuran [534-22-5], 2-methyltetrahydrofuran [96-47-9], or straight chain compounds by hydrogenolysis of the ring. Ethoxylation and propoxylation of furfuryl alcohol provide useful ether alcohols.

The chemistry of furfuryl alcohol polymerization has received much attention over the years. Several recent reviews have been written (5,6,54). Based on the accumulated data, furfuryl alcohol has to be considered a bifunctional monomer in the initial stage and its "normal" reactions give linear chains or oligomers containing essentially two repeating units (**15**,**16**) with (**16**) predominating.

$$-\underset{O}{\overset{}{\bigsqcup}}-CH_2OCH_2- \qquad\qquad -\underset{O}{\overset{}{\bigsqcup}}-CH_2-$$

(**15**)   (**16**)

In 1996, the mechanism of resinification was clearly and elegantly established (55). Development of color (a phenomenon which up until now has never been adequately explained) has been shown to be due to conjugated sequences along linear oligomeric chains (**17**). Branching and crosslinking, which cause gelation and isolubilization are described to take place through the reaction of terminal methylol groups or disubstituted furan rings with nonfuranic double bonds in the oligomeric chain.

(**17**)

Manufacture.   Furfuryl alcohol has been manufactured on an industrial scale by employing both liquid-phase and vapor-phase hydrogenation of furfural (56,57). Copper based catalysts are preferred because they are selective and do not promote hydrogenation of the ring.

Furfuryl alcohol is shipped in bulk or drums. Although not corrosive to metals, it is a powerful solvent and penetrant; containers, tanks, lines, and valves need to be in good condition to avoid potential leakage. Furfuryl alcohol can be stored in containers lined with baked phenolic resin coatings; however, it should not be put in containers that are coated with lacquers, varnishes, or epoxy resins because it is an excellent solvent for many such coatings.

Furfuryl alcohol, on long storage, becomes progressively darker and less water soluble, a change that is also caused by heat, acidity, and exposure to air. The reactions responsible for this change in water solubility may be retarded by the addition in small quantity of an organic or inorganic base. Commercial furfuryl alcohol, however, usually does not contain any additives.

Furfuryl alcohol is comparable to kerosene or No. 1 fuel oil in flammability, the Tag Closed Cup flash point is 170°F. In the presence of concentrated mineral acids or strong organic acids, furfuryl alcohol reacts with explosive violence. Therefore, precautions should be taken to avoid contact of such materials with the alcohol. Caution is also recommended to avoid over-catalysis in the manufacture of furfuryl alcohol resins.

World-wide furfuryl alcohol capacity in 1993 was estimated to be 110,000 metric tons (38). As with furfural, new capacity in developing countries is replacing older capacity in developed countries. China and South Africa have become significant producers of furfuryl alcohol. New plants have been built in Asia and Indonesia as well. Consumption of furfuryl alcohol is spread over the globe; the largest use is in the foundry industry which is increasingly moving away from heavily industrialized countries.

Uses.   Furfuryl alcohol is widely used as a monomer in manufacturing furfuryl alcohol resins, and as a reactive solvent in a variety of synthetic resins and applications. Resins derived from furfuryl alcohol are the most important application for furfuryl alcohol in both utility and volume. The final crosslinked products display outstanding chemical, thermal, and mechanical properties. They are also heat-stable and remarkably resistant to acids, alkalies, and solvents. Many commercial resins of various compositions and properties have been prepared by polymerization of furfuryl alcohol and other co-reactants such as furfural, formaldehyde, glyoxal, resorcinol, phenolic compounds and urea. In 1992, domestic furfuryl alcohol consumption was estimated at 47 million pounds (38).

The polymerization or resinification of furfuryl alcohol is highly exothermic, and requires careful control of the reaction conditions. Proper attention to safety cannot be overemphasized. The addition of too much strong acid will lead to a violent, almost explosive, reaction. The resinification of furfuryl alcohol is autocatalytic in the sense that the rate increases geometrically with temperature. During a resinification reaction, temperature control is generally accomplished with refluxing solvent and/or external cooling. As a safety precaution, to prevent a runaway reaction, provision for emergency neutralization of the catalyst is essential. At temperatures above 250°C, in a closed system, furfuryl alcohol undergoes a strong exothermic reaction which is not controlled by base (18). Furfuryl alcohol should not be heated to above 250°C in a closed system.

In practice, intermediate, liquid resins, capable of further reaction are usually prepared. Polymerization is carried to an established end-point as determined by viscosity or other measurements. When the proper end point has been reached, the reaction is terminated by adjusting the pH of the system to 5–8. Such liquid resins can be stored for six months or longer, then catalyzed and reacted further to obtain the final, desired product.

Furfuryl alcohol is an excellent solvent for many resins. A number of applications are based on its reactive solvent properties. When a resin solution is cured, a hard, rigid, thermoset matrix results. The final, cured product often has many outstanding properties. For example, furfuryl alcohol is a reactive solvent for phenolic resins in the manufacture of refractories for ladles holding molten steel. The furfuryl alcohol provides a low viscosity to allow good mixing of the resin with the refractory particles, then reacts with the phenolic resin during curing. When heated to high temperatures, the matrix carbonizes, producing a strong refractory bond.

The industrial value of furfuryl alcohol is a consequence of its low viscosity, high reactivity, and the outstanding chemical, mechanical, and thermal properties of its polymers, corrosion resistance, nonburning, low smoke emission, and excellent char formation. The reactivity profile of furfuryl alcohol and resins is such that final curing can take place at ambient temperature with strong acids or at elevated temperature with latent acids. Major markets for furfuryl alcohol resins include the production of cores and molds for casting metals, corrosion-resistant fiber-reinforced plastics (FRPs), binders for refractories and corrosion-resistant cements and mortars.

Metal-Casting Cores and Molds. In the foundry industry there is a highly significant binder technology based on furfuryl alcohol resins. These binders are cured at ambient temperature or at elevated temperatures. Microwave curing has been used. Even curing with acidic gases is feasible. Total furfuryl alcohol based binder consumption by the foundry industry in 1992 has been estimated at 35 million pounds, representing 27 million pounds of contained furfuryl alcohol (38).

Furan No-Bake (FNB) resins are used worldwide in competition with several other chemical binder systems. FNB binders are used in the production of sand cores and molds that cure without heat (58–61). The FNB process uses a reactive furan resin binder with strong organic acid or phosphoric acid as catalyst. Based on the requirements of the particular system, slow or rapid curing can be accomplished at ambient temperatures, eliminating the need for long oven-baking cycles. Accelerators are sometimes used to increase curing rate (62–65). The catalyst is added to the sand, followed by the binder, often in a continuous mixer. The characteristics of the acid and the binder as well as the mixing ratio determine curing rate characteristics. Other money saving benefits that accrue include: fewer or no reinforcing rods; less pattern wear; dimensional accuracy of the castings; easy shakeout of core sand after pouring; and minimal clean-up of the finished casting. The used sand can easily be reclaimed by mechanical means. Most important, the process is simple, requiring fewer steps to produce cores, and can be operated by semiskilled personnel.

Furan hot-box resins are used in both ferrous and nonferrous foundries (66,67). In this process, resin and catalyst are intimately mixed with dry sand and then blown into heated metal boxes containing a cavity the shape of the desired core. In seconds, the surface of the sand mass hardens and, as soon as the core has cured sufficiently to be rigid and handleable the box is opened and the core removed. Automotive cores with excellent dimensional accuracy and high strengths are made via this forty-year-old process.

Both FNB and Hot Box applications are mature and declining as new technology is being used more and more in the foundry industry. Technological advances continue to be made, several recent patents describe advanced phenol–formaldehyde–furfuryl alcohohol binder systems (68–70). These systems are free of nitrogen compounds that can be detrimental to metal integrity. Systems with extended bench life have also been developed (71).

One newer application, however, is not yet mature. Warm Box systems, curing at significantly lower temperatures, are becoming more widely used. The warm-box resin is based on 2,5-bis(hydroxymethyl)furan (BHMF) resins. Lower binder levels are reported effective at core-curing temperatures of 149–177°C (72). The binders do not contain any nitrogen compounds. An essential part of the new technology is new latent curing catalysts (73,74). Core production rate using warm-box resins is equivalent to the rate with hot-box resins and provides significant improvement in working environment. Lower emissions, both at core-making and pour off are claimed.

Corrosion Resistant Fiber-Reinforced Plastic (FRP). Fiberglass reinforcement bonded with furfuryl alcohol thermosetting resins provides plastics with unique properties. Excellent resistance to corrosion and heat distortion coupled with low flame spread and low smoke emission are characteristics that make them valuable as laminating resins with fiberglass (75,76). Another valuable property of furan FRP is its strength at elevated temperature. Hand-lay-up, spray-up, and filament-winding techniques are employed to produce an array of corrosion-resistant equipment, pipes, tanks, vats, ducts, scrubbers, stacks, and reaction vessels for industrial applications throughout the world.

Reinforced furan resins have been used for many years in process piping and in underground sewer or waste-disposal systems. With a wide range in pH acceptability and good solvent resistance, furan piping has been a logical choice for many services.

The highly crosslinked thermoset character of cured furan resin provides substantially greater corrosion resistance than other resins used in fiberglass-reinforced plastics. Hence, furan FRP is recommended (77) for high performance chemical process equipment in service with chlorinated aromatics, highly oxygenated organic solvents, and other admixtures of organics and aggressive aqueous media of a nonoxidizing nature. Commercially manufactured corrosion-resistant off-the-shelf pipe using furan resins and fiberglass has been available since 1977 (77).

The inherent fire resistance and low smoke properties of furan resins appear to be related to the high degree of charring that takes place with minimum evolution of volatiles when exposed to fire.

Cements and Mortars. One of the original applications for furan resins was in the jointing of bricks and masonry. Furan resins have been extensively used for formulating mortars, grouts, and "setting beds" for brick linings in structures exposed to corrosive media, especially concentrated acids, and for setting tile in floors exposed to alkaline cleaning solutions or other corrosive chemical media. Mortars and grouts for these applications are usually formulated as two-package systems and mixed just prior to use. Silica fillers are satisfactory for many applications, but carbon flour (ground coke or other nongraphitic, nonactivated carbon) is used where resistance to hydrogen fluoride, fluoride salts, or hot, concentrated alkaline cleaning solutions is required.

High Carbon Yield. Furfuryl alcohol and furfural are reactive solvents (monomers) and are effective in producing high carbon yield (heat induced carbonization in a reducing atmosphere). They function as binders for refractory materials or carbon bodies. Furfuryl alcohol usually requires acidic catalysis and furfural basic catalysis. Mixtures of furfuryl alcohol and furfural are generally catalyzed with acid although some systems may be catalyzed with base.

Furfuryl alcohol alone, or in combination with other crosslinkable binders such as phenolic resins, chemical by-products and pitch, catalyzed with acid, gives carbon yields of 35–56%. Furfural together with cyclohexanone, pitch, or phenolic resins gives, under acid catalysis, yields of 35–55% carbon; under basic catalysis yields of 5–50% are achieved. Furfurylideneacetone resins (**13** and **14**), catalyzed by acid or base, give carbon yields of 48–56 and 30–35%, respectively (78).

Other Uses. Anisotropic and isotropic carbon are produced from furfural-modified systems; glassy carbon is produced primarily from furfuryl alcohol or BHMF resins (78,79).

Furfuryl alcohol is used alone or in combination with other solvents for various cleaning and paint removing operations. The ethylene oxide adduct of furfuryl alcohol is especially useful in this type of application (80–83).

Furan

Physical Properties. Furan, a colorless liquid with a strong ethereal odor, is low-boiling and highly flammable. It is miscible with most common organic solvents but only very slightly soluble in water. The physical properties of furan are listed in Table 1.

Chemical Properties. Furan is a heat-stable compound, although at 670°C in the absence of catalyst, or at 360°C in the presence of nickel, it decomposes to form a mixture consisting mainly of carbon monoxide, hydrogen, and hydrocarbons (84). Substitution and addition reactions can be effected under controlled

storage. In the presence of air, if a stabilizer such as Naugard is added, tetrahydrofurfuryl alcohol remains colorless after protracted periods of storage. Peroxide accumulation is low, not dangerous, and readily dischargeable on redistillation.

The reactions of tetrahydrofurfuryl alcohol are characteristic of its structure, involving primary alcohol and cyclic ether functional groups. As a primary alcohol, it undergoes normal displacement or condensation reactions affording new functional groups, (eg, halides, esters, alkoxylates, ethers, glycidyl ethers, cyanoethyl ethers, amines, etc). As a cyclic ether, it is typically unreactive, but the ring can be forced to open by hydrolysis or hydrogenolysis to give a variety of open-chain compounds, some of which may be recyclized to different heterocyclic species.

All the common monobasic (107) and dibasic esters (108) of tetrahydrofurfuryl alcohol have been prepared by conventional techniques; the dibasic esters and some of the mono esters are effective as primary or secondary plasticizers for vinyl polymers. Tetrahydrofurfuryl acrylate [2399-48-6] and methacrylate [2455-24-5], specialty monomers, have been produced by carbonylation (nickel carbonyl and acetylene) of the alcohol (109) as well as by direct esterification (110–112) and ester interchange (111).

Tetrahydrofurfuryl alcohol reacts with ammonia to give a variety of nitrogen containing compounds depending on the conditions employed. Over a barium hydroxide-promoted skeletal nickel–aluminum catalyst, 2-tetrahydrofurfurylamine [4795-29-3] is produced (113–115). With palladium on alumina catalyst in the vapor phase (250–300°C), pyridine [110-86-1] is the principal product (116–117); pyridine also is formed using Zn and Cr based catalysts (118,119). At low pressure and 200°C over a reduced nickel catalyst, piperidine is obtained in good yield (120,121).

Tetrahydrofurfuryl alcohol is oxidized to 2-tetrahydrofurfural [7681-84-7] in good yield by passage, with oxygen, over silver gauze at 500°C (123–124). With chromate oxidizing agents, lactones are also formed (124,125).

Tetrahydrofurfuryl alcohol is surprisingly resistant to hydrogenolysis; under vigorous conditions, however, cleavage of the ring or side chain occurs (126,127).

When tetrahydrofurfuryl alcohol is passed over a variety of alumina-containing catalysts at temperatures ranging from 200–500°C, it undergoes dehydration and ring expansion to 3,4-dihydro-2H-pyran, a major chemical intermediate for 1,5-difunctional open-chain aliphatic compounds (128–131) as well a hydroxyl group protecting agent. The analogous 3,4-dihydro-2H-thiopyran is similarly prepared by reaction of tetrahydrofurfuryl alcohol with hydrogen sulfide over alumina at 300°C (132,133).

Manufacture.   Tetrahydrofurfuryl alcohol is produced commercially by the vapor-phase catalytic hydrogenation of furfuryl alcohol. Liquid phase reduction is also possible.

Normal precautions for chemicals of mild toxicity are applicable to the safe handling and storage of commercial tetrahydrofurfuryl alcohol. Discoloration in storage rarely occurs if the proper precautions are observed; prevention of exposure to air will prevent autoxidation. The list price of tetrahydrofurfuryl alcohol (1997) is $1.15/lb.

Uses. Tetrahydrofurfuryl alcohol is of interest in chemical and related industries where low toxicity and minimal environmental impact are important (134). For many years tetrahydrofurfuryl alcohol has been used as a specialty organic solvent. The fastest growing applications are in formulations for cleaners (135) and paint strippers (136), often as a replacement for chlorinated solvents (137). Other major applications include formulations for crop sprays, water-based paints, and the dyeing and finishing of textiles and leathers. Tetrahydrofurfuryl alcohol also finds application as an intermediate in pharmaceutical applications.

A major use of tetrahydrofurfuryl alcohol is as an ingredient in proprietary stripping formulations. These formulations are used broadly in the automotive industry to remove protective coatings, paint, and grime prior to final painting and finishing (138). Stripping formulations containing tetrahydrofurfuryl alcohol are also available for the lifting and removal of epoxy coatings (139). Tetrahydrofurfuryl alcohol is being widely used as a replacement for chlorinated solvents, especially in the electronics industry.

Because tetrahydrofurfuryl alcohol is virtually colorless, it is used in lacquer formulations for all colors as well as water-white clear products. More specifically, tetrahydrofurfuryl alcohol is a wetting dispersant for most pigments. It has a high boiling point, high toluene dilution ratio, and good miscibility with oils, eg, linseed and soya, and is an excellent solvent for a wide range of resins.

Tetrahydrofurfuryl alcohol is used as solvent and carrier for industrial and commercial cleaning formulations. Product applications include the commercial cleaning of aircraft, ships, and trucks as well as the industrial cleaning requirements for machine tools and dies. The solvent properties of tetrahydrofurfuryl alcohol and its low rate of evaporation are utilized in biocides and pesticides where it serves as solvent–carrier. In a recent paper, tetrahydrofurfuryl alcohol is compared to other solvents as an agricultural adjuvent (140). Tetrahydrofurfuryl alcohol has been cleared by the EPA for use in formulations sprayed on growing food crops. The fact that tetrahydrofurfuryl alcohol is totally miscible with water and easily biodegradable enhances its use in agricultural sprays used for pest control, weed control, and growth retardation.

Tetrahydrofurfuryl alcohol also has been cleared for use in California under Rule 66. It meets the rigid specifications for uv degradation in environmentally sensitive areas. In addition to domestic use, a substantial amount of tetrahydrofurfuryl alcohol is used as a solvent for agricultural chemicals in Europe.

Tetrahydrofurfuryl alcohol is a solvent and coupling agent for a phosphate-type insecticide used to control the gypsy moth. Esters of tetrahydrofurfuryl alcohol are used in preparations employed as insect repellents. Tetrahydrofurfuryl alcohol is also used as a solvent–carrier for an EPA-approved paper slimicide formulation. In this application, the exceptional solvent action of tetrahydrofurfuryl alcohol prevents separation of the active ingredient, especially at low operating temperatures.

Tetrahydrofurfuryl acrylate [2399-48-6] and methacrylate [2455-24-5], reactive unsaturated monomers, are readily polymerized and easily cross-linked by exposure to heat, peroxide catalysts, or uv radiation.

Tetrahydrofurfuryl acrylate is used as a co-reactive viscosity depressant for vinyl terminated epoxy systems. It is also used in the formulation of uv-curable printing inks, coatings, paints, and adhesives. Tetrahydrofurfuryl methacrylate serves as a reactive curing agent in the peroxide-catalyzed production of nitrile rubber. The monomer functions as plasticizer during processing and polymerizes during cure to yield hard vulcanizates useful in electrical cable coatings.

Tetrahydrofurfuryl alcohol is used in elastomer production. As a solvent for the polymerization initiator, it finds application in the manufacture of chloro-hydrin rubber. Additionally, tetrahydrofurfuryl alcohol is used as a catalyst solvent-activator and reactive diluent in epoxy formulations for a variety of applications. Where exceptional moisture resistance is needed, as for outdoor applications, furfuryl alcohol is used jointly with tetrahydrofurfuryl alcohol in epoxy adhesive formulations.

In dyeing applications, tetrahydrofurfuryl alcohol permits higher dye concentrations and wider latitude in solution temperature because of its high solvency for dyes. Through swelling action on synthetic fibers and polarity with respect to leather, tetrahydrofurfuryl alcohol facilitates deeper penetration of dye into the substrate circumventing mere surface tinting. As a dye leveling agent, tetrahydrofurfuryl alcohol aids in the attainment of evenness in color shade.

Other Furan Derivatives

Other furan compounds, best derived from furfural, are of interest although commercial volumes are considerably less than those of furfural, furfuryl alcohol, furan, or tetrahydrofurfuryl alcohol. Some of these compounds are still in developmental stages. Applications include solvents, resin intermediates, synthetic rubber modifiers, therapeutic uses, as well as general chemical intermediates.

Solvents.   Compounds containing the furan ring are generally excellent solvents. Some are miscible with both water and with hexane. Presence of the ether oxygen adds polarity as well as the potential for hydrogen bonding. Ring-substituted derivatives of furfural and furfuryl alcohol are reactive solvents, similar to the parent compounds. Reaction of the aldehyde group on furfural or the alcohol group on furfuryl alcohol give products that are merely solvents, no longer reactive. Ethers of furfuryl alcohol are easy to prepare; most simple ethers have been synthesized. Ethylene oxide adducts [31692-86-1] are used in paint stripping formulations (136,141–143). The esters of furfuryl alcohol are effective solvents, but must be synthesized carefully owing to the reactivity of furfuryl alcohol with acids. Both esters and ethers of tetrahydrofurfuryl alcohol are excellent solvents and are easily prepared. Since double bonds are no longer present, these compounds are more stable than the corresponding furan derivatives. Tetrahydrofurfuryl alcohol–ethylene oxide adducts [31692-85-0] are also useful solvents for paint stripping formulations (136,141,143). 2-Methylfuran [534-22-5] is a good solvent, but 2-methyltetrahydrofuran [96-47-9] (METHF), is completely saturated and is less reactive, less toxic, and has greater potential for a number of solvent applications. It is a more convenient solvent than tetrahydrofuran for Grignard reagents (144), is higher boiling, and wet METHF is more easily recovered and made anhydrous for recycle and reuse. METHF has also found application as a solvent for other organometallic reagents (145), as well as in lithium batteries (146). Ultimately, when fully commercialized, the cost of METHF should approximate that of tetrahydrofuran.

Resins. As mentioned above, both furfural and furfuryl alcohol are widely used in resin applications. Another resin former, 2,5-furandimethanol [1883-75-6] (BHMF), is prepared from furfuryl alcohol by reaction with formaldehyde. It is usually not isolated because oligomerization occurs simultaneously with formation (competing reaction). Both the monomer and oligomers are very reactive owing to difunctionality, and are used primarily as binders for foundry sand (72) and fiberglass insulation (147,148).

Therapeutics. Compounds containing the furan or tetrahydrofuran ring are biologically active and are present in a number of pharmaceutical products. Furfurylamine [617-89-0] is an intermediate in the diuretic, furosemide. Tetrahydrofurfurylamine [4795-29-3] may also have pharmaceutical applications. 5-(Dimethylaminomethyl)furfuryl alcohol [15433-79-1] is an intermediate in the preparation of ranitidine, which is used for treating ulcers. 2-Acetylfuran [1192-62-7], prepared from acetic anhydride and furan is an intermediate in the synthesis of cefuroxime, a penicillin derivative. 2-Furoic acid is prepared by the oxidation of furfural. Both furoic acid [88-14-2] and furoyl chloride [527-69-5] are used as pharmaceutical intermediates. 2-Tetrahydrofuroic acid [16874-33-2] also finds application.

Rubber Modifiers. Derivatives of furan and tetrahydrofurfuryl alcohol are used in the polymerization of synthetic rubber to control stereoregularity and other properties (149,150).

Intermediates. 3,4-Dihydro-2*H*-pyran [110-87-2] is prepared by a ring-expanding dehydration of tetrahydrofurfuryl alcohol. It is used as a protecting agent for hydroxyl compounds and as an intermediate. 2-Methylfuran is a chemical intermediate for 5-methylfurfural [620-02-0] (151) and 3-acetyl-1-propanol [1071-73-4] (152), and is also used in perfume. α-Methylfurfuryl alcohol [4208-64-4] is prepared by the reaction of furfural with methylmagnesium bromide. It is an intermediate in the synthesis of maltol [118-71-8], which is used as a flavoring agent.

Health and Safety

As with all chemical compounds, the Material Safety Data Sheet (MSDS) for each of the specific furan derivatives should be reviewed before starting to work with these materials. Caution should be the keyword when handling chemicals. Additional information on toxic effects of most of these compounds can be found in RTECS (*Registry of Toxic Effects of Chemicals*), HSDB (*Hazardous Substances Data Bank* from the National Library of Medicine), and standard works on toxicology such as Patty's (153). Toxicology studies are taking place on a continuing basis with many chemicals; furan derivatives are no exception. New data may change the perspective on toxicity of these chemicals.

Over the years, many people have been exposed to low levels of these furan derivatives with no evidence of any significant long term effects. Proper precautions, however, should be taken when working with these compounds as furfural, furfuryl alcohol, and furan are moderately toxic, tetrahydrofurfuryl alcohol is less so. Protective clothing should be worn; skin absorption is a likely route of entry. Eye protection is required. Proper ventilation is needed, especially with highly volatile furan. The odor threshold for these compounds is such that they can be detected at levels below the PEL (Permissible Exposure Level), and appropriate action taken as soon as the odor is detected. Specific recommendations from the manufacturer regarding exposure and protective measures should be followed. Since regulations change from time to time, up-to-date exposure limit recommendations from OSHA (Occupational Safety and Health Agency) or ACGIH (American Council of Governmental Industrial Hygienists) need to be consulted and followed.

Toxicity information is summarized in Table 3.

Table 3. Toxic Effects of Furfural and Main Derivatives[a]

		Value for		
Property	Furfural	Furfuryl alcohol	Furan	Tetrahydrofurfuryl alcohol
oral toxicity	149 mg/kg	451 mg/kg	300 mg/kg	2500 mg/kg
inhalation toxicity	1037 ppm 1 h	692 ppm 1 h	3464 ppm	$LC_{2/3}$ 12650 ppm
dermal toxicity	620 mg/kg	667 mg/kg	not established	>5000 mg/kg
skin irritation	slight	slight	not established	slight
eye irritation	moderate	moderate	not established	moderate
ACGIH carcinogen	not listed	not listed	not listed	not listed
IARC carcinogen	not listed	not listed	not listed	not listed
NIOSH carcinogen	not listed	not listed	not listed	not listed
NTP carcinogen	not listed	not listed	not listed	not listed
OSHA carcinogen	not listed	not listed	not listed	not listed

[a]Ref. 154.

BIBLIOGRAPHY

"Furfural and Other Furan Compounds" in *ECT* 1st ed., Vol. 6, pp. 995–1008, by H. R. Duffey, The Quaker Oats Company, and H. J. Barrett, E. I. du Pont de Nemours & Co., Inc.; "Furfural and Other Furan Compounds" in *ECT* 2nd ed., Vol. 10, pp. 237–250, by A. P. Dunlop, The Quaker Oats Company; "Furan Derivatives" in *ECT* 3rd ed., Vol. 11, pp. 499–527, By W. J. McKillip and E. Sherman, The Quaker Oats Company.

1. A. P. Dunlop and F. N. Peters, *The Furans*, ACS Monograph 119, Reinhold Publishing Corp., New York, 1953.
2. M. J. Cook, A. R. Katritzky, and P. Linda, *Adv. Heterocycl. Chem.* **17**, 255 (1974).
3. C. W. Bird, *Tetrahedron* **48**, 335 (1992).
4. M. V. Sargent and T. M. Cresp "Furans" in D. Barton and W. D. Ollis, eds., *Comprehensive Organic Chemistry; The Synthesis and Reaction of Organic Compounds*, Vol. 4, Pergamon Press Ltd., Oxford, U.K., 1979, pp. 693–744.

5. A. Gandini "Furan Polymers" in J. I. Kroschwitz, ed., *Encyclopedia of Polymer Science and Technology*, 2nd ed., Vol. 7, John Wiley & Sons, Inc., New York, 1967, pp. 454–473.
6. A. Gandini, *Adv. Polym. Sci.* **25**, 47 (1977).
7. F. M. Dean, *Advances in Heterocyclic Chemistry*, **30**, 168; **31**, 237 (1982).
8. Dobereiner, *Ann.* **3**, 141 (1832); quoted in Ref. 1, p. 272.
9. Emmet, in *J. Pract. Chem.* **12**, 120 (1837); *Ann.* **28**, 249 (1939); quoted in Ref. 1, p. 272.
10. Stenhouse, *Phil. Mag.* **18**, 122 (1840); *Ann.* **35**, 301 (1840); quoted in Ref. 1, p. 272.
11. Baeyer, *Ber.* **10**, 355, 695, 1358 (1877); quoted in Ref. 1, p. 4.
12. Markwald, *Ber.* **20**, 2811 (1887); quoted in Ref. 1, p. 5.
13. Markwald, *Ber.* **21**, 1398 (1887); quoted in Ref. 1, p. 5.
14. Harries, *Ber.* **34**, 1488 (1901); quoted in Ref. 1, p. 5.
15. G. H. Mains, *Chem. Met. Eng.* **26**, 779 (1922).
16. R. G. Curtis and H. H. Hatt, *Aust. J. Sci. Res.* **A1**, 213 (1948).
17. A. P. Dunlop and F. N. Peters, Jr., *Ind. Eng. Chem.* **32**, 1639 (1940).
18. The Quaker Oats Company, Unpublished data.
19. T.-S. Li, S.-H. Li, J.-T. Li, and H.-Z. Li, *J. Chem. Res. Synop.*, 26–27, (1997); *Chem. Abst.* **126**, 144440 (1997).
20. V. V. Girdyuk, Yu. K. Kirilenko, L. A. Vol'f, and A. I. Meos, *Zh. Prikl. Khim.* **39**, 2601 (1966).
21. I. V. Kamenskii, S. M. Filimonova, Ngo Van Lam, B. Ya. Eryshev, and N. B. Krovyakova, *Plast. Massy* **5**, 5 (1973); *Chem. Abstr.* **79**, 79683n (1973).
22. Pais, C. G. Godwin; A. Keshavaraja; K. Saravanan, and P. Kumar, *J. Chem. Res. Synop.* **9**, 426 (1996).
23. N. Deka, D. J. Kalita, R. Borah, and J. C. Sarma, *J. Org. Chem.* **62**, 1563–1564 (1997).
24. W. J. Chute and G. F. Wright, *J. Org. Chem.* **10**, 541 (1945).
25. H. Gilman and G. F. Wright, *J. Am. Chem. Soc.* **52**, 2550 (1930).
26. V. I. Sakhnenko, M. V. Sokolov, V. V. Kashmet, A. N. Chernobrovyi, *Khim.-Farm. Zh.* **27**, (3), 53-7 (1993) *Chem. Abst.* **121**, 160125 (1994).
27. H. Li and M. Gao, *Hebei Daxue Xuebao, Ziran Kexueban*, **12**, 78 (1992); *Chem. Abst.* 35220 (1992).
28. D. L. Williams and A. P. Dunlop, *Ind. Eng. Chem.* **40**, 239 (1948).
29. J. Z. Marcusson, *Z. Angew. Chem.* **32**, 113 (1919).
30. G. Illari, *Gazz. Chim. Ital.* **77**, 389 (1947).
31. Y. Nakamura and M. Saito, *Kogyo Kagaku Zasshi* **62**, 1173 (1959).
32. N. Galego, Ph.D. Thesis, University of Havana, 1975.
33. N. Galego and A. Gandini, *Rev. CENIC Cent. Nac. Invest. Cient. I Cienc. Fis.* **6**, 163 (1975); *Chem. Abst.* 84, 18211p (1976).
34. N. Galego and F. Lopez, *Rev. Cubana Quim.* **6**, 107 (1992); *Chem. Abst.*, **119**, 181292 (1993).
35. R. Sanchez, C. Hernandez and G. Keresztury, *Eur. Polym. J.* **30**, 43 (1994); *Chem. Abst.*, **120**, 55768 (1994).
36. R. L. Whistler and E. L. Richards, "Hemicelluloses" in W. Pigman and D. Horton, eds., *The Carbohydrates*, Vol. IIA, Academic Press, Inc., New York, 1970, pp. 447–467.
37. G. O. Aspinall, *Polysaccharides*, Pergamon Press Ltd., Oxford, U.K., 1970, pp. 103–115.
38. *Making and Marketing Furfural; Added Value for Agro-Industrial Wastes*, International Trade Center UNCTAD/GATT, Geneva, 1979.
39. J. E. Stone and M. J. Blundell, *Can. J. Res.* **28B**, 676 (1950).

107. R. T. Wragg, *J. Chem. Soc.*, 7162 (1965).

108. L. H. Brown and J. W. Hill, *J. Chem. Eng. Data* **5**, 56 (1960).

109. W. Reppe and co-workers, *Ann.* **582**, 1 (1953).

110. Jpn. Pat. JP58174346, (Oct. 13, 1983) (to Yokkaichi Chemical Co., Ltd.); *Chem. Abst.* **100**, 121755 (1984).

111. Jpn. Pat. JP50050313 (May 6, 1975), K. Kimura, K. Suzuki, E. Kondo, N. Motoyama, G. Terada, and H. Mishina (to Toa Gosei Chemical Industry Co., Ltd.); *Chem. Abst.* **83**, 115343 (1975).

112. Jpn. Pat. JP50013313 (Feb. 12, 1975), K. Suzuki, K. Kimura, Y. Kato, H. Mishina, and E. Kondo, (to Toa Gosei Chemical Industry Co., Ltd.); *Chem. Abst.* **83**, 79945 (1975).

113. U.S. Pat. 2,636,902 (Apr. 28, 1953), A. W. C. Taylor, P. Davies, and P. W. Reynolds (to Imperial Chemical Industries Ltd.).

114. Jpn. Pat. JP60178877 (Sept. 12, 1985), S. Mori, T. Aoki, R. Hamana and Y. Nomura (to Mitsubishi Petrochemical Co., Ltd.); *Chem. Abst.* **104**, 88416 (1986).

115. L. S. Glebov, G. A. Kliger, A. N. Shuikin, and V. G. Zaikin, *Neftekhimiya* **36**, 344 (1996).

116. U.S. Pat. 3,238,214 (Mar. 1, 1966), D. G. Manly, J. P. O'Halloran, and F. J. Rice (to The Quaker Oats Company).

117. J. H. Choi and W. Y. Lee, *Appl. Catal. A* **98**, 21 (1993); *Chem. Abst.* **118**, 256952 (1993).

118. Soviet Pat. SU1524917 (Nov. 30, 1989), K. M. Akhmerov, D. Yusupov, U. D. Nazirova, and A. B. Kuchkarov, (to Tashkent Polytechnic Institute); *Chem. Abst.* **112**, 198136 (1990).

119. U. D. Nazirova and K. M. Akhmerov, *Dokl. Akad. Nauk. UzSSR* 40 (1988); *Chem. Abst.* **109**, 230746 (1988).

120. U.S. Pat. 3,163,652 (Dec. 29, 1964), D. G. Manly, J. P. O'Halloran, and F. J. Rice (to The Quaker Oats Co.).

121. Jpn. Pat. JP60174776 (Sept. 9, 1985), S. Mori, T. Aoki, R. Hamana and Y. Nomura, (to Mitsubishi Petrochemical Co., Ltd.); *Chem. Abst.* **104**, 88449 (1986).

122. Brit. Pat. GB593,617 (Oct. 21, 1947), J. G. M. Bremner and R. R. Coats (to Imperial Chemical Industries).

123. R. A. Karakhanov, V. Vagabov, N. P. Karzhavina, O. M. Ramazanov, Y. S. Mardashev, and M. M. Vartanyan, *Zh. Prikl. Khim.* (*Leningrad*) **54**, 454 (1981); *Chem. Abst.* **95**, 24893 (1981).

124. S. Baskaran, I. Islam, and S. Chandrasekaran, *J. Chem. Res. Synop.* 290 (1992); *Chem. Abst.* **117**, 251152 (1992).

125. S. Baskaran and S. Chandrasekaran, *Tetrahedron Lett.* **31**, 2775 (1990).

126. E. P. Goodings and C. L. Wilson, *J. Am. Chem. Soc.* **73**, 4801 (1951).

127. S. Landa and J. Mostecky, *Collect. Czech. Chem. Commun.* **20**, 430 (1955).

128. O. W. Cass, *Chemurgic Dig.* **8**(6), 6 (1949).

129. R. Paul, *Bull. Sec. Chim. France* **14**, 158 (1947).

130. O. W. Cass, *Ind. Eng. Chem.* **40**, 21C (1048).

131. U.S. Pat. 3,518,281 (June 30, 1970), D. C. Holtman (to Emery Industries, Inc.).

132. R. F. Naylor, *J. Chem. Soc.*, 2749 (1949).

133. Yu. K. Yur'ev and E. G. Vendel'shtein, *J. Cen. Chem. USSR* **22**, 751 (1952).

134. *QO Tetrahydrofurfuryl Alcohol, Bulletin 206*, QO Chemicals, part of Great Lakes Chemical Corporation, West Lafayette, Ind.

135. U.S. Pat. 5,514,294 (May 7, 1996), G. W. Bohnert and co-workers (to Allied Signal Inc.).

136. U.S. Pat. 4,737,195 (Apr. 12, 1988) C. M. Caranding and R. E. Koch (to Amchem Products, Inc.).

137. U.S. Pat. 5,128,057 (July 7, 1992), M. L. Bixenman and G. C. Wolf (to Kyzen Corporation).
138. U.S. Pat. 4,514,325 (Apr. 30, 1985), A. Russo and T. Mauthner (to J. Hall Co.).
139. U.S. Pat. 5,456,853 (Oct. 10, 1995), M. M. Myers (to Rust-Oleum Corp.).
140. K. J. Doyel and co-workers, C. L. Foy, ed., *Adjuvants Agrichemistry*, CRC, Boca Raton, Fla., 1992, pp. 225–234.
141. U.S. Pat. 5,259,993 (Nov. 9, 1993), S. M. Short (to Cook Composites and Polymers Co.).
142. U.S. Pat. 4,619,706 (Oct. 28, 1986), D. G. Squires, L. Hundley and R. A. Barry (to Texo Corp.).
143. U.S. Pat. 4,666,002 (Dec. 28, 1981), C. M. Carandang (to Amchem Products, Inc.).
144. L. Poppe, K. Recseg and L. Novak, *Syn. Com.* **25**, 3993 (1995).
145. WO. Patent 97/06097, (Feb. 20, 1997), T. Rathman and co-workers (to FMC Corp.).
146. S. Tobishima and co-workers *J. Appl. Electrochem.* **27**, 902 (1997).
147. U.S. Pat. 5,486,557 (Jan. 23, 1996), D. W. Akerberg (to QO Chemicals, Inc.).
148. U.S. Pat. 5,589,536 (Dec. 31, 1996), C. M. Golino, C. A. Rude and co-workers (to QO Chemicals, Inc., and Schuller International, Inc.).
149. U.S. Pat. 4,429,091 (Jan. 31, 1984), J. E. Hall (to Firestone Tire and Rubber Co.).
150. A. F. Halasa and W. L. Hsu, "New Ether Modifiers for Anionic Polymerization of Isoprene," *Polym. Prepr.* **37**(2), (1996).
151. Jap. Pat. 57091982 A2 (Nov. 28, 1980) (to Sumitomo Chemical Co., Ltd.).
152. V. N. Pavlychev and co-workers, *Khim. Prom-st.*, (7), 394 (1980); *Chem. Abst.* **93**, 238781 (1980).
153. G. D. Clayton and F. E. Clayton, eds., *Patty's Industrial Hygiene and Toxicity* 4th ed., John Wiley & Sons, Inc., New York, 1991.
154. Great Lakes Chemical Corp., information summarized from Material Safety Data Sheets.

General References

Refs. 1, 7–9.
M. V. Sargent and T. M. Cresp "Furans" in D. Barton and W. D. Ollis, eds., *Comprehensive Organic Chemistry*: *The Synthesis and Reactions of Organic Compounds*, Vol. 4, Pergamon Press Ltd., Oxford, U.K., 1979, pp. 693–744.

R. H. KOTTKE
Great Lakes Chemical Corporation

HIGH PURITY GASES

High purity industrial gases are routinely delivered in large quantities having purities exceeding 99.999% (>5 nines pure). Other commodity materials, whether in liquid or solid form, much less typically have purity exceeding 99.9%. There are many applications for gases where purity even higher than 99.999% is required. To service these applications, the technology for manufacturing and delivering high purity gases has developed into a multibillion dollar business worldwide. Japan and the United States have the majority of the market.

There is no universally accepted definition of what purity levels correspond to high purity. However, gases having total impurities specified <1 ppm on a molar or volume basis must be manufactured and handled differently from regular gases if that specification is to be maintained. A good working definition of high purity is gases having certain individual impurities held to levels <0.1 ppm.

The technology for manufacturing and delivering high purity gases has largely developed to support the manufacture of advanced semiconductor materials needed as part of the overall process of manufacturing integrated circuits. Purity requirements in some of these high performance materials have reached the point where impurity concentrations in the starting gases of 1 part in 10^{12} have been related to reduced yields and poor performance in the resulting integrated circuits. Depending on a volume, high purity gases can be delivered using either bulk systems, where a plant-wide distribution system is integrated with central gas storage facilities, or cylinders, where a short local distribution system is supplied from a single high pressure cylinder.

Gases used in the manufacture of semiconductor materials fall into three principal areas: the inert gases, used to shield the manufacturing processes and prevent impurities from entering; the source gases, used to supply the molecules and atoms that stay behind and contribute to the final product, and the reactive gases, used to modify the electronic materials without actually contributing atoms or molecules.

Nitrogen [7727-37-9], N_2; oxygen [7782-44-7], O_2; hydrogen [1333-74-0], H_2; argon [7440-37-1], Ar; and helium [7440-59-7], He, are classified as semiconductor bulk gases. The following gases are delivered in cylinders and are classified as semiconductor specialty gases: (1) silicon precursor gases such as silane [7803-62-5], SiH_4; dichlorosilane [4109-96-0], SiH_2Cl_2; trichlorosilane [10025-78-2], $SiHCl_3$; and tetrachlorosilane [10026-04-7], $SiCl_4$; (2) dopant gases such as arsine [7784-42-1], AsH_3; phosphine [7803-51-2], PH_3; diborane [19287-45-7], B_2H_6; and boron trifluoride [7637-07-2], BF_3; (3) etching gases such as nitrogen trifluoride [7783-54-2], NF_3; sulfur hexafluoride [2251-62-4], SF_6; boron trichloride [10294-34-5], BCl_3; hydrogen fluoride [7664-39-3], HF; chlorine [7782-50-5], Cl_2; chlorine trifluoride [7790-91-2], ClF_3; hydrogen bromide [10035-10-6], HBr; tetrafluorosilane [7783-61-1], SiF_4; and other halocarbons; and (4) reactant gases such as ammonia [7664-41-7], NH_3; hydrogen chloride [7647-01-0], HCl; nitrous oxide [10024-97-2], N_2O; tungsten hexafluoride [7783-82-6], WF_6; and carbon dioxide [124-38-9], CO_2.

Production and Purification

The separations processes used for manufacturing high purity gases are generally the same as those used for making lower purity products. Purification by distillation (qv) and adsorption (qv) are often used and, provided basic data are available, can generally be designed using standard engineering practices. Some separation processes that are not economical for the manufacture of lower purity gases do find applications in making high purity gases. Chemical conversion processes, where the impurities are converted into more easily separable forms through a selective chemical reaction, are often employed in point-of-use purifiers which typically process streams having less than five standard liters per minute flow rate.

Whereas the underlying separation or purification technology may be straightforward, the purity achieved is often far less than that which the separation processes are capable of producing. More often than not, recontamination by impurities released by the materials of construction used in the purification, storage, and delivery equipment represents the true limit to the purity that can be achieved in practice.

Generally, the purification methods used for manufacturing high purity gases can be divided according to volume throughput requirements and of course differ according to the chemical and physical properties of the gas being purified. For the manufacture of high purity gases on a large scale, the industry tends to employ a few well-known methods. When production or purification is required on a smaller scale, a greater number of separation methods and combinations of these methods are used in actual practice.

Bulk Gases. The bulk gases are usually characterized by high volume flow requirements in the manufacturing process. Historically these have consisted of nitrogen, oxygen, argon, hydrogen, and to a lesser extent, helium. Because these gases are manufactured on a large scale, small-scale supply schemes tend to be more of a distribution process for product already purified on a large scale. These small-scale distribution systems may, however, incorporate so-called onsite or point-of-use purifiers to ensure that any impurities picked up during distribution are removed.

Generally the bulk gases are also nonreactive with their containment systems. However, this is becoming less true because the need for large quantities of high purity HCl, trichlorosilane, and other corrosive gases has grown pushing these into the bulk gases category.

Nitrogen. Because of numerous applications in semiconductor manufacturing, high purity nitrogen is produced both at high volumes and at some of the lowest impurity levels seen for any of the high purity gases. Streams up to 10,000 nm^3/h can be produced at impurity levels of 1 part in 10^9.

Distillation. All high purity nitrogen is manufactured from air using multistage cryogenic distillation (qv). Depending on the customer's flow rate and purity requirements, the original separation from air is carried out either locally on the customer's property or at a large air separation (qv) unit (ASU) located some distance away.

For volume throughput requirements larger than ~1000 nm^3/h, self-contained and highly automated on-site generators are installed on the user's property. The gaseous nitrogen produced by the generator is then delivered throughout the user's plant using a dedicated high purity distribution system. Impurity contents on the order of 1 part in 10^9 can be achieved in the delivered product using these systems.

For smaller volume requirements, nitrogen is manufactured and liquefied in a large merchant air separation unit owned by the gas supplier. Liquid nitrogen is transported by tanker truck and stored in a large cryogenic tank on the customer's site. When needed by the customer, liquid nitrogen is withdrawn from the tank, vaporized, filtered, and delivered throughout the plant using a dedicated distribution system. Impurity levels on the order of 1 part in 10^8 can be achieved by trucking liquid nitrogen over distances exceeding 1500 km. However, product costs are determined largely by delivery distance and may be uneconomical at that range.

Chemical Conversion. In both on-site and merchant air separation plants, special provisions must be made to remove certain impurities. The main impurity of this type is carbon monoxide, CO, which is difficult to separate from nitrogen using distillation alone. The most common approach for CO removal is chemical conversion to CO_2 using an oxidation catalyst in the feed air to the air separation unit. The additional CO_2 which results, along with the CO_2 from the atmosphere, is then removed by a prepurification unit in the air separation unit.

At throughputs below 500 nm^3/h, a wide variety of inert gas purification processes based on chemical conversion can be used to produce high purity nitrogen. For the most part, these processes employ chemical reactions to convert O_2, H_2O, CO, and CO_2 impurities into nonvolatile products. Typically, the impure nitrogen is passed through a bed of reagents, which may or may not require elevated temperatures to function, where the conversion reactions occur causing the impurities to remain behind in the bed of reagents. These processes require that the bulk of the oxygen in the feed be removed by some other method. For this reason, they are mainly used to convert standard purity grades of nitrogen into high purity ones.

For nitrogen flows above 10 nm^3/h, it is important to minimize any heating or cooling of the process stream because of the energy costs. Heating for regeneration purposes does not carry this economic penalty. There are two types of chemical conversion processes in general use which meet this requirement. Both function effectively at room temperature and below and are capable of removing O_2, H_2O, CO, and CO_2 to levels below 1 part in 10^9.

The first process utilizes a bed of nickel catalyst which has been regenerated with hydrogen to reduce the nickel content to metallic form. The finely divided metal then reacts with impurities and retains them in the bed, probably as nickel oxide in the case of oxygen or as physisorbed compounds for other impurities. Periodically, the bed is regenerated at elevated temperature using hydrogen to restore the metallic content. The nickel process can be used and regenerated indefinitely.

The second process utilizes a bed of high surface-area porous zirconium–iron alloy pellets. The bed is regenerated periodically at elevated temperature to allow impurities on the surface of the pellets to diffuse into the bulk. The renewed surfaces are effective for removing impurities from the gas stream until, with use, they once again become saturated. The zirconium–iron process has the advantage of not requiring hydrogen for regeneration; however, zirconium–iron pellets must be replaced once these have fully reacted.

Combined Distillation and Chemical Conversion. On-site generators using distillation are almost always combined with chemical conversion purifiers in large bulk high purity nitrogen supply systems. Practical issues, such as ensuring a continuous supply of inert nitrogen during an interruption of electrical power or when the generator is being serviced, dictate that liquid nitrogen be stored for use when the generator is unavailable. Often backup nitrogen must be obtained from standard commercial sources that are not high purity. Also, even if high purity liquid nitrogen from the generator can be accumulated for backup purposes, commercial purity liquid nitrogen may be the only source available if the generator is out of service for more than a day or so.

The delivery of high purity nitrogen from commercial purity liquid nitrogen sources is accomplished by using high capacity chemical conversion purifiers which are active only when the backup stream is in use. Even so, the equipment cost is high and every attempt is made to tightly integrate the chemical conversion purifier with the nitrogen generator.

Oxygen. High purity oxygen for use in semiconductor device manufacture is produced in relatively small quantities compared to nitrogen. There are two different purification processes in general use for manufacturing the gas: distillation and chemical conversion plus adsorption.

For applications in semiconductor device manufacture, it is important to remove the chemically nonreactive noble gas impurities such as krypton, xenon, and particularly radon. Through spontaneous decay, radioactive isotopes of these gases in the purified oxygen can cause a wide range of radioactive daughter products to be incorporated into the electronic device materials. Further decay of the radioactive daughter elements provides a continuous source of ionizing radiation inside the device that can lead to a type of defect known as soft errors, particularly in large dynamic memories.

Distillation. As for nitrogen, all high purity oxygen is derived from air through the air separation (qv) process using cryogenic distillation (qv). Generally, air separation units that manufacture commercial purity oxygen also remove nitrogen and other light impurities to levels low enough for high purity applications. Argon is still present at concentrations above 1 part in 10^5, but is not considered an impurity for most applications of high purity oxygen. Heavy impurities, typically hydrocarbons and noble gases, are usually not removed from commercial purity oxygen. To make high purity grades, an additional distillation column is added to the process. Some on-site nitrogen generators are also able to generate a small stream of high purity oxygen.

Chemical Conversion and Adsorption.   Where additional distillation is not practical, hydrocarbons and heavy noble gases can also be removed by combining chemical conversion with adsorption. Commercial purity oxygen is passed through a high temperature bed of oxidation catalyst; usually combinations of platinum and rhodium on a high surface area alumina support. Hydrocarbon impurities are oxidized to CO_2 and H_2O. Temperatures in excess of 700°C are needed in the catalyst bed to ensure complete oxidization of methane. Because the process flow rates are typically small, these temperatures can be maintained without excessive energy costs. The CO_2 and H_2O products from catalytic oxidation of hydrocarbon impurities are removed using a temperature swing adsorption (TSA) process. The adsorbent is typically one of the molecular sieves (qv), such as 13X. If designed appropriately, the adsorption process can also be used to remove the heavy noble gases. Usually, however, only radon removal is necessary. Because radon has a 3.8-d half-life, it is only necessary to slow its transit through the adsorber bed for sufficient time for natural decay to reduce the concentration to an acceptable level.

Argon.   High purity argon has many applications as an inert gas during the manufacture of semiconductor devices. High consumption rates are typical during processes which deposit high purity silicon epitaxial films onto polished wafers. Lower volume applications range from silicon crystal pulling to sputter deposition of metal films. In all these applications, nitrogen is a reactive impurity which, in addition to O_2, H_2O, CO_2, CO, and all hydrocarbons, must be removed from the argon to low levels.

Distillation.   Conventional purity argon is separated from air using a combination of distillation and chemical conversion. High purity argon is made the same way. However, more stages of distillation are designed into the low pressure and pure argon columns in the air separation plant. The improvement in the efficiency of the low pressure column typically reduces the nitrogen impurity to below 1 ppm. Because argon is found at a concentration of only 1 vol % in air, it is not practical to supply argon from on-site generators.

Chemical Conversion.   Except for control of nitrogen impurity levels, the same chemical conversion methods used for nitrogen purification at low flow rates can also be used for argon purification. Although used less commonly for argon purification than for nitrogen purification, these chemical conversion methods are applied in point-of-use purifiers located close to where the gas is consumed.

There are also chemical conversion purifiers which remove nitrogen in addition to other more reactive impurities. These purifiers require elevated temperatures to function and consequently are restricted to small process flow rates of typically a few liters per minute.

Hydrogen and Helium.   Whereas hydrogen and helium are very different chemically, these gases have low boiling points and are normally liquefied during manufacture. Because the boiling points are so low, even very small amounts of trace impurities tend to freeze and form solid deposits. To prevent formation of these deposits, which can eventually plug process lines, trace impurities must be removed prior to liquefaction. Similar methods are used to purify hydrogen and helium prior to liquefaction.

Hydrogen. High purity hydrogen is usually delivered and stored as a cryogenic liquid and vaporized when needed. This vaporized liquid seldom needs any further processing to meet high purity specifications. However, in situations where liquid hydrogen cannot be used, high pressure hydrogen gas can also be delivered and further purified on-site to meet high purity specifications.

On-site purification of gaseous hydrogen is accomplished using combinations of chemical conversion, cryogenic adsorption, and palladium membrane processes depending on the purity and throughput requirements (see MEMBRANE TECHNOLOGY). For small hydrogen streams, stand-alone hydrogen purifiers based on selective permeation through palladium membranes are available. These units take advantage of the high permeability of hydrogen through palladium metal. Generally, the membranes are heated to get higher processing capacity per unit area. However, provisions must then be made to remove hydrogen from the membrane prior to cool-down to prevent formation of small cracks which then allow impurities to leak through the membrane. Small to large hydrogen streams can be purified by adsorption carried out at cryogenic temperatures, usually near the boiling point of liquid nitrogen. Activated charcoal (qv) and silica gel (qv) are both effective adsorbents for purification. However, if any significant quantity of oxygen must be removed, activated charcoal should not be used because of the potential for an explosive chemical reaction with the adsorbed oxygen during regeneration. Regeneration of the adsorbent is accomplished through a process of heating and reverse purging. Large hydrogen streams can be purified using a room-temperature palladium catalyst bed to convert oxygen impurity into water followed by cryogenic adsorption. Because the oxygen impurity has been removed, the higher adsorption capacity of activated charcoal can then be utilized without danger of explosion.

Helium. High purity helium is usually not required in large quantities and is therefore not commonly delivered as a cryogenic liquid. Instead, high pressure cylinders are filled from a liquid helium source by the gas supplier and then transported to the customer. When high purity helium is required, the high pressure gaseous helium is processed through an on-site purifier.

Because helium is not chemically reactive, the same chemical conversion processes used for purification of nitrogen and argon are also applied to helium purification. Additionally, cryogenic adsorption processes are used to remove the nonreactive impurities. As for hydrogen, the same precautions must be taken against possible adsorbent explosions when dealing with significant amounts of oxygen impurity.

Helium is separated from helium-bearing natural gases usually, but not always, in the process of removing nitrogen to improve the fuel value of the natural gas. Thus, in a sense, the principal part of commercial helium is as a by-product. Natural gas components include methane and heavier hydrocarbons, helium, nitrogen, water vapor, carbon dioxide, sometimes hydrogen sulfide, small amounts of argon, and traces of neon and hydrogen. Water, carbon dioxide, and sulfides are first removed by scrubbing with monoethanolamine and diethylene glycol, followed by drying with alumina. Then the natural gas is concentrated in helium as the higher boiling hydrocarbons are liquefied and collected. Crude helium, concentrated to about 70% and containing other permanent gases, undergoes a cryogenic purification where the majority of nitrogen and argon are liquefied. Activated charcoal operating at liquid nitrogen temperatures or below is used to adsorb all remaining nonhelium gases. Helium that exceeds 99.9999% in purity results.

Specialty Gases. The specialty gases are generally more reactive than the bulk gases and usually have low volume flow requirements in their applications. Historically, these have been delivered almost exclusively in standard compressed gas cylinders. However, as the need for increased quantities has arisen, bulk specialty gas supply systems utilizing larger ton-sized containers and tube trailers are being developed.

Purification of specialty gases can be divided into two areas: purification done by the gas supplier on a bulk scale prior to filling the cylinder or other delivery container, and purification carried out by the consumer on a point-of-use scale generally just prior to use.

Bulk Purification. Many specialty gases originate as by-products or low purity intermediate chemicals produced during the course of manufacturing something else. For example, most of the high purity silane produced in the United States originates as an intermediate in the manufacture of metallurgical silicon.

In some cases the chemical manufacturer purifies a portion of this intermediate stream to make a high purity product. In other cases, the chemical manufacturer sells a low purity product to a gas company and the gas company purifies it to make a high purity product. In both bases, purification is done on a continuous basis, rather than cylinder by cylinder. The purification processes tend to utilize standard methods.

Distillation. Processes which utilize either simple liquid vapor flash processes or multistage distillation tend to be used often for purification of bulk specialty gases. The fact that the separation process is well understood goes a long way toward reducing overall risk when the gases are difficult to handle. Distillation has been applied to the manufacture of flammable gases, for example ultrapure silane, as well as many corrosive gases.

Adsorption and Chemical Conversion. In some cases, removal of moisture or oxygen added by small amounts of air contamination is all that is necessary to make a gas suitable for high purity applications. For example, high purity HCl is available in tube trailer quantities but may require removal of moisture inadvertently added during transport before it is distributed to process equipment within a semiconductor plant. With a limited objective, it is usually most effective to use an adsorption process. As long as the total mass of impurity which must be removed is low enough, the process may be designed either with or without the capability for repeated regeneration.

Chemical conversion processes can also be used for moisture and oxygen removal. These tend to be the same ones developed for the smaller point-of-use purifiers. Consequently there is little economy of scale and they are seldom able to be regenerated.

Point-of-Use Purification. For the user of cylinder quantities of reactive specialty gases, there are only a limited number of ways to remove impurities and obtain high purity. Specialized point-of-use purifiers have been developed that purify small streams of many important reactive gases. Whereas these point-of-use purifiers cannot remove all important impurities, they are usually effective for removing the contamination added by the users' gas distribution system, mostly air and moisture.

One important class of point-of-use processes utilizes a porous polymer containing reactive metals. Variations in the metal and polymer chemistry are made to optimize the process for different gas applications. This is an active area of development and purifiers are available for most of the principal specialty gases.

Delivery and Control

Once a gas has been purified, it must be brought to its intended point of use without being degraded by the addition of excessive contamination. This transport or delivery process can be technically challenging. Often, contamination is unintentionally added even before the gas has left the purifier during its original manufacturing process. This is especially true for the corrosive specialty gases, where minimizing contamination by iron and other metallic species arising from the purification system is a primary concern.

Technologies to prevent recontamination during the delivery process have been an intense area of development. Work in the 1970s focused mainly on preventing contamination of high purity inert gases such as N_2 and Ar by atmospheric leaks and the release of atmospheric impurities which had been adsorbed on the inner surfaces of delivery systems during assembly. More recently, work has focused on preventing contamination of corrosive gases, such as HCl and HBr, with contaminants that contribute metallic impurities such as Fe and Ti to the electronic materials.

Delivery methods for high purity gases can be divided both according to chemical reactivity with respect to the containment system and, to a less significant degree, according to volume throughput requirements. Many highly flammable gases such as H_2 and SiH_4 are still inert with respect to the containment systems. Even though special provisions must be made because of flammability, the technology used to deliver flammable gases is similar to that employed for inert gases.

Bulk Gases. Attaining high purity gases where they are used requires a suitable gas distribution system. To achieve a high purity distribution system, there must be an absence of dead zones, external leakage, outgassing, and particulate contamination. Dead zones are obvious sources of carry-over contamination and must be designed out of the system. External leakage can arise from the diffusion of air through small holes, which takes place despite a reverse pressure gradient. Leakage can also occur because of permeation, especially through organic polymers used for gasketed seals. Outgassing is a general term applied to the release of any gaseous contaminants from the materials of construction. It can result from impurities adsorbed on surfaces and dissolved within the bulk material of both system components and particulate contamination.

detection sensitivity (ppb to <ppb), a dedicated, high purity blending system must be utilized to minimize background and produce accurate low ppb calibration gases. To carry out calibration, apims response is measured for a zero gas and at least five blended impurity concentrations. These concentrations are evenly distributed in the range of 0–50 ppb. The ion-current response recorded at each concentration for the individual impurity is normalized against the total ion-current response. The data are expressed as a percentage in terms of relative ion intensities, $R_a = 100\, I_a/I_T$, where R_a is the relative ion intensity of impurity a; I_a, the ion-current response of impurity a; and I_T, the total ion-current response. The relative intensity response data is regressed against blended impurity concentration data to produce the linear model, $R_a = k_a(C_{am} + C_0)$, where C_{am} is the blended impurity concentration of impurity a; C_0, the background impurity level; and k_a, the multiplication constant. Possible sources of background response include instrument noise, sample system outgassing, or interference from other impurity response signals. Proper setup, purging, and operation of the instrument should reduce background levels well below 1 ppb.

Rearranging the above expression yields impurity concentration as a function of relative intensity, $C_a = R_a/k_a$, where C_a represents both sample concentration and any background effects. The stability of the calibration must be confirmed at least every two weeks by analysis of a known mixed impurity standard.

Particulate Impurities.   Particle counters require factory calibration every two years. In addition, the background signal associated with both the instrument and its sampling system (tubing, sampler, sample valves, regulators, and fittings) must be quantified so that it may be subtracted from sample measurements. To accomplish this, an absolute filter (<0.01 μm rating) is employed. The absolute filter removes all particles entering the sampling system, so any particle registered by the counter can be directly attributed to the sampling system or instrument noise.

To measure background, the absolute filter is placed at the beginning of the sampling system and purged for a minimum of one hour at a rate of 100 standard liters per minute (SLPM). After the purge, the sampling system is plumbed up to the particle counter and a minimum of 0.14 standard cubic meters (5 SCF) of gas is analyzed in 0.003 or 0.0003 standard cubic meter (0.1 or 0.01 SCF) increments. The average of these runs is accepted as the system background. It is recommended that the background be checked again at the completion of certification measurements for a particular system. The background varies depending on the matrix being analyzed, therefore when switching matrices, the background must again be quantified.

Sampling and Analysis Guidelines.   As a general safety consideration, all gases should be vented to an external area and whenever possible, inert gases should be used as the test gas for piping systems.

Sample Line Connection Procedure.   The following procedure is used for connecting sample lines to sampling ports. For best results, a one-eighth in. (0.318 cm) diameter 316SS electropolished tubing having one-fourth in. (0.635 cm) end fittings is recommended for sampling gas for both gaseous and particulate impurity analyses. The one-eighth in. diameter tubing allows a larger degree of flexibility than does the one-fourth in. tubing. Butt weld glands that adapt one-fourth in. end fittings to one-eighth in. tubing are available (Swagelok). The actual orbital welding can be performed by fabricators.

The following are mandatory: (1) a new pair of clean, dry, plastic, disposable gloves must be worn during this procedure to prevent the potential contamination of the gas system by oils, dirt, and moisture present on the hands; (2) prior to making a connection, all required materials must be gathered and close at hand, including all tools, gloves, gaskets, adapter fittings, sample line, etc; (3) all sampling components must be precleaned prior to assembly, and these components must be kept clean at all times with ends capped and stored in cleanroom plastic bags; (4) appropriate precautions must be taken if making a connection to a hydrogen system, which includes grounding all components prior to and after making the connection and confirming that there is no ignition source or oxidizer near by (area must be isolated with red "Hazard Area" tapes); (5) sample connections are made to valved ports only; (6) for bulk gases other than hydrogen, the valve is opened slightly to ensure that a flow comes out the port once its cap is cracked open (for hydrogen, it is recommended that the connection be made with the valve closed and the next step omitted); (7) standing to the side of the port (ie, not in front of the line of flow), the analyst should crack open the fitting slightly and ensure that gas is flowing out of the fitting, use the valve to adjust the flow leaking past the fitting to a substantial but manageable level, and if there is no flow with the valve opened, the fitting must be resealed immediately and the cause determined; (8) the cap and old gasket are removed and the sample line with new gasket connected; and (9) prior to connection to the analyzer distribution manifold, the sample line must be purged. The purge flow rate should be as high as possible, given all safety and gas consumption constraints. Sample line purging should be performed for at least 30 minutes. If the connection is made to a hydrogen system, this purging step is eliminated and the sample tubing immediately isolated until such time as it can be connected to the analyzers.

Sample Analysis. The following procedure is used for sample analysis: (1) after sample connections are made and the sample tubing purged for 30 minutes, the tubing is connected to the sampling manifold which serves to distribute sample gas to each of the analyzers, and should analyzers have specific pressure and/or flow requirements, a means of regulating these parameters should be incorporated into the manifold (suggested design: high purity regulator attached to a multivalve manifold having flow meters on each flow-critical leg); (2) the sampling system and analyzers are then purged for at least two hours (if this is the first use of the system, or if it has been shut down for a significant time, longer purge times will likely be required to reach representative level). It is crucial that the analytical system be continuously purged or effectively isolated when moving to a new point to avoid contamination and a mobile analytical cart with a purified, on-board purge is ideal for maintaining high purity conditions; (3) when analyzing multiple locations, it is recommended that separate sample lines be set up purging in advance to reduce analysis cycle time, but this should only be done when all safety considerations and gas consumption factors are taken into account; (4) repeated analyses are then performed for approximately one hour to determine the stability of the analyzers and if the required specifications have been met; (5) it is recommended that five consecutive analyses showing the impurities within specification be completed (average concentrations are recorded on a standard certification form given the following information: customer facility, sampling location, analyst, time and date, impurity specification, measured impurity concentration for each completed analysis, measured impurity concentration mean, and analyzers employed); (6) if any impurity fails to meet specification, the analyst should check for leaks, check the calibration, and/or increase purge flow (slowly) if safe to do so. Another option is to cycle-purge the system by repeatedly closing the vent valve to pressurize and then opening the valve wide to rapidly depressurize. The analyzers must be isolated during this process and care should be taken not to allow the system pressure to fall so low (eg, <69 kPa (<10 psig)) that air can gain access and contaminate the system; (7) after the above six steps have been taken, sampling conditions are returned to normal and the system retested; (8) gas chromatographic analyses are based on a direct comparison of the area or peak heights of the individual impurity against its respective standard calibration curve, however if the concentration of impurity found in the sample is less than the lowest standard evaluated, then an extrapolated linear regression value may be determined; (9) the particle count for a measured gas is the difference between the count taken during the analysis of the sample and that measured for the sampling system background according to standard statistical procedures; and (10) particle measurements are generally accomplished more effectively when performed separately from vapor impurities because a higher level of purging is usually required for particle analysis and a complex sampling system can generate additional particles under these conditions.

Applications of High Purity Gases

The applications of high purity gases are primary in the semiconductor industries. From 1991 to 1995, the North American semiconductor bulk gas sales increased from U.S. $214 to $252 million (annual growth rate of 4.2%) and specialty gas sales increased from U.S. $78 million to $169 million (annual growth rate of 21.4%).

In addition to the microelectronics industry, other applications for high purity oxygen include fiber optics manufacturing, production of pharmaceuticals, and usage as calibration media in research and development laboratories, and in the pollution control field. Applications for high purity hydrogen include oxidation processing and epitaxial growth for both silicon and gallium arsenide.

Argon has become an important commodity owing to its ability to remain inert even under extreme conditions of temperature and pressure. High purity argon is likely to be used in the high technology fields of electronics, fiber optics, research and development, powder metal spraying, and hot isostatic pressing.

Other applications of high purity specialty gases include hydrogen bromide for etching single-crystal silicon, polysilicon, and aluminum; nitrogen trifluoride as a fluorine source for *in situ* cleaning processes for chemical deposition equipment, semiconductor etching and deposition, and high energy chemical lasers; and sulfur hexafluoride as a key etching material in certain semiconductor manufacturing processes because it has a selectively high etch rate of silicon compared to silicon dioxide or silicon nitride.

The primary driving force for high purity gases has been the increasing purity demands from high technology industries such as ceramics, optical fibers, and silicon wafer fabrication. In 1985, the typical impurity requirement in high purity nitrogen for wafer fabrication was on the order of 500 to 100 ppb on a molar or volume basis. The impurity requirement changed to the range of 10 to 1 ppb in 1990 and to 1 to 0.1 ppb in 1995. Future demands for even higher purity gases are expected.

In addition to purification to remove contaminants, accurate analyses of extreme purity gases and selection of storage and delivery equipment to minimize recontamination, all the while keeping costs down, pose serious challenges for the manufacturers of high purity gases. One analytical breakthrough in the 1990s was the development of atmospheric pressure ionization mass spectrometers that monitor < ppb levels of impurities continuously. Further improvement of analytical techniques is a focus of industrial research and development facilities.

BIBLIOGRAPHY

General References

R. DiNapoli and A. M. Sass, in J. J. McKetta, ed., *Encyclopedia of Chemical Processing and Design*, Vol. 31, Marcel Dekker, Inc., New York, 1990, p. 236.
W. H. Whitlock, "The Ultra-High Purity Challenge", in *Separation of Gases, Proceeding of the Fifth BOC Priestley Conference*, Birmingham, U.K., Sept. 19–21, 1989, Royal Society of Chemistry, 1990.
M. Finnergan, *CryoGas Int.*, 13 (Apr. 1996).
H. A. Grieco, *CryoGas Int.*, 20 (Apr. 1996).
J. D. Borkman, W. R. Couch, and M. L. Malczewski, *Microcontamination*, 23 (Mar. 1992).
S. D. Cheung, G. L. Mooney, and D. L. Jenson, *Micro*, 59 (Oct. 1995).
N. M. Chodhury and L. J. Mostowy, *Micro*, 33 (Sept. 1995).
J. M. Davidson and T. P. Rhane, *Microcontamination*, 34 (Mar. 1987).
E. F. Ezell, A. Athalye, and M. Chigrinskiy, *Microcontamination*, 93 (July/Aug. 1996).
J. Hart and A. Paterson, *Microcontamination*, 63 (July 1994).
G. Kasper, H. Y. Wen, and H. C. Wang, *Microcontamination*, 18 (Jan. 1989).
K. Siefering, H. Berger, and W. Whitlock, *Microcontamination*, 31 (Sept. 1992).
Ibid., p. 23 (Nov. 1992).
K. Siefering, W. Whitlock, and A. Athalye, *Microcontamination*, 41 (Mar. 1994).
R. M. Thorogood, A. Schwarz, W. T. McDermott, and C. D. Holcomb, *Microcontamination*, 28 (Aug. 1986).
P. Burggraaf, *Semicon. Int.*, 70 (Mar. 1990).
R. Iscoff, *Semicon. Int.*, 52 (Mar. 1991).
Ibid., p. 42 (Apr. 1992).

WALTER H. WHITLOCK
EDWARD F. EZELL
SHUEN-CHENG HWANG
The BOC Group, Inc.

HORMONES, ADRENAL-CORTICAL

Since the introduction of cortisone (1948) and hydrocortisone (1951), adrenal-cortical hormones have remained an important and unreplaced drug class. Though not without adverse effects, these compounds have continued to be the drug of choice in the treatment of afflictions ranging from the moderate skin rash to severe acute inflammatory disorders, and are included in many other therapeutic regimes.

The adrenal cortex releases both mineralocorticoids (from the zona glomerulosa) and glucocorticoids (from the zona fasciculata/reticularis). In addition, some androgenic and estrogenic steroids are synthesized by the adrenal gland. Once released by the adrenal cortex, the primary endogenous function of glucocorticosteroids is in influencing carbohydrate and protein metabolism. Mineralocorticoids regulate sodium reabsorption in the collecting tubules of the kidney. As with the mineralocorticoids, the generation of glucocorticosteroids is intricately balanced; where the production and regulatory process malfunctions, the result is either an excess (eg, Cushing's syndrome) or a deficiency (eg, Addison's disease) in glucocorticoid levels.

However, corticosteroids and their metabolites (1) were early recognized as possessing powerful antiinflammatory and immunomodulatory properties. Even prior to 1950, reports of the antiarthritic properties of cortisone (**1**) by Hench and co-workers (2) indicated the potential for these compounds to reduce the suffering of patients with inflammatory diseases. This awareness, combined with the first synthesis of naturally occurring glucocorticoids (11-desoxycorticosterone), led not only to the massive increase in research in the area of steroid synthesis and physiology, but to a Nobel prize in 1950 for early steroid pioneers Hench, Reichstein, and Kendall.

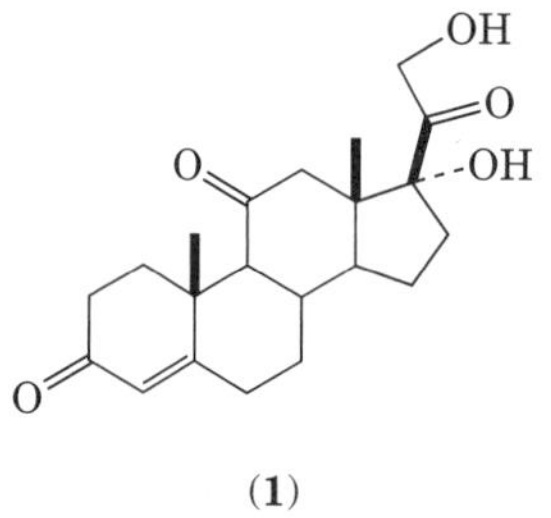

(1) 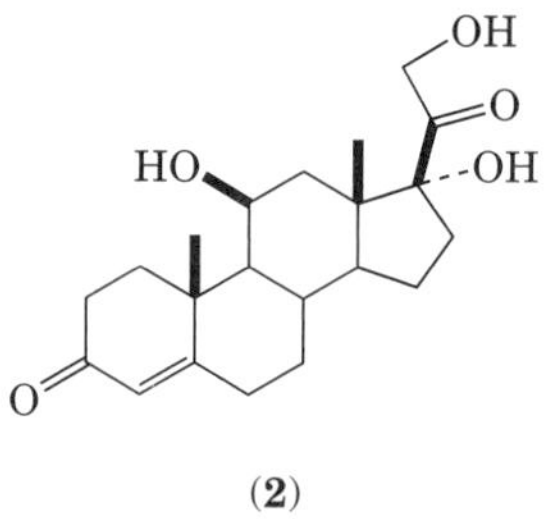   (2) 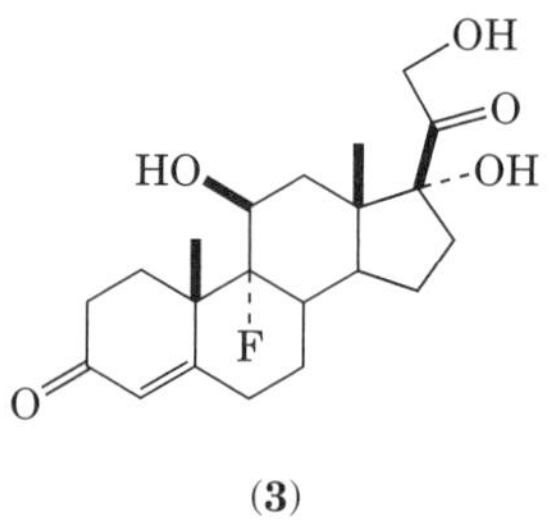   (3)

Progress in the field received further impetus with the development of synthetic steroids exhibiting activities far greater than that of the natural hormones. The synthesis of 9α-fluorocortisol (**3**), described in 1953 by Fried and Sabo (3), opened the way for development of many more highly active antiinflammatory agents, and indeed some of today's most active antiinflammatory agents (ca 1997) bear some resemblance to this 9α-fluorinated steroid.

Comprehensive reviews treating the pharmacological structure–activity relationships of glucocorticosteroids (4–7) and mineralocorticoids (8) have been published, as well as reviews of the mechanism of action (9,10). All natural adrenocorticoids are derivatives of the planar ring system 5α-pregnane (**4**). Substituents lying above the plane of the rings are assigned a β-configuration, indicated by a dotted line. Angular methyl groups at C-10 and C-13 have the β-configuration and are often shown simply by solid bonds. Tertiary hydrogen atoms at C-8, C-9, C-14, and C-17 are usually omitted unless their stereochemistry differs from that shown in (**4**).

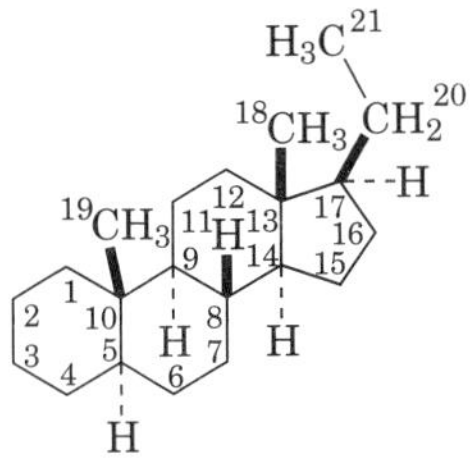

(**4**)

Clinical Use of Adrenal-Cortical Steroids

The predominant clinical use of corticosteroids is a result of their associated antiinflammatory properties. These are commonly used as topicals for the suppression of symptoms, including inflammation, occurring in a particular disease state; these compounds are rarely considered curative in their usage. Many other disease states do, however, respond well symptomatically to treatment with corticosteroid therapy. Some of these (11) are listed below.

Antiinflammatory Activity

Adrenocortical insufficiency
Organ transplants
Liver disease
Adrenogenital syndrome
Nephrotic syndrome
Acute spinal cord injury
Hypercalemia
Hematologic disorders
Myasthenia gravis
Neoplastic disease

Allergic conditions and conjunctivitis
Cerebral edema
GI diseases; ulcerative colitis and anorectal disorders
Bacterial meningitis
Rheumatic disorders and collagen diseases
Ophthalmic, otic, and nasal disorders
Dermatologic diseases
Respiratory diseases

Mineralocorticoid therapy is a less common, though still important aspect of the medicinal use of adrenal-cortical steroids. Very important applications include use in treating adrenocortical insufficiency or in other adrenocoritcoid

Table 1. (Continued)

Substance	Chemical name	Structure number	Level of activity[c]	Degree of potency[d] Very high	High	Medium	Low
fluocinolone acetonide	6α,9-difluoro-11β,21-dihydroxy-16α,17-[(1-methylethylidene)bis(oxy)]-pregna-1,4-diene-3,20-dione		IV		X		
halcinonide	21-chloro-9-fluoro-11β-hydroxy-16α,17-[(1-methylethylidene)bis(oxy)]-pregn-4-ene-3,20-dione		II		X		
mometasone acetonide			III				
triamcinolone acetonide	9-fluoro-11β,21-dihydroxy-16α,17-[(1-methylethylidene)bis(oxy)]-pregna-1,4-diene-3,20-dione	(**49**)	III		X		
β-methasone benzoate	17-(benzoyloxy)-9-fluoro-11β,21-dihydroxy-16β-methylpregna-1,4-diene-3,20-dione		III			X	
clocortolone pivalate	9-chloro-21-(2,2-dimethyl-1-oxopropoxy)-6α-fluoro-11β-hydroxy-16α-methylpregna-1,4-diene-3,20-dione					X	
flurandrenolide	6α-fluoro-11β,21-dihydroxy-16α,17-[(1-methylethylidene)bis(oxy)]-pregn-4-ene-3,20-dione		IV			X	
fluticasone propionate	6α,9-difluoro-11β-hydroxy-16α-methyl-3-oxo-17α-(1-oxopropoxy)-androsta-1,4-diene-17β-carbothioic acid, S-fluoromethyl ester					X	
hydrocortisone butyrate	11β,21-dihydroxy-17-(1-oxobutoxy)-pregn-4-ene-3,20-dione		V			X	

hydrocortisone valerate	11β,21-dihydroxy-17-[(1-oxopentyl)oxy]-pregn-4-ene-3,20-dione	X	V	
prednicarbate mometasone fluroate	9α,21-dichloro-17-[(2-furanyl-carbonyl)oxy]-11β-hydroxy-16α-methyl-pregna-1,4-diene-3,20-dione	X	V	
aclometasone dipropionate	7α-chloro-11β-hydroxy-16α-methyl-17α,21-bis(1-oxopropanoxy)-pregna-1,4-diene-3,20-dione		VI	X
desonide	11β,21-dihydroxy-16α,17-[(1-methylethylidene)bis(oxy)]-pregna-1,4-diene-3,20-dione		VI	X
dexamethasone (74)	9-fluoro-11β,17,21-trihydroxy-16α-methylpregna-1,4-diene-3,20-dione			X
dexamethasone sodium phospate	9-fluoro-11β,21-dihydroxy-16α-methyl-21-(phosphonooxy)-pregna-1,4-diene-3,20-dione, disodium salt			X
hydrocortisone	11β,17,21-trihydroxy-pregn-4-ene-3,20-dione			X
hydrocortisone acetate	21-(acetyloxy)-11β,17-dihydroxy-pregn-4-ene-3,20-dione			X

[a] Ref. 12.
[b] Ref. 13.
[c] Ranging from I (most active) to VI (least active), as specified by the DI (13).
[d] Range in which compounds listed by FC were placed (12).

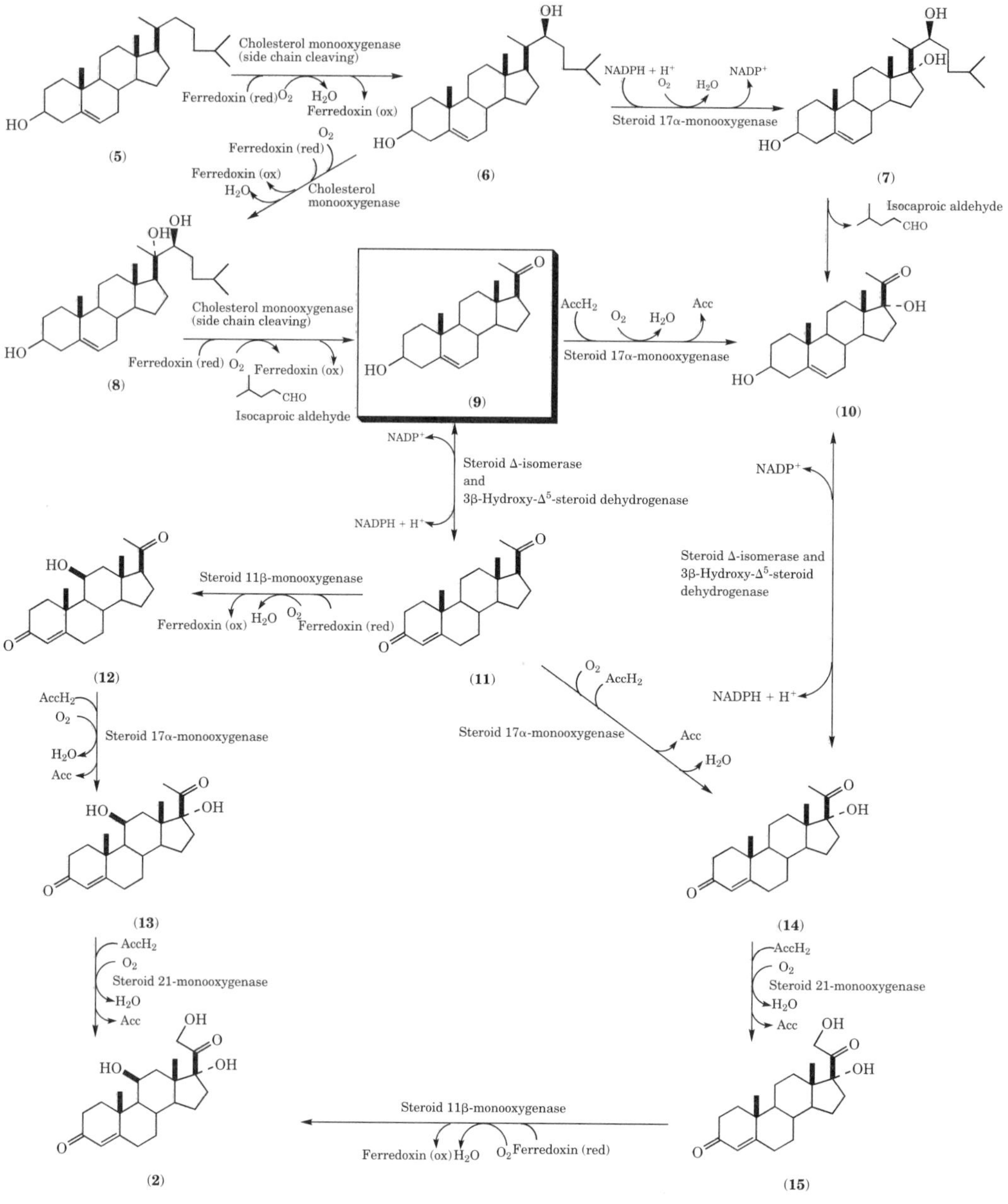

Fig. 1. Biosynthetic pathways for formation of cortisol from cholesterol.

C-20 Reduction: Two stereoisomers can result from this transformation, although cortisol is thought to act primarily with $(R)20\beta$-hydroxysteroid dehydrogenase. This is a first step in the metabolism of corticosterone.

Cleavage of C-17 acyl/alkyl substituents: Resulting primarily in cholan-17-ones, this is a relatively minor metabolic pathway. Corticosterone is not known to undergo this transformation before excretion.

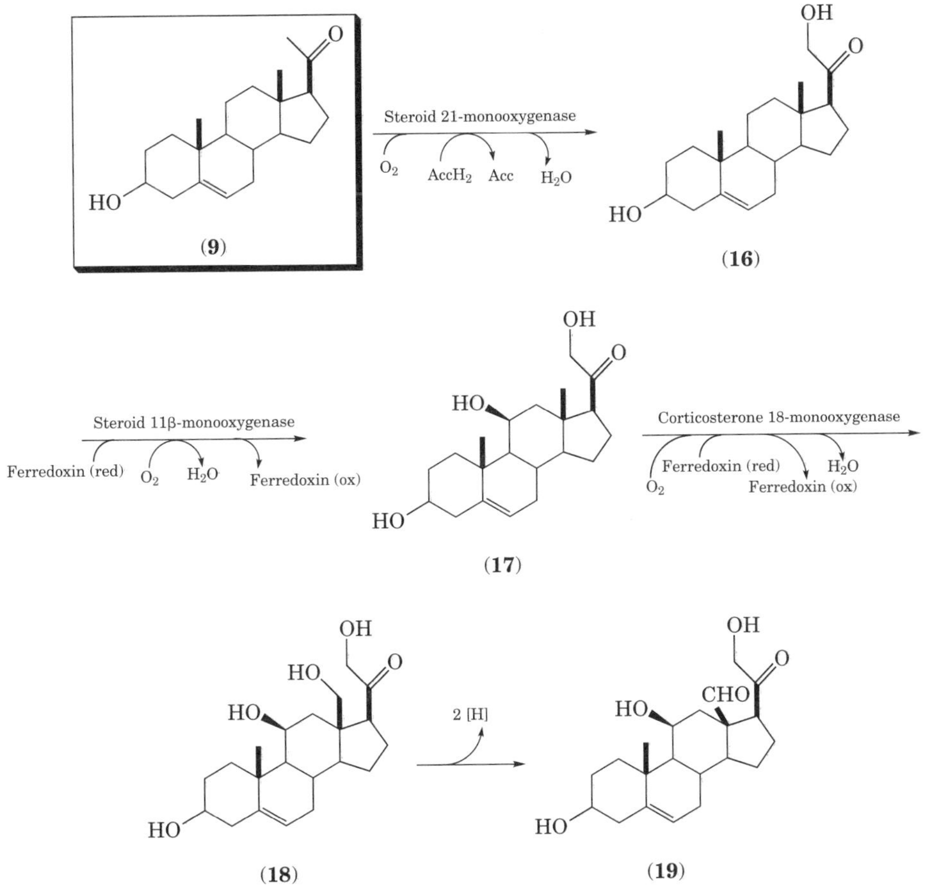

Fig. 2. Key steps in the mineralocorticoid biosynthesis.

C-6 Hydroxylation: This biotransformation is more predominant in infants than in adults, and can prevent other metabolic transformations.

Glucuronidation: Complexation of the steroid to glucuronic acid, most predominantly via the C-3 hydroxyl, leads to a considerable portion of the excreted metabolites of all glucocorticoids. In infants, sulfurylation (formation of a sulfate ester) is also predominant (16).

Other reactions: Most of the metabolites of cortisol are neutral (alcohol or glucuronide complex) compounds. However, oxidation at C-21 to C-21 carboxylic acids (17) accounts for some of the identifiable metabolites of glucocorticoids (18).

Compounds having the 16,17 ketal, eg, budesonide, amcinonide, fluocinonide, halcinonide, triamcinolone acetonide, and flurandrenolide, also undergo metabolism by routes that parallel that of cortisol metabolism. Unsymmetrical acetals such as budesonide are also metabolized by routes not available to the more metabolically stable symmetrical $16\alpha,17\alpha$-isopropylidene-dioxy-substituted compounds (desonide, flunisolide, and triamcinolone acetonide). Isozymes within the cytochrome P450 3A subfamily are thought to catalyze the metabolism of budesonide, resulting in formation of 16α-hydroxyprednisolone and 6β-hydroxybudesonide (19,20) (Fig. 3) in addition to the more common metabolic steps (oxidation via Δ^6, reduction of Δ^3, etc).

Steroids having the greatest number of substituents generally have the slowest rate of metabolism. Groups that appear to activate by effects on metabolism include 2β-methyl, which stabilizes the resulting molecule to the action of the 4,5-reductase (21) and to the action of 20-keto reductase (22). Similarly, 6α-methyl protects the A ring against metabolic destruction (22). Again, the introduction of 16α-hydroxyl, as in triamcinolone (**48**), prolongs the half-life (23). The 16α-methyl and 16β-methyl groups have similar action. Triamcinolone is unusual in that it is metabolized principally to the 6β-hydroxy derivative (23).

Mechanism of Action of Antiinflammatory Steroids

Receptor Structures for Glucocorticoids and Mineralocorticoids. The effects of adrenal-cortical steroids are thought to result from their interaction with intracellular receptors, and a great deal of attention has been given to the task of determining the structure and function of the glucocorticoid receptor (GR) as well as the mineralocorticoid receptor (MR). Though studied in less detail, the primary function of the MR seems to be nearly identical to that of the GR, with the primary differences being the restricted expression of the MR (limited mostly to tissues in the kidney, colon, salivary and sweat glands, and hippocampus), and the different proteins encoded by the activated DNA-receptor complex. The highly homologous GR and MR proteins are also very similar in action, and the description of the more thoroughly examined GR is adequate for the understanding of the MR structure and function.

The primary structure of the GR has been identified as a 795 residue protein. Though there does not yet exist an x-ray structure, a significant amount of 3-D structure elucidation of the GR has been made. The GR is a member of the nuclear receptor superfamily, which includes receptors for other steroids, vitamin D, and thyroid hormones, and some other various proteins (24). A high degree of homology is found between receptors in this class, and each contains a similar domain make-up. These domains (sections of primary structure) each have a functional duty, and generally describe regions of hormone binding, nuclear translocation, dimerization, DNA binding, and transactivation (25).

Within the carboxyl terminus portion of the protein (residues 518–795) is the ligand (hormone) binding domain (HBD) (25). It is here that much of the interaction of the receptor with hsp90 occurs (26) (thought to be residues 595 to 614 of the rat GR (27)). Three of five cystein residues in this area are spaced close together in the binding pocket (28), and two of these are critical to the receptor's ability to bind specifically glucocorticoids (29). Binding of the hormone to the HBD allows the activation of the receptor to the DNA-binding state, and during this process, the conformation of the protein changes, in part the result of the dissociation of hsp90. Antiglucocorticoids and certain other modulators as well as sodium molybdate will interact with this domain and inhibit activation.

The DNA binding domain is highly conserved among species, and changes to the amino acid sequence in this region result in changes in receptor function (30). A structural feature which characterizes the GR-DNA binding domain are the two so-called zinc fingers, in each of which two zinc (+2) ions are held in

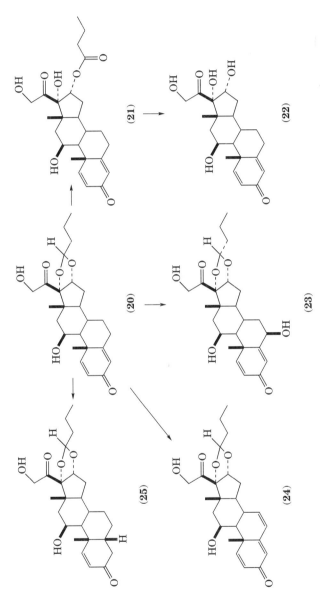

Fig. 3. Metabolism of 22R-budesonide in human liver microsomes.

model (57). This is a result of LC-1 inhibition of phospholipase A_2, which converts membrane phospholipids to arachidonic acid, along with its effect on other cellular components of certain inflammatory responses (10,58). Antibodies raised against LC-1 are able to reverse the antiinflammatory action of LC-1 (59–61). Corticosteroids reduce phospholipase A_2 activity, which results in the diminished release of arachidonic acid, and this subsequently leads to limiting the formation of prostaglandins, thromboxane, and the leukotrienes (9,62).

Glucocorticoids have been shown to inhibit gene transcription of other proteins involved in the inflammatory process, including the key inflammation mediators called cytokines (IL-1, IL3-6, IL8, GM-CSF, TNFα) (10,58,63–65). Steroids have been also shown to suppress the formation of cytokine receptors (10); dexamethasone, in particular, downregulates gene transcription of angiotensin II type 2 receptors (66).

Mineralocorticoids follow a mechanistic route similar to that of glucocorticoids, though differing in the proteins expressed. The activated MR-DNA complex promotes the expression of aldosterone-induced proteins (AIPs), which then act to increase Na^+ conductance of the luminal membrane and concurrently increase Na^+ pump activity of the basolateral membrane. These actions result from a number of AIP-influenced cellular characteristics, including a modification of tight junction permeability, an increase in membranous Na^+ channels and pumps, activation of silent Na^+ channels and pumps, and an increase in mitochondrial ATP-production-related enzymes.

Synthesis of Glucocorticoids

Hydrocortisone and Prednisolone. Following the discovery of the antiinflammatory actions of cortisone (**1**) and cortisol (**2**), there was a need not only to develop highly efficient routes to the corticoids, but to discover novel structures with fewer side effects than those of the corticoids, eg, sodium and water retention, reduced carbohydrate tolerance (steroid diabetes), osteoporosis, and depressed host defense.

A major difficulty in the manufacture of corticosteroids was the lack of an abundant raw material containing an 11-oxygenated function. This problem initially was solved by using 12α-hydroxylated cholic acid (**26**) (obtained from ox bile); later it was converted into cortisone acetate by a 32-stage process (66). A more recent development in procuring readily available starting material was the discovery that perfusion of cortexone (**16**) through the adrenal gland gives corticosterone (**17**), albeit in low yield, thereby demonstrating the existence of enzymes capable of hydroxylating progesterone derivatives at position C-11. Microorganisms capable of introducing an 11β-hydroxyl group into a steroid were discovered. *Rhizopus arrhizus* transformed progesterone (**11**)—which was readily available from diosgenin or from the soybean sterol, stigmasterol (**27**)—into 11α-hydroxyprogesterone (**28**) in excellent yield; even better results were obtained using *Rhizopus nigricans* (67). Other organisms include *Corynebacterium simplex* which converts cortisone (**1**) and cortisol (**2**) into their 1-dehydro derivatives, prednisone (**29**) and prednisolone (**30**) (68), respectively. These steroids surpassed their parent hormones in antirheumatic and antiallergic activity and produced lower mineralocorticoid activity and other side effects. Nearly all corticoids on the market other than cortisone are 1-dehydro steroids.

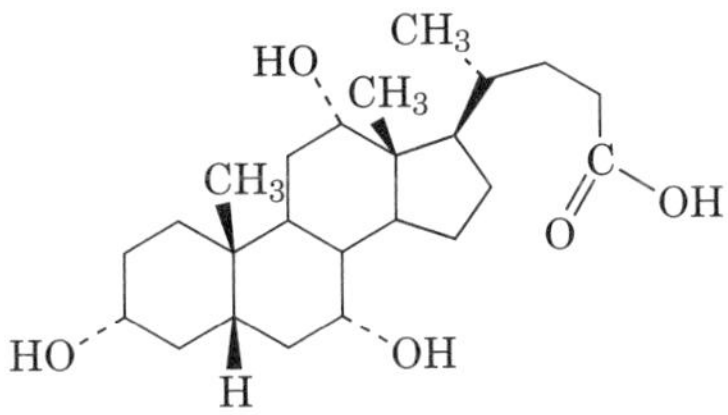

(26)

The classic process of corticosteroid manufacture employs stigmasterol (**27**), obtained from soybean oil, as the raw material. The sterol is first oxidized to stigmastadien-3-one (**31**) by the Oppenauer method. Ozonolysis of the latter in methylene chloride/1% pyridine yields 3-oxobisnorchol-4-enyl aldehyde (**32**) (69). Azeotropic distillation of the aldehyde with piperidine containing a trace of *p*-toluenesulfonic acid (pTS) leads to formation of the enamine (**33**) (70) which, when oxidized with sodium dichromate in anhydrous acetic acid, gives progesterone (**11**) in high yield. Microbiological oxidation of the (resulting) progesterone (**11**) with *Rhizopus nigricans* yields 11α-hydroxyprogesterone (**28**) in yields exceeding 90% (67). The latter is oxidized to 11-oxoprogesterone (**34**), which is condensed with diethyl oxalate in the presence of sodium methoxide to yield the 21-ethoxyoxalyl derivative (**35**). The last compound is treated with bromine giving the 21,21-dibromo-21-carbethoxylate, which is treated without isolation

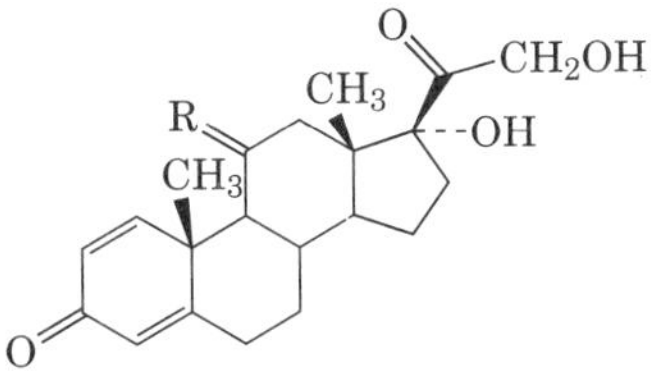

(30) R = $\overset{H}{\underset{}{|}}$ OH

(29) R = O

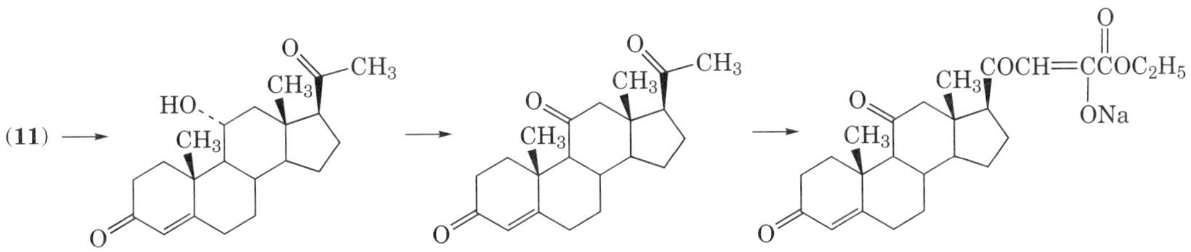

(11) →

(28)   (34)   (35)

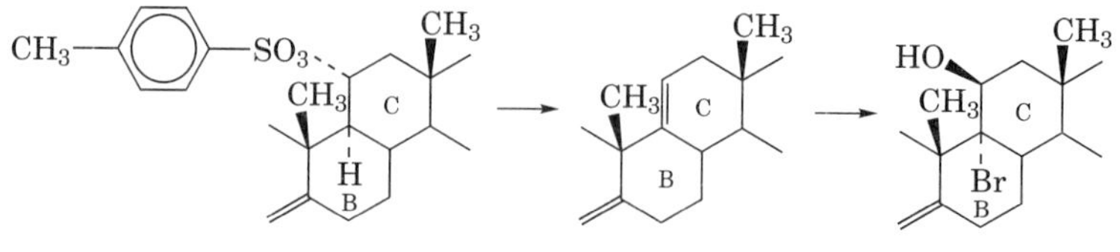

(43)   (44)   (45)

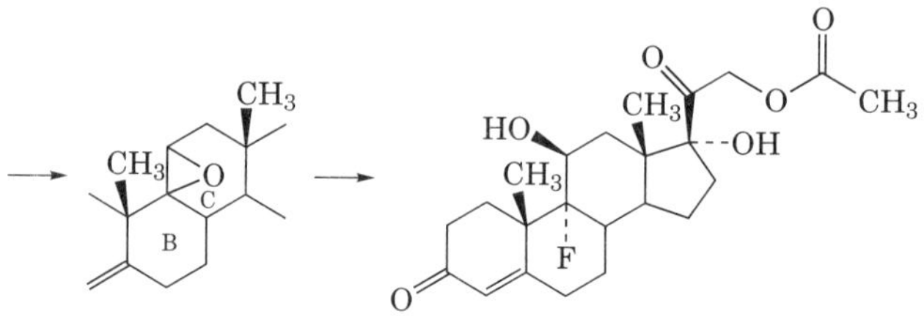

(46)   (47)

activity and, because of the presence of the 16α-hydroxyl group, is almost entirely devoid of salt-retaining activity. It had been shown that steroidal $16\alpha,17\alpha$-
dihydroxy-pregnan-20-ones readily forms acetonides, and application of this
reaction to triamcinolone (**48**) yields triamcinolone acetonide (**53**) (74) which
significantly exceeds the parent steroid in topical antiinflammatory activity.
16α-Hydroxyprednisolone $16\alpha,17\alpha$-acetonide (desonide) (**54**) has been introduced into clinical practice (75).

Methylated Glucocorticoids. The preparation of 2α-methyl-9α-fluoro-
cortisol has been reported (76). This compound shows enhanced glucocorticoid
activity and greatly enhanced mineralocorticoid activity, so much that
it surpasses aldosterone (**19**) in sodium-retaining and potassium-excreting
potency. Attention was then turned to the preparation of 6-methylated corticoids

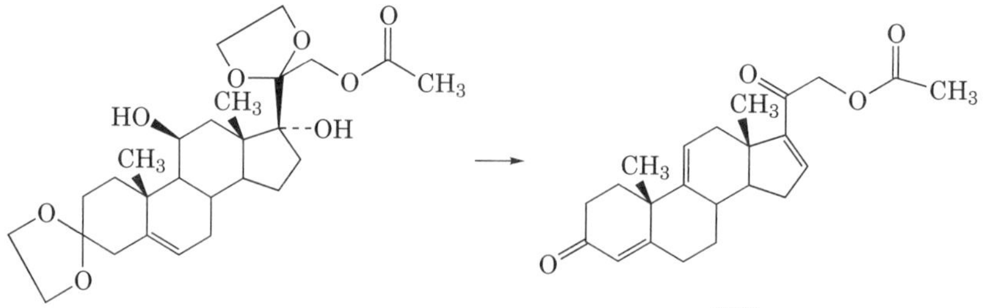

(50)

(49)

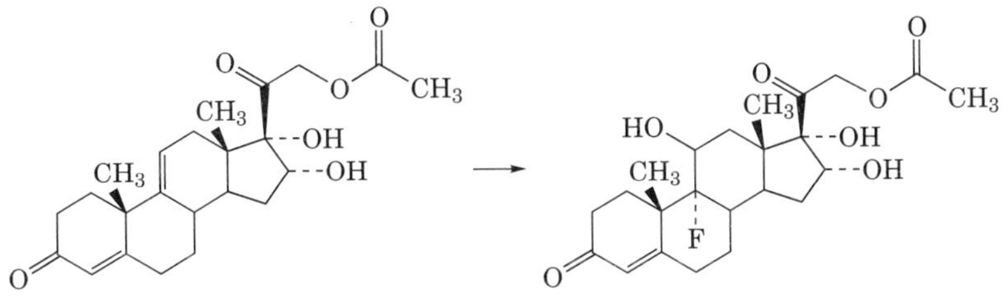

(51) → (52)

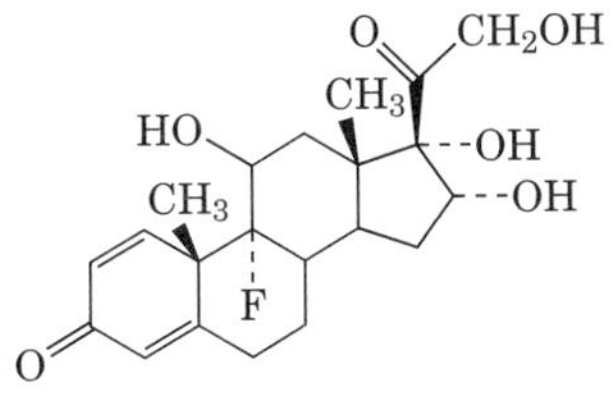

(48)

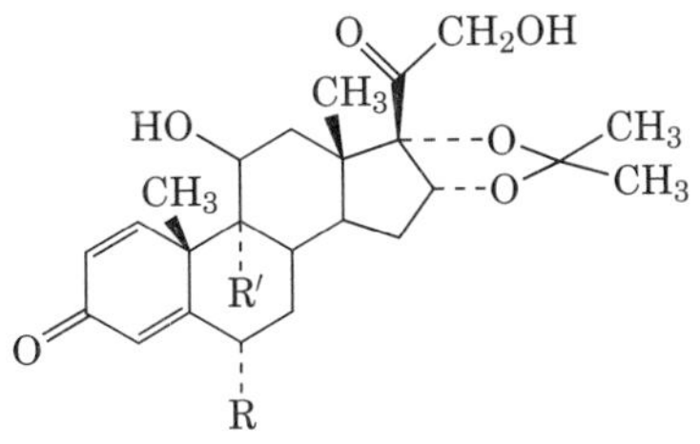

(53) R = H; R′ = F
(54) R = R′ = F
(92) R = F; R′ = H

when potentiation of glucocorticoid activity linked to negligible mineralocorticoid activity resulted: 6α-methylprednisolone (55) proved to be the compound of choice. It was prepared from 11α-hydroxyprogesterone acetate (28-acetate), which was converted into the bis-ethylene ketal and then by reaction with peracetic acid into the 5α,6α-epoxide (56). The latter compound was treated with methyl magnesium bromide (when fission of the epoxide ring occurred) to give the 5α-hydroxy-6β-methyl-bis-ketal (57)

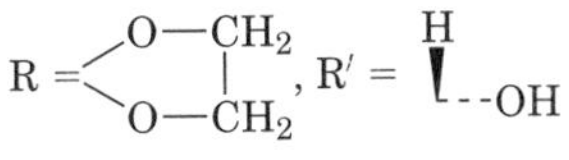

Removal of the ketal protecting groups, followed by oxidation with sodium dichromate in acetic acid gave 5α-hydroxy-6β-methyl-5α-pregnane-3,11,20-trione (57) (R = R′ = O). The latter was submitted to oxalylation, dibromination, Favorski rearrangement, and zinc dust reduction (see (39) → (40)) to yield 3,11-diketo-6α-methylpregna-4,17(20)-dien-21-oic acid (58) (see also (36)). The last compound was converted into the pyrrolidyl enamine (cf (37)) which, on reduction with LiAlH₄ and regeneration of the 3-keto group by treatment with buffered methanol followed by acetylation, gave 11β,21-dihydroxyl-6α-methylpregna-4,17(20)-dien-3-one acetate (59). Oxidation with phenyliodosoacetate in t-butanol/pyridine in the presence of catalytic OsO₄ furnished 6α-methylhydrocortisone, which subsequently was converted into 6α-methylprednisolone by microbiological oxidation with *Septomyxa affinis* (77). The potentiating effect of the 6α-methyl group upon glucocorticoid activity is well illustrated by the observation that $16\alpha,17\alpha$-isopropylidene-6-methylpregna-1,4-diene-3,20-dione (**60**) (77) is an excellent topical antiinflammatory agent even though it lacks the 11β-hydroxyl group characteristic of the natural glucocorticoids (78) and originally believed to be essential for glucocorticoid activity. The 21-desoxycorticoid $11\beta,17\alpha$-dihydroxy-9α-fluoro-6α-methylpregna-1,4-diene-3,20-dione (fluoromethalone) (**61**) also is an excellent topical antiin-flammatory steroid (79).

Methylation of hydrocortisone/prednisolone in positions C-4, C-7, C-12, and C-21 failed to give useful products. Methylation at C-16, in contrast, led to 16α- and 16β-methyl-9α-fluoroprednisolones which were exceptionally useful. Both series were prepared using 3α-acetoxy-5β-pregn-16-ene-11,20-dione derived from desoxycholic acid (80). A much shorter route was subsequently developed from 16-dehydropregnenolone (**62**) which was readily available from diosgenin.

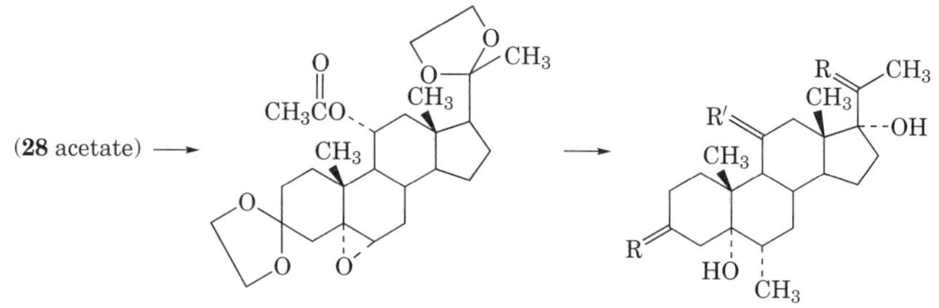

(**28** acetate) ⟶

(**56**)

(**57**)

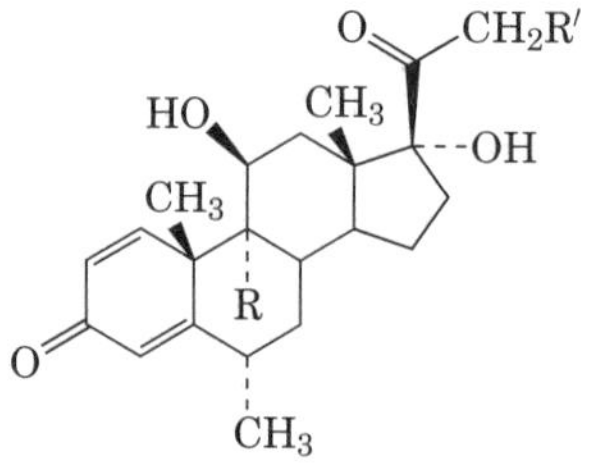

(**55**) R = H, R' = OH
(**91**) R = F, R' = H

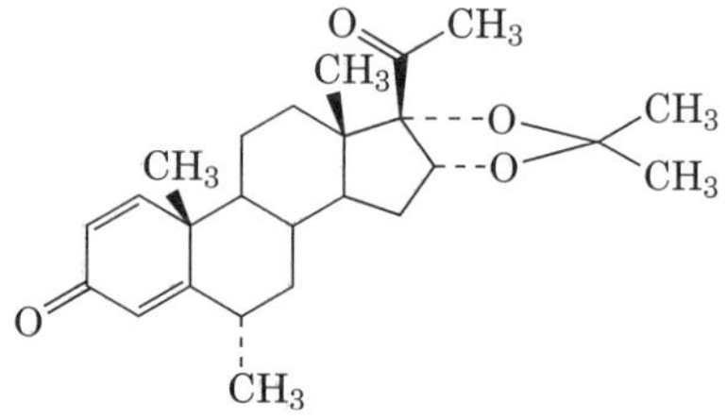

(**60**)

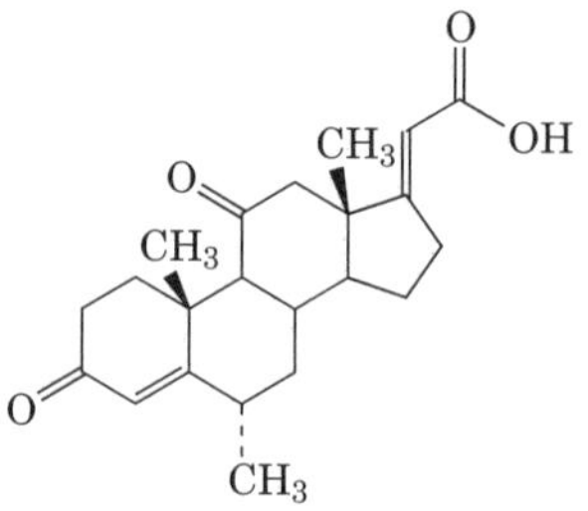

(**58**)

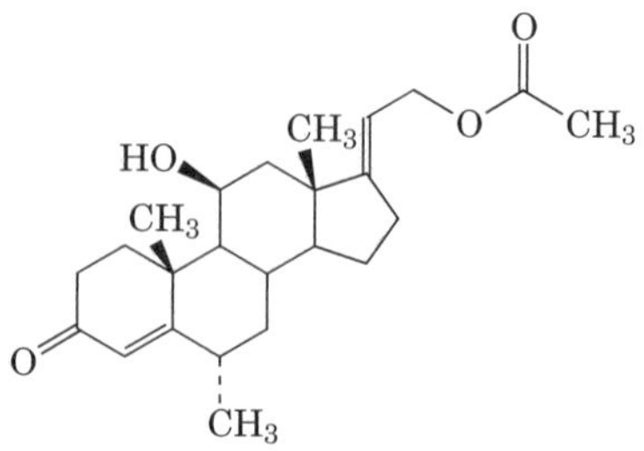

(**59**)

Pregnadienolone (**62**) was converted into 16α-methylpregnenolone (**63**) by reaction with CH_3MgI/Cu_2Cl_2 (81). Catalytic hydrogenation of the last compound gave the 5α-pregnane derivative (**64**) which passed smoothly on enforced acetylation into the enol acetate (**65**). The latter, on treatment with peracetic acid, yielded the epoxide (**66**), which was converted into 17α-hydroxy-16α-methyl-5α-pregnan-20-one (**67**). Bromination of the last compound at C-21 gave (**68**) from which, by metathesis with acetate ion, 21-acetoxy-3β,17α-methyl-5α-pregnan-20-one (**69**) was obtained. Chromic acid oxidation of the last compound yielded the 3-one (**70**), which was converted using the 2,4-dibromo derivatives (**71**) into the 1,4-dien-3-one (**72**). Microbiological hydroxylation furnished the 11-epi 16α-methyl prednisolone (**73**) converted by standard procedures into 9α-fluoro-16α-methyl-11β,17α-trihydroxypregna-1,4-diene-3,20-dione (9α-fluoro-16α-methylprednisolone; dexamethasone) (**74**) (82). The last compound is used in the diagnosis of Cushing's syndrome.

Introduction of a 16β-methyl group into the corticosteroid molecule was effected by a reaction (83) whereby a 16-dehydropregnenolone (**62**) was treated with diazomethane to form the pyrazoline (**75**) which was decomposed with perchloric acid in acetone to give the 16-methylpregn-16-en-20-one derivative (**76**). Catalytic hydrogenation yielded the 16β-methyl intermediate (**77**), which was converted into 9α-fluoro-16β-methyl-11β,17α,21-trihydroxypregna-1,4-diene-3,20-dione (9α-fluoro-16β-methylprednisolone, betamethasone) (**78**) by standard procedures (80).

Hecogenin has been used as starting material for the preparation of betamethasone (**78**); this genin is found in the sisal plant, *Agave sisalana*. The 12-oxo group present in hecogenin (**79**) is transferred to C-11 (84). Bromination of hecogenin in benzene yields the 11α,23a-dibromo derivative (**80**). The latter is treated with sodium hydroxide in aqueous *t*-butanol to yield the crystalline 23-bromo-ketal (**81**) which is acetylated and then debrominated with zinc dust. Reduction with calcium/liquid ammonia yields 11-oxotigogenin (**82**), which is converted into betamethasone (**78**) by standard procedures.

A variant of the 16-methyl group is 9α-fluoro-16-methylene-11β,17α,21-trihydroxypregna-1,4-diene-3,20-dione (fluprednylene) (**83**) (85).

6-Fluorocorticoids. The 6α-fluoro substituent resembles the 6α-methyl group in strongly potentiating glucocorticoid activity. 6α-Fluoro-16α-methyl-prednisolone (paramethasone) (**84**) was prepared from 16α-methylpregnenolone (acetate) (**63**). The 5α,6α-epoxide (**85**) of the last compound was treated with boron trifluoride to yield the 6β-fluoro-5α-hydrin (**86**) which gave the enol triacetate (**87**) with acetic anhydride/acetyl chloride. Reaction with perphthalic acid gave the 17α-hydroxy-20-one (**88**) which was converted into the 21-acetoxy-derivative (**89**) by the standard bromination/acetoxylation procedure. Oxidation

$$R = O\overset{\overset{\displaystyle O}{\|}}{C}CH_3, \quad R' = \underset{\text{--H}}{\overset{\text{OH}}{|}}$$

to the 3-one (**89**), followed by treatment with hydrogen chloride and cautious

$$R = O\overset{\overset{\displaystyle O}{\|}}{C}CH_3, \quad R' = {=\!=}O$$

alkaline hydrolysis gave 17α,21-dihydroxy-6α-fluoro-16α-methylpregn-4-ene-3,20-dione (**90**), converted into 6α-fluoro-16α-methyl-11β,17α,21-trihydroxy-

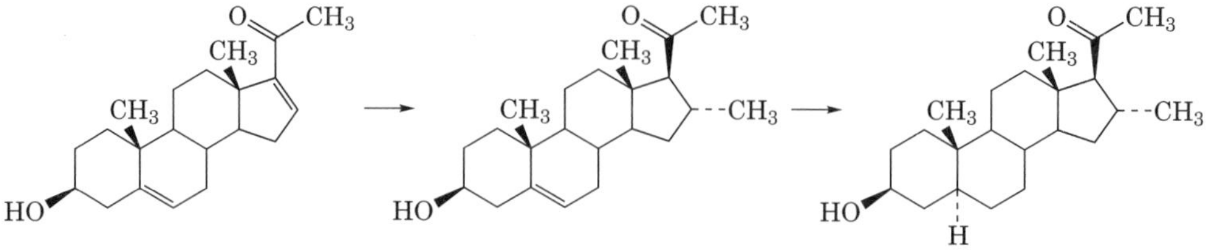

(**62**)   (**63**)   (**64**)

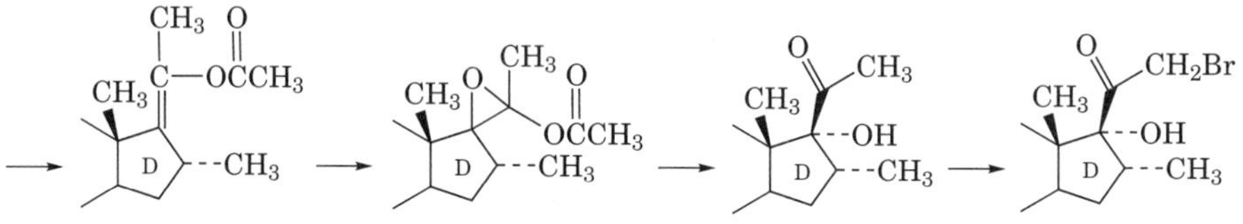

(**65**)   (**66**)   (**67**)   (**68**)

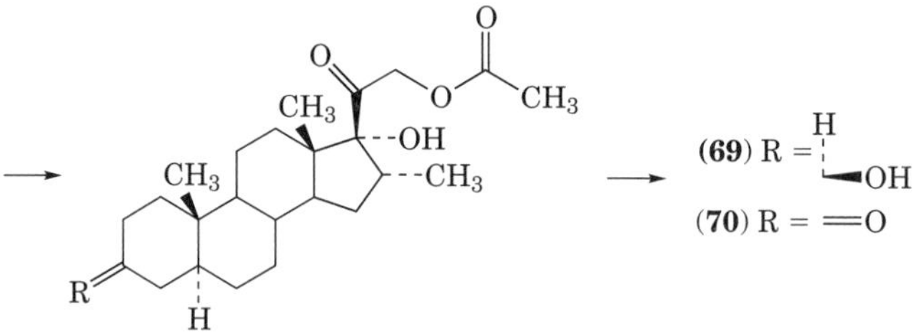

(**69**) R = H, OH
(**70**) R = ═O

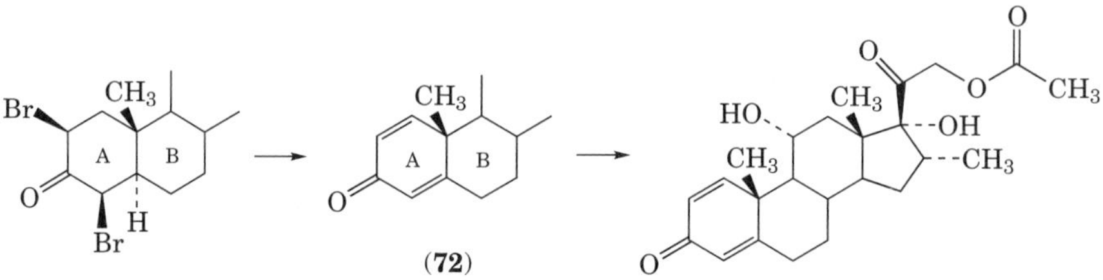

(**71**)   (**72**)   (**73**)

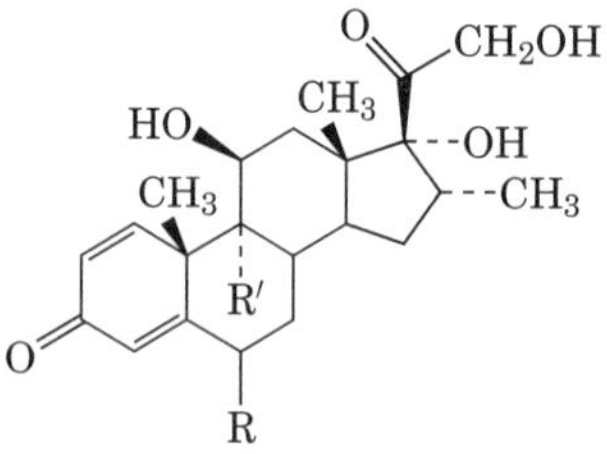

(**74**) R = H, R′ = F
(**84**) R = F, R′ = H
(**61**) R = R′ = F

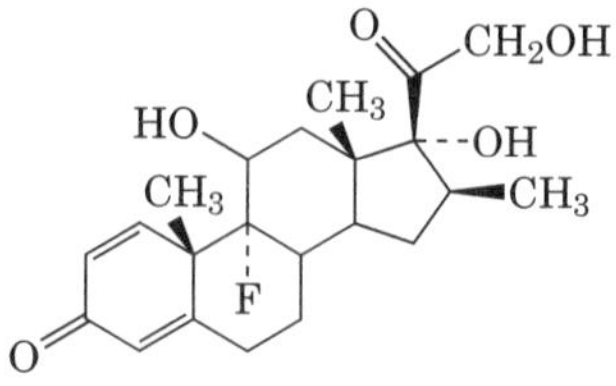

(**78**)

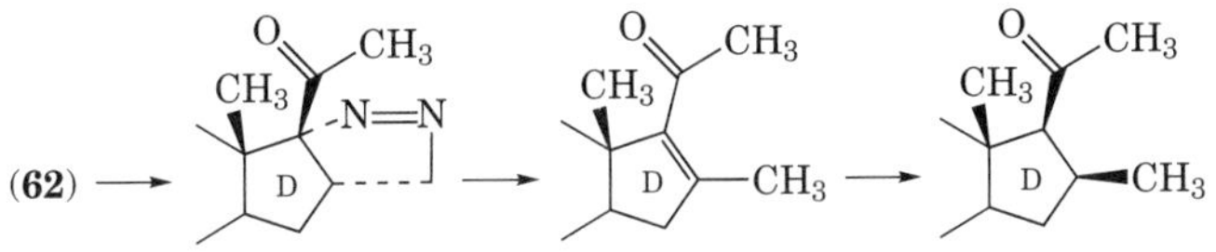

pregna-1,4-diene-3,20-dione (*para*-methasone) (**84**) and into 6α,9α-difluoro-16α-methyl-11β,17α,21-trihydroxypregna-1,4-diene-3,20-dione (flumethasone) (**91**) by standard procedures (86). The same general procedure was used to prepare 6α,9α-difluoro-11β,16α,17α,21-tetrahydroxypregna-1,4-dien-3,20-dione-16,17-acetonide (fluocinolone acetonide) (**54**) (87), which proved to be extremely effective by the topical route and is widely used as a standard for determining the potency of new corticoids. 6α-Fluoro-11β,16α,17α,21-tetrahydroxypregna-1:4-diene-3,20-dione-16α,17α-acetonide (flurandrenolone acetonide) (**92**) also was prepared (88).

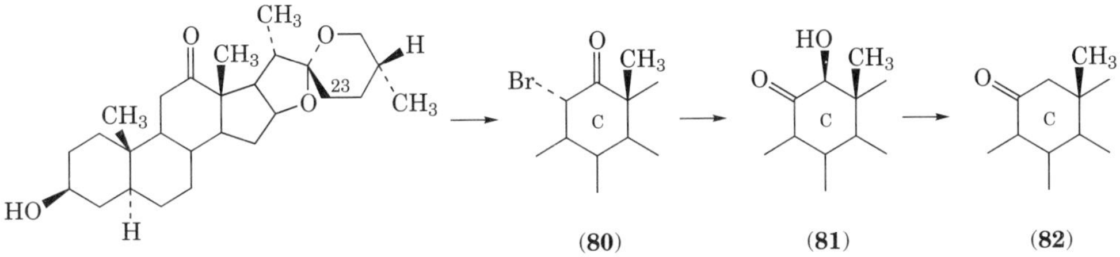

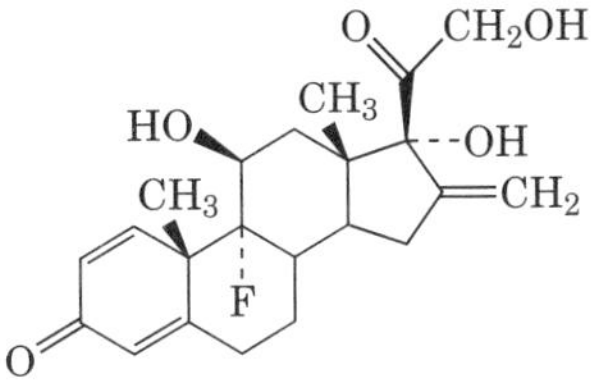

An alternative route to 6α-fluorosteroids was developed by treating the enol acetate (**93**) of a Δ^4-3-one with perchloryl fluoride in aqueous dioxane to obtain a mixture of the 6-fluoro-epimers (**94**) which were converted into the 6α-fluoro-Δ^4-3-ones (**95**) on treatment with H$^+$.

17-Acylated Corticoids. The corticoid side-chain of (**30**) was converted into the cyclic ortho ester (**96**) by reaction with a lower alkyl ortho ester RC(OR')$_3$ in benzene solution in the presence of *para*-toluenesulfonic acid (88). Acid hydrolysis of the product at room temperature led to the formation of the 17-monoesters (**97**) in nearly quantitative yield. The 17-monoesters (**97**) underwent acyl migration to the 21-monoesters (**98**) on careful heating with H$^+$. In this way, prednisolone 17α,21-methylorthovalerate was converted quantitatively into prednisolone 17-valerate, which is a very active antiinflammatory agent (89). The intermediate ortho esters also are active. Thus, 17α,21-(1'-methoxy)-pentylidenedioxy-1,4-pregnadiene-11β-ol-3,20-dione [(**96**), R' = CH$_3$, R = C$_4$H$_9$] is at least 70 times more potent than prednisolone (89). The above conversions

would possess antiinflammatory activity on topical administration but would be devoid of systemic activity when hydrolysis to the free acid occurs followed by

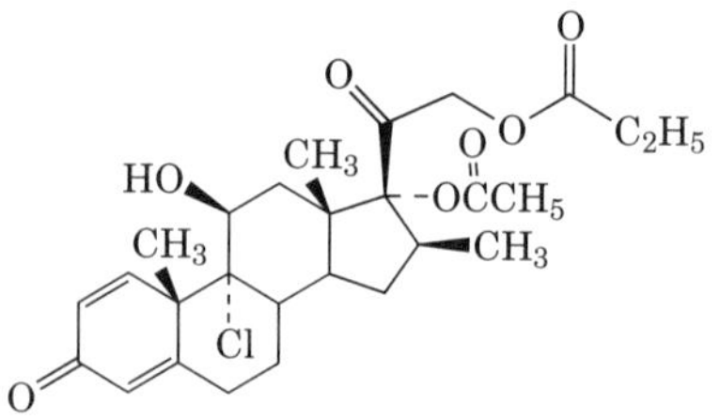

(101)

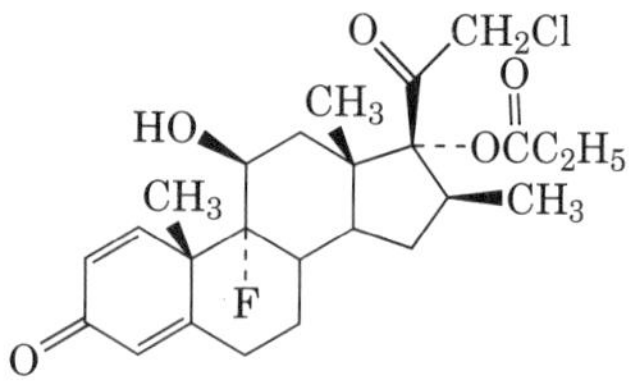

(102)

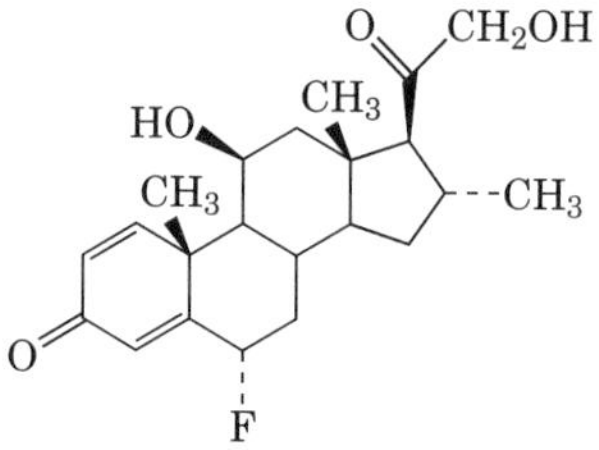

(103)

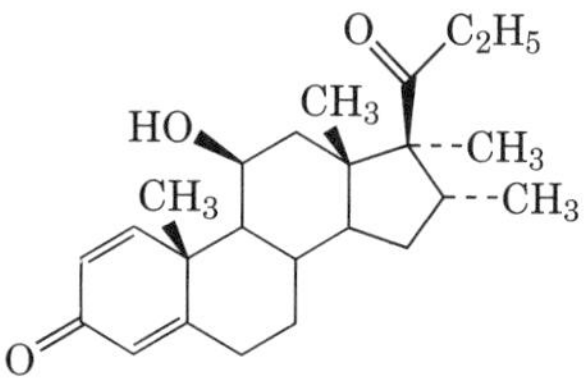

(104)

rapid elimination (100). The required esters (**105**, R = alkyl) were readily prepared by oxidation of fluocortolone (**103**) with alcoholic cupric acetate (101) whereby a mixture of the 20α- and 20β-hydroxypregnanoic esters (**106**) was obtained followed by oxidation with manganese dioxide to give the esters (**105**). Biological studies reveal that the esters (**105**) are highly active in the (topical) vasoconstriction assay but are devoid of systemic antiinflammatory activity as shown by the liver glycogen, paw edema, thymus involution, and other assays, and they were without mineralocorticoid activity. The butyl ester (fluocortin butyl ester) (**105**, R = C_4H_9) was selected for clinical study as a topical corticosteroid devoid of systemic activity (102). Urinary metabolites of prednisolone likewise contain products with β-glycolic and β-glyoxalic side chains. The methyl ester (**106**) of the latter in the paw edema assay is least effective if administered orally or intraperitoneally and most effective if administered topically or subcutaneously (103).

Another series of antiinflammatory carboxylic acids that are derived from cortienic acid (**107**), a minor adrenal metabolite, has been described (104,105). Esterification of both the 17α-hydroxyl group and the carboxylic acid of (**107**) were required to develop a compound of high topical potency with low systemic activity. Peak activity was generally associated with a 17α-propionoxy group and a 17β-fluoromethoxy carbonyl (eg, (**108**)), or 17β-methoxycarbonyl residue.

The D-homo-corticoids represent the last group of antiinflammatory steroids worthy of mention (106).

In 1973 D-homo corticosteroids (**109–112**), eg, D-homo-9α-fluoroprednisolone acetate (**111**) were reported to have antiinflammatory activity (107).

Compounds such as 21-acetoxy-11β-fluoro-9α-chloro-17aα-hydroxy-D-homo-pregn-4-en-3,20-dione (**110**) had especially strong topical activity with weak systemic activity (108). Other preparations of D-homocorticoids included

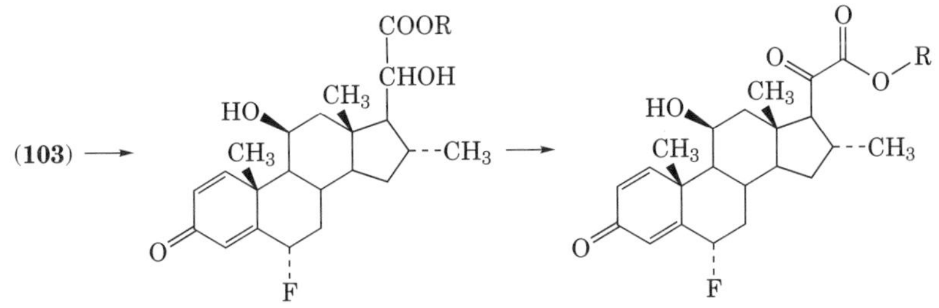

(**106**)   (**105**)

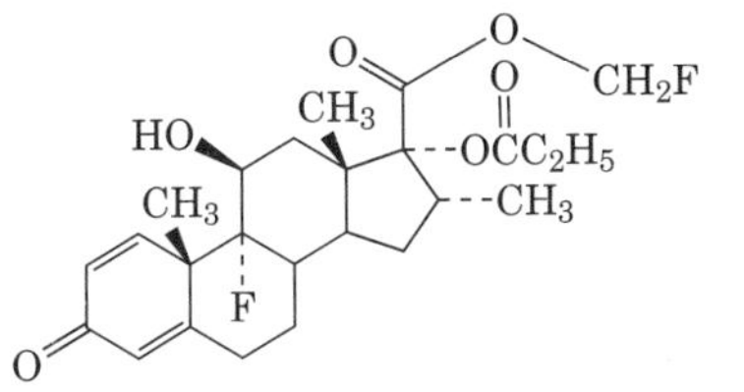

(**108**) 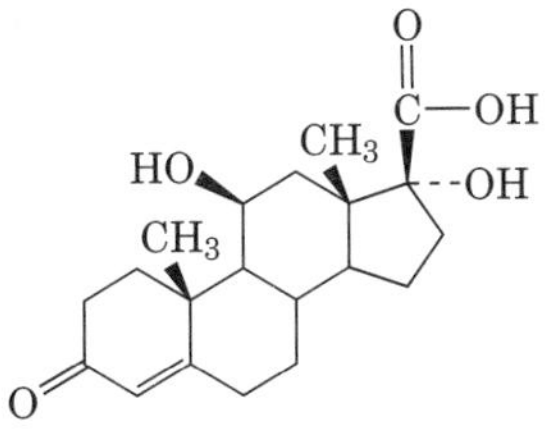   (**107**)

11β-hydroxy-3,20-dioxo-17α-methyl-D-homo-1,4-pregnadien-21-oic acid methyl ester (**111**), which was thought to have high topical antiinflammatory activity with insignificant systemic activity (109). An important result of the D-homocorticoid work was the development of a convenient synthesis of Δ^{17}-D-homopregnenolone (**113**) from the readily available 16α,17α-cyclomethylene pregnenolone (**114**) by acid-catalyzed rearrangement (110). Conversion of (**113**) into D-homocorticoids followed conventional routes. A number of patents covering D-homocorticoids had been issued and covers compounds such as 11β-hydroxy-17aα-propionyloxy-3-oxo-D-homoandrost-4-ene-17aβ-carboxylic acid chloromethyl ester (**112**) (111).

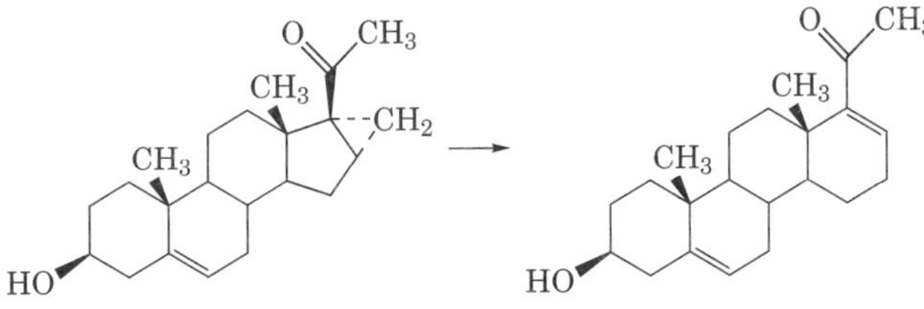

(**114**)   (**113**)

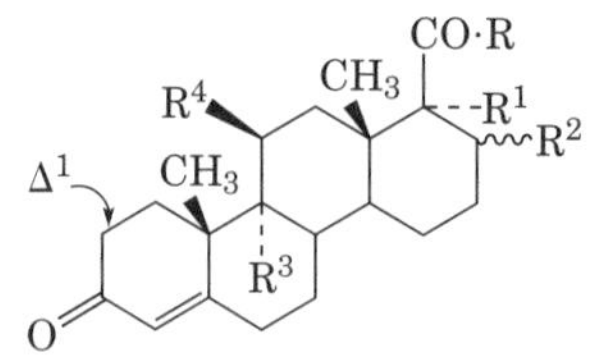

	R	R^1	R^2	R^3	R^4
(109)	Δ^1	CH$_2$OCCH$_3$ (O)	OH	H	F	OH
(110)		CH$_2$OCCH$_3$ (O)	OH	H	Cl	F
(111)	Δ^1	COCH$_3$ (O)	H	α-CH$_3$	F	OH
(112)		OCH$_2$Cl	OCC$_2$H$_5$ (O)	H	H	OH

The Mineralocorticoids

Studies have shown that aldosterone (**19**) (112,113) from adrenal extracts was intensely active in the survival and sodium retention assays in the adrenalectomized rat.

A photochemical partial synthesis of aldosterone (**19**) made the hormone available on an industrial scale for the first time (114). Corticosterone acetate (**51**; acetate) is treated with nitrosyl chloride in pyridine at 20°C to yield the 11-nitrite (**115**). Irradiation of (**115**) leads to rearrangement with formation of the C$_{18}$-oxime (**116**). Removal of the oxime residue with nitrous acid furnishes aldosterone (**19**) in excellent yield.

Hyperaldosteronism is accompanied by elevation of blood pressure (115), and can be treated with an aldosterone antagonist, eg, spironolactone (**117**) which

(17·acetate) → (**115**) → (**116**) → (**19**)

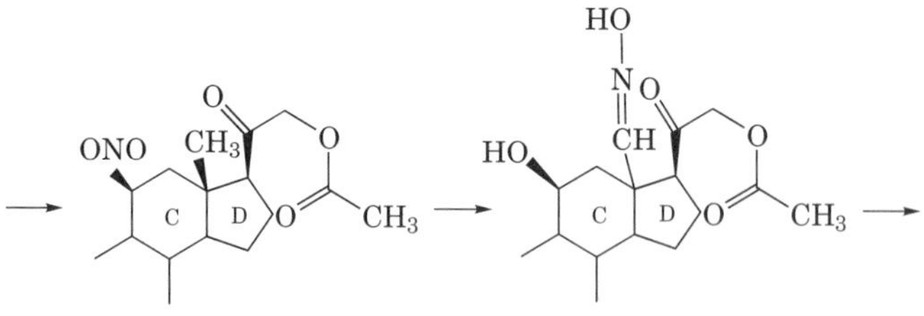

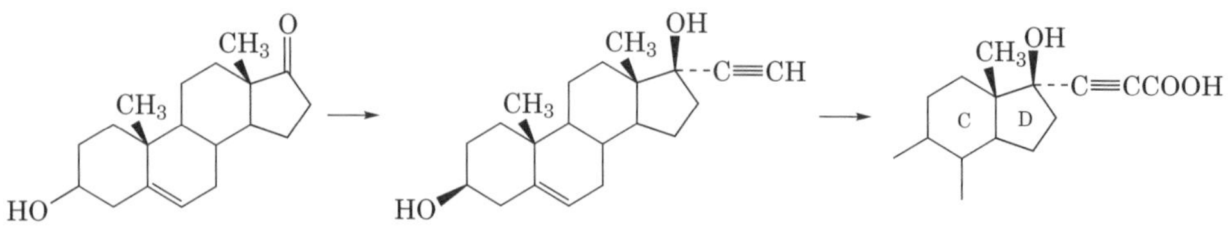

(**118**) → (**119**) → (**120**)

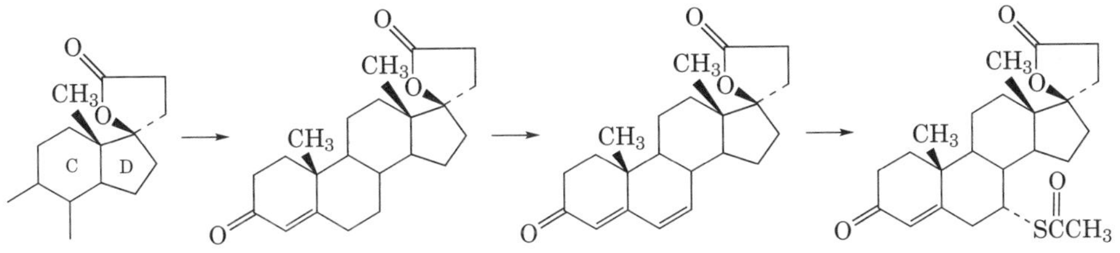

(**121**)   (**122**)   (**123**)   (**117**)

is synthesized from dehydroepiandrosterone (DHA) (**118**) (116). Ethynylation of DHA gives the 17β-hydroxy-17α-ethinyl derivative (**119**), which is converted using its ethynyl Grignard derivative and treatment with CO_2 into the 21-carboxylate (**120**). Hydrogenation leads to formation of the lactone (**121**), which is converted by the Oppenauer method into the 3-oxo-Δ^4-lactone (**122**). The last compound is treated with chloranil in *t*-butanol to give the 4,6-diene (**123**) which adds thioacetic acid to give spironolactone (**117**). The market for mineralocorticoids is extremely modest as their main value lies in the treatment of Addison's disease.

Adrenal-Cortical Steroid Antagonists

Antagonists of glucocorticoid and mineralocorticoid activity have found increased use clinically in the treatment of hypertension, Cushing's disease, and heightened intraocular pressure.

Some general structural features that lead to both observed differences in steroidal conformation (via x-ray structures) and to noticable effects in bioactivity (RBAs, agonism vs antagonism, magnitude of activity, etc) are the following: (*1*) Unsaturation of the A-ring: Alteration of the A ring conformation. Except for the A-ring diazoles, all the GR antagonist known have the 4-ene-3-one structure. As this is shared with the most potent glucocorticoids, it is thought that this structural feature increases binding affinity to the steroid receptor (117). (*2*) The hydroxy groups at C-11, C-17, and C-21, although not necessary for a compound to bind to the GR, seem to be individually or collectively responsible for agonistic activity (though some compounds with hydroxyls at one or more of these positions, (eg, RU-486 (**124**) and dexamethasone oxetanone (**125**), still act as GR antagonist). The absence of any of these hydroxyls may be one factor related to antagonistic activity (117), though not necessarily a determining factor. The lack of a hydroxyl at C-11 is a noticeable factor in the case of some GR antagonists. Where as all antiinflammatory glucocorticoid agonists have a hydroxyl at C-11 (or are readily solvolyzed to a C-11 OH), many antagonists though not all have no hydroxy moiety at C-11. For example, compounds that have shown antagonistic, eg, ZK 98299 (**126**), or mixed agonist activities, eg, RU-486 (**124**), each have substantially bulky groups at C-11.

An interesting comparison between agonist and antagonist to the GR is that of dexamethasone (**74**) (agonist) with dexamethasone oxetanone (**125**) (antagonist). The x-ray structures of these compounds indicate that all four rings overlap very well and thus should fit into the ligand binding site in a similar manner. As with differences in the magnitude of agonist activity, the agonist/antagonist distinction may be the result of chemical factors, including the hydrogen bond-donating or accepting ability of the ligand. In this example, both compounds can accept a hydrogen bond at C-20, though only the agonist is able to donate hydrogen bonds (117).

The most useful GR antagonist clinically is RU-486 (118,119), which also shows significant PR antagonism. This compound has been characterized more completely as a partial antagonist in certain cell lines (120,121), and has thus sometimes been called a type II antagonist. Although type II antagonist-occupied receptor can bind to DNA, it cannot interact productively with the DNA in a manner equivalent to interaction of an agonist-occupied receptor (122). RU-486 blocks the transactivation of the GR and inhibits the production of associated proteins, notably dexamethasone-induced extracellular LC-1 (55).

The defining feature of RU-486 is the 11β-aryl moiety. Most steroidal GR antagonists include 11β-aryl or other 11β substituents. The tight binding affinity of RU-486 to the GR is presumed to be in part the result of a tight fit of this 11β-aryl component in a lipophillic pocket within the GR hormone binding region. ZK 98299 is an 11β-aryl GR antagonist very similar in structure to RU-486, though it is a pure (type I) antagonist. Like RU-486, it also has antiprogestational activity. The 11β-N-methyldihydroindol-5-yl analogue of RU-486 (**127**), another compound with GR antagonistic properties and a strong GR binding affinity, has a rotomer which fits well in the space (11β-pocket) occupied by the 11β-substituent of RU-486.

(**124**)   (**126**)   (**127**)

Another class of glucocorticoid antagonists, the 10β-androstanes, eg, RU-39305 (**128**) and RU-43044 (**129**), also bind tightly to the GR (Table 2). Relative binding affinitys (RBA) to the GR (vs dexamethasone) do not lead to clear SARs in regard to unsaturation at $\Delta 1$ or $\Delta 6$. However, space-filling models indicate a similarity between the 3-D space occupied by the 10β- and 11β-compounds (123), although the aromatic moiety of RU-39305 (**128**) is not in the same orientation as that of RU-486. These compounds have very little affinity for the progesterone receptor (PR).

Research conducted by Simons using antiglucocorticoids, including compounds which covalently bind to the GR (124), eg, dexamethasone 21-mesylate, has better defined the structure and function of the GR. Spiro C-17 oxetanes have shown potent antiglucocorticoid activity in whole cell systems (125,126).

Table 2. Relative Binding Affinities of 10β-Androstanes[a]

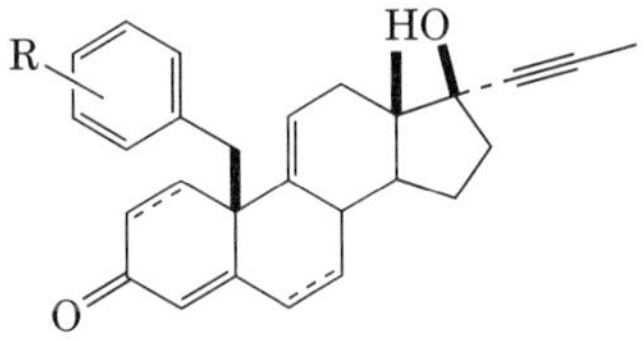

RU Code (Structure number)	Unsaturation	R	Thymus GR RBA[b]	PR RBA	Thymocytes IC$_{50}$ nM
39305		H	57	<0.1	100
43044		p-CH$_3$	130	<0.2	200
43065		p-CH$_3$	17	<0.1	500
46759	Δ1, Δ6	p-CH$_3$	33	<0.1	500
44068	Δ1	p-CH$_3$	15	<0.1	>1000
44427	Δ6	p-CH$_3$	3	<0.1	1000

[a] Adapted from Ref. 123.
[b] Dexamethasone = 100.

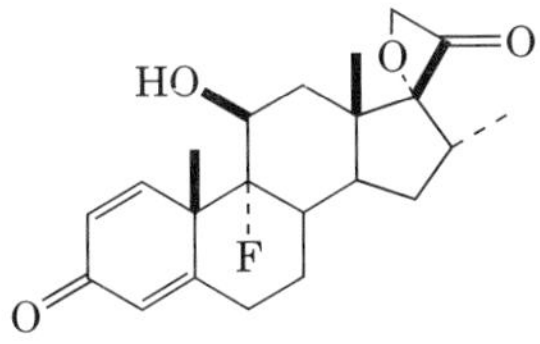

(**125**)

Although both the 10β- and the 11β-group have been found to be important in connection with many GR antagonists, neither is a requirement for antiglucocorticoid activity. Cortexolone (**15**) acts as a GR antagonist in the true sense of the term, blocking GR transactivation (127–129) although it does not include an 11β-substituent.

Spironolactone is the most clinically useful steroidal aldosterone antagonist, and unlike GR antagonists, this compound is utilized much more frequently than aldosterone agonists. Interfering with Na$^+$ reabsorption and K$^+$ secretion in the late distal segment, this compound is predominantly used with other diuretics. Canrenone, an olefinic metabolite of spironolactone, and potassium canrenoate, in which the C-17 lactone has been hydrolyzed open, are also potent mineralocorticoid antagonists.

BIBLIOGRAPHY

"Adrenal Cortical Hormones" under "Hormones" in *ECT* 1st ed., Vol. 7, pp. 495–513, by H. B. MacPhillamy, Ciba Pharmaceutical Products and T. F. Gallagher, Sloan-Kettering

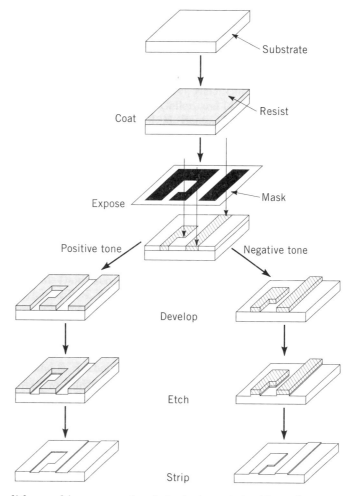

Fig. 1. The lithographic process. A substrate is coated with a photosensitive polymer film called a resist. A mask with transparent and opaque areas directs radiation to preselected regions of the resist film. Depending on resist characteristics, exposed or unexposed portions of the film are removed using a developer solvent. The resulting pattern is then transferred to the substrate surface and the resist is stripped.

Lithographic resists enable a host of technological advances that have had far-ranging impacts. For example, the rapid progress in microelectronics technology in large part stems from refinement of the lithographic techniques used to fabricate computer integrated circuits. The ability to pack ever-increasing numbers of discrete electronic components in a given area of an integrated circuit, when coupled with the economies of scale that result from high volume manufacturing, has led to both improved performance and lower cost with each succeeding generation (5). Today the annual global market for semiconductors far exceeds $100 billion U.S.

The advantages of miniaturization are now being exploited in areas beyond microelectronics. Adaptation of materials and processes originally devised for semiconductor manufacture has allowed fabrication of sensors (for example, pressure meters and accelerometers used in the automotive industry) (6,7), complex optical (8) and micromechanical (6,7,9) assemblies, and devices for medical diagnostics (6,7,10) using lithographic resists.

Essential Attributes of Lithographic Resists

Regardless of the specific application, all resists must display certain fundamental functional properties:

1. The resist composition must form *uniform, defect-free films* on the substrate of interest. To achieve this, the resist components must be soluble in and inert toward the selected coating solvent or medium, the formulation must form glassy films without component precipitation, separation or dewetting, and must be stable during post-application heating steps. A variety of different coating techniques have been applied, including spray-coating, curtain-coating, spin-coating or whirling, dip-coating, dry-film lamination, and most recently, electrodeposition onto a conducting substrate from an aqueous emulsion of resist components.

2. The coated film must display adequate *adhesion* to the substrate after coating, and through the develop and image transfer steps. Adherence is influenced by resist composition, substrate surface composition, preparation and cleanliness, coating conditions, develop and etch process details, and, with negative crosslinking resists, by the exposure dose. Adhesion loss in mild cases is manifested by the lifting of a resist pattern at its perimeter (often visible under microscopic examination as a contrasting band) and in extreme cases by complete detachment of the resist pattern.

3. The resist must have suitable *radiation sensitivity*. Today's exposure tools are so costly that tool throughput is a key measure of performance. The overall time to expose a resist film is the sum of the times to load and position the substrate in the exposure tool, to align the substrate and the mask, to irradiate the film, and to unload the complete part. In the optimum case the resist exhibits sufficient radiation sensitivity so that the fraction of the overall cycle apportioned to irradiate the film does not limit the number of substrates exposed in a given period of time.

4. The resist must provide *high fidelity reproduction* of the mask image on the substrate. This attribute is generally quantified as the resist's resolution, that is, the smallest feature that can be consistently printed, and its contrast, which is a fundamental measure of how the resist response changes with increasing radiation dose. In high resolution lithography, diffraction and imperfections in the optical system cause stray radiation to be scattered into regions of the resist pattern that nominally are unexposed. A resist with high contrast can effectively reject the stray radiation and yield a more accurate replica of the original mask. Resist contrast and resolution are influenced by its composition, by process conditions, and exposure tool characteristics.

5. The patterned resist must provide an effective *protective barrier* during image transfer. The image transfer step frequently includes harsh chemi-

cal or physical treatments that erode or otherwise degrade the resist pattern. A resist's capacity to withstand the more vigorous image transfer processes (eg, plasma etching) can vary widely with its composition.

6. After image transfer, the patterned resist must be readily and completely *removable* without substrate damage. The pattern often can be stripped from the substrate with a mild organic solvent. Proprietary stripper formulations or plasma oxidation treatments are utilized when the imaging chemistry or image transfer process has insolubilized the pattern.

Historical Development of Resist Materials

Most modern lithographic resists are evolved from materials first developed for the printing industry (11). Increasingly specialized resists and processing techniques have been introduced as applications of lithographic technology have grown in scope and sophistication. Figure 2(**a**) presents a time-line noting significant events in this evolution; many of these are described in detail in this section. Figure 2(**b**) shows one representation of Moore's Law (5), which traces the evolution of semiconductor lithographic technology, as characterized by the minimum feature size versus the year of commercial introduction of the corresponding device generation.

Dichromated Resists. The first compositions widely used as photoresists combine a photosensitive dichromate salt (usually ammonium dichromate) with a water-soluble polymer of biologic origin such as gelatin, egg albumin (proteins), or gum arabic (a starch). Later, synthetic polymers such as poly(vinyl alcohol) also were used (11,12). Irradiation with uv light (λ in the range of 360–380 nm using, for example, a carbon arc lamp) leads to photoinitiated oxidation of the polymer and reduction of dichromate to Cr(III). The photoinduced chemistry renders exposed areas insoluble in aqueous developing solutions. The photochemical mechanism of dichromate sensitization of PVA (summarized in Fig. 3) has been studied in detail (13).

Poly(vinyl cinnamate) Resists. Dichromated resists exhibit numerous shortcomings which include lot-to-lot variability of the components, aging of the formulated resists in solution and in coated form, poor process stability (due to a sensitivity to variations in temperature and humidity), and intrinsically low photosensitivity requiring long exposure times for adequate insolubilization.

In the 1950s resist systems with substantially improved processing characteristics were developed. The first commercially available member of this class, KPR, was introduced in 1953 by Eastman Kodak. Originally targeted for printing applications, KPR was used in the fabrication of circuit boards and semiconductor devices later in the decade. This material is a crosslinking system based on the photodimerization of poly(vinyl cinnamate) chains. Pendant carbon–carbon double bonds on adjacent polymer strands undergo photocyclization to form a crosslinked, insoluble network in exposed areas (Fig. 4). Although poly(vinyl cinnamate) is intrinsically photoactive, its electronic absorption peaks near a wavelength of 290 nm, and absorbs only weakly at wavelengths longer than 350 nm, where the lamps commonly used in exposure tooling (eg, a high pressure mercury arc) show strong emission. Its spectral sensitivity can be extended to longer wavelengths by addition of an appropriate absorbing dye. The dye is believed to

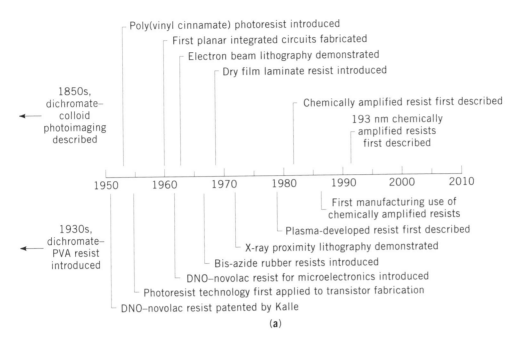

(a)

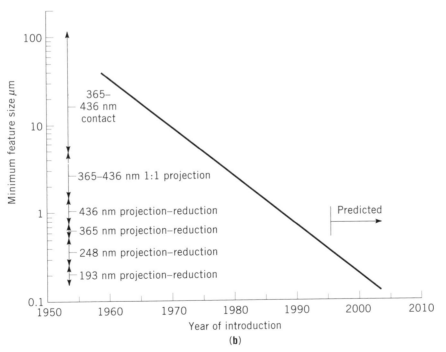

(b)

Fig. 2. Evolution of lithographic technology. (**a**) A time line noting significant events in the evolution of lithographic technology. (**b**) The minimum feature size used in a commercial microelectronic device is plotted against its year of introduction. Also indicated are the exposure tool technologies used to produce the different device generators. In projection lithography, printed resist features can be either the same size as those on the mask (1:1), or can be a factor of 4−10 smaller (reduction).

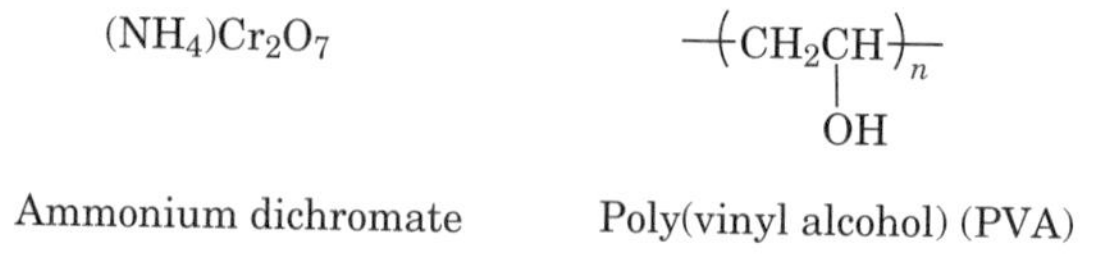

$$(NH_4)Cr_2O_7$$

$$-(CH_2CH)_n-$$
$$|$$
$$OH$$

Ammonium dichromate   Poly(vinyl alcohol) (PVA)

$$Cr^{6+} \xrightarrow{h\nu} [Cr^{6+}]^*$$

$$[Cr^{6+}]^* + PVA \longrightarrow Cr^{5+} + [PVA\cdot]$$

$$Cr^{5+} + [PVA\cdot] \xrightarrow{h\nu} Cr^{3+} + \text{Crosslinked PVA}$$

Fig. 3. Chemistry of dichromated poly(vinyl alcohol) resist. Initially the dichromate ion absorbs light; the light-activated Cr^{6+} species undergoes an electron-transfer reaction with the PVA matrix to form a polymer radical. These undergo further reactions to form Cr^{3+} and a crosslinked matrix (13).

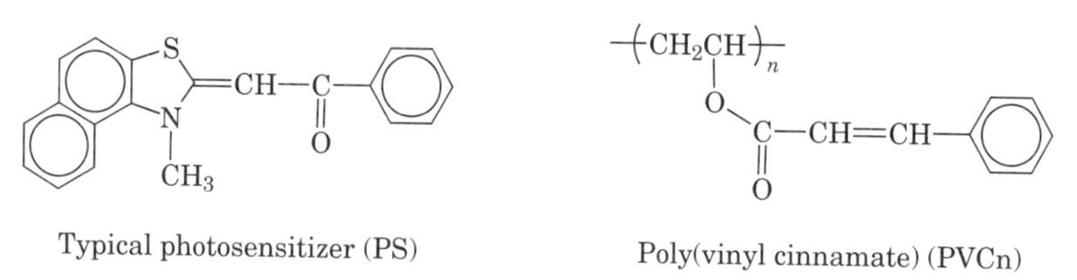

Typical photosensitizer (PS)   Poly(vinyl cinnamate) (PVCn)

$$PS \xrightarrow{h\nu} {}^1[PS]^* \xrightarrow{ISC} {}^3[PS]^*$$

$${}^3[PS]^* + PVCn \longrightarrow PS + {}^3[PVCn]^*$$

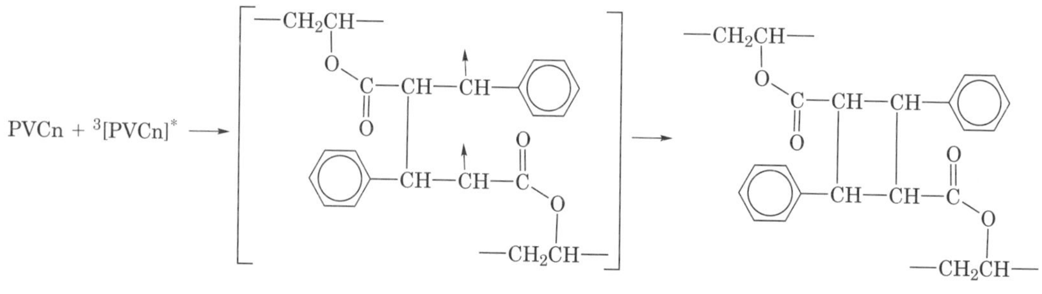

$$PVCn + {}^3[PVCn]^* \longrightarrow$$

Fig. 4. Chemistry of poly(vinyl cinnamate) negative-acting resist. Initial light absorption by the photosensitizer is followed by energy transfer to produce a pendant cinnamate group in a triplet electronic state. This combines with a second cinnamate on another polymer chain, forming a polymer–polymer linkage upon deactivation.

act as a triplet photosensitizer, transferring excitation energy to a cinnamate acceptor. The intermediate triplet cinnamate adds to a ground state cinnamate to form a biradical species, which closes to produce the final crosslinked cyclic structure.

Resists Based on Bis-Azide Crosslinking Photochemistry. Negative-acting resist compositions that combine an unsaturated hydrocarbon polymer derived from polyisoprene with an organic aromatic bis-azide were introduced in the 1960s. These resists exhibited photosensitivity and adhesion significantly improved over available poly(vinyl cinnamate) systems. Numerous commercial formulations based on this chemistry have been marketed. The polymers in these resists are prepared by treating natural rubber derived from latex, or synthetic poly(isoprene), with a metal or acidic catalyst to induce partial ring formation along the chain (Fig. 5(**a**)). The resulting "cyclized rubber" products displays improved adhesion and flow characteristics with a softening temperature of about 65°C, properties suitable for thin-film coating applications (14).

During exposure of the formulated resist, azide groups in the sensitizer undergo sequential loss of molecular nitrogen. Each fragmentation produces a highly reactive nitrene with two unpaired electrons. The nitrene rapidly deactivates by a number of pathways (Fig. 5(**b**)). Crosslinking results when nitrene intermediates add to residual carbon–carbon double bonds or insert into carbon–hydrogen bonds, converting the matrix into an insoluble network.

First introduced for printing applications, bis-azide–rubber resists were soon adopted by the developing microelectronics industry and used extensively for device fabrication through the 1970s. However, as the semiconductor industry

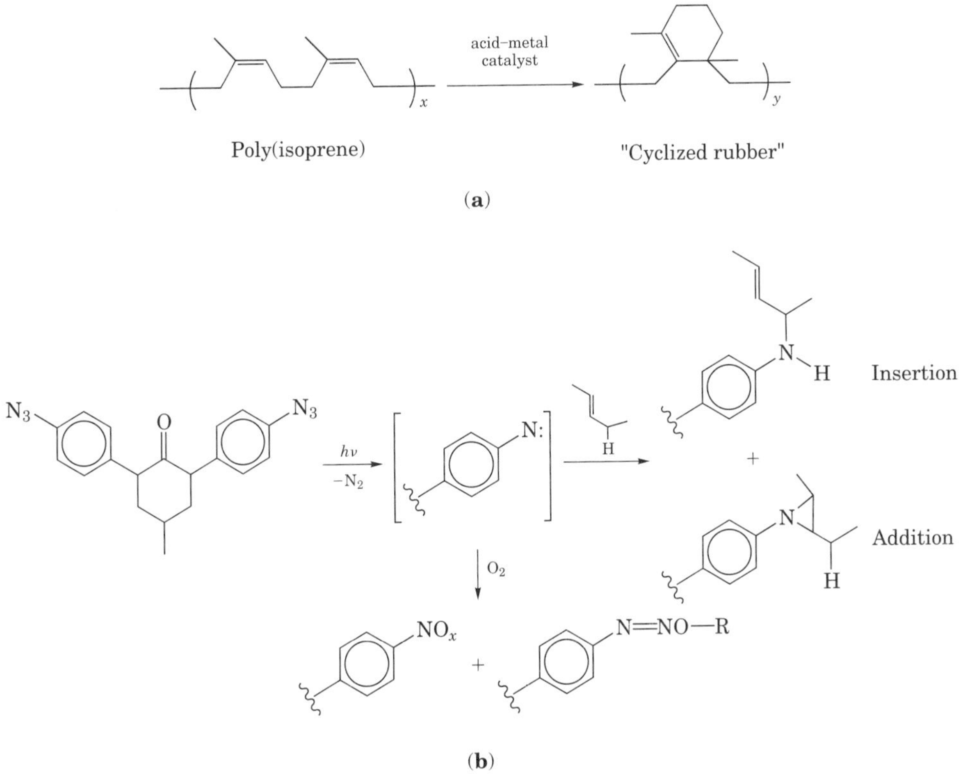

Fig. 5. Chemistry of cyclized rubber–bis-azide negative acting resist. (**a**) Preparation of cyclized rubber resin from polyisoprene; (**b**) photochemistry of aromatic bis-azide sensitizers. The primary photoproduct is a highly reactive nitrene which may combine with molecular oxygen to form oxygenated products, or may react with the resin matrix by addition or insertion to form polymer–polymer linkages.

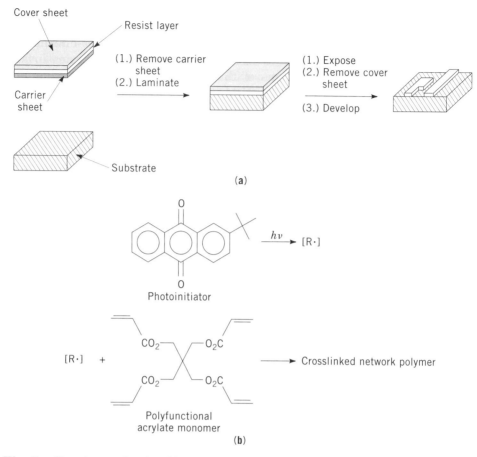

Fig. 7. Chemistry of a dry-film, negative-acting photoresist. (**a**) Polymerizable layer sandwiched between a polyolefin carrier sheet and a polyester cover sheet. (**b**) Chemical structures of typical components.

where plated through-holes serve as electrical conductors to join levels of circuitry (Fig. 8).

As PWB technology is refined to provide greater integration using finer conductor lines, there is renewed interest in liquid resists. The absence of a cover sheet and the ability to apply thinner films both contribute to improved resolution and to an intrinsically lower consumables cost (16,17).

Resist Materials for Imaging with Ionizing Radiation. Although most lithographic processing uses light in the visible to uv wavelength region (known as photolithography), a number of specialized, low volume applications employ high energy radiation and have unique resist requirements as a result. These include very high resolution lithography techniques using beams of electrons or x-rays to induce chemical changes in the resist film. Electron beam or e-beam lithography is used extensively for the fabrication of high resolution photomask patterns, and for direct writing or circuit patterns for low volume, custom microelectronic applications (18). The applicability of proximity-mode x-ray lithography in large-scale device fabrication has been extensively researched (19). To

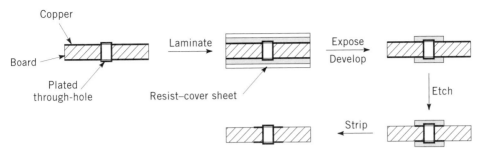

Fig. 8. Protection of a plated through-hole in a PWB during etching by using a tenting dry-film photoresist.

date, however, photolithography has been extended to image resolutions well beyond those first anticipated a decade ago, deferring a conversion to x-ray tooling. At present the principal commercial application of x-ray lithography is the LIGA (a German acronym for lithography–electroplating–molding) process for forming micromechanical parts (6).

In e-beam lithography, a finely focused beam of high energy electrons is scanned in a pattern across a resist-coated substrate. Collisions of impinging (primary) electrons and secondary electrons with atoms of the resist film produced ionic and radical species that undergo further reaction, modifying properties of the surrounding matrix. In x-ray lithography, initial absorption of a photon produces an energetic photoelectron whose collisions with atoms of the resist lead to the same intermediates produced by e-beam radiation. For this reason there tends to be a correlation between the sensitivity of a resist material for electron beam and x-ray lithography. However, the chemistry differs substantially from that caused by nonionizing (ie, long wavelength ultraviolet to visible) radiation, so that there is no assurance that a resist designed for photolithography will be found suitable for these methods.

Many examples of positive- and negative-tone x-ray and e-beam resists have been described (1,20,21). A number of these are single-component systems consisting of a polymer with intrinsic radiation sensitivity. If radiation-induced crosslinking is the predominant reaction, then the resist is negative-acting. Conversely, if chain scission and/or depolymerization are the principal reactions, then the polymer functions as a positive-tone resist. One representative example is a positive-tone resist developed by Bell Laboratories in the early 1970s that has been widely applied to commercial photomask manufacture by electron-beam lithography (22). This resist is the polymer poly(butene-1-sulfone) (PBS), an alternating copolymer of butene and sulfur dioxide (Fig. 9). Electron-beam irradiation of PBS is believed to produce a cation radical center initially that destabilizes the chain. Scission at that site creates radical- and cation-terminated polymer strands which may undergo further fragmentation before deactivation.

Modern Resists for Microlithography

Today the market for photoresists used in the manufacture of integrated circuits is estimated to be $750 million U.S. annually, with diazonaphtho-

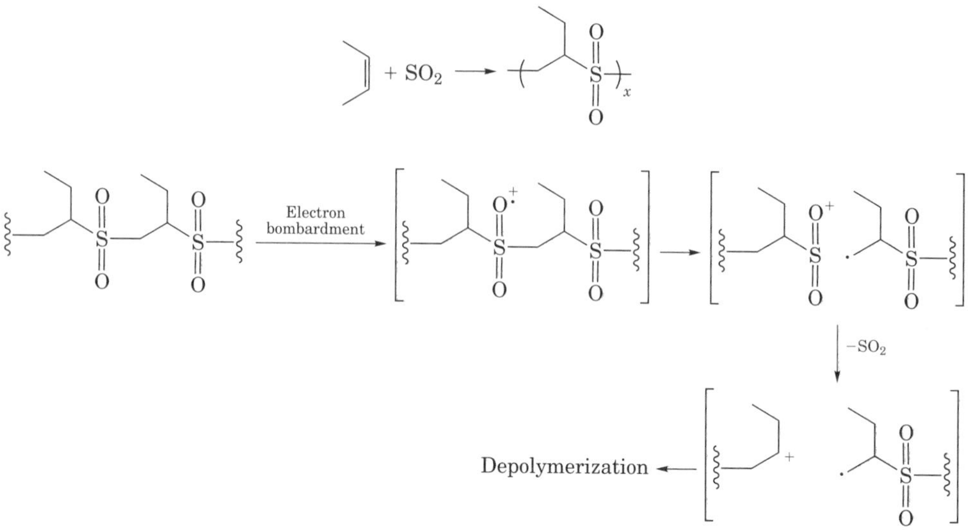

Fig. 9. Chemistry of PBS electron beam resist.

quinone–novolac resist products accounting for almost all of this. Newly introduced commercial resist systems designed for deep-ultraviolet applications are now being used to build leading edge semiconductor products, however, and forecasts suggest that use of such advanced materials will grow several-fold over the next five years.

Positive-Tone Photoresists based on Dissolution Inhibition by Diazon-aphthoquinones. The intrinsic limitations of bis-azide–cyclized rubber resist systems led the semiconductor industry to shift to a class of imaging materials based on diazonaphthoquinone (DNQ) photosensitizers. Both the chemistry and the imaging mechanism of these resists (Fig. 10) differ in fundamental ways from those described thus far (23). The DNQ acts as a dissolution inhibitor for the matrix resin, a low molecular weight condensation product of formaldehyde and cresol isomers known as novolac (24). The phenolic structure renders the novolac polymer weakly acidic, and readily soluble in aqueous alkaline solutions. In admixture with an appropriate DNQ the polymer's dissolution rate is sharply decreased. Photolysis causes the DNQ to undergo a multistep reaction sequence, ultimately forming a base-soluble carboxylic acid which does not inhibit film dissolution. Immersion of a patternwise-exposed film of the resist in an aqueous solution of hydroxide ion leads to rapid dissolution of the exposed areas and only very slow dissolution of unexposed regions. In contrast with crosslinking resists, the film solubility is controlled by *chemical* and *polarity* differences rather than molecular size.

DNQ–novolac resists exhibit several important practical attributes. First, DNQ photochemistry is not inhibited by oxygen, so unlike free-radical based systems the resist can be exposed in noncontact modes with unimpaired imaging properties. Second, films of the resist dissolve in aqueous base by a surface-limited etching reaction, with no evidence of swelling. Presumably this is because only the ionic, deprotonated phenolate form of the polymer is sufficiently polar

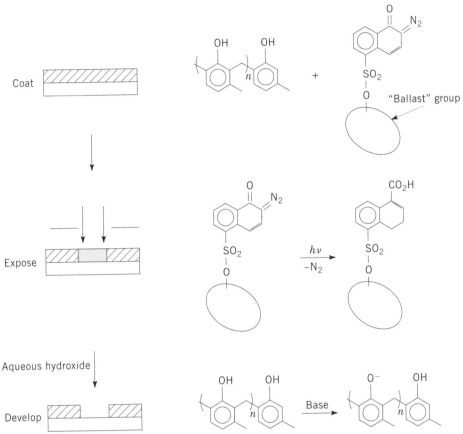

Fig. 10. Chemistry of diazonaphthoquinone–cresol novolac positive-acting resist. During patternwise exposure, the DNQ undergoes photolysis that destroys its inhibitory effect on film dissolution.

to be solubilized by the aqueous developing solvent. Finally, the developing solvent is nonflammable and water-based, an advantage from the viewpoints of workplace safety and materials handling and disposal.

DNQ–novolac resist chemistry has proved to have remarkable flexibility and extendibility. First introduced for printing applications, DNQ–novolac resists have been available since the early 1960s in formulations intended for electronics applications. At present, most semiconductor manufacturing processes employ this resist chemistry. Careful contemporary research and engineering support the continuing refinement of this family of materials.

DNQ Synthesis and Properties.   DNQ photosensitizers are synthesized by base-catalyzed condensation of a diazonaphthoquinone sulfonyl chloride with a mono- or polyhydroxy species to produce a sulfonate ester (Fig. 11). Since the structures of the photolabile diazonaphthoquinone group and the photoinert ballast can be readily and independently changed using this route, many DNQ variants have been prepared. Modification of the diazonaphthoquinone group in large part has been restricted to positioning the sulfonate functionality on either the 4- or 5-location. The location of the sulfonate influences the chromophore's

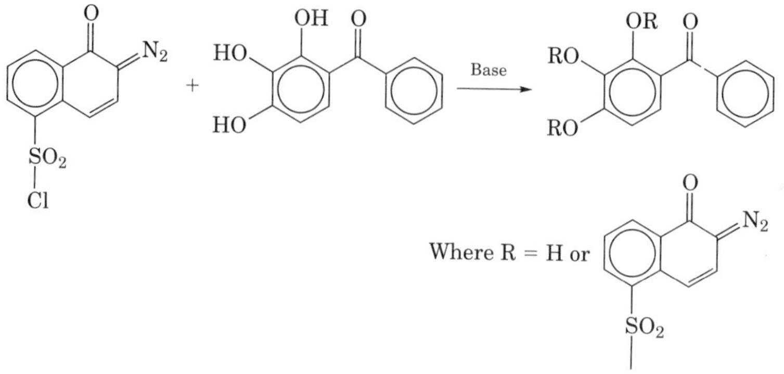

(a)

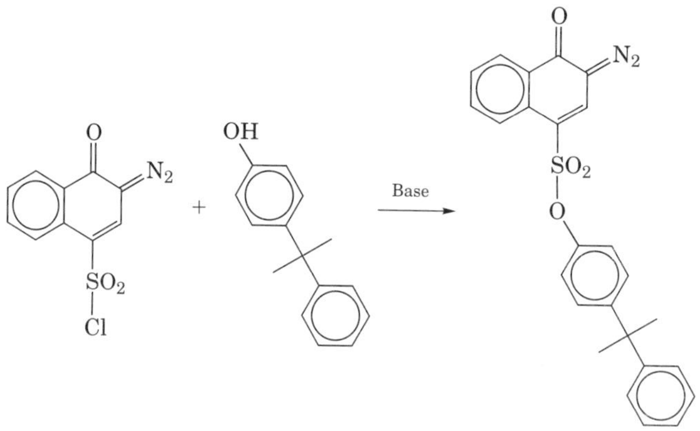

(b)

Fig. 11. Synthesis of DNQ photosensitizers found in commercial resists. (**a**) Condensation of 1,2-diazonaphthoquinone-5-sulfonyl chloride with 1,2,3-trihydroxybenzophenone. Often the reaction is not carried to completion so the product is a mixture of mono-, di- and trisubstituted products. (**b**) Condensation of 1,2-diazonaphthoquinone-4-sulfonyl chloride with *p*-cumylphenol.

absorption properties, with the long-wavelength absorption (λ_{max} of ~380 nm) of the 4-sulfonate blue-shifted in comparison to the 5-sulfonate isomer (λ_{max} of ~400 nm). More significantly, the substitution pattern can influence the course of the photoinduced reaction. Like the 5-sulfonate isomers, a 4-sulfonate DNQ first undergoes a Wolff rearrangement to form a carboxylic acid upon photolysis. This initial product is labile, however, and the 4-sulfonate analogue can undergo further facile hydrolysis, forming a strongly acidic sulfonic acid as the final product (Fig. 12). This has been exploited in the design of an "image-reversing" resist, where sulfonic acid catalyzes an insolubilizing reaction if an exposed film is heated (25).

Structural variation of the ballast or backbone group has been explored in far greater depth. By correlating structure and lithographic properties, researchers have identified numerous factors that can influence resist performance. These include the number of diazonaphthoquinone sulfonate groups per

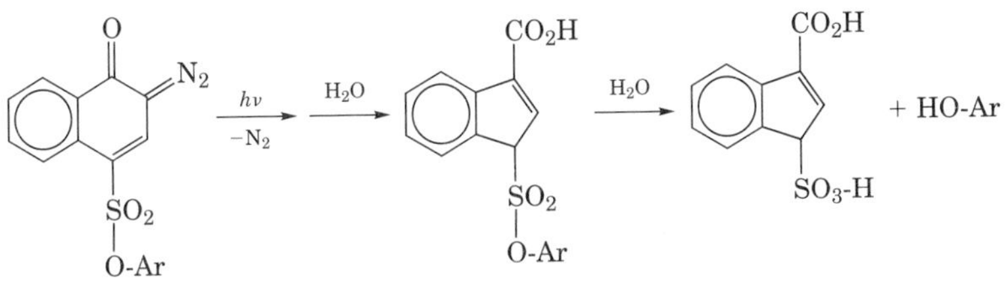

Fig. 12. Photo-induced chemistry of a 4-sulfonyl DNQ. The intermediate species reacts with adventitious water in the resist film to produce a sulfonic acid and a phenol.

sensitizer molecule (26), the spatial proximity of those groups (27), the backbone's size and the degree of esterification of polyfunctional DNQs (28,29), backbone hydrophobicity (28), and DNQ's capacity for hydrogen-bonding with the phenolic matrix resin (30).

    Novolac Synthesis and Properties.   Novolac resins used in DNQ-based photoresists are the most complex, the best-studied, the most highly engineered, and the most widely used polymers in microlithography. Novolacs are condensation products of phenolic monomers (typically cresols or other alkylated phenols) and formaldehyde, formed under acid catalysis. Figure 13 shows the polymerization

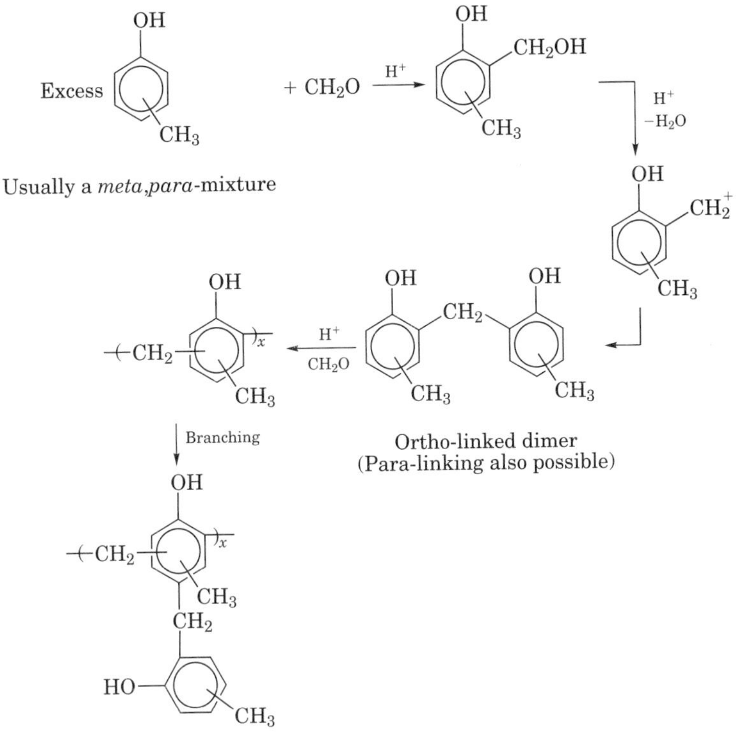

Fig. 13. Polymerization chemistry of phenol–formaldehyde condensation: synthesis of novolac resin. The phenol monomer(s) are used in stoichiometric excess to avoid gellation, although branching invariably occurs due to the multiple reactive sites on the aromatic ring.

chemistry and polymer structure formed in the step growth polymerization (31) of novolac resins.

The polymerization mechanism is thought to involve electrophilic attack on the aromatic ring at the positions ortho and para to the phenolic hydroxyl group (23). Initially, formaldehyde adds to the ring, forming a benzyl alcohol. In the second step, protonation of the alcohol and loss of a water molecule produces a benzyl cation that can undergo electrophilic substitution with another aromatic ring to form a dimeric product. Repeated reaction of the products with formaldehyde initiates a sequence of condensations to generate novolac polymer.

Molecular Weight Effects. In step-growth polymerization, high molecular weight product is formed only when the following two requirements are met: (*1*) strictly difunctional monomers are used in a 1/1 stoichiometric ratio, and (*2*) the extent of conversion of monomer to polymer approaches 100% (31). In such a case, classical step-growth kinetics would predict a most probable distribution of molecular weights with a polydispersity of 2.0. In the case of novolac synthesis, however, the phenolic monomer is not perfectly difunctional, but more likely should be considered trifunctional. One result of this "excess" functionality is that branching occurs early in the polymerization (Fig. 13). If the reaction is brought to a high degree of conversion, gellation and crosslinking result; the product is unusable (and difficult to remove from the reactor, as these polymerizations are often run in the bulk using no polymerization solvent). To avoid this, the polymerization is run with excess phenol and is not brought to high conversion.

The combined effects of branching, which leads to the formation of high molecular weight products, and the low overall extent of conversion, which favors formation of low molecular weight oligomers, yield a polymerization product of some complexity. As synthesized, novolacs have a very low average molecular weight, with a multimodal molecular weight distribution. Figure 14 shows a size-exclusion chromatogram of a typical novolac resin, illustrating the multimodal molecular weight distribution characteristic of this class of polymer.

Figure 15 graphically shows the well-known general relationship between polymer properties and molecular weight. Most polymers in practical use inhabit the plateau region of this curve, where properties change only slightly with

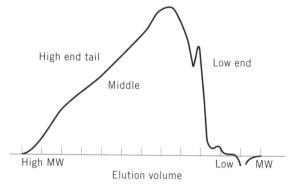

Fig. 14. Molecular weight characteristics of novolac resins. Shown is the size exclusion chromatogram for a typical commercial novolac polymer. The unsymmetrical peak shape reflects the multimodal molecular weight distribution of the polymer.

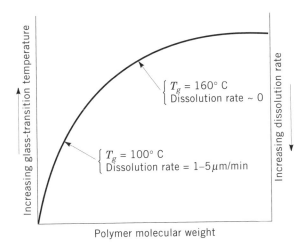

Fig. 15. Properties of novolac fractions are a strong function of molecular weight.

molecular weight. Novolac resins, however, reside on the early portion of this curve, where a slight change in molecular weight strongly influences material properties. The broad molecular weight distribution, coupled with low average molecular weight, ensures that portions of the novolac mixture exhibit the full range of properties. In Figure 14, three main molecular weight regions can be distinguished: a region that contains small oligomers, a high molecular weight tail with $M_n > 4000$, and a midrange portion between these extremes. Isolation and examination of each of these individual components, a difficult task by traditional methods, has recently been achieved using supercritical fluid fractionation (32). This study showed that a typical novolac contains 25–50% high molecular weight polymer that in isolated form is insoluble in aqueous alkaline developer solutions, and 15–25% oligomeric liquid that dissolves essentially instantaneously in alkaline developer. Examination of materials properties of the near-monodisperse novolac fractions demonstrates the pronounced influence of molecular weight. For example, thermal analysis of the separated novolac fractions reveals a substantial change in glass-transition temperature (from below room temperature to above 160°C) depending on whether the low or the high end of the molecular weight distribution is chosen.

As one might anticipate, lithographic properties of DNQ–novolac resists are strongly influenced by polymer molecular weight and molecular weight distribution. For example, in one study, dissolution rates of both exposed (R_e) and unexposed (R_u) resist films were shown to decrease with increasing molecular weight, though the ratio of these rates R_e/R_u, a measure of contrast, was found to be independent of molecular weight (33). This same study also found that molecular weight distribution influences resist contrast, which improves with increasing polydispersity until a maximum contrast is attained. In recent work, a series of experimental DNQ-based resist formulations were prepared by systematically blending solvent-fractionated (34) or supercritical fluid fractionated (32) novolacs. By comparing lithographic imaging results for all the polymer mixtures, the influence of individual fractions on photospeed, contrast, and post-development residue and resistance to thermal flow were established. From this and related work has emerged a set of design principles that facilitate more precise optimization of resin molecular weight in resist formulations.

Structural Effects. The molecular structure of novolac resins can be readily manipulated by varying the mixture of phenolic monomers in the polymerization feed, and by selecting a different polymerization catalyst. Novolacs used in DNQ-based resists often are prepared from a mixture of cresol isomers, in particular a mixture of *para*- and *meta*-cresols, rather than a single component. The polymerization chemistry is sufficiently general and flexible such that a range of phenolic homologues and derivatives have been incorporated into novolacs with the goal of improving lithographic properties (23).

Numerous studies have probed how novolac microstructure influences resist lithographic properties. In one example, a series of resists were formulated from novolacs prepared with varying feed ratios of *para-/meta*-cresol. These researchers found that the dissolution rate decreased, and the resist contrast increased, as the *para-/meta*-cresol feed ratio increased (33). Condensation can only occur at the *ortho* position of *para*-cresol, but can occur at both the *ortho*- and *para*- positions of *meta*-cresol. It is believed that increased steric factors and chain rigidity that accompany increased *para*-cresol content modify the polymer solubility.

Deep-Ultraviolet Chemically Amplified Resists based on Acid Catalysis. In any optical imaging system, the size of the smallest element that can be accurately resolved is related to the wavelength of exposing light: the smaller the wavelength, the finer the feature that can be resolved (35). In the microlithographic arena, improved resolution is achieved by incrementally shifting the exposure wavelength to smaller values as refinements in optics, tooling and process technology permit. The microelectronics industry currently is beginning a transition from exposure tools designed to use monochromatic light at $\lambda = 436$ and 365 nm (two strong emission lines in the spectrum of mercury arc lamps) to more advanced exposure tools that use light at 248 nm (the output wavelength of a krypton fluoride excimer laser), the deep-ultraviolet or duv region. This wavelength shift has profound implications from the viewpoint of resist design. DNQ–novolac resists, now the industry standard, are impractical for duv photolithography for two reasons.

The first can best be illustrated by consideration of resist optical properties. Figure 16 shows the optical absorption spectrum of a typical DNQ–novolac photoresist film before and after exposure with 365 nm light. The absorption band at long wavelengths (300–450 nm) is that of the DNQ sensitizer. As the DNQ is photolyzed this characteristic absorption decreases. The increased transparency facilitates efficient photolytic conversion of the sensitizer throughout the depth of the film. If such a film is exposed at 248 nm, however, the strong nonbleaching absorbance at the wavelength (due in part to the novolac matrix polymer, and in part, to the DNQ and its photoproduct) sharply attenuates the beam as it passes through the film. The result is highly nonuniform photolysis of photosensitizer in the resist film, with insufficient conversion near the substrate interface. Figure 17 illustrates the damaging effect on resist imaging properties.

Second, the overall brightness of light sources available for duv exposure is much less than that of mercury arc lamps used for 365 nm wavelength. Though KrF excimer lasers are generally regarded as powerful sources of uv light, the

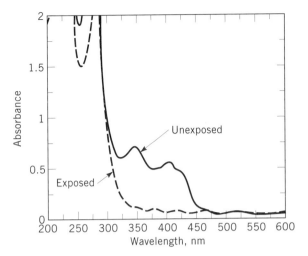

Fig. 16. Optical transmittance spectrum of a typical DNQ–novolac resist. (——), The transmittance of a 1 μm film after coating. (— — —), the transmittance of the film after exposure with light of 365 nm wavelength. Note the sharp increase in transparency at the exposure wavelength upon exposure, and the high opacity of the film at wavelengths below 300 nm.

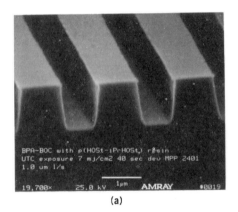

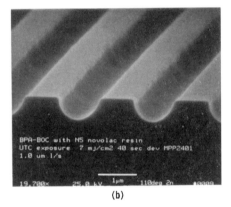

| (a) | (b) |

Fig. 17. Influence of optical absorbance on photoresist imaging properties. Shown are scanning electron micrographs of relief images obtained using two different photoresist formulations, processed under identical conditions. The formulations are functionally identical but differ in the transparency of the film at the exposing wavelength (36). (**a**) Image obtained when the film is weakly absorbing (~65% transmittance). (**b**) Image obtained when the film absorbs strongly (<20% transmittance) at the exposing wavelength. Note the incomplete development to substrate and the sloping sidewalls.

spectral output line is relatively broad. Limitations of available lens materials make correction for chromatic aberration difficult, so the source output beam must span only a very narrow wavelength range. Introduction of optical line narrowing elements into the beam leads to a large overall attenuation. Roughly speaking, a photoresist designed to be used with duv exposure tooling now at hand must be at least tenfold more efficient in its utilization of absorbed photons than a DNQ–novolac resist. Since the quantum yield for DNQ–novolac systems is on the order of ~0.3, a 10× improvement in efficiency is a considerable challenge.

One potential approach extends the idea of chemical amplification introduced in our preceding description of dry-film resists. In 1982, Ito and co-workers (37,38) recognized that if a photosensitizer producing an acidic product is photolyzed in a polymer matrix containing acid-labile groups, the acid will serve as a spatially localized catalyst for the formation or cleavage of chemical bonds.

An early practical example of this concept is shown in Figure 18. The acid-generating sensitizer is an onium salt, a class of compounds that efficiently generate a Bronsted acid upon photolysis. The polymer is poly(*t*-butoxyoxycarbonylstyrene) (PTBOCST). Photolysis forms a small quantity of acid in those areas exposed to radiation. During a later heating step the acid catalyzes thermolysis of pendant *t*-butoxyoxycarbonyl (TBOC) groups, converting the nonpolar PTBOCST into the polar poly(hydroxystyrene) (PHOST) and gaseous products, while regenerating the initial acid. The converted latent image can be developed in either positive tone by selecting a polar solvent, or in negative tone by selecting a nonpolar solvent as the developer. The catalytic chain length for this system, the average number of TBOC groups fragmented by each acid molecule, has been measured to be in the range 800–1100 (39). Therefore only extremely low exposure doses (in the range of 1–5 mJ/cm^2) are

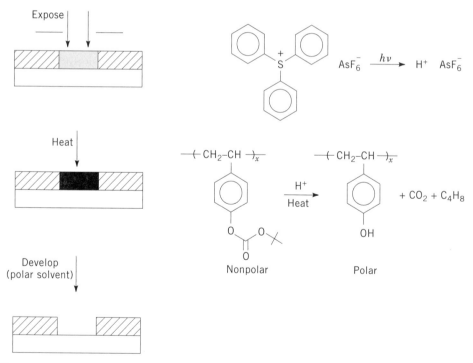

Fig. 18. Chemistry of PTBOCST chemically amplified resist. Patternwise exposure creates small quantities of acid. In a subsequent heating step, pendant TBOC groups are cleaved under acid catalysis. The exposed and unexposed areas can then be differentiated on the basis of solubility.

needed to image this system, which is roughly two orders of magnitude more photosensitive than typical DNQ–novolac resists.

Since the original proof of concept, and a later demonstration of its practical use in semiconductor manufacturing (40), applications and extensions of this concept have proliferated. In the following sections these systems are described in greater detail with emphasis on the resist formulation at a components level (41).

Photoacid Generators. Practical applications of photoacid generators (PAGs) have been actively pursued since onium salts were first reported to serve as photopolymerization catalysts in the 1970s (42). Onium salts are particularly useful as photoinitiators since they generate both Bronsted acids and free radicals and consequently can simultaneously initiate both cationic and radical cures. Owing to the high quantum efficiency for acid production, these ionic PAG's are well-suited for chemically amplified photoresist applications. Onium salts are one of several classes of photoacid generators. A variety of other PAG compounds (both ionic and nonionic) have been developed as researchers have sought to optimize the following key functional properties:

1. Quantum yield of acid generation
2. Photoacid characteristics
    a. Acid strength, pK_a
    b. Acid volatility
    c. Acid diffusion length
3. Wavelength response
4. Solubility
5. Thermal stability
6. Manufacturing costs
7. Toxicity

In the following sections the properties of photogenerators of strong Bronsted acids and their use in microlithography are summarized.

Ionic Photoacid Generators. Ease of synthesis, high thermal stability and good quantum yield have made sulfonium and iodonium salts the most widely used onium salts. Figure 19 depicts some representative examples.

By altering the counter ion in these salts (generally accomplished by a metathesis reaction) the properties of the photoacid can be readily modified, often without significant impact on the photochemical reactivity of the cation. In this way, the acid strength of the photogenerated acid can readily be varied from that of so-called "super acids" to the comparatively weak organic acids (48). Other modifications, such as the use of sterically bulky counterions (eg, substituting camphorsulfonate for methanesulfonate anion), have been studied in attempts to reduce acid volatilization and diffusion (which presumably improves lithographic resolution) (47).

Diffusion of acid is believed to be an important factor determining CA resist performance. At the molecular level, diffusion is necessary to effect acid catalyzed reactions in the resist film. Diffusion also acts to smooth sidewall fine structure (resist standing waves) that results from interference effects within the resist film due to reflections from the film interfaces. (Many DNQ resists,

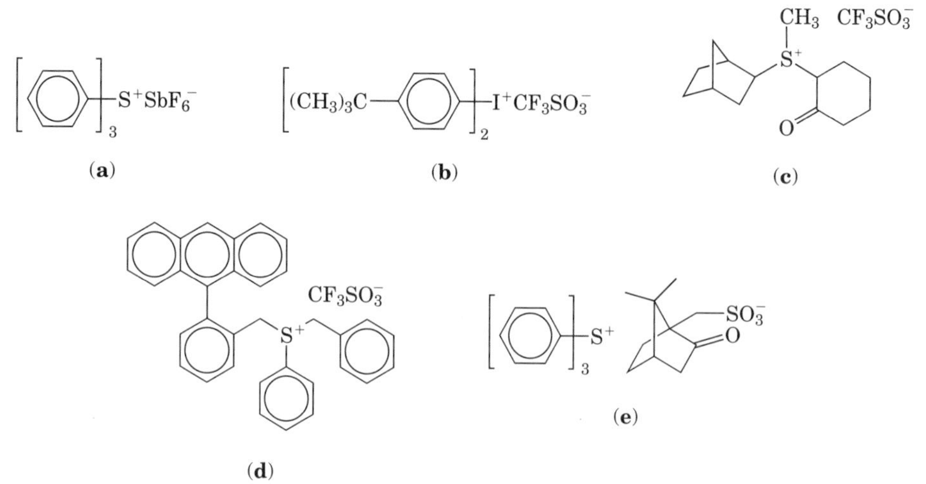

Fig. 19. Structures of selected sulfonium and iodonium salt photoacid generators designed for different lithographic applications. Modifications in the structure of the anion lead to changes in the properties of the photogenerated acid; changes in the structure of the cationic chromophore modify the light absorption properties of the PAG (43–47). (**a**) duv PAG generating inorganic acid; (**b**) duv PAG generating organic acid; (**c**) PAG designed for 193 nm lithography; (**d**) PAG designed for long wavelength activity via intramolecular electron transfer sensitization; and (**e**) PAG designed to generate bulky, weak acid.

which in principle require no thermal processing step after exposure, are heated after exposure to smooth out standing waves). However, if the mean diffusion length is too large the spatial profile of photogenerated acid is blurred and the final developed resist image appears distorted. The problem becomes more acute at finer resolution, and the characterization of acid diffusion and methods for its control in CA resists are active areas of research. The use of sterically bulky acids is one means for controlling photoacid diffusion, however, several other approaches are being pursued and are described in later sections.

The solubility properties of the PAG itself can play an important role in the overall resist performance as well (50). Solubility differences between the neutral onium salt and the acidic photoproducts can be quite high and will affect the resist contrast. In fact onium salts can serve as dissolution inhibitors in novolac polymers, analogous to diazonaphthoquinones, even in the absence of any acid-sensitive chemical function (51).

Simple aryl onium salts (Fig. 19) above absorb in the wavelength range of 200–330 nm, making them well-suited for applications requiring duv exposure. Spectral response can be shifted to longer wavelengths either by varying the substituents (Fig. 19) or by sensitization. Aromatic hydrocarbons with low oxidation potentials have been used to shift the activity of the PAG to wavelengths >500 nm via electron transfer sensitization (52). The absorption properties can also be shifted to shorter wavelengths by employing aliphatic substituents in the PAG (Fig. 19). These latter materials have been used in resists designed for exposure with 193 nm light, where aryl-substituted onium salts are claimed to increase absorption of the formulated resists to unacceptably high values (45).

The chemical pathways leading to acid generation for both direct irradiation and photosensitization (both electron transfer and triplet mechanisms) are complex and at present not fully characterized. Radicals, cations, and radical cations all have been proposed as reactive intermediates, with the latter two species believed to be sources of the photogenerated acid (Fig. 20) (53). In the case of electron-transfer photosensitization, aromatic radical cations (generated from the photosensitizer) are believed to be a proton source as well (54).

Nonionic Photoacid Generators. Examples of such materials are listed in Figure 21, and include a variety of structural types which may undergo several different photochemical rearrangements. As one example, irradiation of the 2,6-dinitrobenzyl ester shown in Figure 21 results in an intramolecular rearrangement (the well known *ortho*-nitrobenzyl rearrangement (55)) ultimately producing toluenesulfonic acid (56). In this case, no radical flux is produced upon photolysis, minimizing the possibility of crosslinking side reactions that can lead in some cases to resist scumming or even negative-tone behavior in positive resist systems (57). In another type of photochemical reaction, electron transfer from a photoexcited phenolic moiety of the polymer matrix to a ground (electronic) state PAG molecule (analogous to that depicted in Fig. 20) leads to the formation of methanesulfonic acid (Fig. 21) (58). Finally, 4-sulfonyl DNQ sulfonic acid

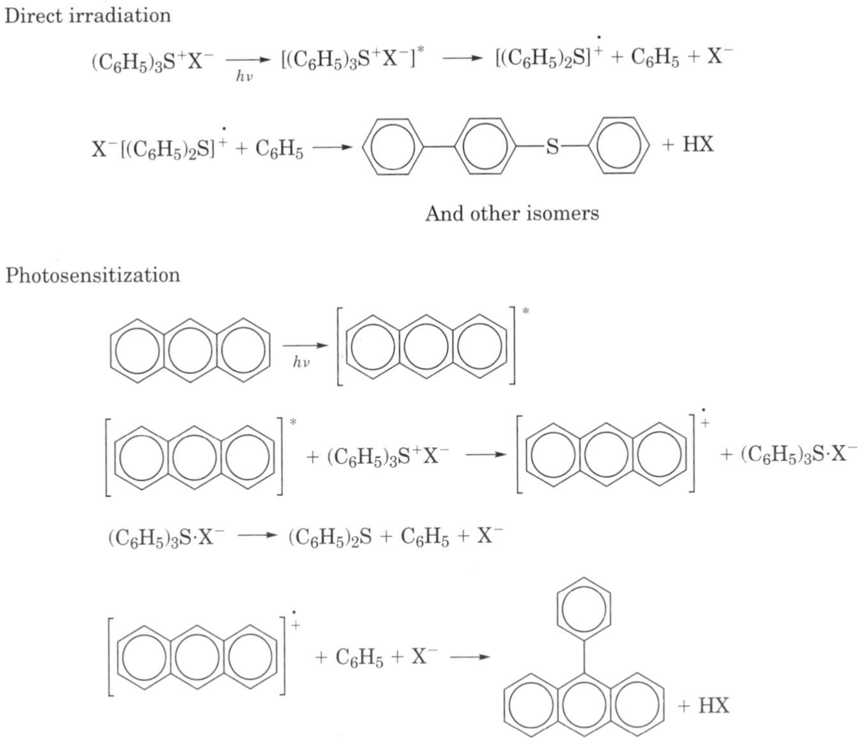

Direct irradiation

$$(C_6H_5)_3S^+X^- \xrightarrow{h\nu} [(C_6H_5)_3S^+X^-]^* \longrightarrow [(C_6H_5)_2S]^{\cdot+} + C_6H_5 + X^-$$

$$X^-[(C_6H_5)_2S]^{\cdot+} + C_6H_5 \longrightarrow \text{(structure)} + HX$$

And other isomers

Photosensitization

$$(C_6H_5)_3S\cdot X^- \longrightarrow (C_6H_5)_2S + C_6H_5 + X^-$$

Fig. 20. Proposed photochemical mechanisms for the generation of acid from sulfonium salt photolysis. Shown are examples illustrating photon absorption by the onium salt (direct irradiation) as well as electron transfer sensitization, initiated by irradiation of an aromatic hydrocarbon.

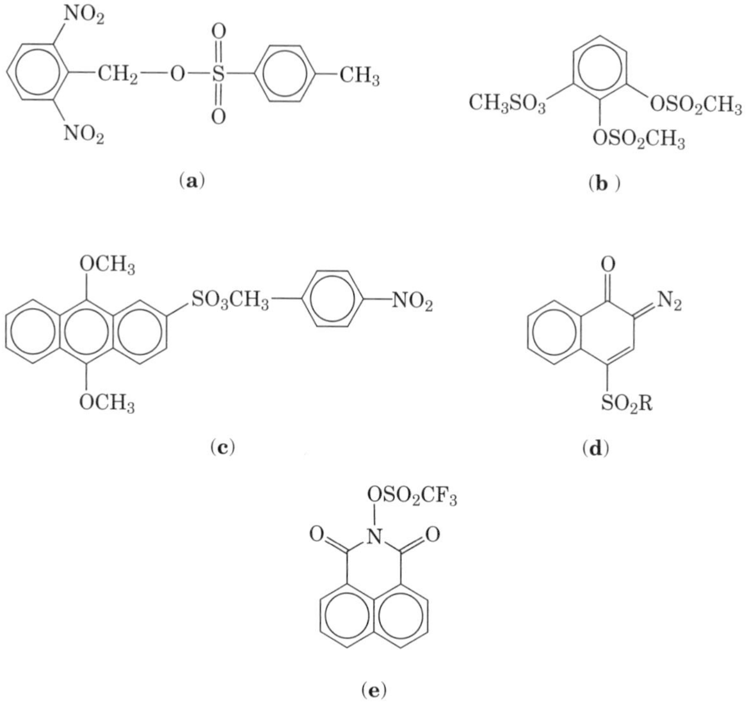

Fig. 21. Representative nonionic photoacid generators. A variety of photochemical mechanisms for acid production are represented. In each case a sulfonic acid derivative is produced (25,56,58–60). (**a**) PAG that generates acid via O-nitrobenzyl rearrangement; (**b**) PAG that generates acid via electron transfer with phenolic matrix; (**c**) PAG that is active at long wavelengths via electron transfer sensitization; (**d**) PAG that generates both carboxylic acid and sulfonic acid; and (**e**) duv PAG generating trifluoromethane sulfonic acid.

generators, introduced earlier in the description of negative-tone novolac-based resists and shown in Figure 12, have been used as PAGs in CA resists, most effectively at exposure wavelengths longer than 300 nm.

To achieve the best overall resist performance, the optimum PAG for a given resist system, whether ionic or nonionic, must balance the functional properties listed earlier in this section. The development of new photoacid generators, and the characterization of their functional properties, are considered key to the design of resists with increased levels of performance.

Acid Catalyzed Chemistry.   Acid catalyzed reactions form the basis for essentially all chemically amplified resist systems for microlithography applications (61). These reactions can be generally classified as either crosslinking (photopolymerization) or deprotection reactions. The latter are used to unmask acidic functionality such as phenolic or pendant carboxylic acid groups, and thus, lend themselves to positive tone resist applications. Acid-catalyzed polymer crosslinking and photopolymerization reactions, on the other hand, find application in negative tone resist systems. Representative examples of each type of chemistry are listed below.

Positive-Tone Photoresists. The ester, carbonate, and ketal acidolysis reactions which form the basis of most positive tone CA resists are thought to proceed under *specific acid catalysis* (62). In this mechanism, illustrated in Figure 22 for the hydrolysis of *tert*-butyl acetate (type $A_{Al}1$) (63), the first step involves a rapid equilibrium where the proton is transferred between the photogenerated acid and the acid labile protecting group:

$$P + HA \underset{}{\overset{K_{eq}}{\rightleftharpoons}} PH^+ + A^- \quad \text{(fast)}$$

$$PH^+ \overset{k}{\longrightarrow} X \quad \text{(slow)}$$

The rate of the reaction in such case is $R = k\,[PH^+]$, where P is the reactant (ie, a repeat unit bearing the acid-labile protecting group).

In resist systems the situation is more complex than that just outlined. The photogenerated acids can be so strongly acidic that they dissociate completely upon formation. In such case, the proton resides on other basic components of the resist film, which can include polymer functionality (ie, ester or ether co-monomers), adventitious impurities such as water or residual casting solvents, as well as additives such as coating aids. The reactant PH^+ will be in equilibrium with all other protonated species in the resist film. The concentration of each protonated species is then determined by the basicity of its unprotonated conjugate. As deprotection proceeds, the concentrations of the basis resist components changes, shifting the equilibria and influencing the overall kinetics. Substitution of an acid of different strength, or introduction of a basic impurity or additive, perturbs the concentration of PH^+, and thereby, the imaging behavior of the resist.

Figure 23 lists representative acid-labile protecting groups that have been incorporated in positive-tone CA resist systems. These groups can be pendant to the matrix polymer chain, can be attached to a monomeric or polymeric additive that acts as a dissolution inhibitor (64–66), or can even be appended to the PAG structure (67). The kinetics of acid-catalyzed deprotection vary significantly with structure. In particular, the activation energy, E_a, which characterizes how the rate of deprotection changes with temperature, has a large impact on CA-resist processing characteristics. For example, removal of the *tert*-butyl group from *tert*-butyl methacrylate has a relatively high E_a (measured to be 35 kcal/mole in experiments in resist films (68)), and resists based on this chemistry require postexposure processing at temperatures 120°C or higher. Since the extent of deprotection (and the related changes in the dimensions of the lithographic image) show a significant dependence on temperature, very precise temperature control is required. In contrast, resists based on low E_a chemistry (eg, the acetal system

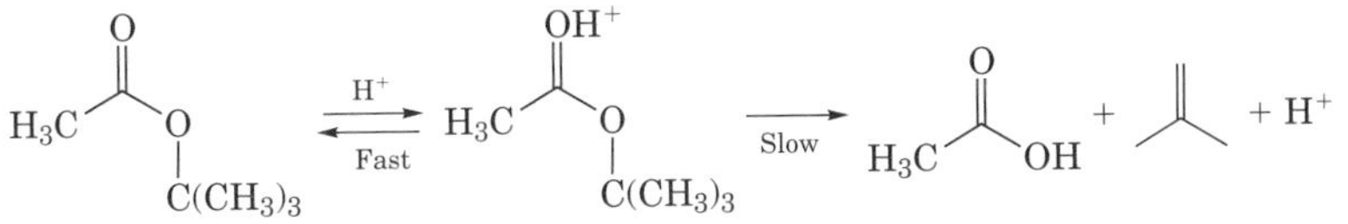

Fig. 22. Mechanism for the hydrolysis of *tert*-butyl acetate under strongly acidic conditions.

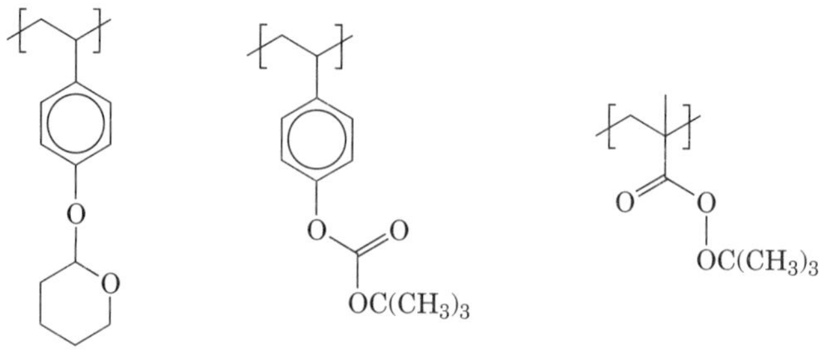

Fig. 23. Representative protecting groups for phenolic and carboxylic acid based systems. (**a**) The polymer-based protecting groups are listed in order of increasing activation energy for acid-catalyzed deprotection. (**b**) Acid-labile monomeric dissolution inhibitors, a bifunctional system based on protected bisphenol A. (**c**) Another system that combines the function of dissolution inhibitor and PAG in a single unit.

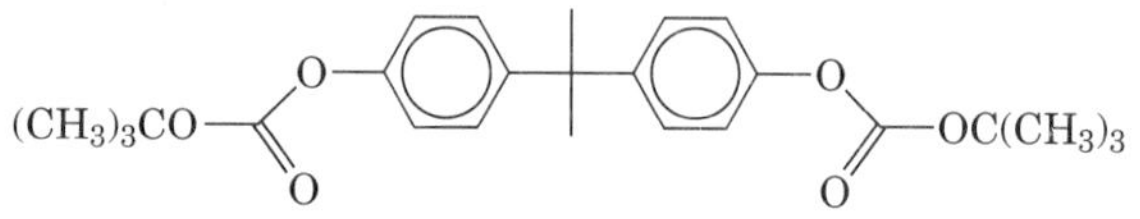

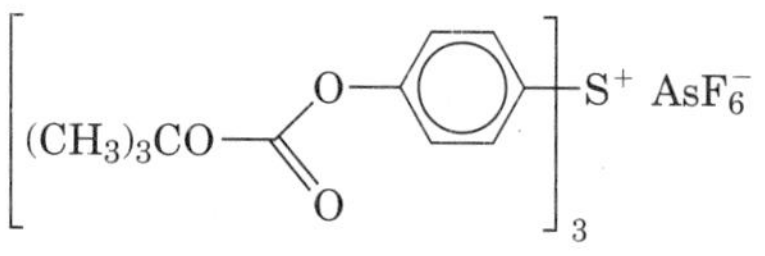

shown in Fig. 23) undergo deprotection at or near room temperature and exhibit little variation in final image linewidth with processing temperature (69).

Another potential advantage of these low activation energy systems is a reduced dependence on the precise timing of process steps (70–72) since there is in effect no delay between the exposure and postbake steps. A potential disadvantage of these systems is that their intrinsically higher reactivity can reduce the stability of the formulated resist solutions during storage (ie, its shelf life). As with PAGs, in choosing a specific protecting group, a number of functional properties must be balanced to optimize overall resist performance.

Negative-Tone Photoresists.   Many negative-tone CA resist employ acid-catalyzed crosslinking to effect insolubilization of the exposed pattern. These systems function in a manner analogous to radical-based crosslinking resists, in some cases exhibiting the same limitations. One of the first CA resist systems described in the chemical literature is based on the acid-catalyzed crosslinking of epoxy groups (Fig. 24(**a**)) (43). Although such resists display excellent photospeed and are immune to inhibition by molecular oxygen, they suffer from the same solvent-induced image distortion during development that is seen with radical crosslinking resists. Careful attention to formulation and process conditions is required to achieve high resolution imaging. These difficulties, coupled with the

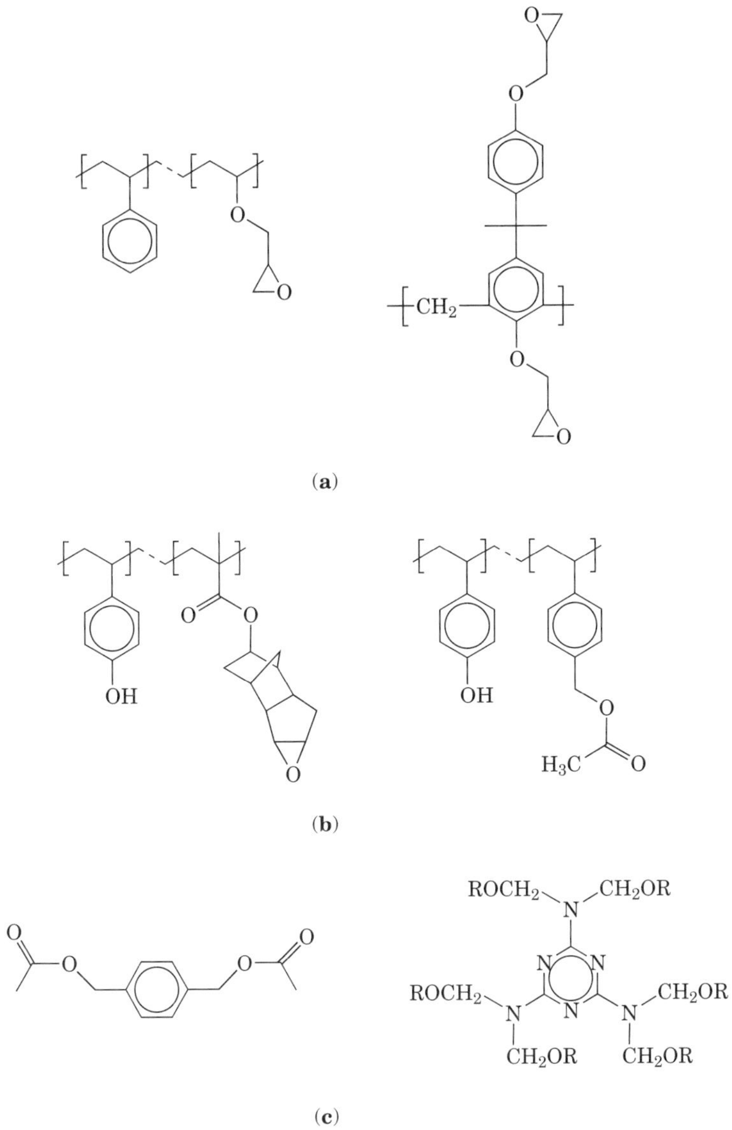

(a)

(b)

(c)

Fig. 24. Representative crosslinking systems employed in negative tone CA resists. (**a**) Epoxy polymers requiring organic solvent development. (**b**) PHOST-based crosslinking systems requiring aqueous development. (**c**) Monomeric crosslinking agents used in PHOST matrix polymers.

practical problems associated with the use of organic solvents in a manufacturing setting, has limited the acceptance of such resists.

Several CA negative-tone photoresists based on polyhydroxystyrene (PHOST) polymers incorporate the crosslinking chemistries shown in Figure 24(**b**) and (**c**). The phenolic structure renders these systems developable using aqueous alkaline solutions. Dissolution properties of this type of crosslinking resist closely resembles the surface limited etching found in DNQ–novolac

Polymers for Chemically Amplified Resists. Derivatives of Polyhydroxystyrene. Earlier it was noted that novolac-based resists are too strongly absorbing at 248 nm to be used for duv exposure. Many derivatives of poly(4-hydroxystyrene) (PHOST) are sufficiently transparent at 248 nm to provide a phenolic platform for resist design. Figure 27 compares the optical absorption characteristics of PHOST and novolac polymers. Resists used in duv lithography are typically copolymers of hydroxystyrene and an acid-labile comonomer, often a protected hydroxystyrene, as depicted in Figure 28. These polymers are almost always produced by free-radical solution polymerization of a derivative of 4-hydroxystyrene (HOST), and thus, are usually medium molecular weight polymers with a relatively narrow polydispersity (between 1.5 and 2.5) (86). As detailed earlier, the unique properties of novolacs stem in part from the unusual condensation polymerization mechanism which produces substantial quantities of oligomeric products and high molecular weight branched polymer. Analogous fractions are not produced in the polymerization of protected hydroxystyrenes.

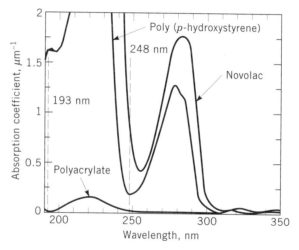

Fig. 27. Optical absorption spectra of thin, 1 μm-films of novolac, polyhydroxystyrene and polyacrylate polymers. The novolac resin is transparent only above 300 nm. While polyhydroxystyrene also absorbs strongly below 300 nm, it exhibits a region of adequate transparency centered near 248 nm. The acrylate polymer is quite transparent over the uv range.

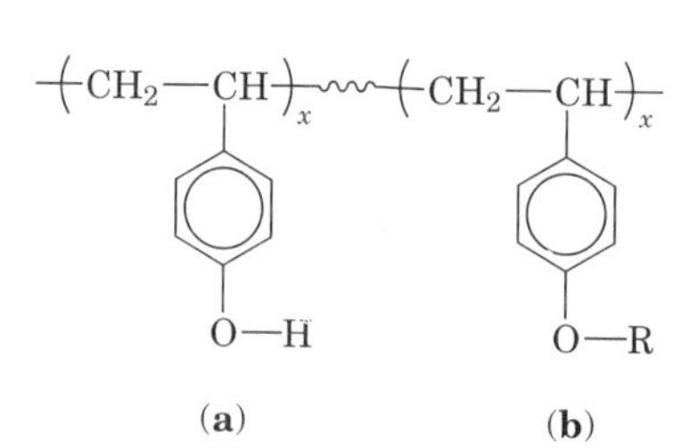

$$\left.\left(CH_2-CH\right)_x \sim \left(CH_2-CH\right)_x\right.$$

(a)   (b)

Fig. 28. Traditional duv-resist design using derivatives of polyhydroxystyrene. Monomer (**a**) contributes hydrophilic character to the polymer, and its acidic phenol group enhances aqueous base solubility; monomer (**b**) provides acid-labile pendant groups.

The functional properties of phenolic resins based on HOST bear little resemblance to those of novolacs, even though the polymers are structurally similar (in fact, isomeric!).

Table 1 compares properties of a "typical novolac" with those of a "typical" HOST homopolymer (PHOST).

PHOST is often prepared by polymerization of 4-acetoxystyrene followed by base-catalyzed hydrolysis (Fig. 29). The acetoxystyrene monomer's stability and polymerization kinetics allow production of PHOST of higher quality than is easily obtained by direct radical polymerization of HOST. The PHOST homopolymer product is then partially or fully derivatized with an acid-cleavable functionality to produce the final resist component.

In some instances, the resist polymer can be prepared in a single step by direct polymerization of the protected monomer(s) (37,88), entirely avoiding the intermediate PHOST. HOST-containing resist polymers have also been prepared by free-radical copolymerization of a latent HOST and a stable, acid-labile monomer, eg, the copolymerization of acetoxystyrene with *tert*-butyl acrylate, followed by selective removal of the acetoxy group (89) (Fig. 30).

Acrylic Polymers.   Although CA resists based on phenolic polymers are suitable for duv applications, the next resist generation is designed for imaging with light of $\lambda = 193$ nm, where all aromatic polymers are opaque (Fig. 27). Acrylic polymers as a class are relatively transparent in this spectral range. Positive-tone CA resists based on acrylic polymers, first designed for PWB fabrication, have been demonstrated to produce high resolution images using 193-nm exposure tooling (90). They are prepared by single-step free-radical solution polymerization with no post-polymerization chemistry required. This class of

Table 1. Comparison of Properties of a Typical Novolac with Those of a Typical HOST Homopolymer

Property	Novolac	PHOST[a]
molecular weight	low	medium
molecular weight distribution	broad	narrow
T_g	100°C	160–180°C
dissolution rate	slow	fast

[a]PHOST represents poly(4-hydroxystyrene). The properties of the isomeric poly(2-hydroxystyrene) and poly(3-hydroxystyrenes) differ substantially (87).

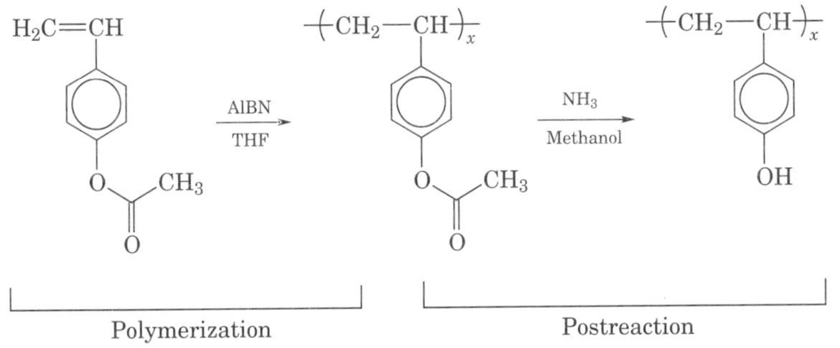

Fig. 29.   Synthesis of polyhydroxystyrene (PHOST).

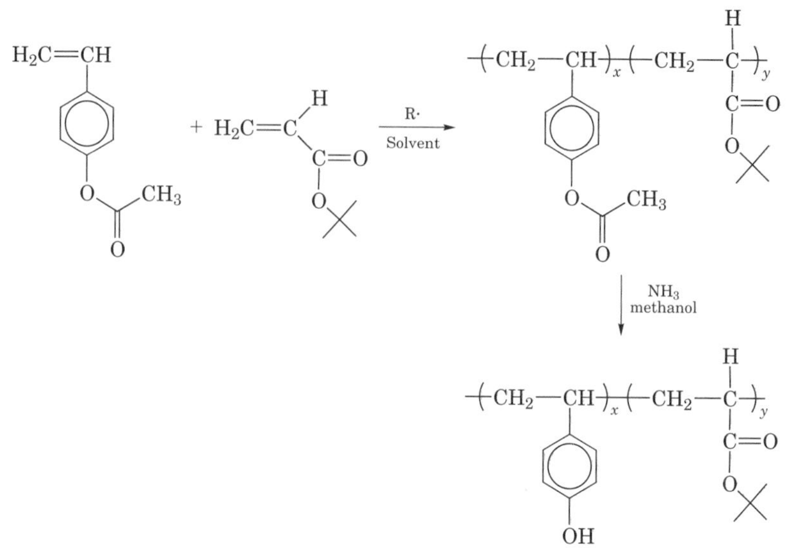

Fig. 30. Synthesis of an acid-labile copolymer by radical copolymerization using a latent HOST, followed by selective deprotection (89).

material is typified by the methacrylate terpolymer shown in Figure 31. Each monomer imparts specific characteristics to the polymer, providing a mechanism for precise balancing of functional properties.

Incorporation of cyclic aliphatic, (alicyclic) side groups markedly improves the plasma etch resistance of acrylic polymers, without reducing optical transparency at 193 nm (91). Figure 32 presents structures of some acrylic polymers currently under study for use in 193-nm CA resists (92–94). Recently, polymers with main-chain alicyclic structures have been described that offer similar properties (95,96).

Polymer Properties and Lithographic Performance in Chemically Amplified Resins. Optical Absorption.   Today, CA resists are being used in advanced duv photolithography applications where the exposure wavelength is between 190 and 250 nm. Most organic compounds absorb strongly in this spectral range, however, and the design of polymers for CA resists that maintain acceptable

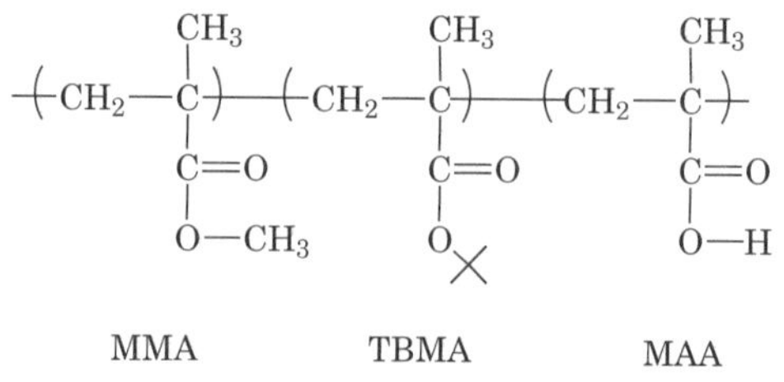

Fig. 31. An acrylic terpolymer designed for chemically amplified resist applications. The properties each monomer contributes to the final polymeric structure are for MMA, PAG solubility, low shrinkage, adhesion and mechanical, strength for TBMA acid catalyzed deprotection; and for MMA, aqueous base solubility and increased T_g.

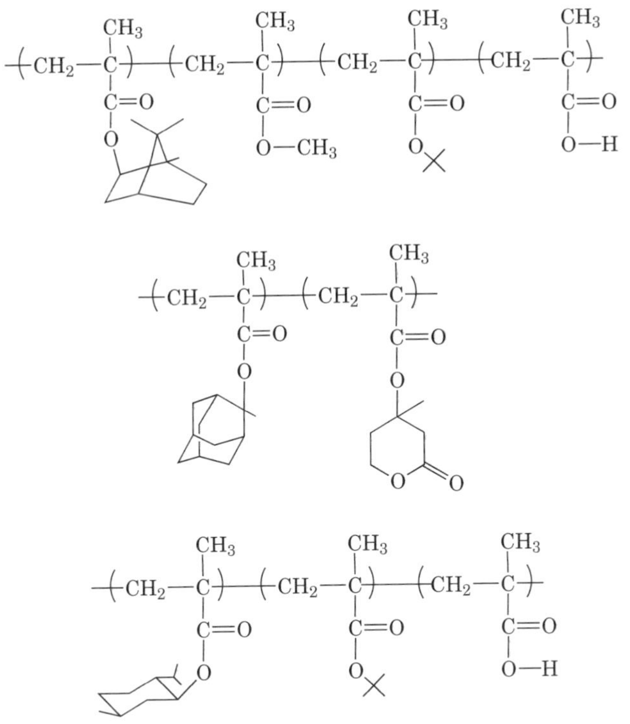

Fig. 32. Acrylic polymers found in CA resists designed for lithographic applications using 193 nm light.

optical transparency while satisfying all other functional requirements is a challenge. The optimum optical absorbance for a nonbleaching photoresist is about 0.4 absorbance units per micron, representing a balance between conflicting requirements for high transparency to provide uniform illumination through the depth of the film, and high absorption to suppress thin-film optical reflectivity effects that reduce process latitude (97,98). To maximize light absorption by the photosensitizer, a rule of thumb is that the absorbance of the polymeric component should not exceed 0.2 per micron at the exposure wavelength.

Dry Etch Resistance. Gas-phase plasma etching image transfer processes (99) are ubiquitous in modern semiconductor fabrication. A polymer's stability in a plasma etch environment is a key consideration in resist design. Plasma etching processes evolved using DNQ–novolac resists so new lithographic materials must exhibit comparable etch resistance. Depending on structure and elemental composition, polymer plasma etch resistance can vary by orders of magnitude (100). A relationship has been established linking a resist's chemical composition with its resistance to a reactive ion etch (RIE) environment. Devised by Ohnishi and co-workers (101), a compositional parameter with good correlation to RIE rate is

$$Ohnishi\ Parameter = N/(N_c - N_o)$$

where N, N_c and N_o are the total number of atoms in the polymer repeat unit, number of carbon atoms, and number of oxygen atoms, respectively. This model serves as a fair predictor of etch rates for polymers under conditions where physical ion bombardment is a significant component, as is the case with RIE. The relation fails for low ion energy plasma conditions (eg, in a downstream glow discharge) where etching mechanisms are mainly chemical in nature. The Ohnishi parameter predicts that polymers having a high effective carbon content (low Ohnishi parameter) will exhibit low etch rates. Incorporation of oxygen and hydrogen into the repeat unit structure increases the etch rate, while an increased carbon content reduces the rate. Aromatic rings are, thus, quite etch resistant while aliphatic structures containing oxygen are etched rapidly. These concepts have been recently refined by introducing an additional parameter based not on composition but on polymer structure, termed the ring parameter (102). The added parameter extends the predictive capability of the relation to materials in plasma processes with low ion energies. This model recognizes that polymer etch rate is closely coupled to the fraction of the resist mass existing as or contained in ring structures, and can accurately predict the etch rate of many new alicyclic materials developed for 193-nm resists.

Thermal Properties. The glass-transition temperature of the polymer can influence CA resist properties in several ways. As with all resists, the final relief image must not exhibit significant thermal flow at the elevated temperatures reached during image transfer. With crosslinking systems, thermal flow is generally not a concern, but with positive-tone systems it can be problematic. Although the T_g of the base resist polymer may be high (eg, the T_g of PHOST homopolymer is 160–180°C), internal plasticization by the acid-labile group and external plasticization by resist additives and residual solvent will lower T_g of the final resist image (103).

Unlike DNQ–novolac resists, CA-resist imaging characteristics are determined to a significant extent by thermally activated bimolecular chemistry taking place during postexposure processing. Since the polymer serves here as the reaction medium, its properties and state influence the course and evolution of chemical change. One particular concern is the impact of diffusion of the photogenerated acid on process latitude and on minimum achievable resolution. Several studies have probed how properties of the polymeric matrix influence diffusion in CA resists.

The impact of CA-resist thermal processing on acid mobility has been investigated, using a novel experimental method for generating sharply defined concentration gradients (104). In this work the mobility of photogenerated acid was seen to increase sharply in films as the temperature of the postexposure bake was increased. The authors concluded that optimum lithographic performance is obtained when the postexposure processing temperature is as low as possible, and in particular is below the T_g of the film. In contrast, acid mobility was shown to increase as the postapply temperature was lowered. This effect was attributed to increased residual solvent content which facilitates acid migration.

In other work, the impact of thermal processing on linewidth variation was examined and interpreted in terms of how the resist's varying viscoelastic properties influence acid diffusion (105). The authors observed two distinct behaviors, above and below the resist film's glass transition. For example, a plot of the rate of deprotection as a function of postexposure processing temperature show a change in slope very close to the T_g of the resist. Process latitude was improved and linewidth variation was minimized when the temperature of postexpose processing was below the film's T_g.

Resistance to Airborne Chemical Contamination. Several methods for attenuating a CA resist's sensitivity to the presence of trace volatile bases were described earlier. Resist process stability in the presence of airborne bases also can be improved by consideration of how this stability is influenced by properties of the resist polymer. Using radiotracer methods to tag the basic contaminant, the kinetics and extent of contaminant uptake can be quantified (106). In one study, a wide range of polymer structures were cast as thin films on silicon wafers, and then stored in air containing a known, very low concentration of ^{14}C-labeled NMP (107). The rate of NMP uptake was found to vary by almost a factor of fifty depending on the polymer structure. The contaminant uptake of each polymer could be related to its physical properties, specifically its solubility parameter and T_g. The correlation between NMP uptake and polymer T_g (Fig. 33) was attributed to a reduction in free volume (a consequence of the nonequilibrium nature of the spin-coating process) due to polymer film annealing during postapply baking, thereby reducing diffusant mobility.

One practical conclusion that can be drawn from this result is that susceptibility to airborne chemical contaminants can be minimized if a CA-resist film is baked after coating at a temperature at or above its T_g. However, at that time few existing resist systems had sufficient thermal stability to allow such treatment (108). By deliberately selecting polymer, protecting groups and PAGs with thermal stability in mind, however, CA resists can be designed that can be heated at or above T_g (109,110). Such resists display sharply improved environmental stability compared to analogous systems processed below T_g.

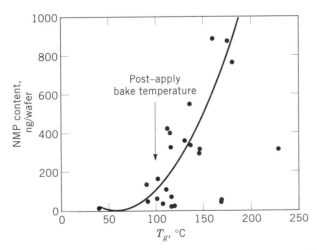

Fig. 33. Correlation of NMP uptake with polymer T_g for a series of 24 different polymeric films. All films were heated after coating at 100°C for 300 seconds (107).

Extending the Chemically Amplified Resist Concept

Microlithography as used in modern semiconductor manufacturing has advanced largely through evolutionary refinement of exposure tool optics and mechanics and resist materials. The switch to duv-CA resists and excimer-laser-based exposure tools now underway in many respects represents a departure from this evolutionary process, and constitutes a considerable technical and financial challenge to the electronics industry. If continued improvements in lithographic performance are to be achieved in the future, it is clear that still more advanced technologies will be introduced (111). Figure 34 (112) lists several lithographic technologies being studied as extensions and replacements for duv technology when the scale of microelectronic devices shrinks to the 0.1 μm range. In parallel with the development of new exposure technologies utilizing high energy

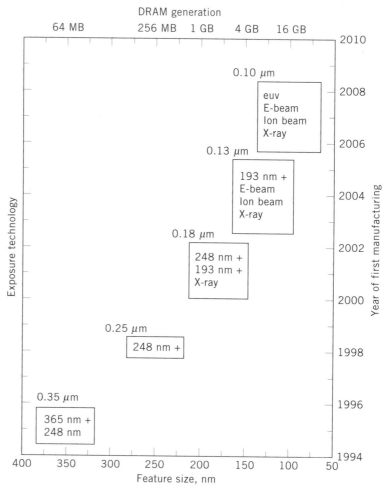

Fig. 34. Lithography approaches under consideration for production of future generations of semiconductor devices. The printed feature size, exposure technology, corresponding generation of dynamic random access memory (DRAM) integrated circuits and the anticipated year of first manufacturing are included. For example, the candidates for manufacturing 4 gigabit DRAM devices at 0.13 μm ground rules (beginning in 2004) are 193-nm lithography with resolution enhancement techniques, or e-beam, ion-beam or x-ray lithography. Optical enhancements or resist modification indicated by + sign. (Figure based on the 1994 SIA roadmap, (112)).

radiation and particle-beam sources, modifications and refinements of duv photolithography (248 and 193 nm indicated by a + sign in Fig. 34) are being pursued. The simultaneous development of these technologies is extremely costly,
so much of the research today is done in industry–government consortia worldwide. There is considerable uncertainty (and in some cases controversy (113,114))
over how future technology investments should be made since it is clear that
the semiconductor industry will not be able to support simultaneously, the in-
frastructures required by all the technologies listed here. Although Figure 34
explicitly specifies only exposure technologies, the availability, properties and
functions of suitable resist materials are key considerations when assessing the
viability of these options. In the following section these new imaging technologies, and their potential impact on resist technology are described.

Thin-Film Imaging Resists. The use of shorter exposure wavelengths and
other refinements in exposure tool optics improve resolution, however, this is
usually accompanied by an increased difficulty in maintaining the image pattern
in focus through the thickness of the resist film; that is, there is a trade-
off between exposure tool resolution and its depth-of-focus. A resist material
that could be used as a very thin film (at a thickness of ca 200 nm or less),
yet still would act as a useful mask throughout subsequent processing steps,
would go far in circumventing the optics trade-off. Thin-film imaging (TFI)
resists incorporating silicon, when coupled with resist development using oxygen-
reactive ion etching (O_2–RIE), are examples of such materials.

Figure 35 shows schematically the process flows of the two primary types
of TFI methods, using bilayer and top surface imaged (TSI) resists. Both methods rely on the differences between the oxygen plasma etch rates of polymers
that contain elements forming refractory oxides (such as silicon) and simple organic polymers comprised only of elements forming volatile oxides. When a thin
film of a resist containing silicon is patterned on top of a polymer underlayer
containing no silicon and then placed in an oxygen plasma, the imaged film
immediately forms a highly etch-resistant layer of SiO_2 while the unprotected
underlayer rapidly erodes. In O_2–RIE the underlayer etch can be performed
anisotropically, transferring the pattern of the top film vertically into the thick
underlayer. The final bilayer pattern constitutes the resist relief image that is to
be used for pattern transfer into the substrate. Since the imaging layer is thin
compared to the overall resist thickness, this approach eases the difficulty of
maintaining focus, and offers other advantages, which include planarization of
underlying substrate topographical features, the potential for reducing substrate
reflectivity using opaque underlayers, and the potential for printing images with
much higher aspect ratio images than is possible in single-layer resists. These
advantages come at the cost of increased process complexity compared to single-
layer resist techniques.

Bilayer Resists. A variety of silicon-containing bilayer resists have been
described, with examples of both CA and conventional imaging chemistries in
negative and positive tone (115). Resist films containing 10–12% silicon by
weight of solids generally show enough O_2–RIE etch resistance to serve as the
imaging layer of a useful bilayer system. The design of such systems is not a
simple matter, however, as the introduction of this proportion of silicon (either as
an additive, as part of the polymer backbone, or in functional groups pendant to

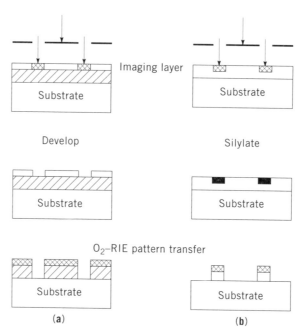

Fig. 35. Process flow for thin-film imaging lithography: (**a**) bilayer process and (**b**) top surface imaging. The bilayer process shown here employs a positive-tone imaging layer. The TSI process illustrated relies on preferential silicon incorporation in the exposed regions of the imaging layer to give a negative-tone pattern.

the polymer chain) can strongly perturb the polymer solubility and/or its T_g, to the detriment of its lithographic properties. Figure 36 provides several examples of this type of resist.

Top-Surface Imaging Resists. TSI resists differ from bilayer systems in that the silicon image is introduced into the polymer after the imaging step (120). The same deprotection chemistry used to unmask phenolic hydroxyl groups in CA resists can also be used to spatially direct their silylation when the film is treated with a silylating reagent (121). In principle, this technique should afford very high silylation contrast since the unexposed regions of the film contain no reactive sites. In a second method, radiation-induced polymer crosslinking slows diffusion of a silylation reagent into a polymer film containing functional groups that can combine with the reagent (122). Both CA or non-CA systems of this type have been described. While patterned films show a preferential incorporation of silicon in the uncrosslinked areas of the resist, there is measurable silylation in the crosslinked areas. The inherently lower silylation contrast must be overcome by careful optimization of the etch processes and silylation reagent.

In each of these approaches, imaging is confined to the top of a single polymeric film by adjusting optical absorption. The penetration depth of the silylation agent and the attendant swelling of the polymer film must also be controlled to avoid distortion of the silylated image. Resists of this type are capable of very high resolution (Fig. 37).

Alternative Exposure Technologies. *Optical Enhancements.* The term wavefront engineering has been coined to describe a set of optical techniques

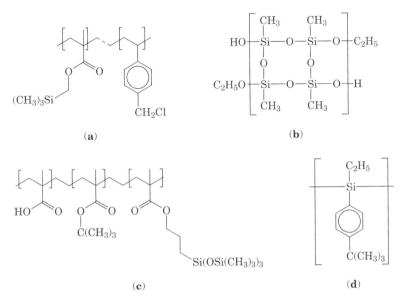

(a)

(b)

(c)

(d)

Fig. 36. Representative bilayer resist systems. Both CA and non-CA approaches are illustrated (116–119). (**a**) Crosslinking E-beam resist, 193-nm thin-film imaging resist; (**b**) acid catalyzed negative-tone crosslinking system; (**c**) positive-tone CA resist designed for 193-nm applications; and (**d**) positive-tone resist based on chain scission.

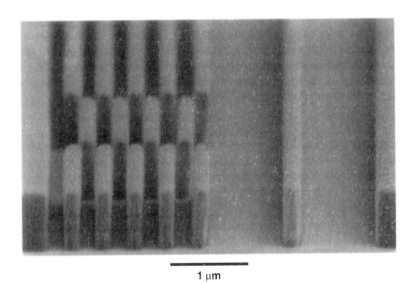

1 µm

Fig. 37. Resist images obtained with a crosslinking monocomponent TSI resist (PHOST polymer), crosslinked by photo-oxidation using light at 193-nm wavelength. After exposure, the film was treated with a vapor of dimethylsilyldimethylamine and then plasma developed using O_2–RIE (122). SEM photomicrography courtesy of MIT Lincoln Laboratory.

271

However, this is accompanied by a number of technical challenges, including the development of new x-ray sources, an all-reflective optical design and a new type of reflective mask suitable for soft x-ray imaging (Fig. 40). Proponents of euv lithography predict resolution scalable to 0.07 microns while maintaining a large depth of focus.

Light absorption at a wavelength of 13 nm is determined by the atomic weights of constituent elements of the absorbing material and its density. In general, organic materials strongly absorb euv light, so researchers anticipate that TFI resists will be called for. Given the low brightness of available soft x-ray sources, and the potential for radiation damage to optical components, it is anticipated that high sensitivity CA resists must be used if euv lithography is to be adopted for full scale device manufacturing.

Proximity X-Ray. The practice wherein improved resolution is achieved by decreasing the wavelength of the imaging radiation reaches its limit with x-ray lithography. A significant investment has been made in the use of synchrotrons (Fig. 41) as light sources for lithography, using emission in the spectral range of 1–2 nm (130). The high transparency of organic films at this wavelength, combined with a large effective depth of focus, enable the fabrication of resist structure with high aspect ratios. Proximity x-ray lithography has been used to create features as small as 30 nm (131).

The construction of suitable masks is one of the most challenging aspects of X-ray lithography (Fig. 42). The mask substrate must be thin to avoid significant attenuation of the exposing beam but must be sufficiently rugged to survive handling and cleaning. The absorber material must be fabricated as a relatively thick film to achieve adequate blocking of the beam. Finally, since proximity x-ray lithography is a simple shadow patterning technique, it requires a mask with the same dimensions as the ultimate feature to be patterned (Optical lithography tools usually project a demagnified image of the mask onto the resist film, simplifying mask fabrication since the mask pattern is four or five times larger than the final resist feature).

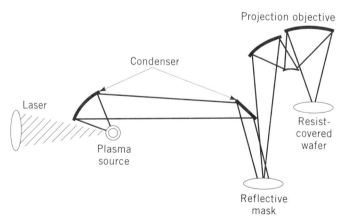

Fig. 40. Schematic of an euv exposure tool. Key features are the excimer laser-driven x-ray source and the reflective optical elements (including the mask) in the imaging system. Courtesy of Sandia National Laboratories.

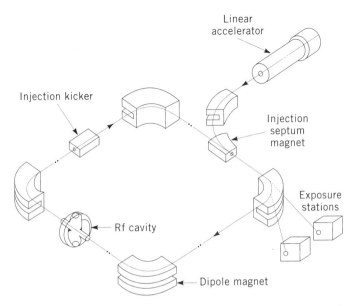

Fig. 41. Diagram of a compact storage ring. Each time the circulating electron beam is bent by one of the dipole magnets it loses energy by emitting x-ray photons which are directed to an exposure station. One ring can serve as the illumination source for many attached exposure stations. A constant ring current is maintained by replacing these energy losses by means of an rf accelerating cavity (130).

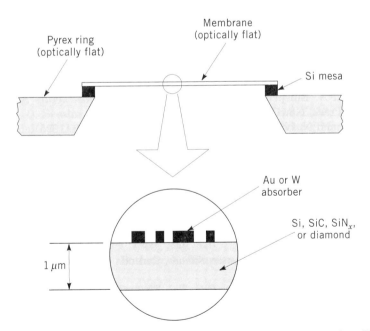

Fig. 42. Diagram of a mask used for 1:1 proximity x-ray lithography. These thin-membrane masks, required for optimum transmission when using patterned metal absorbers, must remain free of distortion to maintain pattern fidelity during exposure.

56. T. Neenan, F. Houlihan, E. Reichmanis, J. Kometani, B. Bachman, and L. Thompson, *Macromolecules* **23**, 145 (1990).

57. N. Hayashi, L. Schlegel, T. Ueno, H. Shiraishi, and T. Iwayanagi, *Proc. Soc. Photo-Opt. Instr. Eng.* **1466**, 377 (1991).

58. L. Schlegel, T. Ueno, H. Shiraishi, N. Hayashi, and T. Iwayanagi, *Chem. Mater.* **2**, 299 (1990).

59. K. Naitoh, K. Yoneyma, and T. Yamaoka, *J. Phys. Chem.* **96**, 238 (1992).

60. H. Ito, G. Breyta, D. Hofer, T. Fischer, and B. Prime, *Proc. Soc. Photo-Opt. Instr. Eng.* **2438**, 53 (1995).

61. Photobase generators also have been applied to a limited extent in microlithographic applications; see M. Shirai and M. Tsunooka, in *Progress in Polymer Science*, Vol. 21, Elsevier, UK, 1996, pp. 1–45.

62. T. H. Lowery and K. S. Richardson, *Mechanism and Theory in Organic Chemistry*, 3rd ed., Harper and Row, New York, 1987, Chapt. 7.

63. Ref. 48, Table 10.14, p. 380.

64. D. McKean, S. MacDonald, N. Clecak, and G. Willson, *Proc. Soc. Photo-Opt. Instr. Eng.*, **1925**, 246 (1993).

65. R. Allen, Q. Ly, G. Wallraff, C. Larson, W. Hinsberg, W. Conley, and K. Muller, *Proc. Soc. Photo-Opt. Instr. Eng.* **1925**, 246 (1993).

66. H. Ito, M. Ueda, and R. Schwalm, *J. Vac. Sci. Technol. B* **6**, 2259 (1988).

67. D. Funhoff, H. Binder, and R. Schwalm, *Proc. Soc. Photo-Opt. Instr. Eng.* **1672**, 46 (1992).

68. G. Wallraff, J. Hutchinson, W. Hinsberg, F. Houle, P. Seidel, R. Johnson, and W. Oldham, *J. Vac. Sci. Technol. B* **12**, 3857 (1994).

69. G. M. Wallraff, J. Hutchinson, W. Hinsberg, F. Houle, and P. Seidel, *Microelectronic Eng.* **27**, 397 (1995).

70. W. Huang, R. Kwong, A. Katani, and M. Khojasteh, *Proc. Soc. Photo-Opt. Instr. Eng.* **2195**, 37 (1994).

71. K. Przybilla, Y. Kinoshita, T. Kudo, S. Masuda, H. Okazaki, M. Padmanaban, G. Pawlowski, H. Roeschert, W. Speiss, and N. Suehiro, *Proc. Soc. Photo-Opt. Instr. Eng.* **1925**, 76 (1993).

72. T. Hattori, A. Imai, R. Yamanaka, T. Ueno, and H. Shiraishi, *J. Photopolym. Sci. Technol.* **9**, 611 (1996).

73. R. Allen, W. Conley, and J. Gelorme, *Proc. Soc. Photo-Opt. Instr. Eng.* **1672**, 513 (1992).

74. W. Feeley, J. Imhof, and C. Stein, *Polym. Eng. Sci.* **26**, 1101 (1986).

75. J. Frechet, S. Matsuszczak, H. Stover, C. Willson, and B. Reck, in E. Reichmanis, S. A. MacDonald and T. Iwayanagi, eds., *Polymers in Microlithography, ACS Symposium Series 412*, American Chemical Society, Washington, D.C., 1989, p. 74.

76. S. MacDonald, N. Clecak, H. Wendt, G. Willson, C. Snyder, C. Knors, N. Deyoe, J. Maltabes, J. Morrow, A. McGuire, and S. Holmes, *Proc. Soc. Photo-Opt. Instr. Eng.* **1466**, 1 (1991).

77. O. Nalamasu, E. Reichmanis, M. Cheng, V. Pol, J. Kometani, F. Houlihan, T. Neenan, M. Bohrer, D. Mixon, and L. Thompson, *Proc. Soc. Photo-Opt. Instr. Eng.* **1466**, 13 (1991).

78. R. Cox, L. Druet, A. Klausner, T. Modro, P. Wan, and K. Yates, *Can. J. Chem.* **59**, 1568 (1981).

79. A. Oikawa, Y. Hatakenaka, Y. Ikeda, Y. Kokubo, M. Tanishima, N. Sanoth, and N. Abe, *Proc. Soc. Photo-Opt. Instr. Eng.* **2438**, 599 (1995).

80. T. Kumeda, Y. Tanaka, A. Ueyama, S. Kubota, H. Koezuka, T. Hanawa, and H. Morimoto, *Proc. Soc. Photo-Opt. Instr. Eng.* **1925**, 31 (1993).

81. A. Oikawa, N. Santoh, S. Miyata, and Y. Hatakenaka, *Proc. Soc. Photo-Opt. Instr. Eng.* **1925**, 92 (1993).

82. Y. Kawai, A. Otaka, A. Tanaka, and T. Matsuda, *Jpn. J. Appl. Phys.* **33**, 7023 (1994).
83. Y. Kawai, A. Otaka, J. Nakamura, A. Tanaka, and T. Matsuda, *J. Photopolym. Sci. Techn.* **8**, 535 (1995).
84. K. Asakawa, T. Ushirogouchi, and M. Nakase, *Proc. Soc. Photo-Opt. Instr. Eng.* **2438**, 563 (1995).
85. S. Satio, N. Kihara, T. Naito, M. Nakase, T. Nakasugi, and Y. Kato, *J. Photopolym. Sci. Technol.* **9**, 677 (1996).
86. J. E. McGrath, *J. Chem. Ed.* **58**(11), 844 (1981).
87. R. Dammel, M. Rahman, P. Lu, A. Canize, and V. Elango, *Proc. Soc. Photo-Opt. Instr. Eng.* **2195**, 542 (1994).
88. O. Nalamasu, A. Timko, M. Cheng, J. Kometani, M. Galvin, S. Heffner, S. Slater, A. Blakeney, N. Munzel, R. Schulz, H. Holzwarth, C. Mertesdorf, and T. Schacht, *Proc. Soc. Photo-Opt. Instr. Eng.* **1925**, 155 (1993).
89. H. Ito, G. Breyta, D. Hofer, R. Sooriyakumaran, K. Petrillo, and D. Seeger, *J. Photopolym. Sci. Tech.* **7**, 433 (1994).
90. R. Allen, G. Wallraff, W. Hinsberg and L. Simpson, *J. Vac. Sci. Technol. B* **9**, 3357 (1991).
91. Y. Kaimoto, K. Nozaki, S. Takechi, and N. Abe, *Proc. Soc. Photo-Opt. Instr. Eng.* **1672**, 66 (1992).
92. R. Allen, G. Wallraff, R. DiPietro, D. Hofer, and R. Kunz, *Proc. Soc. Photo-Opt. Instr. Eng.* **2438**, 474 (1995).
93. K. Nozaki, K. Watanabe, E. Yano, A. Kotachi, S. Takechi, and I. Hanyu, *J. Photopolym. Sci. Technol.* **9**(3), 509 (1996).
94. N. Shida, T. Ushirogouchi, K. Asakawa, and N. Nakase, *J. Photopolym. Sci. Technol.* **9**(3), 457 (1996).
95. R. Allen, R. Sooriyakumaran, J. Opitz, G. Wallraff, R. DiPietro, G. Breyta, D. Hofer, R. Kunz, S. Jayaraman, R. Schick, B. Goodal, U. Okoroanyanwu, and G. Willson, *Proc. Soc. Photo-Opt. Instr. Eng.* **2724**, 334 (1996).
96. T. Wallow, F. Houlihan, O. Nalamasu, E. Chandross, T. Neenan, and E. Reichmanis, *Proc. Soc. Photo-Opt. Instr. Eng.* **2724**, 355 (1996).
97. T. Ohfuju, O. Nalamasu, and D. Stone, *J. Vac. Sci. Technol. B* **11**, 2714 (1993).
98. W. Conley, *Proc. Soc. Photo-Opt. Instr. Eng.* **2438**, 41 (1995).
99. J. Much, D. Hess, and E. Aydil, in ref. 1, pp. 377–493.
100. L. Pedersen, *J. Electrochem. Soc.* **129**, 205 (1982).
101. H. Gokan, S. Esho, and Y. Ohnishi, *J. Electrochem. Soc.* **130**(1), 143 (1983).
102. R. Kunz, S. Palmateer, A. Forte, R. Allen, G. Wallraff, R. DiPietro, and D. Hofer, *Proc. Soc. Photo-Opt. Instr. Eng.* **2724**, 365 (1996).
103. Though often overlooked, in some cases the amount of residual coating solvent in a resist film can be as much as 20% of the total film weight: see, for example, W. Hinsberg, S. MacDonald, C. Snyder, H. Ito, and R. Allen, in *Polymers for Microelectronics, ACS Symposium Series 537*, American Chemical Society, Washington, D.C., 1994, p. 101.
104. L. Schlegel, T. Ueno, N. Hayashi, and T. Iwayanagi, *J. Vac. Sci. Technol. B* **9**, 278 (1991).
105. F. Vinet, N. Buffet, P. Fanton, L. Pain, and P. Paniez, *Proc. Soc. Photo-Opt. Instr. Eng.* **2438**, 202 (1995).
106. W. Hinsberg, S. MacDonald, N. Clecak, and C. Snyder, *Proc. Soc. Photo-Opt. Instr. Eng.* **1672**, 24 (1992).
107. W. Hinsberg, S. MacDonald, N. Clecak, C. Snyder, and H. Ito, *Proc. Soc. Photo-Opt. Instr. Eng.* **1925**, 43 (1993).
108. P. Paniez, C. Rosilio, B. Mouanda, and F. Vinet, *Proc. Soc. Photo-Opt. Instr. Eng.* **2195**, 14 (1994).

109. H. Ito, W. England, N. Clecak, G. Breyta, H. Lee, D. Yoon, R. Sooriyakumaran, and W. Hinsberg, *Proc. Soc. Photo-Opt. Instr. Eng.* **1925**, 65 (1993).
110. G. Breyta, D. Hofer, H. Ito, D. Seeger, K. Petrillo, H. Moritz, and T. Fischer, *J. Photopolym. Sci. Tech.* **7**, 449 (1994).
111. The following publications are among those in which new work in advanced resist chemistry can be located: *Proceedings of the Society of Photo-Optical Instrumentation Engineers, Journal of Vacuum Science and Technology, Journal of Electrochemical Society, Micro/NanoEngineering*, and *Chemistry of Materials*.
112. The SIA 1995 National Technology Roadmap from K. H. Brown, *Proc. Soc. Photo-Opt. Instr. Eng.* **2438**, 33 (1995). A good source of information on the new exposure technologies can be found in the Proceedings of the International Conference on Electron, Ion and Photon Beam Technology and Nanofabrication, published annually in Issue 6 of the *Journal of Vacuum Science and Technology B*.
113. R. DeJule, *Semiconductor International* **19**(9), 85 (1996).
114. P. N. Dunn, *Solid State Technology* **37**(6), 49 (1994).
115. R. D. Miller and G. M. Wallraff, *Advanced Materials for Optics and Electronics* **4**, 95 (1994).
116. B. Smith, D. Mixon, A. Novembre, and S. Butt, *Proc. Soc. Photo-Opt. Instr. Eng.* **2438**, 504 (1995).
117. H. Watanabe, Y. Todokoro, and M. Inoue, *J. Vac. Sci. Technol. B* **9**, 3436 (1991).
118. U. Schaedeli, E. Imguely, A. Blakeney, P. Falcigno, and R. Kunz, *Proc. Soc. Photo-Opt. Instr. Eng.* **2724**, 344 (1996).
119. G. Wallraff, R. Miller, M. Baier, E. Ginsberg, and R. Kunz, *J. Photopolym. Sci. Technol.* **5**, 111 (1992).
120. D. Seeger, D. La Tulipe, R. Kunz, C. Garza, and M. Hanratty, *IBM J. Res. Devel.* **41**, 105 (1997).
121. S. MacDonald, G. Willson, and J. Frechet, *Acc. Chem. Res.* **27**, 151 (1994).
122. S. Palmateer, R. Kunz, M. Horn, A. Foote, and M. Rothschild, *Proc. Soc. Photo-Opt. Instr. Eng.* **2438**, 455 (1995).
123. M. D. Levenson, *Solid State Technol.* **38**(2), 57 (1995).
124. Y. Nakayama, S. Okazaki, N. Saitou, and H. Wakabayashi, *J. Vac. Sci. Technol. B* **8**, 1836 (1990).
125. S. Okazaki, *Proc. Soc. Photo-Opt. Instr. Eng.* **2438**, 18 (1995).
126. L. R. Harriott, S. Berger, C. Biddick, M. Blakey, S. Bowler, K. Brady, R. Camarda, W. Connelly, A. Crorken, J. Custy, R. Dimarco, R. Farrow, J. Felker, L. Fetter, R. Freeman, L. Hopkins, H. Huggins, C. Knurek, J. Kraus, J. Liddle, M. Mkrtychan, A. Novembre, M. Peabody, R. Tarascon, H. Wade, W. Waskiewicz, G. Watson, K. Werder, and D. Windt, *J. Vac. Sci. Technol. B* **14**, 3825 (1996).
127. T. Chang, M. Thompson, E. Kratchmer, H. Kim, M. Yu, Y. Lee, S. Rishton, B. Hussey, and S. Zolgharnain, *J. Vac. Sci. Technol. B* **14**, 3774 (1996).
128. H. Loschner, G. Stengl, A. Chalupka, J. Fegerl, R. Fischer, G. Lammer, L. Malek, R. Nowak, C. Traher, and P. Wolf, *J. Vac. Sci. Technol. B* **11**, 487 (1993).
129. G. Kubiak and D. Kania, eds., *OSA Trends in Optics and Photonics on Extreme Ultraviolet Lithography*, Optical Society of America, Washington, D.C., 1996.
130. M. Wilson, A. Smith, V. Kempson, M. Townsend, J. Schouten, R. Anderson, A. Jorden, V. Suller, and M. Poole, *IBM J. Res. Develop.* **37**, 357 (1993).
131. H. Smith and M. Schattenberg, *IBM J. Res. Develop.* **37**, 319 (1993).

W. D. HINSBERG
G. M. WALLRAFF
R. D. ALLEN
IBM Research Division

MICROBIAL AND VIRAL FILTRATION

Several physicochemical methods exist to ensure the safety of biological and biopharmaceutical products in manufacture. Whereas physical methods such as heat and radiation may be used, these are often not viable options owing to detrimental effects on product quality. For example, in the case of products that are heat labile or biochemically complex, these methods may result in alteration of the chemistry or function of the product, or random adsorption of key components. Filtration (qv) is the separation of particles from a fluid (liquid or gas) by passage of that fluid through a permeable medium. Sterile filtration ensures complete removal of viable organisms. Advances in membrane technology (qv) have resulted in the availability of filtrative devices for the removal of viruses in addition to bacteria. Thus membrane filtration is becoming increasingly the method of choice for sterilization of biologicals, especially when the product is heat labile, because the filtration process is inherently nondestructive.

An overview of the general principles of filtration having specific application to bacterial and viral removal is given herein. The emphasis is on ensuring that the sterility and/or safety of biologicals and biopharmaceuticals be maintained.

Filtration for Bacterial Removal

The introduction of parenteral drugs at the beginning of the twentieth century created a concomitant need for a suitable method to ensure adequate sterilization of these biochemically complex heat-labile products. Several different types of filters were introduced into the industrial arena: porcelain filter cartridges, asbestos–cellulose layers, and membrane filters. The porcelain filters (Chamberlain) were used extensively for the manufacture of antibiotics. Problems associated with cleaning and concerns over cross-contamination of products arose, however, and these filters fell into disuse. The first filter medium to be used on an industrial scale was the Seitz EK Filter (EK = *entkeimung* or germ removal). The asbestos content of that filter limited its applications. The first membrane disks were introduced in 1929 and represented a breakthrough in filtration technology. Four decades later membrane filter cartridges were introduced for processing large batches of parenterals.

The earliest commercially available filters were manufactured in two pore sizes: 0.45 and 0.8 μm. The 0.45 μm-rated membranes were considered to be sterilizing-grade filters and were successfully used in the sterile filtration of pharmaceuticals and parenterals. The membrane filters were qualified using *Serratia marcescens*, a standard bacterium, having dimensions of 0.6 $\times$ 1 μm. However, in the late 1960s it became apparent that the matrix of the 0.45 μm-rated filters could be penetrated by some pseudomonad-like organisms (1). For sterile filtration applications in the 1990s, 0.2 μm-rated membranes are the industry standard in the manufacture of sterile parenterals and pharmaceuticals.

General Principles. *Mechanisms of Filter Retention.* In general, filtrative processes operate via three mechanisms: inertial impaction, diffusional interception, and direct interception (2). Whereas these mechanisms operate concomitantly, the relative importance and role of each may vary.

Direct interception refers to a sieve-type mechanism in which contaminants larger than the filter pore size are directly trapped by the filter. This sieve retention mechanism of particle arrest is the mechanism of choice and occurs owing to geometric or spatial restraint. This type of particle arrest is considered to be absolute, that is, it is independent of filtration conditions.

Inertial impaction involves the removal of contaminants smaller than the pore size. Particles are impacted on the filter through inertia. In practice, because the differential densities of the particles and the fluids are very small, inertial impaction plays a relatively small role in liquid filtration, but can play a major role in gas filtration.

Diffusional interception or Brownian motion, ie, the movement of particles resulting from molecular collisions, increases the probability of particles impacting the filter surface. Diffusional interception also plays a minor role in liquid filtration. The nature of liquid flow is to reduce lateral movement of particles away from the fluid flow lines.

Types of Filters. In general, there are two types of filters used for microbial removal: depth and membrane. The first type removes microorganisms and particles mainly through retention by entrapment or impaction and adherence. These rely on filter matrix depth to achieve particulate contaminant retention. However, using depth filters, microbial cells may sometimes be set free because of high differential pressure, the high flux of fluid passing through the filter, or filter matrix shifting. The primary mechanism of bacterial cell retention by membrane filters is the sieving effect, due to the highly stable, uniform pore matrix, so that trapped cells are not released. Thus membrane filtration is the method of choice for pharmaceutical and biological applications where absolute microbe retention is required. Membrane filters used for sterile filtration applications are typically constructed from, but not limited to, polymers such as cellulose esters, nylon, polyesters, polytetrafluoroethylene (PTFE), poly(vinylidene fluoride) (PVDF), polycarbonate, polypropylene, and polysulfone.

Membrane Filter Ratings. Filters are rated based on the ability to remove particles of a specific size from a fluid. There is, however, no standard on which method is to be used to specify performance. In general, the absolute rating, or cutoff point of a filter refers to the diameter of the largest particle, normally expressed in micrometers which can pass through the filter. The absolute rating is determined under carefully controlled conditions using industry-accepted reference standards, such as silica suspensions, latex beads, or microorganisms. In a filtration system, the actual form of the contaminants is not necessarily spherical, in which case the nominal diameter is generally taken as the largest of the linear dimensions. Many filter manufacturers use a nominal filter rating, which is an arbitrary value determined by the filter manufacturer and expressed in terms of percentage retention of a specific test contaminant (usually latex or glass beads) of a given size distribution. This nominal rating also represents a nominal efficiency figure (~50–95%), or more correctly a degree of filtration. Nominal rating standards are, however, arbitrary and thus comparison of nominally rated filters is imprecise. This rating system is not used to characterize sterilizing-grade filters.

Sterilizing-grade filters require that biological retention capability be evaluated using a microbial challenge test. *Brevundimonas* (*Pseudomonas*) *diminuta* (ATCC 19146) is the standard test microorganism used for the validation of sterilizing-grade filters. By U.S. FDA definition, sterilizing-grade filters refer to filters which can remove a minimum concentration of 10^7 colony forming units per square centimeter (cfu/cm^2) of *Brevundimonas* (*Pseudomonas*) *diminuta* (ATTC 19146) and yield sterile effluent (3).

Filter Selection. A variety of product- and process-related factors govern filter selection. Considerations include the characteristics of the fluid to be filtered, ie, its chemical composition and compatibility with the filtration system (inclusive of the membrane, filter hardware, piping, etc), the level of bioburden present, specifications on effluent quality, the volume of product to be filtered, flow rate, and temperature.

Membrane-Feed Compatibility. The feed stream must be compatible with the membrane selected. The composition of the feed as well as pH and operating temperature must be considered. Materials having excessively low or high pH may not be compatible with certain membrane polymers. The temperature of operation must fall within the membrane manufacturer's recommended temperature range. Most fluids that are sterile filtered are water based and thus compatible with most membrane materials, such as polyamides, PVDF, polysulfone, and cellulose acetate. In addition to the compatibility of the membrane itself, the compatibility of the filter components, such as the cage, core, and supporting materials, must also be considered. For sterile processes, the biological safety of the membrane filter or filter cartridge must be demonstrated by the performance of the USP ⟨88⟩ Class VI (121 C) Plastics Test for Biological Reactivity. Because there is no specific listing for materials in contact with pharmaceutical products, filter materials of construction are often selected based on the listing for food contact in the *Code of Federal Regulations* (4).

Effluent Quality. The criteria to be met by the effluent or filtrate must be clearly defined. For aseptic processes a typical requirement is sterilization through an 0.2 μm-rated sterilizing-grade filter, as defined by ASTM Standard F838-83 (5). The objective of the filtration is to remove contaminants from the

bioburden: the specification for challenge level, a minimum of 1×10^7 cfu/cm^2 of filter area, is far in excess of the bioburden routinely encountered in typical pharmaceutical process streams. Challenge concentrations higher than 1×10^8 cfu/cm^2 are not recommended as caking/plugging from excessive bioburden can occur; (3) process considerations during testing: passage of microorganisms through partially retentive filters may be enhanced by application of high pressure; temperature may affect fluid viscosity as well as viability and growth rates of microorganisms; and (4) solution characteristics of the carrier fluid: the pH, ionic strength, osmolarity, and presence of additives such as surfactants, especially in the case of simulated process streams, may influence microbial retention; it is necessary to ensure that these variables are controlled in the test, to assure reproducibility of test results.

Factors that could potentially affect microbial retention include filter type, eg, structure, base polymer, surface modification chemistry, pore size distribution, and thickness; fluid components, eg, formulation, surfactants, and additives; sterilization conditions, eg, temperature, pressure, and time; fluid properties, eg, pH, viscosity, osmolarity, and ionic strength; and process conditions, eg, temperature, pressure differential, flow rate, and time.

The efficiency of the filter is evaluated in terms of the titer reduction or log reduction value (LRV). The Titer reduction is the ratio of the number of microorganisms in the input suspension to the number of organisms in the effluent. Similarly, the LRV is the $\log_{10}$ of this ratio. The ratio of the difference between the numbers of challenge microorganisms recovered upstream and downstream of the test filter to the average total challenge received by the filter provides an indication of the removal efficiency of the filter, ie,

$$\text{removal efficiency, } \% = \frac{\text{average total challenge} - \text{average total recovery}}{\text{average total challenge}}$$
$$\times 100$$

When the filtrate is sterile, the number 1 is substituted for the average total recovery and the value is expressed as greater than the calculated value.

Although *Brevundimonas* (*Pseudomonas*) *diminuta* (ATCC 19146) is most commonly used for sterilizing-grade filter validation, in certain applications other bacteria are used. For example, when it is necessary to demonstrate removal of mycoplasma in applications involving sera and tissue culture media, membranes having a smaller pore size rating, eg, 0.1 μm, are frequently used. For these membranes, *Acholeplasma laidlawii* may be employed for validation purposes (9).

Integrity Testing. The only test of a filter's ability to remove bacteria is demonstration of its performance characteristics by bacterial-retention testing. However, a bacterial challenge is a destructive test and precludes subsequent use in a filtration operation. Therefore, filter manufacturers provide validation documentation for a filter with correlation of microbial removal to other nondestructive physical integrity tests. Examples of nondestructive tests most commonly used in the industry are bubble-point determinations, pressure hold testing, and forward (diffusive air) flow measurements. These methods have the advantage of serving as in-process checks on the integrity of the sterilizing membrane while ensuring proper pore size. In the preparation of parenterals current practice is to subject the filter assembly to a nondestructive integrity test both prior to and after completion of the filtration operation.

The industry-accepted integrity tests are performed by applying gas, eg, air or nitrogen, pressure to a wetted filter and monitoring the air flow. These tests, applicable to both hydrophilic and hydrophobic membrane filters, include forward flow, bubble point, and pressure hold. The forward flow test quantitatively measures the diffusive flow as well as flow through any open pores in a wetted membrane filter. The test is performed by wetting a membrane filter and applying a predetermined constant pressure. The test pressure is established for a particular membrane filter by the filter manufacturer. The diffusional gas (air or nitrogen) flow rate as well as the flow through any open pores is measured through the wetted membrane. The gas flow is usually defined in units of mL/min. The filter is integral if the forward flow value is lower than the manufacturer's specified value. The bubble point is qualitative and is dependent on the observation downstream of bubbling through the largest pores of a wetted filter. This test is usually employed for small surface area filters such as membrane disks. To perform the visual bubble point test, the filter membrane is first wet using an appropriate solvent, then air or nitrogen pressure is slowly applied until the wetting fluid is expelled from the largest pores and gas bubbles appear from a submerged tube in a downstream collection vessel. The pressure hold is a modified form of upstream forward flow testing and involves the measurement of decay in pressure over a specified time period for a given filter assembly and wetting fluid. Because the test is performed upstream, the downstream sterile connections are not disturbed. The advantage of this test is that it can be performed after sterilization of the filter assembly, as well as pre- and post-filtration. The filter housing is pressurized to the test pressure specified by the filter manufacturer, then the filter is isolated from the pressure source. The diffusion of gas across the wetted membrane is measured as a decay in pressure over a specified period of time. The pressure hold and forward flow tests are related through the ideal gas law (10). Automated filter integrity test instruments are available in order to provide reproducibility.

Validation Considerations.   The need to validate all processes related to the manufacturing of pharmaceutical and biopharmaceuticals has been well established in good manufacturing practices (GMP) regulations and various other guidelines. Filtrative particle removal may be attributable to other mechanisms in addition to direct interception or size exclusion, including, for example, adsorptive particle arrest. It is therefore necessary to validate filter performance as the efficiency of the given filter is dependent on the physical, eg, viscosity and temperature, as well as the chemical, eg, presence of surfactants, composition of the suspending fluid. Microbial retention is required to be demonstrated under simulated pharmaceutical conditions in order to document the performance claims of the filter (3).

Validation of sterile filtration processes is required to be carried out under worst-case conditions. Typically, multiple filter lots (usually three) are challenged with the product under actual or simulated process conditions. A membrane filter having a pore size that allows penetration by the challenge organism is also included as a control. Most commonly, an 0.45 μm-rated filter is included when validation of 0.2 μm-rated sterilizing-grade filters are tested, and should show incomplete retention of *B. diminuta* (1). Essentially, the criteria for the selection of the challenge organism is that it should be small enough to challenge the retentivity of the sterilizing-grade filter and simulate the smallest organism that may occur during production.

If a specific organism has been identified as a routine bioburden, this organism may be substituted for *Brevundimonas* (*Pseudomonas*) *diminuta* (ATCC 19146). Ideally, validation experiments are conducted in the product under conditions that closely simulate process conditions. This is, however, not always possible. For example, prior to any microbial retention study it is necessary to determine viability of the test organism in the test fluid. If the product affects the viability of the test organism, as, for example, in the case of cytotoxic drugs, an appropriate substitution is essential. A placebo, ie, a formulation designed to simulate the actual product having the ionic strength, osmolarity, viscosity, surface tension, and other attributes equivalent to the product, but which does not contain the active drug substance should be used. Alternatively, the native formulation may be modified, ie, the active ingredient is present but the bacteriocidal component removed/neutralized/diluted to ensure no effect on microbial viability. Similarly, if other process conditions affect the viability of the test organism, appropriate modifications are essential. Simulation of other process conditions is also essential. For example, hydraulic process conditions should be simulated during the bacterial challenge to assess any effect on the filter relative to its ability to retain bacteria. Such conditions include maximum differential pressure and pulsing. The pressure differential across the test filter should meet or exceed the maximum pressure differential observed during processing, within the design specifications provided by the filter manufacturer. This serves to validate the filter's ability to retain bacteria in the product and provide a sterile effluent up to or beyond the maximum process pressure differentials.

Sterilization Considerations. In sterile filtration processes the downstream side of the filter must be sterilized and must remain sterile during the entire process. Presterilized (gamma-irradiated) filters may be available or alternatively, filters may be sterilized by the user. The most common method of sterilization is by steam under pressure. The sterilization process must be validated to ensure that sterile conditions are met for a given system. Methods that involve steam are validated through the use of thermocouples and/or biological indicators to ensure sterilization of the system. The filters may be sterilized in an autoclave. Alternatively, the sterilization may be undertaken in-place, called sterilization in-place (SIP) or *in situ* sterilization. Most filter manufacturers provide protocols and recommendations for these procedures. Small-volume systems tend to be autoclaved; larger systems are typically SIP. Minimally, a temperature of 121°C is used for autoclave sterilization; more commonly, a sterilization temperature of 125°C is employed. Temperatures in excess of 125°C may affect the plastics used for filter construction, thereby affecting the physical integrity of the membrane. Standard precautions must be followed during autoclaving: the system must be purged of air to achieve reliable sterilization, and the filter assembly wrapped using a porous barrier to ensure steam penetration. It is critical that excessive differential pressures are not created during the autoclaving which would result in damage to the membrane. These same considerations are all the more relevant during *in situ* sterilization. Additionally, in SIP operations the condensate must be drained throughout the steam cycle to prevent accumulation. This is achieved by keeping the drains and steam traps partially open during the steaming cycle. After the steam valve is closed, a noncondensable gas, such as air or nitrogen, is introduced into the housing to prevent formation of a vacuum on the upstream side of the filter. If it is necessary that the system be completely dry prior to use, the air-nitrogen flow can be continued until the system is dry and cool to operating temperature.

Other methods of sterilization may also be used. Irradiation has the advantage of assuring sterility without any residual gas components. However, several polymers used in filter manufacture may have limited resistance to irradiation sterilization. As with any other process, it has to be validated for sterility. Spores of *Bacillus pumilus* are the indicator organisms for validation of radiation sterilization. Another method is gas sterilization using a gas such as ethylene oxide. For successful sterilization of filters by ethylene oxide, the filters must be dry and wrapped so as to allow penetration of the gas into the matrix or the filter. In addition to the health and safety concerns associated with ethylene oxide gas itself, by-products such as ethylene chlorohydrin and ethylene glycol may be generated which also constitute health hazards. If ethylene oxide is used, appropriate venting is necessary following the sterilization cycle; however, in spite of venting there are concerns that some of the by-products may remain in the filter matrix. The sterilization method must be validated. The indicator organism recommended is *Bacillus subtilis* spores.

Sterile Filtration of Gases.   Primary applications for sterile gas filtration are the sterilization of fermentor inlet air, fermentor vent gas, vents on water for injection tanks, and vacuum break filters during lyophilization. Operational and process considerations apply. Typically, the membrane in gas filtration applications is a hydrophobic membrane, eg, PVDF or PTFE, although there are applications in which the liquid (condensate) in the system is well controlled and hydrophilic membranes may be used. The inherent hydrophobicity of membrane filters used for fermentor air sterilization allows these filters to remove bacteria completely from inlet air, even when exposed to moisture (11). The effluent for gas filtration applications is typically filtered at the 0.2-μm level.

Verification of the microbial retention efficiency of the membrane filters may be undertaken using either liquid or aerosol challenge tests. A liquid challenge test is more stringent. Furthermore, this test can provide retention information for process conditions such as extreme moisture after sterilization or air entrained with water drops. A liquid challenge is performed using a protocol similar to that described for liquid filtration.

Aerosol challenges may be conducted using essentially a test setup composed of three components: the nebulizer, mixing chamber, and a sampling system. An applicable system was first described in 1978 (12). A schematic representation of an aerosol challenge setup is provided in Figure 2. The challenge microorganism is aerosolized using a nebulizer. The aerosol is then mixed with compressed dry air to ensure that the monodispersed microbial challenge to the filter is delivered as a dry aerosol, rather than as microdroplets. Sampling may be done using a vacuum switch device that alternates between the upstream and downstream impingers, and a split-stream liquid impingement method. Any excess air flow not collected by the impingers is vented through an exhaust filter located upstream of the filter. Following the challenge, the buffer from the impingers located upstream and downstream of the test filters are assayed using standard microbiological methods. The dual impingers allow for precise

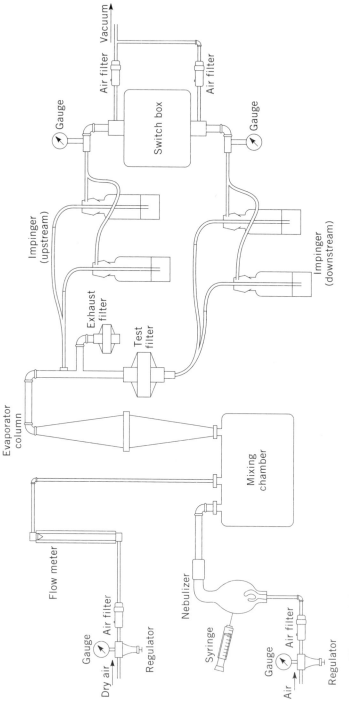

Fig. 2. Schematic of an experimental aerosol bacterial challenge setup for evaluation of bacterial retention.

determination of the actual challenge level for each test filter and calculation of the efficiency of titer reduction of the input challenge level.

Filters for use in sterile gas filtration must conform to standards similar to those mandated for sterile liquid filtration. Nondestructive integrity tests may be applied. The tests are performed by wetting the filter with an appropriate solvent, commonly 60/40 isopropyl alcohol/water for hydrophobic membranes, and applying air or nitrogen gas at a preset pressure.

Filtration of Virus Removal

General Principles.   Filtration was traditionally used for the removal of bacteria and mycoplasma from biologicals that were heat labile. Advances in filtration technology have resulted in the availability of filtration devices for applications involving removal of viruses. The virological safety of biologicals and biopharmaceuticals is a key consideration in their manufacture. Much of the concern regarding viral contaminants in therapeutic agents centers around blood and blood products as well as biopharmaceuticals which have a blood or tissue component to their production. Human viruses of greatest concern have included human immunodeficiency virus (HIV), hepatitis B virus, hepatitis C virus, cytomegalovirus, and parvovirus. Nonhuman viruses such as bovine viral diarrhea virus are of concern if raw materials derived from these animals are incorporated into the production scheme. Viruses represent a diverse group which include enveloped and nonenveloped viruses, and ribonucleic acid (RNA) and deoxyribonucleic acid (DNA) viruses of various sizes. The focus herein is on removal of viruses from fluids.

Methods used to ensure virological safety are briefly classified as either virus inactivation or virus removal methods. The former includes chemical inactivation, pasteurization, uv inactivation, and solvent–detergent and ion-exchange (qv) chromatography (qv) (13–16). Whereas these methods can be very effective depending on the inactivation process and the nature of the product, there are limitations to the application of these methods. Stabilizers, sometimes used to ensure that the biological activity is not compromised during the treatment, need to be removed from the final product. Heat treatment can denature certain proteins (17,18). Processes such as solvent–detergent, chemical treatment, and uv irradiation are often not uniformly effective against all viruses, especially enveloped ones.

The most desirable mechanism for the removal of viral particles using filtration is size exclusion. However, as in the case of bacterial removal by filtration, other mechanisms may also influence virus removal. These factors can include viral adsorption to the filter surface by electrostatic interactions; changes in pore size characteristics during filtration owing to deposits of material on the membrane surface, eg, development of a gel layer; and the filtration conditions, flow rate, pressure, temperature, etc. For example, adsorption can be a complex relationship between fluid pH, membrane chemistry, and the level of organics or protein in the fluid. Thus the removal of viruses by particle size minimizes many of the variables affecting the level of retention and can be a predictable means of sterilization. Removal of viral particles from fluids by size exclusion is preferable. However, concomitant with the requirement for adequate virus

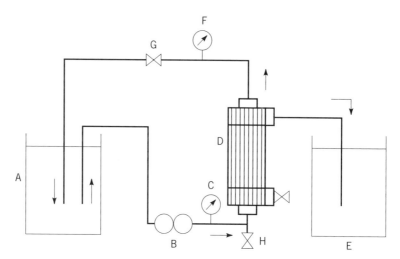

Fig. 4.   Diagram of a hollow-fiber ultrafilter filtration system where A corresponds to the retentate reservoir; B, circulation pump; C, pressure gauge at module inlet; D, ultrafilter module; E, permeate reservoir; F, pressure gauge at module outlet; G, value to control module outlet pressure; and H, drain valve.

be determined, as in the case of bacterial removal, by comparing the virus concentration in the input suspension to the concentration of virus in the effluent.

The plaque assay is desirable because it is very sensitive and only detects infectious viral particles. However, there are viral agents which cannot be supported by cell lines. In these cases other methods must be used. The polymerase chain reaction (PCR), which amplifies DNA or RNA from viral agents, can be used to detect the presence and quantity of viral agents. The amount of RNA or DNA target in the initial sample can be determined by competitive PCR where the quantity of amplified product is compared to a control PCR product where the initial amount of target is known. Quantification is also possible by an end-point dilution method similar to that used to determine a tissue culture infections dose. PCR methods can be very sensitive; however, detection is based on the presence of nucleic acid and consequently the method does not differentiate between an infectious or noninfectious viral particle. In general, the detection of viral agents by either cell culture or PCR assay tends to be virus specific. Therefore multiple assay systems may be needed to detect more than one type of virus.

Effectiveness of Membrane Filtration.   *Microfiltration.*   Various membrane filters have been used to remove viral agents from fluids. In some cases, membranes which have pores larger than the viral particle can be used if the filtration is conducted under conditions which allow for the adsorption of the viral particle to the membrane matrix. These are typically single-pass systems having pore sizes of 0.10–0.22 μm. Under situations which allow optimum adsorption, between 10–10^2 particles of poliovirus (28–30 nm) were removed (34–36). The formation of a cake layer enhanced removal (35). The titer reduction when using 0.10–0.22 μm membrane filters declined under conditions which minimized adsorption. By removal standards, these filters remove viruses at a rate on the low end of the desired titer reduction and the removal efficiency varies with differences in fluid chemistry and surface chemistry of viral agents (26).

Smaller pore size microfilters in single-pass systems which have pore sizes small enough to remove some viruses by size exclusion have been examined (26,37,38). Minimum levels of virus removal can be established for these systems if fluid and process conditions are employed which minimize removal of viral particles by mechanisms other than size selection.

Nylon filter membranes have been tested using 47-mm disks of the filter material (26,37). Influenza A virus (80–120 nm) and HIV (80–100 nm) were removed to below detectable limits in all fluids tested. However, removal of viruses smaller than influenza A virus was not as efficient. Titer reduction results for small (25–50 nm) viruses produced large differences depending on fluid type. The highest titer reductions were observed from high purity water and solutions having low concentrations of protein; smaller titer reductions were observed from solutions containing bovine serum. These results suggest that other factors in addition to size exclusion were enhancing the titer reduction for viruses in the 25–50 nm size range. Diminished adsorptive effects have been observed in the presence of serum or by pretreating a normally adsorbent membrane with serum or gelatin.

A PVDF membrane filter has been shown to remove $>10^6$ particles of virus for viruses >50 nm independent of fluid type (8). Viruses smaller than 50 nm are not removed as efficiently but are removed in a predictable manner which correlates to the virus particle size. The chemistry of the suspending fluid affects titer reduction for viruses <50 nm owing to other removal mechanisms, such as adsorption, coming into play. The effects of these other mechanisms can be minimized by using filtration conditions that minimize adsorption.

Ultrafiltration. Ultrafilters have also been examined for viral removal by size exclusion utilizing tangential flow and hollow-fiber membrane systems (27–29,30–32,35,36). The titer reduction varies depending on virus size and membrane filter pore size distribution. Removal of poliovirus by a 30,000 molecular weight polysulfone ultrafilter removes $>10^4$ particles of poliovirus in water of various qualities. Tangential flow ultrafiltration has been extensively tested using viruses ranging from 28 nm (Phi X174) to 80–100 nm (Murine leukemia virus) in size using a 70 KD PVDF membrane filter (28) (Table 1). There appeared to be a correlation between virus size and titer reduction. In this study, virus removal appeared to be enhanced in the presence of human serum albumin. Regenerated cellulose fiber (BMM) hollow-fiber ultrafilters from 10–80 nm have been tested using human blood borne viruses (30–32) (Table 1). Polysulfone (PS) and polyacrylonitrile (PAN) hollow-fiber ultrafilters have also been tested for their ability to remove viruses between 100–25 nm.

Integrity Testing. As in the case of bacterial removal, it is necessary to carry out an integrity test on the filter, minimally, following filtration to ensure filter performance. Ideally, the integrity test should be performed both pre- and post-use. This is possible when a nondestructive integrity test method is used. Integrity tests used for virus removal filters include a forward-flow test similar to the test done on bacterial removal filters. This test is nondestructive and amenable to use both pre- and post-filtration. The test is correlated to virus removal by the filter manufacturer. Another nondestructive test is the liquid porosimetric integrity test, also correlated to virus removal. Another integrity test includes a gold particle removability test (GPT). This test is destructive and

Table 1. Removal of Viral Particles from Fluids by Ultrafiltration

Filter					
Size reference	Material	Virus (size)	Fluid[a]	Log titer reduction	Reference
500K	PS[b]	MS2 (25 nm)	DI	~1.50	35
300K	ceramic[c]	MS2	DI	4.00	
100K	cellulosic[d]	MS2	DI	>6.00	
			water	>6.00	
30K	PS[e]	polio (28–30 nm)	water	>4.00	
~70 nm	PVDF[f]	Phi X174 (28 nm)	PBS	2.93	28,29
		Phi X174	PBS/HSA	7.42	
		polio (28–30 nm)	PBS	3.1–3.51	
		polio	PBS/HSA	4.2	
		SV-40 (40–45 nm)	PBS	4.89–5.65	
		SV-40	PBS/HSA	>5.7	
		Sindbis (54 nm)	PBS	7.41	
		Reo-3 (78 nm)	PBS	7.18	
		Reo-3	PBS/HSA	>7.6	
		MuLV (85 nm)	PBS/FCS	>6.82	
50 nm	cellulose[g]	HCV (35 nm)	plasma	1.50	32
		HCV	plasma	3.50	
		HCV	plasma	4.00	
		HCV	plasma	>4.00	
105 nm	cellulose[g]	HIV (100 nm)	plasma	>5.77	30
		HIV	plasma	>5.67	
		HIV	plasma	>5.85	
		HIV	plasma	>5.74	
		HIV	plasma	>5.94	
50,000	PAN[h]	polio (28–30 nm)	DMEM-10	4.59	33,36
		phage T1[i]	DMEM-10	6.27	
13,000	PAN[h]	polio (28–30 nm)	DMEM-10	>6.51	
		phage T1[i]	DMEM-10	7.55	
6,000	PAN[h]	phage PP7 (25 nm)	DMEM-10	7.62	
		phage T1[i]	DMEM-10	7.80	
6,000	PS[b]	polio (28–30 nm)	DMEM-10	>6.40	
		phage T1[i]	DMEM-10	7.99	

[a]DI = deionized water; PBS, phosphate-buffered saline; HSA, human serum albumin; FCS, fetal calf serum; and DMEM, Dulbecco's Eagles minimum essential medium + 10% fetal bovine serum. [b]Polysulfone hollow fiber. [c]Tubular. [d]Hollow fiber. [e]Polysulfone flat sheet. [f]Poly(vinylidene fluoride), tangential flow. [g]Regenerated cellulose hollow fiber. [h]Polyacrylonitrile hollow fiber. [i]50-nm head, 150-nm tail.

therefore is applicable only post-use. In general, the integrity test results must correlate with the virus removal claims, as specified by the filter manufacturer.

Validation Considerations. Mechanisms other then size exclusion may be operative in the removal of viruses from biological fluids. Thus virus removal must be validated within the parameters set forth for the production process and using membrane material representative of the product line of the filter.

The validation study for filtrative virus removal essentially involves challenging (spiking) the product using high titers of infectious virus under conditions that simulate process parameters and quantitating virus in pre- and post-treatment samples. Pre-purification treatments and post-purification modification reactions must also be validated. The choice of virus for validation studies is not as clearly defined as in bacterial filtration where there is an industry-accepted standard. No one viral agent can serve as a generic model virus. Thus, validation studies should be conducted using a panel of viruses that includes known contaminants which may represent identifiable and theoretical risks to product contamination, for example, HIV in the case of blood products; suspected contaminants or model viruses resembling suspected contaminants; or a range of viruses of differing properties which are not themselves considered likely contaminants. In some cases, use of surrogate viruses is necessary as some pathogenic viruses, eg, Hepatitis B and Hepatitis C, are not easily propagated in cell culture. Factors affecting virus clearance results include the choice of model viruses, the appropriateness of the scaled-down version, and the search for process variables which may alter the efficacy of virus inactivation–elimination steps, etc.

The virus reduction factor of an individual purification or removal–inactivation step is defined as the $\log_{10}$ of the ratio of the virus load in the pre-purification material divided by the virus load in the post-purification material. A clearance factor for each stage can be calculated and the overall clearance capacity of the production process assessed. Total virus reduction is calculated as the sum of individual log reduction factors. Individual manufacturing steps must possess fundamentally different mechanisms of virus removal or inactivation in order for values to be considered cumulative. Additionally, because viruses vary greatly with regard to inactivation or removal profiles, only data for the same model virus can be cumulative.

A membrane filter which can uniformly remove all viral agents regardless of the size of the viral agent is not available. Part of the difficulty is that the efficient recovery of the biological product diminishes as the size difference between the virus and biological product lessens. Thus a balance needs to be met where virus removal and product recovery are optimized.

Improvements in membrane technology, validation of membrane integrity, and methods to extend filter usage should further improve the performance of membrane filters in removal of viral particles. Methods to improve or extend filter life and increase flow rates by creating more complex flow patterns could possibly be the focus of the next generation of membrane filters designed to remove viral particles.

BIBLIOGRAPHY

1. F. W. Bowman, M. P. Calhoun, and M. White, *J. Pharm. Sci.* **56**, 222 (1967).
2. M. Osumi, N. Yamada, and M. Toya, *J. Pharm. Sci. Technol.* **50**, 30 (1996).
3. *Guidelines on Sterile Drug Products Produced by Aseptic Processing*, Center for Drugs and Biologics and Office of Regulatory Affairs, U.S. FDA, Washington, D.C., June 1987.
4. "Indirect Food Additives Subpart B: Substances for Use as Basic Components of Single and Repeat Use Food Contact Surfaces," *Code of Federal Regulations*, Title 21, Part 177, U.S. Government Printing Office, Washington, D.C., 1994.
5. *Standard Test Method for Determining Bacterial Retention of Membrane Filters Utilized for Liquid Filtration*, ASTM F838-83, American Society for Testing and Materials, Philadelphia, Pa., 1983; reapproved 1988.

6. "Current Good Manufacturing Practice in Manufacturing Processing, Packing or Holding of Drugs: General," *Code of Federal Regulations*, Title 21, Part 210.3 (b) (6), U.S. Government Printing Office, Washington, D.C., 1994.

7. "Current Good Manufacturing Practice for Finished Pharmaceuticals," *Code of Federal Regulations*, Title 21, Part 211.72, U.S. Government Printing Office, Washington, D.C., 1994.

8. H. Aranha and J. Meeker, *PDA J. Pharm. Sci. Technol.* **42**, 67 (1995).

9. J. Meeker, J. B. James, J. M. Martin, and G. Howard, Jr., Technical data, STR RUF08BA, Pall Corp., East Hills, N.Y., 1990.

10. *Validation Guide for the Palltronic™ FFE03-P Filter Integrity Test Instrument*, TR-FFE03-P, Pall Corp., East Hills, N.Y., 1992.

11. C. F. Bruno and L. A. Szabo, *Biotechnol. Bioeng.* **25**, 1223 (1983).

12. R. Duberstein and G. Howard, *J. Paren. Drug. Assoc.* **32**, 192 (1978).

13. E. D. Gomperts, *Lancet* **23**, 295 (1986).

14. M. S. Horowitz, C. Rooks, B. Horowitz, and M. Hilgortner, *Lancet ii*, **186** (1988).

15. P. M. Mannucci and M. Colombo, *Lancet*, 782 (1988).

16. P. Murphy, T. Nowak, S. M. Lemon, and J. Hilfenhaus, *J. Med. Virol.* **41**, 61 (1993).

17. M. Gleeson, L. Herd, and C. Burns, *Ann. Clin. Biochem.* **27**, 592 (1990).

18. A. M. Prince, B. Horowitz, M. S. Horowitz, and E. Zang, *Eur. J. Epidemiol.* **3**, 103 (1987).

19. Centers for Disease Control, *Morbid. Mortal. Weekly Rep.* **45**, 29 (1996).

20. R. D. Aach and R. A. Kahn, *Ann. Intern. Med.* **92**, 539–546 (1980).

21. Y. J. Wang and co-workers, *Vox Sang*, **67**, 187–190 (1994).

22. S. C. Darby, D. W. Ewart, P. L. Giangrande, P. J. Dolin, R. J. D. Spooner, and C. R. Rizza, *Nature*, **377**, 79 (1995).

23. S. Builder, R. Van Reis, N. Paoni, and J. Ogez, in R. W. Spier, J. B. Griffiths, J. Stephenne, and P. J. Crooy, eds., *Advances in Animal Cell Biology and Technology for Bioprocesses*, Butterworths, Stoneham, Mass., 1989, p. 452.

24. "Viral Safety Evaluation of Biotechnology Products Derived from Cell Lines of Human or Animal Origin," ICH Harmonized Tripartite Guideline, IFPMA, Geneva, Switzerland, 1997, pp. 1–27.

25. "Note for Guidance on Plasma Derived Medicinal Products," CPMP/BWP/269/95, Committee for Proprietary Medicinal Products (CPMP), London, Mar. 13, 1996.

26. K. H. Oshima, A. K. Highsmith, and E. G. Ades, *Environ. Topical. Water Qual.* **9**, 165 (1994).

27. A. J. DiLeo, A. E. Allegrezza, and S. E. Builder, *Bio/Technol.* **10**, 182 (1992).

28. A. J. DiLeo, D. A. Vacante, and E. F. Deane, *Biologicals*, **21**, 275 (1993).

29. Ref. 28, p. 287.

30. Y. Hamamoto and co-workers, *Vox Sang*, **56**, 230 (1989).

31. S. Sekiguchi and co-workers, *Membrane*, **14**, 253 (1989).

32. T. Yuasa and co-workers, *J. Gen. Virol.* **72**, 2021 (1991).

33. K. H. Oshima and co-workers, *Can. J. Microbiol.* **41**, 316 (1995).

34. K. Hou, C. P. Gerba, S. M. Goyal, and K. S. Zerda, *Appl. Environ. Microbiol.* **40**, 892 (1980).

35. J. G. Jacangelo, S. S. Adham, and J-M. Laine, *J. AWWA*, **87**, 107 (1995).

36. S. S. Madaeni, A. G. Fane, and G. S. Grohman, *J. Membr. Sci.* **102**, 65 (1955).

37. K. H. Oshima and co-workers, *J. Acquir. Immune Defic. Syndr.* **8**, 64 (1995).

38. K. H. Oshima, T. T. Evans-Strickfaden, and A. K. Highsmith, *Biologicals* **124**, 137 (1996).

General References

H. Aranha-Creado, K. Oshima, S. Jafari, G. Howard, Jr., and H. Brandwein, *Scientific and Technical Report*, STR-PUF-24, Pall Ultrafine Filtration Co., East Hills, N.Y., 1995.

"Bacterial Endotoxins Test ⟨85⟩," "Biological Reactivity Tests, *in vivo* ⟨88⟩," "Particulate Matter in Injections ⟨788⟩," and "Purified Water," *USP 23*, The U.S. Pharmacopeial Convention, Rockville, Md., 1994.

T. J. Leahy and M. J. Sullivan, *Pharm. Technol.* **2** (11).

R. V. Levy and T. J. Leahy, in S. S. Block, *Disinfection, Sterilization, and Preservation*, 4th ed., Lea & Febiger, Philadelphia, Pa., 1991, pp. 527–551.

P. Keating and co-workers, *BioPharm* **5** (1992).

HAZEL ARANHA-CREADO
K. OSHIMA
Pall Corporation

MICROEMULSIONS

There is no official or universally accepted definition of what constitutes a "microemulsion." In fact, for several years, some leading scientists in microemulsion research considered the term to be an unnecessary and even an unfortunate one. Nevertheless (Table 1), during the years from about 1975 to 1980 the word ascended from obscurity to ubiquity. By the end of 1996 there were 13 widely available English-language books (1–9) with the word "Microemulsion" in their titles (10). About 70 more books on surfactants are in print, of which those on industrial applications (9,11–18), and environmental effects (19–21) are of particular interest here.

The concept of microemulsions now holds a central role within the field of surfactant technology. Perhaps the most fundamental fact captured by the term is that, contrary to a popular saying, oil and water can mix.

Table 1. Growth in Microemulsion Publications

Year	Number[a] of publications
1970	12
1975	29
1980	123
1985	237
1990	410

[a] In *Chemical Abstracts*.

Definition of a Microemulsion

The term microemulsion was introduced by Schulman, who studied surfactant solutions as early as 1943 (22). At that time it was widely accepted that "oil and water do not mix," and Schulman understood that an emulsion scatters light because it contains droplets whose diameters are large compared to the wavelength of light (see EMULSIONS). Thus, the term *micro*emulsion implies a components, at the corners of the phase diagrams, and phases, which always contain at least minute concentrations of all components present in the system.)

In Figure 1, the pairs (or triad) of phases that form in the various multiphase regions of the diagram are illustrated by the corresponding test-tube samples. Except in rare cases, the densities of oleic phases are less than the densities of conjugate microemulsions and the densities of microemulsions are less than the densities of conjugate aqueous phases. Thus, for samples whose compositions lie within the oleic phase-microemulsion binodal, the upper phase (ie, layer) is an oleic phase and the lower layer is a microemulsion. For compositions within the aqueous phase-microemulsion binodal, the upper layer is a microemulsion and the lower layer is an aqueous phase. When a sample forms two layers, but the amphiphile concentration is too low for formation of a middle phase, neither layer is a microemulsion. Instead the upper layer is an oleic phase ("oil") and the lower layer is an aqueous phase ("water").

In three-phase systems the top phase, T, is an oleic phase, the middle phase, M, is a microemulsion, and the bottom phase, B, is an aqueous phase. Microemulsions that occur in equilibrium with one or two other phases are sometimes called "limiting microemulsions," because they occur at the limits of the single-phase region.

Rigorous and useful as it is, the definition of limiting microemulsions does not specify which compositions in the single-phase region should also be called microemulsions. In Figure 1 there is critical micelle concentration point for formation of normal micelles on the amphiphile–water side of the phase diagram and an inverse critical micelle concentration point for formation of inverted micelles on the amphiphile–oil side of the diagram (28). (Depending on the compounds and temperature, these CMC points may be easily detected, or exist only in a formal sense.) Geometrically, each of the two CMC points can be connected by a line to a nearby plait point. These lines serve to separate the single-phase area of the diagram formally into (nonlimiting) microemulsion, aqueous, and oleic regions. However, such formalities seldom are needed; and sometimes they may even be counterproductive.

Temperature and Salinity Scans

The locations of the tietriangle and binodal curves in the phase diagram depend on the molecular structures of the amphiphile and oil, on the concentration of cosurfactant and/or electrolyte if either of these components is added, and on the temperature (and, especially for compressible oils such as propane or carbon dioxide, on the pressure (29,30)). Unfortunately for the laboratory worker, only by measuring (or correctly estimating) the compositions of T, M, and B can one be certain whether a certain pair of liquid layers are a microemulsion and conjugate aqueous phase, a microemulsion and oleic phase, or simply a pair of aqueous and oleic phases.

However, often the identities (aqueous, oleic, or microemulsion) of the layers can be deduced reliably by systematic changes of composition or temperature. Thus, without knowing the actual compositions for some amphiphile and oil of points T, M, and B in Figure 1, an experimentalist might prepare a series of samples of constant amphiphile concentration and different oil–water ratios, then find that these samples formed the series (*a*) 1 phase, (*b*) 2 phases, (*c*)

3 phases, (*d*) 2 phases, (*e*) 1 phase as the oil–water ratio increased. As illustrated by Figure 1, it is likely that this sequence of samples constituted (*a*) a "water-continuous" microemulsion (of normal micelles with solubilized oil), (*b*) an upper-phase microemulsion in equilibrium with an excess aqueous phase, (*c*) a middle-phase microemulsion with conjugate top and bottom phases, (*d*) a lower-phase microemulsion in equilibrium with excess oleic phase, and (*e*) an oil-continuous microemulsion (perhaps containing inverted micelles with water cores).

Historically, however, it has been much more common for experimentalists to introduce a new variable into Figure 1, changing either the temperature of one or more samples of fixed composition, or the electrolyte concentration in a series of samples of fixed amphiphile–oil–water ratio. The former constitutes a temperature scan; the latter experiment is widely known as a salinity scan. When the temperature of an amphiphile–oil–water system is varied, the phase diagram can be plotted as a triangular prism (because temperature is an intensive or field variable). When a fourth component (eg, NaCl) is added at constant temperature, tetrahedral coordinates, are appropriate (conjugate phases have different salinities, and the planes of different tietriangles are no longer parallel).

Changes of temperature or salinity cause the phase diagram to evolve through the sequence illustrated by Figure 2. As the temperature is decreased from some temperature between T_{lc} and T_{uc}, the aqueous phase-microemulsion binodal shrinks until it becomes a single point, the lower critical endpoint. Simultaneously, the tietriangle shrinks in height, becoming a lower critical tieline of zero height at $T = T_{lc}$. Conversely, when the temperature is raised, the oleic phase-microemulsion binodal shrinks, until it becomes an upper critical end point at T_{uc}. Below T_{lc} and above T_{uc} the respective binodal curves and plait points disappear, leaving only a single two-phase region. Figure 2 illustrates the phase diagrams of a nonionic amphiphile–oil–water system at $T < T_{lc}$ and $T > T_{uc}$, respectively. Because of the relative densities of the phases, for $T < T_{lc}$ the system is sometimes said to contain an upper phase microemulsion in equilibrium with an aqueous phase, whereas for $T > T_{uc}$ the system may be said to contain a lower phase microemulsion in equilibrium with an oleic phase. However, if the existence of middle-phase microemulsions at intermediate temperatures is unknown, the respective phase pairs are likely to be called simply oil and water.

Of all the characteristic points in the phase diagram, the composition of the middle phase is most sensitive to temperature. Point M moves in an arc between the composition of the bottom phase (point B) at T_{lc} and the composition of the top phase (point T) at T_{uc}, reaching its maximum surfactant concentration near $T = (T_{lc} + T_{uc})/2$. (Points B and T move by much smaller amounts, also.) The complete nonionic-amphiphile–oil–water–temperature phase diagram is illustrated by Figure 3, including the S-shaped curve of T, M, and B compositions; the two lines of plait points, which terminate at the lower and upper critical end points; and the lower and upper critical tielines (at T_{lc} and T_{uc}, respectively).

Figure 4 illustrates the analogous amphiphile–oil–water–electrolyte phase diagram, including a representative tietriangle, the S-shaped curve of T, M, and B compositions, the lower and upper critical end points (R and Q, respectively), and the lower (PR) and upper (QS) critical tielines (31). For

However, as given by group renormalization theory (45), the values of the universal exponents depend on the (thermodynamic) dimensionality of the system. For four dimensions (as required by the phase rule for the existence of tricritical points), the exponents have classical values. This means the values are multiples of 1/2. The dimensions of the volume of tietriangles are (31)

$$h = h_O \left| \frac{T - T_{tc}}{T_{tc}} \right|^{3/2} \tag{4a}$$

$$w = w_O \left| \frac{T - T_{tc}}{T_{tc}} \right|^{2/2} \tag{4b}$$

$$l = l_O \left| \frac{T - T_{tc}}{T_{tc}} \right|^{1/2} \tag{4c}$$

As confirmed by experiment (47), the dependence of the optimal tension on the distance from the tricritical point is given (40) by

$$\sigma = \sigma_O \left| \frac{T - T_{tc}}{T_{tc}} \right|^{4/2} \tag{5}$$

where T_{tc} is the (absolute) tricritical point temperature and the values of h_O, l_O, w_O, and σ_O depend on the chemical system. Thus, there is a simple, universal, thermodynamic relationship between the width of the three-phase region, the composition (ie, amount of solubilized oil or brine) of the middle-phase microemulsion at optimum, and the optimal tensions. Once the values of h_O, l_O, w_O, T_{tc}, and σ_O are known, equations 4 and 5 provide convenient equations for the description of the compositions of phases T, M, and B and the interfacial tensions among them.

Moreover, except for small tradeoffs between h_O and σ_O that may be obtained by changing the surfactant, the goals of simultaneously lowering the tensions and increasing the width of the three-phase region are mutually contradictory. As in the design of heat engines, thermodynamics can save some work in the design of microemulsion formulations (33)!

Microemulsions and Macroemulsions

Operationally, it is not always easy to determine whether a given sample is a microemulsion or macroemulsion. Close to a critical point, a microemulsion may not be transparent, but may be confused with a macroemulsion. An extremely stable macroemulsion may require more than one researcher's lifetime to separate into bulk layers, and thus, appear to be a microemulsion, but it is nevertheless a (macro) emulsion. Such experimental problems added to the early difficulties of developing the microemulsion concept. When the number of components is so large that determination of the phase diagram is impractical, advanced techniques such as quasi-elastic light scattering, small-angle x-ray or neutron diffraction, or nuclear magnetic resonance may be used to determine if the sample is a microemulsion by characterization of its structure (2).

However, the formal differences between microemulsions and macroemulsions are well defined. A microemulsion is a single, thermodynamically stable, equilibrium phase; a macroemulsion is a dispersion of droplets or particles that contains two or more phases, which are liquids or liquid crystals (48).

Nevertheless, possibilities for confusion abound. From the definitions of microemulsions and macroemulsions and from Figure 1, it immediately follows that in many macroemulsions one of the two or three phases is a microemulsion. Until recently (49), it was thought that all nonmultiple emulsions were either oil-in-water (O/W) or water-in-oil (W/O). However, the phase diagram of Figure 1 makes clear that there are six nonmultiple, two-phase morphologies, of which four contain a microemulsion phase. These six two-phase morphologies are oleic-in-aqueous (OL/AQ, or O/W) and aqueous-in-oleic (AQ/OL, or W/O), but also, oleic-in-microemulsion (OL/MI), microemulsion-in-oleic (MI/OL), aqueous-in-microemulsion (AQ/MI), and microemulsion-in-aqueous (MI/AQ) (49).

Although they have not yet all been reported, theoretically there are twelve three-phase emulsion morphologies formed by the top, microemulsion (ie, middle), and bottom phases (50,51): totally engulfing $T/M/B$, $M/T/B$, $M/B/T$, $B/M/T$, $B/T/M$, $T/B/M$; nonengulfing $(T + M)/B$, $(B + M)/T$, and $(B + T)/M$; and partially engulfing $(T + M)/B$, $(B + M)/T$, and $(B + T)/M$. These twelve three-phase emulsion morphologies, in which one of the phases is a middle-phase microemulsion, are illustrated by Figure 7.

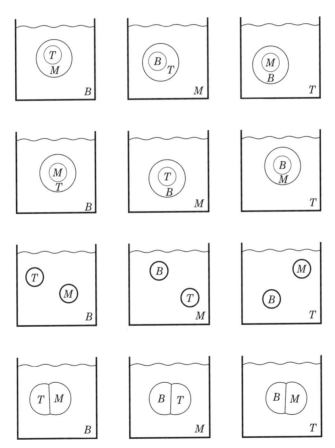

Fig. 7. Cartoon illustration of the twelve theoretical morphologies of three-phase macroemulsions in which one the phases is a middle-phase microemulsion.

Emulsion Morphology Diagrams

Just as phase diagrams clarify what constitutes a microemulsion, emulsion morphology diagrams (52–54) provide maps that clarify the relationships among microemulsions and the various emulsion morphologies in which these microemulsions can appear. The emulsion morphology diagram of Figure 8 illustrates each of the six two-phase emulsion morphology regions and, for the various regions within the tietriangle, shows which phase is the continuous phase. Separating the different morphology regions (eg, OL/MI and MI/OL) are narrow hysteresis regions, within which, depending on its history, the emulsion may be in either (eg, OL/MI or MI/OL) morphology. For two-phase microemulsion emulsions, the hysteresis region forms a cusp that terminates at a plait point. For the T-continuous, B-continuous, and M-continuous regions the morphology (nonengulfing, partially engulfing, or totally engulfing) of the dispersed phases is believed (50,55) to depend on the wettability condition (56) among the three phases.

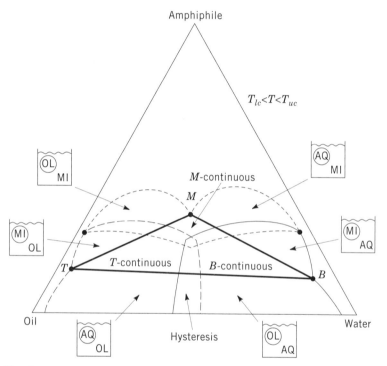

Fig. 8. Emulsion morphology diagram, illustrating where the microemulsion in various macroemulsion morphologies is a continuous phase or dispersed phase. Morphology boundaries: —, aqueous, continuous; − − −, oleic, continuous; - - - -, microemulsion, continuous.

Economic Aspects

There are no statistics available for microemulsion products or their annual values, but data for the surfactant industry can be taken as a guide. Annually updated lists of commercial surfactants and their suppliers are available from several sources (57,58). *Chemical & Engineering News* annually publishes a feature article on "Soaps and Detergents" in its fourth issue (59,60). The market for surfactants is immense. For example, in 1995, the U.S. market for laundry and dishwashing detergents, soaps, and household cleaners was $9.6 billion (59). In 1994 the United States consumed 2.18 billion kg of anionic surfactants, 0.86 billion kg of nonionic surfactants, 0.32 billion kg of cationics, and

0.024 billion kg of amphoteric surfactants, for a total of almost 3.4 billion kg. During the period from 1994 to year 2000, the annual growths in these demands were expected to be 2.6%, 2.9%, 3.5%, and 6.0% respectively. Hence, the projected demands for the year 2000 were 2.54 billion kg for anionics, 1.02 billion kg for nonionics, 0.39 billion kg for cationics, and 0.03 billion kg for amphoterics (59). Demand for many surfactants was expected to grow even faster in other parts of the world than in the U.S. For example, in 1995, North America consumed 13%, Western Europe 15%, Latin America 19%, and the Asia-Pacific region 37% of the 2.8 million tons of linear alkylbenzene surfactants consumed in the world. For the year 2005, these fractions were predicted to be 9%, 8%, 17%, and 45%, respectively, of a total worldwide consumption of 4.0 million tons (59) (see SURFACTANTS).

BIBLIOGRAPHY

1. I. D. Robb, *Microemulsions*, Plenum Press, New York, 1982.
2. V. Degiorgio and M. Corti, *Physics of Amphiphiles: Micelles, Vesicles, and Microemulsions*, Elsevier Science Publishing Co., New York, 1985.
3. D. O. Shah, *Macro and Microemulsions: Theory and Applications*, American Chemical Society, Washington, D.C., 1985.
4. S. E. Friberg and P. Bothorel, *Microemulsions: Structure and Dynamics*, CRC Press, Boca Raton, Fla., 1987.
5. H. L. Rosano and M. Clausse, eds., *Microemulsion Systems*, Marcel Dekker, New York, 1987.
6. M. Bourrel and R. S. Schechter, *Microemulsions and Related Systems: Formulation, Solvency, and Physical Properties*, Marcel Dekker, New York, 1988.
7. S. H. Chen and R. Rajagoplan, *Micellar Solutions and Microemulsions*, Springer-Verlag, New York, 1990.
8. D. Roux, *Micelles, Membranes, Microemulsions, and Monolayers*, Springer-Verlag, New York, 1994.
9. C. Solans and H. Kunieda, eds., *Industrial Applications of Microemulsions*, Marcel Dekker, New York, 1996.
10. *Global Books in Print (CD-ROM)*, R. R. Bowker, New Providence, N.J., March 1997.
11. D. H. Smith in D. H. Smith, ed., *Surfactant-Based Mobility Control: Progress in Miscible-Flood Enhanced Oil Recovery*, American Chemical Society, Washington, D.C., 1988.
12. D. R. Karsa, ed., *Industrial Applications of Surfactants*, Royal Institute of Chemistry, Cambridge, 1990.
13. M. Gratzel and K. Kalyanasundaram, eds., *Kinetics and Catalysis in Microheterogenous Systems*, Marcel Dekker, New York, 1991.
14. D. R. Karsa, ed., *Industrial Applications of Surfactants*, Royal Institute of Chemistry, Cambridge, 1992.
15. D. A. Sabatini and R. C. Knox, eds., *Transport and Remediation of Subsurface Contaminants*, American Chemical Society, Washington, D.C., 1992.
16. T. F. Tadros, ed., *Surfactants in Agrochemicals*, Marcel Dekker, New York, 1992.
17. N. Kosaric, ed., *Biosurfactants: Production, Properties, Applications*, Marcel Dekker, New York, 1993.
18. M. M. Reiger and L. D. Rhein, eds., *Surfactants in Cosmetics*, Marcel Dekker, New York, 1997.
19. S. S. Talmage, ed., *Environmental and Human Safety of Major Surfactants*, Lewis Publishers, Boca Raton, Fla., 1994.
20. D. R. Karsa and M. R. Porter, eds., *Biodegradability of Surfactants*, Blackie Academic and Professional, London, 1995.
21. M. J. Schwuger, ed., *Detergents in the Environment*, Marcel Dekker, New York, 1996.

22. T. P. Hoar and J. H. Schulman, *Nature* **152**, 102 (1943).
23. D. H. Smith, *J. Colloid Interface Sci.* **108**, 471 (1985).
24. S. E. Friberg and I. Lapczynska, *Prog. Colloid Polym. Sci.* **56**, 16 (1975).
25. H. Kunieda, *Bull. Chem. Soc. Jpn.* **56**, 625 (1983).
26. D. H. Smith and G. L. Covatch, *J. Colloid Interface Sci.* **170**, 112 (1995).
27. P. Ekwall in G. H. Brown, ed., *Advances in Liquid Crystals*, Academic Press, New York, 1975.
28. M. Kahlweit, R. Strey, and G. Busse, *J. Phys. Chem.* **94**, 3881 (1990).
29. G. D. Efremova and A. V. Shvarts, *Russ. J. Phys. Chem.* **40**, 486 (1966).
30. J. R. DiAndreth and M. E. Paulaitis in D. H. Smith, ed., *Surfactant-Based Mobility Control: Progress in Miscible-Flood Enhanced Oil Recovery*, American Chemical Society, Washington, D.C., 1988.
31. J. C. Lang, and B. Widom, *Physica* **81A**, 190 (1975).
32. M. Kahlweit, R. Strey, R. Schomaecker, and D. Haase, *Langmuir* **5**, 305 (1989).
33. D. H. Smith, *AOSTRA J. Rsch.* **4**, 245 (1988).
34. R. B. Griffiths, *J. Chem. Phys.* **60**, 195 (1974).
35. D. O. Shah and R. S. Schechter, eds., *Improved Oil Recovery by Surfactant and Polymer Flooding*, Academic Press, New York, 1977.
36. J. J. Taber, *Soc. Petrol. Eng. J.* **9**, 3 (1969).
37. G. L. Stegemeier in D. O. Shah and R. S. Schechter, eds., *Improved Oil Recovery by Surfactant and Polymer Flooding*, Academic Press, New York, 1977.
38. H. Kunieda and K. Shinoda, *Bull. Chem. Soc. Jpn.* **55**, 1777 (1982).
39. P. D. Fleming, J. E. Vinatieri, and G. R. Glinsmann, *J. Phys. Chem.* **84**, 1526 (1980).
40. B. Widom, *J. Chem. Phys.* **62**, 1332 (1975).
41. B. Widom, *Phys. Rev. Lett.* **34**, 999 (1975).
42. W. C. Griffin, *J. Soc. Cosmetic Chemists* **1**, 311 (1949).
43. H. Kunieda and K. Shinoda, *J. Disp. Sci. Technol.* **3**, 233 (1982).
44. K. Shinoda and H. Kunieda in P. Becher, ed., *Encyclopedia of Emulsion Technology*, Vol. 1, Marcel Dekker, New York, 1983.
45. J. S. Rowlinson and B. Widom, *Molecular Theory of Capillarity*, Clarendon Press, Oxford, 1984.
46. P. D. Fleming and J. E. Vinatieri, *J. Colloid Interface Sci.* **81**, 319 (1981).
47. D. H. Smith, *J. Chem. Phys.* **85**, 1545 (1986).
48. *International Union of Pure and Applied Chemistry Manual on Colloid and Surface Science*, Butterworths, London, 1972.
49. D. H. Smith in H. L. Rosano and M. Clausse, eds., *Microemulsion Systems*, Marcel Dekker, New York, 1987.
50. D. H. Smith, G. K. Johnson, Y. C. Wang, and K.-H. Lim, *Langmuir* **10**, 2516 (1994).
51. G. K. Johnson, D. Dadyburjor, and D. H. Smith, *Langmuir* **10**, 2523 (1994).
52. D. H. Smith and K.-H. Lim, *J. Phys. Chem.* **94**, 3746 (1990).
53. D. H. Smith, J. S. Reckley, and G. K. Johnson, *J. Colloid Interface Sci.* **151**, 383 (1992).
54. D. H. Smith and Y.-C. Wang, *J. Phys. Chem.* **98**, 7214 (1994).
55. S. Torza and S. G. Mason, *Science* **163**, 813 (1969).
56. D. H. Smith and G. L. Covatch, *J. Chem. Phys.* **93**, 6870 (1990).
57. *Chemcyclopedia* (Annual Suppl. to *C&E News*), American Chemical Society, Washington, D.C., 1997.
58. *McCutcheon's Emulsifiers & Detergents, International Ed.*, MC Publishing Co., Glen Rock, N.J., 1997.
59. *C&E News* **74**(4), 32 (1996).
60. *C&E News* **75**(4), 30 (1997).

DUANE H. SMITH
Technical Solutions and
West Virginia University

MOLECULAR MODELING

It can be said that science is the art of building models to explain observations and predict new ones. Chemistry, as the central science, utilizes models in virtually every aspect of the discipline. From the first week of a first chemistry course, students use the scientific method to develop models which explain the behavior of the elements. Anyone who studies or uses chemistry has, in fact, practiced some form of molecular modeling.

A useful method of tracing the origins of molecular modeling is to examine how its offshoot, computer-assisted molecular modeling (CAMM), came to be developed. Molecular modeling today represents a convergence of a number of techniques from different disciplines. Basic techniques used to accomplish modeling objectives necessarily draw on these. Specific software systems today importantly assist the researcher in the study of molecular systems and provide help in deriving a rigorous and consistent explanation for the chemical or biological behavior observed or help the researcher to develop a model for predictions.

The literature of molecular modeling is expanding rapidly. The reasons the number of molecular modeling studies continues to grow are that the tools are becoming easier to use, computers get faster, and researchers' awareness of the utility of these tools is growing. Despite this increasing ease to use, however, the development of a good grasp of the origins of the methods and strategies of molecular modeling could usefully draw on the many citations from the literature that are included herein.

Historical Perspective

Molecular modeling has evolved as a synthesis of techniques from a number of disciplines—organic chemistry, medicinal chemistry, physical chemistry, chemical physics, computer science, mathematics, and statistics. With the development of quantum mechanics (1,2) in the early 1900s, the laws of physics necessary to relate molecular electronic structure to observable properties were defined. In a confluence of related developments, engineering and the national defense both played roles in the development of computing machinery itself in the United States (3). This evolution had a direct impact on computing in chemistry, as the newly developed devices could be applied to problems in chemistry, permitting solutions to problems previously considered intractable.

Into the late 1940s, Nobel Laureate Robert S. Mulliken, a physical chemist at the University of Chicago, maintained a skeptical view regarding the future of applying the theories of physics to solving practical problems in chemistry (4,5). Subsequently, Mulliken (5) related that

> [it was] only in the '50s that really substantial progress was made ... A major and indeed crucial step beyond the development of formulas for molecular integrals was the programming for large electronic digital computers of otherwise excessively time-consuming numerical computation of these integrals, and of their combination to obtain desired molecular wave functions and related molecular properties.

Whereas many scientists shared Mulliken's initial skepticism regarding the practical role of theory in solving problems in chemistry and physics, the work of London (6) on dispersion forces in 1930 and Hückel's π-electron theory in 1931 (7) continued to attract the interest of many, including a young scientist named Frank Westheimer who, drawing on the physics of internal motions as detailed by Pitzer (8), first applied the basic concepts of what is now called molecular mechanics to compute the rates of the racemization of *ortho*-dibromobiphenyls. The 1946 publication (9) of these results would lay the foundation for Westheimer's own systematic conformational analysis studies (10) as well as for many others, eg, Hendrickson's (11) and Allinger's (12). These scientists would utilize basic Newtonian mechanics coupled with concepts from spectroscopy (13,14) to develop nonquantum mechanical models of structures, energies, and reactivity.

Researchers in chemistry and chemical physics whose interests were more theoretical were also very active between 1940 and 1965. The Manhattan Project had focused a great deal of public attention on chemistry and physics, garnering as well the energy and interest of both young and accomplished scientists. Coulson's seminal contribution (15) on molecular orbitals, along with studies of the low lying excited states of benzene by Goeppert-Mayer and Sklar (16), laid the groundwork for the subsequent contributions of Dewar (17), Pople (18), and Pariser and Parr (19), whose bodies of work have provided the basis for innumerable citations detailing the application of quantum mechanical methods to organic chemistry since the 1950s, a subject thoroughly treated in several exemplary texts (20–22). Such efforts (17–19) paved the way for Nobel Laureates Woodward and Hoffman (23) to elaborate elegant orbital-based theories on the relationship between reactivity and molecular electronic structure.

Definition of Molecular Modeling and Uses of CAMM

Molecular modeling refers broadly to any study of molecules utilizing physical or theoretical models to explain an observed or predicted behavior. In practice, physical models have expedited the understanding of small molecules, inorganic complexes, and biopolymers, including such molecules as DNA. The pioneering work of Watson and Crick on DNA would certainly not have been possible without the building of actual physical models of the structure derived from experimental observations of x-ray diffraction patterns. However, physical models have limitations, both in terms of being primarily static representations of dynamic systems, and in being only semiquantitative with respect to scale. Whereas chemists prefer having hands-on experience with actual physical models, their parallel quest to quantify structures and energetics makes it impractical to rely only on such models for all aspects of their work. It is here that computer models, or molecular modeling, enter in.

Molecular modeling can be defined as the application of computational techniques, grounded in theory, to predict or explain observable biological or physical chemical properties. Wherever molecular modeling is practiced using a computer, the technique then becomes computer-assisted (aided) molecular modeling, or CAMM. CAMM, is often used synonymously with CAMD, or computer-assisted molecular (materials) design/discovery. CADD refers to computer-assisted drug design/discovery. A computational technique as used herein is a mathematical model derived from principles of chemistry, physics, or statistics which facilitates molecular modeling. An entire branch of chemistry, ie, computational chemistry, is devoted to developing, benchmarking, and applying computational techniques in order that researchers may be able to better understand and predict properties. Computational chemistry serves as an umbrella under which several disciplines converge to promote the evolution of better technologies to enhance the understanding of molecules and their reactivities. Some of the properties which may be calculated either exactly or approximately by computational methods are the following:

boiling points	dipole moments
melting points	quadrupole moments
crystallization energy	octupole moments
heat capacity	infrared spectra/intensities
heat of formation	nmr spectra/chemical shifts
heat of fusion	optical rotary dispersion
heat of sublimation	raman spectra
heat of vaporization	ultraviolet spectra
entropy	
molar refractivity	ionization potentials
molar volume	electron affinities
partition coefficients	protonation energies and pK_as
	ionic strength
radius of gyration	
elasticity	conformational energies
tensile strength	Boltzmann distributions

Molecular properties can be classified according to their end-point observables, such as chemical (reactivity, solubility, acid–base), physical (a function of physical state—gas, liquid, solid; thermodynamic), or biological (ligand or enzyme; agonist or antagonist). These properties reflect macroscopic, or bulk, properties, which exist only for the bulk material, eg, heat of crystallization, or microscopic properties, which exist for an ensemble of the molecule. As use of CAMM methods expands to address a broader horizon of applications beyond those in organic, medicinal, and biological chemistry, calculations on metals, semiconductors, and magnetic systems are becoming more common (24).

To gain a proper perspective of the role of computed physical properties, the relationship between estimated and computed properties needs to be understood. A thorough reading of Horvath (25) permits formulation of the following definitions of estimating or computing properties.

Interpolating properties. A correlation is found between the desired property and another property or characteristic of related molecules; in this case the desired property may be computed from within the range of application of the correlation, and the interpolated property should be accurately estimated.

Extrapolating properties. In this case the correlation does not extend to include the molecule of interest, but by extending the correlation, it is possible to estimate the desired property. Since the validity of the correlation in the extrapolated region is unknown, the accuracy of the extrapolated property is difficult to estimate.

Computing properties. In many cases it is possible to compute a property, directly or indirectly, with varying levels of accuracy. Such computed properties can be quite comparable to experimental accuracy and indeed may substitute for the experiment in cases where the experiment would be difficult or impossible to perform.

This last definition should be carefully applied as either an interpolation or an extrapolation, particularly for empirical computational methods based on diverse observations. It is critical that users of molecular modeling tools understand where it is appropriate to apply a technique and where it is not, and what degree of accuracy can be expected.

The specific process involved in a given molecular modeling study depends significantly on the nature of the primary objective of the task. However, for many small-molecule and even macromolecular studies, a number of authors have diagrammed the individual steps in the process, and the flowchart in Figure 1 summarizes their efforts. An initial step that is critical to any CAMM project is the generation or retrieval of the pertinent structures themselves, a subject which is treated in detail herein. Structures may range from simple organic molecules or monomers of more complex polymers, to full proteins, enzymes, metal surfaces, or zeolites. The modeling process can be influenced by the initial structure and its geometry. Thus, the selection and development of the starting molecular geometry needs to be given particular attention. An important com-

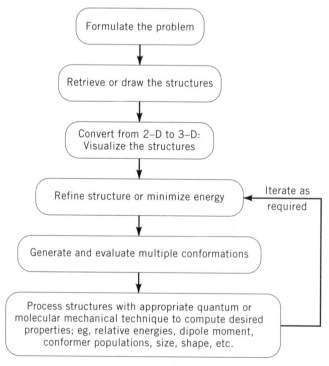

Fig. 1. Flowchart for a typical small molecule modeling project.

ponent of the quality control process, as well as in gaining an understanding of the molecules themselves, is the visual examination of structures involved in a modeling study.

Computer Graphics In Molecular Modeling

The goal of molecular modeling is to define clearly the relationship between chemical constitution, ie, the molecular formula or a topographic representation thereof, its geometric constitution or 3-D topology (the disposition of its atoms in Cartesian space), and its observed (or predicted) properties. The representation and facile manipulation of 3-D arrays of atoms comprise the domain of molecular or computer graphics. From the time of Levinthal's work in the mid-1960s (26), scientists have endeavored to develop and use graphics software and hardware to expedite molecular modeling studies in both the classroom and the research lab. The growth of graphics tools has paralleled the evolution of computing hardware and, indeed, in some circumstances has even given strong impetus to that evolution (27). An example is the development and use of the Evans and Sutherland computer graphics systems in molecular modeling (27). Computing software and hardware systems have had a profound impact on the ability of modelers to compose a modeling study and address all aspects of the work, ranging from generating 2-D drawings of structures to statistical quality control of computed properties via visualization of multidimensional data.

Illustrating this enhancement in the visualization of structure and properties, Figures 2–5 provide increasingly complex and useful structural representations. Figure 2, a stick drawing of the drug Zantac (ranitidine), when viewed in color (not shown here), conveys both 2-D structure data and color-coded electronic charge data. Figure 3 shows the CPK shaded solid surface for ranitidine. In Figure 4, color-coded electronic charge data (color not shown here) have been mapped onto the CPK surfaces. In Figure 5, four layers of the Connolly solvent-accessible dot surface, when color-coded (color not shown here), correspond to the energies of the electrostatic potential. In this figure, the highest charge density, when viewed in color, would be indicated by red dots in areas where there is the strongest attraction to an H^+ atom brought to that point. Conversely, at

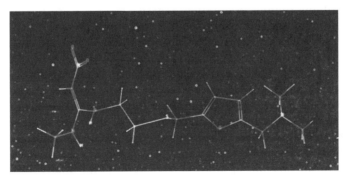

Fig. 2. A stick drawing of the drug Zantac (ranitidine) illustrating the method of showing color-coded electronic charge data (color not shown here), where red would represent the highest charge density, blue the lowest charge density, etc.

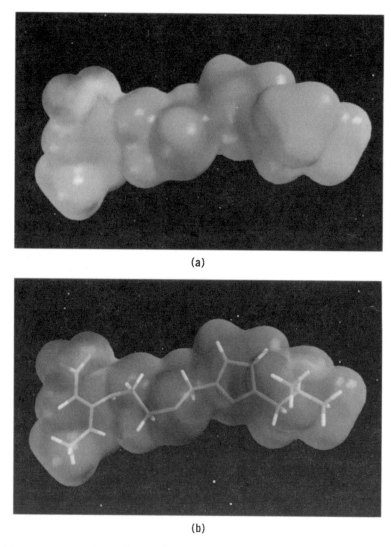

(a)

(b)

Fig. 6. Two representations of ranitidine illustrating the electrostatic potential mapped onto (**a**) a solid and (**b**) translucent constant-density surface (0.001 e); see text for effect of color-coding, not shown here.

on the utilization of AVS in molecular modeling which describes methods for combining orbital descriptions of molecules, as well as models of how the AVS visualization utilities are applied to the understanding of such properties as the electrostatic potential and reactivity indices (35). The utilization of computer graphics tools has also been extended to viewing the chemical literature itself (36) through the introduction of programs and data sources for "kinemages," ie, kinetic images that can be viewed and rotated interactively on a PC or Macintosh. Such capabilities are now more common through utilization of the Internet and via plug-ins to Web browsers.

Computational Methods for Molecular Modeling

Classical and Quantum Mechanics. At the beginning of the twentieth century, a revolution was brewing in the world of physics. For hundreds of years, the Newtonian laws of mechanics had satisfactorily provided explanations and supported experimental observations in the physical sciences. However, the experimentalists of the nineteenth century had begun delving into the world of matter at an atomic level. This led to unsatisfactory explanations of the observed patterns of behavior of electricity, light, and matter, and it was these inconsistencies which led Bohr, Compton, deBroglie, Einstein, Planck, and Schrödinger to seek a new order, another level of theory, ie, quantum theory.

Basically, Newtonian mechanics worked well for problems involving terrestrial and even celestial bodies, providing rational and quantifiable relationships between mass, velocity, acceleration, and force. However, in the realm of optics and electricity, numerous observations seemed to defy Newtonian laws. Phenomena such as diffraction and interference could only be explained if light had both particle and wave properties. Indeed, particles such as electrons and x-rays appeared to have both discrete energy states and momentum, properties similar to those of light. None of the classical, or Newtonian, laws could account for such behavior, and such inadequacies led scientists to search for new concepts in the consideration of the nature of reality.

In 1903, when Planck suggested that the energy emitted from heated bodies, ie, black-body radiation, was not composed of waves, but rather discrete particles or quanta, a long-standing physical anomaly was resolved. Similarly, Planck's theory was applied to the photoelectric effect and was subsequently used by Bohr (37) to develop models of atomic structure. By the time Pauling and Wilson published their treatise on quantum mechanics in 1935 (38), the foundations for a workable quantum theory, explaining black-body radiation, electron distributions around nuclei and in chemical bonds, and the wave–particle duality of photons and electrons, had been detailed by a new generation of physicists (39). The extraordinary progress in the theory of matter made during the first three decades of the twentieth century lead Dirac, one of the pioneers of quantum theory 40, to state, "The underlying physical laws necessary for the mathematical theory of a large part of physics and the whole of chemistry are thus completely known" (41).

At the heart of the revolution in quantum theory is Schrödinger's equation, which, in one dimension, for one electron not interacting with its surroundings, may be written

$$d^2\Psi/dx^2 + 8\pi^2 m/h^2 \, (E - V)\Psi = 0 \tag{1}$$

in which E is the total energy, a constant, and V is the potential energy, which most often is a function of x. The wave function, ψ, provides the solution of this equation and is at the heart of quantum mechanics and its application to problems in chemistry. After expanding the wave equation to three dimensions, and replacing the second derivatives in equation 1 with the Laplacian operator, ∇^2, equation 1 can be rearranged to give

$$\boldsymbol{H} = T + V \tag{2}$$

wherein the energy terms have been on grouped the right, and the Hamiltonian operator, H, is used in classical mechanics to provide a function of momenta and coordinates. For a single electron, the Schrödinger equation is

$$H\Psi = E\Psi \tag{3}$$

in which the H is an operator, the constant E is called an eigenvalue, and the function Ψ is called the eigenfunction. Certain constraints must be met if Ψ is to be physically meaningful, ie, that it be continuous and single-valued over the region of interest. The probability of finding the particle in all space must be unity, ie,

$$\int |\Psi|^2 \, dx \, dy \, dz = 1 \tag{4}$$

The quantity $|\Psi|^2$ represents the probability of finding the electron in the described region. It is also interpreted statistically, in the context of the Heisenberg uncertainly principle, as an expectation value. Streitwieser has eloquently summarized the importance of such concepts in chemical computations (42):

> For organic chemists the importance of quantum mechanics lies not at all in the exact calculations from first principles (*ab initio* calculations), but rather in providing heuristic concepts and insights in establishing qualitative and quantitative semiempirical correlations of experimental data and, especially, in facilitating the application of what has long been the organic chemist's most important tool: reasoning by analogy.

Two important points are made in this statement: first, less important than exact quantification is the development of a heuristic model of chemical behavior; and second, semiempirical correlations are the goal of computations involving applied quantum mechanics. At the time Streitwieser wrote, in 1960–1961, full-scale calculations on systems large enough to be of interest to organic chemists were certainly beyond then-current limits for most levels of theory and the corresponding computer programs. Computing machinery was still quite primitive by today's standards (about 10^6 slower than a good workstation at the time of this writing, ie, early 1997), and a full elaboration of semiempirical quantum mechanical methods was in progress (20). Of particular note is the fact that Streitwieser here suggests that organic chemists are less interested in exact *ab initio* calculation of properties than in developing qualitative concepts of structure–reactivity relationships. As discussed herein, *ab initio* quantum mechanical calculations now have taken their place alongside semiempirical methods in the chemist's computational toolkit for molecular modeling.

Clearly, most chemistry involves interactions of multielectron systems with each other to yield either desired quantities of previously characterized chemicals or new chemical entities (NCEs, in pharmaceutical language). At the atomic level, chemical reactions involve structures which have multiple electrons interacting, perhaps even on multiple atomic centers such that the potential energy of these systems is dependent on the relative positions of the electrons with respect to each other as well as with respect to the nuclei. The Born-Oppenheimer approximation is generally applied; this approximation states that the motion of the nuclei is neglected, and that electron motion is the dominant component of interatomic interactions. Hartree (43) and Fock (44) developed a formalism for reducing the multielectron Schrödinger equation to a sum of single-electron equations, which could be solved to yield what is called a Self-Consistent Field (SCF) Approximate Method. As an interactive method wherein electrons are distributed into shells based on the Aufbau principle, the SCF method made it possible to develop wavefunctions which can be approximated by analytical solutions to provide representations of atomic and molecular orbitals that can be readily visualized. Such visualization facilitates the interpretation of chemical reactivity and chemical reactions.

Numerous methods arose utilizing Hartree-Fock SCF techniques, ranging from the simplest, or Hückel π-electron techniques, to the most complete first principles, or *ab initio* methods. What distinguishes these methods is, in practical terms, which electrons and orbitals are included in the calculations, along with the degree to which the elements of the Fock matrix (representing the operator *H*) are evaluated explicitly, approximated, or neglected completely. These matrix elements, when they are evaluated in the context of the linear combination of atomic orbitals (LCAO) method, ultimately provide a quantitation of the presence and magnitude of the influence of neighboring electrons on each other. For systems in which all σ- and π-electrons are considered, so long as the researcher neglected a selected set of neighboring atom electron–electron interactions LCAO provided the basis for most so-called semiempirical quantum mechanical methods in use around 1960.

The Hückel Molecular Orbital theory (HMO) and its subsequent elaboration, Extended HMO theory (EHT), methods provide the simplest quantum mechanical description of π-electron systems. The development and applications of HMO were reviewed by Streitwieser in his text (20). Whereas the HMO method neglects σ-electrons, it did utilize linear combinations of atomic orbitals (LCAO) for the molecular orbitals. The method worked quite well, particularly for planar conjugated systems and even for certain nonplanar molecules. However, the technique can be best credited as providing the basis for subsequent, more elaborate methods. In Murrell and Harget (22) the development and applications of the Pariser-Pople-Parr (PPP) (19) method is given in a very readable account. It was a particularly useful technique in that adiabatic ionization potentials as well as singlet-triplet separation energies could be computed accurately. Additionally, the PPP method was used subsequently to compute the energetics of the π-electron component of structures (45,46), in conjunction with an adaptation of the Westheimer method to compute gas-phase conformational energies for planar and nonplanar π-electron systems.

The next step in the development of molecular orbital theories involved all-valence-electron methods wherein the concept of zero differential overlap (ZDO) of two-center integrals involved in the wave functions are refined. In these techniques, which introduced drastically reduced numbers of integrals requiring evaluation, investigators found they could incrementally move toward a more complete set of SCF equations without their being computationally intractable.

In 1965, the first group of a series of important papers detailing the Complete Neglect of Differential Overlap (CNDO) and Neglect of Diatomic Differential Overlap (NDDO) methods were published (47). Full computational details of these methods are also available (21,22). The CNDO and NDDO techniques enabled computation of a broad spectrum of geometric features, such as bond lengths, angles, and related properties, including dipole moments, which could be predicted for singly bonded systems for the first time. The methods continue to be used today (ca 1997), although the relatively poor accuracy of the CNDO technique in determining structural and charge distribution properties has led to its being used principally in spectroscopy (48).

Through the 1970s and 1980s investigators continued to improve on the ZDO technique, especially in increasing the accuracy and range of computed quantities, and in taking advantage of the enhanced computing facilities which were becoming available. Advances made included development of a series of programs utilizing the Intermediate Neglect of Differential Overlap (INDO) technique (49). The "Dewar School" continues today to produce semiempirical techniques to permit the computation of properties of organic molecules, organometallics, semiconductors, peptides, and proteins (50,51). Excellent expositions on semiempirical methods derived from NDDO are available (51,52), including a "how to" text (53). However, as Dewar has stated, "MO (Molecular Orbital) Theory is not a description of reality. It is only the embodiment of another molecular model, the MO model" (58).

The ultimate goal of quantum mechanical calculations as applied in molecular modeling is the a priori computation of properties of molecules with the highest possible accuracy (rivaling experiment), but utilizing the fewest approximations in the description of the wavefunction. *Ab initio*, or from first principles, calculations represent the current state of the art in this domain. They are also referred to as nonempirical calculations, although this name is somewhat misleading. *Ab initio* calculations utilize experimental data on atomic systems to facilitate the adjustment of parameters such as the exponents of the Gaussian functions used to describe orbitals within the formalism. Additionally, these Gaussians are a function of the electron–nucleus distance squared, r^2, and represent a significant approximation to the Slater-type or simple exponential functions of r, used to describe electron distributions in elemental quantum mechanics. Whereas Gaussian functions do a creditable job of reproducing experimental properties, they were introduced for pragmatic reasons, namely, to simplify the computation of multicenter integrals. A full and readable exposition and detailing of *ab initio* molecular orbital calculations, including the very important topic of basis sets, is available (54).

The performance of *ab initio* techniques distinguishes them significantly from their predecessors, semiempirical methods. Their consistent reproduction of data from structural, thermodynamic, and reaction sources to a range falling within the error limits of the experimental values provides scientists with an important tool with which to address various modeling problems. Whereas the absolute value of the relative performance of *ab initio* techniques varies for each structural or energetic feature examined, it is not unreasonable to suggest that if the quantity can be computed by both semiempirical and *ab initio* methods, the *ab initio* value will be closer to experiment or an ideal value than any other method. To get a fuller appreciation of the comparison of these two methods, one need only follow the numerous publications of the Pople (55) group detailing the performance of *ab initio* methods in comparison to these published by the Dewar (49) group, which continued the refinement of semiempirical techniques. A mathematically rigorous overview of the utility of *ab initio* calculations is available (56), and many elementary questions about the role of *ab initio* calculations in molecular modeling studies are addressed in a particularly readable account (57) on the basis of which several important points about both semiempirical and *ab initio* calculations are worth summarizing herein.

Ab initio Theory: Caveats and Performance.

Except for approximations made previously, no others are made.

All electrons are treated.

All electron integrals are computed exactly, but with no guarantee of accuracy of the prediction or agreement with experiment.

Basis sets consist of a finite number of Gaussian functions, introducing an inherent limitation on accuracy.

Computing time can be proportional to the fourth power of the number n of basis functions.

Whereas correlation energies can be included, in practice it is even more time-consuming to use them in the determination of molecular geometries (ie, n^6), and determining the correct basis set to use can be difficult.

Geometric properties are quite sensitive to the basis set chosen, including the presence or absence of polarization functions (additional s and p-type functions on H and d-type on heavy atoms).

Geometric and energetic properties are also sensitive to the starting geometries, and to the algorithm used for geometry optimization.

The following conclusions apply to organic molecules of about 25 heavy atoms (~60 atoms total), assuming use of medium-size basis sets (3–21G*):

Geometric properties can be reproduced to within 0.15 ± 0.15 nm (0.015 ± 0.015 Å) for bond lengths, 1–2° for bond angles, and to ±5° for dihedral angles.

Ionization energies can be computed to about ±0.2 eV; rotational barriers to about 0.5 kcal/mol; dipole moments to about ±0.5 D; barriers to inversion to about ±2.5 kcal/mol; infrared frequencies can be computed with about a 15% error (usually too high); and protonation energies are accurate to about 1 pK unit.

Hydrogen bond geometries may be reproduced or predicted fairly well with reasonable, but sometimes underestimated heavy atom–heavy atom distances; radial dependence of the hydrogen bond may be in error.

Semiempirical MO Theory: Caveats and Performance. The same basic theoretical assumptions are made as in *ab initio* theory.

Only valence electrons are considered, and the influence of core shell electrons are accommodated by a nuclear screening factor.

The total number of integrals computed depends greatly on the level of complexity of the method: time cost savings of 2 orders of magnitude can be realized over *ab initio* theory (n^2 vs n^4).

the application of these methods to a full spectrum of studies in structure and energetics.

Molecular Mechanics.   Molecular mechanics (MM), or empirical force field methods (EFF), are so called because they are a model based on equations from Newtonian mechanics. This model assumes that atoms are hard spheres attached by networks of springs, with discrete force constants. The force constants in the equations are adjusted empirically to reproduce experimental observations. The net result is a model which relates the "mechanical" forces within a structure to its properties. Force fields are made up of sets of equations each of which represents an element of the decomposition of the total energy of a system (not a quantum mechanical energy, but a classical mechanical one). The sum of the components is called the force field energy, or steric energy, which also routinely includes the electrostatic energy components. Typically, the steric energy is expressed as

$$E_{Total} = E_{steric} + E_{electrostatic} = E_{bonds} + E_{angles} + E_{vdW}$$
$$+ E_{torsion} + E_{charge/dipole}$$

The overall form of each of these equations is fairly simple, ie, energy = a constant times a displacement. In most cases the focus is on differences in energy, because these are the quantities which help discriminate reactivity among similar structures. The computational requirement for molecular mechanics calculations grows as n^2, where n is the number of atoms, not the number of electrons or basis functions. Immediately it can be seen that these calculations will be much faster than an equivalent quantum mechanical study. The size of the systems which can be studied can also substantially eclipse those studied by quantum mechanics.

In a force field calculation, a molecule in three dimensions is constructed using either Cartesian coordinates x, y and z, or via an internal coordinate matrix consisting of bond distances, bond angles, and dihedral angles to specify the atoms' unique positions. Then the initial structure is evaluated to determine the extent to which each degree of freedom (bonds, angles, etc) deviates from the ideal (the zero-energy value) for the particular element and its hybridization. An energy minimization process follows wherein the energy associated with the distortions from ideal is minimized as the individual atomic positions or degrees of freedom are adjusted. Iteratively, this converges on a "minimum energy" or an "optimized" structure. This structure represents the best attempt of the minimization algorithm to render the smallest deviations in position of each of the atoms such that either the derivatives of the change in energy associated with the deviations are the smallest, or they satisfy either energetic convergence or coordinate change criteria from iteration to iteration. It should be noted that this process is analogous to the geometry optimization process within a quantum mechanical program, except that there the objective is to converge on a structure which yields the smallest energy derivatives and lowest total energy from solution of the SCF equations. Most simple molecular mechanics force fields include terms (Fig. 8) for:

bond stretching: $E_1 = k_1(l_0 - l)$

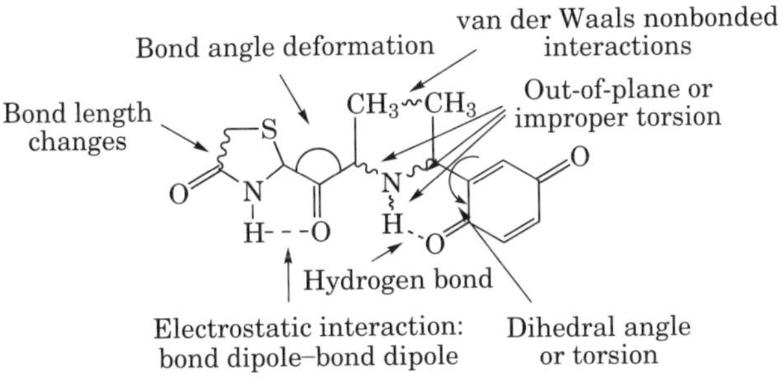

Fig. 8. Structural representation of the energetic components of a typical Molecular Mechanics Force Field.

bond angle distortion (bending): $E_\theta = k_\theta(\theta_0 - \theta)$

dihedral angles: $E_\tau = V_n/2\,(1 \pm \cos n\omega)$

van der Waals nonbonded interactions (1 ... 4 interactions or greater):
$E_{vdW} = C/r_{k_1}^6 - A/r_{k_1}^{12}$,

Coulombic interactions where q_{k_1} can be dipoles or charges $E_{Coul} = q_k\,q_1/r_{k_1}$,

In these equations, k_1, k_θ, V_n, C, and A represent the empirically adjusted constants associated with changes in bond lengths, angles, dihedrals, and nonbonded interactions. These terms plus 1 ... 3 nonbonded interactions, and cross-terms such as stretch–bend are considered in the most complex computational models as well as in the experimental force fields. Practically all of the early molecular mechanics force fields utilized force constants directly from vibrational spectroscopic studies (see background in Refs. 59 and 60). That is, for a particular interaction included in the force field, the force constant applied to an interaction was one which had been experimentally determined. Although this method can be utilized, it is very difficult to develop a generalized force field for a broad spectrum of molecules, because not all experimental force fields are derived to the same level of accuracy, nor are they consistent, ie, having all force constants derived concurrently. For this reason Allinger (60) has suggested:

> ... we must not look at a force field calculation and ask "what interactions are really occurring in the molecule"; rather the question must be "what interactions are really occurring in *our model* of the molecule." Of course the hope is that the answers to the last question will, in fact, converge upon the answers to the first question as force fields improve.

A more detailed account of the MMI force field is also available (61).

Force Fields, Molecular Dynamics, and Vibrational Spectroscopy. The details of the relationship between molecular mechanics force fields and spectroscopic vibrational force fields has been discussed (60). Other fundamental papers on molecular mechanics are available as well (59,62–63,65–67). The link between molecular mechanics and molecular dynamics comes about through the force field itself. In molecular mechanics, the main interest here is in computing the energy of molecules in the gas phase at room temperature in a single, discrete configuration and conformation; time is not a variable in the equations.

periods of time, eg, 500–2000 ps, in order to derive equilibrium properties. It is worthwhile to summarize what properties researchers can expect to evaluate by performing molecular simulations:

Conformational states and energetics

Kinetic properties: rates of reaction and interconversion

Reaction pathways

Solubilities

Diffusion rates

Binding and complexation data

Folding processes

Transition temperatures

Free energies for point mutations

Free energies of binding

As an adjunct to x-ray and nmr for structure refinements

As noted, force fields are a set of equations relating the total energy of the system to its individual interaction components.

Popular force fields for simulations on organic and biomolecules include the following:

Program	Principal author(s)
AMBER	Kollman/Weiner/Singh/Pearlman
CHARMM/CHARMm	Karplus/Brooks/MSI
CFF9X	Hagler/BIOSYM
GROMOS	van Gunsteren
MM2/MM3	Allinger/Welsh
BOSS (Monte Carlo)	Jorgensen/Tirado-Rives

After selection of a force field simulation program which is appropriate to a given problem, the general procedure is as follows:

(*1*) Initial structure equilibration, wherein bad or close contacts are relieved; this may be done with constraints on bonds, eg, to simplify the process (the premise of the SHAKE technique).

(*2*) Structure refinement (locating energy minima) by energy minimization: this step may take a few hundred picoseconds of simulation time if the original configuration is far from a minimum. Note that the researcher may converge to a local minimum which is significantly higher in energy than the global minimum.

(*3*) Techniques used to find global and local energy minima include: sequential simplex, steepest descents, conjugate gradient and variants (BFGS), and the Newton and modified Newton methods (Newton-Raphson).

(*4*) Set up and solve Newton's equation, equation 5, for each atom in the system:

$$F = m_i a_i \, (t) = m_i \cdot \partial^2 r_i \, (t)/\partial t^2 \tag{5}$$

where F = force on atom i at time t, a = acceleration of atom i at time t, and r = position of the atom i at time t.

(5) Evaluate the force F as the negative gradient of the potential energy function:

$$F_i = -\partial v(r_1, r_2, \ldots, rN)/\partial r_i \qquad (6)$$

(6) Compute normal modes. These represent primarily harmonic motions internal to the molecule. There are $3N-6$ displacement eigenvectors, where N is the number of degrees of freedom of the system. The associated eigenvalues are the frequencies.

A variety of techniques have been detailed for handling Newton's equations of motion, equation 5 (86–90). Integration techniques yield atomic positional velocities which are used graphically to display the internal motions and paths of motion computed during the simulation runs. Understanding the behavior of a molecular system as a function of time is a critical element in any simulation, and velocities can be collected into a series snapshots called a trajectory (atomic coordinate sets). From trajectories, researchers can determine the level of cooperation of motions in the folding or conformational processes which a polymer or biopolymer chain might undergo, eg, the interconversion of forms of DNA prior to and during complexation with a protein. Although such simulations represent a full-scale challenge to the simulation technique owing to the wide variety of atom types and associated parameters, a different but no less rigorous type of challenge exists in the simulations of simple polymers of polyethylene. For example, in studies on polyethylene chains of size C_{10} to C_{100} (91), dynamics were used to elucidate the cooperation of motions at neighboring rotational sites, with a finding that for very short time periods the torsional motions of the chains are effectively Brownian in behavior, as had been found earlier. It should be noted that the operational methodology for applying molecular dynamics does vary from application to application, but it is probably reasonable to assume that simulations involving most polymeric systems other than peptides can be addressed similarly, with appropriate modifications for atomic species and periodicity (92). Even peptide systems are addressable by similar methods, despite the breadth of the literature on it that is available suggesting otherwise. The proliferation of dynamics programs for handling peptides and proteins may be the result more of differing perspectives on the method of sampling of excited states, energy refinement algorithms, implementation of constraints, or restraints during the simulation, or force field. The situation is similar to understanding that a car is a vehicle for transporting people and cargo from point A to point B. Whereas the relative comfort and efficiency of the process may vary with the specific vehicle manufacturer, the fact that cars transport people and cargo is common to all manufacturers. All of the subsequent levels of structure and function in peptides and proteins depend intrinsically on the primary structure of the system. Unlike simulations on homopolymers or rigid systems with regular periodic sets of atoms at essentially fixed distances, as in the case of zeolites, the complexity of the simulation problem increases as a function of the variety of the types of structural motifs (helices, sheets, turns, barrels, coils) present, because each motif can behave both as individual atomic entities and as a single collective entity. Numerous investigators have detailed approaches to dealing with the analysis of these systems (73,88–90,93,94). Dynamical characterization of biopolymers and related systems (ca 1997) remains one of the most challenging and stimulating aspects of molecular modeling.

In the context of molecular dynamics, some additional pointers may help give a fuller perspective on the MD simulation operational process:

Initializing the initial kinetic energy and temperature of the system: it is necessary to start the motion at some level, eg, assume a Boltzmann (random) distribution of atomic velocities, at 300 K.

Timesteps for the simulation need to be short enough to capture high frequency motions such as bond stretching, eg, 10^{-15} s.

Simulations need a substantial number of timesteps to sample configurational states such that desired properties are represented well enough to either confirm the experiment or establish valid prediction. Initial equilibrations take 50–100 ps. For systems of >100 atoms, run for 500–2,000 ps if computing resources permit.

Langevin Dynamics: a technique to reduce the total number of equations of motion that are solved. Utilize the Coupled Heat Bath, wherein the method models the solvent effect by incorporating a friction constant into the overall expression for the force.

The SHAKE method for bond constraints reduces the number of degrees of freedom during the initial stages of simulations; it is good for minimizing solvent bath overhead.

Simulated Annealing is a technique which heats up the system and slowly cools it, perhaps to multiple different minima.

Use Nonbonded (NB) truncation methods to reduce size of NB pairlist; it is a dominating term in the calculation! It is important to remember the pairlist increases as N^2, consider truncation of NBs at 100–120 nm (10–12Å), and to experiment with electrostatic cutoffs independently of van der Waals.

Update the NB list every 25–50 timesteps.

Utilize Periodic Boundary Conditions, which permit reduction of the number of nonbonded interactions at greater distances by involving only the "nearest neighbor" atoms from copies of the system which are in different but adjacent cells.

Use a solvent or water bath to capture the influence of solvent on the solute. Use the fewest shells of water/solvent possible, but no fewer than two, if resources are scarce. Use a formal "box" of water, if possible, to reduce the influence of edge effects.

Monte Carlo (MC) techniques for molecular simulations have a long and rich history, and have been used to a great extent in studying the chemical physics of polymers. The majority of molecular modeling studies today do not involve the use of MC methods; however, the sampling capability provided by MC methods has gained some popularity among computational chemists as a result of various studies (95–97). Relevant concepts of MC are summarized herein.

Monte Carlo methods as applied to chemical problems owe their popularity to the work of Metropolis (98), who first utilized them on very early computers to evaluate properties of simple molecular systems. Typically, a set of configurations in a given thermodynamic ensemble is generated by the random sampling of configuration space. By configurations is meant sets of atomic coordinates corresponding to discrete geometries, including those with substantial distortions in bonding or other periodic structural behaviors. In MC calculations, no time relationship exists between successively calculated configurations. Unlike in a molecular dynamics simulation, there is no path or trajectory which the system follows as interconversions occur and states are sampled. Rather, in MC simulations random steps are taken, necessitating that a very large sampling be done to optimize the sampling of the desired configurations. Unfortunately, with the complexity of the potential energy surface of large polymers or proteins, it is not always possible to be assured that one has sampled sufficiently. This open-endedness may be the primary reason that molecular modelers have not embraced Monte Carlo techniques as freely as they have molecular dynamics (MD) simulations, despite the steep computing requirements of the latter.

For many systems the ensemble that is used in an MC simulation refers to the canonical ensemble, (N, V, T). This ensemble permits a rise and fall in the pressure of the system, P, because the temperature and volume are held constant. Thus, the probability that any system of N particles, in a volume V at temperature T is found in a configuration x is proportional to the Boltzmann weighted energy at that state, E_x, and it is given by

$$P_x = \exp(-E_x/kT)/(N!\,Z)$$

where Z is the configurational partition function

$$Z = [(N!)^{-1}] \int \exp(-E_x/kT)\,dx = \exp(-A/kT)$$

and A is the Helmholz free energy. The average value of a property is

$$\langle F \rangle = \int F_x P_x \, dx$$

where F_x is the property evaluated at configuration x. Another popular ensemble used in protein and DNA MC simulations (95) is the (N, P, T) ensemble.

One method used to enhance the efficiency of sampling is biased sampling (98). An algorithm utilizing biased sampling allows low energy configurations to be sampled more often, and is usually more efficient than random sampling at sampling those configurations which contribute more significantly to the true average of a system. For example, to simulate an MC run (92), an algorithm might involve the following steps: (1) calculate the energy of the current configuration, E_1; (2) assign a new configuration, with a new energy, E_2; (3) calculate the weight, Ω, of $\exp(-[E_2 - E_1]/kT)$; (4) compare the result to a uniform, random value in the range of [0,1]; and (5) if the weight Ω is greater than the random value, reset E_1 to E_2, and restart the cycle. The Markov chain that results from this process produces a probability of a given configuration that is proportional to the calculated weight.

The list which follows gives an outline of the properties of a Monte Carlo simulation used in the context of molecular modeling studies for sampling either multiple conformations of smaller, flexible structures or multiple local minima of larger macromolecules or polymers:

Probabilistic, as opposed to deterministic molecular dynamics

Essentially random atomic moves—uses the Metropolis method (98)

Calculate new energy and compare to previous configuration(s)

Keep or discard new configuration

Develop a statistical ensemble of energetically accessible states

Usually sample many millions of states; but not always easy to sample space well

Use Boltzmann or Biased-Sampling Techniques

Umbrella Sampling can give free-energy differences, but not absolute free energies

Usually done in NPT–isothermal/isobaric ensembles, including a water box

Useful for efficient sampling of conformational space in systems such as polymers (92) and peptides (93–95).

Combined Quantum and Molecular Mechanical Simulations. A recently developed technique is one wherein a molecular dynamics simulation includes the treatment of some part of the system with a quantum mechanical technique. This approach, QM/MM, is similar to the coupled quantum and molecular mechanical methods introduced by Warshel and Karplus (45) and at the heart of the MMI, MMP2, and MM3 programs by Allinger (60). These latter programs use quantum mechanical methods to treat the π-systems of the structures in question separately from the sigma framework. The results are combined at the end to render a structure which is optimized and energy-refined to satisfy both SCF and force field energy convergence. Newer QM/MM approaches treat the bulk of the molecule via a molecular mechanical force field to give its dynamical behavior, whereas a selected portion of the structure(s) is treated quantum mechanically to yield information about interactions between the selected segments. This technique is particularly appropriate where applied to systems such as ligands bound to biological receptors or molecular "guests" trapped by "hosts."

In the 1980s the most comprehensive implementation of this method was the program Gaussian80 (UCSF) developed at the University of California at San Francisco (99), which combined the molecular mechanics and dynamics code AMBER (100) and the *ab initio* quantum mechanics program Gaussian80 (101). The program permits researchers to generate structures of interest, then run initial steps of energy minimization to resolve close contacts and significantly distorted bond lengths and angles. The quantum mechanical atoms of the enzyme-substrate complex would be defined next. The dynamics simulation would be run over several hundred picoseconds, with the majority of the atoms dealt with using the MM component of the program; substructures including the quantum mechanical atoms are treated by a full-scale *ab initio* calculation at the STO-3G level, eg, at selected intervals. The process would be captured in a trajectory file, and the quantum mechanical data would permit researchers to investigate reaction mechanisms, proton shuttling, salt-bridge formation, and other intermolecular associations pertinent to the dynamical processes. The general method has been elaborated by a number of researchers, a published survey (102) provides further details and examples of the application of the method.

Treatment of QM/MM calculations would not be complete without mention of the use of a newer quantum mechanical method called density functional techniques. Although density functional theory (DFT) is not new, recent implementations have made it significantly more popular than in the past. In particular, the potential of DFT as the QM component of combined QM/MM efforts has focused more attention of DFT methods; an introduction to DFT and its applications is available (103) which cites the primary literature in DFT, including a classic monograph (104). The comprehensive survey of QM/MM (102) also gives references pertinent to the application of the DFT method to QM/MM studies. Briefly put, DFT methods depend on the use of a functional of the electron density rather than the wavefunction functional, the advantage being that the density is an observable entity whereas the wavefunction is not. Further simplifying the calculations in DFT is the fact that the electron density has three spatial coordinates, regardless of the number of electrons in the chemical system. This makes it possible readily to compute properties directly from the electron density for systems of hundreds of electrons, a feat not easily accomplished routinely within wavefunctional theory. Just as occurs in the case of *ab initio* Hartree-Fock theory, DFT utilizes basis sets, and many of the more popular ones, along with the variants of DFT functionals, are available in commercial QM programs such as the Gaussian series (101) or SPARTAN (105).

Structure Generation, QSAR, and CoMFA Modeling Methods. *Rule-Based Structure Generation Techniques.* Figure 1 summarized the process for small molecule modeling. An important component of that process is the generation of the three-dimensional coordinates of the molecules to be studied. Many modeling program suites have their own 2-D drawing-to-3-D-coordinates conversion routines. However, it was not always so simple or routine to make the 2-D-to-3-D conversions. During the 1980s, significant progress toward simplifying this aspect of modeling was made, particularly in the area of knowledge-based methods. One of the most rigorous, and by far the most successful, of the conversion programs is CONCORD (106), produced at the University of Texas at Austin. CONCORD is a rule-based algorithm which evaluates atomic connectivity tables and maps reasonable bond lengths, angles, and dihedrals to the structures based on connectivity information and other geometric features permissible for a given atom and its normal valence states. The tool has become the de facto standard for the conversion of databases of hundreds to millions of 2-D structure representations into 3-D structures. Because the geometric parameters for structures were derived from AM1 (Austin Model 1, a successor to MINDO) semiempirical quantum mechanical optimized geometries for families of model functionalities, the method is quite extensible and reliable, although it does suffer from some of the same deficiencies which are characteristic of AM1 calculations (eg, underestimated dipole moments, in correct conformations for hydroxyl groups and bioxyl groups). A thorough review of methods for generating 3-D structures is available (107).

Distance Geometry. Another technique which utilizes both rule-based and computational geometry methods for generating 3-D structures is called distance geometry. This technique, which owes its origins to Cayley (108) in 1841, has been popularized as a result of pioneering work (109,110) and by the DGEOM

where $n = 16$, $r = 0.884$, and $s = 0.204$. This equation was derived for a series of fungicide compounds of the type N-substituted aminoacetonitriles, $RNHCH_2CN$, wherein it was found that the potency (ED^{50} = Effective Dose to kill 50%) was dependent on the corrected steric parameter, E_s^c (127). This variant of the Taft steric parameter, E_s, emphasizes the effect of branching in addition to steric bulk of the R group. Another example of a simple linear equation has been derived for the enzymatic hydrolysis of esters by the serine hydrolase, trypsin (128). For a series of esters, $X–Ph–OCOCH_2NHCOPh$, hydrolysis yields equation 8, where X is 4-SO_2NH_2, 4-NH_2, 4-CN, 4-NO_2, 4-$NHCONH_2$, 4-OCH_3, H, 4-CH_3, 4-Cl:

$$\log(1/K_m) = 0.71\ (\pm 0.17)\sigma + 3.31\ (\pm 0.09) \tag{8}$$

where $n = 10$, $r = 0.961$, and $s = 0.100$. These examples are not atypical of the hundreds which can be found in the literature. More exemplary of the relationships involving the log P are given by equation 9:

$$\log(1/C) = a(\log P) + b(\log P)^2 + c \tag{9}$$

where C is the equipotent concentration or dose, and a, b, and c are the coefficients of the linear, quadratic, and constants terms, respectively. Often, the linear term is zero, and a purely parabolic relationship between activity and log P is observed (126). More complex multivariate linear, bilinear, and parabolic equations can be valid for QSARs. However, it is important to perform a critical evaluation of both the biological data and the statistical paradigms before publishing such results. The treatment and testing of both the biological data and the QSAR model equations have been detailed (129–132). More recent studies on this subject can be found in the chapter by Pleiss and Unger (133), and in the journal, *Quantitative Structure–Activity Relationships* (Wiley-VCH).

The range of application of QSAR is extensive. Additional aspects of QSAR which are important and have recently reached the production level of application in industrial QSAR studies (134) include molecular similarity searching, 2-D fingerprinting, 3-D substructure searching, molecular superpositioning, pharmacophore identification, pharmacophore searching, 3-D databases, structural alignments, receptor–ligand binding energetics/modes, 3-D QSAR, molecular shape analysis, 1-D, 2-D and 3-D descriptors, including shape/size, charge, and hydrophobic fields, *de novo* ligand design, and hypothetical active site lattices.

The entire domain of "new-lead" discovery has expanded considerably. This development has affected what have traditionally been divergent approaches, namely QSAR and structure-based design, leading them to become integrated so as to provide a more powerful approach (135). Several recently published comprehensive volumes capture the state of the art and can be consulted to determine precedents relevant to any particular study (134–137).

Comparative Molecular Field Analysis (CoMFA).   Another method for molecular modeling is one which was developed in the mid-1980s but has come into frequent and practical use in the 1990s owing to the rapid advances in workstation computing power and in techniques for aligning diverse sets of structures. This technique, developed by Cramer (138) and frequently referred to as 3-D QSAR, is called Comparative Molecular Field Analysis (CoMFA) (see ENZYME INHIBITORS). CoMFA is a computational technique which attempts to mimic the interaction of a "ligand" with a "receptor" by means of a lattice of points (receptor) within which the molecule of interest (ligand) is placed, and the interactions between the molecule and the grid points are evaluated. The points of the lattice which are inclusive of the volume of the molecule of interest are discarded. The researcher must propose an alignment (superpositioning) of the structures of interest. This can be done by rigid or flexible fitting, in either case an additional QSAR descriptor, the distortion energy of superpositioning, can be added to the classical QSAR regression. All other points are assigned both steric and electrostatic properties, such as Csp^3 carbon van der Waals steric properties, and a $+1$ charge for electrostatics. The interaction energies for both property types are then evaluated for the approximately 2,000 points which remain in the lattice. These data (the xs) are then processed by Wold's technique of partial least squares (PLS) regression analysis (139) against the activity values (the ys) supplied by the researcher. The program produces a model which will both reproduce the training set of data values (bioactivity, or a similar property) and have predictive power as well. The PLS model is cross-validated by successively eliminating observations, rederiving the model, and predicting the eliminated observations. As new, diverse data are added to the training set, the predictive power of the model is enhanced. Additionally, the CoMFA model is visually displayed, indicating regions either where steric bulk is favored or less so, as well as where changes in the electrostatic field enhance or diminish activity. The power of CoMFA is made possible by the PLS technique, because without PLS to reduce the dimensionality of the xs, it would not be possible to derive a believable regression model when presented with a data matrix which is 10–20 rows by 2,000 columns wide. Studies on factor analysis and principal components regression (140), on chemometric tools (141), on the performance of biased regression techniques (142) have helped in understanding the complexity of regression techniques, their pitfalls, and their importance of QSAR. Indeed, a set of recommendations and caveats regarding the use of CoMFA has been published (143), and as with any computational or modeling technique, its capabilities and limitations should become better understood as the frequency of its use grows. In summary, QSAR techniques give researchers a wide range of approaches to the problem of quantitating the relationship between chemical structures, changes in chemical structures, and observed physical and biological properties.

Summary

Clearly, whereas molecular modeling as a practice has its roots in the development of quantum theory at the turn of the twentieth century, it has been the exponential growth in computing power between the mid-1970s and the mid-1990s that has catalyzed the development and application of molecular modeling methods during that period. The spectrum of software systems available (144) covers all aspects of modeling. A sampling is given in Table 1. Although it is

Table 1. Molecular Modeling Software Systems[a]

System(s) acronym(s)	System name or type	System producer
Databases and DB management systems		
		Daylight Chemical Information Systems
MACCS/ISIS		Molecular Design Limited, Inc. (MDLI)
CCDC	Cambridge Crystallographic Data Center X-ray Database	Cambridge Crystallographic Data Center
PDB	Brookhaven Protein Data Bank	Brookhaven National Laboratory
CAST-3D	The Chemical Abstracts Service 3-D Structure DB	Chemical Abstracts Service
UNITY	Module of the SYBYL suite	Tripos
Desktop modeling and data management		
ISIS Draw, SAR, Excel		Molecular Design Limited, Inc.
AccuModel	ISIS Add-ins	Microsimulations, Inc.
SAS, JMP, MINITAB	statistics	SAS
ChemDraw, Chem3-D, ChemOffice		CambridgeSoft
Alchemy 2000		Tripos
CaChe		Oxford Molecular
Standalone software programs (workstation versions)		
NBO, GAUSSIAN, SPARTAN, MOPAC, PROAIM	quantum mechanical	
PEOE Methods for Charges		Gasteiger, Scheraga, Gombar (HDI)
MM3, MM4, MMFF	molecular mechanical	Allinger/Halgren
AMBER, GROMOS, QUANTA/CHARMm	molecular dynamics	(see text)
BOSS	Monte Carlo simulations	Jorgensen
Molecular modeling software suites (workstation)		
Chem-X		Chemical Design, Ltd.
Hyperchem		Hypercube, Inc. (Ostlund and co-workers)
InsightII, Cerius		Molecular Simulations/ Biosym
MacroModel		Columbia University (W. C. Still)
OMG		Oxford Molecular
SYBYL		Tripos
2-D to 3-D converters		
CONCORD	converter/conformation generator	Tripos
CORINA	converter/conformation generator	Gasteiger/OMG

Table 1. *Continued*

System(s) acronym(s)	System name or type	System producer
WIZARD	converter/conformation generator	Dolata & Leach
DGEOM	distance geometry program	QCPE (UCSF/DuPont)
QSAR		
ADAPT	pattern recognition toolkit	P.C. Jurs (MDLI)
APEX	full range of statistical treatments for QSAR	MSI
CHEMEST		Technical Database Services
CLOGP		Biobyte, Inc. (Pomona MedChem Project)
MOLCONN-X	Kier & Hall Indices	L. Hall
QSAR-PC		Biosoft, Inc.
SigmaStat	stat analysis package	Jandel Scientific Software
TOPKAT	toxicology prediction tool	Health Designs Inc. (HDI)
TOPMOST	computes charges, descriptors for QS(APT)R	HDI
Other applications		
AutoDOCK	Monte Carlo docking of ligands to receptors	A. Olson at Scripps Institute
CAVEAT	database generator for new ligands	P. Bartlett at University of California, Berkeley
DISCO	tool for deriving pharmacophore from active compounds	Y. Martin at Abbott Labs; available from Tripos
DOCK	ligand docking/active site probe tool	I. D. Kuntz at University of California, San Francisco
GRIN/GRID	nonbonded force probe of active sites	Molecular Discovery, Ltd. (Goodford)
HINT	adjunct to CoMFA, computes hydrophobic fields	EduSoft, Inc. (Kellog/ Abraham)
LEAPFROG	generates new leads from fragments	Tripos
LUDI	*de novo* ligand design	MSI (Böhm)

[a]Ref. 144.

not possible to duplicate the exhaustive survey of available systems given in
Reference 144, Table 1 furnishes a starting point from which to search for the
right tool to address specific problems.

An understanding of the precedents in both methods development and
applications citations in the literature is thus critical to the researcher working
in fields that employ molecular modeling as a tool. With it, the varied application
and untapped potential of molecular modeling may be used more profitably in
individual researchers' specific fields of interest.

BIBLIOGRAPHY

1. W. Heisenberg, *Z. Phys.* **33**, 879 (1925).
2. E. Schrödinger, *Ann. Phys.* **79**, 361, 489; **80**, 437; **81**, 109 (1926).
3. J. D. Bolcer and R. B. Hermann, in K. B. Lipkowitz and D. B. Boyd, eds., *Reviews in Computational Chemistry*, Vol. 5, VCH Publishers, New York, 1994, pp. 1–63.
4. R. S. Mulliken, in D. A. Ramsey and J. Hinze, eds., *Selected Papers of Robert Mulliken*, University of Chicago Press, Chicago, 1975, pp. 39–42.
5. R. S. Mulliken, *Life of a Scientist*, Springer-Verlag, Berlin, 1989, pp. 136–161.
6. F. London, *Z. Phys.* **63**, 245 (1930).
7. E. Hückel, *Z. Phys.* **70**, 204 (1931).
8. J. D. Kemp and K. S. Pitzer, *J. Chem. Phys.* **14**, 733 (1936).
9. F. H. Westheimer and J. E. Meyer, *J. Chem. Phys.* **14**, 733 (1946).
10. F. H. Westheimer, in M. S. Newman, eds., *Steric Effects in Organic Chemistry*, John Wiley & Sons, Inc., New York, 1956, p. 523.
11. J. B. Hendrickson, *J. Am. Chem. Soc.* **83**, 5537 (1961).
12. N. L. Allinger, J. Allinger, and M. A. DaRooge, *J. Am. Chem. Soc.* **86**, 4061 (1964) and references therein.
13. G. Herzberg, *Molecular Spectra and Molecular Structure. II. Infrared and Raman Spectra of Polyatomic Molecules*, Van Nostrand Reinhold, New York, 1945; E. B. Wilson, Jr., J. C. Decius, and P. C. Cross, *Molecular Vibrations*, McGraw-Hill Book Co., Inc., New York, 1955.
14. T. L. Hill, *J. Chem. Phys.* **14**, 465 (1946).
15. C. A. Coulson and G. S. Rushbrooke, *Proc. Cambridge Phil. Soc. Math. Phys. Sci.* **36**, 193 (1940).
16. M. Goeppert-Mayer and A. I. Sklar, *J. Chem. Phys.* **6**, 645 (1938).
17. M. J. S. Dewar, *J. Chem. Soc.*, 2329 (1950); M. J. S. Dewar, *J. Chem. Soc.*, 3532 (1952).
18. J. A. Pople, *Trans. Faraday Soc.* **49**, 1375 (1953); J. A. Pople, *J. Phys. Chem.* **61**, 6 (1957).
19. R. Pariser and R. Parr, *J. Chem. Phys.* **21**, 466, 767 (1953).
20. A. Streitwieser, Jr., *Molecular Orbital Theory for Organic Chemists*, John Wiley & Sons, Inc. New York, 1961.
21. J. A. Pople and D. L. Beveridge, *Approximate Molecular Orbital Theory*, McGraw-Hill Book Co., Inc., New York, 1970.
22. J. N. Murrell and A. J. Harget, *Semi-empirical Self-consistent-Field Molecular Orbital Theory of Molecules*, Wiley-Interscience, London, England, 1972.
23. R. Hoffman and R. B. Woodward, *Acc. Chem. Res.* **1**, 17 (1968); R. B. Woodward and R. Hoffman, *Conservation of Orbital Symmetry*, Verlag Chemie, Weinheim, Germany, 1970.
24. E. Wimmer, *Chem. Design Auto. News* **8**(4), 1 (1993); R. C. Merkle, *Chem. Design Auto. News* **8**(9/10), 1 (1993); see also *J. Comput.-Aided Mater. Design*.
25. A. L. Horvath, *Molecular Design*, Vol. 75, Elsevier, Amsterdam/New York, 1992.
26. C. Levinthal, *Sci. Am.* **214**, 42 (1966).
27. R. Langridge and T. Klein, in C. A. Ramsden, vol. ed., *Comprehensive Medicinal Chemistry*, Vol. 4, *Quantitative Drug Design*, Pergamon Press, New York, 1990, pp. 413–430.
28. R. E. Hubbard, in N. C. Cohen, ed., *Guidebook on Molecular Modeling in Drug Design*, Academic Press, San Diego, Calif., 1996, pp. 19–54.
29. T. A. Jones, *J. Appl. Crystallogr.* **11**, 268 (1978).
30. D. Sayer, ed., *Computational Crystallography*, Oxford University Press, New York, 1982.
31. G. N. Phillips, Jr., *Biophys. J.* **69**, 1281 (1995).

32. R. Langridge, T. E. Ferrin, I. D. Kuntz, and M. L. Connolly, *Science* **211**, 661 (1981).
33. M. L. Connolly, *Science* **221**, 709 (1983); M. L. Connolly, *J. Appl. Crystallogr.* **16**, 548 (1983).
34. F. M. Richards, *Annu. Rev. Biophys. Bioeng.* **6**, 151–176 (1977).
35. K. Flurchik and L. Bartolotti, *J. Mol. Graphics* **13**, 10 (1995).
36. D. C. Richardson and J. S. Richardson, *Trends in Biochem. Sci.* **19**(3), 135–138 (1994).
37. N. Bohr, *Phil. Mag.* **26**, 476 (1913).
38. L. Pauling and E. B. Wilson, Jr., *Introduction to Quantum Mechanics*, McGraw-Hill Book Co., Inc., New York, 1935.
39. L. deBroglie, *The Revolution in Physics*, Noonday Press, Inc., New York, 1953; H. Eyring, J. Walter, and G. E. Kimball, *Quantum Chemistry*, John Wiley & Sons, Inc., New York, 1944; W. Kauzmann, *Quantum Chemistry*, Academic Press, New York, 1957.
40. P. A. M. Dirac, *The Principles of Quantum Mechanics*, Oxford University Press, New York, 1958.
41. P. A. M. Dirac, *Proc. Roy. Soc. (London)* **A123**, 714 (1929).
42. A. Streitwieser, Jr., *Molecular Orbital Theory for Organic Chemists*, John Wiley & Sons, Inc., New York, 1961, pp. 6–7.
43. D. R. Hartree, *Proc. Cambridge Phil. Soc.* **24**, 89, 111, 426 (1927).
44. V. Fock, *Z. Phys.* **62**, 795 (1930).
45. A. Warshel and M. Karplus, *J. Am. Chem. Soc.* **94**, 5612 (1972).
46. N. L. Allinger and J. T. Sprague, *J. Am. Chem. Soc.* **95**, 3893 (1973).
47. J. A. Pople, D. P. Santry, and G. A. Segal, *J. Chem. Phys.* **43**, S129 (1965); J. A. Pople and G. A. Segal, *J. Chem. Phys.* **43**, S136 (1965).
48. M. Zerner, in K. B. Lipkowitz and D. B. Boyd, eds., *Reviews in Computational Chemistry*, Vol. 2, VCH Publishers, New York, 1991, pp. 313–356.
49. R. C. Bingham, M. J. S. Dewar, and D. H. Lo, *J. Am. Chem. Soc.* **97**, 1285 (1975), and subsequent papers by Dewar and collaborators.
50. M. J. S. Dewar, E. F. Healy, J. J. P. Stewart, and W. Thiel, in Ref. 48.
51. J. J. P. Stewart, *J. Comput.-Aided Mol. Design* **4**, 1 (1990).
52. J. J. P. Stewart, in K. B. Lipkowitz and D. B. Boyd, eds., *Reviews in Computational Chemistry*, Vol. 1, VCH Publishers, New York, 1990, pp. 45–82.
53. T. Clark, *A Handbook of Computational Chemistry*, John Wiley & Sons, Inc., New York, 1985, Chapts. 3 and 4.
54. D. Feller and E. R. Davidson, in Ref. 52, pp. 1–44.
55. W. J. Hehre, L. Radom, P. v.-R. Schleyer, and J. A. Pople, *Ab Initio Molecular Orbital Theory*, Wiley-Interscience, New York, 1986, and references therein.
56. C. E. Dykstra, J. D. Augspurger, B. Kirtman, and D. J. Malik, in Ref. 52, pp. 83–118.
57. D. B Boyd, in Ref. 52, pp. 321–354.
58. M. J. S. Dewar, *J. Am. Chem. Soc.* **106**, 669 (1984).
59. L. S. Bartell, *J. Am. Chem. Soc.* **99**, 1977, 3279 (1977), and references therein.
60. U. Burkert and N. L. Allinger, *Molecular Mechanics*, ACS Monograph Series, Vol. 177, American Chemical Society, Washington, D.C., 1982, pp. 38, 157.
61. N. L. Allinger, *Adv. Phys. Org. Chem.* **13**, 1 (1976).
62. O. Ermer and S. Lifson, *J. Am. Chem. Soc.* **95**, 4121 (1973); O. Ermer, *Struct. Bonding* **27**, 161 (1976).
63. S. Lifson and A. Warshel, *J. Chem. Phys.* **49**, 5116 (1968); A. Warshel and S. Lifson, *Chem. Phys. Lett.* **4**, 255 (1969); A. Warshel and S. Lifson, *J. Chem. Phys.* **53**, 582 (1970); A. T. Hagler and S. Lifson, *Acta Crystallogr. Sect. B* **30**, 619 (1974); A. T. Hagler, E. Huler, and S. Lifson, *J. Am. Chem. Soc.* **96**, 5319 (1974); A. Warshel and M. Levitt, *Nature* **253**, 694 (1975). M Levitt and A. Warshel, *J. Mol. Biol.* **106**, 421 (1976); M. Levitt and A. Warshel, *J. Am. Chem. Soc.* **100**, 2607 (1978).

118. A. J. Leo and C. Hansch, *J. Org. Chem.* **36**, 1539 (1971); C. Hansch, in C. J. Cavallito, eds., *Structure–Activity Relationships*, Vol. 1, Pergamon Press, Oxford, U.K., 1973, p. 150; C. Hansch and A. J. Leo, *Exploring QSAR: Fundamentals and Applications in Chemistry and Biology*, American Chemical Society, Washington, D.C., 1995, Chap. 4 and 5.

119. J. M. McFarland, *J. Med. Chem.* **13**, 1192 (1970).

120. C. Hansch, A. R. Steward, S. M. Anderson, and D. Bentley, *J. Med. Chem.* **11**, 1 (1968); C. Hansch and J. M. Clayton, *J. Pharm. Sci.* **62**, 1 (1973).

121. Cerius², a Software Suite for Molecular Modeling, Version 3.5, Molecular Simulations, Inc., San Diego, Calif., 1997.

122. R. W. Taft, in M. S. Newman, ed., *Steric Effects in Organic Chemistry*, John Wiley & Sons, Inc., New York, 1956, pp. 556–675.

123. P. N. Craig, *J. Med. Chem.* **14**, 680 (1971).

124. C. Hansch, S. H. Unger, and A. B. Forsythe, *J. Med. Chem.* **16**, 1212 (1973).

125. A. Leo, P. Jow, C. Silipi, and C. Hansch, *J. Med. Chem.* **18**, 865 (1975).

126. T. Fujita, in Ref. 27, pp. 497–560; C. Hansch and T. Fujita, eds., *Classical and Three-Dimensional QSAR in Agrochemistry*, ACS Symposium Series No. 606, American Chemical Society, Washington, D.C., 1995.

127. O. Kirino, H. Ohshita, T. Oishi, and T. Kato, *Agric. Biol. Chem.* **44**, 31 (1980).

128. C. D. Selassie, M. Chow, and C. Hansch, *Chem.-Biol. Interact.* **68**, 13 (1988).

129. Y.C. Martin, *Quantitative Drug Design*, Marcel Dekker, New York, 1978.

130. D. Cuthbert and F. S. Wood, *Fitting Equations to Data*, John Wiley & Sons, Inc., New York, 1971.

131. A. Cammarata, R. C. Allen, J. K. Seydel, and E. Wempe, *J. Pharm. Sci.* **59**, 1496 (1970).

132. J. G. Topliss and R. J. Costello, *J. Med. Chem.* **15**, 1066 (1972).

133. M. A. Pleiss and S. H. Unger, in Ref. 27, pp. 561–587.

134. H. Kubinyi, eds., *3-D QSAR in Drug Design: Theory, Methods and Applications*, ESCOM Science Publishers, Leiden, The Netherlands, 1993.

135. A. Itai, Y. Mizutani, Y. Nishibata, and N. Tomioka, in N. C. Cohen, *Guidebook on Molecular Modeling in Drug Design*, Academic Press, San Diego, Calif., 1996, pp. 93–138.

136. P. S. Magee, D. R. Henry, and J. H. Block, *Probing Bioactive Mechanisms*, ACS Symposium Series No. 413, American Chemical Society, Washington, D.C., 1989.

137. C. H. Reynolds, M. K. Hollaway and H. K. Cox, *Computer-Aided Molecular Design: Applications in Agrochemicals, Materials and Pharmaceuticals*, ACS Symposium Series No. 589, American Chemical Society, Washington, D.C., 1995.

138. R. D. Cramer, D. E. Patterson, and J. D. Bunce, *J. Am. Chem. Soc.* **110**, 5959 (1988).

139. S. Wold, A. Ruhe, H. Wold, and W. J. Dunn, *SIAM J. Sci. Stat. Comput.* **5**, 735 (1984).

140. S. Wold, *Technometrics* **20**, 397 (1978).

141. M. Baroni, G. Costantino, G. Cruciani, D. Riganelli, R. Valigi, and S. Clementi, *Quant. Struct.–Act. Relat.* **12**, 9 (1993); M. Baroni, D. Bonelli, S. Clementi, G. Cruciani, C. Ebert, and B. Skagerberg, *Quant. Struct.–Act. Relat.* **9**, 101 (1990).

142. K. G. Kowalski, *Chemometrics Intel. Lab Syst.* **9**, 177 (1990).

143. U. Thibaut, G. Folkers, G. Klebe, H. Kubinyi, A. Merz, and D. Rognan, in Ref. 134, pp. 711–728.

144. D. B. Boyd, in Ref. 102, pp. 303–380.

General References

Historical perspective:

Reference 3 is also a general reference.

R. M. Davis, *Science* **195**, 1099 (1977).

M. S. Tute, in C. A. Ramsden, vol. ed., *Comprehensive Medicinal Chemistry*, Vol. 4, *Quantitative Drug Design*, Pergamon Press, New York, 1990, pp. 1–32.

Definition of Molecular Modeling and Uses of CAMM:

L. M. Balbes, S. W. Mascerella, and D. B. Boyd, in K. B. Lipkowitz and D. B. Boyd, eds., *Reviews in Computational Chemistry*, VCH Publishers, New York, Vol. V, 1994, pp. 337–379.

Chemical Design Automation News, Elsevier Science, Inc., New York, 1985–1997.

N. C. Cohen, ed., *Guidebook on Molecular Modeling in Drug Design*, Academic Press, San Diego, Calif., 1996.

D. Liotta, ed., *Advances in Molecular Modeling*, Vols. 1–2, JAI Press, Inc., Greenwich, Conn., 1988–1990.

Computer Graphics In Molecular Modeling:

References 27 and 28 are also general references.

G. R. Marshall and C. B. Naylor, in C. A. Ramsden, vol. ed., *Comprehensive Medicinal Chemistry*, Vol. 4, *Quantitative Drug Design*, Pergamon Press, New York, 1990, pp. 431–458.

J. M. Blaney and C. Hansch, in C. A. Ramsden, vol. ed., *Comprehensive Medicinal Chemistry*, Vol. 4, *Quantitative Drug Design*, Pergamon Press, New York, 1990, pp. 459–496.

C. Levinthal, *Sci. Amer.* **214**, 42 (1966).

Computational Methods for Molecular Modeling:

L. Pauling and E. B. Wilson, Jr., *Introduction to Quantum Mechanics*, McGraw-Hill Book Company, Inc., New York, 1935.

L. DeBroglie, *The Revolution in Physics*, Noonday Press, New York, 1953.

H. Goldstein, *Classical Mechanics*, Addison-Wesley Publishing Company, Reading, Mass., 1956.

A. Streitwieser, Jr., *Molecular Orbital Theory for Organic Chemists*, John Wiley & Sons, Inc., New York, 1961.

J. Hammer, *The Conceptual Development of Quantum Mechanics*, McGraw-Hill Book Co., Inc., New York, 1966.

J. P. Tollenaere, in N. C. Cohen, ed., *Guidebook on Molecular Modeling in Drug Design*, Academic Press, San Diego, Calif., 1996, pp. 337–356. Contains a superb glossary of terminology.

Molecular Mechanics and Molecular Dynamics:

U. Burkert and N. L. Allinger, *Molecular Mechanics*, ACS Monograph Series, Vol. 177, American Chemical Society, Washington, D.C., 1982.

D. B. Boyd and K. B. Lipkowitz, *J. Chem. Educ.* **59**, 269 (1982). The method and underlying philosophy of molecular mechanics.

K. Rasmussen, in G. Berthier and co-workers, eds., *Lecture Notes in Chemistry*, Vol. 27, *Potential Energy Functions in Conformational Analysis*, Springer-Verlag, Berlin, 1985.

M. Karplus and J. A. McCammon, *Sci. Amer.*, 42 (Apr. 1986).

P. Kollman, *Annu. Rev. Phys. Chem.* **38**, 303 (1987).

W. F. van Gunsteren and P. K. Weiner, eds., *Computer Simulations of Biomolecular Systems: Theoretical and Experimental Applications*, ESCOM, Leiden, the Netherlands, 1989.

T. P. Lybrand, in K. B. Lipkowitz and D. B. Boyd, eds., *Reviews in Computational Chemistry*, Vol. 1, VCH Publishers, New York, 1990, pp. 295–320.

Reviews in Computational Chemistry, K. B. Lipkowitz and D. B. Boyd, eds., Vol. 1–10, VCH Publishers, New York, 1990–1997.

Structure Generation, QSAR, and CoMFA Modeling Methods:

Reference 134 is also a general reference.

C. A. Ramsden, vol. ed., *Quantitative Drug Design*, Vol. 4, *Comprehensive Medicinal Chemistry*, Pergamon Press, New York, 1990.

P. Krogsgaard-Larsen and H. Bungaard, eds., *A Textbook of Drug Design and Development*, Harwood Academic Publishers, Switzerland, 1991.

F. Sanz, J. Giraldo, and F. Manaut, *QSAR and Molecular Modeling: Concepts, Computational Methods and Biological Applications*, J. R. Prous Science Publishers, Barcelona, Spain, 1995.

M. Kuchar, ed., *QSAR in the Design of Bioactive Compounds*, J. R. Prous Science Publishers, Barcelona, Spain, 1992.

J. G. Topliss, *Perspectives in Drug Discovery and Design* **1**, 253 (1993).

SALVATORE PROFETA, JR.
Monsanto

MOLECULAR RECOGNITION

Receptor–Substrate-/Host–Guest-Chemistry

Molecular recognition is a central point. It may be said that without molecular recognition, there would be no life in this world. Vital biochemical processes such as enzyme action, molecular transport, genetic information, processing and protein assembly all involve molecular recognition as an essential action (1–3). Understanding of its principles is still a problem although first elucidation of the rules that govern molecular recognition dates back to the late nineteenth century (4). Strictly speaking, in 1894 Emil Fischer, a farsighted chemist, came up with his brilliant "lock-and-key" idea. In his famous paper (5) he proposed that enzyme and substrate can be compared to lock and key getting selectivity between molecules involved. Moreover, and this occurred even earlier, it was Paul Ehrlich who recognized that molecules do not act if they do not bind, thus, introducing the concept of receptor (6). Finally binding or fixation requires interaction, affinity between the partners that may be related to the idea of coordination introduced by Alfred Werner (7).

According to these basic concepts, molecular recognition implies complementary lock-and-key type fit between molecules. The lock is the molecular receptor and the key is the substrate that is recognized and selected to give a defined receptor–substrate complex, a coordination compound or a supermolecule. Hence molecular recognition is one of the three main pillars, fixation, coordination, and recognition, that lay foundation of what is now called supramolecular chemistry (8–11).

Behind this new direction of supramolecular chemistry, the chemistry beyond the molecule, is a highly interdisciplinary field of science covering the chemical, physical, and biological features of chemical species of greater complexity than molecules themselves that are held together and organized by means of intermolecular (nonbinding) interactions (12). The chemistry of molecular recognition is also the core of host–guest chemistry which is a sub-discipline or a particular aspect of supramolecular chemistry mostly involving inclusion and complex formation (13) (see also INCLUSION COMPOUNDS).

Principles of Receptor Design

Information Storage and Read Out. Picking up the thread of the introduction, molecular recognition is defined by the energy and the information involved in the binding and selection of substrates by a given receptor molecule that may also involve a specific function (14). Mere binding is not recognition, although it is often taken as such. Instead, one may say that recognition is binding with a purpose, like receptors are ligands with a purpose. It implies a pattern recognition process through a structurally well-defined set of intermolecular interactions. Molecular recognition, thus, deals with the molecular storage and supramolecular read out of molecular information (9).

Information may be stored in the architecture of the receptor, in its binding sites, and in the ligand layer surrounding the bound substrate such as specified in Table 1. It is read out at the rate of formation and dissociation of the receptor–substrate complex (14). The success of this approach to molecular recognition lies in establishing a precise complementarity between the associating partners, ie, optimal information content of a receptor with respect to a given substrate.

Complementarity. To a first approximation, complementarity should take two forms (Fig. 1). Firstly, the shape and size of the receptor cavity must complement the form of the substrate. Secondly, there must be a chemical complementarity between the binding groups lining the interior of the cavity and the external chemical features of the substrate (15).

The weak intermolecular forces that are principally involved in stabilizing receptor-substrate interactions and involved in molecular recognition processes (16) are summarized in Table 2. Examples are shown in Figure 1.

Most effective differentiation of the receptor between substrates will occur when multiple interactions are involved in the recognition process. The more binding regions (contact area) present, the stronger and more selective will be the recognition (17). This is the case for receptor molecules that contain intramolecular cavities, clefts or pockets into which the substrate may fit (Fig. 1).

Reorganization and Preorganization. On principle there are two different modes of receptor behavior illustrated in Figure 2. One of them, already evoked, is represented by the so-called lock-and-key image (Fig. 2a), involving complementary fit concept between rigid substrate and rigid receptor or rigid guest and rigid host relating to conformational flexibility of the molecular constituents forming the receptor–substrate (host–guest) complex (4). Receptors of this type

Table 1. Structural Parameters for Storage of Information in a Chemical Receptor

Receptor	Parameter
architecture	size
	shape
	connectivity
	cyclic order
	conformation
	chirality
	dynamics
binding sites	electronic properties
	(charge, polarity, polarisability, van der Waals attraction and repulsion)
	size
	shape
	number
	arrangement
	reactivity
	(protoniziable, deprotonizable, reducible, oxidizable)
surrounding	thickness
ligand layer	overall polarity
	(lipophilic, hydrophilic)
	specific polarity
	(exo/endo-lipo/polarophilic)

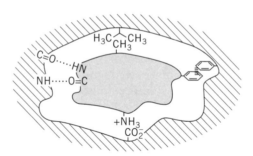

Fig. 1. Schematic representation of a receptor–substrate (host–guest) complex involving cavity inclusion of the substrate and the formation of different types of weak supramolecular interactions between receptor (hatched) and substrate (dotted).

Table 2. Types of Interactions in Molecular Recognition

hydrogen bonding between basic and acidic centers
electrostatic attraction between anionic and cationic centers
metal-ligand interaction
dipole–dipole interaction
π-stacking and charge-transfer interaction between aromatic residues in the receptor
  and delocalized regions of the substrate
van der Waals attraction between hydrophobic regions on the two components
covalent bonds, that can be reversibly formed and broken (eg, disulfides, borate esters).

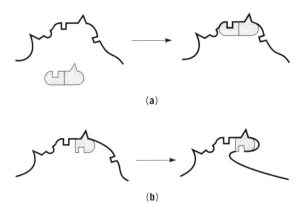

Fig. 2. Principle mechanisms of formation of a receptor–substrate complex: (**a**) Fischer's rigid "lock-and-key" model; (**b**) "induced fit" model showing conformational changes of the receptor (solid line) upon substrate binding.

are expected to present very efficient recognition between complementary partners, ie, both high stability and high selectivity of the receptor–substrate complex. The advantage of complementary receptor preorganization comes from minimizing the unfavorable entropy involved in substrate binding (see below).

However, in most biological system there is a degree of flexibility in the receptor (2). The approach of the substrate leads to conformational changes and an organization of the binding site around it. With this induced fit mechanism of binding (Fig. 2**b**), a higher entropy price is paid but there are several advantages (18). A flexible receptor will permit a more wrap around interception or even complete encapsulation with the substrate involving many more potential binding interactions. This may lead to high selectivity of binding involving the amplification of molecular recognition interactions illustrated in Figure 3 (19).

In case of the rigid lock-and-key type receptor forming five hydrogen bonds plus two extended electrostatic attractions (Fig. 3**a**), one mismatched hydrogen bond will result in only a small reduction in overall binding free energy

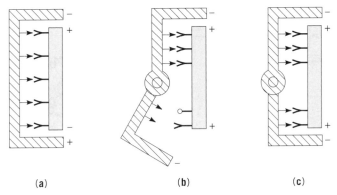

Fig. 3. Schematic approach illustrating amplification of molecular recognition effects of (**a**) matched rigid, (**b**) mismatched rigid, and (**c**) flexible type of receptor–substrate assemblies (19).

Table 3. Comparison of Cation and Cavity Diameters

Cation	Cation diameter,[a] Å	Crown ether	Cavity diameter,[a] Å
Li^+	1.36	12-crown-4 (**1a**)	1.2–1.5
Mg^{2+}	1.56		
Na^+	1.90	15-crown-5 (**1b**)	1.7–2.2
Ca^{2+}	2.12		
Sr^{2+}	2.54	18-crown-6 (**1c**)	2.6–3.2
K^+	2.66		
Ba^{2+}	2.86		
Rb^+	2.98		
Cs^+	3.38	21-crown-7 (**1d**)	3.4–4.3

[a] Å = 0.1 nm.

higher level of organization that is directly related to the mode and efficiency of molecular recognition (14,30).

It has already been stressed that a concave receptor is a favorable case. Under these circumstances the receptor cavity is lined with binding sites directed towards the bound species (Fig. 1). This corresponds to Cram's definition of a receptor (host) molecule providing binding sites that are convergent, as contrasted with the bound substrate (guest) featuring divergent complementary sites, ie, the substrate is more or less completely surrounded by the receptor forming an inclusion complex (20,31). This widely used principle of convergence defines a convergent or endo-supramolecular chemistry (host–guest chemistry) with endo-receptors (endo-hosts) effecting endo-recognition (Fig. 5a) (9).

The opposite procedure consists in making use of an external receptor surface rather than an internal cavity as substrate receiving site. This amounts to the passage from a convergent endo-supramolecular chemistry to a divergent or exo-supramolecular chemistry, and from endo- to exo-receptors (Fig. 5b) (9). Here receptor–substrate binding occurs by surface-to-surface interaction which may be termed affixation as contrasted with inclusion. Exo-recognition with strong and selective binding, in particular, requires a large enough contact area and a sufficient number of complementary interactions along the interface. Such a mode of molecular recognition also finds biological analogies, for instance at the antibody–antigen interface of immunological importance (32). Metallo-exoreceptor aggregation, molecular recognition at organic and inorganic monolayers, films and solid surfaces bearing recognition groups, as well as the design of supramolecular solid architectures and materials, are other important instances of the exo-recognition principle discussed in more detail below.

Apart from the basic classification in convergent endo- and divergent exo-receptors, molecular receptors (hosts) of extremely varied structural types have been developed (30) including the acyclic podands, the macrocyclic crown ethers, coronands or torands, the macropolycyclic cryptands and speleands, the spherands, cavitands and carcerands, the calixarenes and cryptophanes, the clathrands, the helicands and other organization framework species (12). Each of these trivial names refers to a particular aspect of the structure involving the overall receptor topology, that is connectivity, dimensionality and cyclic order. Examples of compounds are illustrated elsewhere in this article (see also

INCLUSION COMPOUNDS) while a selection of possible topologies are shown in Figure 6, from a linear receptor (a) to spherical (i) or cylindrical (j) tricyclic structures following the classification of graphs (14). Moreover, all these types of receptors may possess a single receptor unit to recognize and bind a single substrate, eg, (d) or contain more than one discrete binding subunit (h) being characteristic of monotopic and oligo-/polytopic receptors, respectively. The size and the shape of the binding cavity that they define and their rigidity or flexibility are determined by the nature of the structural subunits making up the branches of the graphical representations in Figure 6. In this respect they serve the same purpose for conveying properties of molecular recognition, mostly based on geometrical discrimination.

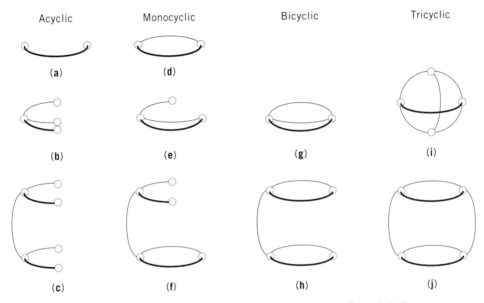

Fig. 6. Topological structure types of receptors (hosts) (14).

Simple Modes of Molecular Recognition

Substrates involved in molecular recognition may feature a particular shape, size, state of charge, chemical affinity or optical specification (19,30,33–36). In general most of these parameters share. Nevertheless there may be dominating features of a certain substrate molecule to be used by a complementary receptor in the recognition process (9).

Size and Shape Dominated Substrate Recognition. Perhaps the simplest recognition process is that of a spherical substrate, in its most elementary form a ball-shaped metal ion of defined diameter. Supramolecular chemistry in itself started with this problem, in particular having the effort for recognition and binding of alkali and alkaline earth cations (37). Numerous studies have been performed that are reported in many papers and summarized in reviews and books (12) showing that three main classes of receptors providing spherical recognition property may be distinguished (38). They are (*1*) macro-cyclic polyethers, the well-known crown ethers and their derivatives (24); (*2*) the macropolycyclic cryptands (25,26); and (*3*) the acyclic analogues of crown compounds and cryptands usually designated as podands (22,23). Prototypical compounds for each substance class are given by compounds (**1**)–(**3**) (Fig. 7). They all possess a spherical or quasispherical negatively polarized cavity pre-pared for the accommodation of alkali- and alkaline earth metal ions that have complementary size, giving rise to a feature termed as spherical recognition (9). Coronates, cryptates or podates are the names of the respective inclusion complexes (39).

The match between crown cavity diameter and cation diameter is obvious from Table 3 showing that, eg, Li^+ and 12-crown-4 (**1a**) or K^+ respectively Ba^{2+} and 18-crown-6 (**1c**) correspond. Similar are the cryptands of gradually increasing cavity size [2.1.1], [2.2.1] and [2.2.2] for Li^+, Na^+ and Sr^{2+} or K^+ and

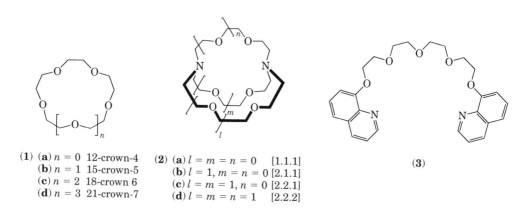

(**1**) (**a**) $n = 0$ 12-crown-4   (**2**) (**a**) $l = m = n = 0$ [1.1.1]
　　　 (**b**) $n = 1$ 15-crown-5　　　　 (**b**) $l = 1, m = n = 0$ [2.1.1]
　　　 (**c**) $n = 2$ 18-crown 6　　　　 (**c**) $l = m = 1, n = 0$ [2.2.1]
　　　 (**d**) $n = 3$ 21-crown-7　　　　 (**d**) $l = m = n = 1$　 [2.2.2]

(**3**)

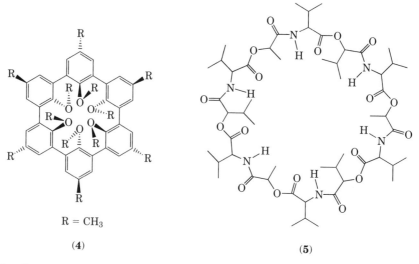

R = CH₃

(**4**)　　　　　　　(**5**)

Fig. 7. Crown type and analogous receptor molecules of different varieties; (**1**) crown ethers; (**2**) cryptands; (**3**) a podand; (**4**) a spherand; and (**5**) the natural depsipeptide valinomycin.

Ba^{2+}, while the small cavity of [1.1.1] fits H$^+$. The example of such a matched cryptate where K$^+$ is accommodated into the cavity of [2.2.2] is illustrated in Figure 8a (25). Although acyclic podands do not provide a permanent cavity, they may create one by encircling a spherical cation with the length of the receptor molecular thread being the controlling parameter (22,40). Nevertheless, from what has already been said, low preorganization and topology of the podands handicap the substrate recognition which is increasingly higher in the circular

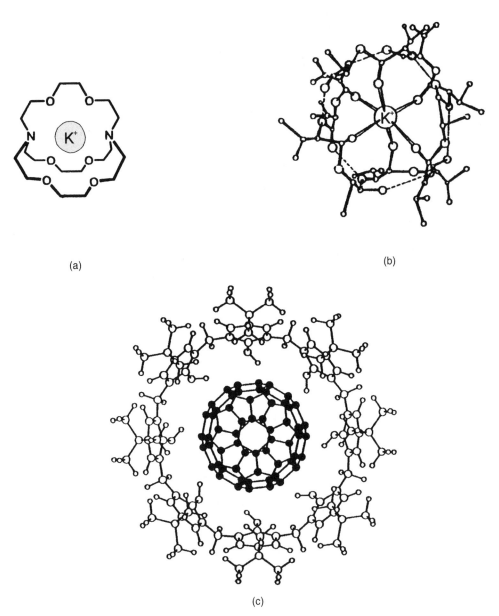

(a)

(b)

(c)

Fig. 8. Spherical recognition: (**a**) K$^+$ cryptate of [2.2.2] (**2d**); (**b**) K$^+$ complex of valinomycin (**5**); and (**c**) inclusion compound of C$_{60}$ into *tert*-butylcalix[8]arene.

crown and spheroidal cryptand case, but is most pronounced for the spherand type of receptor (eg, **4**) also mentioned before.

Alkali and alkaline earth cation recognition is also effected rather efficiently by other macrocyclic receptors that have been synthesized (21) such as the cryptospherands (an amalgamation of cryptand and spherand topology) (20,31) and lariat ethers (41), the latter is characterized by a crown ether ring that has attached extra podand arms for feeling and lock in of the substrate. Moreover, natural macrocycles displaying antibiotic properties are also very efficient in the recognition of alkali metal ions (42). For instance, valinomycin (**5**) gives a strong and selective complex in which a K^+ ion is included in the macrocyclic cavity in octahedral environment of six carbonyl oxygens (Fig. 8**b**) (43).

In a word, all these receptors are more or less able to discriminate against cations that are either smaller or larger than their cavity (44). However, in a strict sense, discrimination of metal-ion spheres does not concern with molecular recognition but selection of the carbon ball C_{60} certainly does. In fact, the fullerene C_{60} has been included into the cavity of octa-*tert*-butylcalix[8]arene (Fig. 8**c**) shutting out C_{70} and making a very convenient and efficient C_{60} purification possible without any expensive apparatus (45).

Recognition of a tetrahedral substrate geometry requires the construction of a receptor molecule with a tetrahedral recognition site (9). This may be realized by positioning four suitable binding sites at the corners of a tetrahedron and incorporating them into a bridged molecular framework such as shown with compound (**6**) (Fig. 9**a**) (25,38,46). In fact, the tetrahedral NH_4^+ cation is very firmly held inside the cavity of the tricyclic cryptand (**6**), forming the

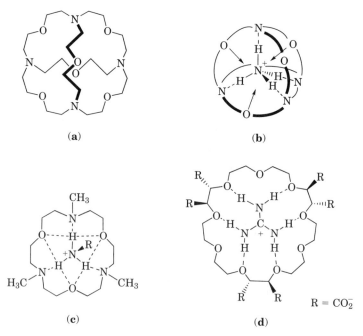

(a)

(b)

(c)

(d)   R = CO_2^-

Fig. 9. Representative examples of (**a, b**) tetrahedral; (**c**) trigonal; and (**d**) circular recognition.

ammonium cryptate (Fig. 9**b**) (47). This cryptate presents a high degree of receptor–substrate complementarity in that the ammonium ion fits into the cavity of (**6**) and is held by a tetrahedral array of hydrogen bonds including also electrostatic interactions with the oxygen atoms (48). Unsymmetrical derivatives of (**6**) display notably perturbed NH_4^+ binding, with a marked loss in recognition behavior (49).

Recognition of a primary ammonium ion, by analogy, is achieved by making use of a symmetrical triazacoronand enabling a trigonal recognition process in its inclusion complex (Fig. 9**c**) (50). A recognition site for secondary ammonium groups is provided by the diaza analogue of 12-crown-4 [cf. (**1a**)] (51). It binds via two hydrogen bonds (52). On the other hand, the derivative of a 27-membered crown compound yielded a particular stable inclusion complex with the guanidinium cation that is bound through an array of six hydrogen bonds suggesting an almost circular recognition mode (Fig. 9**d**) (53,54).

Linear recognition is mainly a subject of molecular length recognition (9). For that reason preferential substrates bear two recognizable functional groups at a distance corresponding to a ditopic receptor molecule (Fig. 10**a**) (55). An example is given by the cylindrical macrotricyclic cryptand that yields cryptates with terminal diammonium cations $H_3N^+\text{-}(CH_2)_n\text{-}N^+H_3$ of matching molecular length (Fig. 10**c**) (56). A complementary substrate, eg, relating to the receptor with R = naphthalene-2,6-diyl in Figure 10**c**, would be the 1,5-pentane

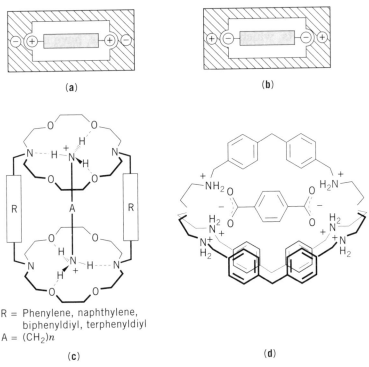

R = Phenylene, naphthylene,
    biphenyldiyl, terphenyldiyl
A = $(CH_2)n$

(a)   (b)   (c)   (d)

Fig. 10. Linear recognition: diagrammatic representation of the recognition of linear dicationic (**a**) and dianionic (**b**) substrates; (**c**, **d**) typical examples of receptor–substrate complexes.

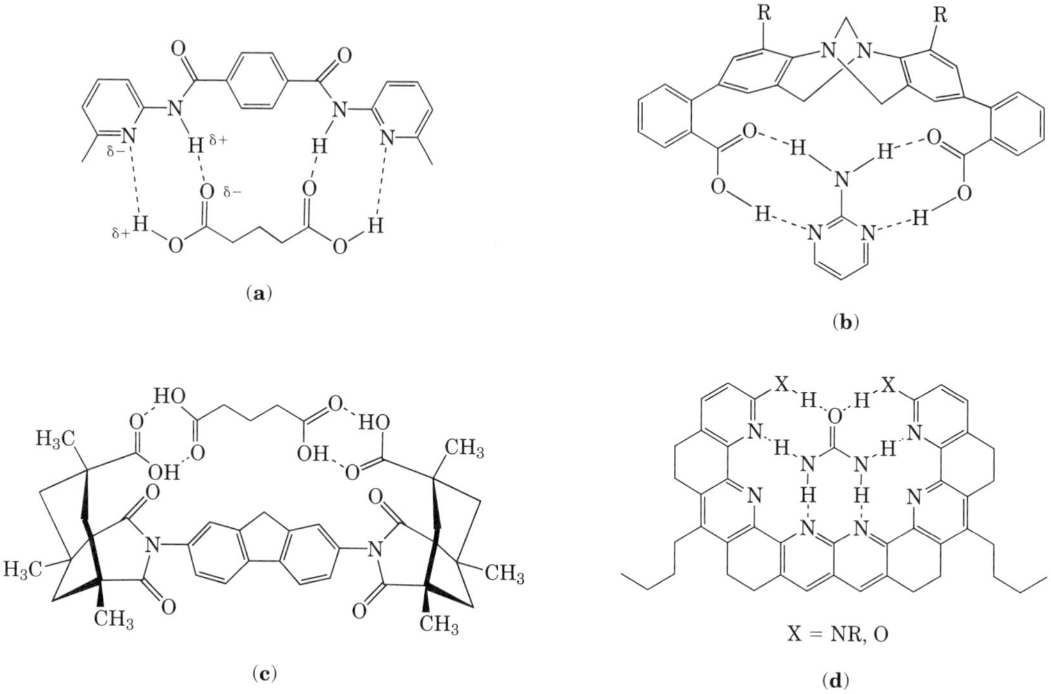

Fig. 13. Hydrogen bond dominated substrate recognition of: (**a**), (**c**) dicarboxylic acids; (**b**) 2-aminopyrimidine; and (**d**) urea.

acid-type receptors are also efficient in the recognition of 2-aminopyridines, respectively of 2-aminopyrimidine (Fig. 13**b**) (80).

An alternative approach to dicarboxylic acid recognition has been developed by using a receptor cleft based on a rigid spacer and two Kemp's triacid binding sites specified in Figure 13**c** (81–83). In this case, the principal binding force is a double carboxylic acid dimer interaction leading to both strong and selective binding to carboxylic acids of complementary shape and size, ie, glutaric acid (84). Another obvious cleft type receptor showing both a rigid molecular framework and complementary orientation of hydrogen bonding (85) to allow, eg, the efficient recognition and binding of urea is illustrated in Figure 13**d** (86). Although the receptor clefts that use carefully designed and rigid components to hold hydrogen bonding groups at a fixed distance apart are very common in this field, macropolycycles comprising similar functions offer an alternative albeit more synthetically challenging solution to the same problem, discussed more detailed below where the recognition of particular substrate molecules is dealt with.

π-Stacking and Charge-Transfer Dominated Substrate Recognition. Nature's strategy for the recognition of substrates featuring a flat aromatic framework (planar recognition) affords another recognition element, namely π-π stacking interactions between aromatic rings, ie, aromatic groups of receptor and substrate that meet a parallel face-to-face orientation, apart from hydrogen bonding being also typical of the nucleotide recognition (77). A very simple example showing the principles of π-π stacking supported substrate recognition is illustrated in Figure 14**a** (87). The flat heterocyclic substrate, uracil derivative, fits in face-to-face mode into the conformationally stepped receptor macroring containing a napththalene π-stacking unit and being bound via a system of extra hydrogen bonds to a diamidopyridine unit. The association constant for the analogous receptor molecule without the π-stacking unit is more than four times lower. The role of the π-stacking subunit in the above receptor clearly indicates that doubling the π-stacking contribution should lead to substantial improvement of recognition behavior and increasement of binding energy. This has led to a molecular tweezer design strategy to flank a hydrogen bonding carboxylic acid by two π-stacking anthracene units corresponding to the receptor–substrate relationship demonstrated in Figure 14**b** (88).

In a way, π-π stacking and charge transfer type of recognition have something in common. Examples making facts obvious are presented in Figure 14**c**, **d**. The macrobicyclic intercaland of Figure 14**c** and related receptors have been found to recognize flat shaped substrates through π-π stacking and bind them to form a molecular cryptate, in particular if electron donating substrate species are involved to allow charge-transfer interaction, such as planar molecular anions

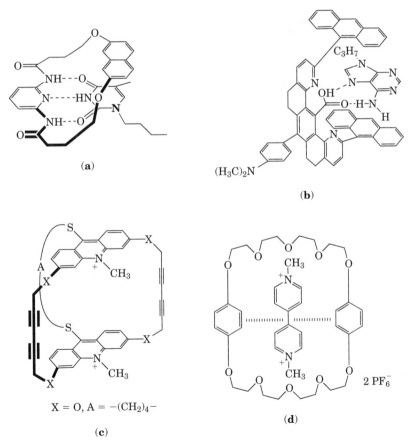

(a)

(H₃C)₂N   (b)

X = O, A = –(CH₂)₄–

(c)

2 PF₆⁻

(d)

Fig. 14. π-Stacking and charge-transfer dominated recognition of flat aromatic–heteroaromatic substrates (formation of intercalates).

or nucleic acids (89). Similarly, large ring electron donor aromatic crown ethers were designed that yield charge-transfer type intercalation complexes with paraquats (Fig. 14**d**) (90). Vica versa, macrocyclic bipyridinoquates form charge-transfer supported intercalation complexes with flat aromatic substrates having electron-donating substituents (91).

Lipophilic Interaction Dominated Substrate Recognition. Making recognition through lipophilic interaction possible require receptors presenting large and more or less rigidly connected architectures of macrocyclic or cage-like nature (92). Here only some illustrative examples can be given (see also INCLUSION COMPOUNDS), referring the reader to specific reviews of this vast subject (8–12).

The naturally occurring cyclodextrins having endo-lipophilic cone-shape are perhaps the most important and also the first receptor molecules whose selective inclusion properties towards lipophilic organic molecules were recognized (93,94). They comprise a family of cyclic oligosaccharides, composed of 6, 7, and 8 glucose units in its most familiar representatives (α, β, and γ-cyclodextrin, respectively) providing endo-lipophilic and exo-hydrophilic cone-shaped molecular cylinders of increasing size (Fig. 15**a**). Cyclodextrins form size and shape selective inclusion compounds with a wide variety of substrates including benzene derivatives, paraffins and noble gases (95).

Calixarenes (from the Latin *calix*) may be understood as artificial receptor analogues of the natural cyclodextrins (96,97). In its prototypical form they feature a macrocyclic metacyclophane framework bearing protonizable hydroxy groups made from condensation of *p*-substituted phenols with formaldehyde (Fig. 15**b**). Dependent on the ring size, benzene derivatives are the substrates most commonly included into the calix cavity (98), but other interesting substrates such as C_{60} have also been accommodated (Fig. 8**c**) (45).

As mentioned, calixarenes fall into the cyclophane-type of compounds that has emerged the central class of synthetic receptors in molecular recognition involving all kinds of ortho-, meta- and para-bridged aromatic macrocycles and oligomacrocycles, ie, pocket, open vessel, and macrocage receptors (99,100). A beautiful construction of this type are the cryptophanes, an example of which is shown in Fig. 16**a** (101,102). Cryptophanes are of much interest in particular for their ability to recognize and bind derivatives of methane that match the cavity. Although substrates are rather cut off outside here, the most extreme case of imprisonment of substrates is provided by the carceplexes (103). They are the inclusion complexes of carcerands (104). A prototype example is illustrated in Figure 16**b**. These containers have a virtually closed molecular surface indicating that carceplexes are formed during shell closure of two hemisphere components (cavitands) templated around the substrate (31,105,106). They also found that high structural recognition is involved in this capture, and the shell closure to give empty carcerands do not occur. Once substrates have been trapped in the cavity they can be released again only by destruction of the carcerand framework. Hemicarcerands are like carcerands except that by conformational modifications they can generate portals which join the inner phase of the receptor with the outer phase making an equilibrium of the substrate between in and out possible (103–106).

While the previous receptors are typically used in organic solvents, except for the cyclodextrins, there are special cases of cyclophane receptors supplied

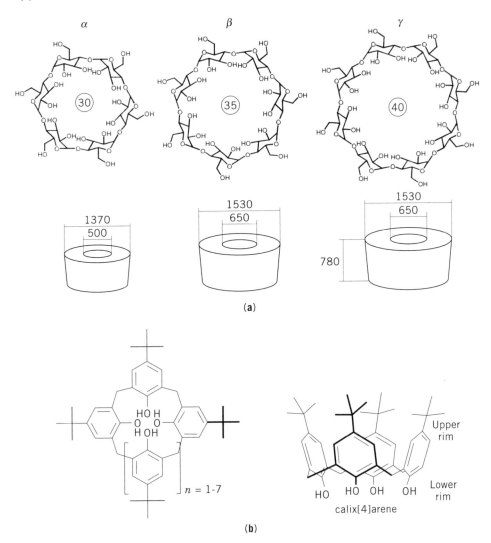

Fig. 15. Prototype examples of (**a**) cyclodextrins and (**b**) calixarenes, showing conformational structures and dimensions.

with peripheral charges (ammonium units) (107–112) or ionizable groups (carboxylate functions) (113,114) (Fig. 17) to allow substrate recognition, as in nature, in an aqueous medium, profiting from the solvophobic effects of water (115).

Multiple and Multisite, Coreceptor- and Coupled-System Substrate Recognition

Once recognition units for specific groups and individual features of a substrate have been identified, one may consider combining several of them within the same receptor. Thus far, though not carefully directed, the previous receptors in many cases, and as pointed out already possess this property of nonindividual

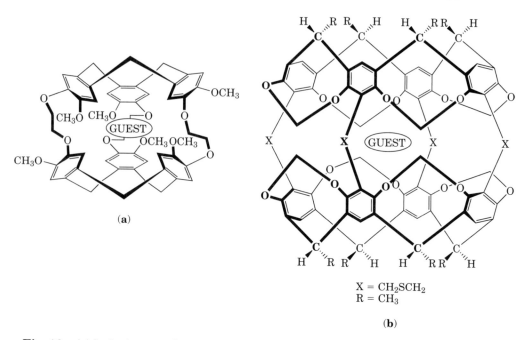

X = CH$_2$SCH$_2$
R = CH$_3$

(b)

Fig. 16. (a) Inclusion complex of a cryptophane; and (b) a carceplex (carcerand inclusion complex).

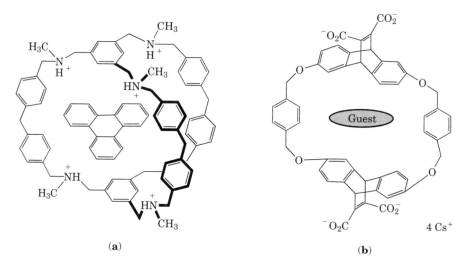

4 Cs$^+$

(a)   (b)

Fig. 17. Inclusion complexes of endo-hydrophobic–exopolarophilic receptors involving charges of different nature.

interaction modes. More carefully directed, this leads to multiple and multisite recognition depending on the design of binding subunits which may cooperate for the simultaneous complexation of several substrates or of a multiply bound polyfunctional species to yield polynuclear complexes (homo- or heteronuclear) and mononuclear polyhapto-type complexes, respectively (116). Moreover, one may distinguish co-receptor systems for which the binding of several substrates is commutative and cascade systems, for which the substrate binding steps are noncommutative but follow a given sequence (9).

A typical example of the multiply binding type of molecular recognition of a polyfunctional species is demonstrated in Figure 18 (117). Owing to the different binding sites that are an electron deficient guandinium group, a negatively polarized macroring and a naphthalene π-stacking unit, the receptor shown in Figure 18a should allow simultaneous interaction with, eg, the carboxylate, the ammonium and the aromatic groups of a substrate that is an aromatic zwitterionic amino acid (tryptophan) such as illustrated in the receptor–substrate complex of Figure 18b (66,118).

Numerous receptors comprising other multiply docking units and binding sites have been synthesized (19,35,77,117–120) including, eg, the speleands that in its frame comprise a negatively polarized macroring and an apolar but rigid shaping component for the size selective recognition of a primary ammonium cation (121).

The macrocyclic hexaimine structure of Figure 19a forms a homodinuclear cryptate with Cu(I) (122), whereas crown ether boron receptors (Fig. 19b) have been applied for the simultaneous and selective recognition of complementary cation–anion species such as potassium and fluoride (123) or ammonium and alkoxide ions (124) to yield a heterodinuclear complex (120).

Metalloreceptors based on a designed cationic inclusion complex as the specific accommodation site for the recognition of an uncharged guest molecule is another useful development (119,125). Metallomacrocycles of salen-type containing a complexed uranyl cation are most common here (126). The triple porphyrin receptor shown in Figure 19c is a more complex example of this strategy that operates in double recognition mode when forming complexes with zinc and 2,4,6-tri-4-pyridyl-s-triazine matching size and geometry of the tritopic metalloreceptor (127).

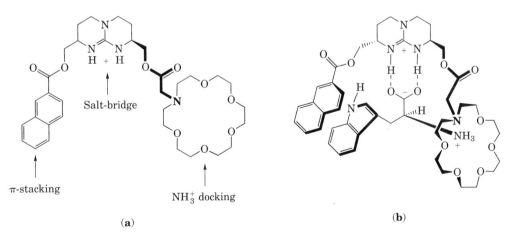

Fig. 18. (a) A bicycloguanidinium receptor; and (b) its three-point association to the zwitterionic α-amino acid trypthophan (118).

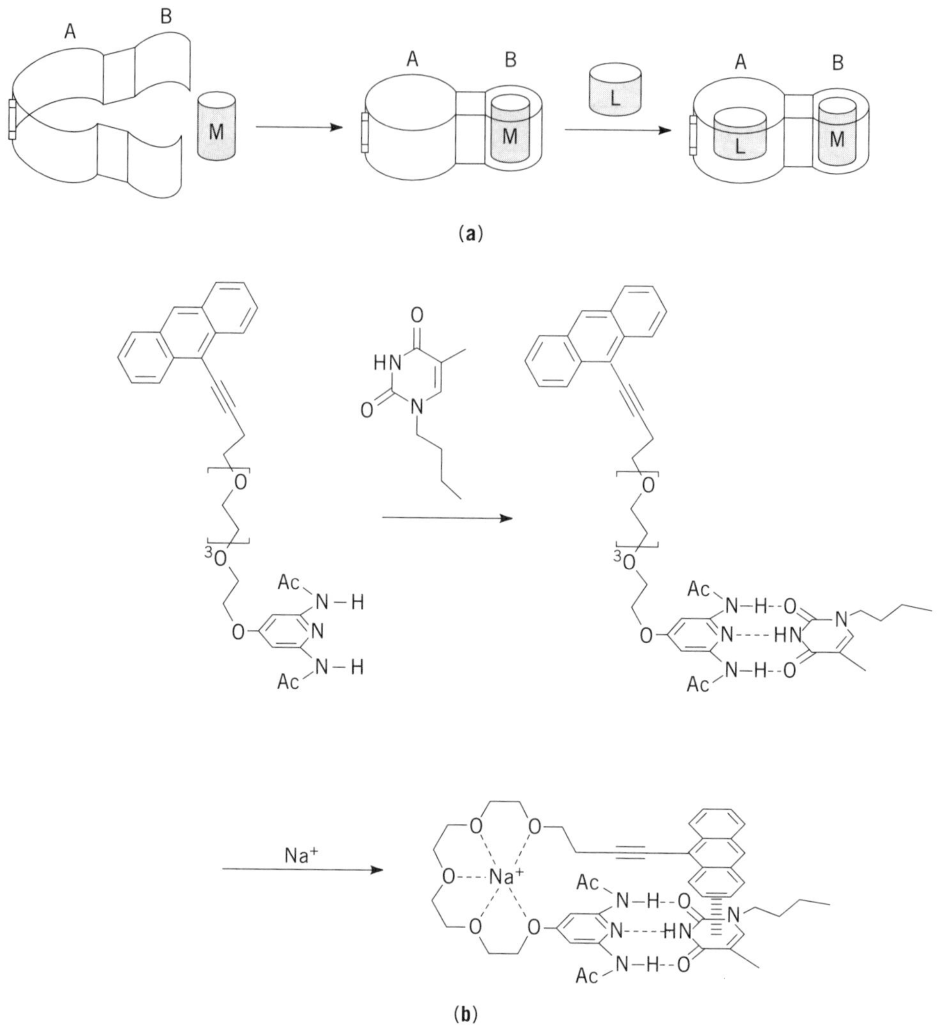

(a)

(b)

Fig. 21. (**a**) Diagrammatical representation of an allosteric receptor mechanism; (**b**) allosteric binding of a thymine derivative promoted on K⁺ uptake (34).

optically resolved receptor structure in order to make possible two diastereomeric receptor–substrate complexes allowing differentiation (Fig. 22**a**) (136).

Following this line, a great variety of optically resolved (optically active) crown compounds were prepared for the resolution of racemic cationic substrates, eg, chiral primary ammonium salts, protonated α-aminoalcohols and α-amino acid derivatives, through complexation (137–139). Among the highest enantiomer recognition properties for chiral ammonium ions were obtained with crown ethers having one or two 1,1-binaphthyl chiral barriers in the framework (Fig. 23**a**) (27,37,140). Others contain a spirobifluorene chiral subunit (141) or are derived from terpenoids, amino acids, and hydroxy acids that make use of the natural pool of chiral compounds (138). Typical examples for the latter classes of

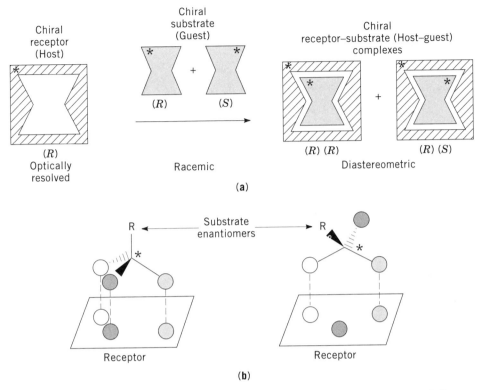

Fig. 22. Principle of chiral receptor-substrate recognition: (**a**) formation of diastereomeric inclusion complexes; (**b**) three-point interaction model.

receptors are shown in Figure 23**b**, **c** where natural α-D-glucose (142) or tartaric acid (143) are the chiral sources. A further important family of chiral receptors derived from natural glucose are the cyclodextrins (Fig. 15**a**) (93–95). Moreover, most of these receptors (cf. Fig. 23**a**–**c**) are carefully designed systems in that they contain at least one C_2 axis of symmetry (dissymmetric compound type), a tactic that makes the receptors nonsided with respect to perching substrates, eg, ammonium guests (136). Beyond that C_3 symmetric receptor molecules have also been used to advantage in chiral recognition such as the cryptophane in Figure 16**a** (101,102) or the basket shaped chiral host presented in Figure 23**d** (144). Although besides chiral ammonium ions amides are the substrate class of compounds to be very efficiently resolved by the majority of receptors, particular enantiomer recognition properties including steroid hormones have also been reported (145). The cryptophane is typical of the chiral resolution of methane derivatives (eg, CHFClBr) (146) and the basket-shaped host of Figure 23**d** exhibits extremely high enantioselectivity for various peptides (144).

Interpretation of these results are in keeping with the complementarity between chiral receptor and chiral substrate as sketched in Figure 22 (notice the orientation of the stars) and visualized in Figure 18**b**. This figure shows that the zwitterionic amino acid tryptophan of natural configuration (S) ideally complements the three sites of the chiral (optically resolved) receptor cleft,

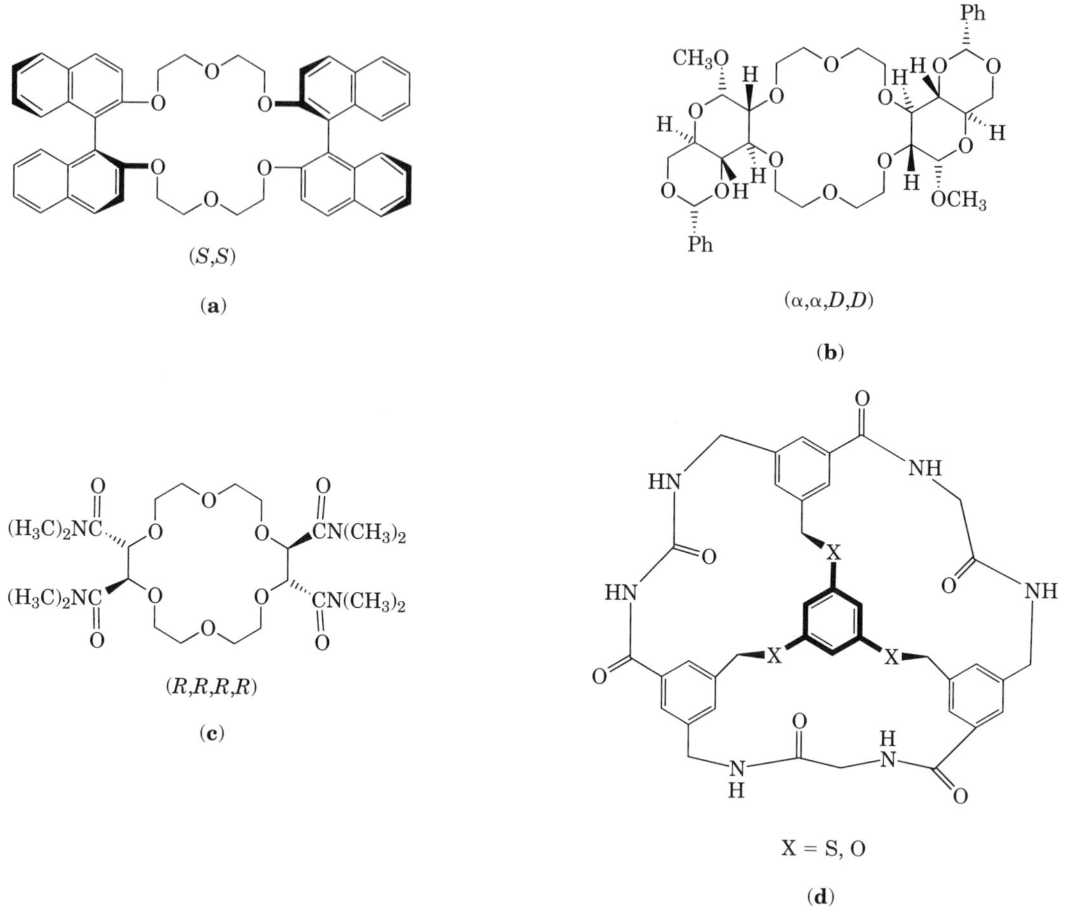

(S,S)

(a)

(α,α,D,D)

(b)

(R,R,R,R)

(c)

X = S, O

(d)

Fig. 23. Prototypical receptor molecules for chiral (enantioselective) substrate recognition.

while the optical antipode of tryptophan (R-configurated enantiomer) is less suited for binding (117). These facts express what is generally called the three-point interaction principle (136,147) illustrating fit or misfit of chiral receptor substrate recognition (Fig. 22b).

For more details on this topic see References 138, 139, 148; for more illustrations on the chiral fit concept see also INCLUSION COMPOUNDS.

Artificial Receptors for Particular Substrate Recognition

The recognition of substrates from many different compound classes has been discussed. Nevertheless some particular substrates remain that are biorelevant species or play central roles as drugs. Barbiturates are such an important family of drugs and are the target for molecular recognition (19,77). According to their structure, the barbiturate moiety essentially fuses two imide groups within a six-membered ring. Thus, two diaminopyridine units correctly positioned in a macrocyclic ring should bind to all six of the accessible hydrogen bonding sites in barbiturates, as shown in Figure 24**a** (149). A crystal structure of a respective receptor–substrate complex has been performed that comes up to the expectations (19,77).

The structural and synthetic relationships shared between barbiturates and urea, which is another substrate of high physiological interest, suggest that the above receptor strategy could be modified for the selective complexation of urea. The designed modification for urea recognition involves replacement of the H-bond donating pyridine-6-amido groups in the previous barbiturate receptor by two H-bond accepting groups that differ by 120° in alignment to

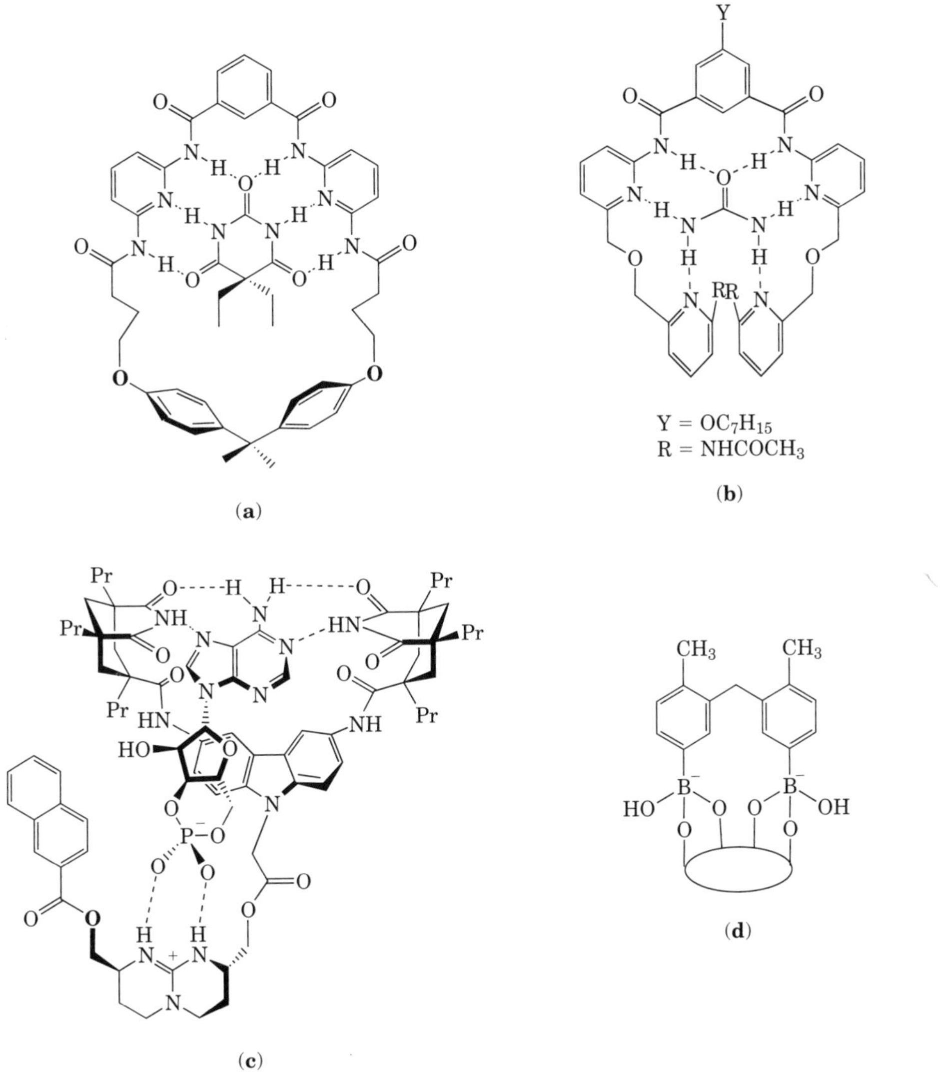

(a)

(b)

$Y = OC_7H_{15}$
$R = NHCOCH_3$

(c)

CH_3 CH_3

(d)

Fig. 24. Receptor substrate complexes involving particular substrate recognition.

the substrate. Such an arrangement of binding groups exists in the receptor molecule for urea illustrated in Figure 24b (19,77). An approach using an acyclic, but rigid hydrogen bond heteroaromatic framework such as the naphthiridine derivative presented in Figure 13d has also been developed (86), and last but not least the above metalloreceptor strategy based on a metal center (eg, UO_2) as an electrophilic binding site for the carbonyl oxygen of urea included in a macrocyclic receptor was successfully applied (119,125,126). A modification of the rigid heterocyclic cleft type receptor mentioned above (cf. Fig. 13d) has also yielded good results in the recognition of uric acid, the key product of the purine metabolism (150).

Nature's strategy for the recognition of nucleic acid bases offers an ideal example of directional and orientational dependence of simultaneous hydrogen bonding and aromatic stacking interactions to make advantage for an artificial receptor design of nucleic acid bases. A taste of it has already been given in Figures 14a (87) and 14b (88) illustrating effective recognition of an uracil and adenine derivative, respectively (19,77). In addition, receptors where two such sites are linked through a spacer element attenting to the recognition of multiple nucleic acid base derivatives have also been carried out successfully (77,151).

Nucleotides, the building blocks of DNA and RNA strands, beside a polyphosphate chain and ribose or deoxyribose also contain a nucleobase. According to this importance, efforts have been directed toward the problem of phosphate recognition and the development of artificial receptors capable of distinguishing nucleotides with respect to both the nucleobase and the phosphate chain (66,118). A tricky solution of the problem is presented in Figure 24c, taking advantage of the phosphate affinities of a bicycloguanidinium cation and the well-established chelating capacity of a molecular cleft featuring two converging imide functions (152).

Boronic acid–diol covalent interactions creating five- or six-membered rings reversibly form in aqueous media, thus, providing an important tool in the recognition of saccharides (153–155). Moreover, many monosaccharides possess at least two binding sites (diol area) which differ from other monosaccharides. Based on this strategy a number of small saccharide selective receptor molecules with conformationally well-defined distance and orientation between two boronic acid functionalities have been designed. An example is given in Figure 24d (156). D-Glucose yields a relatively strong 1:1 complex at pH 11.3, whereas complexes with galactose, talose, maltose, cellobiose, and lactose are weaker. The complex with glucose is believed to involve bonds to the C1-C2 and C4-C6 diols as sketched in Figure 24d. Chiral recognition of saccharides along this line using chirally modified diboronic acid receptors have also been realized (157). A second main category of saccharide receptors are typical of bowl-shaped molecules belonging to the resorcinarenes (158) (see INCLUSION COMPOUNDS) and other more recent examples (153). What is more, there are good expectations for the development of artificial adrenalin (159) and peptide receptors (160).

However, all the receptors hitherto discussed are monomolecular species which possess a monomolecular cavity, pocket, cleft, groove or combination of it including the recognition sites to yield a molecular receptor–substrate complex. They can be assembled and preserved in solution although there are dependences (see below). By way of contrast, molecular recognition demonstrated in the following comes from multimolecular assembly and organization of a nonsolution phase such as polymer materials and crystals.

Molecular Recognition in Polymers and Solids

If a polymer is prepared in the presence of molecules, the "print molecules" which are extracted after polymerization, the remaining polymer may contain cavities, prints, or footprints that can recognize the print molecule (161). Actually, the cast relates to the matrix molecule like lock and key of Emil Fischer's long-known principle (5) (cf. Fig. 1). A scheme of this clever imprinting technique is illustrated in Figure 25**a**, and a relevant example of imprinted polymer is shown in Figure 25**b**. The template monomer of Figure 25**b** is a compound in which two

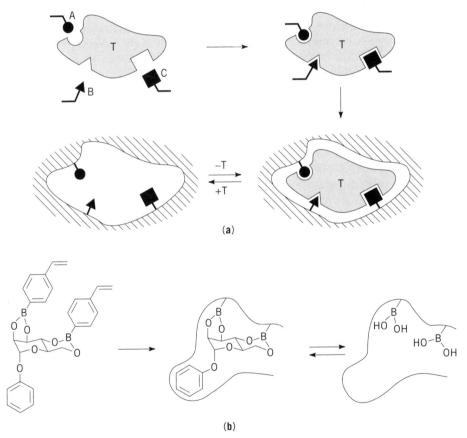

(a)

(b)

Fig. 25. Schematic representation of imprinting: (**a**) crosslinking polymerization in the presence of a template (T) to obtain cavities of specific shape and a defined spatial arrangement of functional groups (binding sites, A–C); (**b**) crosslinked polymer prepared from the template monomer and ethylene dimethacrylate with and without the templates (161).

polymerizable moieties of vinylphenylboronic acid have been bound to phenyl α-D-manopyranoside as the print molecule. A 1:9 ratio of the monomer and ethylene dimethacrylate were copolymerized in an inert solvent to yield a macroporous polymer of which the template is split off providing chiral cavities each bearing a pair of boronic acid groups. Polymers of this type have excellent ability for recognition of the template molecule, also in the given enantiomer configuration when subjected to its racemate.

By analogy, a great many of other functionalized styrenes, including carboxylic acids, amino acids, Schiff bases, or specific compounds, eg, L-DOPA, have successfully been applied as print templates. Moreover, it has also been shown that silica gel can be imprinted with similar templates, and that the resulting gel has specific recognition sites determined by the print molecule (162–164).

In a sense, molecular recognition using hollow organic crystals, in particular clathrate structures (165), is similar although the interactions forming the framework are noncovalent but weak interactions (16). For the same reason typical multinuclear crystalline inclusion compounds and clathrates are not stable in solution, but decompose unlike the monomolecular inclusions and more stable receptor–substrate complexes (165). Nevertheless molecular recognition behavior of crystalline inclusion compounds is both various dependent on the structure that can be cavity-, layer- or channel-type (167), and in many cases highly selective including chiroselectivity (168).

A packing motif giving a general idea of such an efficient chiral recognition machinery that uses the crystalline inclusion phenomenon is illustrated in Figure 26 (169). The optically active receptor molecule (Fig. 26**a**), a bulky derivative of lactic acid, chiroselectively yields a crystalline inclusion compound with (R)-configurated 3-methylcyclohexanone (Fig. 26**b**), refusing the steric mirror image (S-configuration) of the substrate, whereas the correctly configurated substrate (R) ideally matches the intermolecular lattice space (Fig. 26**c**). Along this line a number of bulky crystalline hosts have been designed capable of chiral substrate recognition, eg, of alcohols, phenyloxirane, sulfoxides and lactones, to say nothing of more simple constitutional isomer recognition and of the recognition of other chemically different species (165,167,170). However, it would mean doubling of information to go into details here since this particular topic is extensively covered under the subjects extra molecular cavity and lattice type inclusion compounds (see INCLUSION COMPOUNDS).

Microporous inorganic materials dominated historically by the zeolites and alumosilicates, and the great variety of more recent nonoxide and coordination framework materials should also be mentioned here (171–174) but not discussed in detail (see ZEOLITES etc). This type of molecular recognition is usually known as molecular sieving.

Molecular Recognition at Interfaces and Surface Monolayers

There are three advantages to study molecular recognition on surfaces and interfaces (monolayers, films, membranes or solids) (175): (*1*) rigid receptor sites can be designed; (*2*) the synthetic chemistry may be simplified; (*3*) the

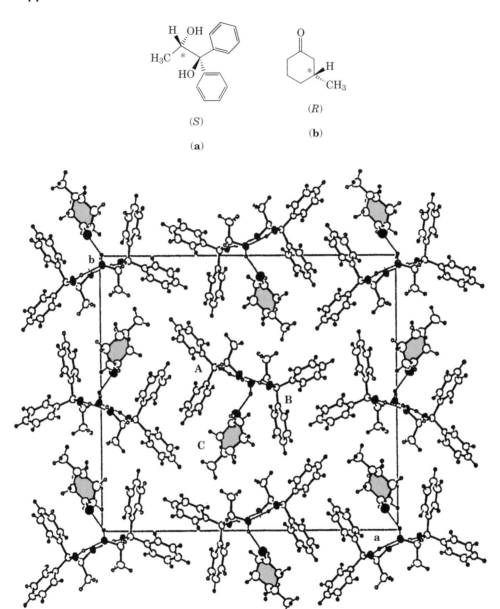

Fig. 26. Clathrate receptor chemistry: (**a**) a chiroselective crystalline host compound (clathrand); (**b**) a typical guest molecule to be included in the specified configuration; and (**c**) the crystal structure of the respective clathrate (A and B denote host and C the guest species) (169).

surface can be attached to transducers which makes analysis easier and may transform the molecular recognition interface to a chemical sensor. And, which is also a typical fact, this kind of molecular recognition involves outside directed interaction sites, ie, exo-receptor function (9) (cf. Fig. 5**b**).

To begin with, molecular recognition of crystal interfaces make possible the control of crystal growth processes in that suitably designed auxiliary molecules act as promoters or inhibitors of crystal nucleation inducing, for instance, the resolution of enantiomers or the crystallization of desired polymorphs and crystal habits (176). As an example (Fig. 27), crystals of achiral glycine, due to their enantiopolar arrangement, may differentiate between (R)- and (S)-amino acids when being used as additives (177). Thus an α-amino acid additive such as alanine of configuration (R) capable of replacing a glycine molecule will block crystal growth only at one of the two enantiopolar crystal faces, the (*pro-R*)-face [(010)-face in Fig. 27**b**]. An (S)-amino acid practises the same effect on the (*pro-S*)-face of the glycine crystal [(01̄0)-face], and addition of racemic (R,S)-amino acid suppresses growth at both enantiopolar faces (Fig. 27**c**). In consequence, the crystal of glycine changes its habit from a symmetric bipyramide to either two asymmetric mirror-image pyramids or a platelet making engineered crystal habits possible (178) (Fig. 27). These impurity crystals having large and hydrophobic planes orientedly swim at the air-water surface which can be applied towards the resolution of further amino acids and for the direct and relative assignment of the absolute configuration of chiral molecules and crystals using stereochemical correlations (179).

Following another direction, it has previously been shown that alkanethiols spontaneously adsorb to Au from dilute solutions of ethanol and other nonaqueous solvents, and that the resulting self-assembling monolayers (SAMs) assume a close-packed overlayer structure on Au (111) and other textured Au

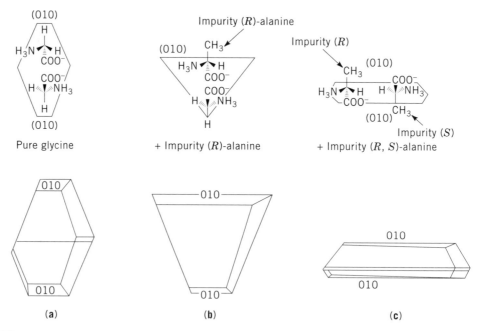

Fig. 27. Recognition at crystal interfaces and its role in the engineering of crystal morphology and configurational assignment of molecules (176,177).

surfaces, being quite robust in aqueous solutions and vapor-phase ambients (180). This mode of self-assembly chemistry has been used to synthesize monolayer assemblies that function as molecular recognition interfaces based on the presence of recognizer end groups (181). Thus one-component SAMs formed of *n*-alkanethiols having extra carboxylic acid functionalized end groups specifically adsorb vapor-phase acid-terminated molecules via H-bonding (Fig. 28) or vapor-phase amine-terminated molecules via proton-transfer interaction, exhibiting chemical complementorship (182). It has also been demonstrated that two-component SAMs, which consist of inert *n*-alkanethiol framework molecules and defect inducing template molecules, can discriminate between solution-phase probe molecules based on their geometrical properties, similar to the imprinting technique discussed before (cf. Fig. 25**a**) but on the two-dimensional level only (182). In order to create analogous molecular recognition property on an oxide surface, a silica overlayer was prepared on tin oxide by the chemical vapor deposition of silicon alkoxide using preadsorbed benzoate anions as template molecules, which were then removed to yield vacancies in the overlayer capable of size recognition of the probe molecules (183).

In an obvious example of molecular recognition at the air–water interface, the receptor consists of an organized monolayer formed from amphiphiles that have complementary binding sites (184). For example, the double-chain triazine amphiphile illustrated in Figure 29 (185) was employed for the formation of a monolayer receptor at the air–water interface which specifically interacts with barbituric acid dissolved in the water subphase, creating a supramolecular strand, in close analogy to solid-state and solution structures formed of the hydrogen-bonded components (186,187). The stabilization conferred on the monolayer by its networking with barbituric acid made its imaging by atomic force microscopy (afm) possible, while the noncomplexed monolayer is destroyed by the scanning tip of the afm (185). Vice versa, amphiphilic derivatives of barbituric acid form a monolayer receptor at the air–water interface interacting with complementary substrates such as 1,4,6-triaminopyridine, melamine and urea (188). The recognition reaction that occurred was investigated by measuring the pressure-surface area isotherms and by using uv-visible spectroscopy. For

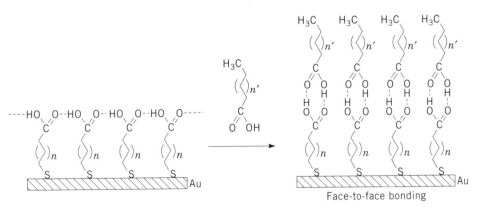

Fig. 28. Self-assembly monolayer to function as molecular recognition interface (182).

is very uncommon for molecules in a crystal structure to be related by rotation axis or mirror planes, because identical parts of molecules avoid one another, except for molecular sites having a so-called self-complementary donor–acceptor group (195). Self-complementary groups such as the carboxylic acid, the amide, the urea function or its combinations form finite, one-dimensional tape, two-dimensional layer, or three-dimensional motifs of organic molecules mostly obtained from hydrogen bonding. Representative examples are given in Figure 31 (196).

In solution, highly ordered structures created via self-recognition and self-assembly of a programmed H-bonding molecular component are also possible (197) such as the hexameric pseudo supermacrocycle of a designed DNA base hybrid shown in Figure 32 (198), to say nothing of the self-assembly of organic Langmuir and Langmuir-Blodgett films (199) where self-recognition at the air–water interface is of vital importance as well (cf. Fig. 29).

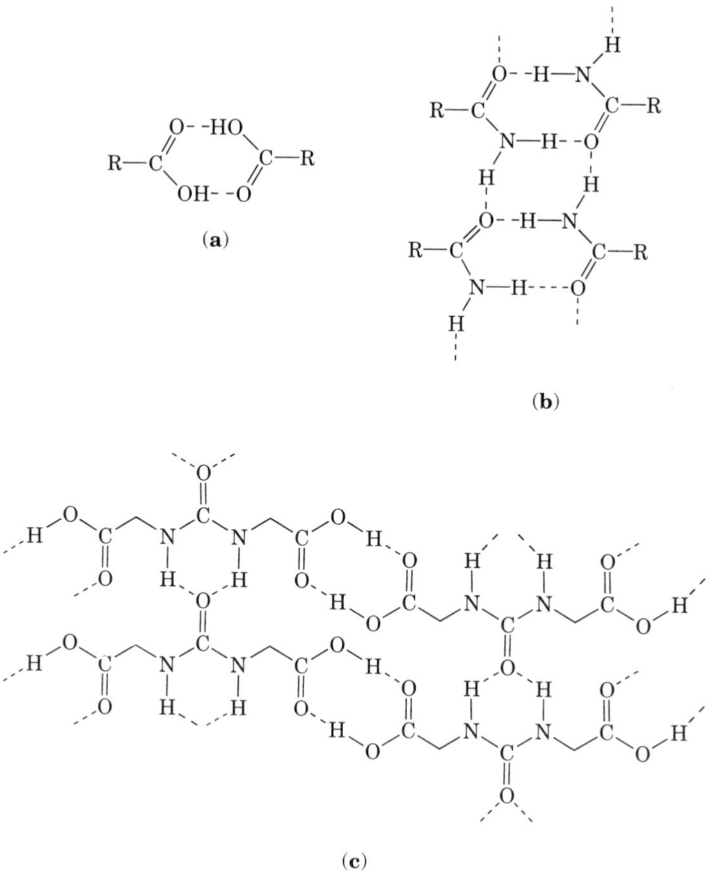

Fig. 31. Supramolecular (hydrogen-bonded) motifs of self-complementary molecules (196).

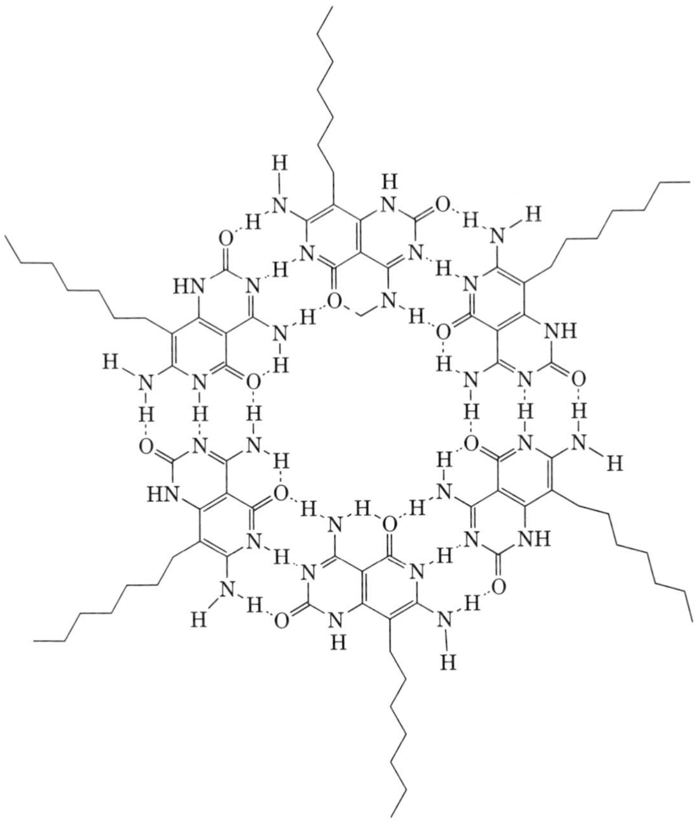

Fig. 32. Hexameric self-assembled supermolecule involving self-recognition (198).

With respect to inorganic self-recognition and self-assembly this would involve preferential binding of like metal ions by like ligands in a mixture of ligands and ions (9). Indeed, selective formation of double-helicates was obtained from mixtures of oligo-bipyridine strands in the presence of suitable metal ions, eg, Cu^+, without significant crossover, ie, the desired helicates are generated with self-recognition (Fig. 33) (200). Similar self-recognized triple-helicates have also been obtained from bis-catechols and Ga(III) (201).

A particular point of interest included in these helical complexes concerns the chirality. The helicates obtained from the achiral strands are a racemic mixture of left- and right-handed double helices (Fig. 34) (202). This special mode of recognition where homochiral supramolecular entities, as a consequence of homochiral self-recognition, result from racemic components is known as optical self-resolution (203). It appears in certain cases from racemic solutions or melts (spontaneous resolution) and is often quoted as one of the possible sources of optical resolution in the biological world. On the other hand, the more commonly found process of heterochiral self-recognition gives rise to a racemic supramolecular assembly of enantio pairs (204).

14. J.-M. Lehn, *Struct. Bonding* **16**, 1 (1973).

15. H.-J. Schneider, T. Blatter, U. Cuber, R. Juneja, T. Schiestel, U. Schneider, I. Theis, and P. Zimmermann, in H.-J. Schneider and H. Dürr, eds., *Frontiers in Supramolecular Chemistry and Photochemistry*, VCH, Weinheim, 1991, p. 29.

16. M. Mascal, *Contempor. Org. Syn.*, 31 (1994); see also G. R. Desiraju, *Angew. Chem.* **107**, 2541 (1995); *Angew. Chem. Int. Ed. Engl.* **34**, 2311 (1995).

17. H.-J. Schneider, *Chem. Soc. Rev.*, 227 (1994).

18. D. E. Koshland, Jr., *Angew. Chem.* **106**, 2468 (1994); *Angew. Chem. Int. Ed. Engl.* **33**, 2375 (1994).

19. D. Hamilton, in H. Dugas, ed., *Bioorganic Chemistry Frontiers*, Vol. 2, Springer, Berlin-Heidelberg, 1991, p. 115.

20. D. J. Cram, *Angew. Chem.* **100**, 1041 (1988); *Angew. Chem. Int. Ed. Engl.* **27**, 1009 (1988).

21. E. Weber in S. Patai and Z. Rappoport, eds., *Crown Ethers and Analogs*, Wiley, Chichester, 1989, p. 305.

22. F. Vögtle and E. Weber, *Angew. Chem.* **91**, 813 (1979); *Angew. Chem. Int. Ed. Engl.* **18**, 753 (1979).

23. G. W. Gokel and O. Murillo, in Ref. 12, Vol. 1, p. 1.

24. G. W. Gokel, *Crown Ethers and Cryptands, Monographs in Supramolecular Chemistry*, Vol. 3, The Royal Society of Chemistry, Cambridge, 1991.

25. J. M. Lehn, *Acc. Chem. Res.* **11**, 49 (1978).

26. B. Dietrich, P. Viout, and J. M. Lehn, *Macrocyclic Chemistry*, VCH, Weinheim, 1993.

27. D. J. Cram and K. N. Trueblood, in F. Vögtle, ed., *Host Guest Complex Chemistry I, Top. Curr. Chem.*, Vol. 98, Springer, Berlin-Heidelberg, 1981, p. 43; see also D. J. Cram and K. N. Trueblood, in Ref. 13, p. 125.

28. D. J. Cram, *Angew. Chem.* **98**, 1941 (1986); *Angew. Chem. Int. Ed. Engl.* **25**, 1039 (1986).

29. J.-M. Lehn, *Angew. Chem.* **100**, 91 (1988); *Angew. Chem. Int. Ed. Engl.* **27**, 89 (1988).

30. E. Weber and F. Vögtle, in Ref. 12, Vol. 2, p. 1.

31. D. J. Cram, *From Design to Discovery*, American Chemical Society, Washington, D.C., 1990.

32. A. G. Amit, R. A. Marinzza, S. E. V. Phillips, and R. J. Poljak, *Science* **233**, 747 (1986).

33. S. Hauptmann and G. Mann, *Stereochemie*, HTB Spektrum Akademischer Verlag, Heidelberg, 1996, p. 291.

34. H.-J. Schneider and A. K. Mohammad-Ali, in Ref. 12, Vol. 2, p. 69.

35. A. R. van Doorn, W. Verboom and D. N. Reinhoudt, in G. W. Gokel, ed., *Advances in Supramolecular Chemistry*, Vol. 3, JAI Press, Greenwich, 1993, p. 159.

36. B. König, *J. Prakt. Chem.* **337**, 339 (1995).

37. C. J. Pedersen, *Angew. Chem.* **100**, 1053 (1988); *Angew. Chem. Int. Ed. Engl.* **27**, 1021 (1988).

38. E. Weber and F. Vögtle, in Ref. 13, p. 1.

39. E. Weber and H.-P. Josel, *J. Incl. Phenom.* **1**, 79 (1983).

40. R. Hilgenfeld and W. Saenger, in F. Vögtle, ed., *Host-Guest Complex Chemistry II, Top. Curr. Chem.* Vol. 101, Springer, Berlin-Heidelberg, 1982, p. 1; see also, R. Hilgenfeld and W. Saenger, in Ref. 13, p. 43.

41. G. W. Gokel and O. F. Schall, in Ref. 12, Vol. 1, p. 97.

42. M. Dobler, *Ionophores and Their Structures*, Wiley, New York, 1981.

43. M. R. Truter, *Struct. Bonding* **16**, 71 (1973).

44. Y. Inoue and G. W. Gokel, eds., *Cation Binding by Macrocycles*, Marcel Dekker, New York, 1990.

45. T. Suzuki, K. Nakashima, and S. Shinkai, *Chem. Lett.*, 699 (1994); see also T. Anderson, K. Nilson, M. Sundahl, G. Wetsman, and O. Wennerström, *J. Chem. Soc., Chem. Commun.*, 604 (1992).

46. J.-M. Lehn, *Pure Appl. Chem.* **49**, 857 (1977).

47. E. Graf and J.-M. Lehn, *J. Am. Chem. Soc.* **97**, 5022 (1975).

48. E. Graf and J.-M. Lehn, *Helv. Chim. Acta* **64**, 1040 (1981).

49. E. Graf, J.-P. Kintzinger, J.-M. Lehn, and J. LeMoigne, *J. Am. Chem. Soc.* **104**, 1672 (1982).

50. J.-M. Lehn and P. Vierling, *Tetrahedron Lett.* **21**, 1323 (1980).

51. J. C. Metcalfe, J. F. Stoddart and G. Jones, *J. Am. Chem. Soc.* **99**, 8317 (1977).

52. J. Krane and O. Aune, *Acta Chem. Scand.* **B 34**, 397 (1980).

53. J.-M. Lehn, P. Vierling, and R. C. Hayward, *J. Chem. Soc., Chem. Commun.*, 296 (1979).

54. J. W. H. M. Uiterwijk, S. Harkema, J. Geevers, and D. N. Reinhoudt, *J. Chem. Soc., Chem. Commun.*, 220 (1982).

55. I. O. Sutherland, in G. W. Gokel, ed., *Advances in Supramolecular Chemistry*, Vol. 1, JAI Press, Greenwich, 1990, p. 65.

56. I. O. Sutherland, *J. Incl. Phenom.* **7**, 213 (1989).

57. C. Pascard, C. Riche, M. Cesario, F. Kotzyba-Hibert, and J.-M. Lehn, *J. Chem. Soc., Chem. Commun.*, 557, (1982).

58. J. S. Bradshaw, K. E. Krakowiak, and R. M. Izatt, *Azo-Crown Macrocycles*, Wiley, New York, 1993.

59. J. S. Bradshaw and J. Y. K. Hui, *J. Heterocycl. Chem.* **11**, 649 (1974).

60. S. R. Cooper, *Acc. Chem. Res.* **21**, 141, (1988).

61. S. R. Cooper, ed., *Crown Compounds: Toward Future Applications*, VCH, Weinheim, 1992.

62. T. L. Ho, *Hard and Soft Acids and Bases Principle in Organic Chemistry*, Academic Press, New York, 1977.

63. K. N. Raymond, *Coord. Chem. Rev.* **105**, 135 (1990).

64. F. Ebmeyer and F. Vögtle, in H. Dugas, ed., *Bioorganic Chemistry Frontiers*, Vol. 1, Springer, Berlin-Heidelberg, 1990, p. 143.

65. K. Gloe, H. Stephan, O. Heitzsch, H. Bukowsky, E. Uhlemann, R. Pollex, and E. Weber, *J. Chem. Soc., Chem. Commun.*, 1955 (1994).

66. C. Seel, A. Galán, and J. de Mendoza, in Ref. 11, p. 101.

67. H. E. Katz, in J. L. Atwood, J. E. D. Davies and D. D. MacNicol, eds., *Inclusion Compounds*, Vol. 4, Oxford University Press, Oxford, 1991, p. 391.

68. F. P. Schmidtchen, in F. Vögtle and E. Weber, eds., *Biomimetic and Bioorganic Chemistry II, Top. Curr. Chem.*, Vol. 130, Springer, Berlin-Heidelberg, 1986, p. 101.

69. E. Graf and J.-M. Lehn, *J. Am. Chem. Soc.* **98**, 6403 (1976).

70. F. Schmidtchen and G. Müller, *J. Chem. Soc., Chem. Commun.*, 1115 (1984).

71. P. D. Beer, N. C. Fletcher, A. Grieve, J. W. Wheeler, C. P. Moore, and T. Wear, *J. Chem. Soc., Perkin Trans 2*, 1545 (1996).

72. B. Dietrich, J. Guilhem, J.-M. Lehn, C. Pascard, and E. Sonveaux, *Helv. Chim. Acta.* **67**, 91 (1984).

73. W. M. Hosseini and J.-M. Lehn, *Helv. Chim. Acta* **69**, 587 (1986); J.-M. Lehn, R. Méric, J.-P. Vigneron, I. Bkouche-Waksman, and C. Pascard, *J. Chem. Soc., Chem. Commun.*, 62 (1991).

74. A. M. Echavarren, A. Galán, J. de Mendoza, A. Salmerón, and J.-M. Lehn, *J. Am. Chem. Soc.* **111**, 4994 (1989).

75. G. A. Jeffrey and W. Saenger, *Hydrogen Bonding in Biological Structures*, Springer, Berlin-Heidelberg, 1991.

76. J. D. Watson and F. H. C. Crick, *Nature* **171**, 737 (1953).

144. J.-I. Hong, S. K. Namgong, A. Bernardi, and W. C. Still, *J. Am. Chem. Soc.* **113**, 5111 (1991).
145. Y. Murakami, O. Hayashida, T. Ito, and Y. Hisaeda, *Chem. Lett.*, 497 (1992).
146. H. Grosenick, V. Schurig, J. Costante, and A. Collet, *Tetrahedron: Asymmetry* **6**, 87 (1995).
147. T. D. Booth, D. Wahnon, and I. Wainer, *Chirality* **9**, 96 (1997); see also, V. Davankov, *Chirality* **9**, 99 (1997).
148. T. H. Webb and C. S. Wilcox, *Chem. Soc. Rev.*, 383 (1993).
149. S. K. Chang and A. D. Hamilton, *J. Am. Chem. Soc.* **110**, 1318 (1988).
150. T. R. Kelly and M. P. Maguire, *J. Am. Chem. Soc.* **109**, 6549 (1987).
151. A. D. Hamilton and D. Little, *J. Chem. Soc., Chem. Commun.*, 297 (1990).
152. G. Deslongchamps, A. Galán, J. de Mendoza, and J. Rebek, Jr., *Angew. Chem.* **104**, 58 (1992); *Angew. Chem. Int. Ed. Engl.* **31**, 61 (1992).
153. J. Haseltine and T. J. Doýle, in *Organic Synthesis: Theory and Applications*, Vol. 3, JAI Press, Greenwich, 1996, p. 85.
154. T. D. James, K. R. A. S. Sandanayake, and S. Shinkai, *Angew. Chem.* **108**, 2039 (1996); *Angew. Chem. Int. Ed. Engl.* **35**, 1910 (1996).
155. T. D. James, P. Linnane, and S. Shinkai, *J. Chem. Soc., Chem. Commun.*, 282 (1996).
156. Y. Shiomi, M. Saisho, K. Tsukagoshi, and S. Shinkai, *J. Chem. Soc., Perkin Trans. 1*, 2111 (1993).
157. T. D. James, K. R. A. S. Sandanayake, and S. Shinkai, *Nature* **374**, 345 (1995).
158. Y. Aoyama, in Ref. 12, Vol. 2, p. 279; see also, K. Kobayashi, Y. Asakawa, Y. Kikuchi, H. Toi, and Y. Aoyama, *J. Am. Chem. Soc.* **115**, 2648 (1993).
159. T. Schrader, *Angew. Chem.* **108**, 2816 (1996); *Angew. Chem. Int. Ed. Engl.* **35**, 2649 (1996).
160. H.-J. Schneider, *Angew. Chem.* **105**, 890 (1993); *Angew. Chem. Int. Ed. Engl.* **32**, 848 (1993).
161. G. Wulff, W. Vesper, R. Grobe-Einsler, and A. Sarhan, *Makromol. Chem.* **178**, 2799 (1977); see also G. Wulff, in J. S. Siegel, ed., *Supramolecular Stereochemistry*, Kluwer, Dordrecht, 1995, p. 13.
162. G. Vlatakis, L. I. Andersson, R. Müller, and K. Moosbach, *Nature* **361**, 645 (1993).
163. J. H. G. Steinke, I. R. Dunkin, and D. C. Sherrington, *Advances in Polymer Science* **123**, 81 (1995).
164. G. Wulff, *Angew. Chem.* **107**, 1958 (1995); *Angew. Chem. Int. Ed. Engl.* **34**, 1812 (1995); see also, G. Wulff, in W. T. Ford, ed., *Polymeric Reagents and Catalysts, ACS Symposium Series*, Vol. 308, American Chemical Society, Washington, D.C., 1986, p. 186.
165. E. Weber, ed., *Molecular Inclusion and Molecular Recognition–Clathrates I and II, Top. Curr. Chem.*, Vols. 140 and 149, Springer, Berlin-Heidelberg, 1987 and 1988.
166. E. Weber, in Ref. 165, Vol. 140, p. 1.
167. J. L. Atwood, J. E. D. Davies and D. D. MacNicol, eds., *Inclusion Compounds, Vol. 1–3*, Academic Press, Inc., London, 1984; Vols. 4–5, Oxford University Press, Oxford, 1991.
168. F. Toda, in Ref. 12, Vol. 6, p. 465; see also, F. Toda, in Ref. 165, Vol. 140, p. 43 and D. Worsch and F. Vögtle, in Ref. 165, Vol. 140, p. 21.
169. E. Weber, C. Wimmer, A. L. Llamas-Saiz, and C. Foces-Foces, *J. Chem. Soc., Chem. Commun.*, 733 (1992); see also, E. Weber and C. Wimmer, *Chirality* **5**, 331 (1993).
170. Ref. 12, Vol. 6.
171. C. L. Bowes and G. A. Ozin, *Adv. Mater.* **8**, 13 (1996).
172. T. Bein, ed., *Supramolecular Architecture, ACS Symposium Series*, Vol. 499, American Chemical Society, Washington, DC, 1992.
173. A. Müller, H. Reuter and S. Dillinger, *Angew. Chem.* **107**, 2505 (1995); *Angew. Chem. Int. Ed. Engl.* **34**, 2311 (1995).

174. I. Dance, in G. R. Desiraju, ed., *The Crystals as a Supramolecular Entity, Perspectives in Supramolecular Chemistry*, Vol. 2, Wiley, Chichester, 1996.

175. J.-H. Fuhrhop and J. Köning, *Membranes and Molecular Assemblies: The Synkinetic Approach, Monographs in Supramolecular Chemistry*, Vol. 5, The Royal Society of Chemistry, Cambridge, 1994.

176. L. Addadi, Z. Berkovitch-Yellin, I. Weissbuch, J. van Mill, L. J. Shimon, M. Lahav, and L. Leiserowitz, *Angew. Chem.* **97**, 476 (1985); *Angew. Chem. Int. Ed. Engl.* **24**, 466 (1985); see also I. Weissbuch, R. Popovitz-Biro, L. Leiserowitz, and M. Lahav, in Ref. 4, p. 173.

177. I. Weissbuch, L. Addadi, Z. Berkovitch-Yellin, E. Gati, S. Weinstein, M. Lahav, and L. Leiserowitz, *J. Am. Chem. Soc.* **105**, 6615 (1983).

178. Z. Berkowitch-Yellin, *J. Am. Chem. Soc.* **107**, 8239 (1985).

179. I. Weissbuch, L. Addadi, L. Leiserowitz, and M. Lahav, *J. Am. Chem. Soc.* **110**, 561 (1988).

180. E. Delamarche, B. Michel, H. A. Biebuyck, and C. Gerber, *Adv. Mater.* **8**, 719 (1996).

181. L. H. Dubois and R. G. Nuzzo, *Annu. Rev. Phys. Chem.* **43**, 437 (1992).

182. R. M. Crooks, O. Chailapakul, C. B. Ross, L. Sun and J. K. Schoer, in T. E. Mallouk and D. J. Harrison, eds., *Interfacial Design and Chemical Sensing, ACS Symposium Series*, Vol. 561, American Chemical Society, Washington, D.C., 1994, p. 104.

183. N. Kodakari, N. Katada, and M. Niwa, *Chem. Vap. Deposition*, **3**, 59 (1997).

184. C. M. Paleos and D. Tsiourvas, *Adv. Mater.* **9**, 695 (1997).

185. H. Koyano, K. Yoshihara, K. Ariga, T. Kunitake, Y. Oishi, O. Kawano, M. Kuramori, and K. Suchiro, *J. Chem. Soc., Chem. Commun.* 1769 (1996).

186. J.-M. Lehn, M. Mascal, A. DeCian, and J. Fischer, *J. Chem. Soc., Chem. Commun.*, 479 (1990).

187. C. T. Seto, J. P. Mathias, and G. M. Whitesides, *J. Am. Chem. Soc.* **115**, 1321 (1993).

188. R. Ahuji, P.-L. Caruso, D. Möbius, W. Paulus, H. Ringsdorf, and G. Wildburg, *Angew. Chem.* **32**, 1082 (1993); *Angew. Chem. Int. Ed. Engl.* **32**, 1033 (1993).

189. K. Kurihara, K. Ohto, Y. Honda, and T. Kunitake, *J. Am. Chem. Soc.* **113**, 5077 (1991).

190. Y. Ikeura, K. Kurihara, and T. Kunitake, *J. Am. Chem. Soc.* **113**, 7342 (1991).

191. C. M. Paleos, Z. Sideratou, and D. Tsiourvas, *J. Phys. Chem.* **100**, 13 898 (1996).

192. J. S. Nowick, J. S. Chen, and G. Noronha, *J. Am. Chem. Soc.* **115**, 7636 (1993).

193. J. D. Dunitz, in G. R. Desiraju, ed., *The Crystal as a Supramolecular Entity, Perspectives in Supramolecular Chemistry*, Vol. 2, p. 1; see also, C. Pascard, in G. Tsoucaris and co-eds., *Crystallography of Supramolecular Compounds*, Kluwer, Dordrecht, 1996, p. 127.

194. A. I. Kitaigorodskii, *Molecular Crystals and Molecules*, Academic Press, New York, 1973; see also, C. P. Brock and J. D. Dunitz, *Chem. Mater.* **6**, 1118 (1994).

195. M. Simard, D. Su and J. D. Wuest, *J. Am. Chem. Soc.* **113**, 4696 (1991); see also, J. Rebek, *Acc. Chem. Res.* **17**, 258 (1984) and E. Fan, C. Vicent, S. J. Geib, and A. D. Hamilton, *Chem. Mater.* **6**, 1113 (1994).

196. J. C. MacDonald and G. M. Whitesides, *Chem. Rev.* **94**, 2383 (1994).

197. J. Rebek, Jr., *Acta Chem. Scand.* **50**, 707 (1996).

198. M. Mascal, N. M. Hecht, R. Warmuth, M. H. Moore, and J. P. Turkenburg, *Angew. Chem.* **108**, 2348 (1996); *Angew. Chem. Int. Ed. Engl.* **35**, 2204 (1996).

199. A. Ulmann, *Ultrathin Organic Films*, Academic Press, Boston, 1991.

200. R. Krämer, J.-M. Lehn, and A. Marquis-Rigault, *Proc. Natl. Acad. Sci. USA* **90**, 5394 (1993).

201. D. L. Caulder and K. N. Raymond, *Angew. Chem.* **109**, 1508 (1997); *Angew. Chem. Int. Ed. Engl.* **36**, 1440 (1997).

202. J.-M. Lehn and A. Rigault, *Angew. Chem.* **100**, 1121 (1988); *Angew. Chem. Int. Ed. Engl.* **27**, 1095 (1988).

203. A. Collet, in Ref. 12, Vol. 10, p. 113.
204. J. Jacques, A. Collet, and S. H. Wilen, *Enantiomers, Racemates, and Resolutions*, Wiley, New York, 1981.
205. J.-M. Lehn, in Ref. 4, p. 307.

EDWIN WEBER
Technische Universität Bergakademie Freiberg
Institut für Organische Chemie

NANOTECHNOLOGY

Molecular nanotechnology (also known as nanotechnology, molecular engineering, or molecular manufacturing) is the production of functional materials and structures in the 0.1 to 100 nm range (the nanoscale) by any of a variety of physical and chemical methods. These methods include nanolithography, direct atomic and molecular manipulation with nanoscale probes, biotechnological selection and production of useful nanomaterials, and chemical synthesis and self-assembly of functional molecules and molecular aggregates (1–3). Various materials produced by such means are being investigated for a wide variety of current technological applications; they are also being examined as functional elements and building blocks of larger structures. Nanoscale structures and devices may be constructed synthetically from their atomic or molecular constituents (the synthetic or "bottom-up" approach, also referred to as nanochemistry) (4–6), or by fabrication techniques that use methods to form small structures from larger ones (the reductive or "top-down" approach) (7,8). Some of the structures and devices that have resulted, and have been envisaged to be forthcoming, from nanotechnology are molecular electronic devices (eg, molecular wires, diodes, transistors, memories), sensors, disease-fighting agents (eg, microbicidal, anticancer), and assemblers (devices that manufacture a particular molecule or assembly of molecules) using nanoscale mechanical devices (eg, nanopumps, motors, gears, and bearings) (2,9–12). A natural analogy to nanotechnology can be found in the biosphere, wherein small and large molecules interact to form complex structures (including many of the structures and devices listed above) necessary for all the functions of living organisms (13,14).

Synthetic vs Reductive Technologies

Many of the devices that have thus far been envisioned as products of nanotechnology (eg, nanoscale environmental sensors, information processors, and actuators) cannot be produced by the large-scale microfabrication techniques currently in use; indeed, these techniques (eg, photolithography) have nearly reached the limits of miniaturization imposed by both practical and fundamental constraints. The commercial production of nanoscale electronic materials is limited by factors such as electron tunneling, quantum mechanical confinement effects, heat dissipation, and difficulty and cost of fabrication (7,15,16). Although there continue to be advances in the development of processing methods that embody the reductive approach, the further development of nanotechnology now hinges on the understanding and manipulation of physical laws and processes at the nanometer level, such as electronic, interatomic, and intermolecular interactions that can be manipulated to allow efficient assembly of nanostructures. These include covalent, electrostatic, van der Waal's, dipole, and π-electron interactions, as well as hydrogen bonding, electron delocalization, and solvation (eg, hydrophobic) and thermal effects (17,18).

Biological systems employ a variety of synthetic strategies and processes that make efficient use of these interactions to form highly ordered molecular (eg, proteins, DNA) and supermolecular units (eg, plasma membranes, ribosomes) that perform a wide variety of functions, including light harvesting, catalysis, environmental sensing, information storage and processing, locomotion, and self-replication (4,19–23). These synthetic processes include covalent bond formation and polymerization, molecular templating, and molecular self-assembly. The distribution and functional success of biological nanostructures in living organisms are an evolutionally-directed balance between the internal stability of the nanostructures (by atomic and molecular interactions) and their responses to the external environment. Thus, biological systems form important existence theorems and design criteria for the production of functional, nanoscale materials through nanotechnology employing the synthetic approach to nanostructures.

In recent years, a wide range of synthetic strategies have been developed that have sought to control molecular and atomic interactions to produce functional materials through nanotechnology using the bottom-up principle. These strategies have ranged in approach from those in which biological processes and synthetic pathways are utilized directly or are mimicked, to those that have no analogy in the natural world. Examples of synthetic strategies that have utilized biological processes directly include the expression of proteins and nucleic acids that are designed for specific function [eg, molecular recognition (24), catalysis (25,26), therapeutics (27–30)] or to exhibit specific structural properties (eg, three-dimensional structure) (14,24). Biomimetic strategies include schemes for formation of functional bilayer membranes (31), thin organic films (32), and vesicles (33) from synthetic precursors, and template-directed crystallization and polymerization (34). Other nonbiological strategies include total synthesis of functional, nonnatural molecules and local probe-assisted construction of atomic and molecular structures (35–39).

An important nature-mimicking methodology involves the use of covalent synthesis followed by molecular self-assembly of the synthesized molecules (4,5). These molecules are generally small mono- or oligomers that interact with each other and with other kinds of mono- or oligomers to form thermodynamically stable, nanoscale structures. Several interactions can play an important role in such assembly, eg, electrostatic, van der Waal's, dipole, and $\pi-\pi$ electron interactions, hydrogen bonding, and hydrophobic effects (18). Biological systems that use these noncovalent interactions to assemble large molecules (eg, proteins, DNA) into compact, functional nanostructures provide the basis for these endeavors.

Related Technologies

Molecular nanotechnology is thus a consequence of current and future research and development efforts involving many broad areas of study, including biotechnology (14,40), protein and nucleic acid structure (41–44), supramolecular chemistry (2,4,6,45,46), molecular recognition (24,40,47), self-assembly (4,5), interface and colloid science (48,49), zeolite and clay chemistry (50), electrochemistry (51,52), electronics micro- and nanofabrication (8,53–55), and local probe microscopy (37–39,56). This article focuses on the current state-of-the-art of methods that produce nanostructures and devices. These methods include nanolithography, atom manipulation by local probes, biotechnology, nanochemical synthesis, rational design, and supramolecular chemistry.

Existence Theorem: Nanobiology

Analogies: Structure and Function.   Biology is replete with complex, functional nanoscale structures formed by directed synthesis and subsequent self-assembly of the component molecules (13,14,57). Initially, the small molecules are produced by covalent synthesis, with the larger, functional structures resulting due to many weak, noncovalent interactions that energetically overcome interactions with the solvent and the entropic advantages of disintegration of the ordered aggregates. The final structure, thus, represents a thermodynamic minimum, and incorrect subunits are rejected in the dynamic, equilibrium assembly. In this process, complementarity in shape and polarity provides the foundation for the association (binding) between components (eg, phospholipids, polypeptide chains, proteins, nucleic acids). Shape-dependent association based on the nonspecific van der Waal's and hydrophobic interactions are made stronger and more specific by hydrogen bonds and electrostatic interactions, as well as, in some cases, covalent bonds (eg, disulfides) (14). Often, positive cooperativity is displayed, ie, conformations of individual subunits change upon binding in such a manner that their affinity for other components of the final structure increases. Moreover, the amount of information required to execute the assembly of a particular structure (eg, a protein) is minimized by use of only a few types of molecules and a limited number of binding interactions (13).

For example, a polypeptide is synthesized as a linear polymer derived from the 20 natural amino acids by translation of a nucleotide sequence present in a messenger RNA (mRNA). The mature protein exists as a well-defined three-dimensional structure. The information necessary to specify the final (tertiary) structure of the protein is present in the molecule itself, in the form of the specific sequence of amino acids that form the protein (57). This information is used in the form of myriad noncovalent interactions (such as those in Table 1) that first form relatively simple local structural motifs (helix and sheet structures associated through networks of hydrogen bonds); these motifs then tend to aggregate in ways that associate hydrophobic regions with one another and out of

Figure 1 schematically depicts the intricate structure of the membrane bound GP Ib-IX-V complex. The complex is formed and stabilized by several of the processes and interactions described above (also see Table 1), including those associated with protein folding and assembly into membranes (hydropho-

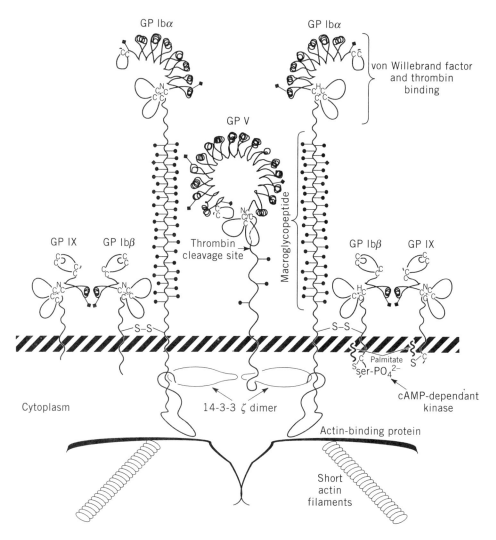

Fig. 1.   The GP Ib-IX-V complex. The complex consists of seven transmembrane polypeptides denoted GP Ibα (mol wt 145,000), GP Ibβ (mol wt 24,000), GPIX (mol wt 17,000) and GP V (mol wt 82,000), in a stoichiometry of 2:2:2:1. The hatched region represents the plasma membrane. The area above the hatched region represents the extracellular space; that below represents the cytoplasm. The complex is a major attachment site between the plasma membrane and the cytoskeleton. Two molecules associated with the cytoplasmic domain are depicted: a 14-3-3 dimer, which may mediate intracellular signaling, and actin-binding protein, which connects the complex to the cortical cytoskeleton and fixes its position and influences its function. The dark circles represent O-linked carbohydrate and the dark squares represent N-linked carbohydrate. Both types of carbohydrate are added after synthesis of the polypeptide chains (64).

bic interactions, hydrogen bonding, disulfide bridging) and those resulting in the quinary structure of the complex (ie, those responsible for the stabilization of the seven-protein structure). The complex is a major attachment site between the plasma membrane and the platelet membrane skeleton. Through its interaction with actin-binding protein, the complex is likely to modulate actin polymerization, and hence, activate the platelet in response to extracellular cues such as binding of plasma proteins (eg, von Willebrand factor, thrombin) and shear.

Biochemical studies of the GP Ib-IX-V complex have identified several distinct functions of particular components of the complex that are related to the platelet's role in hemostasis. These include the binding of von Willebrand factor (vWf) that is immobilized on the vascular subendothelium (ie, the blood vessel wall), and the binding of thrombin. One of the most remarkable features of these interactions is that they are regulated so that they normally occur only when hemostasis is needed, ie, during bleeding. For example, it has been shown that GP Ib will bind vWf when vWf is immobilized on the vessel wall, and not when vWf is in solution in the plasma where it circulates along with the platelets. The ability of GP Ib to discriminate bound versus unbound vWf is thought to be related to the increased shear rates to which the platelet is subjected in flow near stationary surfaces. Thus, in addition to functioning as a molecular sensor and transducer (ie, transducing activation signals after vWf and thrombin binding), the GP Ib-IX-V complex is likely to serve as a mechanical sensor as well, whose molecular recognition capacity is activated only in the proper shear environment.

Another interesting class of membrane proteins that embody an example whereby a nanoscale device is produced by an organism and distributed into its environment for function comprises the pore-forming cytotoxins which are produced by a wide range of bacteria for the purpose of perforating the cells of other bacteria, or of host macro-organisms they are infecting. The gramicidin antibiotics are oligopeptides synthesized and secreted by certain bacteria for the purpose of permeabilizing, and hence killing, other bacteria. The ability of individual gramicidin peptides to assemble within lipid bilayers to form transmembrane pores has been utilized by Cornell and co-workers in synthetic molecular assemblies for biosensing (31).

The sensor (Fig. 2) comprises a supported, functionalized, lipid bilayer separated from a gold electrode by lipids substituted with spacers capable of forming gold–sulfur bonds. The membrane contains two types of gramicidin-based ion channels, one tethered to the gold, the other tethered to a hapten competitor and capable of diffusing laterally within the membrane. The membrane serves three purposes: (1) to anchor the recognition elements (antibody or Fab fragment and hapten competitor); (2) to allow diffusion of the hapten-tethered gramicidin; and (3) to prevent transport of ionic species between the analyte solution and the gold electrode. It may also serve to inhibit nonspecific interactions with other biological molecules in the solution to be analyzed. Upon binding the analyte, the released hapten-bound gramicidin can dimerize with the gold-tethered gramicidin to form a transmembrane pore that allows ionic transport to the gold surface. If a potential is applied, analyte binding is thus detected as a result of ion transport across the membrane. The fact that one analyte binding event can result in the transport of many ionic species across the membrane results in a large amplification effect during transduction (31).

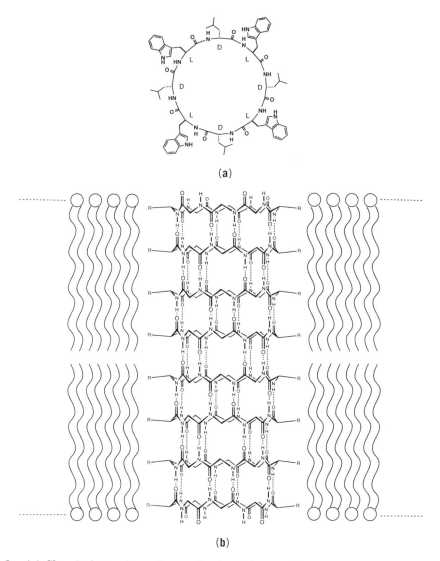

(a)

(b)

Fig. 3. (**a**) Chemical structure of a synthetic cyclic peptide composed of an alternating sequence of D- and L-amino acids. The side chains of the amino acids have been chosen such that the peripheral functional groups of the flat rings are hydrophobic and allow insertion into lipid bilayers. (**b**) Proposed structure of a self-assembled transmembrane pore comprised of hydrogen bonded cyclic peptides. The channel is stabilized by hydrogen bonds between the peptide backbones of the individual molecules. These synthetic pores have been demonstrated to form ion channels in lipid bilayers (71). Courtesy of Dr. M. Reza Ghadiri.

semiconductor industry shows that, as the limits of a particular manufacturing technology are approached and surpassed, the economic consequences have been a spectacular increase in the cost of building new factories. At present, building a new silicon chip fabrication facility needs a capital layout of about $1.5 billion. Although the size of the smallest possible features on the chip have shrunk by 14% every year, the price of lithography equipment has risen at a rate of 28% annually (15). However, the technological barriers faced by the lithographers may not be completely insurmountable, nor may they be prohibitively expensive. In the following sections the limitations of current methods of microfabrication that prohibit their use for nanofabrication (fabrication of nanoscale structures and devices), new advances being made in current technologies that may be used in nanofabrication, and new reductive approaches to micro- and nanofabrication are discussed.

Limitations. The number of transistors present on a chip has doubled approximately every 18 months since the integrated circuit was first developed (a rate of increase predicted by Gordon Moore of Intel Corp. in 1960 that has become known as Moore's Law). The main reason for this continuing decrease in the minimum feature sizes of transistors (and consequent increase in density of transistors on the chip) has been the development of photolithography, the prototypical reductive method (7,15,16,81–83). Photolithography creates patterns in layers of silicon, insulators, and metals to produce the integrated circuit. For the accurate reproduction of features onto the silicon wafer, the wavelength of light used must be at least as small as the smallest feature size (eg, feature resolution varies as the wavelength of light used and is inversely proportional to the aperture of the objective lens and the depth of focus (DOF), requiring innovative solutions in planarization technologies and mask design). Other issues related to further scaling down of integrated circuits include the effects of power supply and threshold voltage of the transistors, short channel lengths, thickness of the gate oxide, high electric fields, fluctuations in the number of dopant atoms, and interconnect delays (7,84). The most important limitation for further size reduction, however, remains the development of new photo- and other lithographic techniques.

Current Advances. New light sources are currently being developed, such as the krypton–fluoride ultraviolet laser (wavelength 0.248 μm for features as small as 0.25 μm) and excimer lasers (wavelength 0.193 μm for features below 0.2 μm) (81). But these technologies still need to overcome several obstacles before they can be implemented by the semiconductor industry. For example, for wavelengths below 0.2 μm, the current photoresists absorb so much light that throughput suffers. Also, the fused silica glass lenses used to demagnify the image absorb light and heat up, resulting in a degradation of the image. Problems associated with DOF become more acute, requiring further innovation in planarization technology. At the present time, these technologies are in the development phase, and it is expected that they will be ready for transfer to manufacturing facilities in the near future (15,81). New photoresist materials, based on the deposition and patterning of monomolecular layers (eg, self-assembled monolayers and Langmuir-Blodgett films), are also being developed to address the afore-mentioned issues of further reduction in the limits of photolithography; these materials and methods are discussed in "Molecular Self-Assembly."

Further reduction in feature size to achieve nanoscale structures by photolithography will necessitate the use of ever smaller wavelengths of light. For example, many manufacturers have explored the use of x-rays (wavelengths less than 1 nm) for the lithographic process for producing structures smaller than 0.1 μm (55,85). Use of x-rays, however, faces a host of obstacles to implementation (55,81,85). For example, commercial patterning tools for x-ray masks constitute a major challenge: they require masks made of materials such as gold or tungsten to absorb the x-rays. The features of gold masks must have very high aspect ratios (eg, 5) in order to absorb x-rays in opaque regions of the pattern that can have widths smaller than 100 nm. For optical and uv lithography, masks are typically four or five times larger than the actual image on a chip. They are focused onto the chip with demagnification to achieve a greater degree of definition in the lithography process. However, despite numerous efforts (86), no commercially feasible way to focus x-rays exists, resulting in the use of masks that are exactly the same size as the circuit. This means that at each of the 20 or more mask levels, an alignment tolerance of only a few tens of nanometers is required. This is a difficult goal to achieve. One approach to circumvent these problems involves using a high-powered laser to bombard a metal target to generate x-rays, which then illuminate a reflective mask. A series of mirrors then demagnify the image to its actual size on the silicon wafer (85).

A new technique developed by Brueck and co-workers for creating submicron feature sizes uses interference effects between two coherent laser beams along with multiple exposures (without the need for a mask) to form highly complex patterns in the photoresist (87,88). Linewidths as small as a quarter of the wavelength of the laser light used are achievable; for a readily available source such as an Ar^+ laser (wavelength 364 nm, close to that of the mercury I-line), a minimum feature size as small as 90 nm is obtainable. Because the interference pattern does not vary with depth (ie, normal to the wafer) for distances of as much as several meters, the issue of DOF ceases to be a problem. It is also possible to project images as large as 30×30 cm^2. Indeed, interferometric lithography can be considered to be the limit reachable by the new mask-oriented methods being developed to extend photolithography for sub-0.2 μm features, such as phase-shift mask and off-axis illumination techniques (87,88).

Focused particle-based lithography, such as ion- (53), neutral atom- (89,90), and electron-beam lithography (91), are capable of achieving very high resolutions. For example, electron-beam lithography employs a focused beam of energetic electrons to draw lines directly on the photoresist (54,91). However, the method of writing each circuit feature separately is serial and inherently slow so that the technology cannot be used for simultaneous fabrication of many chips. This precludes the use of electron-beam lithography for direct writing of circuit patterns for ultra-large scale integration. Instead, the technology is used to create the masks that are subsequently used in photolithography. To speed up the process of electron-beam lithography, methods are being explored to scan a broad electron beam across the entire chip by projecting the beam through an appropriate mask (in mimicry of photolithography) (91).

Even if new methods are developed for denser integrated circuits with smaller features, there are other areas of computing that present formidable challenges. For example, future computers would need data storage and handling capabilities in the hundreds of gigabits to terabit range (15,81). Existing compact disc- or magnetic drive-based storage devices cannot hold such vast amounts of information, nor can they deliver them to the ultra-high speed processors at a commensurate rate. One new approach to address these technical barriers is optical recording using holography (81). Holographic data storage uses lasers to write and read large blocks of data in a photosensitive material (eg, inorganic crystals such as lithium niobate, barium titanate, strontium barium niobate, and organic macromolecules such as bacteriorhodopsin) (81,92). By this method, storage densities of up to 1 terabit of data per cc of crystal are possible.

The most recent approach to reductive nanofabrication that can indeed construct nanoscale structures and devices uses microscopic tools (local probes) that can build the structures atom by atom, or molecule by molecule. Optical methods using laser cooling (optical molasses) are also being developed to manipulate nanoscale structures.

Atomic and Molecular Manipulation

Scanning Probe Microscopy. The scanning tunneling microscope (STM) can image and manipulate matter on the atomic scale (3,37–39,56,93). In general, when a small conducting probe (the tip, consisting of one or a few atoms, or a metal) is placed close (less than 10 nm) to the surface of a conducting substrate, an electronic current results under a suitable bias due to the overlapping of the electronic wave functions of the probe and the surface. Because this tunneling current is exponentially dependent on the separation of the probe from the surface, imaging resolutions of fractions of an angstrom can be obtained. These images reflect both the topography and the electronic structure of the surface. The STM can also be used to modify surfaces locally. As a result, individual atoms and molecules can be manipulated with atomic-scale precision. A variety of atomic manipulation processes have been demonstrated with the STM. They can be broadly classified as parallel processes, in which the atoms and molecules adsorbed on a surface are moved laterally with an STM tip, and perpendicular processes, wherein atoms and molecules are transferred from the tip to the surface or vice versa (3).

In parallel processes, eg, field-assisted diffusion and sliding, the bond between the surface and the adatom is never completely broken. Field-assisted diffusion of an adatom on the surface occurs due to the presence of the intense, inhomogeneous electric field between the probe tip and the surface, which gives rise to a potential gradient. Sliding of adatoms results because the tip of the STM always exerts a force on the adatom and this force can be tuned in terms of magnitude and direction by proper manipulation of the current, potential, or the position of the tip. This process is schematically shown in Figure 4, as employed by Crommie and co-workers (*vide infra*). The adsorbate is first located with the tip (STM in its imaging mode) and then the tip is placed near the adsorbate. The tip-adsorbate interaction is increased by increasing the tunneling current. The tip is then moved laterally across the surface at constant current to the desired destination, pulling the adsorbate along with it. The process is terminated by reducing the tunneling current (returning the STM to its imaging mode) leaving the adsorbate bound to the surface at the new location.

The second class of atomic manipulations, the perpendicular processes, involves transfer of an adsorbate atom or molecule from the STM tip to the surface or vice versa. The tip is moved toward the surface until the adsorption potential wells on the tip and the surface coalesce, with the result that the adsorbate, which was previously bound either to the tip or the surface, may now be considered to be bound to both. For successful transfer, one of the adsorbate bonds (either with the tip or with the surface, depending on the desired direction of transfer) must be broken. The fate of the adsorbate depends on the nature of its interaction with the tip and the surface, and the materials of the tip and surface.

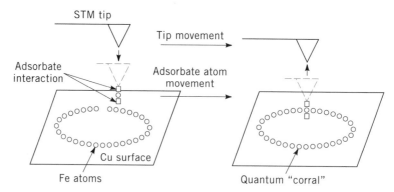

Fig. 4. Atom manipulation by the scanning tunneling microscope (STM). Once the STM tip has located the adsorbate atom, the tip is lowered such that the attractive interaction between the tip and the adsorbate is sufficient to keep the adsorbate "tethered" to the tip. The tip is then moved to the desired location on the surface and withdrawn, leaving the adsorbate atom bound to the surface at a new location. The figure schematically depicts the use of this process in the formation of a "quantum corral" of 48 Fe atoms arranged in a circle of about 14.3 nm diameter on a Cu(111) surface at 4 K. Adapted from Crommie and co-workers (94).

Directional adatom transfer is possible with the application of suitable junction biases. Also, thermally-activated field evaporation of positive or negative ions over the Schottky barrier formed by lowering the potential energy outside a conductor (either the surface or the tip) by the application of an electric field is possible. Electromigration, the migration of minority elements (ie, impurities, defects) through the bulk solid under the influence of current flow, is another process by which an atom may be moved between the surface and the tip of an STM.

Several striking examples demonstrating the atomically precise control exercised by the STM have been reported. A "quantum corral" of Fe atoms has been fabricated by placing 48 atoms in a circle on a flat Cu(111) surface at 4K (Fig. 4) (94). Both STM (under ultrahigh vacuum) and atomic force microscopy (AFM, under ambient conditions) have been employed to fabricate nanoscale magnetic mounds of Fe, Co, Ni, and CoCr on metal and insulator substrates (95). The AFM has also been used to deposit organic material, such as octadecanethiol onto the surface of mica (96). New applications of this type of nanofabrication are being reported at an ever-faster rate (97–99).

Optical Manipulation. Laser beams provide another means of capturing individual atoms or molecules (89). When an atom is irradiated from both sides by laser light at a frequency slightly lower than the frequency at which the atom absorbs photons, then the atom loses some of its momentum. In particular, the laser beam propagating in a direction opposite that of the motion of the atom increases in frequency due to the Doppler effect, resulting in light absorption with subsequent isotropic emission (scattering). The light propagating in the same direction as the atom is not absorbed, so that the atom is pushed in a direction opposite its motion and slows down. By surrounding the atom with three sets of counterpropagating laser beams orthogonal to each other the atom can be cooled (ie, slowed) in all three dimensions (100). The first demonstration of this atom-trapping strategy resulted in the cooling of sodium atoms to 240 μK. Because the light field acts as a viscous drag force, the combination of laser beams is known as optical molasses (89).

While near-resonant light exerts both scattering forces and dipole forces on single atoms, similar forces are also exerted on larger dielectric objects. A single laser beam can attract a dielectric object to its focal point through the dipole force, resulting in optical trapping; this has led to the development of a device called optical tweezers that can trap and manipulate small objects (ranging from large polymers to living cells) (89,101). Optical tweezers can handle live bacteria and other organisms without apparent damage (102). Even organelles within the cell can be manipulated without rupturing the cell wall (103). Optical tweezers are being used to study the mechanical properties of several biomolecular motors, including those associated with bacterial flagella and the contraction of muscles (104–106). On a smaller scale, single molecules of DNA have been manipulated and their elastic properties measured (107). New applications include the use of optical tweezers to measure the binding force of biomolecules to surfaces and to other biomolecules (108,109).

Optical trapping can also be used as a lithographic tool (90). For example, a combination of optical molasses and an optical standing wave have been used to focus a beam of neutral sodium atoms and deposit them in the desired pattern on a suitable substrate (eg, silicon). Pattern resolutions of the order of 40 nm with good contrast (up to 10:1 between the intended features and the surrounding unpatterned areas) and deposition rates of about 20 nm/min were obtained (90).

Despite advances, it seems unlikely that the reductive approaches outlined above can, by themselves, reach the level of control, flexibility, discrimination, and versatility of atomic and molecular manipulation that will be needed to manufacture the molecular and supramolecular nanodevices envisaged to be the products of nanotechnology. Studies of biological nanodevices (eg, proteins) suggest that under proper conditions, atoms and molecules can assemble into functional nanoscale units that can carry out all the functions of life.

Synthetic Approaches

The synthetic (bottom-up) approach offers a level of control over the selection and placement of atoms and molecules that is ultimately much higher than that offered by other methods of large-scale microfabrication (eg, fabrication of integrated circuits) (2,4–6,17,22,46,49,110–112). Such synthesis can employ a variety of chemical methods utilizing some or all of the forces and interactions listed in Table 1 to produce nanoscale molecules. Most chemical synthetic reactions that produce large molecules (eg, collections of atoms that can act as nanodevices) generate polydisperse materials. These materials are mixtures of oligo- or polymeric chains of varying molecular weights. For nanotechnological applications, it is important to have synthetic strategies that yield compounds of uniform length, size, and shape; inhomogeneities can be detrimental to the designated function of the molecule. Thus, the specific objectives of synthesis are to discover and develop rapid and efficient methods for the precise control of composition, molecular weight, stereochemistry, aggregation, and placement of functional molecules. Four strategies are currently in use (either separately or in combination with one another) for the fabrication of large molecules (substances with molecular weights of a few hundred to a few million): biotechnological synthesis (14,113), sequential covalent synthesis (5,111), covalent polymerization (6), and molecular self-assembly (4).

In nature, complex functional molecules such as enzymes are produced by using the chemical synthetic approach outlined above (13). The molecules are first formed by covalent synthesis. For example, proteins are formed under the direction of information contained in mRNA by the action of a variety of enzymes (protein catalysts) and cofactors. In particular, the polypeptide is formed by the directional joining of amino acids (brought to the site of the synthesis, eg, a ribosome, by a transfer RNA (tRNA) molecule) through amide bonds. The primary structure of the amino acids (ie, sequence) during synthesis is specified by the mRNA (a transcript of the original gene encoded in the DNA). Once the polypeptide is produced, it then undergoes many conformational changes (governed almost entirely by the sequence of the constituent amino acids) that reduces its size to a compact, native form that is the functional protein. These conformational changes are termed folding (41,57,59). As illustrated in Figure 5, protein folding proceeds by the formation, first, of local structural motifs (eg, α-helices, β-strands, and β-turns) by self-assembly. These motifs are formed as a result of formation of multiple hydrogen bonds between the side chains of amino acids that are far from each other in the linear sequence, but spatially close to each other. These structural motifs then self-assemble to form structural domains (eg, a bundle of α-helices, a barrel of β-sheets, and combinations of these) that are stabilized by hydrogen bonds and other noncovalent interactions (eg, the hydrophobic effect, electrostatic interactions, van der Waal's interactions). This secondary structure of the polypeptides aggregates further to form the tertiary structure (the final structure for the polypeptide chain) and quaternary structure (when more than one polypeptide chain aggregate to form the final protein structure). Often, covalent disulfide bonds and metal ions are employed to bridge the different structural motifs to provide additional structural stability. For example, a zinc-finger polypeptide coordinates a Zn^{2+} ion to provide the necessary stability to form a small protein, a task that would otherwise be impossible due to the small number of amino acids in the protein which precludes the formation of a large enough hydrophobic core (47). Additional chemical functionality, eg, carbohydrates, can also be incorporated in the protein structure through posttranslational modification (13).

RNA is also capable of folding into specific shapes for ligand recognition and catalysis (44,114). Although there are only four naturally occurring bases available for the formation of tertiary structures through noncovalent interactions (while proteins have twenty different amino acids to choose from), myriad RNA shapes can be produced. The predominant noncovalent interaction involves the formation of hydrogen bonds. More importantly, metal ions have been shown to confer extraordinary stability to RNA and RNA fragments. For example, the overall tertiary structure of catalytic RNA (referred to as ribozyme) derived from the self-splicing intron of *Tetrahymena thermophila* results due to cooperative assembly of the RNA molecule upon uptake of at least three divalent metal ions (115). In particular, Mg^{2+}, Ca^{2+}, and Sr^{2+} can participate in general folding of the ribozyme tertiary structure, and Mg^{2+} is further involved in the

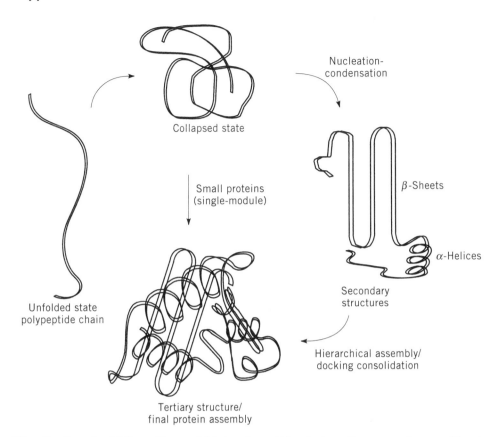

Collapsed state

Nucleation-condensation

Small proteins
(single-module)

β-Sheets

α-Helices

Secondary
structures

Unfolded state
polypeptide chain

Hierarchical assembly/
docking consolidation

Tertiary structure/
final protein assembly

Fig. 5. Protein folding. The unfolded polypeptide chain collapses and assembles to form simple structural motifs such as β-sheets and α-helices by nucleation-condensation mechanisms involving the formation of hydrogen bonds and van der Waal's interactions. Small proteins (eg, chymotrypsin inhibitor 2) attain their final (tertiary) structure in this way. Larger proteins and multiple protein assemblies aggregate by recognition and docking of multiple domains (eg, β-barrels, α-helix bundles), often displaying positive cooperativity. Many noncovalent interactions, including hydrogen bonding, van der Waal's and electrostatic interactions, and the hydrophobic effect are exploited to create the final, compact protein assembly. Further structural consolidation may also occur by covalent bond formation (eg, disulfide bridges).

catalytic activity of the ribozyme. The tRNA for methionine has a tertiary structure more stable than some of its secondary structural motifs in the presence of small concentrations of mono- and divalent cations (eg, Na^+, Mg^{2+}) (44,114). This is primarily due to the creation of negatively charged pockets that have a high affinity for cations. On the other hand, the structural stability of a subunit of ribosomal RNA (rRNA) that binds ribosomal protein L11 and thiazoles is made possible by the presence of either Mg^{2+} or NH_4^+ cations. Such highly specific selectivity for the two ions implies that the rRNA coordinates or forms hydrogen bonds with these ions to form a tertiary structure that is rigid enough to distinguish between these ions and any others. It is important to note that the primary structure of the polynucleotides also determines the final, tertiary structure (similar to the case of polypeptides) (44,114).

Thus, protein and RNA folding studies form a fundamental paradigm for the design and synthesis of new, functional marcomolecules with both final structure and function built into the primary structure of the macromolecule. Such synthetic strategies could be applied to the rational design of functional nanostructures, including drugs, sensing elements, photonic and electronic components, catalysts, and even mechanical devices. The current state of development of these methodologies is addressed in the following sections.

Biotechnological Synthesis.   Molecular manufacturing of functional proteins within living systems follows the well-known path of transcription and translation of gene-coded information present in DNA (13). Typically, a transcript of the genes coding a particular protein is made. This transcript (mRNA) binds to complementary sections of two tRNA molecules. One tRNA contains an amino acid that is next in line to be attached to the growing polypeptide chain (which is attached to the other tRNA). The mRNA then translocates to bind another tRNA (bearing the next amino acid in the sequence) and so on. Biotechnological synthesis of new nanomaterials (eg, proteins) and biotechnological modification of living systems (eg, conferring specific therapeutic properties, new genetic traits) exploit the many ways in which this protein manufacturing machinery can be modified (13). In particular, the DNA of an organism can be altered in a specific manner, resulting in a modified mRNA, and consequently, a modified or new protein. The DNA can be altered by chemical synthetic methods followed by expression of the altered genes in a suitable organism (recombinant DNA technology), or by selection and isolation of natural mutations.

Recombinant DNA technology provides a powerful tool for analysis, synthesis, and alteration of genes and proteins (13). It is based on the ability to rapidly synthesize polynucleotides with any sequence using nucleic acid enzymology (eg, using DNA polymerases, restriction enzymes or endonucleases, DNA ligases). The unique base-pairing attributes of the constituents of the DNA and the ability to express the modified or synthetic DNA in microorganisms and eukaryotic cells result in a powerful tool for the production of synthetic molecules with specific biological functions (13,116,117).

In general, a new DNA chain is made by joining fragments that have complementary cohesive ends produced by the action of a restriction enzyme (13). A DNA ligase then joins the ends of the DNA chains (provided they are within a double helix). Alternatively, tailor-made synthetic genes can be synthesized by sequential addition of activated monomers to a growing chain attached to a solid support, in a method similar to the Merriefield synthesis (*vide infra*) (118). A further method involves the reverse transcription of mRNA to obtain the original DNA chain that encodes the protein that mRNA synthesizes. The DNA chains obtained in these ways are then inserted in vectors to obtain recombinant DNA. Examples of vectors include plasmids (naturally occurring circular DNA that act as accessory chromosomes), λ phages, retroviruses, and adenoviruses (13). The recombinant DNA is introduced into producer cells (eg, *Escheria coli*) by direct uptake from the medium surrounding the cell, by injection into the host cell, or by use of a virus as a vector. The cells that contain the recombinant DNA are selected and cloned to obtain large quantities of cells with the ability to express the genes inserted into their genome. Gene expression in these cells results in the production of large quantities of useful proteins. This technology has now been commercialized to manufacture many naturally-occurring, therapeutic protein drugs, including human insulin, erythropoietin, growth hormone, interferons, tissue plasminogen activator, and granulocyte colony stimulating factor (119).

It is also possible to carry foreign DNA into mammalian cells, for example, by use of retroviruses or by direct injection. For example, the gene for rat growth hormone was injected into fertilized mouse eggs, resulting in the production of giant mice that were almost twice as large by weight as their siblings with unchanged genome (120–122). Tumor-inducing plasmids (carried by the soil bacterium *Agrobacterium tumefaciens*) have been used to deliver foreign genes into plant cells, and represent promising candidates for exploring the genomes of plant cells and for modifying plants to improve their agricultural value (eg, disease resistance, increased yield) (13,123,124). Electroporation (application of intense electric fields that render the cell walls transiently permeable to large molecules), aerosol transport, and liposome delivery are other methods for transporting foreign DNA into cells (125–128).

The development of new, more efficient gene transfer protocols (eg, using retroviruses and adenoviruses) has allowed rapid progress in gene therapy (116, 122,129–142). Here gene transfer involves the introduction of new genetic material into the cells of an intact organism in order to change its phenotype. Applications of gene therapy include substitution of missing proteins (due to genetic disorders, eg, deficiency of adenosine deaminase resulting in severe combined immune deficiency), increased synthesis of naturally occurring proteins (eg, cytokines in tumor cells for immune stimulation), expression of proteins not naturally occurring in man (eg, production of anti-HIV antibody fragments in T cells), and targeted activation or deactivation of protein expression (eg, production of antisense RNA against HIV sequences). Expression of new genes in hematopoietic (primitive) stem cells is the focus of much research because it may provide a powerful tool for the cure of congenital genetic disorders (many cell lineages are derived from primitive stem cells throughout the life of an individual, so that genetic cures introduced into the stem cells can become nearly permanent).

The types of molecules synthesized by biotechnological techniques are restricted to those biomolecules whose structures can be encoded in the DNA of organisms capable of translating them into functional nanomaterials. Other types of molecules and nanomaterials can be synthesized by chemical synthetic approaches, such as covalent syntheses and molecular self-assembly of molecular units.

Covalent Synthesis. The first strategy for chemical synthesis employs elaborate and sophisticated methods for assembling atoms into molecules based on the general strategy of sequential formation of covalent bonds. The atoms can also be assembled into subunits that are then reacted to form more complex, designed molecules (convergent synthesis) (45). Covalent synthesis can be used to form molecules with well-defined composition, connectivity of atoms, and shape. This, however, requires enormous effort and energetic input when applied to very large molecules (or nanostructures), which hinders the use of this strategy alone to fabricate complex nanostructures that can interact with each other, with other nanostructures, and with their macroscopic environment.

Covalent synthesis of complex molecules involves the reactive assembly of many atoms into subunits with aid of reagents and established as well as

pressures (up to 60 MPa), thus acting as nanoscale gas cylinders (171). A single multiwalled carbon nanotube has been attached to the tip of a scanning probe microscope to yield an atomic manipulator of unprecedented dimensions. The nanotube tip is up to 1 μm in length and 5 nm in diameter (97,172). This tip can not only image features in trenches as small as 0.48 μm wide and 0.8 μm deep (which is not possible with the usual pyramidal tip), but also deposit a 40-nm dot of carbon at the bottom of the trench (172).

Template-Assisted Synthesis. Microporous (pore diameter <2 nm) phases with three-dimensional framework structures, eg, zeolites, are routinely synthesized by use of molecular units or hydrated alkali or alkaline earth cations as templates (50,173). Zeolites are crystalline inorganic oxides such as aluminosilicates that have uniform pores (size range 0.2 to 1.0 nm) with regular connectivity (173). Because of the constancy of pore characteristics of particular zeolites, they have been successfully used in commercial catalysis and separation systems, usually based on size and shape selectivity. The recent synthesis of silica-based mesoporous materials by the cooperative assembly of periodic inorganic and surfactant-based structures extends the range of molecular sieves (ie, separation systems based on the size and shape of molecules) into the regime of mesoporous materials (ie, 2 to 50 nm) (174). For example, researchers at the Mobil Oil Company used self-assembled molecular arrays as templates (instead of a single, solvated organic molecule or metal ion) and successfully synthesized the first mesoporous molecular sieves (MCM materials) with controlled pore sizes (between ~1.5 and ~10 nm) (175). These syntheses exploited the formation of micellar arrays of amphiphilic surfactant molecules whose noncovalent interactions with the inorganic polymerizable component (eg, alkoxysilanes such as tetraethoxysilane, TEOS) directed the formation of amorphous silica around them.

Organic ligands covalently bonded to the silica precursor (eg, an alkoxysilane) can be copolymerized with the inorganic component (eg, TEOS) to direct (ie, template) the formation of ceramic materials with very small (<1 nm) pores; unlike zeolites, these materials are amorphous (174,176). Typically, a nonporous hybrid material consisting of an organic template-containing precursor (eg, 3-methacryloxypropyl trimethoxysilane) and an inorganic polymerizable precursor (eg, TEOS) is formed by sol-gel processing (177), followed by the pyrolytic removal of the nonhydrolyzed organic template, leaving behind a material with pores that correspond to the size, connectivity, and perhaps, shape of the organic template ligands (176). The microporous material (pore diameter <1 nm) so obtained can not only exhibit molecular sieving properties similar to those of zeolites, but also allow greater ease and flexibility of processibility. For example, thin defect-free silica membranes can be produced with which sulfur hexachloride (kinetic diameter, KD = 0.55 nm) has been separated from helium (KD = 0.265 nm), nitrogen (KD = 0.364 nm), and propane (KD = 0.45 nm) (176).

The template-assisted synthetic strategies outlined above produce micro- or mesoporous structures in which amorphous or crystalline polymers can form around the organic template ligands (174). Another approach is the use of restricted spaces (eg, pores of membranes, cavities in zeolites, etc.) which direct the formation of functional nanomaterials within their cavities, resulting in the production of ultra-small particles (or dots) and one-dimensional structures (or wires) (178). For example, in the case of polypyrrole and poly(3-methylthiophene), a solution of monomer is separated from a ferric salt polymerization agent by a Nucleopore membrane (linear cylindrical pores with diameter as small as 30 nm) (179–181). Nascent polymer chains adsorb on the pore walls, yielding a thin polymer film which thickens with time to eventually yield a completely filled pore. De-encapsulation by dissolving the membrane in CH_2Cl_2 yields wires wherein the polymer chains in the narrowest fibrils are preferentially oriented parallel to the cylinder axes of the fibrils. These polymeric heterocyclic nanotubes and nanowires have redox properties similar to those of parent, bulk polymers and their electronic conductivity is enhanced with respect to analogous bulk polymers (due to the preferential orientation of the polymer chains along the fibrils).

Nanoscale structures have also been fabricated by template-assisted lithography (eg, micro-contact printing and replica molding) using elastomeric (rather than rigid) organic polymer masters (182,183). The use of elastomers facilitates the easy separation of the master and the replica, thus minimizing damage to the small features on the replica as well as the master. The elastomeric master can also be shaped mechanically, allowing flexible control of the form and size of the features (184). For example, an elastomeric stamp made of poly(dimethylsiloxane) (PDMS) can be generated from an original master (with patterned features of metals, silica, silicon nitride, or photoresists) by cross-linking a viscous prepolymer after it has been poured onto the master. The stamp can be compressed mechanically to obtain patterned features smaller than those on the original master. The elastomeric stamp can then be re-replicated on a rigid organic polymer (eg, polyurethane). Polyurethane replicas with patterned features as small as 30 nm in width and 8 nm in height have been produced (185). Recently, PDMA stamps have also been made incorporating silicone fluid fillers that, when the filler is removed, yield patterned stamps with feature sizes smaller than those obtained without the filler (183). The methodology is potentially amenable to forming nanoscale, patterned organic layers with useful chemical, biochemical, magnetic, optical, and electronic functions.

A similar method of nanoscale patterning involves the use of an elastomeric resist film onto which the desired pattern is generated by compression molding at high temperature (ie, higher than the glass-transition temperature) (186,187). As a result of the compression, the film is thinned in the regions where the master pattern is present. Subsequent to controlled reactive ion etching, the thinned regions of the resist film are removed, leaving behind the thicker regions (which are partially thinned due to etching) and these form the desired pattern on the surface. This pattern can now be used as a resist layer. Poly(methyl methacrylate) (PMMA) is one candidate for this purpose, because it does not shrink or swell over large temperature and pressure ranges and is nonadhesive to silica. Metal (5 nm Ti and 15 nm Au) patterns with feature size of 25 nm have been fabricated by use of these resist templates in conjunction with metal deposition and lift-off (186,187).

Molecular imprinting is a technique by which molecular recognition capabilities are bestowed upon organic or inorganic polymeric systems through templating (174,188). The imprinting imparts structural information ("memory") of a particular molecule, by positioning the functional groups of the polymer (the local environment of the imprint molecule) in a specific geometric configuration that can then recognize the target (imprint) molecule. The polymeric materials are typically copolymers of methacrylates or vinylpyridines. Potential applications of imprinted polymers include catalysis, separations (eg, separation of chiral molecules), and chemical sensing (188).

Molecular Self-Assembly. As seen in "Reductive Approaches," reductive techniques, such as those currently in use in the microelectronics industry, can produce structural features smaller than about 200 nm, although at great economic cost. The use of proximal probes and other nanomanipulative techniques can be considered to be a hybrid of the reductive lithographic techniques and the synthetic strategies of assembling functional nanostructures atom by atom, or molecule by molecule. The organization of nanostructures and devices by the self-assembly of the component atoms and molecules, a ubiquitous phenomenon in biological systems, forms the noncovalent synthetic approach to nanotechnology.

In this approach, well-defined subunits (small molecules similar to, eg, nucleotides) are first formed through covalent synthesis. Second, these subunits aggregate with themselves or with other subunits through covalent or noncovalent (or both) interactions to form large, stable, structurally defined assemblies (eg, large molecules similar to RNA and proteins, and supramolecular assemblies such as DNA, ribosomes). To realize this synthetic approach, it is necessary to understand and control the noncovalent connections within and between molecules and to understand and overcome the possible unfavorable entropy increase involved in bringing many molecules together in a single, ordered aggregate (4,147). For the final supramolecular structure to be stable and to have a well-defined shape, the noncovalent connections must be collectively stable. The strengths of individual noncovalent bonds or interactions are small (eg, the strength of van der Waal's interaction is less than 4 kJ/mol, that of hydrogen bonds is of the order of 2 kJ/mol) relative to typical covalent bonds (150–400 kJ/mol) and comparable to thermal energies (RT = 2.5 kJ/mol at 300 K) (17). Therefore, molecules must be stabilized by many noncovalent interactions (ie, large complementary areas of molecular surface in interacting molecules must be in noncovalent contact).

The following sections contain a review of many of the varied synthetic systems that have been developed to date utilizing noncovalent interactions to form assemblies of molecules. These sections are loosely demarcated according to the most important type of noncovalent interactions utilized in conferring supramolecular order (ie, van der Waal's interactions, electrostatic interactions, and hydrogen bonds). For extensive reviews, see references 1,2,4–6,22,46,49,110–112. Finally, the development of self-assembling, self-replicating synthetic systems is noted.

van der Waal's Interactions. Langmuir mono- and multilayers of amphiphilic molecules (ie, molecules with a long hydrocarbon chain and a hydrophilic head-group) formed at the air–water interface are the prototypical example of a synthetic system based on van der Waal's interactions (189). These layers can form ordered, solidlike phases that can be transferred onto a variety of substrates (by physisorption), including glass, silicon, mica, and metal surfaces. Such films, called Langmuir-Blodgett (LB) films, retain their structural order and can display two-dimensional crystallinity. A variety of amphiphilic molecules have been investigated depending on the projected use of their LB films. In general, the molecules comprise a hydrophilic head group and a long hydrocarbon tail, that may be separated by a central group and/or a linking group that provides the desired properties to the LB film. Common hydrophilic groups include quaternary ammonium, carboxylic acid, alcohol, methyl ester, and ethyl ester. Examples of central groups include biphenyl and azobenzene. LB films have also been fabricated from amphiphilic derivatives of porphyrins, phthalocyanines, polymerizable materials (eg, long chain acids containing vinyl derivatives and diacetylene groups), rodlike polymers (eg, synthetic polyglutamates), preformed polymers (eg, polyimides), and amphiphilic molecules incorporating electrically conductive materials (eg, the electron acceptor tetracyanoquinodimethane, TCNQ, and the electron donor tetrathiafulvalene, TTF) (189).

One of the foremost potential applications of LB films is in the generation of second harmonic radiation at optical frequencies. Noncentrosymmetric films are fabricated by using multilayers containing alternating monolayers. One monolayer contains chromophores with high second-order optical susceptibility and the other having no nonlinear optical properties (190). The LB technique has also been applied to fabricate noncentrosymmetric structures having pyroelectric properties (191). Multilayered LB films incorporating charge-transfer salts (CT salts such as TTF, TCNQ) as central groups have been prepared that display electrical conductivity comparable to that of not only pure crystals of the CT salts, but also metals (192). Even superconductivity has been observed in LB films containing the central group metal-bis-(4,5-dimercapto-1,3-dithiole-2-thione), where the metal is Ni or Au. LB film-based field-effect transistors have been fabricated that exhibit current-voltage characteristics typical of inorganic (eg, Si-based) transistors (192).

A second class of monolayers based on van der Waal's interactions within the monolayer and chemisorption (in contrast with physisorption in the case of LB films) on a solid substrate are self-assembled monolayers (SAMs). SAMs are well-ordered layers, one molecule thick, that form spontaneously by the reaction of molecules, typically substituted-alkyl chains, with the surface of solid materials (193–195). A wide variety of SAM-based supramolecular structures have been generated and used as functional components of materials systems in a wide range of technological applications ranging from nanolithography (196,197) to chemical sensing (198–201).

A prototypical class of SAMs that has been widely investigated is derived from the reaction of one or more substituted alkanethiols ($HS(CH_2)_nR$, where typically $n = 8$–22 and R represents one of a variety of diverse functional groups) with the surface of gold and other coinage metals (ie, silver, copper) (194,195). Figure **6a** illustrates schematically the salient aspects of the structure of this type of SAMs, as inferred from the results of several investigations employing different analytical techniques such as x-ray photoelectron spectroscopy (202), polarized external infrared spectrometry (203), electron and helium diffraction (204), and x-ray reflectivity (205). These SAMs are easily prepared from solutions or vapors of the alkanethiols and are stable in common solvent environments, structurally well-ordered, and easy to modify synthetically (194,195). Gold supports for these SAMs can be easily fabricated by evaporation on a variety of

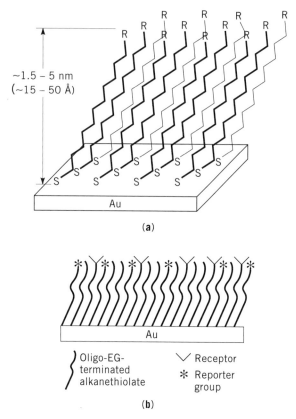

Fig. 6. (**a**) Schematic representation of the structure of a SAM derived by chemisorption of $HS(CH_2)_{11}R$ from solution onto a gold(111) surface (202,204,205,218). For simple terminal functional groups (eg, R = CH_3) the alkyl components of these SAMs are generally found to adopt a fully transextended configuration that is tilted from the surface normal by ~30°. Note that this structure results in a surface whose chemistry is determined primarily by the identity of the terminal functional groups, R. (**b**) Schematic depiction of a modular SAM for use in biosensing applications. The illustration shows a three component mixed SAM of alkanethiolates terminated with a receptor moiety, a reporter moiety, and an oligo(ethylene glycol) moiety. A reagentless biosensor is formed by the specific interaction (binding) of an analyte (eg, a toxin, metabolite, or physiological indicator) with the receptor moiety (eg, a protein fragment, nucleotide, or synthetic receptor) which perturbs an externally-measurable chemical or physical property of the reporter group (eg, a fluorophore). The oligo(ethylene glycol) moieties serve to eliminate nonspecific interactions between the sensor surface and other components (eg, proteins or cells other than the analyte) of the solution to be analyzed.

substrates and can be thin enough to allow transmission optical microscopy, optical spectroscopy, and surface acoustic wave spectroscopy (201,206). These SAMs on gold also lend themselves to investigation by electrochemical methods and surface plasmon resonance (SPR) spectroscopy and microscopy and hence, have been exploited as platforms for affinity-based chemical sensors (198,199,201).

A driving force for the development of this class of molecular assembly has been the ease with which these SAMs can be used to obtain different interfacial chemistries by variation of the terminal functional group, R, and by forming multicomponent SAMs (eg, with two types of terminal functional groups R and R′) (207). This property makes them ideal candidates as model systems for the investigation of interfacial chemical and physical phenomena. Further versatility in function can be obtained by forming patterned SAMs, in which particular areas of the surface contain one type of SAM (eg, with a particular terminal functional group, R) and other areas of the surface contain a different type of SAM (with R′) or no SAM at all. Several methods have been developed for fabricating microscopic and nanoscopic patterns in SAMs of alkanethiolates on gold including: (1) use of polymeric "rubber stamps" to deliver reactive alkanethiols to many different areas of a gold surface simultaneously by micro-contact printing (208); (2) use of uv-lithography to oxidize thiolate groups to noncoordinating sulfonate-species that can be displaced by other alkanethiols (209); (3) use of micropens to deliver and to derivatize alkanethiols directly in specific areas of a gold surface (210); (4) micromachining of the gold to remove a pre-existing SAM followed by formation of a different type of SAM in the newly exposed gold regions (211); (5) use of local probes such as those used in scanning tunneling microscopy and atomic force microscopy to remove pre-existing SAMs (212); and (6) electrochemical stripping of SAMs and blocking of SAM formation on gold microelectrodes (200). SAMs of alkanethiolates on gold or other substrates can also be patterned by electron- or ion-beam lithography (213). These methods for patterning SAMs have been used in a variety of experimental applications for micro- and nanofabrication and in the investigation of interfacial phenomena (see also LITHOGRAPHIC RESISTS).

From the standpoint of the use of SAMs as prototypical materials and functional elements in nanotechnological applications, both the ability to use these structures to control surface chemical properties precisely and the ability to form well-packed molecular layers, are of prime importance. Experimental systems based on SAMs have been devised to study phenomena in which the molecular constitution of the surface governs macroscopic to nanoscopic behavior of the surface. These phenomena include wettability (210,214), acid–base interactions (201,215), adhesion (216,217), corrosion (218), protein adsorption (209,219), biomolecular recognition, and cellular attachment and growth (208,220). Other nanotechnological applications have been in the area of material and nanostructures synthesis, and examples include the use of SAMs as resists for micro- and nanofabrication (196,197) and in templating the growth of nanoclusters, crystals and thin films (34,221). Of particular importance to nanostructure fabrication is the use of SAMs as ultrathin resist materials. Future reductive technologies requiring convenient resist materials for the production of materials with physical features of nanoscopic scale are likely to require nanoscopically thin resists for efficacy. In such cases, it may be that the use of self-assembling molecular films, whose thickness can be easily controlled by appropriate choice of the molecular constituents, and which have an inherent tendency to cover a surface conformally (and thus avoid defects), will allow the generation of resist materials for the high definition production of nanoscopically sculpted materials.

A number of efforts are underway to form SAMs for chemical sensing applications which mimic biological nanostructures (eg, membrane protein receptors) in that they exhibit modular function in which a recognition moiety (a receptor) is colocalized with a transducer moiety (a reporter) that signals the binding of an

and acceptor, and their orientation relative to each other. For example, guanosine monophosphate spontaneously assembles *in vivo* to produce planar rings that can stack themselves in a cylindrical fashion (147). Each planar disc comprises four hydrogen-bonded guanine bases. This observation has led to numerous reports of similar molecules that can assemble by hydrogen bonding into planar and columnar aggregates (eg, folic acid, isoguanosine, dipyridones) (147).

The amine functional group has been exploited to synthesize a variety of molecules that can form hydrogen bonds. For example, Whitesides and co-workers have demonstrated that cyanuric acid and melamine and their derivatives can assemble into myriad supramolecular structures (237,238). As shown in Figure 7, three melamine subunits can be attached to a central hub; the layer of melamines is repeated to obtain a three-tiered molecule **A**. The geometry of the molecule is such that it can bind 9 molecules of the cyanuric acid derivative **B** to form a triply-stacked supramolecular aggregate (molecular weight ~6 kDa) stabilized by 54 hydrogen bonds. Similar to the amide functionality, carboxylic acids can also be employed to form hydrogen-bonded networks in aprotic solvents (147). For example, trimesic acid (1,3,5-benzenetricarboxylic acid) self-assembles into a sheet of hexagonal arrays. Two dimers of biphenyl-3,3'-dicarboxylic acid (which exists as two conformers syn and anti) can self-assemble to produce a cyclic aggregate that looks similar to the number eight. A porphyrin derivative containing two alkyluracil groups has been shown to coordinate a triaminopyrimidine derivative to give a supramolecular cage. Metallorganic compounds are another class of molecules that can form extended hydrogen-bonded complexes (147). For example, a 3×3 planar grid has been generated by the complexation by Ag(I) of 3,6-di-2-pyridylpyridazine (229).

Cyclodextrins and crown ethers also have the ability to form supramolecular aggregates with various shapes (eg, cyclic or helical) (148,239). For example, a β-cyclodextrin derivative containing four lipophilic arms was incorporated into a lecithin bilayer. The compound dimerizes in the bilayer to form an ion channel that enhances the rate of transport of Co^{2+} ions (by an order of magnitude) into the interior of the liposome, compared to an ion channel formed by 18-azacrown-6 (147). Synthetic cyclic peptides containing alternating D- and L-amino acids can self-assemble by hydrogen bonding into stacked rings with a central pore that can act as an ion channel (see Fig. 3). Ghadiri and co-workers have produced such an ion channel from a cyclic octapeptide with a pore size of 0.75 nm (71). The ion channel so-formed has transport properties similar to those of gramicidin A; indeed, the synthetic ion channel (nanotube) transports cations three times faster than gramicidin A. The peptide nanotube also displays selectivity for K^+ versus Na^+. Recently, peptide nanotubes with 1.0 nm and 1.3 nm diameter pores capable of glucose transport were prepared (240).

Helical arrays are another structural motif that noncovalent interactions can generate (eg, DNA in living systems). Along with hydrogen bonds, the hydrophobic effect and metal ion coordination are important contributors to the formation and stability of such supramolecular complexes. For example, tartaric acid-based derivatives have been synthesized that can form either left- or right-handed helical superstructures through hydrogen bonding between complementary uracil–pyridine base pairs; the handedness of the helix is determined by that of the tartaric acid derivatives (147). Double helices have also been demon-

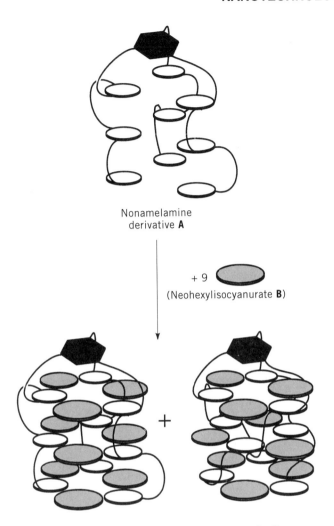

Nonamelamine
derivative **A**

+ 9

(Neohexylisocyanurate **B**)

Fig. 7. Formation of a supramolecular aggregate composed of a compound containing nine melamine rings (the three-layered nonamelamine derivative **A**) and nine molecules of neohexylisocyanurate (**B**). Of the 16 possible conformers that can result, two are shown: The first has the nine molecules of **B** arranged in three rosettes of three molecules each, stacked atop each other; in the second, the rosettes are staggered with respect to each other such that the rosettes in the first and third layers of **A** are aligned with each other, but not with the rosette in the second layer. The supramolecular assembly is stabilized by 54 hydrogen bonds. Adapted from Mathias and co-workers (238).

strated. For example, two molecules of quinquepyridine (or sexipyridine) intertwine upon exposure to metal ions to produce a double helix. Ghadiri and co-workers have synthesized polypeptides that can self-assemble into a variety of helices upon coordination by metal ions (eg, Ru^{2+}) to produce three- and four bundle helices (241–244).

The self-assembly of synthetic DNA molecules also offers control of the structure of matter on the nanometer scale (245). The natural, linear, double-stranded DNA assembly invariably forms a right-handed double helix in compliance with the complementary rule that adenine always binds to thymine and that guanine always pairs with cytosine. Because the hydrogen bonding motif is well-known and easily varied, the possibility of designing synthetic sequences of nucleotides that self-assemble to form branched DNA molecules allows DNA to be engineered in three dimensions and serve, perhaps, as a molecular scaffold for placement of other functional molecules. This has been successfully demonstrated by the preparation of synthetic, branched DNA molecules that can assemble into two-dimensional lattices, three-dimensional cubes and truncated octahedra (245). Such DNA arrays and objects may be useful as macromolecular hosts, as scaffolds for molecular electronic components and for tethering catalysts, proteins, or drugs (245). The isomerization reactions of these DNA lattices (eg, from the normal right-handed structure to a left-handed structure) results in a torsional transition that, with proper choice of the parts of the array that are isomerized, can lead to controlled molecular motion, a key goal of nanotechnology.

Self-Replicating Systems. Recently, molecules have been synthesized that can catalyze covalent bond-making reactions by forming a noncovalently bonded superstructure, a maneuver that converts an intermolecular reaction into an intramolecular one. In general, in such systems, two reaction channels exist: an uncatalyzed bimolecular reaction and a catalyzed ternary reaction. When the reaction product can serve as a template for its own formation and is capable of conserving and expressing its structure, a self-replicating process results (246–249). For example, the aminoadenosine derivative X is recognized by the pentafluorophenyl ester Y, and forms the complex XY (Fig. 8) (250). This complex is stabilized by the hydrogen bonds between the imide unit of Y and the adenine residue of X, and probably also by the $\pi - \pi$ stacking interactions between the adenine base of X and the naphthalene moiety of Y. This arrangement allows the formation of a *cis*-amide bond between the amino function of X and the activated ester of Y. The cis-conformer of the product Z (*cis*-Z) rapidly isomerizes to the more stable *trans*-amide Z, opening up two new binding sites capable of binding X and Y. The binding of X and Y by Z is analogous to the formation of a ternary complex containing the reactants and the bisubstrate reaction template; in this case, moreover, the product of the reaction is Z. The ternary complex facilitates the reaction of X and Y to form the hydrogen-bonded complex ZZ. This dimer can dissociate to form two separate molecules of Z (typically, the noncovalently bonded dimer ZZ is in thermodynamic equilibrium with free Z). Thus, Z is self-replicating. In these and other systems, competition, reciprocal template effects, and mutation have also been observed (246–248).

Similar to the presumed earliest forms of single-celled life, nano- and microstructures have been created that point to a possible approach to the construction of a synthetic, living cell (251,252). This approach involves the use of self-assembled, self-reproducing vesicles of surfactant molecules that incorporate within their cavities the reactants and catalysts necessary for the self-replication of the molecules bearing the information (in their structure) for the replication of themselves as well as the encircling vesicles. The first steps of these syntheses have been demonstrated. At the present time, fragments of catalytic RNA have been synthesized that can self-replicate as well as evolve *in vitro* (153,253). Synthetic polypeptides have also been designed that can self-replicate (254). Surfactant vesicles have been synthesized that can catalyze the formation of

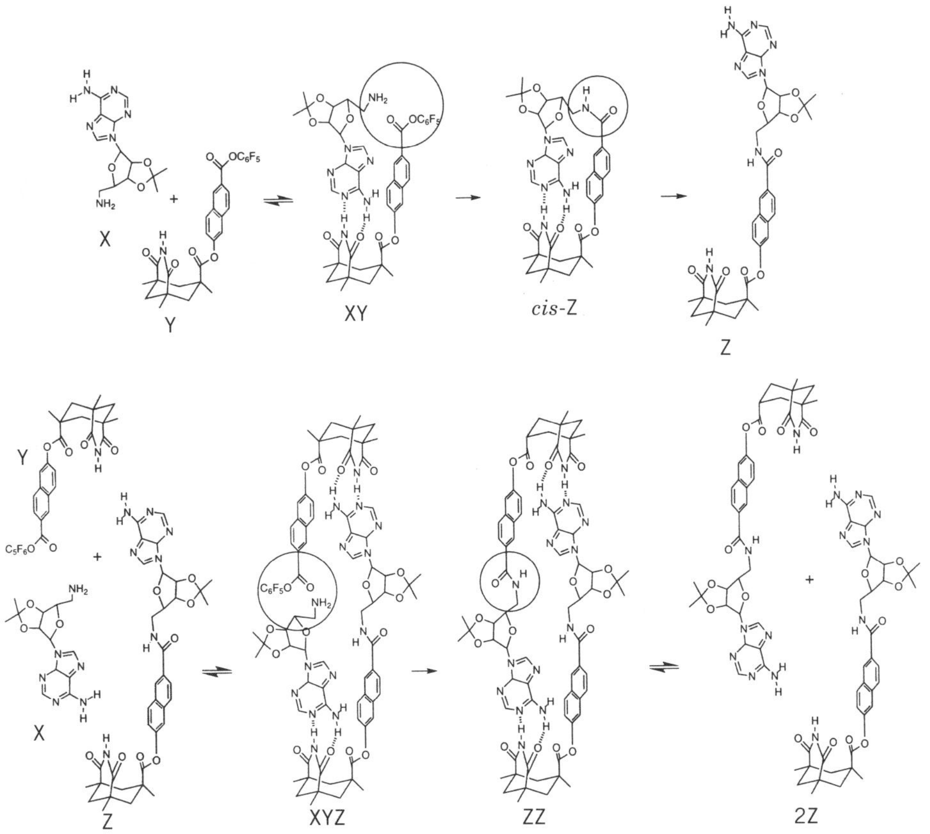

Fig. 8. Replication. The aminoadenosine X and the pentafluorophenyl ester Y form a hydrogen-bonded dimer XY, prior to reaction between the amine and the activated ester groups (shown in the circle). The reaction product is a *cis*-amide conformer *cis*-Z that isomerizes to the more stable *trans*-amide Z. The replicative process is catalyzed by the reaction product Z (also referred to as the template). First, a termolecular complex XYZ is formed from X, Y, and Z. The amide formation reaction (highlighted by the circle) leads to the production of a hydrogen-bonded dimer (ZZ) of the reaction product Z with the template Z. The dimer is in thermodynamic equilibrium with free template in the reaction medium. Adapted from Nowick and co-workers (250).

more surfactant molecules from their precursors and incorporate them into the growing vesicles or form new vesicles (thus earning the term, self-reproducing vesicles) (33,255–258). Recently, it has been shown that enzymatic reactions can be carried out within the vesicles to form the constituents of the vesicle itself. The new vesicles share the reactants of the reaction being carried out within them (similar to cell division) (259).

The studies reviewed above demonstrate that chemists have begun to mimic some of the strategies prevalent in biological systems. The design of intelligent substrates, ie, molecular units with all the information necessary for forming supramolecular structures encoded in their primary structure, has already begun. The production of complex, synthetic, functional nanoscale devices by molecular self-assembly of constituent parts is, thus, an important paradigm in the development of nanotechnology.

Conclusions and Outlook

Nobel-laureate Richard Feynman once said that the principles of physics do not preclude the possibility of maneuvering things atom by atom (260). Recent developments in the fields of physics, chemistry, and biology (briefly described in the previous sections) bear those words out. The invention and development of scanning probe microscopy has enabled the isolation and manipulation of individual atoms and molecules. Research in protein and nucleic acid structure have given rise to powerful tools in the establishment of rational synthetic protocols for the production of new medicinal drugs, sensing elements, catalysts, and electronic materials.

Much of the current progress in nanotechnology is confined to understanding and harnessing pre-existing, exquisitely-evolved nanomachinery, ie, natural living systems. Many of these developments, eg, biotechnolgical synthesis of functional proteins (eg, enzymes, antibodies), have progressed to commercial production. Table 2 summarizes the various reductive and synthetic strategies employed to date. The growing scientific and technical progress in nanotechnology will, in the next few decades, continue to produce many new biotechnology-based as well as biomimetic and nonnatural (ie, synthetic) nanodevices. The manufacture of this broad class of products will play a vital role in the world economy of the twenty-first century and the future of humanity.

Table 2. Some Approaches in Reductive (Top Down) and Synthetic (Bottom Up) Nanofabrication

Approach	Examples	References
reductive (also known as top-down)	conventional microfabrication	
	photolithography	16
	x-ray lithography	85
	E-beam lithography	91
	imprint lithography	186,187
	local probe lithography	38,261–263
synthetic (also known as bottom-up)	nanochemical synthesis	5,35,36,145
	biotechnological expression	13,14,264
	molecular templating	178,180–182,188,265
	molecular self-assembly	2,4,17,73
hybrid (reductive/synthetic)	local probe-assisted synthesis	37,38,56,266,267
	microcontact printing	183,268
	optical–laser manipulation	102,269–271

ACKNOWLEDGEMENT

Support for this work was provided by ONR Multidisciplinary University Research Initiative Grant N00014-95-1-1315, by ONR Grant N00014-95-1-0901, and by NSF Grant HRD-9450475.

BIBLIOGRAPHY

1. P. Ball, *Designing the Molecular World: Chemistry at the Frontier*, Princeton University Press, Princeton, 1994.
2. W. M. Tolles, in G. M. Chow and K. E. Gonsalves, eds., *210th National Meeting of the ACS*, ACS, Chicago, Ill., 1996, pp. 1–15.
3. J. A. Stroscio and D. M. Eigler, *Science* **254**, 1319–1326 (1991).
4. G. M. Whitesides and co-workers, *Acc. Chem. Res.* **28**, 37–44 (1995).
5. G. A. Ozin, *Adv. Mater.* **4**, 612–649 (1992).
6. D. Philp and J. F. Stoddart, *Angew. Chem. Int. Ed. Engl.* **35**, 1154–1196 (1996).
7. S. Asai and Y. Wada, *Proc. IEEE* **85**, 505–520 (1997).
8. K. D. Wise and K. Najfi, *Science* **254**, 1335–1342 (1991).
9. L. Fabrizzi and A. Poggi, *Chem. Soc. Rev.* **24**, 197–202 (1995).
10. K. E. Drexler, *Engines of Creation* Anchor Press/Doubleday, Garden City, 1986, pp. 1–298.
11. M. Krummenacker and J. Lewis, *Prospects in Nanotechnology: Toward Molecular Manufacturing*, John Wiley & Sons, Inc., New York, 1995, p. 297.
12. E. Regis, *Nano: The Emerging Science of Nanotechnology* Little, Brown and Company, Boston, 1995, pp. 1–325.
13. L. Stryer, *Biochemistry*, W. H. Freeman & Co., New York, 1988.
14. M. Edelstein, in M. Krummenacker and J. Lewis, eds., *Prospects in Nanotechnology: Toward Molecular Manufacturing*, John Wiley & Sons, Inc., New York, 1995, pp. 67–91.
15. G. D. Hutcheson and J. D. Hutcheson, *Sci. Am.* **274**, 54–62 (1996).
16. R. F. Service, *Science* **274**, 1834–1836 (1996).
17. G. M. Whitesides, J. P. Mathias, and C. T. Seto, *Science* **254**, 1312–1319 (1991).
18. B. D. Ratner, *J. Biomed. Mater. Res.* **27**, 837–850 (1993).
19. S. M. Block, *Nature* **386**, 217–219 (1997).
20. R. R. Birge, *Computer* **25**, 56–67 (1992).
21. I. Petersen, *Sci. News* **150**, 26 (1996).
22. J.-M. Lehn, *Angew. Chem. Intl. Ed. Engl.* **29**, 1304–1309 (1990).
23. P. Hogeweg, *Ber. Bunsenges, Phys. Chem.* **98**, 1135–1139 (1994).
24. F. Toda, *Bioorg. Chem.* **19**, 157–168 (1991).
25. K. Johnson, R. K. Allemann, H. Widmer, and S. A. Benner, *Nature* **365**, 530–532 (1993).
26. S. Borman, *C&EN* **75**, 8–9 (1997).
27. M. S. Goodman, V. Jubian, B. Linton, and A. D. Hamilton, *J. Am. Chem. Soc.* **117**, 11610–11611 (1995).
28. E. M. Gordon, M. A. Gallop, and D. V. Patel, *Acc. Chem. Res.* **29**, 144–154 (1996).
29. D. R. Burton, *Acc. Chem. Res.* **26**, 405–411 (1993).
30. J. C. Hogan, *Nature Biotech.* **15**, 328–330 (1997).
31. B. A. Cornell and co-workers, *Nature* **387**, 580–583 (1997).
32. A. Ulman, *Chem. Rev.* **96**, 1533–1554 (1996).
33. S. A. Walker, M. Kennedy, and J. A. Zasadzinski, *Biophys. J.* **72**, MP248 (1997).
34. P. Calvert and P. Rieke, *Chem. Mater.* **8**, 1715–1727 (1996).
35. R. W. Armstrong and co-workers, *J. Am. Chem. Soc.* **111**, 7525–7530 (1989).
36. R. B. Woodward, *Pure Appl. Chem.* **33**, 145–177 (1973).
37. P. Avouris, *Acc. Chem. Res.* **28**, 95–102 (1995).
38. P. M. Campbell, E. S. Snow, and P. J. McMarr, *Surf. Sci.* **361/362**, 870–873 (1996).
39. E. S. Snow, P. M. Campbell, and F. K. Perkins, *Proc. IEEE* **85**, 601–611 (1997).
40. M. Pique, in M. Krummenacker and J. Lewis, eds., *Prospects in Nanotechnology: Toward Molecular Manufacturing*, John Wiley & Sons, Inc., New York, 1995, pp. 93–112.

the nitrile. For example, cyanomethane (methyl cyanide) is named acetonitrile [75-05-8] because hydrolysis of its cyano group yields acetic acid. Nitriles which contain additional functional groups are typically named as cyano-substituted compounds, (eg, cyanoacetic acid). Nitriles which contain a hydroxy (-OH) group on the carbon atom that is bonded to the cyano moiety are known as cyanohydrins (qv). According to *Chemical Abstracts*, aliphatic nitriles are named as derivatives of the longest carbon chain and the carbon of the nitrile is included.

General Preparations and Chemical Properties

Nitriles may be prepared by several methods (1). The first nitrile to be prepared was propionitrile, which was obtained in 1834 by distilling barium ethyl sulfate with potassium cyanide. This is a general preparation of nitriles from sulfonate salts and is referred to as the Pelouze reaction (2). Although not commonly practiced today, dehydration of amides has been widely used to produce nitriles and was the first commercial synthesis of a nitrile. The reaction of alkyl halides with sodium cyanide to produce nitriles (eq. 1) also is a general reaction with wide applicability:

$$RX + NaCN \longrightarrow RCN + NaX \tag{1}$$

where X = Cl, Br, or I. If dimethyl sulfoxide is used as solvent, high yields of nitriles can be obtained with both primary and secondary alkyl chlorides (see SULFOXIDES). This method also may be used for preparing dinitriles. Reaction times usually are less than one hour (3).

Ammoxidation, a vapor-phase reaction of hydrocarbon with ammonia and oxygen (air) (eq. 2), can be used to produce hydrogen cyanide (HCN), acrylonitrile, acetonitrile (as a by-product of acrylonitrile manufacture), methacrylonitrile, benzonitrile, and toluinitriles from methane, propylene, butylene, toluene, and xylenes, respectively (4).

$$RCH_3 + NH_3 + O_2 \xrightarrow{\text{catalyst}} RCN + H_2O \tag{2}$$

Processes have been developed whereby the oxygen is supplied from the crystal lattice of a metal-oxide catalyst (5) (see ACRYLONITRILE; METHACRYLIC ACID AND DERIVATIVES).

Addition of HCN to unsaturated compounds is often the easiest and most economical method of making organonitriles. An early synthesis of acrylonitrile involved the addition of HCN to acetylene. The addition of HCN to aldehydes and ketones is readily accomplished with simple base catalysis, as is the addition of HCN to activated olefins (Michael addition). However, the addition of HCN to unactivated olefins and the regioselective addition to dienes is best accomplished with a transition metal catalyst, as illustrated by DuPont's adiponitrile process (6–9).

Chemistry and Uses of Nitriles

As a class of compounds, nitriles have broad commercial utility that includes their use as solvents, feedstocks, pharmaceuticals, catalysts, and pesticides. The versatile reactivity of organonitriles arises both from the reactivity of the C≡N bond, and from the ability of the cyano substituent to activate adjacent bonds, especially C—H bonds. Nitriles can be used to prepare amines, amides, amidines, carboxylic acids and esters, aldehydes, ketones, large-ring cyclic ketones, imines, heterocycles, orthoesters, and other compounds. Some of the more common transformations involve hydrolysis or alcoholysis to produce amides, acids and esters, and hydrogenation to produce amines, which are intermediates for the production of polyurethanes and polyamides. An extensive review on hydrogenation of nitriles has been recently published (10).

Acrylonitrile [107-13-1] (qv), the largest volume organonitrile, is an important monomer both for plastics and synthetic fibers. Acetonitrile, a by-product of acrylonitrile manufacture, is commercially important for solvent extraction, reaction media, and as an intermediate in the preparation of pharmaceuticals (qv) and other organic chemicals (see EXTRACTION, LIQUID-LIQUID EXTRACTION). Propionitrile [107-12-03], a by-product of the electrodimerization of acrylonitrile to adiponitrile, is used as a chemical intermediate. Hydrogenation of organonitriles to amines provides important intermediates both for polyurethanes (by way of isocyanates) and polyamides (nylons); adiponitrile is used almost exclusively by the manufacturers in the production of 1,6-diaminohexane (hexamethylenediamine), an intermediate for nylon 6,6. Other nitriles that are produced in thousands of metric tons per year include acetone cyanohydrin, 2-amino-2-methylpropionitrile, and fatty acid nitriles. Acetone cyanohydrin is an intermediate for the preparation of methyl methacrylate and arcylic resins, (eg, lucite and plexiglas) and for 5,5-dimethylhydantoin, which is used to make commercial water treatment chemicals. 2-Amino-2-methylpropionitrile is an intermediate for the preparation of azobis(isobutyronitrile), which is a widely used polymerization initiator, (eg, Vazo 64) and in the production of some agrichemicals. Other aminonitriles are unisolated intermediates in the production of chelants such as ethylenediaminetetraacetate (EDTA) and nitrilotriacetate (NTA). The fatty acid nitriles are intermediates in the production of a large variety of commercial amines and amides. The nitriles that are commercially available in the United States (1994) and their manufacturers are listed in Table 1 (11). The U.S. Department of Commerce estimates that more than 2.5 million metric tons of nitrile compounds were produced in the United States in 1994. Global production of nitriles has increased significantly in the last twenty years. Excellent resources for identifying producers of organonitriles worldwide include Chem Sources International and SRI International's *Directory of Chemical Producers.*

General Health and Safety Factors

As a class of compounds, the two main toxicity concerns for nitriles are acute lethality and osteolathyrsm. A comprehensive review of the toxicity of nitriles, including detailed discussion of biochemical mechanisms of toxicity and structure-activity relationships, has recently become available (12). Nitriles vary broadly

Table 2. Some Physical Properties of Acetonitrile

Property	Value
CAS Registry Number	[75-05-8]
mol wt	41.05
bp (at 101.3 kPa = 1 atm), °C	81.6
freezing pt, °C	−45.7
density (at 20°C), g/cm^3	0.786
n_D^{20}	1.3441
viscosity (at 20°C), mPa·s(=cP)	0.35
heat of vaporization (at 80°C), J/kg^a	72.7×10^4
heat of fusion (at −45.7°C), J/kg^a	21.8×10^4
heat of combustion (at 25°C), J/kg^a	31.03×10^6
heat capacity (liquid at 20°C), J/(kg·K)^a	22.59×10^2
surface tension, mN/m (dyn/cm)	29.3
coefficient of expansion (at 20°C per °C)	1.37×10^{-3}
specific conductance (at 25°C), S	$(5-9) \times 10^{-8}$
dipole moment, C·m^b	10.675×10^{-30}
dielectric constant	
at 0°C	42.0
at 20°C	38.8
at 81.6°C	26.2
evaporation rate (butyl acetate = 100)	579
flash pt (COC^c), °C	6
flammable limits (in air), vol %	
lower	4.4
upper	16.0

^{a}To convert J to cal, divide by 4.184.
^{b}To convert C·m to D, divide by 3.336×10^{-30}.
^{c}COC = Cleveland open cup.

Table 3. Acetonitrile Binary Azeotropes^a

Component	Bp (at 101.3 kPa = 1 atm)	Acetonitrile, wt %
benzene	73	34
carbon tetrachoride	65	17
1,2-dichloroethane	79	79
ethanol	73	44
ethyl acetate	75	23
methanol	63	81
water	77	84

^{a}Ref. 14.

heterocycles, eg, pyridines (23), and imidazolines (24). It can be trimerized to s-trimethyltriazine (25) and has been telomerized with ethylene (26) and copolymerized with α-epoxides (27).

Most, if not all, of the acetonitrile that was produced commercially in the United States in 1995 was isolated as a by-product from the manufacture of acrylonitrile by propylene ammoxidation. The amount of acetonitrile produced in an acrylonitrile plant depends on the ammoxidation catalyst that is used, but the ratio of acetonitrile:acrylonitrile usually is ca 2–3:100. The acetonitrile is recovered as the water azeotrope, dried, and purified by distillation (28). U.S. capacity (1994) is ca 23,000 t/yr.

Specifications for commercial acetonitrile are given in Table 4. The principal organic impurity in commercial acetonitrile is propionitrile; although small amounts of allyl alcohol also may be present.

Shipping and Storage. The DOT/IMO classification for acetonitrile is class 3 hazard, UN No. 1648. It requires a FLAMMABLE LIQUID label on all containers and is in packing group II. For storage and piping at normal temperatures and pressures, aluminum or carbon steel may be used. Centrifugal pumps are preferred because the solvency of acetonitrile may affect the lubricant in positive-displacement pumps. All tanks, piping valves, and pumps should be electrically grounded. Fire protection equipment call be water spray, alcohol foam, CO_2, or dry chemical.

Health and Safety Factors. The following toxicities for acetonitrile have been reported: oral LD_{50} (rats), 3030–6500 mg/kg; skin LD_{50} (rabbits), 3884–7850 mg/kg; and inhalation LC_{50} (rats), 7500–17,000 ppm (29). Humans can detect the odor of acetonitrile at 40 ppm. Exposure for 4 h at up to 80 ppm has not produced adverse effects. However, exposure for 4 h at 160 ppm results in reddening of the face and some temporary bronchial tightness.

Although acetonitrile has a low order of acute toxicity by ingestion, inhalation, and skin absorption, it can cause severe eye burns. In case of eye contact, eyes should be immediately flushed with water for at least 15 min and a physician should be consulted. In the event of a spill or leak, the spill should be contained, flooded with water, and disposed of according to local regulations. Acetonitrile is flammable (see Table 2) and must be kept away from excessive heat, sparks, and open flame. Associated fires can be extinguished using water spray, alcohol foam, CO_2, or dry chemical extinguishers. OSHA requires that an employee's exposure to acetonitrile in any 8-h shift does not exceed a time-weighted average of 40 ppm (70 mg/m^3) in air (30).

Uses. Because of its good solvency and relatively low boiling point, acetonitrile is used widely as a recoverable reaction medium, particularly for the preparation of pharmaceuticals. Its largest use is for the separation of butadiene from C_4 hydrocarbons by extractive distillation (see AZEOTROPIC AND EXTRACTIVE

Table 4. Specifications for Commercial Acetonitrile

Specification	Value
sp gr, at 20°C	0.783–0.787
distillation range, °C	
initial min	80.5
end pt, max	82.5
purity (min), wt %	99.0
acidity (max), wt %	0.05
copper (max), ppm	0.5
iron (max), ppm	0.5
water (max), wt %	0.3
color (max), Pt–Co	15

DISTILLATION) (31). It also has been proposed for the separation of other olefins, eg, propylene, isoprene, allene, and methylacetylene from hydrocarbon streams (32–34). It is a superior solvent for polymers and can be used as a solvent for spinning fibers and for casting and molding plastics. It is used widely in spectrophotometry and electrochemistry. Since pure acetonitrile does not absorb uv light, it is commonly used as a solvent in high pressure liquid chromatography (hplc) for the detection of materials, eg, residual pesticides, in the ppb range; highly purified hplc grade acetonitrile is routinely available from suppliers.

Acetonitrile also is used as a catalyst and as an ingredient in transition-metal complex catalysts (35,36). There are many uses for it in the photographic industry and for the extraction and refining of copper and by-product ammonium sulfate (37–39). It also is used for dyeing textiles and in coating compositions (40,41). It is an effective stabilizer for chlorinated solvents, particularly in the presence of aluminum, and it has some application in the manufacture of perfumes (qv) (42,43). It also is used as a reagent for the preparation of a wide variety of compounds.

Adiponitrile

Adiponitrile (hexanedinitrile, dicyanobutane, ADN), $NC(CH_2)_4CN$, is manufactured principally for use as an intermediate for hexamethylenediamine (1,6-diaminohexane), which is a principal ingredient for nylon-6,6. However, in 1996, BASF announced the development of a process to make caprolactam from adiponitrile (44,45). Caprolactam is used to produce nylon-6. The implementation of this technology could increase the demand for adiponitrile dramatically.

Pure adiponitrile is a colorless liquid and has no distinctive odor; some properties are shown in Table 5. It is soluble in methanol, ethanol, chloroalkanes, and aromatics but has low solubility in carbon disulfide, ethyl ether, and aliphatic hydrocarbons. At 20°C, the solubility of adiponitrile in water is ca 8 wt %; the solubility increases to 35 wt % at 100°C. At 20°C, adiponitrile dissolves ca 5 wt % water.

Adiponitrile undergoes the typical nitrile reactions, eg, hydrolysis to adipamide and adipic acid and alcoholysis to substituted amides and esters. The most important industrial reaction is the catalytic hydrogenation to hexamethylenediamine. A variety of catalysts are used for this reduction including cobalt–nickel (46), cobalt manganese (47), cobalt boride (48), copper cobalt (49), and iron oxide (50), and Raney nickel (51). An extensive review on the hydrogenation of nitriles has been recently published (10).

Adiponitrile is made commercially by several different processes utilizing different feedstocks. The original process, utilizing adipic acid (qv) as a feedstock, was first commercialized by DuPont in the late 1930s and was the basis for a number of adiponitrile plants. However, the adipic acid process was abandoned by DuPont in favor of two processes based on butadiene (qv). During the 1960s, Monsanto and Asahi developed routes to adiponitrile by the electrodimerization of acrylonitrile (qv).

The reaction of adipic acid with ammonia in either liquid or vapor phase produces adipamide as an intermediate which is subsequently dehydrated to adiponitrile. The most widely used catalysts are based on phosphorus-containing

Table 5. Some Physical Properties of Adiponitrile

Property	Value
CAS Registry Number	[111-69-33]
mol wt	108.14
bp, °C	
at 101.7 kPa[a]	295
at 1.3 kPa[a]	154
freezing pt, °C	2.49
density (at 20°C), g/cm^3	0.965
n_D^{20}	1.4343
viscosity, mPa·s (=cP)	
at 20°C	9.1
at 70°C	2.6
heat of vaporization, J/kg[b]	70.4×10^4
heat of fusion (at 1°C), J/kg[b]	21.3×10^4
heat of combustion, J/kg[b]	40.4×10^6
critical temperature, °C	507
critical pressure, MPa[c]	2.8
specific conductance, S	3.5×10^{-8}
flash pt (closed cup), °C	159
autoignition temperature, °C	550
flammable limits (in air), vol %	
lower	1.7
upper	5.0

[a]To convert kPa to mm Hg, multiply by 7.5.
[b]To convert J to cal, divide by 4.184.
[c]To convert MPa to atm, divide by 0.101.

compounds, but boron compounds and silica gel also have been patented for this use (52–56). Vapor-phase processes involve the use of fixed catalyst beds; whereas, in liquid–gas processes, the catalyst is added to the feed. The reaction temperature of the liquid-phase processes is ca 300°C and most vapor-phase processes run at 350–400°C. Both operate at atmospheric pressure. Yields of adipic acid to adiponitrile are as high as 95% (57).

In the now-obsolete furfural process, furfural was decarboxylated to furan which was then hydrogenated to tetrahydrofuran (THF). Reaction of THF with hydrogen chloride produced dichlorobutene. Adiponitrile was produced by the reaction of sodium cyanide with the dichlorobutene. The overall yield from furfural to adiponitrile was around 75% (see FURAN DERIVATIVES).

In a related process, 1,4-dichlorobutene was produced by direct vapor-phase chlorination of butadiene at 160–250°C. The 1,4-dichlorobutenes reacted with aqueous sodium cyanide in the presence of copper catalysts to produce the isomeric 1,4-dicyanobutenes; yields were as high as 95% (58). The by-product NaCl could be recovered for reconversion to Na and Cl$_2$ via electrolysis. Adiponitrile was produced by the hydrogenation of the dicyanobutenes over a palladium catalyst in either the vapor phase or the liquid phase (59,60). The yield in either case was 95% or better. This process is no longer practiced by DuPont in favor of the more economically attractive process described below.

Dupont currently practices a butadiene-to-adiponitrile route based on direct addition of HCN to butadiene (6–9). It was first commercialized in 1971. All reactions are catalyzed by soluble, air and moisture sensitive, triarylphosphite-nickel(0) complexes.

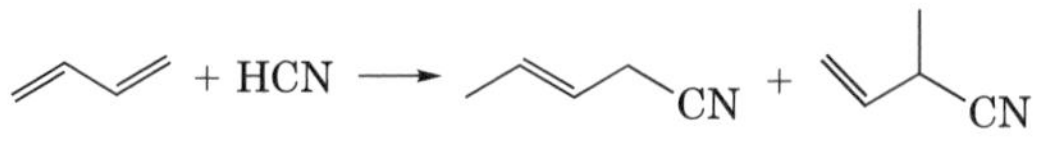

$$\text{(3)}$$

3PN   2M3BN

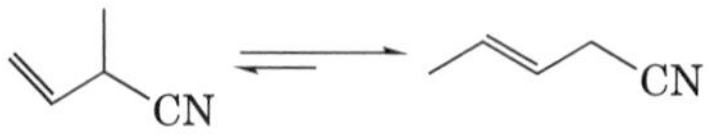

$$\text{(4)}$$

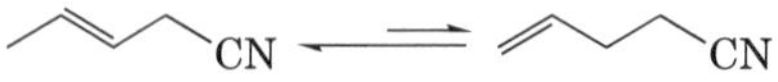

$$\text{(5)}$$

4PN

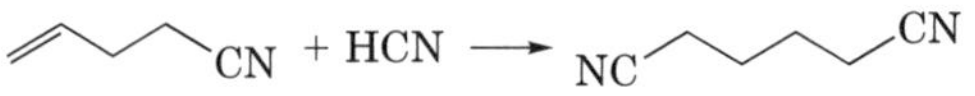

$$\text{(6)}$$

ADN

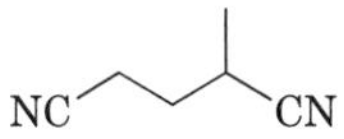

MGN 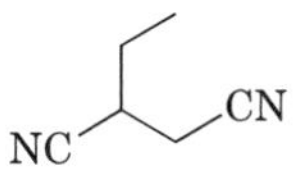   ESN 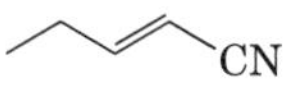   2PN

The first HCN addition (eq. 3) occurs at practical rates above 70°C under sufficient pressure to keep butadiene condensed in solution and produces the 1,4- and 1,2-addition products (3-pentenenitrile [4635-87-4], 3PN, and 2-methyl-3-butenenitrile [16529-56-9], 2M3BN) in a 2 to 1 ratio. Fortunately thermodynamics favors 3PN (about 20:1) and 2M3BN may be isomerized to 3PN (eq. 4) in the presence of a nickel catalyst.

The selective addition of the second HCN to provide ADN requires the concurrent isomerization of 3PN to 4-pentenenitrile [592-51-8], 4PN (eq. 5), and HCN addition to 4PN (eq. 6). A Lewis acid promoter is added to control selectivity and increase rate in these latter steps. Temperatures in the second addition are significantly lower and practical rates may be achieved above 20°C at atmospheric pressure. A key to the success of this homogeneous catalytic process is the ability to recover the nickel catalyst from product mixture by extraction with a hydrocarbon solvent. 2-Methylglutaronitrile [4553-62-2], MGN, ethylsuccinonitrile [17611-82-4], ESN, and 2-pentenenitrile [25899-50-7], 2PN, are by-products of this process and are separated from adiponitrile by distillation.

Recent patent activity suggests that DuPont is developing a new generation of chelating-diphosphite-nickel catalysts for this technology which are significantly more active than the monodentate phosphite based catalyst system used for the last two decades (61–64).

The Monsanto adiponitrile process, first commercialized in 1965 (65–67), involves the dimerization of acrylonitrile at the cathode in an electrolytic cell (eq. 7):

$$2\ CH_2{=}CHCN + 2\ H^+ + 2\ e \longrightarrow NC(CH_2)_4CN \qquad \text{(7)}$$

Small amounts of propionitrile and bis(cyanoethyl) ether are formed as by-products. The hydrogen ions are formed from water at the anode and pass to the cathode through a membrane. The catholyte that is continuously recirculated in the cell consists of a mixture of acrylonitrile, water, and a tetraalkylammonium salt; the anolyte is recirculated aqueous sulfuric acid. A quantity of catholyte is continuously removed for recovery of adiponitrile and unreacted acrylonitrile; the latter is fed back to the catholyte with fresh acrylonitrile. Oxygen that is produced at the anodes is vented and water is added to the circulating anolyte to replace the water that is lost through electrolysis. The operating temperature of the cell is ca 50–60°C. Current densities are 0.25–1.5 A/cm^2 (see ELECTROCHEMICAL PROCESSING).

A typical composition of the catholyte is adiponitrile, 15 wt %; acrylonitrile, 15 wt %; quaternary ammonium salt, 39 wt %; water, 29 wt %; and by-products, 2 wt %. Such a solution is extracted with acrylonitrile and water, which separates the organics from the salt that can be returned to the cell. The acrylonitrile is distilled from the extract and the resultant residue consists of ca 91 wt % adiponitrile, which is purified further by distillation. The overall yield of acrylonitrile to adiponitrile is 92–95%.

Production and Shipment. Estimated adiponitrile production capacities in the U.S. in 1992 were about 625 thousand metric tons and worldwide capacity was in excess of 10^6 metric tons. The DOT/IMO classification for adiponitrile is class 6.1 hazard, UN No. 2205. It requires a POISON label on all containers and is in packing group III. Approved materials of construction for shipping, storage, and associated transportation equipment are carbon steel and type 316 stainless steel. Either centrifugal or positive displacement pumps may be used. Carbon dioxide or chemical-foam fire extinguishers should be used. There are no specifications for commercial adiponitrile. The typical composition is 99.5 wt % adiponitrile. Impurities that may be present depend on the method of manufacture, and thus, vary depending on the source.

Health and Safety Factors. See "General Health and Safety Factors." The following toxicities for adiponitrile have been reported: oral LD$_{50}$ (rats), 300 mg/kg; dermal LD$_{50}$ (rabbits), 2,134 mg/kg; and inhalation 4-h LC$_{50}$ (rats), 1.7 mg. NIOSH has proposed an exposure limit of 4 ppm as a TWA (68).

Uses. The principal use of adiponitrile is for hydrogenation to hexamethylene diamine leading to nylon 6,6. However, as a result of BASF's new adiponitrile-to-caprolactam process, a significant fraction of ADN produced may find its way into nylon-6 production. Adipoquanamine, which is prepared by the reaction of adiponitrile with dicyandiamide [461-58-5] (cyanoguanidine), may have uses in melamine–urea amino resins (qv) (see "Benzonitrile, Uses"). Its typical liquid nitrile properties suggest its use as an extractant for aromatic hydrocarbons.

α-Aminonitriles

α-Aminonitriles are compounds containing both cyano and amine substituents attached to the same carbon atom. They are versatile synthetic intermediates that are used to make aminoacids, agrichemicals, chelants, radical initiators, and water-treatment chemicals. In some cases, aminonitriles produced as intermediates are not isolated, but immediately further reacted, for example by hydrolysis, as is the case in producing ethylenediaminetetraacetate (EDTA) or nitrilotriacetate (NTA). Isolated and commercially available aminonitriles include 2-amino-2-methylpropanenitrile (aminoisobutyronitrile, AN-64) [19355-69-2], 2-amino-2-methylbutanenitrile (AN-67) [4475-95-0], 2-amino-2,4-dimethylpentanenitrile (AN-52) [26842-43-3], and 1-aminocyclohexane carbonitrile (AN-88) [5496-10-6]. The designation in parentheses arise from their identity as intermediates in the production of azo radical initiators (see below).

Historically these compounds have been made in two-step processes. For smaller volumes, reaction of an appropriate ketone or aldehyde with a cyanide salt followed by treatment with an ammonium salt proves satisfactory (Strecker synthesis). For larger volumes, treatment of the ketone or aldehyde with HCN to produce a cyanohydrin, followed by treatment with ammonia has been practiced. However, in 1990, DuPont began practicing a new one-step process (69) in which the a ketone is treated simultaneously with both HCN and ammonia at 40–60°C. The new process, based on the understanding that a cyanohydrin is not on the reaction pathway (Fig. 1), is both faster and more selective. The uncatalyzed reactions are driven to completion by the presence of excess ammonia; water is a by-product. If excess ammonia is removed without separating the by-product water, the reaction may be reversed to produce the reagents. Therefore, these products are often stored and shipped in the presence of excess ammonia.

Physical Properties. α-Aminonitriles are stable at modest temperatures ($<70°C$) in the absence of water; in the presence of water, they can degrade to their original constituents, ie, ketone (aldehyde), ammonia and hydrogen cyanide if insufficient ammonia is present. For this reason they are frequently stored in the presence of excess ammonia. Even in the presence of ammonia, aminonitriles begin to degrade at temperatures above 70°C, depending on the pressure of ammonia. The aminonitriles based on ketones described here are clear colorless liquids, but sometimes appear yellow to brown depending on the synthetic procedure and the amount of decomposition. They are soluble in polar organic solvents and in aromatic solvents. AN-64, AN-67, and AN-88 are soluble in water; AN-52 is insoluble in water. They have an ammonialike odor. Vapor pressure of pure aminonitriles are AN-64, 4 kPa (30 torr) at 66°C; AN-67, 1.9 kPa (14 torr) at 68°C; AN-52, 66 Pa (0.5 torr) at 70°C and 267 Pa (2 torr) at 84°C; AN-88, 200–400 Pa (1.5–3 torr) at 82°C. Specific gravity is about 0.9 for AN-64, -67, and -52, and 1.03 for AN-88.

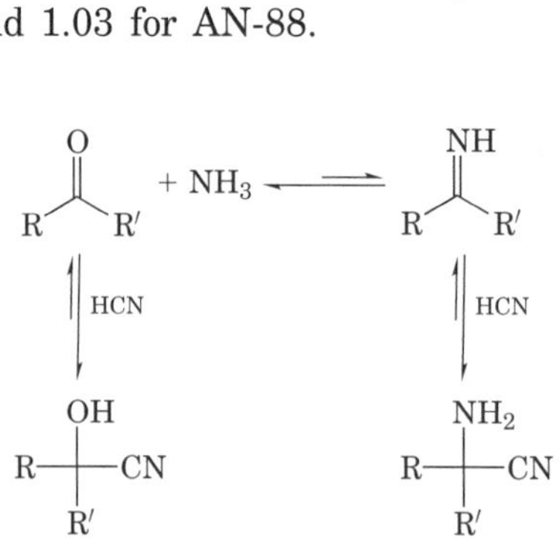

Fig. 1.   Reactive pathway for α-aminonitriles synthesis.

Shipment. The DOT/IMO shipping information is shown in Table 6. Approved materials of construction for shipping, storage, and associated transportation equipment are lined carbon steel (DOT spec. 105 S 500W) and type 316 stainless steel. Water spray, carbon dioxide, chemical-foam, or dry-chemical fire extinguishers may be used.

Health and Safety Factors. See "General Health and Safety Factors." As a class, alpha-aminonitriles are exceedingly acutely lethal from oral, inhalation, dermal, and occular exposure. Some are known to cause osteolathyritic effects in experimental animals. The following toxicology data have been reported for the aminonitriles reported here. AN-64: oral LD_{50} (rats) 45 mg/kg, inhalation LC_{50} (rats) 112 ppm; AN-67: inhalation LC_{50} (rats) 111 ppm; AN-52: inhalation LC_{50} (rats) 100 ppm, dermal LD_{50} (rabbits) 90 mg/kg; AN-88: oral LD_{50} (rats) 200 mg/kg, inhalation LC_{50} (rats) 161 ppm, dermal LD_{50} (rabbits) <200 mg/kg. Additional data for these and other aminonitriles are available (12). These compounds are frequently stabilized with excess ammonia and thermal decomposition leads to the evolution of both ammonia and hydrogen cyanide. Accordingly, first-aid treatment should be consistent with both ammonia and cyanide exposure. In case of ingestion, drink two glasses of water and induce vomiting and call a physician (DO NOT give Syrup of Ipecac). Spills should be contained and treated with dry sodium bicarbonate ($NaHCO_3$) to absorb the spilled aminonitrile and make a dry solid at a ratio of three (3) pounds of $NaHCO_3$ per pound of aminonitrile. Transfer aminonitrile/$NaHCO_3$ solids to plastic or metal drums for disposal. Firefighters may use water, carbon dioxide, dry-chemical or chemical foam extinguishers and should wear self-contained breathing apparatus.

Uses. α-Aminonitriles may be hydrolyzed to aminoacids, such as is done in producing ethylenediaminetetracetate (EDTA) or nitrilotriacetate (NTA). In these cases, formaldehyde is utilized in place of a ketone in the synthesis. The principal use of the ketone-based aminonitriles described above is in the production of azobisnitrile radical initiators (see below). AN-64 is also used as an intermediate in the synthesis of the herbicide Bladex. Aminonitriles are also excellent intermediates for the synthesis of substituted hydantoins by reaction with carbon dioxide; however, this is not currently commercially practiced.

Table 6. Aminonitrile Shipping Information

DOT/IMO	AN-64[a]	AN-67[a]	AN-52[a]	AN-88[a]
hazard class	6.1	6.1	6.1	3
subs. hazard class	3	3	3	6.1
UN No.	2929	2929	2929	1992
label	POISON, FLAMMABLE LIQUID			
packing group	I	I	II	II

[a]Shipping name:
  AN-64, Toxic Liquids, Flammable, Organic, N.O.S. (2-amino-2-methylpropanenitrile).
  AN-67, Toxic Liquids, Flammable, Organic, N.O.S. (2-amino-2-methylbutanenitrile).
  AN-52, Toxic Liquids, Flammable, Organic, N.O.S. (2-amino-2,4-dimethylpentanenitrile).
  AN-88, Flammable Liquids, Toxic, N.O.S. (1-amino cyclohexane carbonitrile).

Azobisnitriles

Azobisnitriles are efficient sources of free radicals for vinyl polymerizations and chain reactions, eg, chlorinations (see INITIATORS). These compounds decompose in a variety of solvents at nearly first-order rates to give free radicals with no evidence of induced chain decomposition. They can be used in bulk, solution, and suspension polymerizations, and because no oxygenated residues are produced, they are suitable for use in pigmented or dyed systems that may be susceptible to oxidative degradation.

DuPont (D) and/or Wako (W) produce several members of this class of compounds: 2,2'-azobis(isobutyronitrile) [78-67-1] (Vazo 64 (D); V-60 (W)); 2,2'-azobis(2-methylbutanenitrile) [13472-08-7] (Vazo 67 (D); V-59 (W)); 2,2'-azobis(2,4-dimethylpentanenitrile) [4419-11-8] [Vazo 52 (D); V-65 (W)]; 1,1'-azobis(cyanocyclohexane) [2094-98-6] [Vazo 88 (D)]; and 2,2'-azobis(4-methoxy-2,4-dimethylpentanenitrile [15545-97-8] [V-70 (W)]. They are crystalline solids that are produced by hypochlorite oxidation of α-aminonitriles (70). Physical properties are listed in Table 7 (71,72).

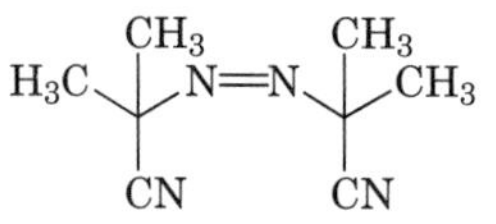

2,2'-Azobis(isobutyronitrile)

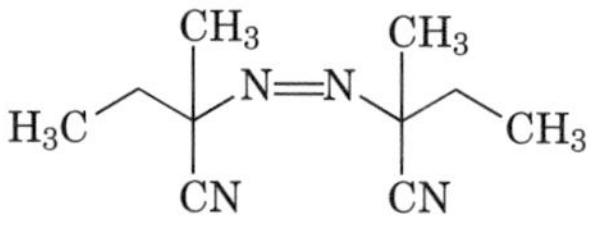

2,2'-Azobis(2-methylbutanenitrile)

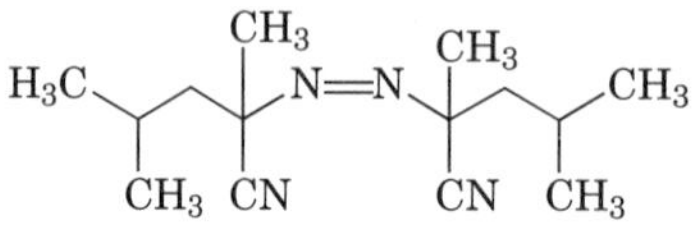

2,2'-Azobis(2,4-dimethylpentanenitrile)

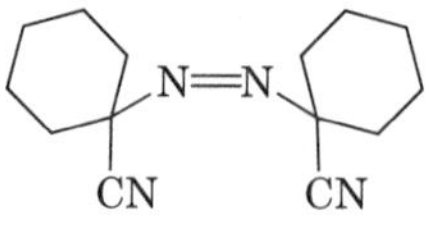

1,1'-Azobis(cyanocyclohexane)

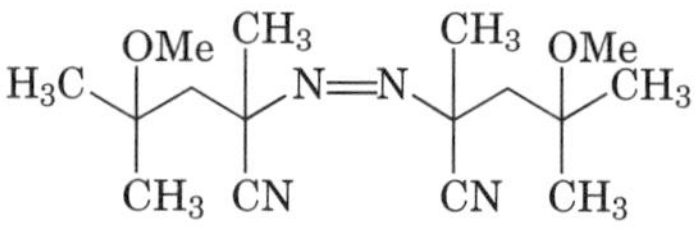

2,2'-Azobis(4-methoxy-2,4-dimethylpentanenitrile)

These compounds are essentially insoluble in water, sparingly soluble in aliphatic hydrocarbons, and soluble in functional compounds and aromatic hydrocarbons.

In solution, the azobisnitriles decompose on heating to form two free radicals with the liberation of nitrogen (eq. 8):

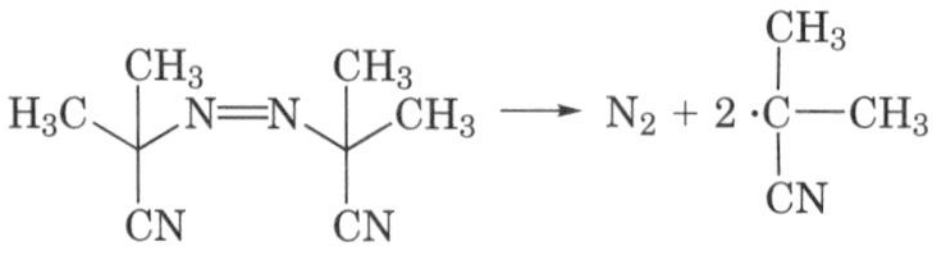

$$\text{(CH}_3\text{)}_2\text{C(CN)-N=N-C(CN)(CH}_3\text{)}_2 \longrightarrow N_2 + 2\, \cdot\text{C(CH}_3\text{)}_2\text{CN} \tag{8}$$

Table 7. Properties of Azobisnitriles

Property	V-70[b]	Vazo 52[a] V-65[b]	Vazo 64[a] V-60[b]	Vazo 67[a] V-59[b]	Vazo 88[a]
CAS Registry Number	[15545-97-8]	[4419-11-8]	[78-67-1]	[13472-08-7]	[2094-98-6]
mol wt	308.4	248.4	164.2	192.3	244.3
mp, °C	50–96 d	45–70	100–103	48–52	113–115
specific gravity			1.128		
bulk density, kg/m^{3c}		~450	~265	~450	~425
specific heat, kJ/kg·K^d		~1.7	~1.7	~1.7	~1.7
heat of combustion, kJ/kg·mol^d			5.05		
10-h half-life decomp temp, °C^e	30	52	64	67	88
half-life ($t_{1/2}$), min $\log(t_{1/2}) = A(1/T) - B$					
A		6767	7142	7492	7660
B		18.04	18.36	19.22	18.39
max storage temp, °C		10	24	24	35

See text for structures and chemical names.
^{a}Registered trademark of DuPont.
^{b}Wako name.
^{c}In form of white noodles.
^{d}To convert J to cal, divide by 4.184.
^{e}In toluene.

Because the decomposition is first order, the rate of free-radical formation can be controlled by regulating the temperature; equations relating half-life to temperature are provided in Table 7. These decomposition rates are essentially independent of the solvent (73).

Shipping and Storage. DOT/IMO shipping information for Vazo products are provided in Table 8. Vazo polymerization initiators must be stored out of the sun and away from heat in a cool, dry place. Since decomposition produces nitrogen and, consequentially, a pressure increase, these compounds should not be stored in glass or in any tightly closed containers other than the original shipping container. The maximum storage temperatures are provided in Table 7.

Table 8. Azobisnitrile Shipping Information

DOT/IMO	Vazo 64[a]	Vazo 67[a]	Vazo 52[a]	Vazo 88[a]
hazard class	4.1	4.1	4.1	4.1
UN No.	3234	3236	3236	3226
label	----------- FLAMMABLE SOLID ------------			
packing group	II	II	II	II

^{a}Shipping name (see text for structure):
  Vazo-64, Self-Reactive Solid Type C, Temperature Controlled (2,2'-azodi(iso-butyronitrile)).
  Vazo-67, Self-Reactive Solid Type D, Temperature Controlled (2,2'-azodi(2-methylbutyronitrile)).
  Vazo-52, Self-Reactive Solid Type D, Temperature Controlled (2,2'-azodi(2,4-dimethylvaleronitrile)).
  Vazo-88, Self-Reactive Solid Type D (1,1'-azodi(hexahydrobenzonitrile)).

two-thirds, compared with the vapor-phase process used by most producers. The process involves the reaction of benzoic acid (or substituted benzoic acid) with urea at 220–240°C in the presence of a metallic catalyst (78).

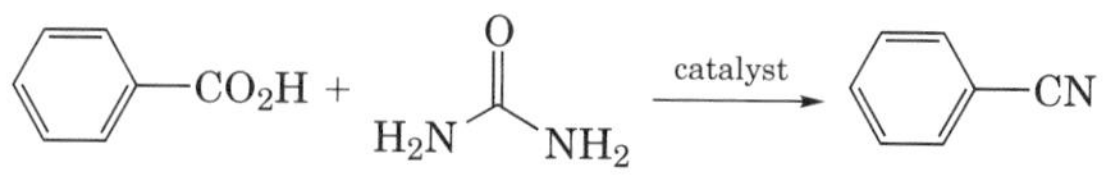

A method for making benzonitrile by dehydrogenation of the Diels-Alder adduct of butadiene and acrylonitrile also has been described (79). Benzonitrile also can be made on a small scale by the dehydration of benzamide in an inert solvent with phosphorus oxychloride or benzenesulfonyl chloride and an organic amine (80,81).

Shipping and Storage. The DOT hazard classification for BENZONI-TRILE is "Flammable", UN No. 2224. Carbon-steel drums and tanks may be used for storage.

Health and Safety Factors. See "General Health and Safety Factors." The following toxicities for benzonitrile have been reported: oral LD_{LO} (rats), 720 mg/kg; skin LD_{50} (rats), 1200 mg/kg; and inhalation LC_{50} (rats), 950 ppm/8 h.

Uses. The most important commercial use for benzonitrile is the synthesis of benzoguanamine, which is a derivative of melamine and is used in protective coatings and molding resins (see AMINO RESINS; CYANAMIDES). Other uses for benzonitrile are as an additive in nickel-plating baths, for separating naphthalene and alkylnaphthalenes from nonaromatics by azeotropic distillation (qv), as a jet-fuel additive, in cotton bleaching baths, as a drying additive for acrylic fibers, and in the removal of titanium tetrachloride and vanadium oxychloride from silicon tetrachloride.

Cyanoacetic Acid and Esters

The physical properties of cyanoacetic acid [372-09-8] and two of its ester derivatives are listed in Table 11 (82). The parent acid is a strong organic acid with a dissociation constant at 25°C of 3.36×10^3. It is prepared by the reaction of chloroacetic acid with sodium cyanide. It is hygroscopic and highly soluble in alcohols and diethyl ether but insoluble in both aromatic and aliphatic hydrocarbons. It undergoes typical nitrile and acid reactions but the presence of the nitrile and the carboxylic acid on the same carbon cause the hydrogens on C-2 to be readily replaced. The resulting malonic acid derivative decarboxylates to a substituted acrylonitrile:

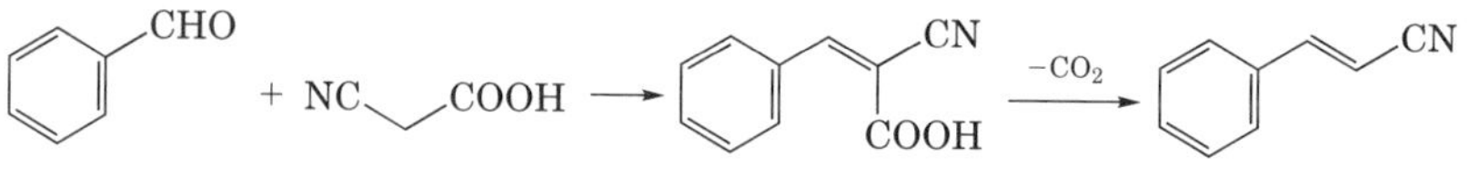

cinnamonitrile
[4360-47-8]

Table 11. Some Physical Properties of Cyanoacetic Acid and Methyl and Ethyl Esters[a]

	Cyanoacetic acid, NCCH$_2$COOH	Methyl ester, NCCH$_2$COOCH$_3$	Ethyl ester, NCCH$_2$COOC$_2$H$_5$
mol wt	85.06	99.09	113.11
bp, °C	108[b]	206	208–210
mp, °C	67		−22.5
n_D^{20}		1.419	1.4177
flash pt, °C	107	110	110

[a]Ref. 70.
[b]At 2 kPa (15 mm Hg).

The methyl and ethyl esters of cyanoacetic acid are slightly soluble in water but are completely miscible in most common organic solvents including aromatic hydrocarbons. The esters, like the parent acid, are highly reactive, particularly in reactions involving the central carbon atom; however, the esters tend not to decarboxylate. They are prepared by esterification of cyanoacetic acid and are used principally as chemical intermediates.

Health and Safety Factors.   The following toxicities have been reported for cyanoacetic acid: oral LD$_{50}$ (rat) 1500 mg/kg; subcutaneous LD$_{LO}$ (rabbit), 1900 mg/kg; and subcutaneous LD$_{LO}$ (frog); 1300 mg/kg (29). For ethyl cyanoacetate the following toxicities have been reported: interperitoneal LD$_{50}$ (mice), 750 mg/kg; subcutaneous LD$_{LO}$ (rabbits), 1500 mg/kg; and subcutaneous LD$_{LO}$ (frogs), 4000 mg/kg.

Uses.   Although cyanoacetic acid can be used in applications requiring strong organic acids, its principal use is in the preparation of malonic esters and other reagents used in the manufacture of pharmaceuticals, eg, barbital, caffeine, and B vitamins (see ALKALOIDS; HYPNOTICS; VITAMINS). Cyanoacetic acid can be used for the preparation of heterocylic ketones.

Isophthalonitrile

Isophthalonitrile [626-17-5] (1,3-dicyanobenzene, IPN), is a white solid which melts at 161°C and sublimes at 265°C. It is slightly soluble in water but readily dissolves in dimethylformamide, N-methylpyrrolidinone and hot aromatic solvents. IPN undergoes the reactions expected of an aromatic nitrile, eg, hydrogenation of both nitrile groups and aromatic ring and chlorination to tetrachloroisophthalonitrile. IPN is prepared by vapor-phase ammoxidation of *meta*-xylene. The oral LD$_{50}$ for rats is 860 mg/kg. It's principal use appears to be as an intermediate to amines. As a reagent, IPN can be used to convert aromatic acids to nitriles in near quantitative yields (83).

2-Methylglutaronitrile

Methylglutaronitrile (2,3-dicyanobutane) [4553-62-2], MGN, is a by-product of DuPont's adiponitrile process. The oral LD$_{50}$ (rats) is 400 mg/kg (29). Some physical properties are listed in Table 12.

Shipping and Storage.   The DOT shipping name for MGN is "Toxic liquid, Organic, N. O. S. (Methylglutaronitrile)" and is in the hazard class 6.1, packing

6. U.S. Pat. 3,496,215 (Feb. 17, 1970), W. C. Drinkard, Jr. and R. V. Lindsey, Jr. (to DuPont).

7. U.S. Pat. 3,536,748 (Oct. 27, 1970), W. C. Drinkard, Jr. and R. V. Lindsey, Jr. (to DuPont).

8. C. Y. Wu and H. E. Swift, *179th American Chemical Society Meeting, Div. of Petroleum Chemistry Symposia Preprint*, Houston, Texas, Mar. 1980, p. 372.

9. C. A. Tolman, R. J. McKinney, W. C. Seidel, J. D. Druliner, and W. R. Stevens, *Adv. Catal.* **33**, 1 (1985).

10. C. De Bellefon and P. Flouilloux, *Catalysis Reviews*, **36**, 459–506 (1994).

11. *United States International Trade Commission Publication* 2933, Nov. 1995.

12. S. C. DeVito, in S. C. DeVito and R. L. Garrett, eds., *Designing Safer Chemicals: Green Chemistry for Pollution Prevention*, American Chemical Society Symposium Series 640, American Chemical Society, Washington, D.C., pp. 194–223.

13. U.S. Pat. 3,362,889 (Jan. 9, 1968) J. F. Hannan (to Du Pont).

14. *Advances in Chemistry Series, Azeotropic Data, No. 6 and No. 35*, ACS, Washington, D.C., 1952 and 1962.

15. U.S. Pat. 3,822,313 (July 2, 1974), J. R. Norell (to Phillips Petroleum Co.).

16. A. Buzas and co-workers, *Rev. Chim.* **22**, 656 (1971).

17. N. S. Kozlov and co-workers, *Dokl. Akad. Nauk BSSR* **2**(4), 326 (1977).

18. G. W. Gokel and co-workers, *Tetrahedron Lett.* **39**, 3495 (1976).

19. U.S. Pat. 3,810,935 (May 14, 1974), W. Leimgruber and M. Weigele (to Hoffmann-LaRoche Inc.).

20. U.S. Pat. 3,825,581 (July 23, 1974), W. A. Gay and D. F. Gavin (to Olin Corp.).

21. H. Felkin and G. Roussi, *C. R. Acd. Sci. Paris ser. C* **266**, 1552 (1968).

22. U.S. Pat. 3,920,718 (Nov. 18, 1975), J. E. Nottke (to DuPont).

23. U.S. Pat. 3,829,429 (Aug. 13, 1974), R. A. Clement (to DuPont).

24. Fr. Pat. 2,121,106 (Sept. 22, 1972), (to Ajinomoto Co., Inc.).

25. T. Cairns and co-workers, *J. Am. Soc.* **74**, 5633 (1952).

26. I. A. Gunevich and co-workers, *Tr. Mosk. Khim. Technol. Inst.* **86**, 5 (1975).

27. A. A. Durgarryan and co-workers, *Arm. Khim. Zh.* **25**, 401 (1972).

28. U.S. Pat. 3,281,450 (Oct. 24, 1966), P. J. Horvath (to B.F. Goodrich).

29. Registry of Toxic Effects of Chemical Substances (RTECS) On-Line Database. National Library of Medicine, Bethesda, Maryland.

30. *OSHA Standard 1910*, June 27, 1974.

31. H. D. Evans and D. H. Sarno, *7th World Petrol. Congr. Proc.* **5**, 259 (1967).

32. E. Ger. Pat. 91,480 (July 20, 1972), G. Hauthal and co-workers.

33. M. Enomotto and co-workers, *Kavgaku Kogaku* **35**, 437 (1971).

34. USSR Pat. 439,143 (Mar. 15, 1976), M. E. Aerov and co-workers.

35. A. Krause and T. Weiman, *Rocznki Chem.* **40**, 1173 (1967).

36. Ger. Offen. 2,237,704 (Feb. 15, 1973), L. W. Gosser and C. A. Tolman (to DuPont).

37. Brit. Pat. 1,312,573 (Apr. 4, 1973), N. T. Notley.

38. D. M. Muir and A. J. Parker, *Adv. Entr. Metall. 3rd., Int. Symp.* 191 (1977).

39. U.S. Pat. 3,607,136 (Sept. 21, 1971), R. A. Smiley and J. A. Vernon (to DuPont).

40. L. I. Primak and co-workers, *Tekst. Promst. (Moscow)* (2), 60–61 (1977).

41. Fr. Pat. 1,471,321 (Mar. 3, 1967), (to Dunlop Co., Ltd.).

42. Jpn. Kokai 76 04,107 (Jan. 14, 1976), T. Kita and M Ishii.

43. W. S. Brud and co-workers, *6th Int. Cong. Essent. Oils*, 73 (1974).

44. *Chem. Week*, **158**, 10 (April 24, 1996).

45. U.S. Pat. 5,495,016 (Feb. 27, 1996) G. Achhammer and E. Fuchs (to BASF).

46. Ger. Pat. 848,498 (Sept. 4, 1952), K. Adam and co-workers (to BASF).

47. Fr. Pat. 1,483,300 (June 2, 1967), K. Adam and co-workers (to BASF).

48. Belg. Pat. 763,109 (Aug. 17, 1971) (to ICI).

49. U.S. Pat. 2,284,525 (May 26, 1942), A. W. Larchar and co-workers (to DuPont).

50. U.S. Pat. 3,696,153 (Oct. 3, 1972), B. J. Kershaw and co-workers (to DuPont).

51. U.S. Pat. 3,821,305 (June 28, 1974) G. Bartalini and M. Gioggioli (to Montedison Fibre).

52. Brit. Pat. 568,941 (Apr. 26, 1945), (to DuPont Co.).

53. U.S. Pat. 3,299,116 (Jan. 17, 1967), R. Romani and M. Ferri (to Societa Rodiatoce).

54. U.S. Pat. 3,153,084 (Oct. 13, 1964), T. M. Veazey and co-workers (to DuPont).

55. Brit. Pat. 893,709 (Apr. 11, 1962), (to Chemstrand).

56. U.S. Pat. 3,242,204 (Mar. 22, 1966), W. A. Lozier (to DuPont).

57. J. Szymanowski and A. Sobczynska, *Przem. Chem.* **66**, 373–377 (1987).

58. U.S. Pat. 2,680,761 (June 8, 1954), R. H. Hallwell (to DuPont).

59. U.S. Pat. 2,749,359 (June 5, 1956), W. H. Calkins and co-workers (to DuPont).

60. U.S. 2,532,312 (Dec. 5, 1950), L. E. Romilly (to DuPont).

61. U.S. Pats. 5,440,067 (August 8, 1995) and 5,449,807 (Sept. 12, 1995), J. D. Druliner (to DuPont).

62. U.S. Pat. 5,512,695 (April 30, 1996) and WO9528228 (Oct. 26, 1995), K. Kreutzer and W. Tam (to DuPont).

63. U.S. Pat. 5,512,696 (April 30, 1996), K. Kreutzer and W. Tam (to DuPont).

64. WO9514659 (June 1, 1995), K. Kreutzer, R. J. McKinney, and W. Tam (to DuPont).

65. M. M. Baizer, *J. Electrochem. Soc.* **111**, 215 (1964).

66. M. M. Baizer, *Chemtech*, 161 (1980).

67. D. E. Danly, *Chemtech*, 302 (1980).

68. R. Hartung, in G. D. Clayton and F. E. Clayton, eds., *Patty's Industrial Hygiene and Toxicology, Volume II, Part D*, John Wiley & Sons, New York, 1994, pp. 3119–3172.

69. U.S. Pat. 5,194,657 (March 16, 1993) R. J. McKinney and R. N. McGill (to DuPont).

70. U.S. Pat. 2,711,405 (June 21, 1955), A. W. Anderson (to DuPont).

71. *VAZO*, Product information bulletin, DuPont.

72. *Azo Polymerization Initiators*, Product information Bulletin, Wako Pure Chemical Industries, LTD.

73. C. Walling, *J. Polym. Sci.* **14**, 214 (1954).

74. U.S. Pat. 2,598,811 (June 3, 1952), J. E. Mahan and S. D. Turk (to Phillips Petroleum).

75. Can. Pat. 545,630 (Sept. 3, 1957), R. T. Corkum and J. M. Salsburg (to American Cyanamid).

76. German Offen. 2,314,151 (Oct. 18, 1973), R. K. Grasselli and J. L. Callahan (to Standard Oil, Ohio) (for example).

77. W. I. Denton and co-workers, *Ind. Eng. Chem.* **42**, 796 (1950).

78. *Comline Chem. Mater.* 21 (Feb. 9, 1987).

79. Brit. Pat. 968,752 (Sept. 2, 1964), M. H. Richmond (to Distillers Co.).

80. Can. Pat. 722,712 (Nov. 30, 1965) C. Herschmann (to Lonza Ltd.).

81. Fr. Pat. 1,170,116 (Jan. 9, 1959), C. R. Stephens (to Charles Pfizer & Co.).

82. *Cyanoacetic Esters*, Technical Brochure, Kay-Fries, Inc. Montvale, N.J.

83. W. G. Toland and L. L. Ferstandig, *J. Org. Chem.* **23**, 1350 (1958).

RONALD J. MCKINNEY
E. I. du Pont de Nemours & Co., Inc.

STEPHEN C. DEVITO
U.S. Environmental Protection Agency

RECYCLING, PLASTICS

Of the 200 million tons of municipal solid waste collected in the United States in 1993 (1), 22% was recycled while 62% was placed in landfills and 16% incinerated (2). Plastics comprised 9.3% of these materials. The number of U.S. residential collection programs increased from 1,000 in 1988 to more than 7,000 involving more than 100 million people in 1993 (2). Approximate 1994 U.S. recycling rates are given in Table 1.

Although the recycling rate for plastic appears low, U.S. plastics production in 1993 was 70 billion pounds, while 700 million pounds was recycled. Other estimates of the U.S. recycling rate have been as much as 3.5% (2,4). Plastics recycling is a world-wide phenomenon. Aided by mandatory recycling laws, Germany's plastics packaging recycling has increased from about 90 million pounds in 1992 to 1.1 billion pounds in 1995 (5). Japan has enacted its own Container and Packaging Law to promote plastics and paper recycling (6). Great Britain has enacted a mandatory target plastics recycling rate of 15% to be achieved in 2001 (7). Export of recovered plastics from the United States to Asia is a significant business for plastics brokers (8).

Table 1. U.S. Recycling Rates of Various Materials[a]

Material	Recycling rate, %
plastics	1
paper and paper board	40
ferrous metals (iron and steel)	37
aluminum	30
glass	7
rubber and leather	3

[a]Ref. 3.

Both economic factors and governmental regulations are driving recycling (9). Energy costs associated with recycling are almost always less than in manufacture of products from virgin materials. Plastics recycling takes only 10–15% of the energy needed to refine petroleum and manufacture virgin resins. Incineration of plastics is a less efficient means of saving energy. For example, 100 lb (45.4 kg) of high density polyethylene has a fuel value of 20×10^6 Btu (19 kJ). Recycling saves twice this, 40×10^6 Btu (38 kJ). Life cycle analysis has been used to determine the most economically and environmentally acceptable method of using recovered plastics: mechanical recycling of the plastic, depolymerization to produce monomers, or incineration to produce energy (10). The most acceptable technology depends on the type of polymer and local business and environmental conditions.

Both economic and environmental factors have led to government regulations designed to promote recycling. In some areas, the number of landfill sites is becoming limited. Although the number of landfills in the United States is declining, the remaining sites are large, modern facilities. Concerns about landfill disposal costs are becoming less of a factor in promoting North American recycling. However, the effect of plastic wastes on the environment is a growing concern.

Separation of Commingled Materials

Random mixing of plastics leads to a significant adverse effect on properties. For example, mixing a few percent polypropylene in polyethylene leads to a significant reduction in tensile strength due to the formation of two immiscible phases having little adhesion (11). Hence different types of plastics must be separated from each other. Solid wastes, particularly from residential curbside collection programs, arrive at material recovery facilities (MRF) as a complex mixture. MRFs are typically built to process 100 to 500 tons of waste per day (1). Unit operations are summarized in Figure 1. The wastes are dumped on a tipping floor. There paper products are separated from metals and plastics. Metals and plastics, mostly containers, are pushed onto a conveyer belt. Two types of magnetic separators remove steel and aluminum from plastics and glass. Density differences or manual sorting are used to separate glass from plastics. The glass containers are hand sorted by color. The plastics are separated into individual polymer types of the MRF or in separate reclaiming facilities (1). Plastic bottles are classified into: clear poly(ethylene terephthalate) (PET) soft drink bottles; green PET soft drink bottles; translucent high density polyethylene milk, water, and juice bottles; pigmented high density polyethylene detergent bottles, poly(vinyl chloride) water bottles, and food containers such as polypropylene ketchup bottles (1). Processing equipment capable of separating pigmented bottles from clear ones has been installed in some facilities (12).

When processing municipal solid wastes, an eddy current separation unit is often used to separate aluminum and other nonferrous metals from the waste stream. This is done after removal of the ferrous metals (Fig. 1). The eddy current separator produces an electromagnetic field through which the waste passes. The nonferrous metals produce currents having a magnetic moment that is phased

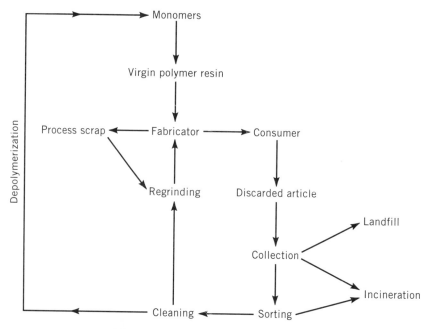

Fig. 1. Process steps in recycling.

to repel the moment of the applied magnetic field. This repulsion causes the nonferrous metals to be thrown out of the process stream away from nonmetallic objects (13).

Another separation device that may be used is the mineral jig. This unit produces a loose vibrating bed of particles in a liquid medium. The vibrations segregate the solids into layers of density. The dense nonferrous metals, primarily lead, zinc, and copper are at the bottom while organics are at the top. The middle layer is primarily glass.

Separation of Impurities.   After separation of the plastics, a number of impurities may still be present. This include inks used to print information and label onto plastics, other types of labels, wood and dirt accumulated during use and disposal of the plastic. Washing technology has been used to remove inks, labels, and encrusted dirt from plastics, particularly bottles (14). A number of technologies have been used to separate other materials from plastics. Froth flotation has been used to separate PVC from PET despite the similar densities of these polymers (15). In laboratory tests, froth flotation separated PVC, polycarbonate (PC), polyacetal, and poly(phenylene ether (PPE)) from each other. Wetting agents such as lignosulfonates, tannic acid, and saponin were required to promote the separation (16). Plastics may be separated from mixed bulk materials based on particle size (17). To separate poly(vinyl butyral) (PVB) in window glass from impurities, the polymer is melted and allowed to flow into supercritical carbon dioxide (18). Another technique to remove impurities from melted polymers is filtration (19).

Plastics

In May 1992, the U.S. Food and Drug Administration established the following guidelines to help assure the consumer safety of plastics recycling processes (20). Primary recycling is the recycling of plastics that are plant scrap and have not been sold for consumer use. Secondary recycling is the physical cleaning and processing of postconsumer plastic products. Tertiary recycling is the chemical treatment of polymers. This treatment is usually depolymerization to produce monomers which are purified and then polymerized to produce new polymer. Using tertiary recycling, materials such as fillers and fibers can be physically removed from the monomer. The monomers can also be purified by distillation and other processes prior to polymerization. The leading example of tertiary recycling is poly(ethylene terephthalate). Tertiary recycling also has been suggested for nylon from discarded carpets (21).

Residential collection programs indicate high collection rates for easily recognized types of containers (Table 2).

Sorted plastic packaging materials are shipped, usually in bales, to processing plants to be converted to polymer resins. The bales are broken and the bottles sorted to ensure that only one type of polymer is further processed. Processing consists of chopping and grinding the bottles into flakes. These flakes are washed. Processing steps such as flotation are used to remove polymeric contaminants from the flakes (15,16). The flakes are melted and converted into pellets.

For high value food packaging applications, minimal migration of contaminants into food products is critical. Currently the FDA requirement is a maximum 0.5 parts per billion (ppb) of noncarcinogenic compounds by dietary exposure (22).

Poly(ethylene terephthalate). About 1.6 billion pounds of PET are used in food packaging applications annually in the U.S. (23). About 42% of produced PET is recycled, mainly soft drink bottles (2). Cleaning of the recovered plastic comprises washing, rinsing, and drying. PVC is a common impurity in PET. Melting PET containing PVC will produce black spots due to charring of the PVC during processing to produce new bottles (24,25). Poly(vinyl chloride) (PVC) and PET have very similar density values; both will sink to the bottom of the water bath during rinsing. Therefore, it is difficult to separate the two polymers after the bottles have been ground into small particles (24). However, froth flotation has been shown to be an effective means of separating these two polymers (15).

For food applications, improved cleaning of PET produced by secondary recycling is needed. Supercritical fluid extraction using carbon dioxide (20) and

Table 2. Residential Recovery Rate by Package Type[a]

Package type	Recovery, %
beverage bottles	65
liquid detergent bottles	50
other rigid containers	10
packaging film	5
average of all plastics	30

[a] Ref. 1.

Contaminants are more easily removed before this densification step than after. A demonstration plant chemically decomposes polystyrene to produce monomer (46). Polystyrene may be cracked at 130°C over sulfated zirconia to produce benzene (46).

Granulated polystyrene foam has been used as an additive in light-weight cement or as a soil additive to retain moisture and minimize compaction (46). Specially designed cup shredding machines for use with vending machines dispensing drinks in PS cups have been commercialized (7). However, recovery rates for other PS packaging products is significantly less than for easily recognized PS foam consumer product packaging (46).

Polystyrene has a high heating value, 46,000 kJ/kg compared to heating oil, 44,000 kJ/kg (46). Thus, incineration for its energy value is another possible application for recovered polystyrene.

Other Plastics. A relatively small amount of poly(vinyl chloride) goes into packaging applications and appears in municipal solid waste (25). The greatest concern with PVC is as a contaminant in other polymers being recycled, particularly PET. Approximately 12 million pounds of PVC was recycled in 1993, about half from packaging (25). Applications for recycled PVC include as an inner layer sandwiched between two virgin PVC layers in pipe and sheet for blister packaging and other packaging applications.

Polyurethane is pulverized to increase its bulk density, mixed with 30–80% of a thermoplastic molding material, gelled, and then granulated to give coated urethane foam particles 0.1 to 0.15 mm in size (48). The particle bulk density is three times that of the polyurethane, while the volume is 15% less. This material may be injection molded or extrusion molded into products (49). Other technologies for recycling polyurethanes have also been reported.

The recycling of engineering thermoplastics such as polyamides, ABS, and PTFE have been discussed (50). Property degradation as a result of use, recovery, and recycling is a concern.

Commingled Plastic Wastes. Owing to the property deteriorations that usually occur on polymer mixing (11), commingled plastics as useful and economic only for low value applications in which mechanical properties are not demanding. Such applications include park benches and parking barriers. Plastics in municipal solid waste streams are often contaminated with paper, which is difficult to separate from the plastic materials. If the cellulose fibers have a sufficient length, 0.2–2 mm (51), they can improve the mechanical properties of the plastic. Usually a reactive compatibilizer is required to improve the compatibility of the polymer phases and promote bonding of the cellulose to the plastics (52). One example cited is the addition of 30% cellulose fiber to a 70:30 mixture of LDPE and high impact polystyrene. An addition level of 30% maleic anhydride grafted styrene/ethylene–butylene–styrene block copolymer was used as the compatibilizer. This additive level was too high to be economic. However, the compatibilizer level was not optimized.

The supply of commingled plastics is much greater than the demand (53). Therefore, a critical issue in recycling commingled plastic wastes is the identification and separation of the plastics that are present. Near infrared (900–1700 nm) spectroscopy has been proposed to identify polyethylene, poly(ethylene terephthalate), polypropylene, polystyrene, and poly(vinyl chloride) (PVC). A spectrograph with an InGaAs-array detector has been developed to record spectra from post-consumer packaging materials located on conveyer belts (54). Atomic absorption spectroscopy can be used when one of the polymers has a different atomic compositions than other polymers in a mixture (53). An example is the separation of poly(vinyl chloride) from polymers not containing chlorine.

The economics of recycling PET are more favorable than recycling HDPE. To increase the recycling of HDPE, the separation of bottles made of these two plastics could be omitted and a mixture processed. Coarse, light-colored powders of the two polymers have been prepared by an experimental solid state shear extrusion pulverization process (55). The powder has been successfully injection molded without pelletization.

The composition of nonmetal residues produced in shredding automobiles is summarized in Table 3.

Each vehicle generates 500–800 pounds of residue. The annual U.S. total is about 3.5 million tons or about 1.3% of the municipal solid waste generated annually (3). The mixture is too complex to separate and recycle. Depending on the amount of glass, water, metal, and dirt present, the residue has a heating value of 4,800–6,800 Btus per pound (3,56). Incineration reduces residue weight by 50% and volume by 80% (4).

The thermal degradation of mixtures of the common automotive plastics polypropylene, ABS, PVC, and polyurethane can produce low molecular weight chemicals (57). Composition of the blend affected reaction rates. Sequential thermolysis and gasification of commingled plastics found in other waste streams to produce a syngas containing primarily carbon monoxide and hydrogen has been reported (58).

One alternative to identifying and separating different types of plastics is using commingled plastics directly. Since the composition and physical properties of commingled plastics can vary from day to day, applications are limited. One such product is a building material containing portland cement as the binder, a filler (sand, gravel, or stone), and a plastic with a maximum particle size of 5–10 mm (59). The ratio of binder to aggregate is 1:4–8 while the volume ratio of binder + filler to plastic is 3–9:1. Concrete made using this material performs as well as standard concrete not containing plastic. In Germany, mixed plastics have been blended with pulverized fuel ash and the mixed used to manufacture fencing and posts (60).

**Table 3. Representative Composition
of Automobile Shredder Residue[a]**

Material	Wt, %
plastics	27
rubber	7
glass	16
textiles	12
fluids	17
other	21

[a]Ref. 3.

Styrene block copolymers have been used as compatibilizers for mixed plastics to permit their processing for applications such as those outlined earlier (52,61).

Laboratory tests indicated that gamma radiation treatment and crosslinking using triallylcyanurate or acetylene produced a flexible recycled plastic from mixtures of polyethylene, polypropylene, general purpose polystyrene, and high impact grade PS (62).

Another alternative to separating commingled plastics is advanced waste recycling. This is the high temperature–high pressure conversion of plastic wastes to form petrochemical process streams. Research is in progress to determine the conditions that will favor conversion of commingled plastic wastes to certain types of chemical feedstocks including synthesis gas (hydrogen + carbon monoxide), hydrogen, crude pyrolysis oil (containing benzene, toluene, and xylene), olefins, and oxygenates such as methanol, esters, and methyl formate (63). This technology has also been evaluated for producing fuels: medium BTU gas for boilers, and liquid fuels such as diesel oil. None of these processes is currently economic.

Thermal cracking of commingled plastics can produce an excellent feed for steam crackers and catalytic crackers. During steam cracking, feed from commingled plastics produced higher yields of ethylene (34% vs 28%), propylene (17% vs 15%), and butylene (12% vs 7%) than did the usual naphtha feed (64). During catalytic cracking, feed generated from plastics provided an 86% yield of naphtha grade product compared to a 62% yield from vacuum oil.

The Conrad recycling process utilizes an auger kiln to apply heat to plastics in the absence of oxygen (65). Feed preparation using the Conrad process requires minimal plastic particle size reduction, washing, and removal of nonplastic contaminants (66). Granulated plastics are introduced into a retort in the absence of oxygen using a rotary air lock. If the plastic particles are in their original form, they are introduced into the retort using a ram feeder. The plastics melt after entering the hot retort. An auger keeps the molten mass moving. Thermally promoted carbon–carbon bond cleavage occurs. As depolymerization proceeds, volatile products are produced and swept out of the reactor. Interestingly, mixing polymers seems to improve thermal cracking of results. A 1:1 mixture of polypropylene and polyethylene cracked at a lower temperature than did polyethylene alone and provided a narrower mixture of products than either polymer did separately (67).

Results obtained for two mixed plastics are summarized in Table 4. A balance exists between process temperature, plastics feed rate, and product yields (67). For example, lower temperatures increase wax formation due to incomplete depolymerization. Slower feed rates and increased residence times reduce wax formation and increase the yield of liquids. The data summarized in Table 4 illustrate that the addition of PET to a HDPE:PP:PS mixture changes the performance of the Conrad process. Compared to the reference HDPE:PP:PS mixture, increased amounts of solids are formed. These are 95% terephthalic acid and 5% mono- and bis-hydroxyethyl esters. At higher temperatures, apparently enough water remains to promote decarboxylation. In contrast, the addition of LDPE or PS to the mixture had little effect on its behavior in the Conrad process.

Table 4. Gas Chromatographic Analysis of the Results of Cracking Polymer Mixtures Using a Conrad Unit[a,b]

Analysis	60:20:20 HDPE:PP:PS	20:48:16:16 PET:HDPE:PP:PS
liquid yield, wt %	73	55
gas yield, wt %	27	38
solids yield, wt %		7
Partial oil analysis, wt %		
aliphatics, carbon number		
$\leq$10	9.4	21.3
11–15	16.6	17.6
16–20	9.9	9.9
21–25	3.5	4.7
26–30	1.7	2.2
31–40+	1.5	2.2
aromatics		
benzene	2.3	1.4
toluene	11.0	3.8
ethylbenzene	5.7	2.6
xylenes		1.1
other alkyl benzenes		8.4
styrene	17.0	8.2
naphthalenes		3.7
unidentified		12.3

[a]Ref. 67
[b]Oven temperature = 649°C; Auger temperature = 527°C for the HDPE:PP:PS mixture, 479°C for the PET:HDPE:PP:PS mixture.

Fiber-reinforced Plastics and Composites. It is usually too expensive to separate fillers and fibers from recovered polymers. Hence, the recycled use of these polymers must tolerate the presence of fillers or fibers (53). Thermoset matrix composites are ground and used as filler for polymers. Remolding is usually by injection or compression molding (68). Fiber-reinforced plastics are recycled primarily from old automobiles and electrical equipment (casings and various plastic parts.) Glass fiber-reinforced plastics have been made into sheet-molding compounds and bulk-molding compounds (69). However, the economical recycling of fiber-reinforced plastics remains a challenge. Dynamic mechanical thermal analysis is said to determine heat resistance, impact resistance, and stiffness of glass-reinforced plastic before and after recycling (70). Thus, it could serve as a tool to determine the suitability of a glass-reinforced plastic for recycling.

Economics and Statistics

Costs of various waste disposal methods are summarized in Table 5.

Polymer recycling process costs for various operations are summarized in Table 6.

Of course, the benefits gained by recycling are also important. These are summarized in Table 7 for PET and HDPE.

Table 5. Estimated U.S. Processing Costs of Waste Disposal Method[a]

Method	Cost estimate, $/t
landfilling	30[b]
incineration for energy recovery	100
recycling of PET and HDPE bottles	100–150

[a]Ref. 71.
[b]Cost varies with the location of the landfill.

Table 6. Approximate Polymer Recycling Costs[a]

| Process step | Costs | |
	$ per pound	%
collection	0.10	27
sorting	0.12	32
subtotal cost	*0.22*	*59*
grinding–cleaning	0.15	41
Total	*0.37*	*100*

[a]Ref. 1.

Table 7. Economic Benefits of Recycling[a]

| Polymer | Value per ton, $ | |
	Incineration and energy recovery[b]	Recycling
PET	28–140	470
HDPE	312	120

[a]Based on $0.14 per kW/h of energy.

U.S. plastics production in 1993 was about 70 billion pounds (1). Plastics recycling continues to increase from its current value of 3.5% in the U.S. (2,4). However, recycling rates of some packaging products are much higher than this. The recycling rate of PET from soft drink bottles was 42% in 1993 while that of HDPE from milk and water jugs was 24% (2). This translates to 448 million pounds of PET and 450 million pounds of HDPE (1). For example, the process cost of recycling PET and HDPE bottles has been given as U.S.$100–150 per ton (71). The value of the PET and HDPE produced from recycled materials is U.S.$470 and U.S.$120 per ton, respectively. These recycling processes can be profitable. The favorable economics of PET recycling have been attributed in part to forward integrated PET recyclers consuming their own product to make bottle resin (24).

Process costs for the methanolysis and glycolysis of PET to produce monomers are similar. Cost estimates are summarized in Table 8.

By comparison, the 1993–1994 cost (per pound of PET) of dimethyl terephthalate and ethylene glycol made from petroleum was $0.35.

However, recycling of many other plastics remains uneconomic (1). This is reflected in a number of companies closing plastics recycling operations in the mid-1990s (73). The costs of recycling commingled plastics has been estimated at U.S.$1700 per ton (72). This is ten times more expensive than recycling

Table 8. Estimated Costs of PET Methanolysis Based on 1993–1994 Data[a]

Item	Cost per pound of PET, $
feedstock	0.30
conversion and handling	0.20
capital costs	0.15

[a]Ref. 72.

easily separated homogeneous products such as PET and HDPE. In Germany, federal mandates require recycling of 60% of all plastic packaging. German projects to recycle over one billion pounds a year of plastics have been announced (74,75). These processes will use high temperature–high pressure processes to depolymerize commingled plastics to produce petrochemical feedstocks. These processes are not economic (65,76).

Few of the products in which polypropylene are used can be recovered in commercial qualities. An exception is battery casings. In 1994, U.S. capacity for recycling polypropylene from battery casings was 265 million pounds annually (45). About 75% of the recovered polypropylene is used in new battery cases which have a recycled PP content of about 50%. Total polypropylene recycling capacity was 350 million pounds annually for an operating rate estimated at 71–90% (45).

In 1993, over 41 million pounds of polystyrene was recycled into new plastics products in the U.S. (77). For commingled plastics, gasification comes closest to competing with low cost landfilling (57).

Price swings, particularly in the PET and HDPE markets have contributed to a retrenchment in the U.S. plastics recycling industry in 1995–1997 (78). In 1994 in the U.S., recycled PET, HDPE, LDPE, and PS had a 16–46% cost advantage (4). This cost advantage largely disappeared by 1996. Bureau of Labor Statistics data indicate U.S. plastics prices in mid-1997 are seven percent below those of mid-1995 after being more than ten percent less in 1996 (79). These lower prices make it more difficult for recycled plastics to compete with virgin resins in the absence of legislation mandating plastics recycling.

BIBLIOGRAPHY

"Recycling, Plastics" in *ECT* 3rd ed., Vol. 19, pp. 993–1002, by Harvey Alter, Chamber of Commerce of the United States.

1. R. G. Saba and W. E. Pearson, in C. P. Rader, S. D. Baldwin, D. P. Cornell, G. B. Sadler, and R. F. Stockel, eds., *Plastics, Rubber and Paper Recycling: A Pragmatic Approach*, American Chemical Society, Washington, D.C., 1995, Chapt. 2, pp. 11–26.
2. Franklin Associates Limited, "Characterization of Municipal Solid Waste in the United States, 1994 Update," Report No. EPA 530-94-042, Nov. 1994.
3. R. A. Pett, A. Golovny, and S. S. Labana, in Ref. 1, pp. 47–61.
4. C. P. Rader and R. F. Stockel, in Ref. 1, Chapt. 1, p. 3.
5. D. Castle, *Packag. Week* **12**(19), 3 (Oct. 10, 1996).
6. *Jpn. Chem. Week* **37** (1898), 1–2 (Oct. 24, 1996).
7. *Plast. Rubb. Wkly* **1659**, 8 (Oct. 25, 1996).

S

SEMICONDUCTORS, ORGANIC

Semiconductors are materials that are characterized by resistivities intermediate between those of metals and of insulators. The study of organic semiconductors has grown from research on conductivity mechanisms and structure–property relationships in solids to include applications-based research on working semiconductor junction devices. Organic materials are now used in transistors, photochromic devices, and commercially viable light-emitting diodes, and the utility of organic semiconductors continues to increase.

The study of organic semiconductors and conductors is highly interdisciplinary, involving the fields of chemistry, solid-state physics, engineering, and biology. This article provides a treatment of the theoretical aspects of organic semiconductors as well as an overview of recent advances in the field and the uses of these materials based on their conductive and optical properties.

Theory

The theory of conduction in organic semiconductors conveniently begins with a discussion of bonding in extended solids, since the nature of conduction is intimately related to the extent of delocalization of orbitals in the material to be studied. For inorganic conducting solids, the orbital overlap of the atoms in the crystal lattice results in the atomic energy levels spreading out to form energy bands rather than discrete levels. These energy bands allow for motion of charge carriers over many lattice sites without interruption by trapping events. The interactions between molecules in a molecular organic solid similarly leads to the formation of energy bands, with the extent of interaction between molecules (the extent of overlap of interacting orbitals) determining the width of the bands.

474

The distinctions between metals, semiconductors, and insulators are based on the band structure of the materials as well as on the electron occupancy of these bands. Figure 1 is a diagram of energy bands and occupancies for various classes of solids. As can be seen in Figure 1, insulators have a filled band formed from the valence orbitals (a "valence band") with a higher-lying unfilled band formed from higher-energy orbitals (the "conduction" band). The region between these has no allowed states, and the energy difference required to promote an electron from the valence to the conduction band is termed the band gap of the material. For an insulator, the band gap is large, typically >4 eV. Since all of the levels of the valence band are filled, no electrons are able to carry current.

Metallic behavior is observed for those solids that have partially filled bands (Fig. 1b), that is, for materials that have their Fermi level within a band. Since the energy bands are delocalized throughout the crystal, electrons in partially filled bands are free to move in the presence of an electric field, and large conductivity results. Conduction in metals shows a decrease in conductivity at higher temperatures, since scattering mechanisms (lattice phonons, etc) are frozen out at lower temperatures, but become more important as the temperature is raised.

Semiconductors show thermally activated conductivity. The reason can be seen from Figure 1c. The band gap of a semiconductor is small relative to that of an insulator. Similar to insulators, pure intrinsic semiconductors have a filled valence band and unfilled conduction band. However, this is strictly true only at absolute zero. At higher temperatures, the energy of the surroundings kT can become large enough to thermally excite electrons from the valence to the conduction band, with the result that semiconductors have partially unfilled orbitals, giving rise to conduction. Typical bandgaps for semiconductors are ~1–2 eV. Increasing temperature causes increased electrical conductivity by providing more carriers in the conduction band.

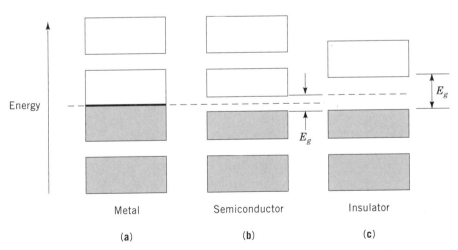

Fig. 1. Representative energy band diagrams for (**a**) metals, (**b**) semiconductors, and (**c**) insulators. The dashed line represents the Fermi Level, and the shaded areas represent filled states of the bands. E_g denotes the band gap of the material.

A variety of organic materials demonstrate reasonable conductivity, and a great deal of study has been devoted to conduction in organic materials. Initial study of conduction in organic materials was focused on molecular crystals (such as anthracene) that have weak intermolecular van der Waals interactions and low conductivities. However, the area of research has broadened and also includes charge-transfer compounds and polymeric materials. Research on new materials has led to synthesis of organic compounds with much larger conductivities than typical molecular crystals. Conducting and superconducting organic solids are now relatively commonplace. A listing of some compounds and associated conductivities are shown in Figure 2.

Certain one-dimensional, chainlike organic semiconductors are found to display metallic properties (decreasing conductivity with increasing temperature) above a certain temperature (1). The phase transition that occurs in these stacked materials is termed a metal-to-insulator transition and is the result of the highly anisotropic nature of these materials. A stack of equally spaced atoms or molecules is subject to a reorganization to a lower symmetry configuration. This reorganization, termed a Peierls distortion (2), is the condensed matter counterpart of the molecular Jahn-Teller distortion. The Peierls deformation results in pairing of molecules along the chain with the formation of short and long intermolecular spacings. This nuclear deformation causes changes in the electronic distribution of the lattice. The result of this redistribution is the opening of an energy gap at the Fermi level, with resulting semiconductivity. Reviews of Peierls distortions can be found in the literature (3).

A common example of the Peierls distortion is the linear polyene, polyacetylene. A simple molecular orbital approach would predict sp^2 hybridization at each carbon and metallic behavior as a result of a half-filled delocalized π-orbital along the chain. Uniform bond lengths would be expected (as in benzene) as a result of the delocalization. However, a Peierls distortion leads to

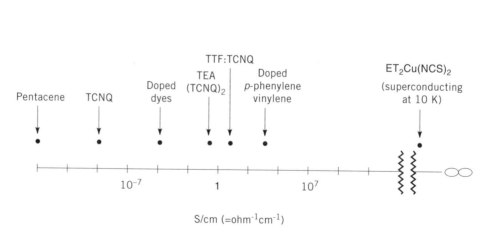

Fig. 2. Conductivity scale for different classes of organic materials.

alternating single and double bonds (Fig. 3) and the opening up of a band gap. As a result, undoped polyacetylene is a semiconductor.

Interaction of molecules to form charge-transfer complexes occurs via transfer of electron density from one molecule (an electron donor, D) to another (an electron acceptor, A), resulting in a partially ionic ground state. The amount of charge transfer is large when the ionization potential of the donor is low and the electron affinity of the acceptor is high. Orbital descriptions and experimental aspects of charge-transfer complexes have been studied extensively, especially for weak complexes (4). In a charge-transfer salt, the intermolecular interactions give rise to periodic arrays of donors and acceptors.

Typical charge-transfer salts form as stacks of planar D and A molecules, though the ratio of D:A need not be 1:1, as the interaction can be spread over more than two molecules. The amount of charge transfer (δ) per $D^{\delta+}A^{\delta-}$ unit in the solid may be less than unity, with partial charges residing on the respective stacks. Interactions between stacks are usually weak, and the observed conductivity is, thus, anisotropic, with the preferred direction being along the stacking axis. Anion (A^-) stacks result in electrical conductivity, while cation (D^+) stacks conduct holes. Strongly bound charge-transfer systems are of primary interest for organic semiconducting applications. A listing of some common strong donors and acceptors is given in Table 1.

Some common single-carrier semiconducting salts are tetracyanoquinodimethane (TCNQ) radicals with alkali or other cations. The study of semiconductivity in these materials led to synthesis of more highly conductive charge-transfer salts based on the strong donor tetrathiafulvalene (TTF) with halogen or other acceptors. Organic materials have been developed with even larger conductivity and superconductivity above liquid-helium temperatures.

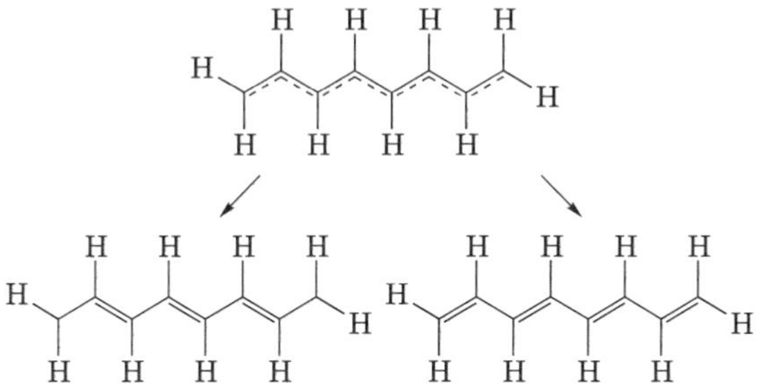

Fig. 3. Representation of the Peierls distortion in *trans*-polyethylene.

Properties

Physical Properties: Electrical. Electrical properties have been the main focus of study of organic semiconductors, and conductivity studies on organic materials have led to the development of materials with extremely low resistivities and large anisotropies. A discussion of conductivity behaviors for various classes of compounds follows.

Charge-Transfer Salts. Most organic molecular crystals have only weak interactions between molecular units and have resulting low conductivities, with

Table 1. Common Donor and Acceptor Molecules Used in Organic Semiconductor and Devices

Compound	CAS Registry Number	Structure
Donors		
2,2',5,5'-tetrathiafulvalene (TTF)	[31366-25-3]	
N,N,N',N'-tetramethyl-*p*-phenyl-enediamine (TMPD)	[100-22-1]	
5,6;11,12-tetratellurotetracene	[64479-92-1]	
3,3',4,4'-tetramethyl-2,2',5,5'-tetraselenafulvalene (TMTSF)	[55259-49-9]	
phthalocyanines (PC), M = Cu	[147-14-8]	
N-methylphenazinium (NMP)	[7432-06-6]	
Acceptors		
7,7,8,8-tetracyanoquinodimethane	[1518-16-7]	
tetracyanoethylene (TCNE)	[670-54-2]	
9,9,10,10-tetracyano-2,6-naph-thylidenedimethane (TNAP)	[6251-01-4]	

Table 1. (Continued)

Compound	CAS Registry Number	Structure
chloranil	[118-75-2]	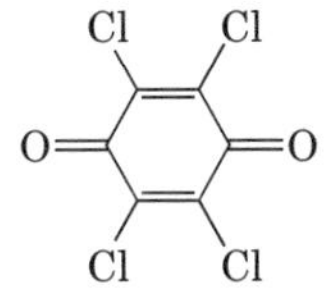
trinitrofluorenone (TNF)	[129-79-3]	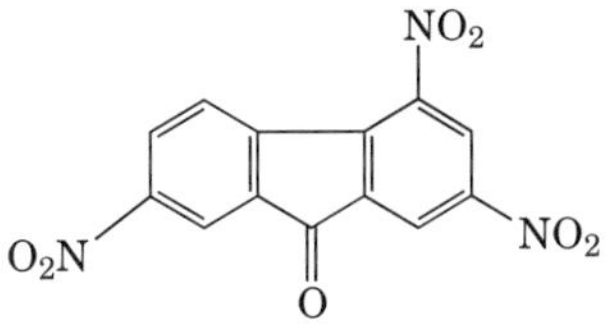

$S < 10^{-10}\Omega^{-1}\mathrm{cm}^{-1}$. Charge-transfer solids tend to have higher conductivities than the molecular compounds from which they are composed, and semiconducting charge-transfer salts of TCNQ were first synthesized in the early 1960s. The simple metal-ion radical salts of the composition $M^+(TCNQ)^-$ have conductivities of $\sim 10^{-9}$–$10^{-4}\Omega^{-1}\mathrm{cm}^{-1}$ at room temperature, intermediate between conductivity values for insulators and typical metals. A variety of other systems show thermally activated conductivity in the range 10^{-10}–$1\ \Omega^{-1}\mathrm{cm}^{-1}$, including phthalocyanines and chloranil or tetracyanoethylene-based charge-transfer complexes. Conductivity in these semiconductive compounds obeys an Arrhenius expression:

$$\sigma = Ae^{-\Delta E/kT}$$

where A is a preexponential factor, k is Boltzmann's constant and ΔE is the activation energy for the thermally activated conductivity.

Proposed conductivity mechanisms and the importance of band motion vs hopping motion in semiconducting organic solids have generated a great deal of controversy. However, it is plausible that for many organic molecular semiconducting materials, the description of the conductivity lies somewhere between the extremes of localized hopping and delocalized band models (5).

The highly conductive class of solids based on TTF–TCNQ have less than complete charge transfer (~ 0.6 electrons/unit for TTF–TCNQ) and display metallic behavior above a certain temperature. However, these solids undergo a metal-to-insulator transition and behave as organic semiconductors at lower temperatures. The change from a metallic to semiconducting state in these chain-like one-dimensional (1D) systems is a result of a Peierls instability. Although for true one-dimensional systems this transition should take place at 0 Kelvin, interchain interactions lead to effective non-1D behavior and inhibit the onset of the transition (6).

The temperature of the metal-to-insulator transition in TTF–TCNQ is 53 K. For systems with increased interchain coupling, the transition temperature for the onset of metallic conduction increases roughly as the square of the interaction between the chains. This behavior is true as long as the coupling between chains remains relatively weak. For compounds with strong interactions between stacks, the material loses its quasi-1D behavior. Thus, the Peierls distortion does not occur even at low temperatures, and the materials remain conductive.

The temperature range of metallic conductivity in organic materials can sometimes be increased by compaction of the compound under large pressures. For example, bis(tetramethyl tetraselenafulvalenium) hexafluorophosphate ((TMTSF)$_2$-PF$_6$) undergoes a metal-to-insulator transition at 12 K at atmospheric pressure. Under high pressures, however, the material stays conducting to low temperatures and goes superconducting at 0.9 K (7). A variety of organic-based materials have now been developed that have strong interchain interactions and possess superconductivity at low temperatures. Examples include bis(tetramethyl tetraselenafulvanenium) perchlorate (TMTSF)$_2$ClO$_4$ and compounds based on the donor molecule bis(ethylenedithio)tetrathiafulvalene (BEDT-TTF) with a variety of inorganic and organic anions (8).

Polymers. The use of polymers in semiconducting applications is growing rapidly. In the undoped form, most of these π-conjugated systems are semiconducting, with a small bandgap separating filled valence and conduction bands. A listing of polymers commonly used in semiconducting applications is given in Table 2. The dark (nonphotoinduced) conductivity of these systems is typically low and increases with increasing temperature, as a result of thermalization of charge carriers. Doping the material leads to large increases in conductivity, with increases of several orders of magnitude possible. Doping removes electrons from the valence band or adds them to the conduction band. However, the charges are not delocalized along the polymer chain. An extra electron in the conduction band, for example, becomes localized as a result of a local geometry change caused by the excess charge. The result of the charge localization is termed a polaron, which as a result of the local energy minimum gives a state in the bandgap of the material. Conduction can then occur by motion of the polaron through the chain.

Conductivities of polymers of technological interest such as polypyrrole and polythiophene are typically ~100Ω^{-1}cm^{-1} in the doped state, and the conductivity can be tuned by reversibly doping and undoping the polymer. Derivatives of these and other polymers have achieved even higher conductivities.

One-dimensional polymers such as polyacetylene which have degenerate ground states are special cases. Conduction in polyacetylene has been extensively studied (9). The nature of conduction in pure, *trans*-polyacetylene is a result of the alternating single bond–double bond structure that results from the Peierls instability. The energy of the system is the same if the alternating pattern of the bonds is reversed. Conductivity has been described for this system in terms of solitons (mobile kinks in the polymer chain that link alternating bond patterns), excitations that are unique to systems with degenerate ground states (10,11). These neutral solitons are mobile, and extend over several bond lengths along the chain. As usually prepared, undoped polyacetylene has *p*-type conductivity as a result of acceptor impurities in the material. The conductivity increases with doping, reaching a plateau value at high doping levels (12). Highly doped *trans*-polyacetylene can display conductivities as high as 10$^5\Omega^{-1}$cm^{-1}.

Table 2. Representative Polymers Used for Organic Semiconductors and Metals

Compound	CAS Registry Number	Structure
Polymers		
polyvinylcarbazole (PVK)	[25067-59-8]	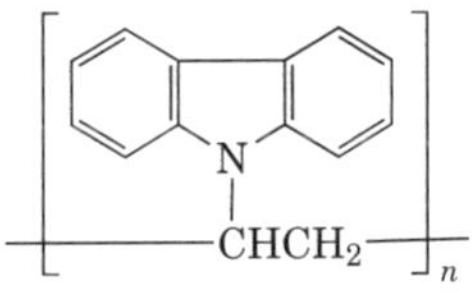
trans-polyacetylene	[25067-58-7]	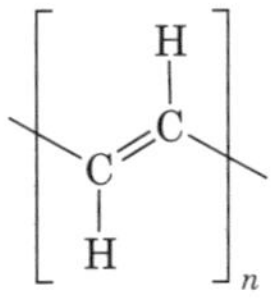
polythiophene	[25233-34-5]	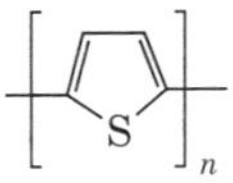
poly(*p*-phenylene)	[25190-62-9]	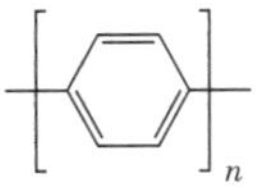
polyaniline (PANI)	[25233-30-1]	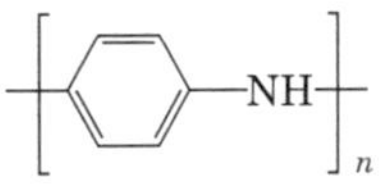
poly(*p*-phenylene vinylene) (PPV)	[26009-24-5]	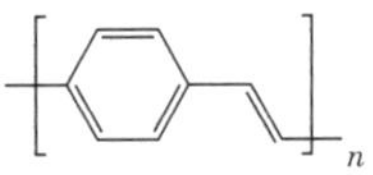

Optical Properties. *Charge-Transfer Salts.* New strong absorbance bands (charge-transfer bands) are often observed in spectra of donor–acceptor solids. These bands are the result of excitation from a ground state that is predominantly donor in character to an excited state that resides mainly on the acceptor. For a given acceptor, the frequency of this band varies linearly (over a small range) with the ionization potential of the donor molecule. This absorption band can occur at lower energy than absorptions of the isolated D and A molecules, and is often in the infrared region. By a suitable choice of donor and acceptor, it is possible to tune the frequency of this absorption band (and the associated fluorescence) throughout the visible region. For compounds such as the simple charge-transfer salts of TCNQ (formula M^+TCNQ^-) the charge transfer is complete, and these salts in solution exhibit spectra characteristic of the anion. In the solid state, an additional absorption band is observed as a result of charge transfer between adjacent $TCNQ^-$ molecules: one acting as a π-donor and the other as a π-acceptor. Highly conducting salts such as those of the formula $M^+TCNQ^-(TCNQ)$ have less than a complete negative charge per TCNQ molecule in the stack. These compounds exhibit an additional transition as a result of promotion of an electron from an anionic to neutral TCNQ site.

Polymers. Polymers used in semiconducting applications such as light emitting diode displays are π-conjugated systems with small bandgaps of 1–3.5 eV. In the undoped state they give strong π-π^* absorption bands in the visible or ultraviolet region, corresponding to transitions from the valence band to the conduction band levels. By tailoring the nature of the electron-donating and withdrawing substituents on the polymer backbone, absorbance and the resulting emission spectra can be shifted throughout the visible region. Conformational changes due to heating can drastically shift the absorption, suggesting possible uses as molecular switches.

Doping of the polymers brings about absorbances in addition to the main bandgap transition. These absorbances occur at lower energies, and grow in at the expense of the band-to-band transition. The new absorbances are due to transitions between states that lie in the bandgap of the material (13). These midgap states are typically polaron states that arise due to the injected charge-carriers being localized by lattice distortions. As the doping levels are increased, absorption at lower energies continues to increase and the original band edge becomes obscured. At high doping levels the polymers show absorption throughout the ir and visible, consistent with metallic behavior.

Chemical Properties: Electrochemistry. Since charge-transfer and mobility of carriers are important features for organic semiconductors and devices, the energies of the highest occupied molecular orbital (HOMO) and lowest unoccupied molecular orbital (LUMO) are of interest. The energies of these levels are related to the efficiency of charge injection into a material as well as the degree of charge-transfer between donors and acceptors. Electrochemical measurements provide a convenient means for determining HOMO and LUMO levels by measuring the potentials required for oxidation and reduction of the material, respectively. Redox potentials also provide a quick method for checking energy gap values obtained from conductivity measurements.

Electron donor molecules are oxidized in solution easily. For example, $E_{1/2}^{ox}$ for TTF is 0.33V vs SCE in acetonitrile. Similarly, electron acceptors such as TCNQ are reduced easily. TCNQ exhibits a reduction wave at $E_{1/2}^{red} = 0.06$V vs SCE in acetonitrile. The redox potentials can be adjusted by derivatizing the donor and acceptor molecules, and this tuning of HOMO and LUMO levels can be used to tailor charge-transfer and conductivity properties of the material. Knowledge of HOMO and LUMO levels can also be used to choose materials for efficient charge injection from metallic electrodes.

Reversible oxidation and reduction of polymers is commonly used to increase conductivity in these systems. Ions from the electrolyte are usually incorporated into the polymer as part of this process (see ELECTRICALLY CONDUCTING POLYMERS).

Stability. *Thermal Stability.* The materials used in organic semiconducting applications are thermally labile upon exposure to high temperatures. For example, many of the compounds used in fabrication of organic light emitting diodes (LEDs) are vapor deposited by resistive heating at relatively low temperatures in vacuum. Compounds such as the hole transporter N,N'-biphenyl-N,N'-bis(3-methylphenyl)-1,1' biphenyl-4,4' diamine (TPD) have glass-transition temperatures in the range of 150°C.

As a result of the organic nature of the materials, chemical oxidation is a problem when heating in atmosphere. For example, TTF melts at 119°C, with the formation of sulfoxides as a result of oxidation of the sulfur atoms in the molecule (14). Polymeric compounds can also oxidize at low temperatures. Polyacetylene is degraded at room temperature by atmospheric oxygen and water vapor. Although other conjugated polymers are typically more thermally stable, most exhibit degradation and decreased conductivity after heating in atmosphere.

Photostability and Atmospheric Stability. The presence of oxygen and water vapor has a profound effect upon the photostability of many organic semiconducting materials. For example, singlet oxygen has been found to be a highly reactive intermediate in the photooxidation of poly(phenylene vinylene) derivatives (15). Oxygen and water vapor have also been implicated in dark spot formation in the degradation of light emitting diodes. As a result, performance lifetimes of these devices can be significantly prolonged by packaging under inert atmosphere. An exception to the general behavior of organic semiconducting systems is observed for *p-n* solar cells fabricated with the donor chloroaluminum phthalocyanine. These cells exhibit improved performance after exposure to water and oxygen, due to formation of a crystalline hydrate (16). This provides further evidence for the need for control over impurity and humidity levels during device production.

Structure

A discussion of structure in organic semiconductors includes chemical structure of individual molecules or unit cells, as well as three-dimensional structure of the bulk material. The structure of individual molecules can be tailored by synthetic methods, with molecules designed for specific applications. For example, rigid, planar structures are often used for dyes in photovoltaics, since this increases exciton mobility. Molecular packing and the structure of thin films can be dependent upon the method of fabrication and the structure of individual units. The nature of molecular and polymeric thin films for semiconducting applications ranges from highly crystalline to amorphous. Challenges ahead are to design better adhesion between organic films and substrates. Better interfacing and orientation are needed for many applications, since these can influence charge injection and transport. Though the structure of organic semiconductors is wide ranging, some common classes are discussed below.

Single-Stack Acceptor. Simple charge-transfer salts formed from the planar acceptor TCNQ have a stacked arrangement with the $TCNQ^-$ units facing each other (intermolecular distances of ca 0.3 nm ($\sim$3Å). Complex salts of TCNQ such as $TEA(TCNQ)_2$ consist of stacks of parallel TCNQ molecules, with cation sites between the stacks (17). The interatomic distance between TCNQ units is not always uniform in these salts, and formation of TCNQ dimers (as in $TEA(TCNQ)_2$) and trimers (as in $Cs_2(TCNQ)_3$) can lead to complex crystal structures for the chainlike salts.

Single-Stack Donor. Ion-radical salts can also be formed from electron donors such as tetrathiafulvalene (TTF) or TMPD (*N,N,N',N'*-tetramethyl-*p*-phenylene diamine) with inorganic acceptors such as halogens. The resulting structure of compounds such as TTF(A)$_x$ (A = acceptor) is a linear chain of parallel stacked TTF molecules. The TTF molecules are registered directly above one another, with anions residing in sites between the TTF chains. Unlike the situation found for the TCNQ salts, the TTF molecules in these solids are equally spaced along the chain. Substantial π-overlap exists between the TTF molecules in the chain, providing for conduction along the stacking direction (see Fig. 4).

A large family of compounds is based on combination of TTF-type donors and TCNQ-type acceptors. The resulting $D^{\delta+}A^{\delta-}$ solids form segregated stacks of D and A molecules with uniform spacings along the stacks. Molecules are tilted with respect to the stacking axis, and the π overlap of the molecules on each chain results in the formation of bands and large conductivity along the stacking direction.

A number of other donor–acceptor molecular solids such as TMPD$^{\delta+}$-TCNQ$^{\delta-}$ or TMPD$^{\delta+}$chloranil$^{\delta-}$ crystallize as mixed stacks of alternating D and A molecules. These compounds typically have much higher resistivities than the segregated salts because the alternating -DADA- sequence leaves no continuous channel for conduction.

Polymers. The individual units in a polymer are well-defined, but bulk polymer usually lacks the periodic structure characteristic of inorganic or small molecule crystalline solids. The structure of semiconducting and conducting polymers is determined by the units of the polymer backbone as well as pendant side-groups (18). Typical conducting polymers are conjugated systems that provide for motion of the electrons through the π-system, and both linear chain (poly-acetylenes) and aryl chain poly(p-phenylene) polymers are commonly used. As chain length increases during polymer growth, the long-chain polymers become

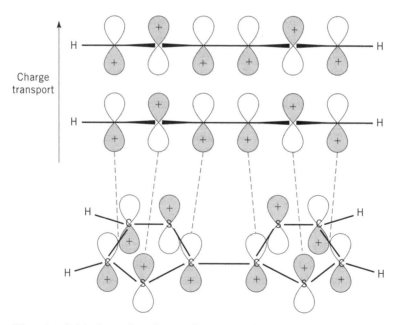

Fig. 4. Orbital overlap for conduction molecular orbitals and molecular stacking in tetrathiafulvalene (TTF).

insoluble and precipitate out during synthesis. The chain length in polymeric solids is thus, influenced by the nature of the solvent and the synthetic process. Although most polymers are fairly amorphous, chemical substitution has been used to design self-assembled oligomeric films with highly ordered structures (19). Film morphology for linear chain polymers can be affected by stretching of the film to orient the chains, resulting in higher crystallinity. The morphology of electrochemically grown conducting polymer thin films has also been shown to be dependent on the nature of the substrate (20).

The conducting polymer poly(sulfur nitride) is unusual in that it is crystalline, consisting of chains of sulfur and nitrogen packed in parallel. Individual chains are planar with alternating cis–trans conformation along the chain, and the interaction with neighboring chains leads to a highly ordered matrix.

Although its use as a transparent electrode in diodes has made it one of the most useful of the conductive polymers, the intractable nature of the material made the structure of polyaniline difficult to determine. Recent studies of polyphenyleneamineimines have conclusively shown that the structure of PANI is an exclusively para-linked system (21).

Synthesis and Manufacture

Donors. Common selenium and sulfur-containing donors such as tetrathiafulvalene [31366-25-3] tetramethyltetraselenafulvalene [55259-49-9] and bis-(ethylenedithio)tetrathiafulvalene [66946-47-3] are commercially available, as are other common donors such as tetramethyl-*p*-phenylenediamine [100-22-1]. Tellurium-containing donors can be prepared using a variety of synthetic routes. Tetratellurafulvalene, for example, is prepared in good yield using tin–lithium exchange (22) (Fig. 5).

Acceptors. Most common acceptor molecules such as tetracyanoethylene or tetracyanoquinodimethane are commercially available. However, TCNQ can be synthesized in high yield by a two-step synthesis involving a condensation of malononitrile with 1,4 cyclohexanedione followed by treatment with an oxidizing agent such as bromine or *N*-bromosuccinamide in pyridine solvent (23) (Figure 6).

Charge-Transfer Salts. Most charge-transfer salts can be prepared by direct mixing of donors and acceptors in solution. Semiconducting salts of TCNQ have been prepared with a variety of both organic and inorganic counterions.

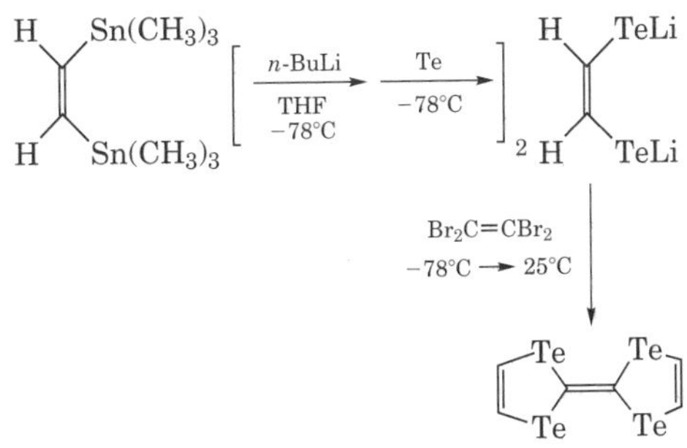

Fig. 5. Synthesis of the common donor tetratellurafulvalene.

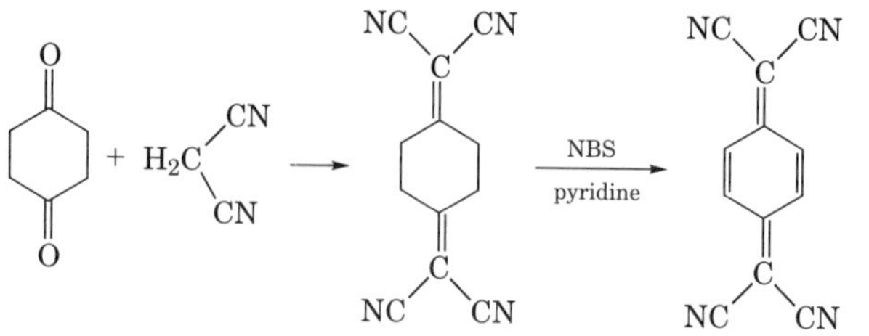

Fig. 6. Synthesis of the acceptor molecule tetracyanoquinodimethane (TCNQ).

Simple salts of the type M^+TCNQ^- can be obtained by direct reaction of a metal such as copper or silver with TCNQ in solution. Solutions of metal iodides can be used in place of the metals, and precipitation of the TCNQ salt occur directly (24).

More highly conducting TCNQ salts of the form $M^+(TCNQ)(TCNQ^-)$ are also easily synthesized, where M^+ can be any one of a variety of organic cations. These salts can be prepared by a number of different methods, though a general method involves mixing of a simple salt with TCNQ in solution. The precipitated salt can then be obtained by filtration.

Compounds containing the donor and acceptor moieties on the same molecule have also been prepared. A class of these compounds contains the D and A parts separated by sulfur atoms (a D–σ-A linkage), with the acceptor portion based on the TCNQ structure (25). Syntheses of these molecules can provide for the design of substituent groups to tailor packing and conducting properties in the solid state. Linkages of this D–σ-A type have garnered interest for possible use as unimolecular rectifiers (26).

Polymers. The π-conjugated polymers used in semiconducting applications are usually insulating, with semiconducting or metallic properties induced by doping (see ELECTRICALLY CONDUCTIVE POLYMERS). Most of the polymers of this type can be prepared by standard methods. The increasing use of polymers in devices in the last decade has led to a great deal of study to improve the processability of thin films of commonly used polymers.

Polyacetylene is the most studied of all of the polymers used in semiconducting applications. Early syntheses of polyacetylene were based on direct polymerization of acetylene using a Ziegler-Natta catalyst (usually titanium tetrabutoxide and triethylammonium) coated onto a glass substrate (27). The resulting polyacetylene films from these direct polymerizations were highly crystalline, insoluble, unstable with respect to atmosphere, and of low tensile strength. Other preparations from polymerization of acetylene have resulted in better film properties and higher conductivities, however, the films are not easily processable. To improve processability and to generate new morphologies of polyacetylene that possess superior properties, precursor methods have been used (28). Polyacetylene has been prepared by ring-opening polymerization of polybenzvalene (29) as well as by polymerization involving addition of catalysts to 1,3,5,7 cyclooctatetraene (COT) (30). A variety of derivatives of polyacetylene can be generated with substituted precursor derivatives.

Polyaniline (PANI) can be formed by electrochemical oxidation of aniline in aqueous acid, or by polymerization of aniline using an aqueous solution of ammonium thiosulfate and hydrochloric acid. This polymer is finding increasing use as a "transparent electrode" in semiconducting devices. To improve processability, a large number of substituted polyanilines have been prepared. The sulfonated form of PANI is water soluble, and can be prepared by treatment of PANI with fuming sulfuric acid (31). A variety of other soluble substituted N-alkylsulfonic acid self-doped derivatives have been synthesized that possess moderate conductivity and allow facile preparation of spincoated thin films (32).

Polythiophene can be synthesized by electrochemical polymerization or chemical oxidation of the monomer. A large number of substituted polythiophenes have been prepared, with the properties of the polymer depending on the nature of the substituent group. Oligomers of polythiophene such as (α-sexithienyl thiophene) can be prepared by oxidative linking of smaller thiophene units (33). These oligomers can be sublimed in vacuum to create polymer thin films for use in organic-based transistors.

Health and Safety Factors

Selenium is an essential element and is beneficial at low concentrations, serving as an antioxidant. Lack of selenium affects thyroid function, and selenium deficiencies have been linked to Keshan Disease (34). Selenium at high levels, however, is toxic. Hydrogen selenide (which is used in semiconductor manufacturing) is extremely toxic, affecting the mucous membranes and respiratory system. However, the toxicity of most organoselenium compounds used as donor compounds for organic semiconductors is not well studied.

Tellurium is not an essential element, and tellurium compounds are in general more toxic than their selenium counterparts. Metallic tellurium is known to have a teratogenic effect in rats, though no studies have been done on the toxicity of tellurium donor compounds (35).

The common acceptor molecule tetracyanoethylene is a poison, and sublimes at relatively low temperature (120°C). The toxicological effects of most cyano-type acceptors have not been fully investigated.

Uses

The performance of early organic semiconductor-based devices was far below that of their Si-based counterparts. In addition, the poor stability and reliability of many organic-based devices slowed their practical use. However, many problems have been overcome and these devices are becoming increasingly in a variety of applications. Organic semiconductors offer many advantages over inorganic-based materials for use in electronic and optoelectronic devices. Among the advantages are the possibility of tailoring properties through synthetic chemical modifications, the relative ease of generating large-area films, and low cost device manufacture. The following is a summary of some current uses and areas of research for potential applications.

Diodes. The first reports of electroluminescence (EL) from organic materials appeared in the 1960s (36–38), but it was not until 1987 that Tang and VanSlyke reported an Organic Light Emitting Diode (OLED) device with reasonable efficiency (39). The key to preparing an efficient device is to use a number of different metal–organic or organic thin-films materials, each serving a different function in the working device. Early devices consisted of two organic layers, one a tertiary amine and the other an aluminum coordination complex (aluminum-tris-8-hydroxyquinoline), the latter of which is responsible for bright green light emission. Over the last several years research on these novel devices has led to new organic and metal–organic materials with emission that covers the entire visible spectrum (40–46). Quantum efficiencies for these devices are comparable to LEDs based on inorganic semiconductors, typically ranging from 1–2.5% (photons/electrons) (41,47). OLEDs can achieve very high brightnesses, of greater than 15,000 cd/m^2 (43,48) and have operational lifetimes greater than 10,000 hours when driven at video brightness (100 cd/m^2) (49).

OLEDs are based on amorphous (glassy) organic films, which can be deposited on virtually any substrate, ranging from rigid supports such as glass or silicon to highly flexible polymer supports (50). This is in contrast to inorganic LEDs, which require crystalline substrates for epitaxial growth of the active materials. An additional feature of OLEDs is that the organic layers are very thin (ca 50–100 nm, 500–1000 Å), and combined with large Franck-Condon shift between absorption and emission, makes them transparent to their own emission (51,52). This transparency may open the door to novel display architectures and applications. Taking these qualities together it is clear that OLEDs have unique characteristics as compared with other potential flat panel display (FPD) technologies. Although technical obstacles must be overcome, OLEDs have potential for certain display applications. This section is an overview of device structure and performance.

Device Structure and Operation.   Efficient generation of electroluminescence in OLEDs depends upon several factors. Injection of carriers (electrons and holes) from the respective electrodes, transport of carriers through the device, and radiative recombination of carriers must all be optimized. A simple OLED consisting of a thin organic or metal–organic film between an indium–tin–oxide (ITO) anode and a metal cathode is shown in Figure 7a. When a potential is applied to the device, the material is oxidized at the anode and reduced at the cathode, leading to the injection of holes and electrons, respectively, into the thin film. The holes and electrons injected into the thin film drift in the presence of the applied field via a hopping mechanism until they are removed at the opposite electrode or encounter an oppositely charged carrier within the film. In the latter case, electron-hole recombination results in the formation of an exciton and resultant emission of light.

Single layer OLEDs have been fabricated with a variety of emitter molecules and conjugated polymers such as poly(phenylene vinylene) (PPV). These single-layer devices are typically not very efficient. To ensure high injection efficiency of both electrons and holes, multiple organic layers are used with each layer optimized for its particular role as a carrier injector or light emitter. The first efficient EL device was a single heterostructure (Fig. 7b), with a tertiary amine hole-transport layer (HTL) and an aluminum complex serving both as the electron-transport layer (ETL) and emissive material (EM) (39). In this device, the preferred conductivity of holes in the HTL and electrons in the ETL leads to a build up of carriers at the HTL–ETL interface. The excitons are formed at the interface, however, and hence are not spatially confined. As a result, it

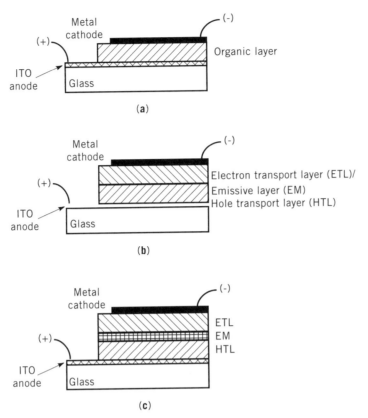

Fig. 7. Schematic of light emitting diodes: (**a**) single-layer device (**b**) single heterostructure (**c**) double heterostructure.

is important that the thickness of the ETL–EM layer emitter be chosen such that the majority of the excitons radiatively decay prior to reaching the adjacent electrode. The choice of ITO for the anode, and a suitable metal as the cathodes is based on the energies of the filled and vacant states of the organic materials relative to the work functions of these contacts. ITO is well matched to inject holes into the HTL HOMO, and low work function metals such as Mg, Al, and Ca can inject electrons into the ETL LUMO. Under forward bias (ITO positive) the injection of holes into the HTL and electrons into the ETL occurs at low (~5 V) potentials, while reversing the bias requires much higher potentials to inject the same current density. Hence, OLEDs show strongly rectifying current–voltage behavior.

    An example "double heterostructure" OLED shown in Figure 7**c** uses an ITO coated glass substrate, upon which a hole transporting layer, typically composed of a tertiary amine (eg, N,N'-biphenyl-N,N'-bis(3-methylphenyl)1-1'biphenyl-4,4'diamine, abbreviated TPD), a thin film of an emissive material such as aluminum-8-hydroxyquinoline(Alq$_3$) and an electron-transporting layer (often an oxidiazole derivative) are sequentially deposited in vacuum (Fig. 8). This molecular multilayer is then capped with a cathode to complete the device. The HTL and ETL have high mobilities for holes and electrons, respectively, but

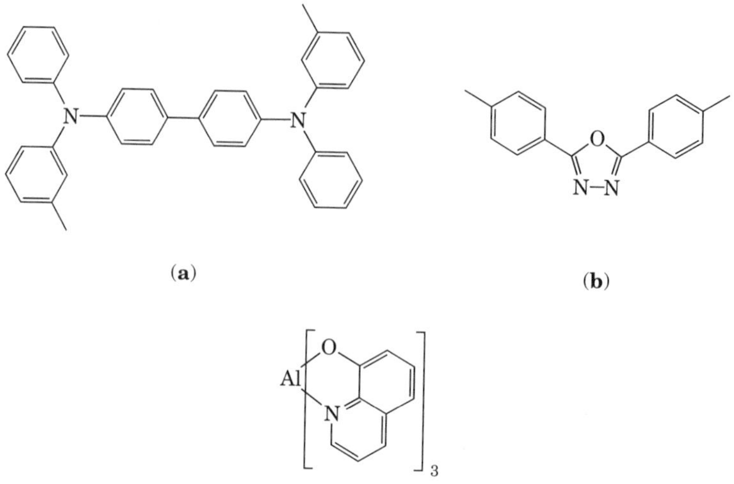

(a)

(b)

(c)

Fig. 8. Structures of (**a**) hole transporter molecule N,N'-biphenyl-N,N'-bis(3-methyl-phenyl)1-1'biphenyl-4,4'diamine (TPD); (**b**) an oxidiazole derivative; and (**c**) the common emitter, 8-hydroxyquinoline.

very low mobilities for the oppositely charged carrier. When an electric field is applied to this device, holes will move through the HTL to the HTL–EM interface and electrons will move through the ETL to the ETL–EM interface. These carries then form a bound state exciton, and subsequently recombine in the EM, leading to efficient light emission. For efficient recombination at the emitter site, the emitter molecule should have a HOMO higher than that of the HTL and a LUMO lower than that of the ETL (53).

The picture presented above for confinement of the excitons within the device is for the EM layer sandwiched between the HTL and ETL. The EM need not be a discrete layer in the OLED, however, for exciton confinement to occur. Alternatively, the EM can consist of a luminescent molecule doped (~1%) into a polymeric or molecular host material (40,41,54,55). So long as the energy gap (or band gap) of the host is higher than that of the EM dopant, excitons will be effectively trapped or confined on the dopant molecules leading to improved EL efficiency. An example of such a dopant-based device involves tetraphenylporphyrin doped into the Alq_3 layer of a TPD–Alq_3 OLED. Even at the dopant concentrations typically used (0.5–1%), the luminescence is entirely due to the tetraphenylporphyrin (54). A wide range of fluorescent and laser dyes have been developed with emission colors spanning the visible spectrum. These materials have very high luminescent quantum efficiencies in dilute solution, making them excellent candidates as emitter dopants in OLEDs, and indeed many have already been incorporated into efficient dopant-based OLEDs.

It has recently been shown that in some cases, the HTL and ETL materials do not need to be segregated into individual layers to achieve high device efficiency. For example, a homogeneous blend can be prepared with a polymeric hole transport (HT) material, and either a molecular or polymeric light emitter and electron transport (ET) material (53,56–61). An OLED prepared in this manner consists of a single organic film sandwiched between an ITO anode and a metal cathode. As in conventional devices, holes are injected from ITO and electrons from the metal cathode, with recombination occurring within the thin polymer film. Polymers and monomeric metal complexes have been used successfully in these blended devices to give colors that span the visible spectrum, with external quantum efficiencies >1% (56,57).

Emission. Generally, the shape of the electroluminescent spectrum is identical to that observed in photoluminescence. The reason for this can be seen by considering molecular LEDs. For closed shell molecules, the ground state S_0 has two electrons in the highest occupied molecular orbital (HOMO), and a completely vacant orbital for the lowest unoccupied molecular orbital (LUMO). In the S_1 excited state, both the HOMO and LUMO are singly occupied. Consider a molecular OLED with a thin organic film between an ITO anode and metal cathode. Since these films consist of weakly interacting molecules, the oxidized and reduced versions of the molecule correspond to the case where holes and electrons, respectively, are located at that molecular site. The holes and electrons move through the device, towards opposite electrodes. When the electron and hole combine, electron transfer from the reduced molecule (electron carrier) to the oxidized molecule (hole carrier) occurs, leading to a molecule to its ground electronic state, S_0, and an adjacent one to the S_1 or Frenkel state (Fig. 9). The S_1 state then relaxes exactly as for photoluminescence. The close relationship between electroluminescence and photoluminescence is not surprising since the emission comes from the same excited state (or Frenkel exciton) in both processes. It should be noted that low levels of impurities can cause quenching of the emission or emission from the impurity due to exciton trapping. As a result, materials used in OLEDs are typically purified by repeated sublimation.

Single emitter devices can be made in a variety of colors, with good efficiency. The color of the device emission can be altered in several ways. One of the most common ways is the use of dopants, as mentioned earlier. Förster energy transfer from the host to the dopant is maximized when the overlap of the host emission spectrum and the dopant absorption spectrum is favorable. By the use of suitable dopants, emission can be tuned and narrowed. In both molecular

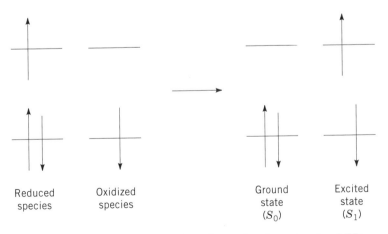

Reduced species   Oxidized species   Ground state (S_0)   Excited state (S_1)

Fig. 9. Representation of exciton formation in molecular LEDs.

and polymeric OLEDs, synthetic methods can be used to tailor the emission. For example, replacing the CH in the 4 position of the quinolate ligand of Alq with a heterocyclic N leads to a 60 nm red shift in luminescence relative to Alq (62). Narrowing and shifting of the emission can also be effected through resonant cavity design. By tailoring the thickness and dielectric constants of the layers, the profile of broad emission from a device can be improved (63).

Device Stability. Stability of OLEDs has been a significant problem, since many organic materials are susceptible to oxidation. The mechanism of degradation in PPV-based devices has been shown to be related to the formation of singlet oxygen (64). Similarly, metal quinolate devices degrade quickly upon exposure to atmosphere water vapor and oxygen. Packaging devices under anaerobic conditions has been shown to extend device lifetime considerably, and devices with operating lifetimes of over 10,000 hours have been reported (49). The role of impurities in device degradation has not been fully established, and the behavior of the organic–contact interface remains an area of intense study. The thermal properties of the organic materials also limit their ultimate lifetime. Increasing the glass-transition temperatures without lowering the carrier mobilities of organic materials is a very active area of research, which may lead to more stable devices or devices that can function well at elevated temperatures.

Efficiency. Efficiency of a device can be reported in terms of an internal quantum efficiency (photons generated/electrons injected). The external quantum efficiency often reported is lower, since this counts only those photons that escape the device. Typically only a fraction of photons escape, due to refraction and waveguiding of light at the glass interface (65). The external efficiency can be increased through the use of shaped substrates (60).

Regardless of device design, for maximum quantum efficiency it is important to have a highly fluorescent emissive center. Many emitters are therefore highly absorbing and rigid, qualities that lead to large fluorescence. Because of spin selection rules governing recombination of the electron and hole, only one quarter of recombinations will lead to singlet excited states. The rest will result in triplet states, which are usually not efficient. This limits the internal quantum efficiency to a nominal maximum of 25%. However, other processes such as triplet–triplet annihilation may increase this value, and the question of the maximum possible efficiency is still not resolved. Though efficiencies for devices have climbed significantly in recent years, further improvements in efficiency are possible.

Outlook. OLEDs in their current form meet the requirements for several low-resolution applications, and with improvements in reliability, efficiency and color tuning they should soon find use in a number of display applications. One area of application will be flexible displays. Organic OLED devices (both molecular and polymer based) on polymer supports have remarkable stability even after repeated bending at sharp angles (67,68). This high degree of substrate flexibility and device stability make these polymer-supported devices attractive for a range of applications, including portable roll-up displays and conformable displays for attachment to windows, windshields, or instrument panels. Aside from bendable devices, transparent OLEDs (TOLEDs) can be created by using a semitransparent metal cathode (69). These devices are >71% transparent when turned off, with light emitted from both top and bottom device surfaces when voltage is applied.

There are technical problems to be overcome before OLEDs can become competitive with liquid crystal technology for use in full color, high definition displays such as computer screens. Improvements in both color saturation and reliability are needed. A review of the prospects of OLEDs for a variety of uses has been published (70). Improvements in color saturation may be realized through the use of microcavities, with layer design used to select red, green, or blue emission from a single broad-emitter material (71,72). Recent development of stacked, independently addressable OLEDs may also lead to improvements in resolution, with three devices (R,G,B) in a vertical configuration providing minimal pixel size (73). Further improvements in the technology should lead to increasing use of OLEDs in a variety of applications.

Photovoltaic Devices. For many inorganic semiconductors, absorption of light can be used to create free electrons and holes. In an organic semiconducting solid, however, absorption of a photon leads to the formation of a bound electron–hole pair. Separation of this pair in an electric field can occur due to differing mobilities of the electron and hole, and this separation of the charge particles gives rise to a photovoltage (see PHOTOVOLTAIC CELLS). This photovoltaic effect is of great interest for use in solar cells, since the conversion of light energy from the sun to electrical energy provides a nonpolluting renewable energy source. The photovoltaic effect is also of interest for photodetectors and for memory storage devices. Photocurrents are of great importance in biological systems as well, and there is interest in light-harvesting devices based on protein pigments (74).

Solar cells and other photovoltaic devices have typically been based on silicon or other inorganic materials, which can provide high power conversion efficiencies (>25%). However, organic materials can also be used to generate a photovoltage. Photovoltaic effects have been studied in a wide variety of materials including solvent-cast films of Buckminsterfullerene (C_{60}) (75) and self-assembled films of alkyl-derivatized metaloporphyrins (76). Many of these systems produce a photovoltage, however, the efficiency is lower than that found in inorganic-based devices.

Device Structure. In both inorganic and organic materials, absorption of light generates electrons and holes, but these will recombine and lead to a loss in device efficiency unless they can be separated and made to flow through an external circuit. There are many types of junctions that are used to separate the electron and hole. In single-crystal inorganic devices, a *p-n* junction is created by adjoining precisely doped *p* and *n*-type regions. For organic devices, a *p-n* junction can be fabricated by sequential deposition of thin films of *n*-type (electron accepting) and *p*-type (electron donating) organic materials. Schottky junctions, formed at semiconductor–metal interfaces, are often used in simple organic photovoltaic cells. In all cases, the potential across the junction is used to sweep the photogenerated carriers towards the opposite electrodes.

A typical organic-based photovoltaic cell (Fig. 10) consists of thin films of an organic semiconducting material sandwiched between conductive surfaces, at least one of which must be transparent (such as glass–indium tin oxide). The organic layer is thin, typically ca 100 nm thick, and can be deposited by solvent

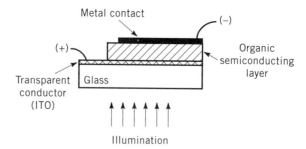

Fig. 10. Typical design for single-layer photovoltaic device.

evaporation or by sublimation under vacuum (77). The organic layer must be relatively thin to minimize resistance in the device. It must also be uniform and defect free (since defects can serve as traps for the charge carriers) and must be also free of pinholes which would lead to short circuiting of the device. In order to maximize efficiency, highly absorbing species are often incorporated into the device. Unlike inorganic devices, organic photovoltaic cells offer the possibility of modifying device properties such as spectral sensitivity through changing of chemical substituents of molecules in the device.

Device Performance. Early sandwich type devices using metal-free phthalocyanines (molar absorptivities of $\sim 10^5$) between gold electrodes were reported to have quantum efficiencies of 0.16% (power conversion efficiencies of 0.009%) (78). Single-dye, Schottky type sandwich cells have been fabricated from both n-type dyes such as rhodamine B and p-type dyes such as merocyanine (79) with appropriate metal contacts. Although single-layer cells have been developed with improved efficiency compared to early devices, cells based on p-n heterojunctions promise superior performance. These cells possess advantages over single-dye devices, including a broader spectral range. Heterojunction solar cells typically consist of two films of organic dyes between indium tin oxide (ITO) and metallic electrodes. To facilitate charge separation, the dyes have large differences in Fermi levels. Highly efficient organic based cells have been fabricated utilizing thermally robust phtalocyanines and perylene derivatives (16). These p-n type devices provide absorption throughout the visible spectrum and monochromatic internal quantum efficiencies of up to 20%.

Aside from dye-based cells, a good deal of study has been devoted to photovoltaic effects in polymer-based devices. Single-layer sandwich-type devices composed of aluminum–polyphenylenevinylene–ITO have been fabricated with solution-cast polymer layers of $\sim 0.5\ \mu$. These devices display quantum efficiencies of up to 5% (power efficiencies of 0.07%), though the efficiency decreases at high illumination levels (80). Other single-layer polymer devices fabricated with polyacetylene show similar efficiencies (81). To obtain optimum efficiency in a photovoltaic device, it is necessary to dissociate the exciton before the charges recombine to generate a photon, and a resulting loss in device efficiency. Multilayer devices consisting of separate donor and acceptor layers generate substantial improvement in device efficiency by facilitating dissociation of the electron-hole pair. In these two-layer devices, excitons that diffuse to the D-A interface are separated (the electron going to the acceptor), and the charges can then be swept on to the electrodes through their respective layers.

However, not all excitons have sufficiently long lifetimes to reach the interface before recombining. To circumvent this problem and increase device efficiency, heterostructure devices have been fabricated. In these devices, donors and acceptors are mixed together to create a network that provides many internal interfaces where charge separation can occur. Heterostructure devices made from the donor polymer poly(2-methoxy-5-(2'-ethyl)-hexyloxy-p-phenylenevinylene) (MEH-PPV) mixed with the acceptor polymer CN-PPV (a cyano-substituted derivative of PPV) showed improved performance (82).

Outlook. Organic materials could provide much cheaper solar cells, since crystalline or epitaxially grown inorganic materials are costly. Also, deposition of organic compounds in vacuum could provide for large-area or custom shaped cells. Recent advances have greatly improved the efficiency of polymer-based solar cells, though they still suffer from a lower efficiency than inorganic-based devices. Also, the polymers suffer from photostability problems that must be overcome. However, they still could see near-term use in applications which do not require exposure to powerful light sources.

Transistors. The use of molecular materials for transistors is attractive because of the facile processability and low cost of organic thin films. Early organic thin-film transistors (TFTs) based on lutetium bis-phthalocyanine (83), polyacetylene (84), or polythiophene derivatives (85) as the active component were shown to have source-drain currents that could be modulated with applied gate voltage. However, performance of early devices was poor compared to inorganic-based devices, and turn-on times were slow. Mobilities for phthalocyanine devices of $\mu = 10^{-4}$ cm^2/V·s and $\mu = 10^{-5}$ cm^2/V·s for polythiophene devices are some four to five orders of magnitude below the mobilities for Si-based devices. Though stable field-effect transistors based on metal phthalocyanines have been developed with mobilities of 10^{-3} cm^2/V·s, much of the research on organic field-effect transistors has focused on semiconducting polymeric materials. Early devices typically generated the polymer electrochemically under inert atmosphere (86) or by spincoating; devices with enhanced performance were fabricated using vacuum deposition of oligomers. Devices from α-conjugated oligomers of polythiophene (α-sexithienyl, α-6T) showed mobility enhancement after heat treatment, with maximum $\mu = 10^{-3}$ cm^2/V·s reported (87). These p-type polycrystalline thin films (thicknesses of 100–200 nm) give on/off current ratios of 10^2–10^3.

"All-organic" transistors based on α-6T were designed with organic materials serving as the semiconductor, insulator, and support (the source and drain contacts were metallic) (88). Aside from being flexible as a result of the organic support, these devices showed very high mobilities. The reported field-effect mobility (4.6×10^{-1}) and transit times are similar to those of amorphous-silicon based transistors. The performance of these devices was noted to be highly dependent on the nature of the semiconductor–insulator interface. Later refinement of α-6T based devices has improved the ratio of source-drain currents in the on and off states, with ratios of $>10^6$ obtained (89). Aside from the p-type polymer, hole transporting devices, n-type devices have been fabricated. In addition, devices have been developed that can function in either n or p type modes.

to the metamagnet family of materials. The $[Fe(C_5(CH_3)_5)]^+[TCNQ]^-$ complex was unique because it was the first member of the metamagnets to exist without extending covalent bonding. Also, the changing of TCNE acceptor to TCNQ increased the spin density in the acceptor portion of the salt, thus, enhancing spin coupling and stabilizing the ferromagnetic behavior of the material. In other studies, bulkier substituent groups were added to the cyclopentadiene ring and other metal ions were substituted in place of iron to enhance the magnetic properties of these organometallic-based magnets.

Electrorheological Materials

Electrorheological materials (fluids) have been known for over a hundred years. An electrorheological (ER) fluid, is one whose viscous properties are modified by applying an electric field (14,15). There are a wide assortment of electrorheological fluids, which are usually a uniform dispersion or suspension of particles within a fluid. A typical example of an electrorheological fluid is a mixture of corn starch in a silicone oil. The mechanism of how electrorheological fluids work is simple. In an applied electric field the particles orient themselves in fiberlike structures (fibrils). When the electric field is off, the fibrils disorient themselves. The damping characteristics of the system can be changed (flexible to rigid). A means of viewing how ER fluids work can be found in a study of the use of electrorheological fluids as a smart space material. Electrorheological fluids were investigated using a single-link flexible-beam test bed. The beam was a sandwich configuration with ER fluids distributed along its length. When the beam is rapidly moved back and forth, the ER fluid is used to provide flexibility during the transient response period of the maneuver (for speed) and made rigid at the end point of the maneuver (for stability). A more practical means of viewing ER fluids is to compared their actions to fly fishing.

Electrorheological fluids are non-Newtonian fluids, that is, the relationship between shear stress and shear strain rate is nonlinear. The changes in viscous properties of electrorheological fluids are only obtained at relatively high electric fields in the order of 1 kV/mm. The practical applications of electrorheological fluids center around their abilities to transferring shear stresses and of acting as a variable damping material in an electric field.

Magnetorheological Materials

Magnetorheological materials (fluids) are the magnetic equivalent of electrorheological fluids. In this case, the particles are either ferromagnetic or ferrimagnetic solids that are either dispersed or suspended within a liquid and the applied field is magnetic (14).

An interesting adaptation of magnetorheological fluids is a series of elastomeric matrix composites embedded with magnetic particles such as iron. During the thermal cure of the elastomer, a strong magnetic field was applied to align the iron particles into chains. These chains of iron particles were locked into place within the composite through the cross-linked structure of the cured elastomer. The resistance of the composite to changes in modulus or deformation was controlled by an external magnetic field. When stimulated by a compressive force, the composite was 60% more resistant to deformation in a magnetic field. When subjected to a shear force, the magnetic-field induced modulus of the composite was an order of magnitude higher in comparison to its modulus measured in a zero magnetic field (19).

Thermoresponsive Materials

Polymeric materials are unique owing to the presence of a glass-transition temperature. At the glass-transition temperatures, the specific volume of the material and its rate of change changes, thus, affecting a multitude of physical properties. Numerous types of devices could be developed based upon this type of stimuli−response behavior, however, this technology is beyond the scope of this article.

Materials that typify thermoresponsive behavior are polyethylene− poly(ethylene glycol) copolymers that are used to functionalize the surfaces of polyethylene films (smart surfaces) (20). When the copolymer is immersed in water, the poly(ethylene glycol) functionalities at the surfaces have solvation behavior similar to poly(ethylene glycol) itself. The ability to design a smart surface in these cases is based on the observed behavior of inverse temperature-dependent solubility of poly(alkene oxide)s in water. The behavior is used to produce surface-modified polymers that reversibly change their hydrophilicity and solvation with changes in temperatures. Similar behaviors have been observed as a function of changes in pH (21−24).

Other examples of materials that respond smartly to changes in temperature are the poly(ethylene glycol)s-modified cottons, polyesters, and polyamide−polyurethanes fabrics. Coupling the thermoresponsiveness of these fabrics with a sensitivity to moisture resulted in a family of fabrics that have properties of thermal adaptability and reversible shrinkage. An interesting application of these materials is as smart pressure bandages. These fabrics, when exposed to an aqueous medium such as blood, contract, thus, applying pressure to the wound to stop the bleeding. As long as the fabric is wet, it applies pressure; dry, the fabric releases the pressure (19).

pH-Sensitive Materials

By far the most widely known classes of pH-sensitive materials are the classes of chemical compounds that include the acids, bases, and indicators. The most interesting of these are the indicators. These materials change colors as a function of pH and usually are totally reversible (see HYDROGEN-ION CONCENTRATION).

In addition to acting as a means of observing changes in pH in titrations and in chemical reactions, indicators have been used in the development of novel chemical indicating devices. During the 1970s, a series of these devices were developed based upon the permeability of organic vapors through polymeric films or through a porous polymeric plug. The devices contained a double-polymer-film pouch containing a glass vial of either a volatile liquid acid or base. Two films were used. One film was to control the diffusion of the organic vapor and the second was a microporous film to protect the system when it was activated. The rest of the device consisted of a paper strip or a porous plug impregnated with a base or an acid (the opposite chemistry as what was in the vial) and an indicator. The system is activated by crushing the vial. The volatile component permeates through the film, it wets and reacts with the chemistry on the paper or in the plug, and in the presence of an indicator, a color change is observed. At higher temperatures, the color change proceeds faster, and at lower temperatures, slower. A family of devices was developed with different time and temperature profiles. These devices were developed to monitor the shipments of time and/or temperature-sensitive items such as pharmaceutics, foods, other perishables, and tasks, such as the changing of intravenous fluids in a hospital.

Other examples of pH-sensitive materials such as the smart hydrogels and smart polymers have been included in other sections of this article.

Light-Sensitive Materials

There are several different types of material families that exhibit different types of response to a light stimuli (14,15). One type is materials that exhibit electro-chromism. This behavior is typically a change in color as a function of an electrical field. Other types of behaviors include thermochromism (color change with heat), photochromic materials (reversible light-sensitive materials), photographic materials (irreversible light-sensitive materials), and photostrictive materials (shape changes due to light usually caused by changes in electronic structure).

An interesting material with both electro- and thermochromism behavior, Li_xVO_2, was evaluated for a "smart window" application (25). Films of Li_xVO_2 were prepared by reactive sputtering and annealing an electrolyte of $LiClO_4$ and propylene carbonate.

Materials are being developed to exhibit both photochromic and photographic behaviors; one such system is based upon a substituted indolinospiro-benzopyran embedded in a polystyrene matrix (26). This system acts as a photochromic system at low exposure in the uv range and at high exposure it functions as a photographic system. The image can be devisualized by heat and can be restored many times with uv irradiation.

Smart Polymers

Even though smart polymers have been used in all types of applications and can exhibit all types of stimuli–response behaviors, the term, smart polymers, has been used as a separate category describing smart materials. The distinction in smartness between materials and polymers can be confusing at times. In medicine and biotechnology, smart polymer systems usually pertain to aqueous polymer solutions, interfaces, and hydrogels. They refer to polymeric systems that are capable of responding strongly to slight changes in the external medium; a first-order transition, accompanied by a sharp decrease in the specific volume of the system. The presence of a poor solvent is one of the main conditions for this phenomenon in swollen polymer networks or linear polymers to occur. A poor solvent causes the forces of attraction between the polymer chain segments to overcome the repulsion forces associated with the extended volume, thus, leading to the collapse of the polymer chain (30). Owing to the recent flurry of activities with smart gels (hydrogels), they are treated separately in this article.

Smart polymers can response to environmental stimuli such as temperature, pH, ions, solvents, reactants, light or uv radiation, stress, recognition, electric fields, and magnetic fields. These stimuli once acted upon, result in changes in phases, shape, optics, mechanics, electric fields, surface energies, recognition, reaction rates, and permeation rates. The polymers that fit into this category include the natural occurring polymers, acrylic polymers and copolymers, and polymers based on combining acid monomers with basic monomers (27,28).

In addition to the polymer systems mentioned elsewhere, one polymer that has been used for a great number of smart applications is poly(vinylidene fluoride), (PVDF). The interesting features of this polymer include its structure, morphology, extraordinarily high piezoelectric and pyroelectric activities, and optical properties (29,30). A smart application of PVDF is in the area of smart aperture antennas. A recent study developed a series of antennas capable of variable directivity and power density. The concept of actuating these antennas was based upon a voltage drop across a film structure of PVDF bonded to metallized Mylar. The change in voltage across the PVDF–Mylar caused the material to expand or contract. This movement caused a moment to develop within the structure, thus, in turn, causing a change in shape (31).

Smart Gels (Hydrogels)

Smart (intelligent) gels (or hydrogels) is not new technology. They are finally reaching commercialization after thirty years of research and development. The concept of smart gels is also more complex than the simple concept of solvent-swollen polymer networks. It is the behavior of the solvent-swollen polymer networks in conjunction with the material being able to response to other types of stimuli; such as temperature, pH, and concentrations of solvents (32–34). The phenomenon was first observed in swollen clear polyacrylamide gels. Upon cooling, the gels would cloud up and became opaque. Warming the gels would regain its clarity. Further investigations into this type of behavior showed that some systems could expand to hundreds of times its volume or could collapse to expel up to 90% of its fluid content with a stimulus of a 1°C change in temperature. These types of behaviors have lead to the recent development of gel-based actuators, valves, sensors, controlled-released systems for drugs and other substances, artificial muscles for robotic devices, chemical memories, optical shutters, molecular separation systems, and of course, toys. Other potential systems for development with smart gels or smart hydrogels include paints, protective coatings, adhesives, recyclable absorbents, bioreactors, bioassay systems, and display devices.

An example of a smart gel chemical composition consists of an entangled network of two polymers; one is a poly(acrylic acid) (PAA), and the second, a tri-block copolymer containing poly(propylene oxide) (PPO) and poly(ethylene oxide) (PEO) in a PEO–PPO–PEO sequence. The PPA portion of the smart gel system is a bioadhesive and is pH responsive. The PPO segments are hydrophobic that help solubilize lipophilic substances in medical applications and the PPO segments tend to aggregate, thus, resulting in gelation at body temperatures. Other types of polymers involved in smart gels include citosan, a hydrolyzed derivative of chitin (a polymer of N-acetylglucosamine which is found in shrimp and crab shells), a copolymer of poly(N-isopropylacrylamide) and poly(acrylic acid), and a graft copolymer of poly(methacrylic acid) and poly(ethylene glycol). These smart gels are being developed for controlled delivery of insulin in diabetic patients and they can be custom-engineered to be temperature, pH, and glucose responsive. Other smart gels are being tailored to be responsive to changes in ionic strength, pressure, stress, light intensity, electric fields, and magnetic fields. Smart hydrogels containing acid groups swell in a weakly alkaline medium, but collapse in an acid medium. On the other hand, smart hydrogels with basic functional groups behave in an opposite manner; ie, swell in an acidic medium and collapse in a basic medium. Polyampholytic smart hydrogels swell to the maximum extent at neutral pH values. When such gels are subjected to either acidic or basic media they undergo rapid dehydration (35). These smart hydrogens are copolymers of methacrylic acid and 2-(N,N-dimethylamino)ethyl methacrylate.

Smart Catalysts

The development of smart catalysts is a relatively new field of investigation. One class of smart catalysts is based on homogeneous rhodium-based poly(alkene oxide)s, in particular those with a poly(ethylene oxide) backbone. Traditionally chemical catalyzed reactions proceed in a manner in which the catalysts becomes more soluble and active as the temperature is raised. This can lead to exothermal runaways, thus, posing both safety and yield problems. These smart catalysts behave differently. As the temperature increases, they become less soluble, thus, precipitating out of solution and inactive. As the reaction mixture cools down, a smart catalyst redissolves and becomes active again (19). Other smart catalysts are being developed that dissociates at high temperatures (less active) and recombines at low temperatures (more active) (36).

Smart Memory Alloys

Some materials undergo a thermomechanical changes as it passes from one phase to another. The crystalline structure of materials such as alloys based on nickel and titanium enters the martensitic phase as the alloy is cooled below a critical temperature. In this stage, the alloy is easily manipulated through large strains with a little change in stress. As the temperature of the alloy is increased above the critical (transformation) temperature it changes into the austentic phase. In the austentic phase, the alloy regains its high strength and high modulus. It behaves like a "normal" metal. The alloy shrinks during the transformation from the martensitic to austentic phase (14,15).

The use of shape memory alloys as actuators depends on their use in the plastic martensitic phase that has been constrained within the structural device. As stress is generated within the shape memory alloy, it is transmitted to the device, thus, triggering an action. The behavior of shape memory alloys is very complex. The material changes its modulus coupled with a change in mechanical losses, acts as a variable damping material in its martensite phase, and has a large energy storing capacity. The properties of shape memory alloys are further complicated by issues of the chemical composition of the alloys and past histories.

Shape memory alloys (SMAs) can be divided into three functional groups; one-way SMAs, two-way SMAs, and magnetically controlled SMAs (37). The magnetically controlled SMAs show great potential as actuator materials for smart structures because they could provide rapid strokes with large amplitudes under precise control. The most extensively used conventional shape memory alloys are the nickel–titanium- and copper-based alloys. Due to their low cost, iron-based shape memory alloys are becoming more popular in smart structure applications. Iron–manganese–silicon steels, iron–manganese–silicon steels alloyed with chromium, nickel, and cobalt, and iron–manganese–silicon steels alloyed with nitrogen all fit into this category.

As previously mentioned, the nickel–titanium alloys have been the most widely used shape memory alloys. This family of nickel–titanium alloys is known as Nitinol (Nickel Titanium Naval Ordnance Laboratory in honor of the place where this material behavior was first observed). Nitinol have been used for military, medical, safety, and robotics applications. Specific usages include hydraulic lines capable of F-14 fighter planes, medical tweezers, anchors for attaching tendons to bones, eyeglass frames, underwire brassieres, and antiscalding valves used in water faucets and shower heads (38,39). Nitinol can be used in robotics actuators and micromanipulators that simulate human muscle motion. The ability of Nitinol to exert a smooth, controlled force when activated is a mass advantage of this material family (5).

Elastorestrictive Materials

This class of smart materials is the mechanical equivalent of electrostrictive and magnetostrictive materials. Elastorestrictive materials exhibit high hysteresis between strain and stress (14,15). This hysteresis can be caused by motion of ferroelastic domain walls. This behavior is more complicated and complex near a martensitic phase transformation. At this transformation, both crystal structural changes induced by mechanical stress and by domain wall motion occur. Martensitic shape memory alloys have broad, diffuse phase transformations and coexisting high and low temperature phases. The domain wall movements disappear with fully transformation to the high temperature austentic (paraelastic) phase.

Materials with Unusual Behaviors or Unusual Materials

Only a few materials fit into this category, unusually they can be categorized into one of the above material classes. Water fits into the category of materials with unusual behavior. Water is one of the few materials that expands upon freezing. It changes volume by approximately 8% transiting from the liquid to the solid state. Indicating devices were produced based upon the freezing and expansion of water. A glass vial was filled completely with water and nucleating agents and hermetically sealed. The vial was placed on a piece of paper impregnated on the opposite side with a water soluble dye. Then the assembly was packaged together. The device was placed in various type of perishables, such as frozen

If the acoustic pressure amplitude of a propagating acoustic wave is relatively large (greater than ≈ 0.5 MPa), local inhomogeneities in the liquid (eg, gas-filled crevices in particulates) can give rise to the explosive growth of a nucleation site into a cavity of macroscopic dimensions, primarily filled with vapor. Such a bubble is inherently unstable, and its subsequent collapse can result in an enormous concentration of energy (Fig. 1). This violent cavitation event has been termed "transient cavitation" (23). A normal consequence of this unstable growth and subsequent collapse is that the cavitation bubble itself is destroyed. Gas-filled remnants from the collapse, however, may give rise to reinitiation of the process.

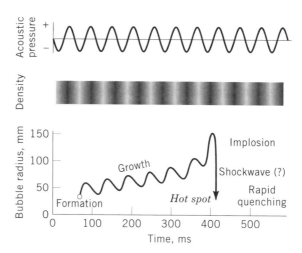

Fig. 1.   Transient acoustic cavitation: the origin of sonochemistry and sonoluminescence.

The generally accepted explanation for the origin of sonochemistry and sonoluminescence is the hot-spot theory, in which the potential energy given the bubble as it expands to maximum size is concentrated into a heated gas core as the bubble implodes. The oscillations of a gas bubble driven by an acoustic field are generally described by Rayleigh-Plesset equation; one form of which, called the Gilmore equation (9,23), can be expressed a second-order nonlinear differential equation given as

$$R\left(1 - \frac{U}{C}\right)\frac{d^2R}{dt^2} + \frac{3}{2}\left(1 - \frac{U}{3C}\right)\left(\frac{dR}{dt}\right)^2 - \left(1 + \frac{U}{C}\right)H - \frac{R}{C}\left(1 - \frac{U}{C}\right)\frac{dH}{dt} = 0$$

$$(1)$$

The radius and velocity of the bubble wall are given by R and U respectively. The values for H, the enthalpy at the bubble wall, and C, the local sound speed, may be expressed as follows, using the Tait equation of state for the liquid.

$$H = \frac{n}{n-1}\frac{A^{1/n}}{\rho_o}\left[(P(R) + B)^{n-1/n} - (P_\infty(t) + B)^{n-1/n}\right] \qquad (2)$$

and

$$C = [c_o^2 + (n - 1)H]$$ (3)

The linear speed of sound in the liquid is c_o. A, B, and n are constants that should be set to the appropriate values for water. Any acoustic forcing function is included in the pressure at infinity term, $P_\infty(t)$. The pressure at the bubble wall, $P(R)$, is given by

$$P(R) = \left(P_o + \frac{2\sigma}{R}\right)\left(\frac{R_o}{R}\right)^{3\gamma} - \frac{2\sigma}{R} - \frac{4\mu U}{R}$$ (4)

where the initial radius of the bubble at time zero is R_o. The ambient pressure of the liquid is P_o, the surface tension σ, the shear viscosity μ, and the polytropic exponent γ.

The validity of the Gilmore equation to compute the behavior of a single, isolated cavitating bubble has been experimentally confirmed. For example, using a light scattering technique, various researchers have obtained measurements of the radius-time curve for single collapsing bubbles, simultaneous with optical emission from sonoluminescence (see below). The single-bubble sonoluminescent emission is seen as a sharp spike, appearing at the final stages of bubble collapse, and the general shape of the theoretical radius-time curve is observed (24–26).

Two-Site Model of Sonochemical Reactivity

The transient nature of the cavitation event precludes conventional measurement of the conditions generated during bubble collapse. Chemical reactions themselves, however, can be used to probe reaction conditions. The effective temperature realized by the collapse of clouds of cavitating bubbles can be determined by the use of competing unimolecular reactions whose rate dependencies on temperature have already been measured. This technique of comparative-rate chemical thermometry was used by Suslick, Hammerton, and Cline to first determine the effective temperature reached during cavity collapse (6). The sonochemical ligand substitutions of volatile metal carbonyls were used as these comparative rate probes (eq. 5, where the symbol $\xrightarrow{)))}$ represents ultrasonic irradiation of a solution, and L represents a substituting ligand). These kinetic studies revealed that there were in fact

$$M(CO)_x \xrightarrow{)))} M(CO)_{x-n} + n\ CO \xrightarrow{L} M(CO)_{x-n}(L)_n$$

where M = Fe, Cr, Mo, W (5)

two sonochemical reaction sites: the first (and dominant site) is the bubble's interior gas-phase while the second is an *initially* liquid phase. The latter corresponds either to heating of a shell of liquid around the collapsing bubble or to droplets of liquid ejected into the hot-spot by surface wave distortions of the collapsing bubble, as shown schematically in Figure 2.

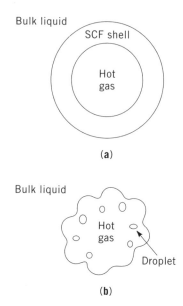

Fig. 2. Two-site models of the sonochemical reactions sites. (**a**) Thermal diffusion shell model; (**b**) surface wave droplet model.

The effective local temperatures in both sites were determined. By combining the relative sonochemical reaction rates for equation 5 with the known temperature behavior of these reactions, the conditions present during cavity collapse could then be calculated. The effective temperature of these hotspots was measured at $\approx$5200 K in the gas-phase reaction zone and $\approx$1900 K in the initially liquid zone (6). Of course, the comparative rate data represent only a composite temperature: during the collapse, the temperature has a highly dynamic profile, as well as a spatial temperature gradient. This two-site model has been confirmed with other reactions (27,28) and alternative measurements of local temperatures by sonoluminescence are consistent (7), as discussed later.

Microjet Formation during Cavitation at Liquid–Solid Interfaces

A very different phenomenon arises when cavitation occurs near extended liquid–solid interfaces. There are two proposed mechanisms for the effects of cavitation near surfaces: microjet impact and shockwave damage. Whenever a cavitation bubble is produced near a boundary, the asymmetry of the liquid particle motion during cavity collapse can induce a strong deformation in the cavity (9). The potential energy of the expanded bubble is converted into kinetic energy of a liquid jet that extends through the bubble's interior and penetrates the opposite bubble wall. Because most of the available energy is transferred to the accelerating jet, rather than the bubble wall itself, this jet can reach velocities of hundreds of meters per second. Because of the induced asymmetry, the jet often impacts the solid boundary and can deposit enormous energy densities at the site of impact. Such energy concentration can result in severe damage to the boundary surface. Figure 3 is a photograph of a jet developed in a collapsing cavity. The second mechanism of cavitation-induced surface damage invokes shockwaves created by cavity collapse in the liquid. The impingement of microjets and shockwaves on the surface creates the localized erosion responsible for much of ultrasonic cleaning and many of the sonochemical effects on heterogeneous reactions. In this process, the erosion of metals by cavitation generates newly exposed, highly heated surfaces that are highly reactive.

A solid surface several times larger than the resonance bubble size is necessary to induce distortions during bubble collapse. For ultrasound of $\approx$20 kHz, damage associated with jet formation cannot occur if the solid particles are smaller than $\approx$200 μm. In these cases, however, the shockwaves created by homogeneous cavitation can create high velocity interparticle collisions (10,11). Suslick and co-workers have found that the turbulent flow and shockwaves produced by intense ultrasound can drive metal particles together at sufficiently high speeds to induce effective melting in direct collisions (Fig. 4) and the abrasion of surface crystallites in glancing impacts (Fig. 5). A series of transition metal powders were used to probe the maximum temperatures and speeds reached during interparticle collisions. Using the irradiation of Cr, Mo, and W powders in decane at 20 kHz and 50 W/cm^2, agglomeration and essentially a localized melting occurs for the first two metals, but not the third (Fig. 6). On the basis of the melting points of these metals, the effective transient temperature reached at the point of impact during interparticle collisions is roughly 3000°C (which is unrelated to the temperature inside the hot-spot of a collapsing bubble). From the volume of the melted region of impact, the amount of energy generated during collision was determined. From this, a lower estimate of the velocity of impact is roughly one half the speed of sound (10). These are precisely the effects expected on suspended particulates from cavitation-induced shockwaves in the liquid.

Fig. 3. Liquid jet produced during collapse of a cavitation bubble near a solid surface. The width of the bubble is about 1 mm. Reproduced with permission (8).

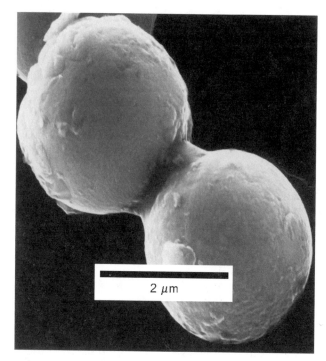

Fig. 4. Scanning electron micrograph of 5-μm diameter Zn powder. Neck formation from localized melting is caused by high-velocity interparticle collisions. Similar micrographs and elemental composition maps (by Auger electron spectroscopy) of mixed metal collisions have also been made. Reproduced with permission (10).

Sonoluminescence

Types of Sonoluminescence. In addition to driving chemical reactions, ultrasonic irradiation of liquids can also produce light. Sonoluminescence was first observed from water in 1934 by Frenzel and Schultes (29). As with sonochemistry, sonoluminescence derives from acoustic cavitation. It is now generally thought that there are two separate forms of sonoluminescence: multiple-bubble sonoluminescence (MBSL) and single-bubble sonoluminescence (SBSL) (17,24,30). Since cavitation is a nucleated process and liquids generally contain large numbers particulates that serve as nuclei, the cavitation field generated by propagating or standing acoustic wave typically consists of very large numbers of interacting bubbles, distributed over an extended region of the liquid. If this cavitation is sufficiently intense to produce sonoluminescence, then this phenomenon is called multiple-bubble sonoluminescence (MBSL) (2,17).

Under the appropriate conditions, the acoustic force on a bubble can be used to balance against its buoyancy, holding the single bubble isolated in the liquid by acoustic levitation. This permits examination of the dynamic characteristics of the bubble can in considerable detail, from both a theoretical and an experimental perspective. Such a bubble is typically quite small, compared to an acoustic wavelength (eg, at 20 kHz, the resonance size is approximately 150 μm). It was

(a)

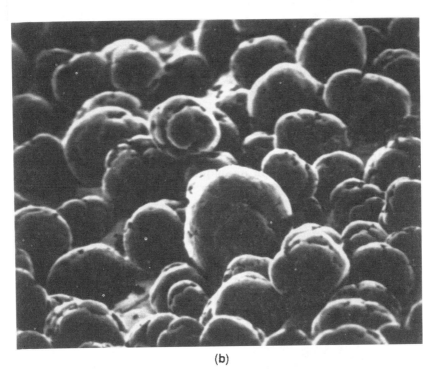

(b)

Fig. 5. The effect of ultrasonic irradiation on the surface morphology and particle size of Ni powder. Initial particle diameters (**a**) before ultrasound were ≈160 μm; (**b**) after ultrasound, ≈80 μm. High velocity interparticle collisions caused by ultrasonic irradiation of slurries are responsible for the smoothing and removal of passivating oxide coating. Reproduced with permission (11).

speculations on the possibilities of inertial confinement (*hot*) fusion have been made (40,41).

Spectroscopic Probes of Cavitation Conditions. Determination of the temperatures reached in a cavitating bubble has remained a difficult experimental problem. As a spectroscopic probe of the cavitation event, MBSL provides a solution. High resolution MBSL spectra from silicone oil under Ar have been reported and analyzed (7). The observed emission comes from excited state C_2 and has been modeled with synthetic spectra as a function of rotational and vibrational temperatures, as shown in Figure 7. From comparison of synthetic to observed spectra, the effective cavitation temperature is 5050 ± 150 K. The excellence of the match between the observed MBSL and the synthetic spectra provides definitive proof that the sonoluminescence event is a thermal, chemiluminescence process. The agreement between this spectroscopic determination of the cavitation temperature and that made by comparative rate thermometry of sonochemical reactions is surprisingly close (6).

The interpretation of the spectroscopy of SBSL is much less clear. At this writing, SBSL has been observed primarily in aqueous fluids, and the spectra obtained are surprisingly featureless. Some very interesting effects are observed when the gas contents of the bubble are changed (39,42). Furthermore, the spectra show practically no evidence of OH emissions, and when He and Ar bubbles are considered, continue to increase in intensity even into the deep ultraviolet. These spectra are reminiscent of black body emission with temperatures *considerably* in excess of 5000 K and lend some support to the concept of an imploding shock wave (41). Several other alternative explanations for SBSL have been presented, and there exists considerable theoretical activity in this particular aspect of SBSL.

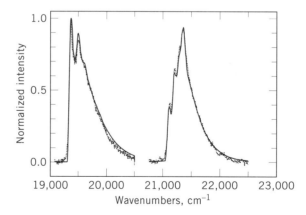

Fig. 7. Sonoluminescence of excited state C_2. Emission from the $\Delta v = +1$ manifold of the $d^3\Pi_g - a^3\Pi_u$ transition (Swan band) of C_2. Reproduced with permission (7). $\cdots\cdots$, Observed sonoluminescence from polydimethylsiloxane silicone oil under Ar at 0°C; ———, best fit synthetic spectrum, with $T_v = T_r = 4900$ K.

Sonochemistry

In a fundamental sense, chemistry is the interaction of energy and matter. Chemical reactions require energy in one form or another to proceed: chemistry stops as the temperature approaches absolute zero. One has only limited control, however, over the nature of this interaction. In large part, the properties of a specific energy source determines the course of a chemical reaction. Ultrasonic irradiation differs from traditional energy sources (such as heat, light, or ionizing radiation) in duration, pressure, and energy per molecule. The immense local temperatures and pressures and the extraordinary heating and cooling rates generated by cavitation bubble collapse mean that ultrasound provides an unusual mechanism for generating high energy chemistry. Like photochemistry, very large amounts of energy are introduced in a short period of time, but it is thermal, not electronic, excitation. As in flash pyrolysis, high thermal temperatures are reached, but the duration is very much shorter (by $>10^4$) and the temperatures are even higher (by five- to ten-fold). Similar to shock-tube chemistry or multiphoton infrared laser photolysis, cavitation heating is very short lived, but occurs within condensed phases. Furthermore, sonochemistry has a high-pressure component, which suggests that one might be able to produce on a microscopic scale the same macroscopic conditions of high temperature–pressure "bomb" reactions or explosive shockwave synthesis in solids. Figure 8 presents an interesting comparison of the parameters that control chemical reactivity (time, pressure, and energy) for various forms of chemistry.

Experimental Design.   A variety of devices have been used for ultrasonic irradiation of solutions. There are three general designs in use presently: the ultrasonic cleaning bath, the direct immersion ultrasonic horn, and flow reactors.

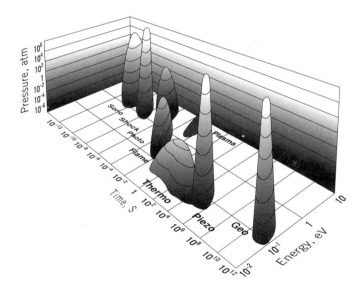

Fig. 8.   Chemistry: the interaction of energy and matter. To convert atm to Pa, multiply by 1.013×10^5.

The originating source of the ultrasound is generally a piezoelectric material, usually a lead zirconate titanate ceramic (PZT), which is subjected to a high a-c voltage with an ultrasonic frequency (typically 15 to 50 kHz). For industrial use, the more robust magnetostrictive metal alloys (usually of Ni) can be used as the core of a solenoid generating an alternating magnetic field with an ultrasonic frequency. The vibrating source is attached to the wall of a cleaning bath, to an amplifying horn, or to the outer surfaces of a flow-through tube or diaphragm.

The ultrasonic cleaning bath is clearly the most accessible source of laboratory ultrasound and has been used successfully for a variety of liquid–solid heterogeneous sonochemical studies. Lower acoustic intensities can often be used in liquid–solid heterogeneous systems, because of the reduced liquid tensile strength at the liquid–solid interface. For such reactions, a common ultrasonic cleaning bath will therefore often suffice. The low intensity available in these devices (≈ 1 W/cm^2), however, can prove limiting. In addition, the standing wave patterns in ultrasonic cleaners require accurate positioning of the reaction vessel. On the other hand, ultrasonic cleaning baths are easily accessible, relatively inexpensive, and usable on moderately large scale. Even in the case of heterogeneous sonochemistry, however, the ultrasonic cleaning bath must be viewed as an apparatus of limited capability.

The most intense and reliable source of ultrasound generally used in the chemical laboratory is the direct immersion ultrasonic horn (50 to 500 W/cm^2), as shown in Figure 9, which can be used for work under either inert or reactive atmospheres or at moderate pressures (<10 atmospheres). These devices are available from several manufacturers at modest cost and are often used by biochemists for cell disruption. A variety of sizes of power supplies and titanium horns are available, thus allowing flexibility in sample size. Commercially available flow-through reaction chambers which will attach to these horns allow the processing of multiliter volumes. The acoustic intensities are easily and reproducibly variable; the acoustic frequency is well controlled, albeit fixed (typically at 20 kHz). Since power levels are quite high, counter-cooling of the reaction solution is essential to provide temperature control. Erosion of the titanium tip is a potential disadvantage, especially in corrosive media. Such erosion is generally a very slow process without chemical consequences (given the high tensile strength and low reactivity of Ti metal) and can be avoided by using the horn to irradiate through a cooling solution into a reaction solution held in a glass container (a so-called cup-horn).

Large-scale ultrasonic generation in flow-through configurations is a well-established technology (43–46). Liquid processing rates as high as 200 L/min are routinely accessible from a variety of modular, in-line designs with acoustic power of ≈ 20 kW per unit. The industrial uses of these units include (1) degassing of liquids, (2) dispersion of solids into liquids, (3) emulsification of immiscible liquids and (4) large-scale cell disruption (45,46).

Homogeneous sonochemistry typically is not a very energy efficient process (although it can be more efficient than photochemistry), whereas heterogeneous sonochemistry is several orders of magnitude better. Unlike photochemistry, whose energy inefficiency is inherent in the production of photons, ultrasound

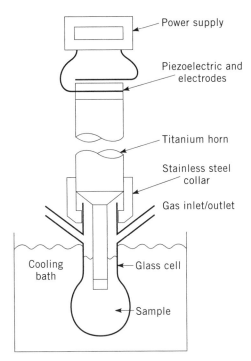

Power supply

Piezoelectric and electrodes

Titanium horn

Stainless steel collar

Gas inlet/outlet

Cooling bath

Glass cell

Sample

Fig. 9. A typical sonochemical apparatus with direct immersion ultrasonic horn. Ultrasound can be easily introduced into a chemical reaction with good control of temperature and ambient atmosphere. The usual piezoelectric ceramic is PZT, a lead zirconate titanate ceramic. Similar designs for sealed stainless steel cells can operate at pressures above 10 bar.

can be produced with nearly perfect efficiency from electric power. A primary limitation of sonochemistry remains the small fraction of the acoustic power actually involved in the cavitation events. This might be significantly improved, however, if a more efficient means of coupling the sound field to generate cavitation can be found.

Sonochemistry is strongly affected by a variety of external variables, including acoustic frequency, acoustic intensity, bulk temperature, static pressure, ambient gas, and solvent (47). These are the important parameters which need consideration in the effective application of ultrasound to chemical reactions. The origin of these influences is easily understood in terms of the hot-spot mechanism of sonochemistry.

The frequency of the sound field is not a commonly altered variable in most sonochemistry. Changing sonic frequency alters the resonant size of the cavitation event and to some extent, the lifetime of the bubble collapse, but the overall process remains unchanged. Sonochemistry is therefore less influenced over the range where cavitation can occur (from tens of Hz to a few MHz). The observed sonochemical rates may change, but well-controlled comparisons of efficiency are lacking at this time and will prove difficult. Subtle differences in product distributions from homogeneous reactions have been occasionally reported (48). At very high frequencies (above a few MHz), cavitation ceases, and sonochemistry is generally not observed.

Acoustic intensity has a dramatic influence on the observed rates of sonochemical reactions. Below a threshold value, the amplitude of the sound field is too small to induce nucleation or bubble growth. Above the cavitation threshold, increased intensity of irradiation (from an immersion horn, for example) will increase the effective volume of the zone of liquid which will cavitate, and thus, increase the observed sonochemical rate. Furthermore, as the acoustic pressures increase, the range of bubble sizes which will undergo transient cavitation increases; this too will increase the observed sonochemical rate. It is often observed experimentally, however, that as one continues to increase acoustic amplitude, eventually rates begin to diminish. At high intensities, the cavitation of the liquid near the radiating surface becomes so intense as to produce a shroud of bubbles which will diminish the penetration of the sound into the liquid. In addition, bubble growth may become so rapid that the bubble grows beyond the size range of transient cavitation before implosive collapse can occur.

The effect of the bulk solution temperature lies primarily in its influence on the bubble content before collapse. With increasing temperature, in general, sonochemical reaction rates are *slower*! This reflects the dramatic influence which solvent vapor pressure has on the cavitation event: the greater the solvent vapor pressure found within a bubble prior to collapse, the less effective the collapse. There is generally a linear correlation of the log of the sonochemical rate and the total solvent vapor pressure (49). When secondary reactions are being monitored (as in chemical reactions occurring after initial acoustic erosion of a passivated surface), temperature will play its usual role in thermally activated chemical reactions. This explains the common observation that rates of heterogeneous sonochemistry often have an optimal reaction temperature: below this temperature, cavitational processes are improved, but secondary chemical reactions are slowed, and at higher temperatures, vice versa.

Increases in the applied static pressure increase the acoustic intensity necessary for cavitation, but if equal number of cavitation events occur, the collapse should be more intense. In contrast, as the ambient pressure is reduced, eventually the gas-filled crevices of particulate matter which serve as nucleation sites for the formation of cavitation in even "pure" liquids, will be deactivated, and therefore the observed sonochemistry will be diminished.

The choice of ambient gas will also have a major impact on sonochemical reactivity. The maximum temperature reached during cavitation is strongly dependent on the polytropic ratio ($\gamma = C_p/C_v$) of the ambient gas, which defines the amount of heat released during the adiabatic compression of that gas. Monatomic gases give much more heating than diatomic, which are much better than polyatomic gases (including solvent vapor). Sonochemical rates are also significantly influenced by the thermal conductivity of the ambient, so even the noble gases affect cavitation differently: He is generally much worse than Ar and Xe is the best; Ar is often the most cost-effective choice. In addition, sonochemical reactions will often involve the gases present in the cavitation event.

The choice of the solvent also has a profound influence on the observed sonochemistry. The effect of vapor pressure has already been mentioned. Other liquid properties, such as surface tension and viscosity, will alter the threshold of cavitation, but this is generally a minor concern. The chemical reactivity of the solvent is often much more important. No solvent is inert under the high temperature conditions of cavitation (50). One may minimize this problem, however, by using robust solvents that have low vapor pressures so as to minimize their concentration in the vapor phase of the cavitation event. Alternatively, one may wish to take advantage of such secondary reactions, for example, by using halocarbons for sonochemical halogenations. With ultrasonic irradiations in water, the observed aqueous sonochemistry is dominated by secondary reactions of OH· and H· formed from the sonolysis of water vapor in the cavitation zone (51–53).

Control of sonochemical reactions is subject to the same limitation that any thermal process has: the Boltzmann energy distribution means that the energy per individual molecule will vary widely. One does have easy control, however, over the energetics of cavitation through the parameters of acoustic intensity, temperature, ambient gas, and solvent choice. The thermal conductivity of the ambient gas (eg, a variable He/Ar atmosphere) and the overall solvent vapor pressure provide easy methods for the experimental control of the peak temperatures generated during the cavitational collapse.

Homogeneous Sonochemistry: Bond Breaking and Radical Formation. The chemical effect of ultrasound on aqueous solutions have been studied for many years. The primary products are H_2 and H_2O_2; there is strong evidence for various high-energy intermediates, including HO_2, H·, OH·, and perhaps $e_{(aq)}^-$. The elegant work of Riesz and collaborators used electron paramagnetic resonance with chemical spin-traps to demonstrate definitively the generation of H· and OH· during ultrasonic irradiation, even with clinical sources of ultrasound (51–53). The extensive work in Henglein's laboratory involving aqueous sonochemistry of dissolved gases has established clear analogies to combustion processes (27,28). As one would expect, the sonolysis of water, which produces both strong reductants and oxidants, is capable of causing secondary oxidation and reduction reactions, as often observed by Margulis and co-workers (54). Most recently there has been strong interest shown in the use of ultrasound for remediation of low levels of organic contamination of water (47,55,56). The OH· radicals produced from the sonolysis of water are able to attack essentially all organic compounds (including halocarbons, pesticides, and nitroaromatics) and through a series of reactions oxidize them fully. The desirability of sonolysis for such remediation lies in its low maintenance requirements and the low energy efficiency of alternative methods (eg, ozonolysis, uv photolysis).

In contrast, the ultrasonic irradiation of organic liquids has been less studied. Suslick and co-workers established that virtually all organic liquids will generate free radicals upon ultrasonic irradiation, as long as the total vapor pressure is low enough to allow effective bubble collapse (49). The sonolysis of simple hydrocarbons (for example, n-alkanes) creates the same kinds of products associated with very high temperature pyrolysis (50). Most of these products (H_2, CH_4, and the smaller 1-alkenes) derive from a well-understood radical chain mechanism.

The sonochemistry of solutes dissolved in organic liquids also remains largely unexplored. The sonochemistry of metal carbonyl compounds is an exception (57). Detailed studies of these systems led to important mechanistic understandings of the nature of sonochemistry. A variety of unusual reactivity patterns have been observed during ultrasonic irradiation, including multiple ligand dissociation, novel metal cluster formation, and the initiation of homogeneous catalysis at low ambient temperature (57).

Applications of Sonochemistry to Materials Synthesis. Of special interest is the recent development of sonochemistry as a synthetic tool for the creation of unusual inorganic materials (58,59). As one example, the recent discovery of a simple sonochemical synthesis of amorphous iron (Fig. 10) helped settle the longstanding controversy over its magnetic properties (60,61). More generally, ultrasound has proved extremely useful in the synthesis of a

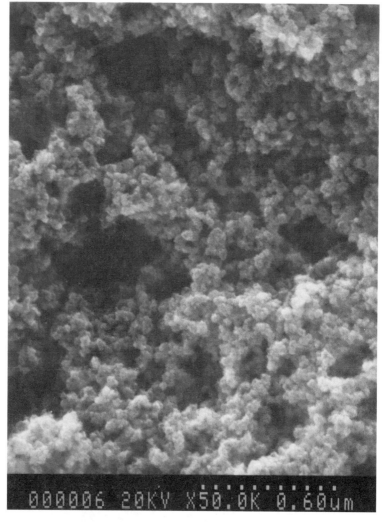

Fig. 10. Scanning electron micrograph of amorphous nanostructured iron powder produced from the ultrasonic irradiation of Fe(CO)$_5$. Reproduced with permission (60).

wide range of nanostructured materials, including high surface area transition metals, alloys, carbides, oxides and colloids (62,64). Sonochemical decomposition of volatile organometallic precursors in high boiling solvents produces nano-structured materials in various forms with high catalytic activities. Nanometer colloids, nanoporous high surface area aggregates, and nanostructured oxide supported catalysts can all be prepared by this general route, as shown schematically in Figure 11. For example, sonication of iron pentacarbonyl with silica generated an amorphous nanostructured $Fe-SiO_2$ supported catalyst. This catalyst showed higher catalytic activity for the Fischer-Tropsch synthesis compared to the conventional Fe–silica catalyst prepared by the traditional incipient wetness method. Sonochemical synthesis of high surface area alloys can be accomplished by the sonolysis of $Fe(CO)_5$ and $Co(CO)_3(NO)$. As another example, ultrasonic irradiation of $Mo(CO)_6$ produces aggregates of nanometer-sized clusters of face centered cubic molybdenum carbide. The extremely porous material had a high surface area and consisted of aggregates of $\approx$2-nm sized particles. The catalytic properties showed that the molybdenum carbide generated by ultrasound is an active and highly selective dehydrogenation catalyst comparable to commercial ultrafine platinum powder.

Sonochemistry is also proving to have important applications with polymeric materials. Substantial work has been accomplished in the sonochemical initiation of polymerization and in the modification of polymers after synthesis (3,5). The use of sonolysis to create radicals which function as radical initiators has been well explored. Similarly the use of sonochemically prepared radicals and other reactive species to modify the surface properties of polymers is being developed, particularly by G. Price. Other effects of ultrasound on long chain polymers tend to be mechanical cleavage, which produces relatively uniform size distributions of shorter chain lengths.

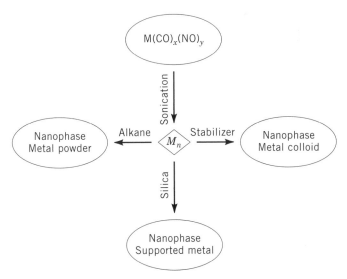

Fig. 11. Sonochemical synthesis of various forms of nanostructured materials. $n = 100-1000$.

Another important application has been the sonochemical preparation of biomaterials, most notably protein microspheres (65–68). Using high intensity ultrasound and simple protein solutions, a remarkably easy method to make both air-filled microbubbles and nonaqueous liquid-filled microcapsules has been developed. Figure 12 shows an electron micrograph of sonochemically prepared microspheres. These microspheres are stable for months, and being slightly smaller than erythrocytes, can be intravenously injected to pass unimpeded through the circulatory system. The mechanism responsible for microsphere formation is a combination of *two* acoustic phenomena: emulsification and cavitation. Ultrasonic emulsification creates the microscopic dispersion of the protein solution necessary to form the proteinaceous microspheres. Alone, however, emulsification is insufficient to produce long-lived microspheres. The long life of these microspheres comes from a sonochemical cross-linking of the protein shell. Protein cysteine residues are oxidized during microsphere formation by sonochemically produced superoxide. These protein microspheres, have a wide range of biomedical applications, including their use as echo contrast agents for sonography, magnetic-resonance-imaging contrast enhancement, drug delivery, among others.

Heterogeneous Sonochemistry: Reactions of Solids with Liquids. The use of ultrasound to accelerate chemical reactions in heterogeneous systems has

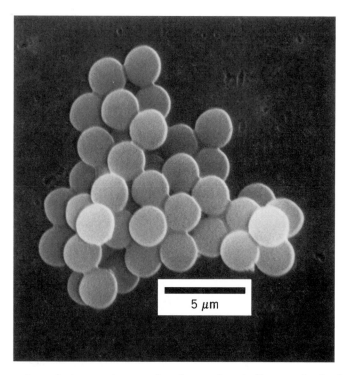

Fig. 12. Scanning electron micrograph of sonochemically synthesized hemoglobin microspheres.

become increasingly widespread. The physical phenomena which are responsible include the creation of emulsions at liquid–liquid interfaces, the generation of cavitational erosion and cleaning at liquid–solid interfaces, the production of shock wave damage and deformation of solid surfaces, the enhancement in surface area from fragmentation of friable solids, and the improvement of mass transport from turbulent mixing and acoustic streaming.

The use of high-intensity ultrasound to enhance the reactivity of reactive metals as stoichiometric reagents has become an especially routine synthetic technique for many heterogeneous organic and organometallic reactions (11–15), particularly those involving reactive metals, such as Mg, Li or Zn. This development originated from the early work of Renaud and the more recent breakthroughs of Luche (12,13). The effects are quite general and apply to reactive inorganic salts and to main group reagents as well (69). Less work has been done with unreactive metals (eg, V, Nb, Mo, W), but results here are promising as well (11). Rate enhancements of more than tenfold are common, yields are often substantially improved, and by-products avoided. A wide range of synthetically useful heterogeneous sonochemical reactions have been listed in Table 1. The applications of sonochemistry to organic synthesis have been reviewed recently in great detail (13).

The mechanism of the sonochemical rate enhancements in both stoichiometric and catalytic reactions of metals is associated with dramatic changes in morphology of both large extended surfaces and of powders. As discussed earlier, these changes originate from microjet impact on large surfaces and high-velocity interparticle collisions in slurries. Surface composition studies by Auger electron spectroscopy and sputtered neutral mass spectrometry reveal that ultrasonic irradiation effectively removes surface oxide and other contaminating coatings (11). The removal of such passivating coatings can dramatically improve reaction rates. The reactivity of clean metal surfaces also appears to be responsible for the greater tendency for heterogeneous sonochemical reactions to involve single electron transfer rather than acid–base chemistry (70).

Applications of ultrasound to electrochemistry have also seen substantial recent progress. Beneficial effects of ultrasound on electroplating and on organic synthetic applications of organic electrochemistry (71) have been known for quite some time. More recent studies have focused on the underlying physical theory of enhanced mass transport near electrode surfaces (72,73). Another important application for sonoelectrochemistry has been developed by J. Reisse and co-workers for the electroreductive synthesis of sub-micrometer powders of transition metals (74).

Sonocatalysis. Ultrasound has potentially important applications in both homogeneous and heterogeneous catalytic systems. The inherent advantages of sonocatalysis include (1) the use of low ambient temperatures to preserve thermally sensitive substrates and to enhance selectivity; (2) the ability to generate high energy species difficult to obtain from photolysis or simple pyrolysis; and (3) the mimicry of high temperature and pressure conditions on a microscopic scale.

Homogeneous catalysis of various reactions often uses organometallic compounds. The starting organometallic compound, however, is often catalytically inactive until loss of metal-bonded ligands (such as carbon monoxide) from the

calculations serve to demonstrate the importance of using high energy beams for physical imaging on the atomic scale in electron microscopy.

Electron microscopy is done in one of two ways: scanning electron microscopy (sem) or transmission electron microscopy (tem). Scanning transmission electron microscopy (stem) is simply tem carried out in a scanning mode. Sem is the most common and well-known electron microscopy method for the physical imaging of surfaces. This technique is based on the interaction with a surface of a primary beam of electrons with energy typically in the range of 0.5–40 keV. This primary electron beam is first demagnified by a condenser lens and then focused onto the sample surface using a series of objective lenses where it is rastered across the surface using a series of scanning coils. Sem must be done in vacuum so that the electrons can travel unimpeded for adequate distances. The typical surface magnification realized is on the order of 10–10,000 X depending on the energy of the primary electron beam.

In most sem analyses, secondary electrons created near the position of the impinging primary beam are detected. However, surface images can also be obtained through collection of backscattered primary electrons. The secondary electrons, which result from ionization of the sample atoms in the near-surface region by the incident or backscattered primary electrons, have much smaller energies of ca 0–20 eV. Various mechanisms result in the ejection of these secondary electrons from the surface. However, due to their very low energies, secondary electrons have escape depths of only a few nm. Therefore, sem is a highly surface sensitive tool even though the primary beam electrons may penetrate into the sample to depths of 0.5–5 μm.

Secondary electrons can usually be collected with very high efficiencies. The efficiency of secondary electron creation, therefore, becomes important in sensitivity. The yield of secondary electrons per primary electron is dictated primarily by two factors, the primary beam energy and the angle of collection. The secondary electron yield actually shows a slight negative dependence on primary beam energy since the higher the primary beam energy, the deeper its penetration and the less efficient is the secondary electron escape from the surface. Thus, the secondary electron yield varies as ca $E^{-0.8}$ for primary beam energy E. In addition, the secondary electron yield depends on angle of collection, generally increasing as the angle from the surface normal increases.

Given the option to obtain images by collection of either secondary electrons or backscattered primary electrons, one might reasonably ask which is the better strategy for surface imaging. Both approaches have strengths and weaknesses depending on the information desired. Generally, images from secondary electrons have higher resolution, since they depend only on the primary beam diameter. Secondary electron images also have higher contrast than backscattered electron images. However, backscattered electron images have their own set of advantages. Chief among these is that backscattered electron images are generally more sensitive to topography than secondary electron images, since backscattering increases significantly at sharp edges on a surface. Backscattering efficiency is a function of the atomic number (Z) of the atom from which the backscattering occurs. Thus, backscattered images have a pronounced dependence on elemental composition of the sample surface. Secondary electron images actually have a slight dependence on elemental composition as well, since some of the secondary electrons are generated by backscattered primary electrons. When acquiring backscattered electron images, either of these attributes can be maximally exploited by the manner in which the collected backscattered electrons are processed. Backscattered electrons are detected with a two-half annular detector. Images sensitive to elemental composition differences across the surface are obtained when the electron intensities in each half of the detector are added. In contrast, images sensitive to surface topography are obtained by subtracting the electron intensities in each half of the detector.

A block diagram of a scanning electron microscope is shown in Figure 1. A vacuum on the order of ca 10^{-5} Pa is required in order to ensure adequate electron pathlengths. This level of vacuum is readily attainable with diffusion pumps trapped with liquid nitrogen.

The electron sources used in most sems are thermionic sources in which electrons are emitted from very hot filaments made of either tungsten (W) or lanthanum boride (LaB_6). W sources are typically heated to ca $2500-3000$ K in order to achieve an adequate electron brightness. LaB_6 sources require lower temperatures to achieve the same brightness, although they need a better vacuum than W sources. Once created, these primary electrons are accelerated to some desired energy with an energy spread (which ultimately determines lateral resolution) on the order of ca 1.5 eV.

One important sem source that is not based on thermionic emission is the field emission (fe) source. Fe-sem systems typically give images of much higher resolution than conventional sems due to the much narrower energy distribution (on the order of 0.25 eV) of the primary electron beam. A fe source is a pointed W tip from which electrons tunnel under the influence of a large electric field.

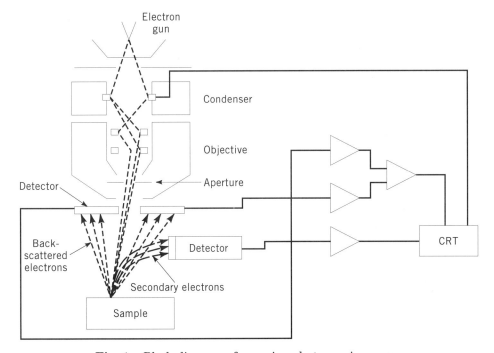

Fig. 1. Block diagram of scanning electron microscope.

This different mechanism of electron generation also results in a brightness comparable to a conventional thermionic source with much less current.

Once the primary electron beam is created, it must be demagnified with condenser lenses and then focused onto the sample with objective lenses. These electron lenses are electromagnetic in nature and use electric and magnetic fields to steer the electrons. Such lenses are subject to severe spherical and chromatic aberrations. Therefore, a point primary beam source is blurred into a primary beam disk to an extent dependent on the energy and energy spread of the primary electrons. In addition, these lenses are also subject to astigmatism. All three of these effects ultimately limit the primary beam spot size and hence, the lateral resolution achievable with sem.

The detectors used in sem are of two types depending on whether secondary or backscattered electrons are being detected. As mentioned above, backscattered electrons are detected with an annular detector placed above the sample such that electrons backscattered along the surface normal can be detected. Secondary electrons are detected with a detector that is positively charged to attract these electrons. This detector is placed at large angles with respect to the surface normal to most efficiently detect secondary electrons whose paths are predominantly away from the surface normal as discussed above. The absolute efficiencies of both detectors are relatively low; therefore, the signals must be amplified before being sent to an image intensifier (CRT) for viewing. For sem, the electron scan rate in the CRT is set to match the raster rate of the primary electron beam across the surface so that an image in real time is observed.

A scanning electron microscope can also be equipped with additional instrumentation for electron-excited x-ray analysis (9). In many systems, this is performed in the mode known as energy dispersive x-ray analysis (edx). Other common acronyms for this method are eds for energy dispersive spectroscopy or edax for energy dispersive analysis of x-rays.

Edx is based on the emission of x-rays with energies characteristic of the atom from which they originate in lieu of secondary electron emission. Thus, this technique can be used to provide elemental information about the sample. In the sem, this process is stimulated by the incident primary beam of electrons. As will be discussed below, this process is also the basis of essentially the same technique but performed in an electron spectrometer. When carried out this way, the technique is known as electron microprobe analysis (ema).

Sem/edx typically occurs in a volume of sample larger than that from which backscattered electrons are observed. Thus, sem/edx samples the surface to a greater depth than does sem imaging. Signals typically result from the upper several μm of the near-surface region. The intensity of the x-rays emanating from a sample at a given energy is proportional to the number density of atoms generating these x-rays. Thus, sem/edx can be used semiquantitatively. The fundamental principles and instrumentation for this approach are discussed in greater detail below in the section describing techniques for elemental composition.

A second important mode of electron microscopy is transmission electron microscopy (tem) (11). An image of a sample in tem is obtained using the transmission of electrons by a sample in a method analogous to optical microscopy using photons. Thus, this method provides a magnified image of a transparent sample with an electron beam using objective and projector lenses as shown in Figure 2. Obviously, the requirement for an electron beam-transparent sample is the most important limitation of this approach. For most materials, this means that the sample thickness must be on the order of ca 20–200 nm. Given the extremely small thickness of tem samples, the surface area-to-bulk volume is very large; therefore, tem provides information which is essentially surface in nature.

Instrumentation for tem is somewhat similar to that for sem; however, because of the need to keep the sample surface as clean as possible throughout the analysis to avoid imaging surface contamination as opposed to the sample surface itself, ultrahigh vacuum conditions (ca 10^{-7}–10^{-8} Pa) are needed in the sample area of the microscope. Electron sources in tem are similar to those used in sem, although primary electron beam energies needed for effective tem are higher, typically on the order of ca 100 keV.

Several lenses are used in a transmission electron microscope. The condenser lenses provide uniform illumination of the sample over the area of interest. The objective lens provides the primary image and therefore, determines the lateral resolution of the image. The objective lens aperture is important in controlling the contrast of the image. The final magnification of the image is performed by one or more projector lenses. The final image is typically recorded on a fluorescent or phosphorescent screen where it can be captured by a video camera for viewing. As noted above, all of these lenses are subject to serious aberrations which ultimately limit the resolution of the microscope to greater than the diffraction limit (the theoretical resolution limit for this approach.) Moreover, these lens aberrations restrict the angular range of the electron beam resulting in the need for very tall instruments. Despite these shortcomings, tem is a very powerful surface imaging tool with atomic resolution in some cases, providing sample magnifications between 100–500,000 X.

Scanning Probe Microscopies. A relatively new group of techniques for surface imaging is the scanning probe microscopies (14,15). The most common of these tools are scanning tunneling microscopy (stm) and atomic force microscopy (afm). These techniques were developed in the 1980s and have undergone explosive growth and acceptance since their initial discovery. Stm, the first scanning probe microscopy method to be developed, was discovered by Binnig and Rohrer (16) in 1981 at IBM Zurich. These researchers received the 1986 Nobel Prize in Physics for their work. The significance of this discovery is a readily accessible technique capable of generating real-space images of a surface with atomic-scale resolution.

Stm is based on the tunneling of electrons between a very sharp, electrically conductive tip (hopefully terminating in a single atom) and a conductive sample when the tip is brought very close (<1 nm) to the sample. A schematic for this technique is shown in Figure 3. The magnitude of the tunneling current will vary in a well-defined exponential manner with tip-to-sample distance. Thus, the magnitude of the tunneling current can be used to generate an image of the surface. Stm requires that both the tip and the sample be electrically conductive. Therefore, the samples that can be studied with stm are restricted to either conductors or semiconductors. The exponential dependence of tunneling current on distance between the tip and the sample gives outstanding (ie, sub-nm)

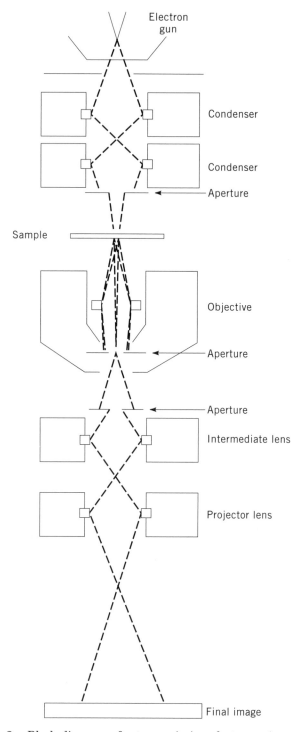

Fig. 2. Block diagram of a transmission electron microscope.

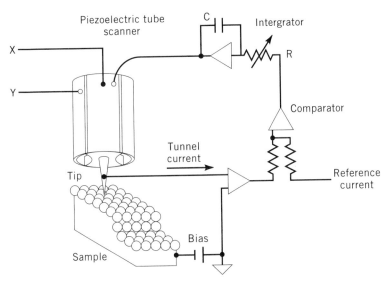

Fig. 3. Block diagram of a scanning tunneling microscope (17).

vertical resolution. Lateral resolution is not quite as good and is limited by the size of the tip.

Stm experiments can be performed in one of two common operating modes as shown in Figure 4. In the constant-current mode of operation, the instrument uses electronic feedback to keep the tunneling current constant by moving the tip closer or further away from the surface as the tip is scanned across the sample surface. Therefore, in the simplest analysis, the height of the tip is a direct measure of the sample surface topography. This method of operation is somewhat more time-consuming than the constant height mode described below, because of the time that it takes to move the tip up and down. However, this approach is excellent for imaging irregular surfaces with high precision.

The constant height mode of operation results in a faster measurement. In this analysis, the tip height is maintained at a constant level above the surface and differences in tunneling current are measured as the tip is scanned across the surface. This approach is not as sensitive to surface irregularities as the constant current mode, but it does work well for relatively smooth surfaces.

The simplest interpretation of stm images is in terms of surface topography. However, care must be exercised in this interpretation, since in reality, tunneling probability is really measured. The many subtleties of stm data interpretation are beyond the scope of this chapter. The interested reader is referred to references 14 and 15 for a more detailed discussion of these issues.

A second scanning probe microscopy that is perhaps more widely used for surface imaging today is afm. This technique is related to stm, but provides a direct topographical image of a surface without the complications arising from the electronic surface structure that cloud image interpretation in stm. In this device, a sharp tip several microns long terminating in a tip <10 nm in diameter is attached to a cantilever several hundred microns in length. In the original configuration of this method shown in Figure 5, the tip is dragged across the

reference material containing A and B that is close in suspected composition to the unknown. In this case,

$$\left(\frac{N_A}{N_B}\right)_{UNK} = \left(\frac{I_A}{I_B}\right)_{UNK} \frac{T_B \; D_B \; \sigma_B \; L_B \; \lambda_B}{T_A \; D_A \; \sigma_A \; L_A \; \lambda_A} \qquad (11)$$

and

$$\left(\frac{N_A}{N_B}\right)_{STD} = \left(\frac{I_A}{I_B}\right)_{STD} \frac{T_B \; D_B \; \sigma_B \; L_B \; \lambda_B}{T_A \; D_A \; \sigma_A \; L_A \; \lambda_A} \qquad (12)$$

Therefore, the relative atomic ratio of the unknown material can be determined from

$$\left(\frac{N_A}{N_B}\right)_{UNK} = \frac{(I_A/I_B)_{UNK}(N_A/N_B)_{STD}}{(I_A/I_B)_{STD}} \qquad (13)$$

This approach is the most useful for routine quantitative xps analysis.

Auger Electron Spectroscopy. Auger electron spectroscopy (aes) is also based on an electron ejection process like xps, but the electrons that are monitored in aes are secondary electrons (19). These secondary or Auger electrons arise from a process shown schematically in Figure 14. The process occurs after primary electron emission such that a core level hole exists. Incident electrons are conventionally used in aes to stimulate primary electron emission, although incident x-rays can also be used as in xps. The presence of this core level electron hole results in electron relaxation from a valence level to fill the core level hole. When this relaxation occurs, the excess energy that is released stimulates

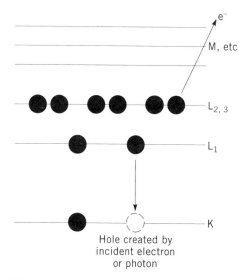

Fig. 14. Schematic of the Auger electron emission process induced by creation of a K level electron hole.

ejection of a secondary or Auger electron from another valence electron energy level. The final state of the atom after Auger electron emission is thus doubly ionized; however, because these vacancies are in valence electron levels further away from the nucleus, this final doubly ionized state is more stable than the initial singly ionized state. The energy of this Auger electron is *independent* of the energy of the incident electron beam and depends only on the energies of the core and valence levels involved. The kinetic energy of the Auger electron is given by the difference in energy between the core level and the valence level from which the electron comes that fills the core level vacancy minus the energy that it takes to remove the Auger electron from the singly ionized atom and the spectrometer work function. Thus, for a KLL Auger electron

$$E_{KL(1)L(2,3)} = E_K - E_{L(1)} - E^*_{L(2,3)} \qquad (14)$$

where $E_{KL(1)L(2,3)}$ is the kinetic energy of the Auger electron, E_K is the energy of the K core level, $E_{L(1)}$ is the energy of the valence level from which the electron comes to fill the K level core vacancy, and $E^*_{L(2,3)}$ is the binding energy of the $L_{(2,3)}$ level *in the presence of a hole in the $L_{(1)}$ level*. Auger electrons typically range in energy from ca 10 to 2000 eV.

Any three subshells within an atom can be involved in the Auger process as long as the final state is significantly more stable than the initial state. The Auger transition is identified by a three-letter label in which the first letter indicates the core level in which the initial electron vacancy resides, the second letter represents the valence level from which the electron comes that fills the initial vacancy, and the third letter indicates the valence level from which the ejected Auger electron comes. KLL, LMM, and MNN Auger processes are common. Any atom has many possible Auger transitions, but the probability of each of these varies with the energy levels involved. Thus, the sensitivity of aes to different atoms varies. A plot of relative sensitivity factors for different atoms normalized to the Ag 351 eV MNN Auger transition is shown in Figure 15 (28).

An alternative mechanism of excess energy release when electron relaxation occurs is through x-ray fluorescence. In fact, x-ray fluorescence favorably competes with Auger electron emission for atoms with large atomic numbers. Figure 16 shows a plot of the relative yields of these two processes as a function of atomic number for atoms with initial K level holes. The cross-over point between the two processes generally occurs at an atomic number of 30. Thus, aes has much greater sensitivity to low Z elements than x-ray fluorescence.

Auger spectra are plots of the intensity of Auger electrons as a function of kinetic energy. Figure 17**a** shows a plot of direct electron intensity as a function of kinetic energy for a copper surface covered with contaminants (19). The Auger electrons of interest are superimposed on a very large background of elastic and inelastically scattered primary electrons used to excite the Auger process. At the lowest kinetic energies, the background increases sharply in an electron "cascade" of inelastically scattered electrons. The intensity of the Auger electrons in relation to this electron cascade is so small as to almost be invisible on this large electron background intensity. Thus, Auger spectra are typically plotted as derivative spectra in which the large background contributions are minimized.

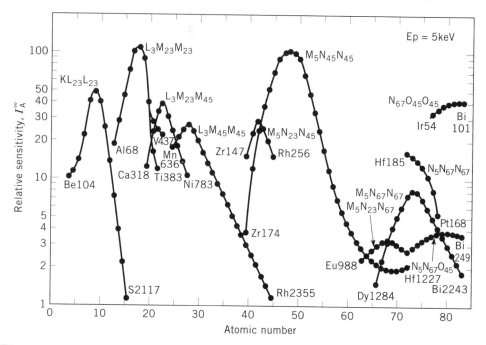

Fig. 15. Auger sensitivity factors relative to the Ag MNN Auger transition as a function of atomic number (19).

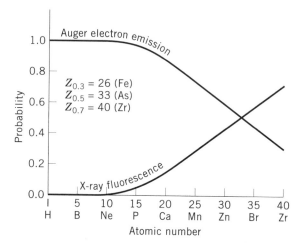

Fig. 16. Relative probabilities of Auger electron emission and x-ray fluorescence for initial K level electron hole as a function of atomic number (19).

An example of such a derivative spectrum is shown in Figure 17**b**. The Auger peaks are clearly much more visible in this derivative spectrum than they are in the direct intensity spectrum.

Quantitation of Auger spectra follows the same lines of reasoning as discussed above for quantitation of xps data (19,27). Thus, the current due to Auger

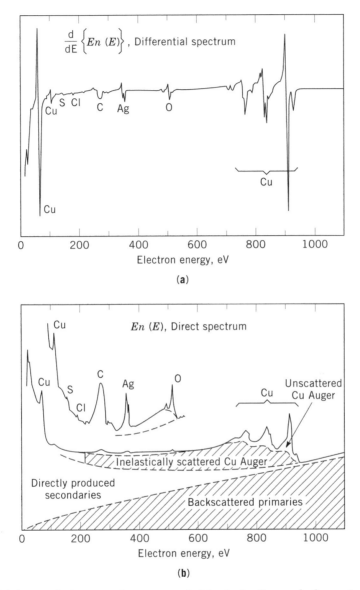

Fig. 17. (a) Auger electron spectrum presented in derivative mode for contaminated Cu surface. (b) Auger electron spectrum presented in direct intensity mode for same sample. Contributions to background intensities indicated (19).

electrons at a particular kinetic energy, I, is given by

$$I = I_0 \, T \, D \, \sigma \, B \int_0^\infty N \, \exp\!\left(\frac{-z}{\lambda \, \cos \, \theta}\right) dz \tag{15}$$

where I_0 is the primary electron beam current, σ is the Auger cross-section, T and D are instrumental efficiencies as defined as above, N is the atom number density, λ is the inelastic mean free path, and B is the backscattering factor. This latter parameter takes on values greater than one and corrects for the creation of Auger electrons by backscattered primary electrons.

Equation 15 is not used directly for quantitation of aes data due to the same limitations discussed above for xps. Particularly troubling for aes is the inability to determine B for a given matrix. Thus, the analyst is left with comparing aes data from an unknown with those of known materials in an attempt to estimate relative atom ratios. This process proceeds along the same lines as for xps and allows quantitation of aes data to a precision of typically better than ca 20%.

Aes is often combined with surface ion sputtering using a high energy ion beam to achieve profiling of the elemental composition as a function of depth into the solid (3,19). This approach is typically implemented using a beam of inert gas ions such as Ar^+ accelerated to 200–5000 eV to sequentially sputter away the surface layer by layer. Simultaneous with this sputtering process, aes can be used to probe the elemental composition in the sputtering crater that is created.

An example of how such depth profiling information is presented is shown in Figure 18 for a hypothetical system composed of a substrate of element B covered by a thin film of material containing element A. This Figure shows plots of aes intensity as a function of sputtering time for the ideal response and the real response expected from this system. For a given sputtering rate, the sputtering time is related to depth into the sample. If depth profiling of a chemical system with a sharp interface were ideal, and hence sputtering of the material occurred perfectly uniformly, the intensity-sputtering time profiles would mimic the physical sharpness of the interface between A and B. In reality, this does not happen due to a variety of experimental limitations. First, the sputtering process is physically destructive and can lead to significant sputtering-induced roughness which blurs the interface, since parts of B can be exposed before all of A is sputtered away. In addition, the sputtering process can

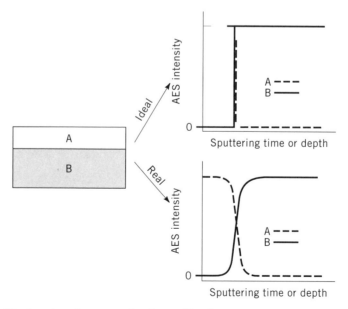

Fig. 18. Ideal and real sputter-depth profiles for thin film of A and substrate B.

cause implantation of atoms of A into the B substrate, such that their signals are still observed beyond the depth of the original interface. The situation is further complicated by the fact that materials have different rates of sputtering. This aspect of the process can lead to further loss of the real interfacial boundary in the data. All of these features lead to the broadening of the interface in the real depth profiles shown in Figure 18.

Several other aspects of depth profiling by a combination of ion sputtering and aes are noteworthy. It is important that the sputtering ions be chemically nonreactive such that the surface chemistry does not change during the sputtering process. The analyst must also remain cognizant of the possibility of the sputtering process altering the Auger electron emission. As noted above, sputtering is an abrasive process which can lead to the undesirable production of secondary electrons, some of which may be Auger electrons. This problem can be overcome through the use of a much smaller ion beam current density for sputtering than the electron beam current density used to excite aes. The disadvantage of this approach is slower sputtering.

An additional consideration when depth profiling by simultaneous sputtering/aes is where within the sputtering crater the aes electron beam is positioned. Due to the inherent current density profile of the ion beam, the sputtering crater that is produced is bowl-shaped. Ideally, therefore, sampling in the center bottom of the crater for aes produces information over the shortest depth range. If the sides of the crater are sampled, a much greater range of depths contribute to the aes intensities, further blurring the interface.

Finally, it is difficult to calibrate the depth scale in a depth profile. This situation is made more complicated by different sputtering rates of materials. Despite these shortcomings, depth profiling by simultaneous ion sputtering/aes is commonly employed, because it is one of the few techniques that can provide information about buried interfaces, albeit in a destructive manner.

Xps and Aes Instrumentation. The instrumentation required to perform xps and aes analyses is generally sophisticated and expensive (19). The need for UHV conditions in order to retain surface cleanliness for a tractable period of time was mentioned above. Beyond this requirement (and the hardware that accompanies it), the most important components of an electron spectrometer system are the source, the electron energy analyzer, and the electron detector. These will be discussed in turn below.

Xps requires a source that can provide a single x-ray line reasonably narrow in energy. The absolute energy requirement for this x-ray line is that it must be energetic enough to generate photoelectrons from core levels of a majority of the elements with reasonable resolution. Of the many possible x-ray sources, those which best meet this requirement are the Al $K_{\alpha(1,2)}$ x-ray line at 1486.6 eV and the Mg $K_{\alpha(1,2)}$ x-ray line at 1253.6 eV. These lines have full-widths-at-half-maxima (FWHM) that are ca 0.85 and 0.7 eV, respectively.

X-rays are produced from these materials by electron bombardment using electrons whose kinetic energies are higher than the x-ray energies produced. Once produced, the x-rays leave the source through an x-ray-transparent window typically made of Al. To avoid melting of the x-ray producing material, a thin film (ca 10 μm) of Al or Mg is coated onto a cooled Cu block acting as the anode. Electrons are generated at the cathode. Figure 19 shows how this is commonly the thermionic sources, because the electron emission is concentrated to the small area of the field emission tip. Focusing in both of these sources is done by electrostatic lenses. Today's thermionic sources typically produce spot sizes on the order of 0.2–0.5 μm with beam currents of 10^{-8} A at 10 keV. If field emission sources are used, spot sizes down to ca 10–50 nm can be achieved.

The heart of the electron spectrometer system is the electron energy analyzer (19). The electrons to be measured in xps and aes have relatively low kinetic energies; therefore, the analyzers rely on electrostatic focusing in which electron trajectories are manipulated using electric fields. The figure-of-merit for analyzers is energy resolution. Analysts refer to analyzer resolution in one of two ways: absolute energy resolution (ie, the FWHM of an electron spectroscopic peak or ΔE) and relative resolution, R, which is given by

$$R = \frac{\Delta E}{E_K} \tag{17}$$

where E_K is the kinetic energy at the peak position. Relative resolution is often referred to in terms of the spectrometer resolving power, ρ, which is the inverse of the relative resolution.

In most electron spectroscopic analyses, the kinetic energies of the electrons entering the analyzer are retarded to either a constant energy or by a constant factor. These approaches lead to two modes of operation: the constant analyzer energy (CAE) mode and the constant retard ratio (CRR) mode. In the CAE mode, electrons are retarded to a constant pass energy, E_p, that is fixed over the entire spectrum. This mode results in constant absolute resolution throughout the spectrum and is the most commonly employed mode for xps. This retardation results in an improvement in relative resolution by a factor equal to E_p/E_K compared to the resolution before retardation. This mode is tantamount to providing constant analyzer transmission throughout a spectral scan in xps and makes the spectra easier to quantify, since ΔE is constant. The trade-off, however, is that at low kinetic energies, the S/N of the signal degrades considerably due to the large increase in background electron intensity.

The CRR mode involves retarding the electron kinetic energies to a constant ratio of E_K/E_p where E_p is the energy passed by the analyzer. Thus, the energies are retarded by a constant factor. Spectra acquired in this mode are less easy to quantify, but small peaks at low kinetic energies are readily detected. This mode of operation results in spectra of constant relative resolution throughout. The relative resolution is improved in this mode by a factor of E_p.

These modes of operation are used in conjunction with the two most popular energy analyzers, the cylindrical mirror analyzer (CMA) and the concentric hemispherical analyzer (CHA). The most common form of the CMA used today is the double-pass version diagramed in Figure 21. This device consists of two perfectly coaxial cylinders of radii r_o and r_i. The outer cylinder is held at a potential of $(-V_o)$ and the inner cylinder is held at ground. The spatial separation of electron energies in this device can be understood by consideration of the behavior in the first stage of the CMA shown in Figure 21. Electrons of kinetic energy E_K emitted from a source (ie, the sample) which lies exactly on the axis of the CMA are collected by the analyzer, usually at an angle, α, of 42° with respect

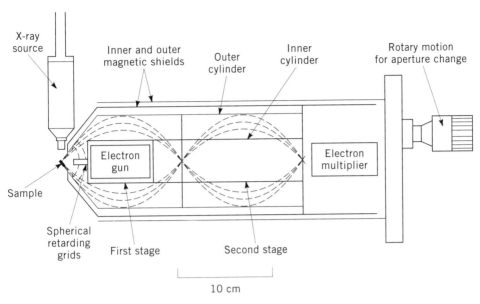

Fig. 21. Double-pass cylindrical mirror analyzer (19).

to the analyzer axis. The kinetic energy that is transmitted by the analyzer is given by

$$E_K = \frac{K\,e\,V_o}{\ln\!\left(\frac{r_o}{r_i}\right)} = \frac{1.31\,e\,V_o}{\ln\!\left(\frac{r_o}{r_i}\right)} \tag{18}$$

where the constant K is 1.31 for $\alpha = 42°$. Obviously, a range of acceptance angles, $\Delta\alpha$, must be used for efficient collection. E_K is systematically varied by sweeping V_o for the scan of a spectrum. For a slit width of w, the energy resolution for the CMA used at this acceptance angle is given by

$$\frac{\Delta E}{E_K} = \left(\frac{0.18\,w}{r_i}\right) + 1.39\,(\Delta\alpha)^3 \tag{19}$$

In xps in which small chemical shifts are employed to ascertain chemical information, energy resolution of the analyzer is an important consideration. Although resolution can be increased by decreasing $\Delta\alpha$, this is done at the expense of sensitivity as dictated by the analyzer luminosity (the product of the analyzer transmission and the acceptance area). Thus, in effectively using the CMA, the analyst is faced with the need to optimize both resolution and luminosity. The second stage of a double-pass CMA becomes critical for achieving these divergent goals. In typical use of a double-pass CMA for xps, the first stage is set to control resolution through $\Delta\alpha$ and the second stage to control luminosity by maximizing transmission of the system to the point at which resolution just begins to be degraded. The optimized situation is most commonly achieved in the CAE mode of operation for xps.

where I_λ is the Bremsstrahlung intensity at an energy E_λ, k is a constant, and Z is the atomic number of the x-ray producing material. The result is that for high Z materials, over 90% of the x-ray fluorescence excitation arising from K ionization processes comes from the Bremsstrahlung radiation as opposed to primary electron beam excitation.

X-rays are collected and analyzed in ema in one of two ways. In wds, x-rays are dispersed by Bragg diffraction at a crystal and refocused onto a detector sitting on a Rowland circle. This arrangement is similar to the production of monochromatized x-rays for xps described above. In the other approach, edx, x-rays are all collected at the same time in a detector whose output scales with the energy of the x-ray (and hence, Z of the material which produces the x-ray.) Detectors used for ema today are almost exclusively Li-drifted Si solid-state detectors.

Wds provides higher energy resolution than edx, but at the expense of spectrometer transmission. However, since in wds only a single energy increment is measured at one time, and hence the background only at that one energy increment is measured and not the entire background distribution, the signal-to-noise ratio for wds is usually greater than edx despite the advantage of edx in terms of spectrometer transmission. The trade-off for this better sensitivity for wds is oftentimes sample stability. Prolonged electron beam exposure times are needed to overcome the limited transmission of a wds spectrometer, and samples can thermally degrade or become altered by thermally-induced atom migration within the sample.

Ema data can be quantitated to provide elemental concentrations, but several corrections are necessary to account for matrix effects adequately. One well-known method for matrix correction is the zaf method (7,31). This approach is based on calculated corrections for major matrix-dependent effects which alter the intensity of x-rays observed at a particular energy after being emitted from the corresponding atoms. The zaf method corrects for differences between elements in electron stopping power and backscattering (the z correction), self-absorption of x-rays by the matrix (the a correction), and the excitation of x-rays from one element by x-rays emitted from a different element, or in other words, secondary fluorescence (the f correction). Once these corrections are made, quantitative ema can be used to determine concentrations of elements in the near-surface region (ca 1 μm) of samples of interest.

Analysis of Surface Molecular Composition. Information about the molecular composition of the surface or interface may also be of interest. A variety of methods for elucidating the nature of the molecules that exist on a surface or within an interface exist. Techniques based on vibrational spectroscopy of molecules are the most common and include the electron-based method of high resolution electron energy loss spectroscopy (hreels), and the optical methods of ftir and Raman spectroscopy. These tools are tremendously powerful methods of analysis because not only does a molecule possess vibrational modes which are signatures of that molecule, but the energies of molecular vibrations are extremely sensitive to the chemical environment in which a molecule is found. Thus, these methods directly provide information about the chemistry of the surface or interface through the vibrations of molecules contained on the surface or within the interface.

The most common and readily accessible methods are those based on ftir spectroscopy. Several of the more common of these are discussed here.

Transmission Fourier Transform Infrared Spectroscopy. The most straight-forward method for the acquisition of ir spectra of surface layers is standard transmission spectroscopy (35,36). This approach can only be used for samples which are partially ir transparent or which can be diluted with an ir transparent medium such as KBr and pressed into a transmissive pellet. The extent to which the ir spectral region (typically ca $600-4000$ cm^{-1}) is available for study depends on the ir absorption characteristics of the solid support material. Transmission ftir spectroscopy is most often used to study surface species on metal oxides. These solids leave reasonably large spectral windows within which the spectral behavior of the surface species can be viewed.

Pellets of such oxides are usually prepared by mixing $5-10$ wt % of the oxide with KBr and pressing a pellet that is several mm thick. Although the preparation of such systems is not difficult, often the pellet must be pressed for long periods of time to allow the KBr to fuse around the oxide particulates. Once prepared, ir spectra of these pellets can be acquired in standard KBr pellet holders.

The advantage of this approach for characterization of surface species is the ease of sample preparation and straightforward data interpretation. Spectra of such samples resemble normal transmission ir spectra and can be quantitated using conventional methods. In addition, this approach is reasonably sensitive with surface coverages down to ca 10% of a monolayer easily observed on oxides such as silica and alumina. One important point to note is the spectral interference that water adsorbed on an oxide surface can produce. In some cases, the spectrum from surface-confined water can swamp out the spectrum of other surface species of interest. In these cases, the transmission signal will not be strong enough to be of much use. This approach also does not work well for solids which are inherently strong ir absorbers for similar reasons. In these cases, approaches other than transmission ftir spectroscopy must be used to acquire surface vibrational spectra.

Diffuse Reflectance Infrared Fourier Transform Spectroscopy. An alternative approach to the acquisition of surface ir spectra for particulate or powdered samples is diffuse reflectance infrared Fourier transform spectroscopy (drifts) (35–37). This technique was originally developed for the ir characterization of powdered organic and inorganic samples. Its use was later extended to the study of surface species and interfacial films on particulate samples that were difficult to study by other vibrational spectroscopic methods. The technique is based on the diffuse reflectance of radiation that occurs when it is directed onto a surface with a matte finish or a sample comprised of a powder. This diffuse reflectance is different than specularly reflected radiation in that it penetrates and interacts with a sample before emerging. While the radiation is within the sample, light scattering occurs such that the diffusely reflected light emerges from the sample at all angles. Therefore, efficient collection requires collection of the diffuse reflectance over the entire 2π solid angle above the sample surface. In acquiring diffuse reflectance spectra of bulk materials, the sample is often mixed with a nonabsorbing matrix such as KBr. This matrix can have the effect of increasing the amount of ir radiation that is diffusely reflected. This practice might also be done for the characterization of surface-confined films or molecules, although such samples are more often examined undiluted by such a matrix to increase sensitivity.

The amount of diffusely reflected light cannot be measured directly; instead, it is typically measured relative to a nonabsorbing reference material such as a powdered alkali halide to allow adequate correction for scattering characteristics of the powdered sample. In acquiring drift spectra, both the sample and the reference must be held in sample cells that are sufficiently deep that a further increase in sample cell depth would not change the magnitude of the signal measured. Once this sample cell depth is reached, the sample is said to be sampled at infinite depth. This depth is generally ca 3 mm for most materials. The signal that is measured is the ratio of the diffuse reflectance of the sample to that of a nonabsorbing reference material. This ratio is given the symbol R_∞. Measurement in this way allows quantitation according to Kubelka-Munk theory (38,39) which is defined in terms of the Kubelka-Munk function as

$$f(R_\infty) = \frac{(1 - R_\infty)^2}{2R_\infty} = \frac{k}{s} \tag{24}$$

where R_∞ is defined above, k is the absorption coefficient, and s is the scattering coefficient. The assumptions made in this derivation are that the specularly reflected component of the signal is small and that s is constant. The first assumption is valid for most commercially available drifts attachments. The second assumption relies on the uniformity in particle size between the sample and the reference material, which can also be achieved with care in sample preparation. Importantly, Kubelka-Munk theory predicts a linear relationship between the Kubelka-Munk function and concentration for dilute concentrations.

Several factors affect the bandshapes observed in drifts of bulk materials, and hence the magnitude of the diffuse reflectance response. Particle size is extremely important, since as particle size decreases, spectral bandwidths generally decrease. Therefore, it is desirable to uniformly grind the samples to particle sizes of <50 μm. Sample homogeneity is also important as is the need for dilute concentrations in the nonabsorbing matrix.

For surface systems, the considerations described above change very little. Most surfaces are sampled directly without dilution for maximum sensitivity. For such studies, R_∞ is usually defined as the diffuse reflectance measured from the solid sample containing the surface molecules or surface film ratioed to the diffuse reflectance measured from the same solid material in the absence of the surface molecules or surface film. Using this approach, the Kubelka-Munk function is linear with surface coverage as long as the surface coverages are kept low. At higher surface coverages, the response can become nonlinear. Using this approach for the study of surface systems such as catalysts, extremely low limits of detection can be realized with sensitivities of 1% of a monolayer or less common. Thus, drifts is a straightforward method to implement, since standard attachments to most commercial ftir spectrometers are readily available. Its ease of implementation makes it a valuable addition to the surface analyst's repertoire for powdered samples.

Attenuated Total Reflectance Fourier Transform Infrared Spectroscopy. Attenuated total reflectance (atr) ftir spectroscopy is based on the principle of total internal reflection (40). Methods based on internal reflection in the uv and visible regions of the spectrum are also common in addition to those in the ir region. The implementation of internal reflection in the ir region of the spectrum provides a means of obtaining ir spectra of surfaces or interfaces, thus providing molecularly-specific vibrational information.

Internal reflection starts by consideration of an interface between two media, a denser transparent medium with refractive index n_1, and a rarer medium with a complex refractive index $\hat{n}_2$ ($= n_2 - ik_2$ where k_2 is the absorption coefficient of the medium) as shown in Figure 23. If k_2 of the rarer medium is zero, then the rarer medium is also transparent. For two transparent phases, total internal reflection occurs when radiation of wavelength λ_1 ($= \lambda/n_1$ where λ is the vacuum wavelength) is incident on the interface at an angle θ_i greater than the critical angle θ_c ($= \sin^{-1}[n_2/n_1]$). Although no net power is transmitted into the rarer medium, an evanescent standing wave is set up parallel to the interface which decays in amplitude exponentially on both sides of the interface. This evanescent wave has an electric field associated with it which samples the near-interface region of the rarer medium and is the basis of spectroscopies based on internal reflection. Thus, this evanescent field is inherently interface-selective because it only samples the interface region and not the bulk.

The penetration depth of this evanescent field, d_p (defined to be the depth at which the evanescent field decays to $1/e$ of its original value,) is given by

$$d_p = \frac{\lambda_1}{2\pi \left[\sin^2 \theta_i - (n_2/n_1)^2 \right]^{1/2}} \tag{25}$$

The electric field associated with this electromagnetic wave has two notable characteristics which distinguish it from the incident beam and make it useful in the analysis of surfaces and interfaces. First, the electric field intensity (and hence, light intensity) of the evanescent wave is enhanced relative to the incident beam. Secondly, the evanescent field has amplitude in all three directions in contrast to the incident traveling wave which has amplitude in only one direction.

Attenuated total reflection, on which atr–ftir is based, occurs when the rarer medium is absorbing and is characterized by a complex refractive index $\hat{n}_2 = n_2 - ik_2$ (40). The absorbing characteristics of this medium allow coupling

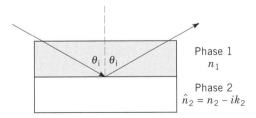

Fig. 23. Total internal reflection occurs at boundary between denser, transparent phase (characterized by n_1) and rarer, absorbing phase (characterized by n_2) when θ_i is greater than the critical angle, θ_c.

to the evanescent field such that this field is attenuated to an extent dependent on k_2. The critical angle in the case of attenuated total reflection loses its meaning, but internal reflection still occurs. Thus, if the internally reflected beam is monitored, its intensity will reflect the loss associated with the internal reflection process at the interface with an absorbing medium.

The extent of the interaction between the evanescent field and the absorbing medium is formally described by the effective thickness, d_e, of this medium. This effective thickness is the thickness of the absorbing phase that would have to be passed through by the incident beam in a transmission experiment to give the same energy loss as in the attenuated total reflection experiment. The exact expressions for effective thickness can be very complex. However, for a single attenuated total reflection of an incident beam of radiation of electric field E_i that occurs at an interface between two bulk phases (ie, phase 2 is not a thin film), d_e is given by

$$d_e = \frac{(n_2/n_1)}{\cos \theta_i} \int_0^\infty |E_i|^2 \, dz \tag{26}$$

where $E = E_i \exp(-z/d_p)$. This expression simplifies to

$$d_e = \frac{(n_2/n_1) \, E_i^2 \, d_p}{\cos \theta_i} \tag{27}$$

when k_2 is very small (ie, $\ll 1$). It is important to note in the expression that the strength of coupling of the incident beam with medium 2 decreases (ie, smaller d_e) as θ_i increases. From this expression for effective sampling thickness, one can develop a Beer's Law expression for internal reflection based on the relationship between absorbance, A, and reflectance, R (the fraction of light reflected at an interface) as given by

$$A = -\log R = \frac{\alpha_2 \, n_2 \, d_p \, I_0}{2.303 \, n_1 \, \cos \theta_i} \tag{28}$$

where $I_0 = E_i^2$ and $\alpha_2 = 4\pi k_2/\lambda_1$.

The real utility of d_e comes in the analysis of thin films. Consider a substrate of refractive index n_3 supporting a thin film of thickness d and refractive index n_2 in contact with an internal reflection element (the prism) of refractive index n_1 as shown in Figure 24. In this case, d_e depends on the polarization of the incident light beam and is given by

$$d_{e,\perp} = \frac{4 \, (n_2/n_1) \, d \, \cos \theta_i}{\left(1 - (n_3/n_1)^2\right)}$$

$$d_{e,\parallel} = \frac{4 \, (n_2/n_1) \, d \, \cos \theta_i\left[\left(1 + (n_3/n_2)^4 \, \sin^2 \theta_i - (n_3/n_1)^2\right)\right]}{\left(1 - (n_3/n_1)^2\right)\left[\left(1 + (n_3/n_1)^2 \, \sin^2 \theta_i - (n_3/n_1)^2\right)\right]} \tag{29}$$

The film thickness d is what would be sampled in a simple ir transmission experiment. Thus, if d_e is the same as d, then the internal reflection experiment

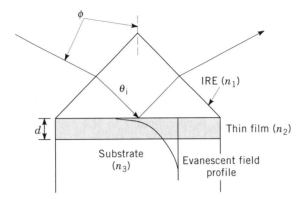

Fig. 24. Attenuated total reflectance of thin film of thickness d and refractive index n_2 on a substrate of refractive index n_3 at the surface of internal reflection element (IRE) of refractive index n_1. Decay of evanescent field beyond thickness of thin film indicated.

samples the same amount of the thin film as the transmission experiment, and no advantage accrues to the use of internal reflection. If, on the other hand, d_e is greater than d, then the effective sampling of the thin film in an internal reflection experiment is greater than in a transmission measurement, and the internal reflection experiment offers greater sensitivity. Conversely, if d_e is less than d, then a transmission measurement provides the greater sensitivity of the two measurement strategies. It should be noted that, according to the above expressions, optimization of d_e is afforded to a certain extent by alteration of θ_i and n_3.

The above discussion was based on the premise of a single internal reflection. An increase in sensitivity can be realized through the use of multiple internal reflections. In the case of multiple reflections, the absorbance A becomes

$$A = -\log R^N = -\log (1 - \alpha \, d_e)^N \tag{30}$$

for N reflections.

Atr–ftir can be readily performed on most commercial ftir spectrometers through the use of an attachment for atr spectroscopy. These devices provide ir-transparent internal reflection elements that are typically made of Ge, KRS-5, ZnSe, or ZnS. These internal reflection elements are made of materials that are of extremely high purity to avoid losses from absorption by impurities in these devices. Coupling of a thin film or surface sample to one of these reflection elements is accomplished by pressing the sample against the element while acquiring the spectrum.

Infrared Reflection-Absorption Spectroscopy. For adsorbed surface species or thin films on ir reflective surfaces such as metals, an alternative method is infrared reflection-absorption spectroscopy (irras) (41). This technique is based on the external reflection on an infrared beam of light at such surfaces, the characteristics of which are highly polarization dependent. Upon reflection of a polarized beam of light at a surface, phase shifts occur that can be quite significant in magnitude (42,43). Given that the electric field experienced by a

Thermographic Sensitivity.   The noise equivalent temperature difference (sensitivity to scene temperature variations in degrees C) may be expressed in terms of the NEP:

$$NE\Delta T = \frac{\pi}{\Omega A_p n_p \epsilon_0} \left[\int_{\lambda 1}^{\lambda 2} \frac{dJ_s}{dT} d\lambda \right]^{-1} (NEP) \tag{3}$$

where ϵ_0 = the optical efficiency of the sensor and is typically 85%
A_p = the detector element (pixel) active area, typically 1.5 E-5 cm^2
Ω = the collection solid angle of the optical system (= $\pi \sin^2(\tan^{-1} 1/2f_n)$, typically 0.15 str
f_n = optical f number, typically 2.2 or less. For fast optics f_n = 1.0
J_s = spectral power density in watts/cm^{2}cmK emitted by an element of the scene.

The integral of the temperature gradient of the spectral power density from wavelength $\lambda 1$ to $\lambda 2$, is readily calculable using the Planck radiation law (5). Constant emissivity is assumed for equation 3.

Thermographic sensitivity for representative sensors may be calculated with equation 3. The results are summarized in Table 1 for both cooled and uncooled focal planes and for scanned and staring type sensors. Scanning sensors use a larger area detector to achieve higher sensitivity and virtual overlap of the detector geometry to suppress aliasing. The optical speed is slower (f_n = 2 vs f_n = 1) to lower the cost of optics. The $D*$ values are typical and refer to the cooled platinum silicide, InSb, AlGaAs superlattice or HgCdTe focal planes and the uncooled pyroelectric barium strontium titanate or semiconductor bolometer focal planes (1,2,6,7).

In all cases, the imaging is conducted in real-time with a near standard television read-out display. This results in a 40,000 Hz bandwidth for scanning an a 100 Hz bandwidth for staring. It is possible to do frame addition and effectively reduce the bandwidth to achieve more sensitivity. The display can be repeated at the normal 60 Hz rate to avoid the appearance of flicker. Satellite infrared weather imaging thermography has a low bandwidth, resulting in a distinctive stepping motion to cloud patterns as displayed during a TV weather report. Advanced staring sensors will greatly improve weather imagery within the next decade yielding better sensitivity, resolution and continuous movement of cloud patterns.

Thermographic imagery is considered useful when the minimum detectable temperature ($NE\Delta T$) is less than 200 mK (8). Thus the uncooled focal plane is not used in the scanning mode. The cooled area focal plane can achieve less than 5 mK but this sensitivity can not be taken advantage of because of sensor instability and limited dynamic range. Stability can be improved by periodically imaging a surface of uniform infrared irradiance, as a "reference" and electronically computing for each pixel a factor for gain correction and a difference value for offset correction. For the scanned sensor these corrections are programmed to occur at the end of each scan. For the staring sensor a low rate chopper blade is a useful, if awkward, device to achieve gain and offset

Table 1. Thermographic Sensitivity For Scanned and Staring Sensors[a]

Figure of merit→	Cooled focal plane				Uncooled focal plane			
	D^*	NEP	$NE\Delta T$ 3–5 μm	$NE\Delta T$ 8–12 μm	D^*	NEP	$NE\Delta T$ 3–5 μm	$NE\Delta T$ 8–12 μm
Sensor type↓	cm Hz$^{1/2}$/W	pW	mK	mK	cm Hz$^{1/2}$/W	pW	mK	mK
scanned linear array $A = 2.2E\text{-}5$ cm^2 $B = 40$ KHz $f_n = 2.0$	5E10	19	750	80	2E9	470	4700	500
staring area array $A = 1.5E\text{-}5$ cm^2 $B = 100$ Hz $f_n = 1.0$	1E10	100	250	25	1E9	40	560	60

[a] $\epsilon_0 = 0.85$.

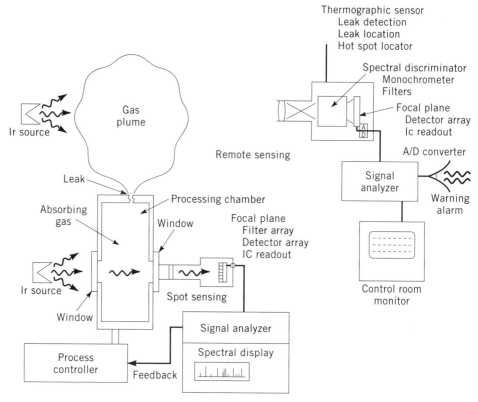

Fig. 2.   Thermography combined with precise spectral discrimination provides for spatial spectroscopy for remote spectral sensing as well as "spot sensing" for process control and safety.

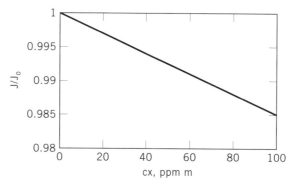

Fig. 3.   Attenuation of infrared as a function of normalized concentration caused by CO absorption in the 4.6 μm band. G_o = 6.6E3 ppm m, spectral bandwidth = 0.2 μm.

constant in the range where the absorption is less than 1%, G_o is given by the slope of the curve. It is clear that the thermographic spectrometer must have high sensitivity to detect small amounts of CO in a reasonable path length of less than 1 meter.

The detection of a specific gas (10) is accomplished by comparing the signal of the detector that is constrained to the preselected spectral band pass with a reference detector having all conditions the same except that its preselected spectral band is not affected by the presence of the gas to be detected. Possible interference by other gases must be taken into account. It may be necessary to have multiple channels or spectral discrimination over an extended spectral region to make identification highly probable. Except for covert surveillance most detection scenarios are highly controlled and identification is not too difficult.

The sensitivity equation can be developed by differentiating equation 4 with respect to J. Since the signal is proportional to J and detection is defined as to when the measured signal to noise ratio equals 3, the gas detection sensitivity (ppm) in terms of the NEP for gas detection becomes:

$$\Delta C = \frac{3G_o\ NEP\ \exp\left(\dfrac{xC}{G_o}\right)}{J_o A_p (n_s n_p)^{1/2} x} \tag{5}$$

This expression includes the use of detector arrays of n_p detectors with additive signals and sample addition of n_s samples to improve sensitivity. Typical sensor parameters are $J_o = 1\ \text{mW/cm}^2$, $NEP = 30\ \text{pW}$, $A_p = 1.5\text{E-5 cm}^2$, $n_s = 60$, $n_p = 1$ for imaging and ca 600 for nonimaging gas detection.

For the case of CO detection at 4.65 μm wavelength, where $G_o = 6.6\ \text{E5}$ ppm-cm, the sensitivity as a function of x and C is shown in Figure 4. The number of signal samples n_s can be taken at a 60 Hz rate giving in this case a measuring time of 1 second. For an absorption tube length of only 50 cm the sensitivity of detection is 0.4 ppm. The specified margin of 3 in the signal to noise ratio rather than 1 improves the probability of identification from 50% to 98%.

An imaging thermographic sensor may be modified for spectral imaging by using a monochrometer or putting a narrow band filter over the sensor aperture

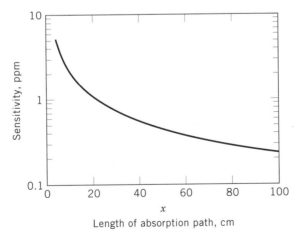

Fig. 4. Sensitivity for the detection of CO using spectral thermography as a function of absorbing path length. Detector NEP = 30 pW, J = 1.0 mW/cm^2, A = 1.5E-5 cm^2, samples = 60, number of detectors = 600, G_o = 6.6E5 ppm-cm; center wavelength = 4.65 μm; spectral bandwidth = 0.2 μm.

(11). The filter may have a constant wavelength or tapered wavelength in the form of a wedge or wheel. A schematic of this arrangement is shown in Figure 2 for the applications of industrial process control and safety. When high spectral resolution (hyperspectral) is required an interferometer spectrometer is utilized in conjunction with the thermographic sensor. At the high dispersion necessary for accurate substance identification remote sensing is very difficult because the signal level is very low. A very narrow spectral bandwidth results in a high value of G_o for gas detection. This can be alleviated somewhat by sensing two or more portions of the spectrum where absorption is expected (multispectral) and performing signal analysis for identification. The presence of several chemicals in any typical scene complicates the entire process.

Applications

Thermal Imaging. Thermography as a night vision tool in the nonmilitary sector is being used by law enforcement (3) agencies to detect the movement of suspects in the complete dark. In fact contraband discarded by suspects in flight have been detected and recovered using their latent heat signature acquired by body contact. A multipurpose commercial thermographic sensor is shown in Figure 5. Sensors are finding use by rescue services looking for survivors floating in the water at night. Since the average adult body even at rest has the energy consumption of a 100 W light bulb and emits 30% of its radiation in the

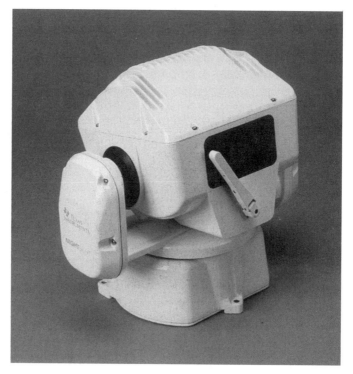

Fig. 5. The Night Sight commercial thermograph made for all-weather operation uses the uncooled pyroelectric focal plane. Courtesy of Texas Instruments Inc.

8 to 12 μm spectral band, the lightly clothed human is a bright thermographic source. All mammals are easy to detect in relatively clear weather making it easier to count livestock at night than in daylight.

Thermal imaging is under development as nighttime vision enhancement for drivers of trucks and automobiles. When perfected the driver will have a display projected on the windshield (heads-up display) showing the forward view as seen by the infrared camera. Distant vision will be similar to daylight viewing. Humans and animals approaching the roadway will be readily seen and there will be no glare from street light or even oncoming headlights. Thermal vision in rain and fog will be better than unaided, but infrared absorption and scattering by water droplets will still greatly reduce penetration. Automobile companies project this option to be available by 2001 and will cost 5% of the average car price. Although thermal imaging is common in military aircraft, commercial airlines have been slow to introduce it because of the high cost of the airborne version. It is being used by a few airlines to help with collision avoidance and landing in light fog.

Efforts have been made to evaluate human trauma caused by disease and injury using thermal imaging. Investigations continue by medical schools to make thermography an early screening tool for the detection of kidney disease and breast cancer (12). A cancer in its early stages grows quickly creating a small amount of excess heat which appears as a bright spot on a thermograph display. Since only one breast is likely to be affected at the early stage, the radiologist looks for thermal pattern asymmetry in comparison of one breast to the other. However complications caused by blood vessels near the skin and weak response to deep lesions, have led to unacceptably high false-positive and false-negative detection as determined by the medical community. A medical thermograph is shown in Figure 6.

Industrial use of thermal imaging typically is the detection of thermal anomalies (13,14) such as leaking pipes and valves, overheating boilers, transformers and power lines, and friction generated heat in bearings. A thermal image of a transformer station is shown in Figure 7. These are simple heat detection applications but the imaging process quickly locates the problem in the three-dimensional industrial environment. The ir camera is scanned like a TV camera and the output signal is digitized for evaluation by a programmed computer than compares the output of each frame with a calibrated reference frame. An alarm declares when a thermal anomaly is detected. Defects in materials (15) and circuit boards (16) can be detected by the associated discontinuities in thermal conduction.

Chemical Gas Detection. Spectral identification of gases in industrial processing and atmospheric contamination is becoming an important tool for process control and monitoring of air quality. The present optical method uses the ftir (Fourier transform infrared) interference spectrometer having high resolution (<1 cm^{-1}) capability and excellent sensitivity (few ppb) with the use of cooled MCT (mercury cadmium telluride) (2) detectors. The spectral data is massive and requires normalization to remove the effects of interfering gases. The ftir equipment is expensive (>100K\$) and difficult to move around. Low cost (<5K\$), small size (<10 kg) and low power spectral thermography is under development. However, its lower spectral discrimination (>20 cm^{-1}) is an issue

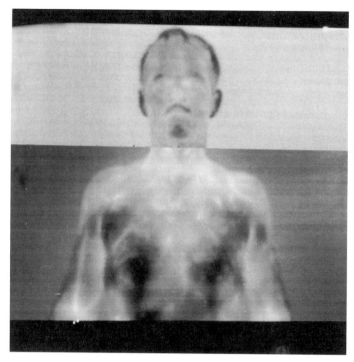

Fig. 6. Thermograph of a person in the 3 to 5 μm spectral band. Spatial resolution is 1 mm and temperature sensitivity is 0.1°C.

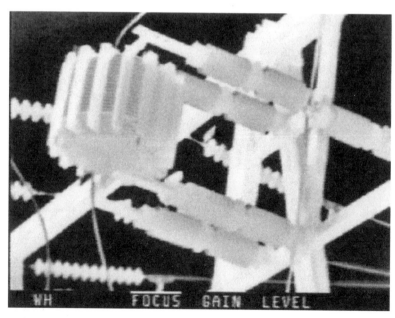

Fig. 7. Thermographic image of a power distribution station and transformer. Spatial resolution is 0.3 m-radians and temperature sensitivity is 0.08 C. Courtesy of Texas Instruments Inc.

and will limit the use of the spectral thermograph to situations where the gases to be detected are known but their concentrations must be monitored. This will result in the use of well defined spectral filtering and signal processing.

Spectral monitoring for chemical process control (17) has been demonstrated for semiconductor integrated circuit manufacture (18,19) and detection of contamination of processing gases and liquids (20). Open path or remote ir sensing (see Fig. 2) is used to monitor workplace gas and vapor exposures, emissions from hazardous waste sites and to track emissions along fence lines (21). Aerosol analysis by infrared spectroscopy are used to set mass detection limits and spectral quality (22). The removal of NO_x has been monitored by analysis of the by-products of the process using ir spectroscopy (23). Ir spectroscopy will be an important process control technique when the system cost is significantly reduced (24). Air quality monitoring increases in importance and U.S. Government agencies are beginning to create Guidance Documents (25,26) for the development and utilization of spectral thermographs.

BIBLIOGRAPHY

1. M. C. Dudzik, ed., *Electro-Optical Systems Design, Analysis and Testing*, Vol. 4 of J. Accetta and D. Shumaker, eds., *The Infrared and Electro-Optical Systems Handbook*, Environmental Research Institute of Michigan, 1993.
2. H. Kaplan, *Photonics Spectra*, 46 (March 1995).
3. P. W. Kruse, *Photonics Spectra*, 103 (March 1995).
4. S. B. Campana, ed., *Passive Electro-Optical Systems*, Vol. 5 of ref. 1.
5. R. B. Emmons, *Infrared Physics* **17**, 415 (1977).
6. K. H. Herrmann, *Measurement* **8**(1), 17 (Jan.–March 1990).
7. H. Kaplan, *Photonics Spectra*, 52 (June 1996).
8. B. F. Andresen, ed., *Proceedings of SPIE—The International Society for Optical Engineering* **2269**, 804 (1994).
9. F. G. Smith, ed., *Atmospheric Propagation of Radiation*, Vol. 2 of Ref. 1.
10. M. D. Schaeberle and P. J. Treado, *Proc. Microscopy Society of America*, 156 (1994).
11. P. J. Treado, I. W. Levin, and E. N. Lewis, *Appl. Spectros.* **48**(5), 607–615 (May 1994).
12. V. A. Kalugin, A. I. Goshenko, V. S. Vetoshnikov, M. E. Belov, and P. M. Grigorishin, *Biomed. Inst. Biomed. Eng.* **23**(4), 155 (July–Aug. 1989).
13. R. J. Murry and B. F. Mitchell, *Proceedings of Annual Reliability and Maintainability Symposium (RAMS)*, 24–27 Jan. 1994, Anaheim, Calif.
14. G. J. Weil and R. J. Graf, *Second Natl. Spec. Conf. Civ. Eng. Appl. Remote Sens. Georg. Inf. Syst.*, ASCE, New York, pp. 189–198 (1991).
15. M. P. Luong, *Nucl. Eng. Des.* **158**(2–3), 363–376 (Sept. 1995).
16. S. Zannoli, *Elettronica-Oggi* (148), 98–100, 106 (Oct. 15, 1992).
17. I. D. Aggarwal, S. Farquharson, and E. Koglin, eds., *Proceedings of SPIE—The International Society for Optical Engineering* **2367**, 243p (1995).
18. C. J. Gamsky, P. M. Dentinger, G. R. Howes, and J. W. Taylor, *Proceedings of SPIE—The International Society for Optical Engineering* **2438**, 143–152 (1995).
19. K. Nishikawa, K. Ono, M. Tuda, T. Oomori, and K. Namba, *Jpn. J. Appl. Physics* **34**(7A), 3731–3736 (July 1995).
20. *Symposium on On-Line Monitoring Institute of Environmental Sciences—Proceedings*, Annual Technical Meeting 1995, Institute of Environmental Sciences, Mount Prospect, Ill., 1995, 106p.

Table 1. Properties of Urea

Property	Value
melting point, °C	135
index of refraction, n_D^{20}	1.484, 1.602
density, d_4^{20}, g/cm^3	1.3230
crystalline form and habit	tetragonal, needles or prisms
free energy of formation, at 25°C, J/mol[a]	−197.150
heat of fusion, J/g[a]	251[b]
heat of solution in water, J/g[a]	243[b]
heat of crystallization, 70% aqueous urea solution, J/g[a]	460[b]
bulk density, g/cm^3	0.74
specific heat, J/(kg·K)[a]	
at 0°C	1.439
50	1.661
100	1.887
150	2.109

[a]To convert J to cal, divide by 4.184.
[b]Endothermic.
[c]Exothermic.

Table 2. Properties of Saturated Aqueous Solutions of Urea

Temperature, °C	Solubility in water, g/100 g solution	Density, g/cm^3	Viscosity mPa·s (=cP)	H$_2$O vapor pressure, kPa[a]
0	41.0	1.120	2.63	0.53
20	51.6	1.147	1.96	1.73
40	62.2	1.167	1.72	5.33
60	72.2	1.184	1.72	12.00
80	80.6	1.198	1.93	21.33
100	88.3	1.210	2.35	29.33
120	95.5	1.221	2.93	18.00
130	99.2	1.226	3.25	0.93

[a]To convert kPa to mm Hg, multiply by 7.5.

Table 3. Properties of Saturated Solutions in Urea in Ammonia[a]

Temperature, °C	Urea in solution, wt %	Vapor pressure of solution, kPa[b]
0	36	405
20	49	709
40	68	952
60	79	1094
80	84	1348
100	90	1267
120	96	507

[a]Ref. 2.
[b]To convert kPa to atm, divide by 101.3.

Table 4. Properties of Saturated Solutions of Urea in Methanol and Ethanol[a]

Temperature, °C	Methanol		Ethanol	
	Urea, wt %	Density, g/cm^3	Urea, wt %	Density, g/cm^3
20	22	0.869	5.4	0.804
40	35	0.890	9.3	0.804
60	63	0.930	15.0	0.805

[a]Ref. 3.

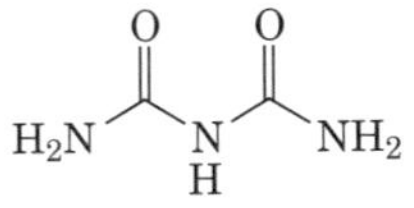

(1)
biuret

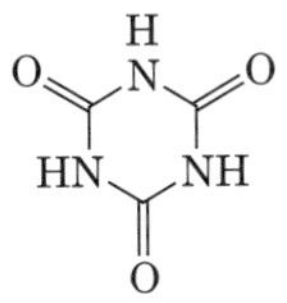

(2)
cyanuric acid

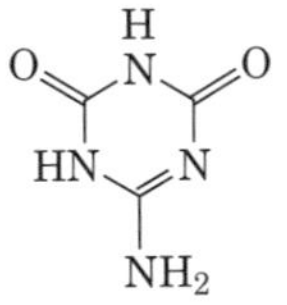

(3)
ammelide

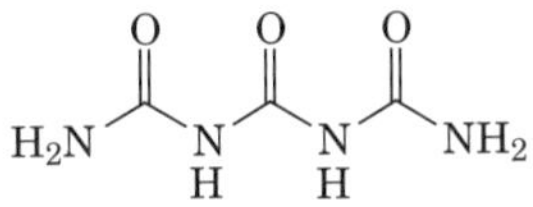

(4)
triuret

Urea acts as a monobasic substance and forms salts with acids (4). With nitric acid, it forms urea nitrate, $CO(NH_2)_2 \cdot HNO_3$, which decomposes explosively when heated. Solid urea is stable at room temperature and atmospheric pressure. Heated under vacuum at its melting point, it sublimes without change. At 180–190°C under vacuum at its melting point, it sublimes without change. At 180–190°C under vacuum, urea sublimes and is converted to ammonium cyanate, NH_4OCN (5). When solid urea is rapidly heated in a stream of gaseous ammonia at elevated temperature and at a pressure of several hundred kPa (several atm), it sublimes completely and decomposes partially to cyanic acid, HNCO, and ammonium cyanate. Solid urea dissolves in liquid ammonia and forms the unstable compound urea–ammonia, $CO(NH_2)_2 NH_3$, which decomposes above 45°C (2). Urea–ammonia forms salts with alkali metals, eg, NH_2CONHM or $CO(NHM)_2$. The conversion of urea is biuret is promoted by low pressure, high temperature, and prolonged heating. At 10–20 MPa (100–200 atm), biuret gives urea when heated with ammonia (6–7).

Urea reacts with silver nitrate, $AgNO_3$, in the presence of sodium hydroxide, NaOH, and forms a diargentic derivative (5) of a pale-yellow color. Sodium hydroxide promotes the change of urea into the imidol form (6):

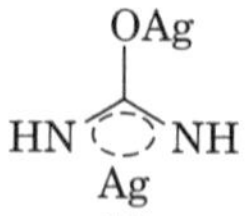

(5) 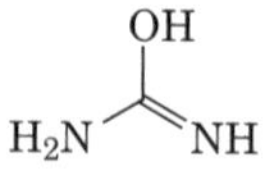   (6)

which then reacts with silver nitrate. Oxidizing agents in the presence of sodium hydroxide convert urea to nitrogen and carbon dioxide. The latter reacts with sodium hydroxide to form sodium carbonate (8):

$$NH_2CONH_2 + 2\,NaOH + 3\,NaOBr \longrightarrow N_2 + 3\,NaBr + Na_2CO_3 + 3\,H_2O$$

The reaction of urea with alcohols yields carbamic acid esters, commonly called urethanes (see URETHANE POLYMERS):

$$\underset{NH_2\overset{\text{O}}{\overset{\|}{C}}NH_2}{} + ROH \longrightarrow \underset{NH_2\overset{\text{O}}{\overset{\|}{C}}OR}{} + NH_3$$

Urea reacts with formaldehyde and forms compounds such as monomethylolurea, $NH_2CONHCH_2OH$, dimethylolurea, $HOCH_2NHCONHCH_2OH$, and others, depending upon the mol ratio of formaldehyde, to urea and upon the pH of the solution. Hydrogen peroxide and urea give a white crystalline powder, urea peroxide, $CO(NH_2)_2 \cdot H_2O_2$, known under the trade name of Hypersol, an oxidizing agent.

Urea and malonic acid give barbituric acid (7), a key compound in medicinal chemistry (see also HYPNOTICS, SEDATIVES, AND ANTICONVULSANTS):

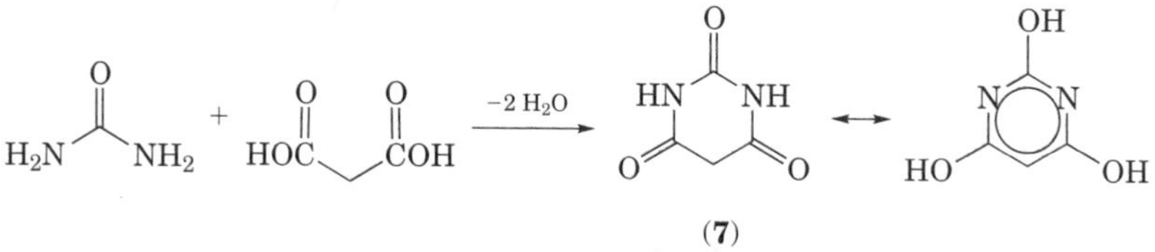

(7)
malonyl urea or
barbituric acid

Manufacture

Urea is produced from liquid NH_3 and gaseous CO_2 at high pressure and temperature; both reactants are obtained from an ammonia-synthesis plant. The latter is a by-product stream, vented from the CO_2 removal section of the ammonia-synthesis plant. The two feed components are delivered to the high pressure urea reactor, usually at a mol ratio >2.5:1. Depending upon the feed mol ratio, more or less carbamate is converted to urea and water per pass through the reactor.

The formation of ammonium carbamate and the dehydration to urea take place simultaneously, for all practical purposes:

$$2\ NH_3 + CO_2 \rightleftharpoons NH_2\overset{\overset{\displaystyle O}{\|}}{C}ONH_4 \tag{1}$$
ammonium
carbamate

$$NH_2\overset{\overset{\displaystyle O}{\|}}{C}ONH_4 \rightleftharpoons NH_2\overset{\overset{\displaystyle O}{\|}}{C}NH_2 + H_2O \tag{2}$$
urea

Reaction 1 is highly exothermic. The heat of reaction at 25°C and 101.3 kPa (1 atm) is in the range of 159 kJ/mol (38 kcal/mol) of solid carbamate (9). The excess heat must be removed from the reaction. The rate and the equilibrium of reaction 1 depend greatly upon pressure and temperature, because large volume changes take place. This reaction may only occur at a pressure that is below the pressure of ammonium carbamate at which dissociation begins or, conversely, the operating pressure of the reactor must be maintained above the vapor pressure of ammonium carbamate. Reaction 2 is endothermic by ca 31.4 kJ/mol (7.5 kcal/mol) of urea formed. It takes place mainly in the liquid phase; the rate in the solid phase is much slower with minor variations in volume.

The dissociation pressure of pure carbamate has been investigated extensively (10–12) and the average values are shown in Table 5.

Ammonium Carbamate. Ammonium carbamate is a white crystalline solid which is soluble in water (2). It forms at room temperature by passing ammonia gas over dry ice. In an aqueous solution at room temperature, it is slowly converted to ammonium carbonate, $(NH_4)_2CO_3$, by the addition of one mol of water. Above 60°C, the ammonium carbonate solution reverts to carbamate solution, and at 100°C, only carbamate is present in the solution. Above

Table 5. Vapor Pressure of Pure Ammonium Carbamate at which Dissociation Begins

Temperature, °C	kPa[a]
40	31
60	106
80	314
100	861
120	2,130
140	4,660
160	9,930
180	15,200; 19,300[b]
200	20,300; 36,500[b]

[a]To convert kPa to atm, divide by 101.3.
[b]The value has been extrapolated because, at temperatures above 170°C, the rate of reaction 2 rapidly increases and it is difficult to determine the carbamate vapor pressure owing to the formation of water and urea and the consequent lowering of the partial pressure of ammonium carbamate.

150°C, ammonium carbamate loses a mol of water and forms urea. The specific heat of solid ammonium carbamate is given in Table 6. Ammonium carbamate melts at ca 150°C, and has a heat of fusion of ca 16.74 kJ/mol (4.0 kcal/mol). The conversion of carbamate to urea begins at ≤100°C. To obtain an appreciable amount of urea at 100°C requires 20–30 h. The rate of conversion increases with increasing temperature (13–15); at 185°C, ca 50% of the ammonium carbamate is converted to urea in ca 30 min.

Conversion at Equilibrium. The maximum urea conversion at equilibrium attainable at 185°C is ca 53% at infinite heating time. The conversion at equilibrium can be increased either by raising the reactor temperature or by dehydrating ammonium carbamate in the presence of excess ammonia. Excess ammonia shifts the reaction to the right side of the overall equation:

$$2 NH_3 + CO_2 \rightleftharpoons NH_2CONH_2 + H_2O$$

Water, however, has the opposite effect. Actual equilibrium constants at various temperatures are given in Table 7. A detailed study of the effect of pressure on urea conversion is given in ref. 17.

Table 6. Specific Heat of Solid Ammonium Carbamate

Temperature, °C	$J/(g{\cdot}K)^a$
20	1.67
60	1.92
100	2.18
140	2.43
180	2.59

^{a}To convert J to cal, divide by 4.184.

Table 7. Reaction Equilibrium Constant^a

Temperature, °C	Reaction equilibrium constant, K
140	0.695
150	0.850
160	1.075
170	1.375
180	1.800
190	2.380
200	3.180

^{a}Ref. 16.

Processing

At this time over 95% of all new urea plants are licensed by Snamprogetti, Stamicarbon, or Toyo Engineering. SNAM utilizes thermal stripping while STAC (Stamicarbon) and Toyo use CO_2 stripping. Only these three processes are, therefore, covered in detail. Process flow sheets are included for others at the end of this section.

As of the end of 1996, about 70 SNAM plants, 125 STAC, and 7 Toyo stripping plants have been built. Currently STAC will design plants for over 3000 MTD, SNAM about 2800 MTD, and Toyo about 2300 MTD as single train units.

STAC, with their current new design (pool reactor), feel the only size limitation will be vessel size. Unless both the vessel fabricator and the intended plant site are "on water", a 4-m-diameter is the maximum that can be transported.

The urea produced is normally either prilled or granulated. In some countries there is a market for liquid urea–ammonium nitrate solutions (32% N). In this case, a partial-recycle stripping process is the best and cheapest system. The unconverted NH_3 coming from the stripped urea solution and the reactor off-gas is neutralized with nitric acid. The ammonium nitrate solution formed and the urea solution from the stripper bottom are mixed, resulting in a 32–35 wt % solution. This system drastically reduces investment costs as evaporation, finishing, (prill or granulation), and wastewater treatment are not required.

Snamprogetti Thermal Stripping Process. The Snamprogetti process is outlined in Figure 1. Initially SNAM utilized NH_3 as the stripping agent. Owing to the high solubility of NH_3 in the synthesis liquid, an overload of NH_3 occurred in the downstream recirculation sections of the plant. At this time only heat is supplied to the stripper to remove unreacted NH_3 and CO_2. Because of the high NH_3 ratio, it is still necessary to have two recirculation sections.

The synthesis recycle loop has the stripped gas going to two high pressure carbamate condensers in series and to a high pressure separator and then back to the reactor. The flow is maintained by using an NH_3-driven liquid–liquid ejector. The reactor is operated at 15 MPa (150 bar) with a NH_3–CO_2 molar feed ratio of 3.5. The stripper is a falling-film type and since high temperatures (200–210°C) are required for efficient thermal stripping, stainless steel tubing is not suitable. Titanium was initially used, but it also was not satisfactory because of erosion near the bottom. At this time a bimetallic tube of zirconium and 25-22-2 stainless steel is used. The zirconium is corrosion-free and the only problem is the difficulty in getting proper welds and separation of the two layers at the bottom ends. Fabrication mistakes have been the only source of problems with this vessel to date.

The stripper offgas going to the high pressure carbamate condensers also contains the carbamate recovered in the medium and low pressure recirculation sections. Both of these systems are similar to those shown in the total-recycle process.

The plant is designed with an excellent heat-exchange system to keep overall steam required to a minimum.

The plant wastewater containing NH_3 and urea is subjected to a desorption–hydrolysis operation to recover almost all the NH_3 and urea. In some plants, this water can then be used for boiler feed water.

The urea solution is evaporated in a two-stage system (99.8%) if the final product is prills, and a single-stage system (+95%) if granules are to be provided.

Stamicarbon CO_2 Stripping Process. In the early 1960s, Stamicarbon introduced the first stripping process. One of the main improvements was the

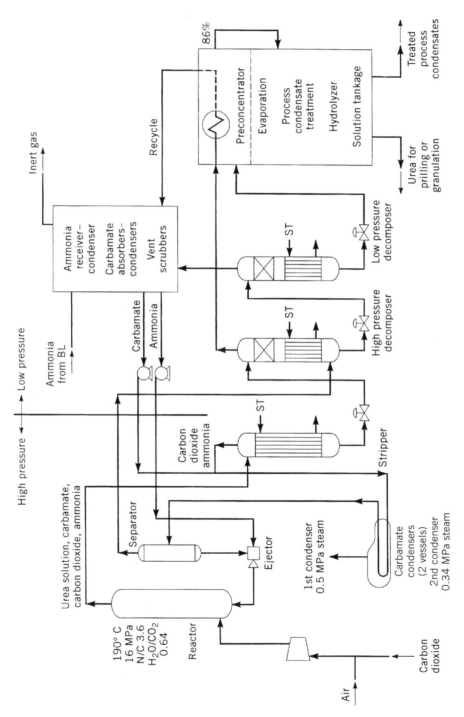

Fig. 1. Snamprogretti thermal stripping urea process. BL = battery limits.

reduction of steam required pre ton or urea to 0.8–0.95 from the 1.8 ton required in the conventional total-recycle process (Fig. 2). Steam usage and electric power required can be varied depending on the final design.

The Stamicarbon process is described in Figures 3–7. The synthesis section of the plant consists of the reactor, stripper, high pressure carbamate condenser, and a high pressure reactor off-gas scrubber. In order to obtain a maximum urea yield per pass through the reactor, a pressure of 14 MPa (140 bar) and a 2.95/1 NH_3–CO_2 molar ratio is maintained. The reactor effluent is distributed over the stripper tubes (falling-film type shell and tube exchanger) and contacted by the CO_2, countercurrently. This causes the partial NH_3 pressure to decrease and the carbamate to decompose.

The urea solution out of the stripper bottom flows to a single-stage low pressure recirculation section (0.4 MPa, 4 bar). The stripper off-gas is sent to the carbamate condenser.

In this condenser, part of the stripper off-gases are condensed (the heat of condensation is used to generate low pressure steam). The carbamate formed and noncondensed NH_3 and CO_2 are put into the reactor bottom and conversion of the carbamate into urea takes place. The reactor is sized to allow enough residence time for the reaction to approach equilibrium. The heat required for the urea reaction and for heating the solution is supplied by additional condensation of NH_3 and CO_2. The reactor which is lined with 316 L stainless steel, contains sieve trays to provide good contact between the gas and liquid phases and to prevent back-mixing. The stripper tubes are 25-22-2 stainless steel. Some strippers are still in service after almost 30 years of operation.

The noncondensable gases introduced in the CO_2 (ie, passivation air) and part of the unreacted NH_3 and CO_2 goes to the high pressure scrubber, which consists of a shell and tube exchanger in the bottom portion and a packed bed in the upper part. In the lower part, most of the NH_3 and CO_2 is condensed, the heat of condensation is dissipated in tempered cooling water. In the upper part, the gases leaving the bottom section are countercurrently contacted with carbamate solution returned from recirculation. The scrubber off-gas, containing nitrogen, oxygen, and very small amounts of NH_3 and CO_2 are vented to the atmosphere after passing through an absorber.

The carbamate solution from the scrubber flows to a high pressure ejector. The NH_3 feed pressure induces enough head to convey the carbamate solution from the scrubber to the carbamate condenser.

As mentioned before, because of design and operating conditions (ie, NH_3/CO_2 ratio, pressure, temperature, reactor volume), only one recirculation stage is required (0.4 MPa) (4 bar). On expansion, a large portion of the carbamate left in the urea solution from the stripper decomposes. The remaining solution passes through a rectifying column, heater, and separator. The gases formed go to a carbamate condenser and are pumped via the high pressure carbamate pump (either reciprocating or centrifugal) to the high pressure scrubber.

The urea solution is then evaporated to 99.8% for prilling (2 stages) or plus 95% for granulation (1 stage).

The Stamicarbon wastewater system consisting of two desorbers, hydrolyzer, hydrolyzer heater, reflux condenser, desorber heat exchanger, and a wastewater cooler is very efficient. Also, in many plants, as the water contains

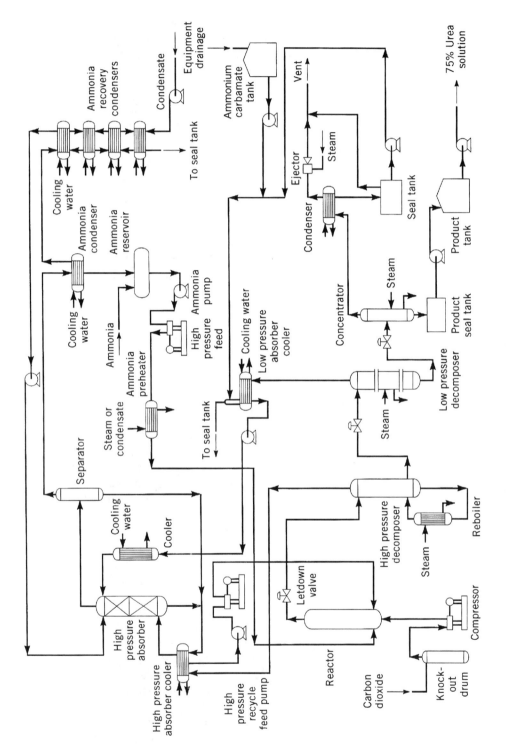

Fig. 2. Typical total-recycle urea process.

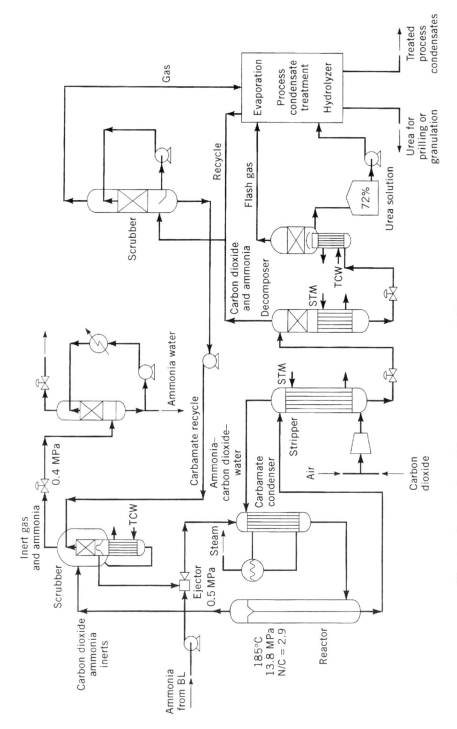

Fig. 3. Stamicarbon CO₂ stripping process. TCW = tempered cooling water.

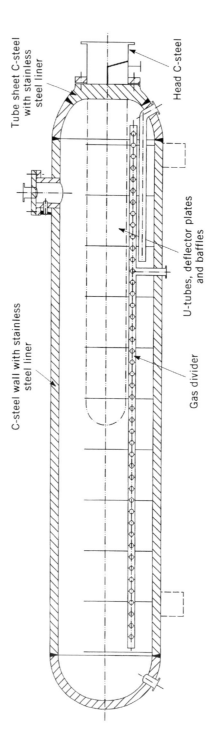

Fig. 7. Pool reactor vessel for Stamicarbon CO_2 stripping process.

Tube sheet C-steel with stainless steel liner

Head C-steel

U-tubes, deflector plates and baffles

C-steel wall with stainless steel liner

Gas divider

610

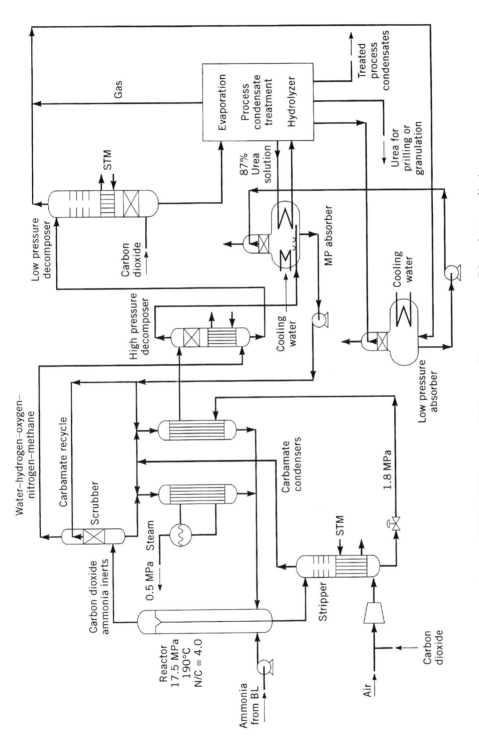

Fig. 8. TEC ACES process. MP = medium pressure; BL = battery limits.

the other to heat the urea solution from the stripper bottoms. The gas–liquid stream from the condenser is recycled back to the reactor by gravity flow. The inert stream from the top of the reactor is purged to the scrubber for recovery of the NH_3 and CO_2.

The urea solution leaving the stripper bottom contains about 12 wt% of NH_3 and is further purified in the 1.8 MPa (18 bar) and 0.2 MPa (2 bar) recovery sections of the plant. The resultant NH_3 and CO_2 separated in the decomposers is absorbed and returned to the synthesis section by the high pressure centrifugal carbamate pump.

The urea solution stream is then fed to the vacuum concentrator unit which operates at 17.3 kPa (130 mm Hg abs) and produces 88.7 wt % urea. It then goes to either two-stage evaporators if prills are made, or a single-stage unit for granule production.

As in all the processes, the process condensate and all other sources of waste urea–NH_3–water contamination go to a waste recovery unit which includes a urea hydrolyzer. The final water discharge is then below 3–5 ppm of NH_3 and urea.

Other Processes.   Flow sheets for typical partial-recycle process and typical once-through urea process are given in Figures 9 and 10, respectively.

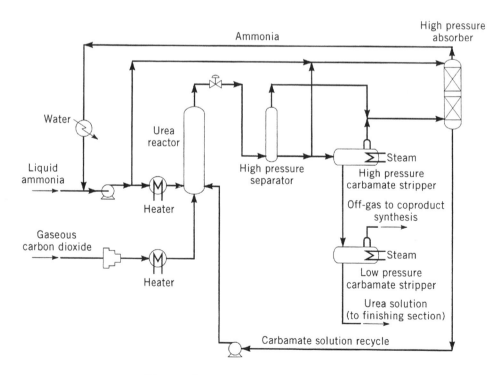

Fig. 9.   Typical partial-recycle process.

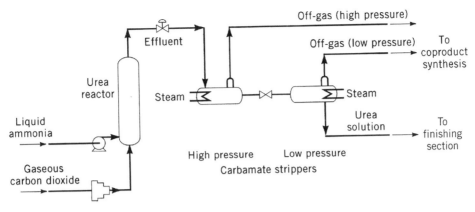

Fig. 10. Typical once-through urea process.

Finishing Processes

Urea processes provide an aqueous solution containing 70–87% urea. This solution can be used directly for nitrogen-fertilizer suspensions or solutions such as urea–ammonium nitrate solution, which has grown in popularity recently (18). Urea solution can be concentrated by evaporation or crystallization for the preparation of granular compound fertilizers and other products. Concentrated urea is solidified in essentially pure form as prills, granules, flakes, or crystals. Solid urea can be shipped, stored, distributed, and used more economically than in solution. Furthermore, in the solid form, urea is more stable and biuret formation less likely.

Prilling. The manufacture of prills is rapidly decreasing owing to both environmental problems and product quality as compared to granules.

In a prilling plant the urea solution from the recovery section is evaporated in two stages to +99.8% strength. It is then pumped to the top of a 50-60m cylindrical concrete tower where it is fed into a spinning bucket containing many (+2000) small holes. The emerging small liquid droplets solidify as they fall and are cooled by a forced or induced draft air flow. The very fine dust that is formed and exits at the top of the tower with the air flow is an environmental problem. Many recovery systems have been tried (ie, water baths, dust collection, etc), but none have been very successful. The quality problem is that prill size must necessarily be small in order to obtain proper solidification and cooling in the fall height that is practical (50–60). Generally, both the crushing and impact strength of the prill is much less than for a granule. This causes many problems in handling both at the plant and in shipping. Stamicarbon has introduced a "seeding" system that has improved prill strength. Also, formaldehyde can be added as it not only improves the crushing strength, but is suppresses the caking tendency in storage. If it were not for environmental considerations, prilling would still be a cheaper option than granulation in a small-scale marketing area (ie, not on a global scale).

Granulation. Almost all new plants produce granules and the Hydro-Agri process is used in the majority of plants (Fig. 11). This process was developed by NSM of Holland many years ago. They have no plant size limitation and

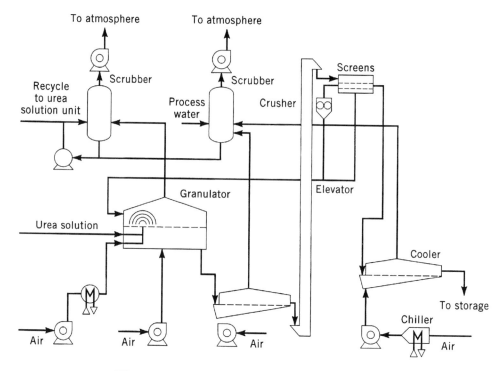

Fig. 11. Hydro-Agri urea granulation process.

will design a single-train unit for production over 3000 MTD. The C&I Girdler drum system has been very successful, but cannot compete in today's market because of restrictions in the train size. Toyo has successfully developed a spout-fluid technology (Figs. 12–14). Three plants are in operation and others are in either the design or construction stage. Stamicarbon also will license a fluid-bed large-scale single-train plant that is somewhat similar to the Hydro-Agri design (Fig. 15).

Wastewater Treatment

Under the pressure of progressively more stringent government regulations with regard to permissible levels of residual NH_3 and urea content in wastewaters, the fertilizer industry made an effort to improve wastewater treatment (see also WATER, SEWAGE).

For each mol of urea produced in a total-recycle urea process, one mol of water is formed. It is usually discharged from the urea concentration and evaporation section of the plant. For example, a 1200 t/d plant discharges a minimum of 360 t/d of wastewater. With a barometric condenser in the vacuum section of the evaporation unit, the amount of wastewater is even higher. Small amounts of urea are usually found in wastewaters because of entrainment carry-over.

The problem in reducing the NH_3 and urea content in the wastewaters to below 100 ppm is because it is difficult to remove one in the presence of the other. The wastewater can be treated with caustic soda to volatilize NH_3. However, in

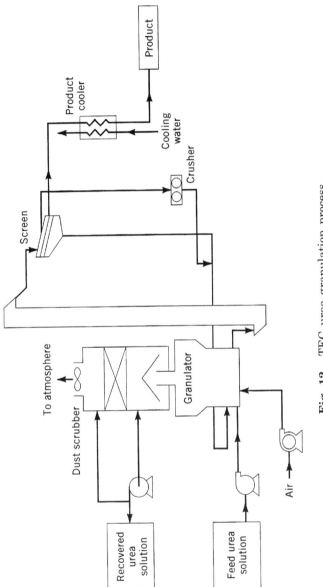

Fig. 12. TEC urea granulation process.

615

WATER-SOLUBLE POLYMERS

Water-soluble polymers find application in a wide variety of areas that include polymers as food sources, plasma substitutes, and as diluents in medical prescriptions. Other areas of importance for water-soluble polymers include detergents, cosmetics, sewage treatment, stabilizing agents in the production of commodity plastics, rheology modifiers in the various processes for petroleum, textile, paper, and latex coatings production. In this article the general features of polymers that influence their water solubility and solution properties will be discussed, beginning with a consideration of carbohydrate polymers provided by natural processes. Carbohydrates exhibit a broad spectrum of properties arising from structural variations of essentially one type of monomer. The discussion of synthetic water-soluble polymers will emphasize differences influenced by variation in the chemical composition of the polymer and the mechanism of polymerization. This article is intended to be a synopsis of water-soluble polymers that have significant commercial impact. A more detailed discussion of water-soluble polymers is available (1–4).

Hydrophilic Groups. Water solubility can be achieved through hydrophilic units in the backbone of a polymer, such as O and N atoms that supply lone-pair electrons for hydrogen bonding to water. Solubility in water is also achieved with hydrophilic side groups (eg, OH, NH_2, CO_2^-, SO_3^-). These groups are found in the various polymers discussed here. Truly unique in its ability to interact and promote water solubility is the $-O-CH_2-CH_2-$ group. The interactions of these groups with water and their placement in the polymer structure influence the water solubility of the polymer and its hydrodynamic volume.

Viscosity Efficiency. A majority of the applications of water-soluble polymers revolve around their role in increasing the viscosity of water solutions.

The hydrodynamic volume of the polymer influences its viscosity efficiency, and thereby its ability to modify the rheology of an application formulation. The primary parameter influencing the hydrodynamic volume is the polymer's molecular weight. The second is in the conformational rigidity or extension of the polymer chain in solution. For example, carbohydrate polymers contain repeating ring structures that facilitate a more rigid structure than observed in nonionic synthetic polymers. At a given molecular weight this effects a greater viscosity efficiency. The rigidity also can be increased by inclusion of charged groups in both synthetic and carbohydrate polymers. Such groups lead to electrostatic repulsions and extension of the chain in deionized aqueous solutions and an increase in the hydrodynamic volume of the polymer. Greater rigidity also can be achieved in carbohydrate polymers if helical conformations are realized. This occurs when certain interunit positional bonding patterns are coupled with pyranosyl ring branches to the main chain.

Carbohydrate Polymers

An anhydroglucose ring or glucopyranosyl unit (Fig. 1a) is the structural unit on which human metabolism is dependent. The anhydroglucose unit provides, through the four hydroxyl units, a diversity of polymer structures that can be formed through variations in positional bonding between the rings (ie, $1 \rightarrow 2$, $1 \rightarrow 3$, $1 \rightarrow 4$, and $1 \rightarrow 6$) (5). Examples of different positional bonding in carbohydrate polymers are illustrated in Figure 2.

Symmetrical $1 \rightarrow 4$ bonding provides the world's most abundant polymer, cellulose (Fig. 2a). The bonds linking hydroxyl units in the plane of the puckered ring are referred to as equatorial bonds. The glucopyranosyl unit is also present in amylose (Fig. 1b). The variance between cellulose and amylose (the latter is a component of starch (qv)) is in the interunit bonding between anhydroglucose rings. In cellulose, the rings are connected through equatorial–equatorial bonding (beta-linkage); in amylose the $1 \rightarrow 4$ inter-ring bonding is equatorial–axial (alpha-linkage; axial denotes a bond perpendicular to the general plane of the ring). The beta-linkage in cellulose (qv), complemented by the intrahydrogen bonding among rings facilitates a linear projection of the polymer. This, complemented by interhydrogen bonding among polymer chains, provides crystallinity

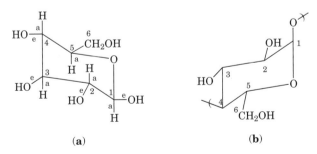

(a)   (b)

Fig. 1. (a) The glucopyranosyl basic unit, where e = equatorial and a = axial bond; (b) amylose with equatorial–axial interunit bonding (the C–H axial bonds have been omitted).

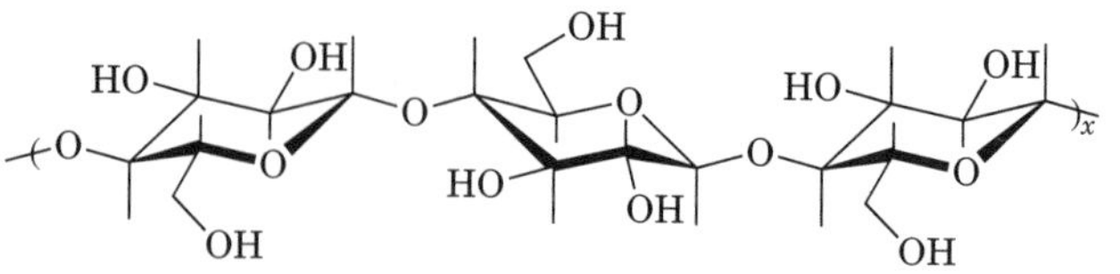

(a)

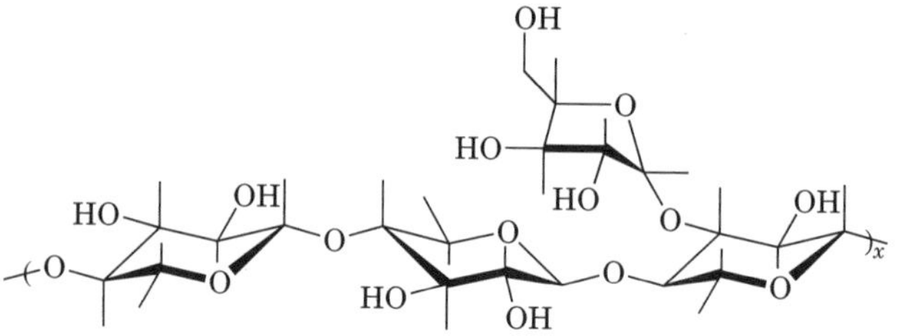

(b)

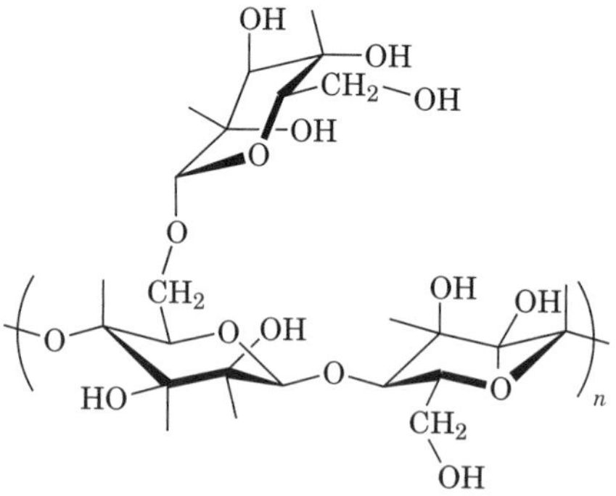

(c)

Fig. 2. Examples of carbohydrate polymers with interunit and branch position differences: (**a**) cellulose; (**b**) flaxseed gum; (**c**) guaran; (**d**) C-6 hydroxyl; and (**e**) *Sclerotium glucanium* polysaccharide (SGPS).

and a rigid structure. Thus, the beta-linkage serves as a structural unit in plants and higher species. To transform the world's most abundant polymer, cellulose, into a water-soluble species requires replacement of some of the hydroxyls to disrupt the extensive hydrogen bonding.

Amylose (Fig. 1b), with alpha-interconnecting linkages, is soluble in hot water, (unlike cellulose), but it retrogrades in low temperature aqueous solutions into a helical conformation and precipitates. Glucopyranose branching from the C-6 hydroxyl provides solubility at low temperatures. Thus this anomeric bonding difference (ie, alpha-$(1 \rightarrow 4)$ instead of the beta $(1 \rightarrow 4)$ linkage in cellulose) provides a readily assessible source of energy and is designated at a storage source in nature. The principal starch product from potato, corn, and other vegetable sources is the branched derivative of amylose (to inhibit precipitation at lower temperatures) known as amylopectin. In higher animal forms it is stored as a more branched derivative (ie, glycogen) to minimize solution viscosities.

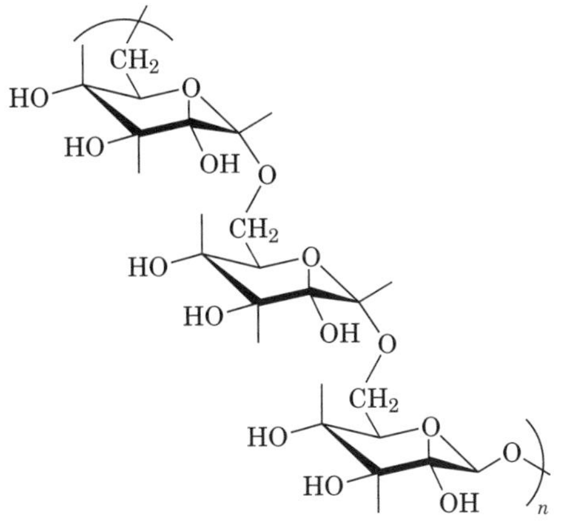

(d)

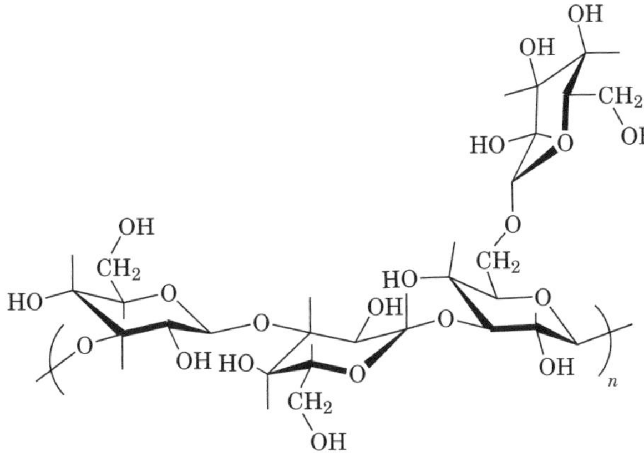

(e)

Fig. 2. *Continued*

The anomeric difference (ie, the alpha- and beta-linkages) between cellulose and amylose is important. Of greater importance in determining the aqueous solution properties of water-soluble polymers is the interunit bonding patterns between rings. When the bonding is between the $1 \rightarrow 3$ positions of the repeating glucose units (rather than $1 \rightarrow 4$ as in cellulose), the polymer is again water insoluble, but the addition of caustic swells the beta-$(1 \rightarrow 3)$ linked polymer. As with amylose, branches from the C-6 position promote solubility in water. *Sclerotium glucanium* polysaccharide (SGPS) is $1 \rightarrow 3$ linked and has $1 \rightarrow 6$ side-chains as shown in Figure 2e. The $1 \rightarrow 3$ bonding with a branch on the C-6 position of every third unit facilitates a triple helix conformation (6,7) and high thickening efficiency. Two naturally occurring carbohydrate polymers with structural main chain similarity to cellulose (ie, in the beta-$(1 \rightarrow 4)$ linkage), but with water solubility without derivatization due to side branch units, are the nonionic fraction of flaxseed gum and guaran (Fig. 2b and 2c, respectively). In Figure 2c the repeating unit is a C-2 epimer of glucopyranose (ie, the C-2 hydroxyl is axial rather than equatorial) and the main chain repeating unit is referred to as mannopyranose. The polymer in Figure 2c is known as guaran or guar gum. The cis stereochemistry in guar gum provides a source of cross-linking with borates and chromium ions that is extensively used in fracturing fluids in petroleum recovery applications (8).

When the interbonding position between repeating rings is through the C-6 hydroxyl (Fig. 2d), there is an additional methylene group between rings. This promotes an extra degree of freedom (9) and water solubility, without derivatization. The primary product with this bonding is known as dextran. The flexibility afforded with this interunit bonding also provides less thickening efficiency than the more segmentally rigid polymers containing interunit bonding through the alicyclic secondary ring hydroxyls. In medical applications in which dextrans are used this is an advantage. In most other application areas this would be a disadvantage. For a more detailed discussion of carbohydrate polymers (see CARBOHYDRATES; CELLULOSE; and STARCH) (10).

CELLULOSE

Commercial Derivatization of Cellulose. Cellulose, the world's most abundant polymer, is derivatized for use in a variety of markets. Two important classes are cellulose esters (qv) and cellulose ethers (qv). Cellulose esters are not water soluble and are not discussed here; cellulose ethers are an important segment of water-soluble polymers.

Commercial derivatization of cellulose begins with the addition of sodium hydroxide to form alkali cellulose (AC) (eq. 1, R = carbohydrate).

$$ROH + NaOH \rightleftharpoons RO^-Na^+ + H_2O \tag{1}$$

The AC may react with methyl chloride or alpha-chloroacetic acid via a direct displacement reaction (eq. 2). The derivatives would be methyl cellulose or carboxymethyl cellulose.

$$RO^-Na^+ + R'X \longrightarrow ROR' + NaX \tag{2}$$

Alternatively, the AC may react with oxiranes (eg, ethylene oxide ($R'' = H$) or propylene oxide ($R'' = CH_3$)) (eq. 3); this is a catalyzed addition and requires a much lower caustic-to-cellulose ratio than is used in direct displacement (eq. 2).

$$RO^-Na^+ + CH_2 \overset{O}{\diagup\!\!\!\diagdown} \underset{\underset{R''}{|}}{CH} \longrightarrow RO-CH_2- \underset{\underset{R''}{|}}{\overset{\overset{O^-Na^+}{|}}{CH}} \tag{3}$$

The derivatives are hydroxyethyl and hydroxypropyl cellulose. All four derivatives find numerous applications and there are other reactants that can be added to cellulose, including the mixed addition of reactants leading to adducts of commercial significance. In the commercial production of mixed ethers there are economic factors to consider that include the efficiency of adduct additions (ca 40%), waste product disposal, and the method of product recovery and drying on a commercial scale. The products produced by equation 2 require heat and produce NaCl, a corrosive by-product, with each mole of adduct added. These products are produced by a paste process and require corrosion-resistant production units. The oxirane additions (eq. 3) are exothermic, and with the explosive nature of the oxiranes, require a dispersion diluent in their synthesis (see CELLULOSE ETHERS).

Biosynthesis. Although cellulose can be derivatized, such materials do not provide the optimum properties desired in many applications, such as retention of viscosity at higher solution temperatures, greater mechanical stability, and greater thickening efficiency. These properties can be approached with carbohydrate polymers in helical conformations, which can be achieved in some carbohydrate polymers prepared by fermentation processes (eg, SGPS; see Fig. 2e) prepared by a yeast (qv) fermentation process. Yeasts have an advantage over fermentation processes using bacteria, because the fermentation can be conducted in low pH media. Yeast are also larger and easier to remove by filtration (qv). However, the most successfully commercial fermentation polymer is XCPS, synthesized by a bacteria, *Xanthomonas campestris*. In this polymer the main chain is simply cellulose, but with three pyranosyl rings branched from the C-3 position (Fig. 3) of every other repeating ring. This arrangement promotes a helical conformation.

With the proper ratio of nutrients and oxygen feed, a water-soluble polymer is produced and accompanied by growth in the microorganism population. Both contribute to the viscosity of the medium and this limits the production process. Fermentation processes require more strenuous mixing and control conditions.

In the carbohydrate polymers discussed so far there is variation in positional or anomeric bonding between different types of polymers, but not within a given polymer chain. Of course such variations would provide different isomers and unique carbohydrate polymer properties. For example, the polysaccharide produced by a yeast *Pullularia pullulans* (Fig. 4) is different from amylose/amylopectin products; this polymer developed by the Japanese is an effective barrier polymer (qv) (11).

Biosynthesis studies of carbohydrate polymers were numerous over the decades that followed the synthesis of SGPS and XCPS; however, these studies did not lead to polymers with the molecular weight and helical characteristics

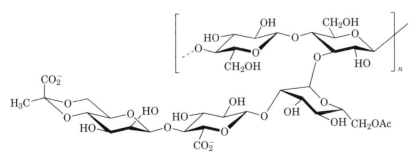

Fig. 3. Structure of *Xanthomonas campestris* polysaccharide.

academic studies, alkaline hydrolysis of the methyl ester requires a lower time than acid hydrolysis of the polymeric ester, and can lead to oxidative degradation of the polymer (38). Poly(methacrylic acid) (PMAA) (**9**) is prepared only by the direct polymerization of the acid monomer; it is not readily obtained by the hydrolysis of methyl methacrylate.

$$
\begin{array}{cc}
\text{\textbf{-}(CH}_2\text{--CH)\textbf{-}} &
\begin{array}{c}
\quad\quad\;\; \text{CH}_3 \\
\quad\quad\;\; | \\
\text{\textbf{-}(CH}_2\text{--C)\textbf{-}} \\
\quad\quad\;\; |
\end{array} \\
\quad\;\; | & \\
\quad\;\; \text{C=O} & \quad\quad\;\; \text{C=O} \\
\quad\;\; | & \quad\quad\;\; | \\
\quad\;\; \text{OH} & \quad\quad\;\; \text{OH} \\
\\
(\textbf{8}) & \quad (\textbf{9})
\end{array}
$$

Free-radical polymerization of the monomer is a commonly used route to prepare PMAA. Over the years this has been done for both PAA and PMAA in aqueous solution, using peroxy initiators such as hydrogen peroxide or persulfate ions. Hydrogen peroxide has the advantage over other initiators in that it does not leave any organic or ionic impurities in the system. Redox systems also have been used, such as persulfate or thiosulfate and ferrous ions plus hydrogen peroxide (39,40); however, the latter system has the disadvantage of leading to an appreciable iron content for the final polymer, which can lead to polymer degradation during applications (41). Because of this the movement has been toward the use of water-soluble organic free-radical initiators to avoid the transition metals present in application formulations. Both PAA and PMAA have multiple pK_as. This property has facilitated many application uses such as dispersing agents for pigments. Carboxylate groups have strong chelating tendencies with cations, particularly divalent ones. This may lead to precipitation of polyelectrolytes, such as PAA and PMAA. To take advantage of this while maintaining solubility, maleic anhydride is copolymerized with styrene. The copolymerization involves an electron–donor acceptor complex that produces an alternating copolymer (42). The anhydride units are hydrolyzed to vicinal carboxylates which enhances the chelating ability over 1,3-carboxylate units, and the aromatic units are sulfonated to maintain solubility in high salinity environments. This type of product has found utility in boiler applications. Unfortunately, the degree of sulfonation of the aromatic unit has not been above 10% and this has limited its market potential.

CATIONIC WATER-SOLUBLE POLYMERS

Cationic monomers are used to enhance adsorption on waste solids and facilitate flocculation (31). One of the first used in water treatment processes (**10**) is obtained by the cyclization of dimethyldiallylammonium chloride in 60–70 wt % aqueous solution (43) (see WATER). Another cationic water-soluble polymer, poly(dimethylamine-*co*-epichlorohydrin) (**11**), prepared by the step-growth

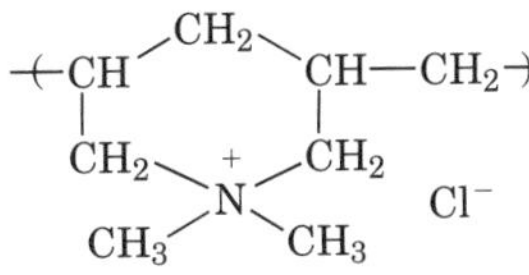

$$\begin{matrix} & CH_2 & \\ +CH & \diagdown CH-CH_2+ \\ | & & | \\ CH_2 & \diagdown \; _+ \; \diagup & CH_2 \\ & N & \\ CH_3 \diagup & \diagdown CH_3 & Cl^- \end{matrix}$$

(10)

$$\begin{matrix} OH & CH_3 \\ | & | \\ +CH_2-CH-CH_2-N+ \\ & | \; _+ \quad Cl^- \\ & CH_3 \end{matrix}$$

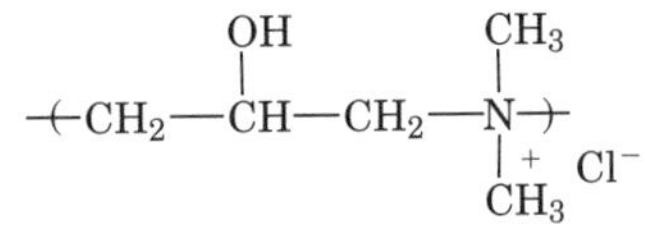

(11)

polymerization of dimethyl amine and epichlorohydrin, is also used for adsorption on clays (qv). This polymer stabilizes shale in drilling fluid formulations in petroleum applications (44). In addition to these cationic polymers used in commodity applications, cationics find use in specialty products such as cosmetic applications. An example of this is the addition of an epichlorohydrin containing an amine group to hydroxyethylcellulose. At a molar substitution of 0.2 to 0.4 per repeating glucopyranosyl ring (see Fig. 1a) the pendent amine provides substantivity to skin and hair (45) and has proven profitability comparable with Carbopol 940.

The literature on the preparation and application of numerous synthetic cationic quaternary water-soluble polymers has been reviewed in detail (46).

INORGANIC WATER-SOLUBLE POLYMERS

Two inorganic water-soluble polymers, both polyelectrolytes in their sodium salt forms, have been known for some time: poly(phosphoric acid) (12) and poly(silicic acid) (13). A more exciting inorganic water-soluble polymer with nonionic

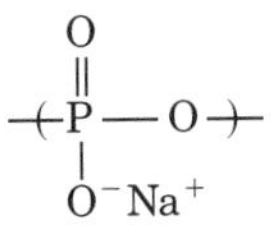

$$\begin{matrix} O & & O \\ \| & & | \\ +P-O+ & & +Si-O+ \\ | & & | \\ O^-Na^+ & & O^-Na^+ \end{matrix}$$

(12) 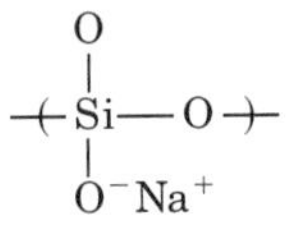   (13)

characteristics has been reported (47). This family of phosphazene polymers is prepared by the ring-opening polymerization of a heterocyclic monomer (14) followed by replacement of the chlorine atoms in the resultant polymer. The

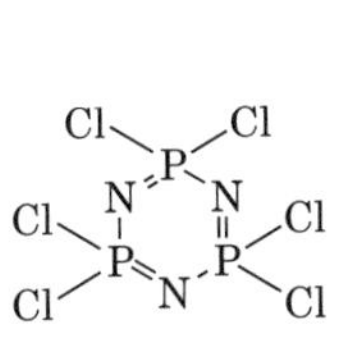

$$\begin{matrix} Cl & & Cl \\ \diagdown & P & \diagup \\ & \diagup \diagdown & \\ Cl & N \; \; N & Cl \\ \diagdown | \; \; \| \diagup \\ Cl & P=P & Cl \\ \diagup & N & \diagdown \\ Cl & & Cl \end{matrix}$$

(14)

chlorine replacement step can be carried out either to introduce only one type of side group or, by simultaneous or sequential substitution, to introduce two or more different types of side groups (see INORGANIC HIGH POLYMERS).

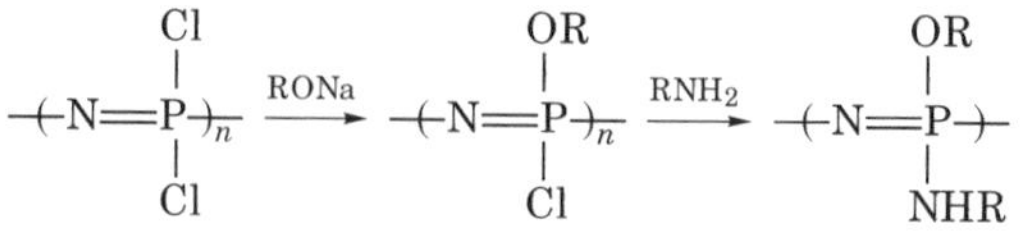

$$
-\!\!\left(N\!\!=\!\!P\right)_{\overline{n}} \;\xrightarrow{\text{RONa}}\; -\!\!\left(N\!\!=\!\!P\right)_{\overline{n}} \;\xrightarrow{\text{RNH}_2}\; -\!\!\left(N\!\!=\!\!P\right)\!\!-
\tag{5}
$$

with Cl/Cl, OR/Cl, OR/NHR substituents respectively.

Qualitative direct displacements on a polymer chain are rare, but possible for $n = 15,000$ in the example shown in equation 5, due to the high reactivity of the P–Cl bond. Some of the organic groups can be placed on the heterocyclic monomer before the ring-opening polymerization if their size is limited to avoid steric influences on the polymerization. The hydrophobic inorganic backbone thus provides an easy route to variable water-solubilizing side nonionic or ionizing groups (Fig. 8). These are stable to hydrolysis at room temperature. The glucosyl (**15**) and glyceryl (**16**) species are sensitive to hydrolysis in neutral pH water at 100°C; they hydrolyze slowly to phosphate, small amounts of ammonia, and glucose or glycerol when the pH is changed. The methylamino-substituted (**17**) and the poly(bis(methoxyethyoxy)phosphazene) (MEEP) (**18**) are stable to water at neutral and basic pH but are sensitive to strong acids.

One of the main advantages of water-soluble polyphosphazenes is the ease with which water-solubilizing side groups can sufficiently cross-link in a stable matrix with high energy radiation such as x-rays, gamma-rays, electron beams, or ultraviolet light. The mechanism of this reaction involves abstraction of hydrogen radicals from the methylene units contiguous to the repeating oxygen bonds

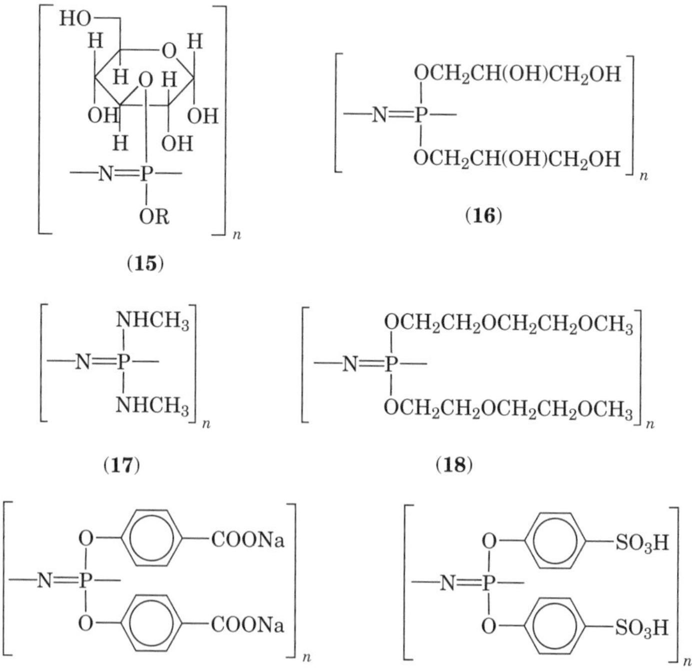

(**15**)

(**16**)

(**17**)

(**18**)

Fig. 8. Water-soluble polyphosphazenes.

in MEEP or to amine groups in other derivatives, followed by cross-combination of the resultant carbon radicals (Fig. 9). In MEEP there are 22 C–H bonds available per repeating unit.

The versatility of water-soluble polyphosphazenes is in the variations in the structures that can be prepared. Structures with a low glass-transition temperature backbone can be modified with a variety of versatile side units. These may find use in solid polymeric ionic conductors, as a means to entrap and immobilize enzymes with retention of enzymic activity, and in biological functions as hydrogels with the capability of exhibiting biocompatibility and bioerodibility properties.

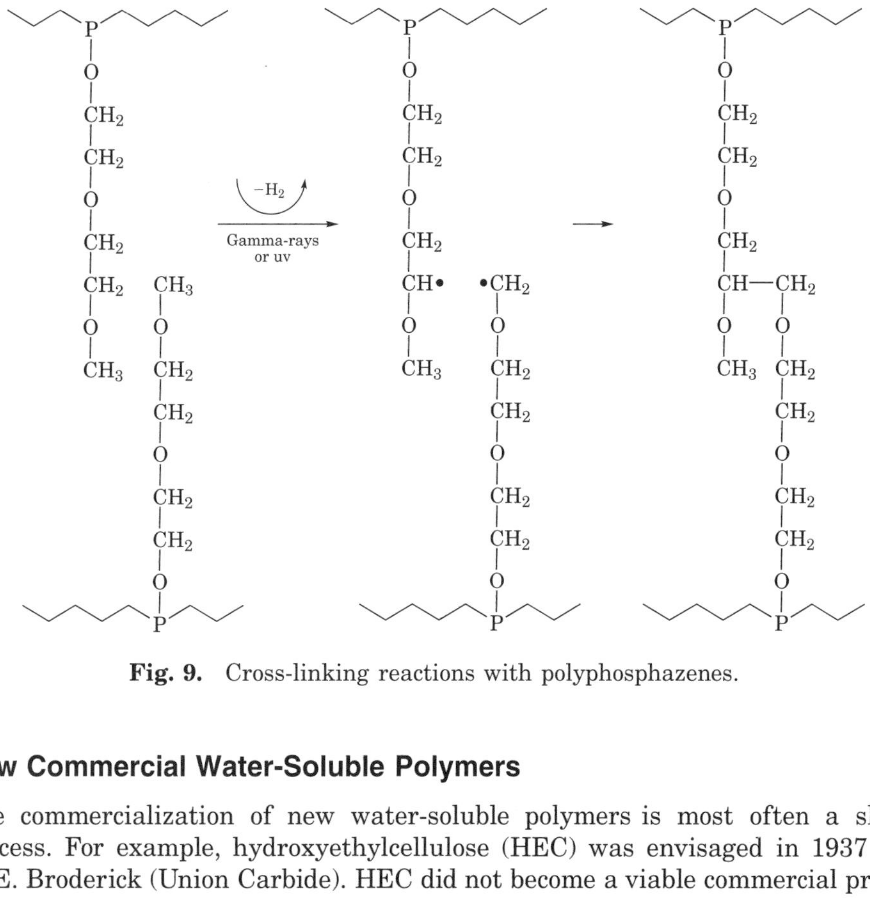

Fig. 9.   Cross-linking reactions with polyphosphazenes.

New Commercial Water-Soluble Polymers

The commercialization of new water-soluble polymers is most often a slow process. For example, hydroxyethylcellulose (HEC) was envisaged in 1937 by A. E. Broderick (Union Carbide). HEC did not become a viable commercial product until the early 1960s. In addition to the general production problems and market development costs, new products face a variety of environmental controls in the 1990s that add more constraints to market development. None the less two more recently developed water-soluble polymers have achieved limited market acceptance and are described below.

One product is poly(2-ethyl-2-oxazoline) (PEOX). It is prepared by the ring-opening polymerization of 2-ethyl-2-oxazoline (**19**) with a cationic initiator (48) eq. 6.

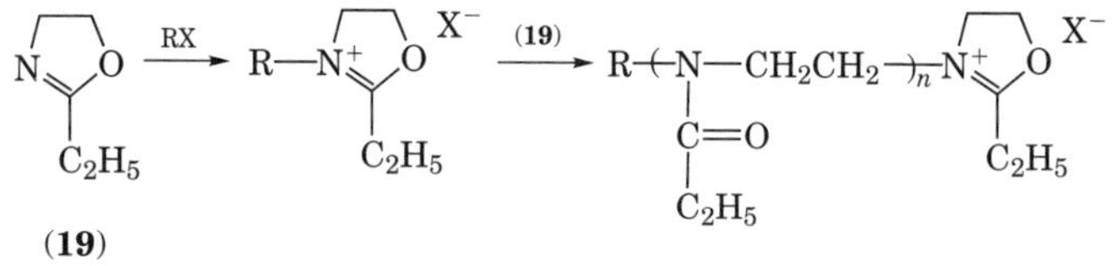

$$(6)$$

Most of the polymer's characteristics stem from its molecular structure, which like POE, promotes solubility in a variety of solvents in addition to water. It exhibits Newtonian rheology and is mechanically stable relative to other thermoplastics. It also forms miscible blends with a variety of other polymers. The water solubility and hot meltable characteristics promote adhesion in a number of applications. PEOX has been observed to promote adhesion comparable with PVP and PVA on aluminum foil, cellophane, nylon, poly(methyl methacrylate), and poly(ethylene terephthalate), and in composite systems improved tensile strength and Izod impact properties have been noted.

Other fairly recent commercial products, poly(vinyl amine) and poly(vinyl amine co-vinyl alcohol), have addressed the need for primary amines and their selective reactivity. Prior efforts to synthesize poly(vinyl amine) have been limited because of the difficulty hydrolyzing the intermediate polymers. The current product is prepared from N-ethenylformamide (**20**) formed from the reaction of acetaldehyde and formamide. The vinyl amide is polymerized with a free-radical initiator, then hydrolyzed (eq. 7).

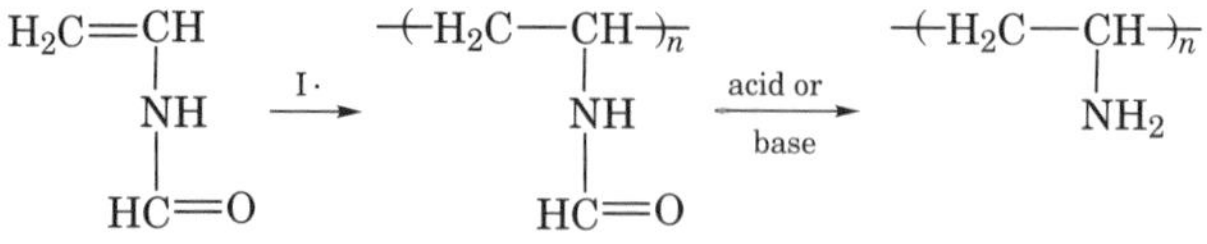

$$(7)$$

The protonated form of poly(vinyl amine) (PVAm–HCl) has two advantages over many cationic polymers: high cationic charge densities are possible and the pendent primary amines have high reactivity. It has been applied in water treatment, paper making, and textiles (qv). The protonated forms modified with low molecular weight aldehydes are useful as fines and filler retention agents and are in use with recycled fibers. As with all new products, unexpected applications, such as in clear antiperspirants, have been found. It is useful in many metal complexation applications (49).

Hydrophobe-Modification of Water-Soluble Polymers

Although many of the new water-soluble polymers discussed above have not achieved large-scale commercial acceptance, there is a class that has achieved outstanding success since the early 1980s: hydrophobically modified water-soluble polymers (HM-WSPs). They have filled certain voids in a number of applications that include cosmetic, paper, architectural, and original equipment manufacturing (OEM) coating areas and have found unsuspected application in the airplane de-icers market. The driving force for the development of HMWSPs is threefold in most application areas:

1. The achievement of high viscosities at low shear rates without high molecular weights, and therefore regain of high viscosities at low shear rates, when the application deformation rates are released. High mol wt polymers are susceptible to mechanical degradation at high deformation rates and permanent loss of high viscosity at low deformation rates.
2. Minimization of the elastic behavior of the fluid at high deformation rates that are present when high molecular weight water-soluble polymers are used to obtain cost-efficient viscosities at low shear rates.
3. Providing colloidal stability to disperse phases in aqueous media, not achievable with traditional water-soluble polymers.

The first criterion was associated with improved secondary and tertiary petroleum recovery processes. This is the justification for the patent applications issued to the Dow (50) and Exxon (51) corporations. The additional costs of production and the increased adsorption of such modified water-soluble polymers are detrimental to the commercial application of such polymers and even the academic studies in this area have decreased in recent years.

The second driving force is most evident in the application of coatings by roll (52) or spray (53). The higher extensional viscosity (see Fig. 7) imparted by higher molecular weight thickeners results in spatter to coatings during roll applications and inhibits misting of spray-applied coatings. Lower molecular weight associative thickeners provide low shear rate viscosities without imparting high extensional viscosities.

The third driving force for acceptance lies in the ability of hydrophobically modified, water-soluble polymers to stabilize the disperse phases of a coating formulation. For example, nonassociative thickeners (those that are not hydrophobically modified) tend to flocculate the disperse phases and this increases the viscosity at low shear rates. This is incorrectly related to the poor flow and leveling, and poor film gloss of an applied coating film. Adsorption and osmotic stabilization of the film-forming polymer colloid and the pigments in a coating formulation are the mechanisms by which the associative thickeners contribute to the improved properties of a coating, and this does not always parallel the viscosities at low shear rates.

The mechanism of adsorption onto the film-forming organic phase occurs through direct hydrophobe adsorption on the disperse phase. Aqueous polyurethane dispersions, used in OEM coatings, are prepared by a step-growth polymerization without external surfactant. Hydrophobically modified, ethoxylated urethane (HEUR) associative thickeners directly adsorb on this disperse phase and thickening is achieved without phase separation (54). In traditional latices used in architectural coatings, the type and concentration of surfactant used in the latices synthesis and added with other components in the coating compete with the hydrophobes of the thickener in the adsorption process (55). This type of competition makes associative thickeners sensitive to variations in other component changes in a coating formulation.

71. S. K. Wolk, G. Swift, Y. H. Paik, K. M. Yocom, R. L. Smith, and E. S. Simon, *Macromol.* **27**, 7613 (1994).

J. Edward Glass
North Dakota State University

WATTLE BARL. See Leather.

WEIGHING AND PROPORTIONING

Weighing is the operation of determining the mass of any material as represented by one or more objects or by a quantity of bulk material. Proportioning is the control, by weighing, of relative quantities of two or more ingredients according to a specific recipe in order to make a mixed product, or to prepare the ingredients for use in a chemical process.

It is likely that volumetric measures were used for quantity determination when commodities were first bartered; however, it has been established with certainty that weighing scales or balances have been in use for at least 7,000 years (1). Measuring by weight instead of by volume eliminates some very considerable inaccuracies from, for example, changes in specific gravity of liquids with temperature, or changes in density of solids owing to voids.

Mass is a fundamental physical property of matter; the mass of an object does not change with its location on the earth. Mass is what is measured in the act of weighing, and scales are calibrated to read in units of mass, eg, kilograms. The terms weight and weighing are used somewhat loosely in this connection, in that weight is a measure of the gravitational force of the earth acting on a mass. Thus, weight depends on the location of the scale relative to the equator and on its altitude. It is common practice to calibrate scales *in situ* using mass standards (test weights); hence, for practical purposes scales actually determine mass.

The terms balance and scale are currently used to describe weighing machines of various forms. The word balance comes from the Latin term *libra bilanx*. *Libra* means scale and *bilanx* means two pans; thus *libra bilanx* means scale with two pans (2). Over the centuries *bilanx* evolved into balance. Balance still refers to sensitive weighing devices with two pans (Fig. 1), although it is also used in a more general sense to refer to any very sensitive laboratory weighing device. The word scale comes from the old Norse word *skal*, which means bowl, and it was originally used to describe the pan of a balance (2). Scale today (ca 1997) is used to describe general-purpose weighing machines found in industrial and retail applications.

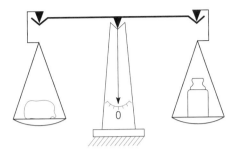

Fig. 1. Balance.

Some common industrial weighing applications include the following:

Verifying quantities of incoming raw materials

Controlling ingredients to the proper proportions

Putting the product into packages of uniform weight, either by packaging directly on a scale or by using a scale to check the performance of filling equipment

Weighing outgoing shipments for purposes of billing and determination of transportation charges

Weighing interdepartmental material transfers for accounting purposes

Controlling the rate of material flow to various pieces of equipment such as grinders or kilns

Weighing Principles

There are many types of scales using many different principles of operation. There are, however, three distinct elements which can be identified regardless of the principle of operation (Fig. 2).

The load-receiving element supports the load during the weighing operation. It may take the form of a scoop, platter, deck, rail, hopper, belt conveyor, or any other configuration appropriate to the material being weighed.

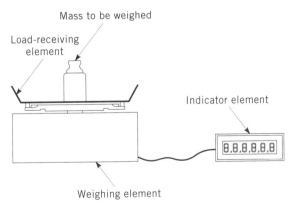

Fig. 2. Basic elements of a scale.

of operation is dependent on temperature, ie, as a result of variations in the viscosity of the hydraulic fluid; and the fact that there is a danger of contamination from leaking hydraulic fluid.

Strain-Gauge Load Cells. The majority of industrial scales today use strain-gauge load cells as the weighing element. The strain-gauge load cell is a device which, when a force is applied to it, gives an electrical output proportional to the applied load.

Figure 5 shows a typical metallic foil strain gauge. It consists of an etched grid of very thin foil attached to a thin insulating backing material. The gauges are bonded firmly to a surface that prevents buckling. When stretched or compressed along the grid lines, the resistance measured between the ends of the grid (solder pads) increases in the case of the former and decreases in the case of the latter.

The central mechanism of a load cell is the spring element that supports the load, and it is designed to have areas of both tensile and compressive strain suitable for application of strain gauges. Spring elements come in a wide variety of designs, such that choice among them is often dictated by their capacity or the method used in supporting the scale. Figure 6 shows a typical spring element for a moment-insensitive load cell that would be used in a small platform scale. Typically, four gauges are applied in such a way that two are in compression and two in tension (C and T, respectively, in Fig. 6) as the spring element is loaded.

The four gauges are wired together to form a Wheatstone bridge (3), as shown in Figure 7. An input voltage (typically 10 V dc) is applied as shown, and the resulting output voltage is measured across points A and B. When no load is applied to the cell, all gauge resistances are the same; the bridge is said to be balanced, and the voltage difference between points A and B is zero. As load is applied to the cell, the resistance of the tension gauges increases, whereas

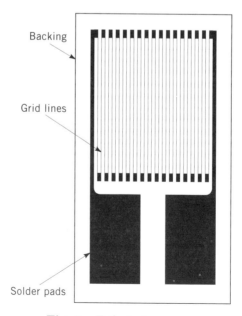

Fig. 5. Foil strain gauge.

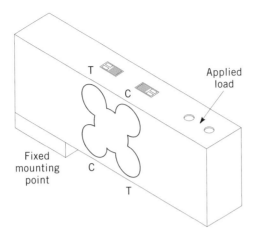

Fig. 6. Load cell spring element. C, gauges in compression; T, gauges in tension.

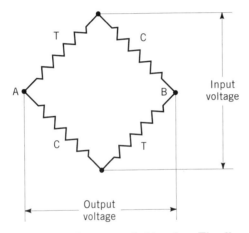

Fig. 7. Wheatstone bridge (see Fig. 6).

that of the compression gauges decreases. The bridge becomes unbalanced, and a voltage difference appears across points A and B. As load continues to increase, the output increases linearly to 20 mV typically at rated capacity.

Figure 8 shows the basic components of an electronic scale based on strain-gauge load cells. A power supply (not shown, but usually housed in the indicator) provides operating power for the indicator and the input voltage for the load cell. The analogue signal produced by the load cell is sensed by the analogue-to-digital (A/D) converter, which sends digital weight information to the digital computer. The digital computer comprises the microprocessor, the working storage (RAM), a permanent memory (EEPROM) for storing calibration data, and the program memory (ROM). The computer filters the raw digital signal and processes it into calibrated weight for the display. The weight is also available at the data interface, from which it can be transmitted to any external device. Also, the indicator can be remotely controlled via the data interface; this is an important feature in systems applications.

ous way to specify the accuracy for an industrial or retail scale is to specify an accuracy class and the number of divisions, eg, Class III, 5,000 divisions. It must be noted that this is not the same as 1 part in 5,000, which is another method commonly used to specify accuracy; eg, a Class III 5,000 d scale is allowed a tolerance which varies from ±0.5 d at zero to ±2.5 d at 5,000 divisions. Calibration curves are typically plotted as in Figure 12, which shows a typical 5,000-division Class III scale. The error tunnel (stepped lines, top and bottom) is defined by the acceptance tolerances listed in Table 1. The three calibration curves belong to the same scale tested at three different temperatures. Performance must remain within the error tunnel under the combined effect of nonlinearity, hysteresis, and temperature effect on span. Other specifications, including those for temperature effect on zero, nonrepeatability, shift error, and creep may be found in *Handbook 44* (5). The acceptance tolerances in Table 1 apply to new or reconditioned equipment tested within 30 days of being put into service. After that, maintenance tolerances apply; they are twice the values listed in Table 1.

Regulation Outside the United States.   Each country establishes its own weights and measures requirements. The majority of these are based on the recommendations of the Organisation Internationale de Métrologie Légale (OIML), in Paris. *R76-1* is the OIML equivalent of *Handbook 44*; it uses accuracy classes and an acceptance tolerance structure similar in many ways to Table 1 (8).

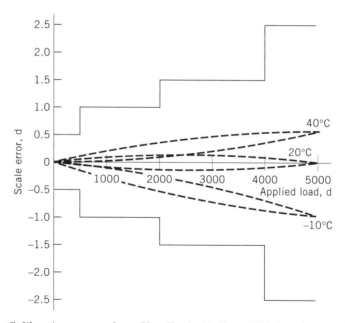

Fig. 12. Calibration curves for a *Handbook 44* Class III 5,000 d scale. See text.

Factors Affecting Weighing Accuracy

Variations in the Force Due to Gravity. The mass of an object is the quantity of matter in the object. It is a fundamental quantity that is fixed, and does not change with time, temperature, location, etc. The standard for mass is a platinum–iridium cylinder, called the International Kilogram, maintained at the International Bureau of Weights and Measures, in Sèvres, France. The mass of this cylinder is 1 kg by definition (9). All national mass standards are traceable to this artifact standard.

It is easiest to appreciate what is meant by mass indirectly, by observing the influence of forces on objects, eg, by picking up an object and sensing the effect of the earth's gravitational force acting on it, and hence "feeling" its weight.

Newton's law of gravitation states that if two particles are a distance r apart, the mutual attraction force between them can be expressed as follows (10):

$$F = \frac{6.673 \times (10^{-11})m_0 m}{r^2} \tag{1}$$

where F is the mutual attraction force in N, m_0 and m are the masses of the particles in kg, and r is the distance between their centers in meters. This equation can be used to calculate the attractive force between the earth and a mass, m, resting on its surface (mass of the earth is 5.976×10^{24} kg and its mean radius is $6,371 \times 10^3$ meters) as follows:

$$F = \frac{6.673 \times (10^{-11})\, 5.976 \times (10^{24})\,(m)}{(6,371 \times 10^3)^2} = 9.82(m) \tag{2}$$

A mass, m, resting on the surface of the earth is attracted to it with a force of $9.82(m)$ N. Put another way, the force per kg experienced by a mass resting on the earth is 9.82 N/kg. The proportionality factor 9.82 is referred to as gravity, or g; it is sometimes referred to as acceleration due to gravity with the equivalent units of m/s^2 (11).

The factor g was calculated using the earth's mean radius; g actually varies with changes in location and altitude. The earth is not spherical; it bulges at the equator; ie, an object at either of the poles is actually nearer to the center of the earth. Also, the earth is rotating on its axis, and an object at the equator is subject to a small centrifugal force that is not present at the poles. For both of these reasons, g increases with increasing latitude from the equator to the poles. In addition, g decreases with increasing altitude above the earth's surface. All of these factors are taken into account in Helmert's equation for g:

$$g = 9.80616 - 0.025928 \cos 2\phi + 0.000069 \cos^2 2\phi - 3.086 \times 10^{-6}h \tag{3}$$

where ϕ is the latitude in degrees and h is the altitude in meters. Table 2 gives the value of g at various locations.

Using an equal-arm balance (Fig. 1), the unknown mass of an object can be determined by placing it in one pan, and adding test weights to the other until the beam balances. The result will be the same regardless of location because the object and the test weights are subject to the same value of g. Any scales that measure an unknown mass by comparing it with a known mass (with or without a lever system), will be unaffected by variations in g. However, this is not a very convenient method of weighing, and the majority of scales in use today determine

Table 2. Values for Gravity, _g_, at Various Locations

Location	Gravity, g, N/kg	Location	Gravity, g, N/kg
United States		_Other countries and areas_	
St. Michael, Alaska	9.82192	Arctic Red River, Canada	9.82434
Quiet Harbor, Alaska	9.81624	Vancouver, Canada	9.80949
Los Angeles	9.79595	Canal Zone, Panama	9.78243
San Francisco	9.79965	Greenwich, U.K.	9.81188
Denver	9.79609	Königsberg, Germany	9.81477
Hartford, Conn.	9.80336	Rome, Italy	9.80367
Washington, D.C.	9.80095	Monrovia, Liberia	9.78165
Key West	9.78970	Bergen, Norway	9.81922
Miami	9.79053	Stockholm, Sweden	9.81843
Honolulu	9.78591	Balia, Brazil	9.78331
St. Louis	9.80001	Kingston, Jamaica	9.78591
New York	9.80267	North or South Pole, sea level	9.83217
El Paso	9.79124	equator, sea level	9.78039
Seattle	9.80733	equator, 500 m above sea level	9.76496

the mass of an object indirectly by measuring the gravitational force acting on it. If a scale is calibrated at one location, and is used at another, the reading would change by a factor of g/g_c, where g and g_c are gravity at the point of use and calibration, respectively. If a mass weighed 5,000 kg on a force-measuring scale in Miami, Florida, it would weigh 5016 kg (5,000 kg $\times$ 9.82192/9.79053) if weighed on the same scale in St. Michael, Alaska, a change of 0.3%. In the extreme case, a variation of 0.7% could be expected. Force-measuring scales should be recalibrated when relocated.

Some scales have a software feature whereby a certain geographical region is identified on initial configuration, and the software then makes the appropriate correction. Some high performance balances have built-in test weights for initial (and periodic) calibration.

Scales are sensitive to force applied in one direction only, eg, a scale with a horizontal platform is sensitive to forces applied perpendicular to the platform. Scales should be leveled before calibration and whenever they are moved; portable scales generally have a bubble level to facilitate leveling.

Slight changes in altitude can significantly affect precision balances of the force-measuring type (note that Helmert's equation for g accounts for the effects of altitude). For example, if an analytical balance is relocated from the basement to the 12th floor of the same building (a change of 48 m in altitude), and is used to weigh a mass of 250 g, the reading will be lighter by 3.8 mg on the 12th floor. Again, recalibration is required.

Moisture Content. In making very fine measurements, it might be necessary to protect a sample from the possibility of losing or absorbing moisture during the weighing operation. This is particularly true if the weighing chamber is at a different temperature or humidity compared to the sample's storage location. In weighing highly volatile liquids, evaporation during the weighing process may cause a steady decline in the weight reading. In either case, the sample should be weighed in the smallest practical covered container.

Materials subject to variable moisture content are frequently treated on a dry weight basis, or on some other standard moisture content basis. In such cases, the moisture content of the material being weighed must be determined, and the actual weight converted to the desired basis. Moisture analyzers are available that incorporate a balance and heating element in one unit. They monitor a sample's weight while the temperature is elevated to expel moisture. Typically, such units operate automatically and display the moisture content directly as a percentage of the original weight.

Temperature Effects.   The output of a strain-gauge load cell changes with temperature because of, for example, changes in gauge resistance and because the gauge and spring element have different coefficients of thermal expansion. Also, with EMFC balances the strength of the permanent magnet changes with temperature, changing the amount of current necessary to balance the load. Good design and careful manufacturing can compensate for most of these effects, but they will not eliminate them completely.

When power is applied to an electronic scale, it warms up and is somewhat unstable until thermal equilibrium is reestablished. The manufacturer's recommendations with regard to warm-up period should be followed.

Strain-gauge load cells are sensitive to temperature gradients induced by, for example, radiant heat from the sun or resulting from high temperature wash down. Load cells should be shielded from such effects or given time to stabilize before use.

Scales will always perform better in a temperature-controlled environment; this is particularly important for high precision balances. Those designed to automatically recalibrate as conditions change will perform better in less-than-ideal conditions.

Where there is a temperature difference between the object to be weighed and the surrounding air, air currents will be induced close to the object's surface (12). These can be significant if extreme accuracy is required. Objects should be allowed to reach thermal equilibrium in the laboratory before weighing. Just as important, the balance should be designed to minimize the temperature rise inside the weighing chamber. In extreme cases, the object should be placed inside the chamber until it reaches thermal equilibrium before weighing. Needless to say, drafts must be avoided.

Buoyant Effect of Air.   Weighing operations performed *in vacuo* are not affected by buoyancy forces. An object in air, however, is subject to a buoyancy force that is equal and opposite to the gravitational force on the mass of air the object displaces (10). If the equal arm balance of Figure 1 is in balance with a test weight of mass, m', in one pan, and material of mass, m, in the other, $m = m'$ if they have the same density. If the densities are different, then the buoyancy forces acting on each pan affect the result. Taking moments about the center pivot point gives

$$g(m - \rho_{\text{air}}V_m) = (m' - \rho_{\text{air}}V_w)g \tag{4}$$

where ρ_{air} is the density of air (1.2 kg/m^3 on average), V_m and V_w are the volumes of the material and test weight, respectively, in m^3. By expressing V_m

and V_w in terms of mass and density, this equation can be expressed as

$$m = m'(1 - \rho_{\text{air}}/\rho_w)/(1 - \rho_{\text{air}}/\rho_m) \tag{5}$$

where ρ_m and ρ_w are the densities of the material and test weight, respectively, in kg/m^3. This expression can be used to calculate the actual mass, m, from the apparent mass, m', indicated by a scale (true for all scales except nuclear). From equation 5, it can be seen that if $\rho_m = \rho_w$, then $m = m'$, as would be expected.

If a 1-kg stainless weight ($m' = 1,000$g, $\rho_w = 8,000$ kg/m^3) is added to one pan of the balance in Figure 1, and material with a density of 1,000 kg/m^3 is added to the other until equilibrium is reached, the amount of the material needed is 1001.05 g, using equation 5. Thus, it takes 1001.05 g of this material to counterbalance 1,000 g of stainless steel, because of the buoyancy effects on the dissimilar volumes.

Buoyancy has no effect in weighing material of the same density as the test weight (used for direct comparison, or used to calibrate the balance), or in proportioning materials of approximately the same density. In general, the recipes used in proportioning are derived empirically, and account for the effects of buoyancy. Buoyancy effects can be significant when the absolute mass must be known accurately, eg, in a laboratory environment, and corrections can be made using equation 5. In industrial and commercial weighing, the effect of buoyancy is generally not considered. Note that variations in air density, eg, due to variations in temperature, can also have a slight effect on weighing repeatability.

Electrostatic and Magnetic Effects.   These two effects are generally small but may be significant in laboratory weighing.

If an electrostatically charged object is weighed close to a charged or conducting surface not being weighed, then weighing accuracy is likely to be affected by the forces which result. Similar problems arise as a result of magnetic effects. Some materials can be permanently magnetized, and others temporarily exhibit magnetic behavior when placed in a magnetic field. When weighed, these materials can seem heavier or lighter if they are in close proximity to a magnetically permeable material, or if they are surrounded by a magnetic field, eg, the earth's magnetic field, or one produced by external equipment.

To avoid electrostatic problems, the air in the laboratory should be maintained at 40% relative humidity or higher, or the air should be ionized. Shielding the sample is also effective against electrostatic and magnetic influences. If the item is electrostatically charged or in a magnetized state, it should be weighed in a metal container. The container can be any conductive metal for electrostatic problems, but it should be magnetically permeable (such as soft iron) for magnetized items. Also, an item that is strongly magnetic should be placed on a nonpermeable spacer such as plastic to lift it away from the scale.

A slightly different arrangement is called for if the balance becomes electrostatically charged, which may happen to the glass panels of the draft shield in low humidity environments, for example, or if it is subjected to an external magnetic field. In this case a shield of the materials described above should be inserted between the item on the balance and the external influence. The shield, eg, a bell-shaped container, must not be supported by the pan of the balance. To facilitate shielding, some balances have an opening which allows a weighing pan to be suspended at any distance below it.

Types of Scales

Scales are available in a variety of designs and configurations to facilitate different weighing operations (7). The two principal categories are industrial and retail scales, and precision scales and balances.

Industrial and Retail Scales. Scales using strain-gauge load cells predominate in this market segment, although mechanical and hydraulic scales are also used to some extent. Many are used in commercial applications, and are typically of accuracy class III or III L with up to 10,000 divisions. The following are descriptions of some industrial and retail scales.

Bench scales are small and light enough to be placed on a bench or table. They generally have a platform on which small quantities can be placed by hand. Alternatively, some have a roller conveyor platform on which the item is moved across the scale. Capacities range from 5 kg to 250 kg.

Retail scales in various configurations are used in stores and supermarkets to determine price. They range from simple weigh-only scales to ones having touch screens and enough memory to store and print data on thousands of products, including the data necessary to print nutritional labels. They range in capacities up to 25 kg.

Portable scales generally have a platform near the floor with the indicator mounted on a column for ease of operation. These scales may be on wheels, and many are battery-powered. Capacities are generally less than 600 kg. Bench scales can be placed on a wheeled table for similar use.

Counting scales are a form of bench scale that can display not only weight, but also the number of parts on the weighing platform. In operation, a small number of parts is put on the scale, which then automatically establishes the average piece weight; if a large quantity is placed on the scale, the scale divides the total weight by the average piece weight and displays the number of parts.

Floor scales are platform scales which are permanently installed in a pit, flush with the floor, or which sit on the floor. These scales typically range in capacity from 500 kg to 30 t, and in size up to that required to accommodate a fork-lift truck.

Truck scales weigh highway vehicles; they may be installed in a pit or above-ground with inclined approach ramps. A capacity of 100-t is typical. For monitoring axle weights, the scale deck can be broken into three scales, which are monitored independently; the individual axle weights or total vehicle weight can be displayed by the indicator.

Railroad-track scales are installed in a pit and support the appropriate length of rail to weigh railroad cars. For static weighing of cars, the rails are generally long enough to support the entire car; capacities range up to 350-t. Some railroad weighing is done with the cars coupled and in motion (CIM). In this case the rails of the scale may be long enough to weigh the entire car or just each truck. Some scales have a flat deck so that highway vehicles can be weighed in addition to railroad cars.

directly (assuming the object is not packaged). Bulk materials may be handled in anything from small containers to railroad cars; the weighing operation consists of measuring the gross weight and subtracting the container's tare weight to arrive at the material's net weight. A truck in–out operation provides a good illustration of this form of weighing; when a truck arrives at a terminal, it pulls onto the scale and its weight (gross or tare, depending on whether it is delivering or picking up a load) is stored in memory by the scale. The truck is then filled or emptied and returns to the scale before exiting the terminal; the scale now has the truck's gross and tare weights and can automatically display the net. Typically the scale prints a ticket listing gross, tare, and net weights, along with the time and date. The information may also be transmitted to a computer for accounting or inventory control.

Tank and hopper scales can weigh in one draft or in multiple drafts. A hopper scale used for multiple-draft weighing of material in transit is commonly called a bulk weigher (Fig. 13). An accumulating or surge hopper with a feed

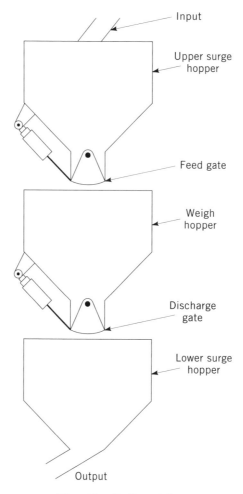

Fig. 13. Bulk weigher.

gate is installed above the scale hopper; a surge hopper is also installed below the scale hopper. In operation, material flows into the scale until the feed gate is closed by the controller; the actual load on the scale is recorded and the scale hopper's discharge gate is opened temporarily to empty the scale. Any material left in the scale is weighed and subtracted from the stored weight to arrive at the net weight for that draft. This operation is repeated until the quantity to be weighed has passed through the scale. This system can be used to weigh a definite amount, such as that contained in a railroad car or a ship, with an error of <0.1% of net weight. If the input is from a car or ship and the output is to storage, it functions as a receiving scale; if the input is from storage and the output to a car or ship, it functions as a shipping scale. As a receiving scale, it weighs a predefined quantity; as a shipping scale, it may be used to weigh out a particular quantity to fill a car or ship optimally.

Some of the various methods of weighing the material in a railroad car, arranged approximately in increasing order of accuracy, are:

Weighing the loaded car on a coupled-in-motion track scale. An entire train, typically, is hauled at slow speed over the scale. The net weight is determined by subtracting the tare weight stenciled on the car

Weighing the material on a belt-conveyor scale as it is put into, or removed from, the car

Weighing the gross weight of the car on a static railroad track scale and subtracting the actual tare weight of the car, determined by weighing the empty car before loading or after emptying, as appropriate

Using a bulk weigher as described above to measure the material either being placed in, or removed from, the car

Weighing the complete load delivered to, or received from, the car as a single draft in a hopper scale.

Controlling Weight. Filling containers to predetermined weights is the most common example of controlling weight. One method is the net-weighing process, in which material is fed into a hopper or tank scale until the desired weight is reached; it is then discharged into the container. This method is often used in bag filling. In the gross-weighing method, the container is placed directly on the scale and filled to the required weight. Drum fillers are an example of the use of this method. Either filling operation may be controlled manually by an operator observing the scale display; more typically, the filling is completely automated for better consistency and greater speed.

Continuous Weighing and Controlling. Scales can be used to determine the weight per unit length of material in sheet or strip form as it is being manufactured or transported. This weight information can be used to control moisture content, addition of sizing or other materials, or to control an extrusion process. The material can pass over a roll that is weighed, or over a belt-conveyor scale. Direct mass measurement with a nuclear scale can also be used for this application.

Process industries frequently need to weigh and control the flow rate of bulk material for optimum performance of such devices as grinders or pulverizers, or for controlling additives, eg, to water supplies. A scale can be installed in a belt conveyor, or a short belt feeder can be mounted on a platform scale. Either can be equipped with controls to maintain the feed rate within limits by controlling the operation of the device feeding the material to the conveyor. Direct mass measurement with a nuclear scale can also be used to measure and control such a continuous stream of material.

Two weigh hoppers can be arranged in parallel to provide accurate quantities of material without interruption (Fig. 14). In operation, weigh hopper A discharges while weigh hopper B is being filled to the appropriate weight. When hopper A empties completely, hopper B begins discharging; hopper A begins filling again, ready to discharge when hopper B is empty. Through correct sequencing, a continuous stream of material can be supplied.

The most accurate flow-rate control can be achieved by using the loss-in-weight method. The total amount of material required for a downstream process is first added to a tank or hopper scale. As the material is discharged, the loss-in-weight is monitored and used to modulate the discharge valve or gate to achieve the desired flow rate.

Material Proportioning. In proportioning operations, two or more materials are weighed and mixed according to a recipe for use in a chemical process or to make a mixed product, such as animal feed. Proportioning can be performed manually on a bench scale or automatically in tank and hopper scales employed with elaborate material handling systems and controls. Proportioning can be done on a continuous basis or, more typically, in batches, which yield greater accuracy.

Where the manufacturing process is continuous, the equipment described herein on continuous weighing can be used. For low accuracy systems, two or more belt-conveyor scales can be run in parallel, supplying different materials at appropriate feed rates. For higher accuracy, two or more parallel weigh hopper systems (Fig. 14) can be sized appropriately for each material and run in parallel

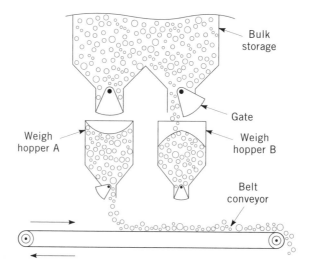

Fig. 14. Parallel weigh hoppers.

to provide continuous proportioning. Multiple loss-in-weight systems may also be used. In some processes, the flow of the primary ingredient is measured but not controlled, and the addition of minor ingredients is modulated accordingly for correct proportioning, using a belt-conveyor scale or loss-in-weight feeder.

The three principal methods of batch proportioning are accumulative, sequential, and simultaneous; the best method for a given application depends on the processing equipment available and the accuracy required.

In the case of accumulative proportioning, a single-weigh hopper having adequate capacity for the entire batch is used (Fig. 15). Each ingredient is weighed in turn and accumulated in the hopper scale until completion of the batch. This system has the advantage of lower cost and space requirements; the disadvantage is that the accuracy of minor ingredients suffers if a small amount of these materials is weighed relative to the scale's capacity. Also, in multiple-recipe systems there is the danger of cross contamination if some materials are not used in all recipes.

In sequential proportioning, the physical configuration is the same as that shown in Figure 15, but each material is weighed and discharged before weighing of the next ingredient begins. If all ingredients weigh about the same, higher accuracy can be achieved with this system by matching the weigh hopper capacity with the weight of material to be weighed. This system is relatively slow because of the multiple-discharge cycles.

In the case of simultaneous proportioning, a weigh hopper is provided for each material, as shown in Figure 16. This system has several advantages. It

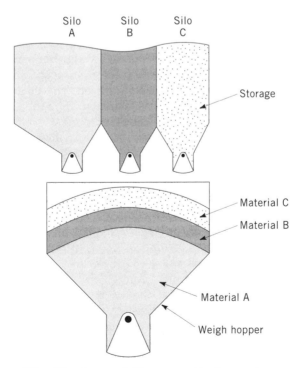

Fig. 15. Accumulative proportioning system.

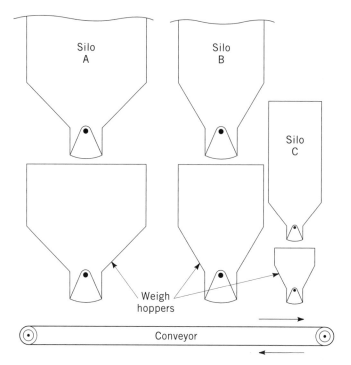

Fig. 16. Simultaneous proportioning system.

is fast because all materials are weighed and discharged simultaneously. Before discharging, the weight of each ingredient can be compared to the recipe and adjusted if necessary. There is less danger of contamination if some materials are not used in all recipes. The capacity of the weigh hoppers can be matched closely to the weight of each ingredient. By simultaneous discharge onto the belt, a certain amount of premixing can be achieved. The disadvantages are the relatively high cost and the amount of space required. Simultaneous systems can be designed to handle up to 150 tons per hour or more, and are often used to feed several mixers or processes. Combination systems are becoming more common: in this arrangement, an accumulative hopper is included in a simultaneous proportioning system. Individual weigh hoppers are provided for the high volume and critical materials, whereas several of the lower volume or less critical materials are proportioned in the accumulative weigh hopper. Assuming proper sizing of such a system, cost is reduced without compromising speed. Regardless of the system used, critical ingredients such as food additives or dyes are often weighed on an appropriate balance and added manually.

Scale Functionality

Mechanical scales indicate the weight applied to the scale through the manipu-lation of a poise weight on a beam, or automatically using a pointer and dial. Simple control functions can be accomplished by adding position sensors to the indicator or lever system. However, many mechanical scales have been converted to electromechanical by inserting a load cell in the steelyard rod; this allows the scale to take advantage of the wide range of functionality available from electronic indicators. In new installations, electronic scales are typically installed if any form of data collection or control is required.

Electronic scales are microprocessor-based devices that provide a range of features for various applications. At its most basic, an electronic scale provides a digital display of the applied weight; a zero button is usually provided so that slight deviations from the zero indication can be corrected periodically. Many simpler scales also provide tare and gross/net buttons. The tare button is used to store an empty container weight in memory for subtraction from subsequent weighments; the gross/net button allows the display to toggle between the gross and net weights. Where it is equipped with numeric or alphanumeric keypads and various function keys, an electronic scale can perform a number of functions directly or indirectly related to weighing. The following examples illustrate some possible uses.

Some indicators are designed for truck in–out operations. The indicator can store tare weights for several hundred vehicles, linking them to the trucks' registration numbers. Regardless of the sequence in which the trucks arrive and depart, the scale can provide a printout of gross, tare, net, and time and date for each transaction with just a single weighing of each vehicle. Net weights can also be accumulated in user-specified categories to provide totals and subtotals for billing or inventory control.

Counting scales display the number of parts on the scale as well as the total weight. They establish the average piece weight (APW) of the parts to be counted, which can be stored in memory along with the part number and an associated tare weight. In the case of cycle counting operations, for example, a bin of parts is placed on the scale and the APW and container weight is recalled by keying in the part number; the scale then calculates the net weight of the parts and divides by the APW to arrive at the parts count. This information can be displayed and stored in memory for automatic uploading to the main computer for correction of inventory records. Counting scales can also be used with bar code scanners to eliminate the need for keying part numbers. It is common practice to store all APW and tare weights on a main computer from which it is downloaded to the scale daily, as an indication of what needs to be cycle counted that day.

Over/under scales are often used for manually filling containers to a desired weight; they are often found in the meat and poultry industry. In addition to the normal weight display, these scales typically have lights or a bar graph to indicate to the operator when the weight is under, within, or over the desired weight range. Employing a part or product look-up (PLU) number, various information can be stored for several items, eg, the target weight, the allowable tolerance values around this target, and an associated tare weight. This information can be programmed at the scale or downloaded from a central computer. Groups of these scales are often networked together with a central computer which can be used to download setup parameters to the scales for fast production line changeovers. The central computer can also collect weight information from each scale for production monitoring and statistical process control.

Scales can be connected with printers, remote displays, bar code scanners, keyboards, relays, computers, and various controllers such as programmable logic controllers (PLCs). Scales typically have one or more simplex or duplex

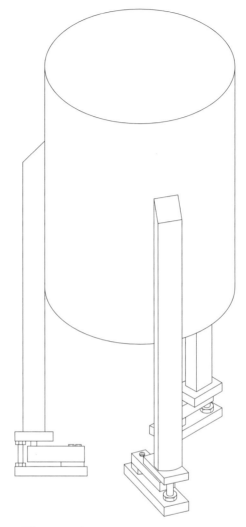

Fig. 18. Compression configuration.

modules with capacities from 25 kg to 100 t and more are available. Most designs
are self-checking and some include protection against tipping; both make the
installation easier. On the other hand, care must be taken to level and align the
modules carefully, and the floor-level modules can be subject to flooding, severe
washdown, and debris accumulation, which can adversely affect accuracy and
cause corrosion.

  Weigh Module Capacity.   The number of weigh modules to be used is often
dictated by the geometry of the vessel. Low capacity tanks can be hung from a
single load cell, but most vessels are supported by three or four modules. In the
case of horizontal tanks or upright vessels of square or rectangular cross section,
it is often most convenient to use four weigh modules. In the case of vertical
vessels of circular cross section, it is common to use three modules symmetrically
arranged. Three-module arrangements are preferred; when more than three are

used, it is usually necessary to shim some of the modules to achieve correct load distribution. Systems using more than 8 modules are rare for this reason.

Weigh module capacity can be calculated from the following:

$$\text{weigh module capacity} = \frac{K \text{ (live load + dead load)}}{N}$$

where K is a safety factor usually between 1.25 and 2, and N is the number of supports, usually the same as the number of weigh modules. The live load is the total weight of material to be weighed on the scale; it should be conservative and take into account variations in material density and moisture content. The dead load is the total weight of the empty vessel including all equipment mounted on the vessel, heating and cooling fluids in a jacketed vessel, the weight of piping supported by the vessel, etc. The factor K is to some extent dictated by the capacities of available weigh modules. If the scale capacity and dead load are known with certainty, and the scale will be loaded without shock, it may be acceptable to use $K = 1.25$. Where the capacity and dead load are based on estimates, where the vessel is subjected to vibration or material surges, or where the weigh modules are not loaded equally, then K values closer to 2 should be used. If the material is in large chunks and falls from a height onto the scale, there is the danger of shock damage to the load cells, and a K factor larger than 2 may be necessary. It must be noted that all weigh modules on a particular vessel must be of the same capacity.

Mechanical Considerations. The support structure for a weigh vessel needs to be rigid so as to minimize undesirable forces on the weigh modules (14). This rigidity also has the beneficial effect of increasing the natural frequency of the scale, which is important where accuracy and speed are required. Support points need to be equally rigid to prevent the transfer of load between weigh modules; also, vessels with long legs need bracing to prevent lateral deflection of the legs as the load is applied. Where modules are used, as shown in Figure 17, gussets should be used to support the horizontal bracket; the vessel wall may need reinforcement also. For highest weighing accuracy, the vessel should be located where it is free of vibrations from rotating equipment or vehicular traffic. Where more than one vessel is supported on a common structure, care must be taken to avoid deflection of one vessel as another is being filled or emptied.

As a vessel is loaded, it moves downward because of deflection of the load cells and support structure. Pipes rigidly attached to a vessel restrict its free movement and assume some portion of the load that cannot be measured by the load cells. This is very detrimental to scale accuracy. Deflection of the load cell is unavoidable; deflection of the vessel support structure should be minimized. Anything which increases vessel deflection, eg, rubber pads used for shock protection, must be avoided. The total number of pipes should be minimized and be of the smallest-diameter, thinnest wall possible. Pipe runs to weigh vessels must be horizontal and the first pipe support should be as far as possible from the vessel. Alternatively, a section of rubber hose or flexible bellows should be used to make the final connection to the vessel. The scale should be calibrated using weights, not by means of an electrical simulation method, which cannot account for the effects of the piping or test the correct functioning of the scale.

If a weigh vessel is located outdoors, the effect of wind forces on the vessel's stability should be considered, particularly if it is a tall slender vessel in an exposed location. An empty vessel may be in danger of tipping; wind forces transfer weight from one module to another and, when full, may cause load-cell damage. In many areas the effects of earthquakes must be considered in the design of the scale and its supporting structure. Building codes exist for the design of structures subjected to wind or seismic loading (15,16). In many instances a weigh module can be selected which can withstand these effects; however, in some cases it may be necessary to add additional horizontal or vertical checking. Vessel stability can be greatly improved if compression mounts are applied in a horizontal plane close to the vessel's center of gravity; this arrangement is convenient if the vessel passes through a floor, for example. All vessels should have safety supports that can hold the vessel if failure of the primary support could lead to loss or injury. This safety backup could be provided by loosely fitting chains or check rods.

It may be necessary to contain dust by enclosing a weigh hopper and using dust seals or flexible connections to seal openings. Figure 19 shows an arrangement where the top of the hopper is fixed to the structure, and the hopper must have an effective vent which minimizes even transient pressure surges; otherwise, unwanted vertical forces will be applied to the scale.

Figure 20 shows an arrangement which is unaffected by air pressure fluctuations, because any force applied to the material is canceled by an equal and opposite force applied to the inside top surface of the hopper. It may be desirable or necessary to vent the hopper for efficient material handling.

Calibration.   The greatest accuracy will be achieved by calibrating with test weights equivalent to scale capacity. Provision should be made for test weights when designing a weigh vessel by providing, for example, a shelf on which the weights can be placed, or eyes from which the weights can be hung. Scales used in commercial applications must be calibrated using test weights or material substitution methods using a specified minimum amount of test weights (5). The substitution method can be used where the amount of test weights available is small compared to scale capacity. The weights (15% of scale

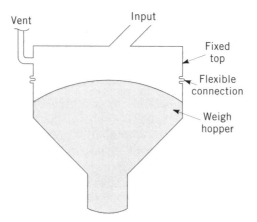

Fig. 19.   Hopper top fixed to structure.

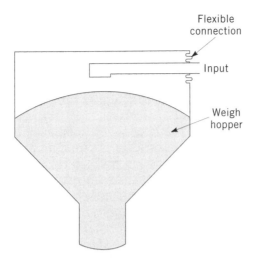

Fig. 20. Top attached to hopper.

capacity, for example) are used to calibrate the scale roughly initially and the scale reading is noted; the weights are removed and any available material is then added until the scale reads the noted value. The test weights are added to the scale along with the material added previously and the new reading is noted; the weights are then removed and more material added until the scale reads the second value noted. This process is repeated until a load close to scale capacity has been added and the final calibration can be performed with what is now a known quantity of material and test weights.

Calibration can also be accomplished using material weighed on another scale. The accuracy of this method depends on the accuracy of the other scale, and care must be taken not to lose any of the weighed material. Scales can also be calibrated electrically using a load cell simulator if the load cells' rated outputs are known accurately. This method does not test the mechanical functioning of the scale and is not very accurate, particularly if it has attached piping that restricts its vertical movement.

Feeding Equipment. For consistency in filling or proportioning operations, the design of the feeding equipment is critical. A front-end loader can be used to fill a hopper scale; however, it will be very difficult to achieve a specific weight repeatedly. If a belt-conveyor is used for filling, the results can vary because of lumps and variations in the quantity of material on the belt. The problem will be compounded if the conveyor coasts after power is removed. The scale and material-handling equipment must be designed as a system to achieve the required accuracy.

The choice of feeder is often dictated by the material properties. For free-flowing materials that do not tend to bridge, rathole, or flood, electric vibrating feeders, belt feeders, screw feeders, and power-operated gates are used. For material that is predominantly in large lumps, recommended types include electric vibrating feeders, belt or slat feeders, and screw feeders of proper design if lumps are not too large. For finely pulverized materials that tend to flood, screw feeders are recommended, preferably double-flight or half-pitch (an auxiliary gate may be required), and rotary vane or star feeders. For some applications it may be desirable to use a rotary vane feeder in conjunction with a vibrating feeder; this combination prevents flooding, and the desired smooth control of flow is achieved because of the vibrating feeder.

The highest throughput can be achieved by filling a hopper directly from a silo or upper surge hopper (see Fig. 13) equipped with a feed gate. An upper surge hopper is also useful for eliminating the erratic flow associated with devices such as belt-conveyors; it also allows the incoming material to be conveyed continuously despite the batch nature of the weighing process.

Where the scale activates an output, indicating that the desired weight has been achieved, the feeding equipment must stop operating immediately. Some feed devices may need a brake to eliminate coasting. Even if a feeder stops instantly, there will be a certain amount of *in-flight* material added to the scale after the desired weight has been achieved. The average amount of in-flight material can be determined and the scale can be programmed to *pre-act*, ie, it activates the stop output signal when the weight on the scale has reached the desired weight minus the in-flight material. For this to be effective, the rate of material flow must be consistent. Best results are achieved by minimizing the amount of in-flight material; hence the height of the feeder over the scale should be minimized and the rate of fill should be very slow. Where the latter is not practical, a *two-speed fill* can be used. Figure 21 shows a storage hopper with a three-position simplex gate. At one extreme it is fully open, allowing fast (*bulk*) fill; at the other extreme it is fully closed, and in the intermediate position (illustrated) it feeds slowly (*dribble*) through a small notch in the gate. Figure 22 illustrates a similar arrangement for filling liquids; in this case, bulk fill is achieved by opening both solenoid valves, and dribble fill is achieved by closing the larger one. Motor-driven feeders can perform bulk and dribble fill conveniently by varying the motor speed. The ratio of bulk to dribble volume

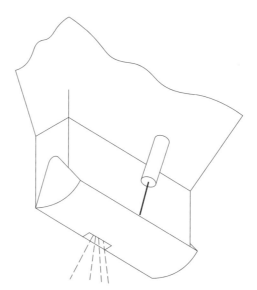

Fig. 21.   Hopper with bulk and dribble fill.

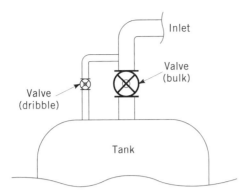

Fig. 22.   Tank with bulk and dribble fill.

can be as much as 1,000 to 1; dribble is often used for the last 5–10% of the material filled. A further refinement is an *autojog* feature. A pre-act is set to ensure that the filled weight is always light; after filling and settling, the weight is compared to the required value. If it is low, the scale jogs the feeder until the batch is within tolerance.

Discharging the Scale.   Where a gate is used to discharge a hopper scale, very little control can be exercised over the rate of discharge. In some cases, particularly where the various scales of a simultaneous batching system discharge onto a single conveyor belt, it is desirable to control the discharge rate so as to premix the materials. This can be accomplished by using a suitable feeding device in place of the gate. Also, using a controllable feeder as the discharge device, it is not necessary to empty the hopper or tank completely since material can be accurately weighed out of the hopper, as well as into it. In the case of materials that cling to the sides of a hopper, use of this device provides a definite advantage. In the case of liquids, more rapid discharge of a definite amount can be accomplished because the last material does not have to be discharged fully with diminishing pressure head; this method is also very useful in the case of very viscous liquids. Of course, it cannot be used when different materials are accumulated in a scale hopper because the composition of the material remaining in the hopper would vary.

Indicators and Controllers.   A batching scale can be controlled manually by an operator observing the weight display and operating the feed and discharge devices at the appropriate time. Some scale indicators are designed specifically for automatic control of batching systems, and the recipe and batching routine can be programmed at the indicator. These indicators can perform functions such as two-speed fill, pre-act, and autojog. They also have digital I/O capability; the inputs can, for example, be used to start and stop the batching process, whereas the outputs are used to control the filling and discharge equipment. This system works well in automating a standalone batching system, because fast reaction times and repeatability can be achieved. The disadvantages are that it is difficult to integrate into a larger system and that the programming can be slow and error-prone.

Another approach to the control of batching systems is to use a relatively simple scale indicator which does nothing but supply weight data to a controller

WELDING

Welding comprises a group of processes whereby the localized coalescence of materials is achieved through application of heat and/or pressure (1–3). It represents a fabrication technology of extremely broad scope; welding principles are used in nearly every product fabricated, including metals and, increasingly, nonmetals. Obvious examples are bridges, buildings, ships, space vehicles, automobiles, offshore platforms, pipelines, pressure vessels, and consumer appliances; less evident examples include plastic cases for computer disks and the very large number of miniature welds used in microcircuits. This field of great technical complexity utilizes robots, computer-controlled machines, and high concentrations of energy in the form of plasma arcs, lasers, and electron beams. Welding involves very rapid processes, including heat transfer, chemical reactions, and metallurgical reactions, and requires complex methods of analysis to control stress, distortion, and fracture.

The principal welding processes are of comparatively recent origin. Forge welding, soldering, and brazing can be traced to ancient times, but the modern processes of arc welding, resistance welding, and gas welding (as well as thermit welding) were discovered in the 1880s. The use of welding in production and repair increased slowly through the early 1900s and received a boost by successes in emergency ship repair during World War I. Welding saw increased use in building construction during the 1930s, and new welding processes, eg, submerged arc welding, were discovered. World War II provided tremendous impetus for further development, including spectacular uses in all-welded ships and for military tanks and development of new processes for welding aluminum. Since the 1940s many new welding techniques have appeared, including electron-beam, laser, and ultrasonic processes (4).

Thus, from the origins of modern welding in the 1880s many new processes have evolved, and today the American Welding Society recognizes nearly seventy methods of welding and more than twenty allied processes employing thermal cutting, thermal spraying, and adhesive bonding (1), some of which are identified in Table 1.

Arc-Welding Processes

In arc welding, the coalescence of metals is achieved through the intense heat of an electric arc, which is established between the base metal and an electrode. The processes listed in Table 1 are differentiated by various means of shielding the arc from the atmosphere (1–3).

In arc processes, either d-c or a-c current is used to establish and maintain an arc between the base metal and an electrode. The electrode itself may be consumed by melting and thus become part of the weld, acting as a filler metal, or it may be a nonconsumable material, eg, tungsten. In the latter case, the heat of the electric arc may simply be used to fuse adjacent base metal (autogenous welding), or a separate filler metal may be added. It is essential that the molten pool of material under the arc, as well as the adjacent solidified but still high temperature metal, be protected from oxygen, nitrogen, and other elements of the atmosphere, since these react with the metal to form oxides and other products

Table 1. Welding Processes[a]

Process	Abbreviation	Process	Abbreviation
arc welding		resistance welding	RW
shielded-metal arc welding	SMAW	resistance spot welding	RSW
		resistance seam welding	RSEW
gas–tungsten arc welding	GTAW	projection welding	RPW
		flash welding	FW
plasma arc welding	PAW	solid-state welding	SSW
gas–metal arc welding	GMAW	diffusion welding	DFW
flux-cored arc welding	FCAW	explosion welding	EXW
stud arc welding	SW	forge welding	FOW
submerged arc welding	SAW	friction welding	FRW
oxyfuel–gas welding	OFW	ultrasonic welding	USW
oxyacetylene welding	OAW	other welding processes	
oxyhydrogen welding	OHW	electron beam welding	EBW
pressure-gas welding	PGW	laser beam welding	LBW
brazing	B	electroslag welding	ESW
resistance brazing	RB	thermit welding	TW
furnace brazing	FB	allied processes	
induction brazing	IB	thermal spraying	THS
soldering	S	adhesive bonding	ABD
dip soldering	DS	thermal cutting	TC
wave soldering	WS		
torch soldering	TS		

[a]Ref. 1.

that reduce the strength and toughness of a weld. Consequently, various forms of shielding are provided around the arc in the different processes.

Shielded-Metal Arc Welding. The essential features of the SMAW process, commonly called stick-electrode welding, are shown in Figure 1(**a**). The arc acts between the consumable electrode wire and the base metal, and droplets of molten filler metal are transferred to the weld pool. The unique feature of this widely used process is the role of the electrode coating. This coating is decomposed by the heat of the arc and provides both a necessary shielding atmosphere for the arc as well as a slag coating over the weld metal, thus affording still further protection from the atmosphere.

In addition to these functions, the coating introduces fluxing agents to the weld pool, assists in establishing the electrical characteristics of the arc, and can be used to provide additional filler metal to the weld. Over a dozen different materials may be used in a single electrode covering, depending on the desired welding characteristics. Common ingredients include cellulose and calcium carbonate to provide shielding gas, titanium dioxide and silica to provide slag, sodium silicate to act as a binder, and ferrosilicon to act as a deoxidizer.

Deposition rates using this process are limited by the fact that each electrode contains a finite amount of filler metal. The time required both to change electrodes and to remove the slag coating between each weld pass lowers the overall productivity of the process.

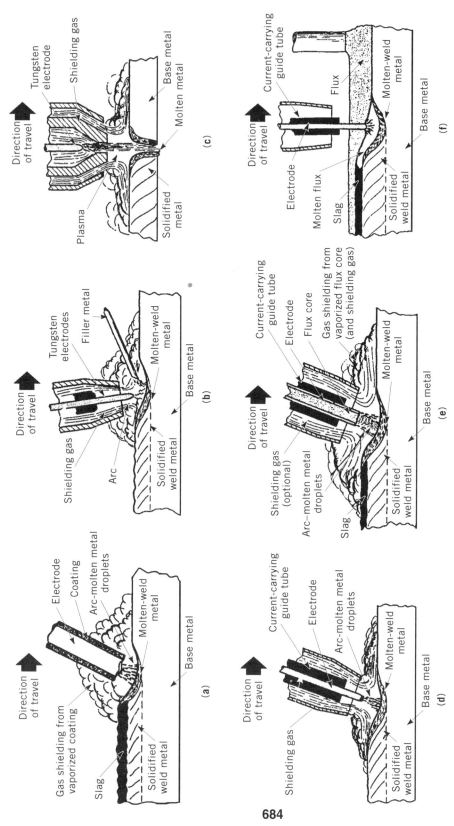

Fig. 1. (a) The shielded-metal arc-welding (SMAW) process. (b) The gas–tungsten arc-welding (GTAW) process. (c) The plasma arc-welding (PAW) process. (d) The gas–metal arc-welding (GMAW) process. (e) The gas shielding from vaporized flux core (and shielding gas) flux-cored arc-welding (FCAW) process. (f) The submerged arc-welding (SAW) process.

684

Gas–Tungsten Arc Welding. The GTAW process, often called the tungsten–inert gas or TIG process, is shown in Figure 1**b**. Here the electrode is nonconsumable and shielding is provided by the flow of inert gas through the welding-torch nozzle. Argon and helium are commonly used, as well as argon–helium mixtures; traces of hydrogen are sometimes present. The arc may be used alone to fuse base metal together or, as shown in Figure 1**b**, separately added filler rod can be fed into the weld pool.

Plasma Arc Welding. In the transferred-arc mode of the PAW process, shown in Figure 1**c**, the arc is between a nonconsumable electrode and the base metal, in a manner similar to the GTAW process. The unique feature is the flow of inert gas around the electrode and through a restricted orifice, which constricts the arc to form a plasma jet. A second, outer stream of shielding gas protects the molten metal from atmospheric contamination. In the nontransferred arc mode of the PAW process, the arc is between the electrode and the constricting orifice. This mode is used for plasma spraying and for heating nonmetals.

Gas–Metal Arc Welding. The features of the GMAW process, also called the metal–inert gas or MIG process, are shown in Figure 1**d**. A consumable bare-wire electrode is fed continuously through the welding torch, and a flow of inert gas through the nozzle provides shielding. This continuous-wire feed leads to higher deposition rates compared to the SMAW process. Common shielding gases include 100% argon, 75–80% argon with the remainder carbon dioxide, and 100% carbon dioxide.

Flux-Cored Arc Welding. The FCAW process is illustrated in Figure 1**e**. The characteristic feature is the hollow, flux-filled, consumable electrode, which is fed continuously through the welding torch. The decomposition of the flux provides both arc shielding and a slag blanket over the weld. In a variation of the process, a shielding gas flows through the welding nozzle to provide additional protection for the arc.

Submerged-Arc Welding. In the SAW process, shown in Figure 1**f**, a loose flux is blanketed over the region of the arc. A consumable, bare-wire electrode is fed continuously through the torch into the weld. The molten- and granular-flux blankets provide the necessary arc shielding and slag cover for the solidified weld. This process, often used with operating currents of over 1000 A, provides the highest deposition rate of all the arc-welding processes. The presence of a loose flux makes vertical and overhead welding impractical with the SAW process.

Arc-Welding Systems

The various welding processes result in systems of varying complexity. They include at least the electrode and a device for holding or feeding it, the work piece, the power source, and heavy-duty cabling to provide a complete electrical circuit. Provisions for supply and control of gas and control of wire feed and movement of the electrode assembly are required, depending on process type and degree of automation.

Welding systems are generally classified as manual, semiautomatic, mechanized, and automatic (2). In manual welding, the operator must maintain the arc, feed in filler metal, and provide travel and guidance along the joint. In semiautomatic welding, the welding machine maintains the arc and feeds filler metal, and the operator controls joint travel and provides guidance. In mechanized welding, the welding machine maintains the arc, feeds filler metal, and moves a mounted torch along the joint; the operator manually adjusts the welding parameters based on observation of the process. In automatic welding, the machine assumes all of the preceding functions. Automatic processes may be further classified, depending on the degree of feedback control used in controlling the welding variables. Automated systems may be dedicated to a specific type of production, or they may be flexible, programmable robot systems.

The power source is the core of any welding system. Electrical power is provided by direct-line power or a generator; the latter is driven by an engine or an electric motor. In the case of welding with a-c current, the welding power source may be in the form of a transformer working from line power, or it may be an engine-driven a-c generator. Welding with d-c power may require a transformer–rectifier working from single- or three-phase line current, or a motor-driven generator. Welding power sources are further differentiated by controls imposed on the output current or voltage. One type of power source, used primarily for manual welding, is capable of providing essentially constant current. Another type, used for semiautomatic and automatic processes, is capable of providing essentially constant voltage.

The introduction of inverter power sources with solid-state electronic components has resulted in weight and size reductions of up to 75% compared to traditional power supplies. The efficiency and performance of these power supplies is also increased. Variable-frequency, pulsed-current power supplies for GMAW and GTAW, wherein a base current level pulses to high currents, depend on inverter technology. Pulsed GMAW, with pulsing rates of 60 to 200 times per second, provides spray metal transfer at lower average currents and with less fume and spatter; the ability to weld thinner material and to make out-of-position welds is also increased. Pulsed GTAW, with pulsing rates of 1 to 20 times per second, provides greater penetration for a given average current and provides for thin-section welding at lower average currents. High (ca 20 kHz) frequency, pulsed GTAW can be used to provide a stiffer, more directional arc.

The system for shielded-metal arc welding, shown in Figure 2**a**, is the simplest system. It consists of the power source, electrode and holder, the base metal, and the electrical cables or leads. When the arc is struck, a complete electrical circuit is provided. With d-c welding, the electrode may be either negative (straight polarity) or positive (reverse polarity). Shielded metal arc welding is only used manually.

A gas–tungsten arc-welding system is more complex. In addition to the components of the shielded-metal arc system, provisions must be made for the inert gas supply and water or air cooling of the welding torch. GTAW systems may range from manual to automatic.

A semiautomatic flux-cored arc system is shown in Figure 3. Controls and drive motor are required for continuous feed of the welding wire to the torch. If a shielding gas is used, provisions for control of this gas supply are needed, and the torch configuration is different. The flux-cored system may be used for gas–metal arc welding (GMAW) by using the shielding gas and the welding torch

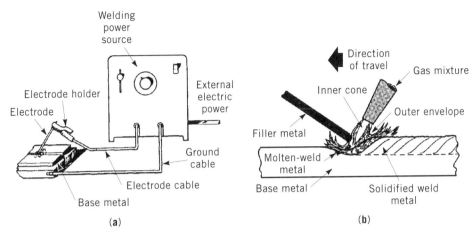

Fig. 2. (**a**) The shielded-metal arc-welding system. (**b**) The oxyacetylene welding process.

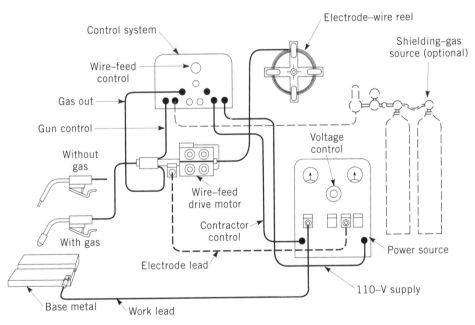

Fig. 3. The flux-cored arc-welding (FCAW) system (2).

configuration for use with gas. Both FCAW and GMAW systems can be used in automatic operations.

A mechanized, submerged-arc-welding system requires a power source, control system, and wire-feed provisions. Granulated flux is fed into the weld joint from a flux hopper, which travels ahead of the welding arc as part of the electrode carriage. Control of the welding torch along the seam of the base metal can be accomplished by a motorized electrode carriage moving along a straight beam, or by a positioner capable of placing the welding torch at any position over a base metal that is itself continually changing position.

and pressure, and free-machining alloys, the inclusion content of which can lead to difficulties as the parts are spun and forged together.

Another variant of the friction welding process, linear friction welding, uses servo-hydraulics pumps to vibrate parts back and forth against each other. Bond areas of approximately 1000 mm^2 can be joined; the attachment of turbine blades to rotors is a prevalent application of this technology.

Brazing and Soldering. In brazing and soldering processes, a molten filler metal flows by capillary action into the closely fit joint between the base metals at a temperature below the melting point of the base metal. Bonding is accomplished by metal-to-metal adhesion, which may involve the formation of intermetallic compounds at the interface between the base and filler metals. Although these characteristics are common features of brazing and soldering, the two processes are differentiated by the melting temperature of the filler metal. In brazing processes, the filler metal melts at a temperature above 450°C, whereas in soldering, the melting point of the filler metal is below 450°C (7). Common solder alloys contain lead with amounts of tin varying from less than 5 wt % to around 60 wt %; the eutectic composition of 61.9 wt % tin and 38.1 wt % lead melts at 183°C. Common applications for lead–tin solders including plumbing and electronics. Current (ca 1997) efforts to achieve lead-free solders are concentrating on alloys such as tin with 3.5 wt % silver.

Transient liquid-phase bonding is a technique that can form a brazing or soldering joint at low temperature that can withstand service at higher temperature, sometimes approaching the melting temperature of the base metal. A boron-rich brazing filler paste is used for the repair of gas-turbine engine components. Boron acts as a melting-point depressant and allows the paste to melt and fill defects. The small, highly mobile boron atoms rapidly diffuse into the parent metal at brazing temperatures; the remaining filler metal has a melting point that approaches that of the base metal. Transient liquid-phase soldering for tooling and microelectronics assemblies depends on the dissolution of a high melting-point metal into a solder to form an intermetallic compound which has a higher melting point than the original solder.

Wetting, the ability of the filler metal to spread through the joint, is a critical factor in these processes. Any oxides or other film must be cleaned from the joint area, generally by a fluxing agent, to ensure good wetting. Uniform heating of the joint area on large parts may best be accomplished in a furnace. Brazing can be used to join ceramics as well as metals. Metal-to-ceramic brazing is often accomplished with the addition of reactive metal, such as titanium or zirconium, to the braze material.

Joining of Polymers and Adhesive Bonding

Polymers are characterized as thermosetting and thermoplastic with respect to the methods by which they are joined. Thermosetting polymers are permanently hard and do not soften upon the application of heat; they are joined by mechanical fasteners and adhesives. Several methods have been devised to join thermoplastic polymers, as well as thermoplastic composite materials, which soften upon heating.

Hot Plate, Infrared, and Hot Gas Welding. These processes involve external means to heat thermoplastic polymers to a viscous state in which the interdiffusion of polymer chain molecules can occur with the application of pressure. In hot plate welding, the two surfaces to be joined are forced against a platen heated to a desired temperature based on the composition of the polymer. After heating occurs, the platen is removed and the two polymer faces are forged together to make a weld. The platen is replaced as a heat source by infrared light in the infrared welding process; the forging step after heating remains the same. In hot gas welding, a stream of heated gas or air is directed at both the joint surfaces and a polymer filler material. In a process similar to the arc welding of metals, the polymer filler metal is fed into the joint to complete the weld.

Friction and Ultrasonic Welding. Both rotational and linear friction can be used to melt the interface between two thermoplastics. The parts are then aligned, and a weld is formed as the interface solidifies. Linear friction welding, also called vibration welding, employs frequencies in the 100–500 Hz range; welding can be accomplished in less than one second but is limited to bond areas approximately 400 mm^2. Ultrasonic welding of polymers involves oscillations of 10–50 kHz, which are dissipated at the bond line to produce heat through both friction and hysteresis. The surfaces to be joined are held together as the sound energy generated by the welding machine is transferred through the parts at right angles to the contacting surfaces. Like the friction process, ultrasonic welding is limited to smaller part sizes.

Adhesive Bonding. As one of the processes allied to welding, adhesive bonding is used particularly in applications where welding or mechanical fastening is either impossible or undesirable (Table 1). Adhesives, which are derived from polymers, are not structural materials and act only to distribute stress over the bond area; as such, proper joint design and minimization of bondline thickness are essential. The selection of an adhesive should take into account the expected service environment, because temperature and atmosphere can degrade adhesives. Stress mode is also a prime consideration in both joint design and adhesive selection. Pure compression, shear, and tension are preferred over mixed stress modes such as peel or cleavage.

Physics and Metallurgy

In most welding processes, the local regions of the adjoining base metals and any added filler metal are melted and resolidified. These features are present, for example, in the arc, gas, electroslag, electron-beam, and laser-welding processes. The analogy of this action to the casting process is often made, where the molten base and filler metals are cast in the mold formed by the nonmolten parts of the base metals. This analogy is of some use because a range of phase transformations is involved as the alloy system solidifies. However, welding differs from conventional casting by the speed of the solidification process. Only a local region of material is melted in welding, and the surrounding, low temperature base metal acts as a large heat sink, producing rapid heat flow from, and the solidification of, the weld zone. This situation produces a complex metallurgical microstructure and physical properties not typical of casting.

Heat Flow. Certain basic features of welding heat flow are illustrated by Figure 5a, which represents isotherms resulting from a welding heat source moving along the joint between two base metals. The central region, under the heat source, is molten. The surrounding elliptical curves are isotherms, which are clustered toward the direction of travel of the heat source and show the temperature distribution at a given instant during welding. Different welding processes, plate thicknesses, and welding speeds modify the details of the clustering and magnitudes of the isotherms, without changing the principles of the behavior shown.

An important aspect of welding heat flow is the thermal cycle at a given location in the material. The nature of the cycle depends on the intensity of the heat source, the speed of welding, the thermal characteristics of the material, and the location in the material. The general behavior is shown in Figure 5b. Curve 1, representing point 1 of Figure 5a, is adjacent to the weld and actually reaches the melting point, T_m, as the arc passes. At points 2–5, two characteristics are evident. First, the peak temperatures that occur, plotted along the line t_p to t_p^*, are progressively lower because the heat spreads over larger regions. Second, there is a progressively greater time delay in reaching peak temperature because of the time required for heat conduction. The weld thermal cycle, involving peak temperatures achieved and the speed of heating and cooling, and possibly preheating and postheating, accounts for many of the subsequent complexities of welding metallurgy.

Solidification. The heat of the electric arc melts a portion of the base metal and any added filler metal. The force of the arc produces localized flows within the weld pools, thus providing a stirring effect, which mixes the filler metal and that portion of the melted base metal into a fairly homogeneous weld metal. There is a very rapid transfer of heat away from the weld to the adjacent, low temperature base metal, and solidification begins nearly instantaneously as the welding heat source moves past a given location.

Solidification begins as atoms of weld metal attach themselves to the solid metal grains at the weld pool edge; this initial growth is epitaxial, ie, the atomic orientation of the base-metal grains continues into the weld pool. Weld metal

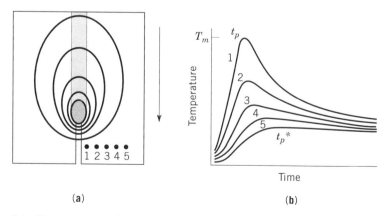

(a)   (b)

Fig. 5. (a) Temperature isotherms in the region of a moving welding arc. (b) Heat-affected-zone thermal cycles at various locations in the base plate (see text).

grain growth continues in the direction of the maximum temperature gradient, which is initially perpendicular to the edge of the weld pool and tends to remain so for fast welding speeds, but which turns toward the arc in a direction parallel to the weld axis for a slower welding speeds (Fig. 6). Cellular and cellular dendritic are the most common weld solidification modes; the mode itself is affected by temperature gradient, speed of solidification, and solute content of the weld metal.

As the weld metal solidifies, impurity elements are rejected into the molten weld pool, eg, sulfur and phosphorus in steel welds (Fig. 7) (8). The final weld metal to solidify, located along the weld centerline at the surface of the weld, has increased levels of these elements, which act to lower the solidification temperature and, in conjunction with the shrinkage stress resulting from solidification,

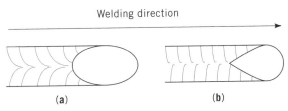

Fig. 6. Weld pool shape and resultant weld–metal solidification direction. (**a**) Slow welding speed. (**b**) Rapid welding speed.

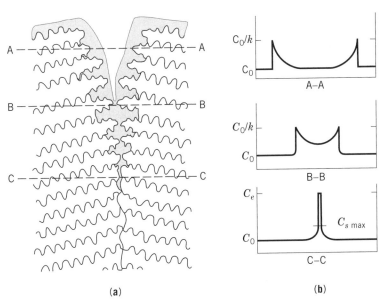

Fig. 7. (**a**) Impurity elements are rejected into the liquid between the dendritic solidification fronts. (**b**) Corresponding impurity concentration profiles. C_0, weld metal composition; k, impurity partitioning coefficient in the liquid; $C_{s\ max}$, maximum impurity solid solubility; C_e, eutectic composition at grain boundary. © NEMJET, The Ohio State University.

increase the susceptibility of a weld to solidification cracking at the centerline. This increased risk of cracking is particularly true for fast welding speeds, which promote solidification toward the weld centerline and segregate impurities there.

The thermal cycle of the welding and solidification process has caused additional complex effects to occur. Differing peak temperatures reached at different locations from the weld significantly affect the microstructure of the base metal in the region surrounding the weld. This region, know as the heat-affected zone, contains metal in various types of transformations, including grain growth and recrystallization. The most pronounced of these occurs at the fusion boundary, where large grain sizes characteristic of the adjacent weld metal are formed. The effects diminish the farther away from the weld one moves until, at some distance, there is no discernible change in the base metal.

Metallurgy.   Welding metallurgy deals with the interactions of the base and filler metals and the interactions of these materials with various chemicals injected into the weld via gases, electrode coverings, fluxing and slagging agents, and surface contaminants. For example, a number of gas–metal reactions are possible where the weld is still at high temperature. Oxygen, nitrogen, water vapor, and carbon dioxide are gases that react with ferrous metals to yield products harmful to the metallurgical properties of a weld. The nature of the slag–metal reactions that occur in the molten state strongly depends on the composition of the flux or the electrode coating. Flux chemistry may be altered to control removal of specific weld-metal impurities, such as the addition of manganese or silicon to provide strong deoxidizing action, or additions to enhance slag removal, with all of these additions influencing the final metallurgical characteristics of the weld.

The microstructure of a weld is the overall arrangement of grains, grain boundaries, and phases that exist once solidification occurs. An important metallurgical tool for understanding weld microstructure is the phase diagram, which, for a given alloy composition, relates material phase and temperature. For example, the iron–carbon, Fe–C, phase diagram shows that a steel of 0.25-wt % carbon melts above 1520°C; at a slightly lower temperature it consists of a mixture of molten metal and delta iron; below 1500°C it is transformed to a mixture of liquid metal and austenite; at 1480°C, it solidifies completely to austenite; and below 815°C, a mixture of austenite and ferrite exists, which is transformed to ferrite and cementite, Fe_3C, at 727°C.

However, phase diagrams represent equilibrium conditions that do not prevail in welding. These nonequilibrium conditions result in changes in the temperatures of phase transformations and in microstructures that have solidified before attaining equilibrium. For example, a very rapid cooling of 0.25-wt % carbon steel results in a martensitic microstructure, a material of greater hardness than is achieved under slow cooling. Cooling rates as well as composition are essential to interpreting the microstructures of deposited weld metal and the surrounding heat-affected zone.

Thus, the metallurgy of welds, comprising the weld metal and surrounding heat-affected zone, is influenced not only by the composition of the materials involved, but also by the welding process, the specific procedures for applying the process, and the heat-transfer characteristics (determined by material, mass, and geometry) of the welded joint (9–12).

Material Properties. The properties of materials are ultimately determined by the physics of their microstructure. For engineering applications, however, materials are characterized by various macroscopic physical and mechanical properties. Among the former, the thermal properties of materials, including melting temperature, thermal conductivity, specific heat, and coefficient of thermal expansion, are particularly important in welding. The last property named greatly influences structural distortion that can occur in welding. The electrical conductivity of a material is important in any welding process where base or filler metal is part of the welding electrical circuit.

The response of materials to force is characterized by mechanical properties, eg, elastic modulus, yield stress, tensile strength, ductility, hardness, and impact or fracture strength. Fatigue strength, which is the ability of a material to withstand cyclic loading, is of particular importance to welded structures. Welded joints exhibit greatly diminished fatigue strength, compared to unwelded base metal. The various mechanical properties should be known over a temperature range that covers the expected service temperature. The fracture resistance of most materials, for example, is temperature-sensitive.

Base and Filler Metals

There is hardly a metal that cannot, or has not, been joined by some welding process. From a practical standpoint, however, the range of alloy systems that may be welded is more restricted. The term weldability specifies the capacity of a metal, or combination of metals, to be welded under fabrication conditions into a suitable structure that provides satisfactory service. It is not a precisely defined concept, but encompasses a range of conditions, eg, base- and filler-metal combinations, type of process, procedures, surface conditions, and joint geometries of the base metals (12). A number of tests have been developed to measure weldability. These tests generally are intended to determine the susceptibility of welds to cracking.

Base Metals. *Carbon Steels.* In addition to iron, these steels contain only carbon and manganese in appreciable quantities as alloying elements. They are mainly used for structural purposes at ordinary temperatures, eg, beams, columns, and storage tanks. Carbon steels are the most easily welded metals. The heat of welding has a metallurgical effect on the base metal that is in accordance with its composition. It can reduce the strength or corrosion resistance of a metal or otherwise change its properties, but this general effect is lower in carbon steels than in other steels. Steels containing up to 0.30 wt % carbon and 1.00 wt % manganese can be readily welded in thicknesses up to ca 50 mm without special techniques. All welding processes discussed above are used to weld carbon steels.

Low Allow Steels. These alloys are carbon steels to which other elements have been deliberately added to impart a particular property. Common alloying elements include nickel to improve low temperature mechanical properties; chromium, molybdenum, and vanadium to improve elevated-temperature properties; and silicon to improve properties at ordinary temperatures. Low alloy steels are not used where corrosion is a prime factor and are usually considered separately from stainless steels.

welding. Alpha titanium alloys, which have a hexagonal close-packed atomic structure, have good weldability and are generally welded in the annealed state. Metastable beta titanium alloys, which have a body-centered-cubic atomic structure, can be welded in the annealed or heat-treated condition. The ductility of welded joints in beta alloys can be low after post-weld aging; therefore these alloys are generally used in the as-welded condition, having good ductility but lower strength. The mechanical properties of alpha + beta titanium alloys can be greatly altered by the welding thermal cycle; weld-metal ductility is generally low. Unalloyed or alpha-titanium filler metal is used to improve ductility in these alloys. T-6Al-4V (6 wt % aluminum, 4 wt % vanadium) has the best weldability among the alpha–beta alloys and accounts for half of industrial titanium usage (3).

Copper and Copper Alloys. The coppers are divided into oxygen-bearing and oxygen-free coppers. Numerous copper alloys are of commercial importance, including those alloys with zinc (brasses), with tin (phosphor bronzes), and with aluminum (aluminum bronzes); all are weldable. In welding copper itself, the copper must be free of oxygen if the joint strength is required to be equal to that of the base metal. Copper alloys and can be welded with the shielded-metal arc, gas–metal arc, and gas–tungsten arc process.

Reactive and Refractory Metals. The reactive and refractory metals, originally used in the aerospace industry, are now welded for many applications. The refractory metals, ie, tungsten, molybdenum, tantalum, and niobium, have extremely high melting points, relatively high density, and high thermal conductivity. The reactive metals, ie, zirconium, titanium, and beryllium, have lower melting points and densities, and, except for zirconium, have higher coefficients of thermal expansion. The metals of both groups are difficult to weld. Their high affinity for oxygen and other gases at elevated temperature excludes those processes that utilize fluxes or those in which heated metal is exposed to the atmosphere. Special care must be taken to maintain a protective atmosphere during welding. Small amounts of impurities cause brittleness. The surfaces must be well prepared and very clean to maintain a contamination-free environment during welding and cooling. Beryllium, because of its toxicity, requires special precautions. Gas–tungsten arc welding may be used for all these metals and gas–metal arc welding for some. Electron-beam and laser welding may also be used.

Filler Metals. Filler metals are added to a weld by melting a consumable electrode or a separate wire fed into the weld pool. In the first category, the filler metal is part of the welding electrical circuit and may be in the form of short lengths of covered wire, as in shielded-metal arc welding, or in the form of continuous reels of wire used in semiautomatic, mechanized, and automatic welding processes. Solid wire is used in the gas–metal and submerged-arc welding processes, whereas a hollow, flux-filled wire is used in flux-cored arc welding. More filler metal in the form of iron powder is sometimes added to the electrode coating or flux. In the second category, the filler metal may be in the form of short lengths of bare solid wire, as used in gas welding or manual gas–tungsten arc welding, or in continuous reel form, used in automatic gas–tungsten arc welding.

Filler metals are manufactured in many special forms for welding the commercial alloy systems described herein. The American Welding Society (AWS) has issued specifications covering the various filler-metal systems and processes (2), eg, AWS A5.28 which applies to low alloy steel filler metals for gas-shielded arc welding. A typical specification covers classification of relevant filler metals, chemical composition, mechanical properties, testing procedures, and matters related to manufacture, eg, packaging, identification, and dimensional tolerances. New specifications are issued occasionally, in addition to ca 30 established specifications. Filler-metal specifications are also issued by the ASME and the Department of Defense (DOD). These specifications are usually similar to the AWS specification, but should be specifically consulted where they apply.

Design

Welded Joints.   The weld joint is the geometric arrangement between two pieces of base metal brought together for purposes of welding (13). There are only five recognized weld-joint configurations: corner, butt, tee, lap, and edge joints (Fig. 8). Thus, a butt joint is located between two members in approximately the same plane, whereas a corner joint brings the edges together with an included angle between the planes of the two parts. A tee joints brings the edge of one part onto the planar surface of the second part with a 90° angle between the plates. In a lap joint, one member overlaps another, and an edge joint brings the edges of two parts together without an included angle.

Weld Types.   A weld may be applied to the various joints in different ways, as shown in (Fig. 9). The fillet weld joins corners and tees. In the plug weld, the two pieces are joined by weld metal deposited in a prepared hole in the overlying piece. For a spot weld, heat is applied to the overlying plate, creating fusion at the interface. Resistance welding is generally used for the spot weld and the seam

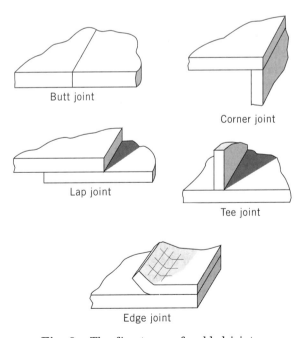

Fig. 8.   The five types of welded joints.

Because steel comprises such a large percentage of welded structures, hydrogen-related discontinuities, which are unique to steel, are listed as a separate category. Fabrication discontinuities occur as a result of the welding process and are described herein. Liquid weld metal can dissolve more gas than solid metal; thus, solidifying weld metal evolves gas bubbles, which may become trapped in the solid metal as porosity. Welding processes that employ a flux are subject to trapped slag inclusions if slag is not completely removed between weld passes. Lack of fusion occurs when weld metal does not properly fuse with either the side of the joint or a previous weld pass. Weld spatter consists of small droplets of electrode material that land beside the weld and adhere to the base metal. Undercut results where a groove, melted at the edge of a weld, is not filled with weld metal. Where the size of a discontinuity exceeds the tolerance limits of a specified welding code, the discontinuity is termed a defect and must be removed, eg, by grinding, and rewelded.

Testing

The integrity of welded structures depends on the integrity of the welds, and much attention is given to testing methods, such as destructive tests, nondestructive tests, and general weld inspection. An objective of many tests is to determine whether welds contain specific defects, such as porosity, slag inclusions, cracks, or lack of fusion (14,15).

Destructive tests destroy the specimen or a portion of the production under examination but provide direct information on properties such as tensile strength, impact strength, ductility, and corrosion resistance. Standard destructive tests include determination of the chemical composition of base or weld metal; corrosion-resistance tests; metallographic tests that employ microscopic examination of polished and etched specimens; and hardness tests across the base metal, heat-affected zone, fusion boundary, and weld metal. Mechanical tests include tensile, impact, and guided bend tests. Destructive tests may be used to evaluate the suitability of welding processes or specific welding procedures for a given application. Simple bend or break tests are used for welder qualification. Test requirements are called out in relevant welding and fabrication codes.

Nondestructive evaluation, also termed nondestructive testing or nondestructive inspection, is extensively used in weld testing (14). Nondestructive tests do no impair the serviceability of the material or component under stress. The most widely used tests for evaluation of welds are liquid penetrant, magnetic particle, ultrasonics, and radiography. Acoustic-emission tests are increasingly used. Nondestructive tests detect and characterize, in terms of size, shape, and location, the various types of weld discontinuities that can occur.

Weld inspection duties of personnel responsible for judging the quality of welding with regard to specifications have been treated (15). Some of these duties involve the visual inspection of welds to determine if they are of the proper size, location, and type and are free of defects. Specifications of materials used must be checked, as must equipment and procedures.

Economic Aspects

More than 1250 companies are involved in the direct manufacture of products associated with welding, such as power sources, electrodes, and fluxes, as well as a wide range of accessories, eg, protective clothing, positioners, and manipulators (16). An extensive distributor network is involved in sales of these welding products. The Standard Industrial Classification (SIC) system of the United States Government lists 14 welding-related industrial groups (17). A search of the *Thomas Register of American Manufacturers* (18) reveals 2455 companies that offer welding services, ranging from arc and resistance welding to the welding of plastics. *Moody's Industrial Manual* indicates that goods using welding in some stage of fabrication account for at least 30% of the gross national product (19).

The U.S. Bureau of Labor Statistics (20) has listed 416,000 persons employed as welders, cutters, and welding machine operators, with 90% employed in the fields of manufacturing, services, construction, and wholesale trades. The same report projects a decline in employment for welders; job prospects remain good, however, as the number of qualified workers entering the market is expected to balance workers leaving the field.

Health and Safety Factors

Welding is carried out in a wide range of industrial environments, such as field construction sites, factory production floors, and small shops. Welders, therefore are subject to the same hazards as all other workers in the metalworking trades. Specific additional hazards of welding include electrical shock, arc radiation, fumes and gases, fires and explosions, compressed gases, cutting and chipping operations, and high noise levels. Many of these hazards also involve workers who share the welding environment. The American National Standards Institute publishes ANSI Standard Z49.1, which covers safe practices for welding and cutting operations (21).

Fumes and gases associated with welding continue to be an area of concern (22–24). Fumes emanate from a number of sources, including electrodes, wires, base metals, coatings and contaminants, ozone produced by the ultraviolet radiation from the welding arc, and gases produced from the heat of the arc. These fumes may lead to a number of health problems, which include acute poisoning under severe cases of ozone and nitrogen oxide concentration, chronic respiratory disease; a condition known as metal-fume fever involving zinc-containing fumes; skin disorders; and disorders of the nervous system, which can be caused by lead or manganese present in welding fumes. A detailed treatment of health and safety in the welding field is found in Reference 25.

BIBLIOGRAPHY

"Welding" in *ECT* 1st ed., Vol. 15, pp. 34–44, by S. A. Greenberg, American Welding Society; in *ECT* 2nd ed., Vol. 22, pp. 241–252, by E. A. Fenton, American Welding Society; in *ECT* 3rd ed., Vol. 24, pp. 502–521, by K. F. Graff, The Ohio State University.

1. *Welding Handbook*, 8th ed., Vol. 1, American Welding Society, Miami, Fla., 1987; see also other volumes in this series.

2. H. B. Cary, *Modern Welding Technology*, Prentice-Hall, Inc., Englewood Cliffs, N.J., 1994.

3. *Welding and Brazing*, Vol. 6 of *Metals Handbook*, 10th ed., ASM International, Materials Park, Ohio, 1993.

4. D. Simonson, *The History of Welding*, Monticello Books, Inc., 1969.

5. *Resistance Welding Manual*, 4th ed., Resistance Welder Manufacturers' Association, Philadelphia, Pa., 1989.

6. M. Schwartz, *Metals Joining Manual*, McGraw-Hill Book Company, Inc., New York, 1979.

7. H. Manko, *Solders and Soldering*, 2nd ed., McGraw-Hill Book Company, Inc., New York, 1979.

8. R. J. Bowers and J. C. Lippold, *Introduction to Materials Behavior*, NEMJET, The Ohio State University, Columbus, Ohio, 1996, p. 13.

9. S. Kou, *Welding Metallurgy*, John Wiley & Sons, Inc., New York, 1987.

10. E. Linnert, *Welding Metallurgy*, 4th ed., Vol. 1, American Welding Society, Miami, Fla., 1994.

11. F. Lancaster, *Metallurgy of Welding*, 5th ed., Chapman & Hall, New York, 1993.

12. D. Stout and W. D. Doty, *Weldability of Steels*, 4th ed., Welding Research Council, New York, 1987.

13. W. Blodgett, *Design of Weldments*, The James F. Lincoln Arc Welding Foundation, Cleveland, Ohio, 1963.

14. *Nondestructive Evaluation and Quality Control*, Vol. 17 of *Metals Handbook*, 9th ed., ASM International, Materials Park, Ohio, 1989.

15. *Welding Inspection*, 2nd ed., American Welding Society, Miami, Fla., 1980.

16. *Welding Design Fabricat.*, 17th Annual Welding and Fabricating Buyers Issue (Jan. 1995).

17. *Standard Industrial Classification Manual*, National Technical Information Service, Springfield, Va., 1987.

18. *Thomas Register of American Manufacturers*, Thomas Publishing Company, http://thomasregister.com, 1996.

19. *Moody's Industrial Manual (American and Foreign)*, Moody's Investors Service, Inc., New York, 1981.

20. *Occupational Outlook Handbook*, Bureau of Labor Statistics, http://stats.bls.gov/oco/ocos226.htm, Apr. 8, 1996.

21. *Safety in Welding and Cutting*, ANSI Z49.1-1988, American Welding Society, Miami, Fla., 1988.

22. Y. Speight and H. C. Campbell, eds., *Fumes and Gases in the Welding Environment*, American Welding Society, Miami, Fla., 1979.

23. *Effects of Welding on Health*, Vols. 1–9, American Welding Society, Miami, Fla., through 1991.

24. *The Facts About Fume*, The Welding Institute, Abington, U.K., 1976.

25. N. Balchin and H. Castner, *Health and Safety in Welding and Allied Processes*, 4th ed., McGraw-Hill Book Co., Inc., New York, 1993.

RANDY J. BOWERS
KARL F. GRAFF
Edison Welding Institute

WETTING AGENTS. See SURFACTANTS; DETERGENCY.

WHEAT AND OTHER CEREAL GRAINS

Origins and History

The word cereal derives from the name of the Roman goddess, Ceres, in whose honor a spring festival, The Cerealis, was celebrated. This indicates the antiquity of these foods and the reverence with which they were esteemed. The latter probably reflected the important role cereals played in the diets of ancient peoples. Cereals are still an important dietary ingredient. As the world population continues to grow, cereals will become an increasing fraction of the diets for more and more people. That is recognized by the use of an ear of wheat as the symbol of the FAO. Below the wheat is the Latin inscription, *Fiat panis* (Let there be bread) (1).

Exactly where and when cereal grains were first cultivated has not been established. Most reports suggest that, for wheat, cultivation first occurred somewhere in the "fertile crescent" which extends from Egypt through Syria to and including the Tigris-Euphrates valley (2). It was in that region that the likely progenitors of wheat, emmer and einkorn, were found. These wild grains still grow in the area extending eastward from Asia Minor to Iran and Afghanistan.

The origins of the other cereals are less certain. Rice may have originated in Africa or Asia but probably was first cultivated somewhere between the southern People's Republic of China and southern Vietnam (3). Corn or maize originated in the Western Hemisphere. The Spanish conquistadors came in contact with corn during their early explorations of Central and South America (4). Corn originated perhaps in the lowlands of South America; others place its origin in Mexico or Guatemala (5). Wherever it was developed, by the time Columbus made his first voyage of exploration, corn as a food crop had spread over much of the Americas as well as to the West Indies (5). By 1492, the Indians had developed corn culture to such a high state that it is thought to have then ranked highest among cereals in efficiency of food production (6). The origin of rye is more difficult to ascertain. There appears to be no reference to it prior to the Christian era. It is still a principal ingredient in the diet of some peoples in northern Europe as it was in the UK until the late eighteenth century (7). Since barley grows wild in the Syria–Palestine area, it is likely to have originated there (7). Millet may have originated in the Sudan where pearl millet is still widely cultivated (3).

Botanically, cereals are simple, one-seed fruits. They include wheat, rice, corn (also called maize in some parts of the world), rye, barley, oats, sorghum, and millet. All cereals contain large amounts of starch and little fat; the fat is associated primarily with the germ and scutellum (single cotyledon or the first leaf). In most cases, the lipids in cereals contain a high concentration of unsaturated fatty acids, which are protected from oxidation by the presence of tocopherols. Actually, the oil from wheat germ was the starting material for the isolation of α-tocopherol (see VITAMINS, VITAMIN E). As long as the antioxidants are in proximity to the lipids, oxidation of the lipids and the accompanying rancidity is minimized. This is one reason why wheat is stored in the kernel stage until it is milled. Generally, wheat is milled only when there is an order for flour.

However, even in the intact kernel, the lipids are subject to hydrolytic rancidity (8), which together with oxidative rancidity can be estimated by determining the free fatty acid content. That is done by determining the amount in milligrams of potassium hydroxide required to neutralize the free fatty acids from 100 g of moisture-free grain (9). This is called the fat acidity value. One of the main purposes of milling, in addition to producing an acceptable flour, is to remove the wheat germ in as intact a state as possible. Thereby, the concentration of the lipids is reduced, the development of oxidative rancidity decreased, and the shelf life of the flour extended.

Cereals were among the earliest plants cultivated. They were related to some of the wild grasses indigenous to those parts of the world where civilizations had their origins. All of them are ideally suited for use as food under both primitive, and advanced conditions. For one thing, they can be stored for long periods under many conditions, thus providing a reserve against food shortages. This was done in ancient Egypt when Joseph advised the Pharaoh to store grain during the seven "plenteous" years to be used in the following seven "lean" years. In addition, cereals are nutritious foods that can be used in many ways, thus facilitating their incorporation in the diet at high levels over long periods of time. Even today, there are some areas, such as rural Iran, where wheat products, especially in the form of the flat breads indigenous to that region, provide as much as 70–90% of the daily caloric intake. Most of the cereals also respond well to primitive methods of agriculture with good yields and, with advanced technology and improved varieties, large yields per worker can be secured. Much more food is secured from fields planted in grains than can be obtained from cattle or other animals on the same land. The importance of this became evident during World War I. In the early stages of that conflagration, the German High Command made a conscious decision to continue the prewar level of meat, milk, and egg production to provide adequate nutrition for the men in the armed services and for the civilians who would be called on for arduous work in connection with the war. Had they decided instead on conversion of meadows and pastures to wheat fields and the direct consumption of the grain by human beings, the yield of food for the German people would have been far greater than it was when the land was given over to raising cattle, pigs, sheep, and poultry. This led a group of Scottish physicians to suggest that this mistake probably did more than any army general to lose the war (10). Finally, cereals grow in a wide variety of climatic and soil conditions, and they successfully compete with weeds for the limited amounts of nutrients and water where these plant factors are in short supply. This is an important reason why cereals have played, and continue to play, such an important role in the development of the human race.

Nutritional Value of Cereals

Deficiency Diseases. Not only did cereals make an important contribution to improving the general status of humankind, but they also were important dietary components of some groups of people who showed certain nutritional deficiencies. This observation led to the discovery of some of the vitamins. These deficiency diseases have been most prominently associated with use of rice, corn, and wheat.

Beriberi, Thiamine Deficiency. The recognition of vitamins and their importance to the health of human beings came about when Eijkman, a Dutch pathologist, was sent to Java in an attempt to cure an epidemic of beriberi that had appeared in one of the hospitals. Eijkman kept a flock of chickens on the hospital grounds to assist in discovering the disease agent he assumed was involved in the etiology of beriberi. These chickens were fed the scraps from the plates of the hospital patients—primarily polished rice, the common food in that part of the world (11).

Although Eijkman recognized the condition in his polyneuritic chickens as the analogue of beriberi, he misinterpreted its cause. He suggested on the basis of bacteriological concepts then dominating the medical field that beriberi resulted from the ingestion of a toxic substance associated with the starchy part of the rice. According to that theory, rice polishings contained an antitoxin which neutralized the toxin present in the polished rice. It was Eijkman's successor who established that polished rice lacked some substance essential for normal physiological functioning and that this substance occurred in rice polishings, beans, meat, and other foods (12). The critical ingredient is now known to be thiamine (vitamin B_1). The absence from the diet of similar substances is responsible for the development of scurvy, pellagra, rickets, etc (13).

Pellagra, Niacin Deficiency. It was 220 years after the first description of pellagra that nicotinic acid was discovered to be the cure for black tongue in dogs (14), a condition suggested by a veterinarian in North Carolina to be similar to human pellagra (15).

The contrast between a high incidence of pellagra among the inhabitants of the southeastern United States and its absence among the corn-eating people south of the Rio Grande was a puzzle for a number of years. It finally became evident that the lime water in which the corn kernels were steeped prior to being made into tortillas liberated nicotinic acid from the bound form (16). Untreated kernels of corn contain nicotinic acid as a complex from which it cannot be made available by human or animal digestive processes. However, a weak alkali, such as lime water, releases nicotinic acid from its complex and makes it available for absorption. Furthermore, the usual diet of the corn eaters south of the Rio Grande includes a large complement of beans. Beans are a rich source of tryptophan which can be converted to nicotinic acid by enzymes in the human body (see VITAMINS, NIACIN, NICOTINIC ACID, AND NICOTINAMIDE).

Zinc Deficiency. A nutritional problem associated with consumption of large amounts of whole wheat products is the unavailability of dietary zinc, first observed in a patient with immature development and dwarfism in southern Iran (17). The patient showed marked improvement when placed on a well-balanced, nutritious diet for a year. In 1962, similar patients were observed in Egyptian villages. A deficiency of zinc was identified as the primary reason for the development of this condition. This deficiency of zinc results from the binding of that metal by the phytates present in whole wheat (18). Even when an excess of zinc is present in the diet, if conditions are right, that zinc may be complexed with the phytate and thus rendered unavailable to the body (see MINERAL NUTRIENTS). Thus, a zinc deficiency may develop when its dietary level appears adequate but the diet contains large amounts of a food, such as whole wheat, that is high in phytates.

The high fructose syrups are produced from corn starch after the latter has been hydrolyzed to glucose by either acid or α-amylase. Treatment of the resulting fluid with an isomerase converts ca 42% of the glucose to fructose. The remaining solids consist almost exclusively of glucose and other saccharides (34). After clarification, this product is concentrated to a base of 71% solids, in which state it is bulk-distributed. The enzymes involved in the isomerization are of bacterial origin, most of them belonging to the *Streptomyces* group. The enzyme is immobilized by fixation on cellulose (see ENZYMES, IMMOBILIZED).

A solution is available with a 90% fructose content. Because of the sweetness of fructose, this syrup imparts the same sweetness to a product as a larger amount of sucrose (see SWEETENERS). Since on a weight basis the monosaccharides have the same number of food calories as sucrose, this gives a means of reducing the caloric content of the food. Reduced caloric foods, such as jellies and preserves, salad dressing, and table syrup, are now being manufactured with 90% fructose corn syrup and are recognized as having quality improvements compared to the use of other sweeteners (34).

Health Problems Associated with the Consumption of Cereals

Celiac Disease.   A disturbance of the lower gastrointestinal tract, celiac disease is a chronic disease characterized by loss of appetite and weight, depression and irritability, and diarrhea frequently followed by constipation (35). One of the more disturbing features of celiac disease is the large, frothy, foul-smelling stools. The disease may develop in childhood or later in life. Frequently, the patients who develop the disease in adulthood report having had some of the symptoms during childhood.

This disturbance was recognized shortly after World War II as being related to the ingestion of wheat. A group of physicians in the Netherlands was impressed by the fact that during the war they saw many cases of celiac disease. During that time, wheat was the primary staple of the diet. However, at the end of the war, other foods again became available and the number of children who developed celiac disease decreased. One Dutch group of investigators had a seven-year-old female patient who displayed extreme fluctuations in her symptoms. These changes were shown to be associated with the presence or absence of bread in her diet. Using that patient as the test subject, the group soon learned that her symptoms worsened shortly after she consumed foods containing wheat gluten (36). Although considerable work has been done attempting to identify the mechanism whereby gluten causes the disturbance characterizing celiac disease, little more can be said beyond that this appears to be an allergic reaction. For victims of celiac, the primary therapy involves the elimination of any dietary product that contains gluten. Since that protein occurs not only in wheat but also in rye and barley, the complete elimination of gluten from the diet becomes an onerous task.

Disturbances Associated with Flour-Aging Agents.   Bread made with flour freshly milled from wheat has a different texture and appearance than that associated with bread. Only after the flour has been aged by storage for a few months does the gluten develop the toughness that permits the bread to retain the shape it had when it was put in the oven. The aging of the flour likely involves a mild reaction of oxygen and some part of the gluten molecule. Additional evidence for this is the destruction of the yellowish pigment present in the flour prior to its storage. To hasten the reactions that occur during storage, chlorine gas and oxides of nitrogen were tried. Although they improved the freshly milled flour, they were not completely satisfactory. In 1921, nitrogen trichloride was introduced as a flour agenizing substance (37).

Although no untoward effects were observed that could be attributed to the long-term ingestion of agenized flour by people in the United States and the UK, and despite the absence of any symptoms in human subjects given large amounts of flour heavily treated with nitrogen trichloride prior to incorporation into foods, the use of nitrogen trichloride by the milling industry in the United States was prohibited after Aug. 1, 1949 (38). That order recommended the use of chlorine dioxide for agenizing flour. Like nitrogen trichloride, chlorine dioxide is unstable and must be passed into the flour immediately after being generated. The chlorine dioxide is formed by passing chlorine mixed with air into a solution of sodium chlorite (39) (see CHLORINE OXYGEN ACIDS AND SALTS).

Ergot.   Ergotism is only indirectly related to health hazards associated with the consumption of cereals because the alkaloid which is responsible for ergotism is produced by a fungus (*Claviceps purpurea*) that grows primarily on rye. That the fungus may grow on other cereals than rye is implied in a report of this disturbance among the members of a British family who consumed no rye bread (40) (see ALKALOIDS).

The disease takes two different forms depending apparently upon whether ergotoxin, the alkaloid in ergot, attacks predominantly the nervous or the circulatory system. The former condition is characterized by severe convulsive seizures; the latter produces an intense burning and itching of the skin called St. Anthony's Fire.

Abortion is one of the prominent characteristics of ergot poisoning among both women and cattle. This is due to the action of ergotoxin in contracting the placental muscles. For that reason, small, regulated doses of ergotoxin are used to control bleeding following childbirth (see also HORMONES, POSTERIOR PITUITARY HORMONES).

Colonic Cancer.   Another area in which cereals may play an important role in maintaining the health of human beings is as a source of dietary fiber (qv). During the past decade, evidence has been collected that certain diseases, to which people in the more-developed countries are prone, owe their origin to the almost complete removal on the fibrous parts of cereals consumed in the diet. There is an almost complete absence of a variety of tumors in the lower gastrointestinal tract among the natives of African villages, whereas the incidence of these colonic tumors among those of European extraction living in the area is as high as in northern Europe (41). The primary, visible difference between these two groups of people is the nature of their diets. The native Africans consume a diet composed largely of unrefined foods and consequently high in fiber. On the other hand, the neighboring people of European origin rely on foods similar to those available in their countries of origin. The foods eaten by the latter group are either devoid of, or very low in, dietary fiber.

Prior to this work, dietary fiber, of which cellulose is one of the more important constituents, was considered important primarily as a means of preventing or overcoming constipation. Otherwise, dietary fiber was considered to be a metabolically inert substance. A large variety of diseases such as appendicitis, hiatus hernia, gallstones, ischemic heart disease, diabetes, obesity, dental caries, and duodenal ulcers are now suspected to be associated with the consumption of a highly refined diet (42).

Diabetes.   Fiber has also been shown to be an aid in treating diabetes. Diabetic patients who required 20 or less units of insulin per day could do without their insulin injections provided they increased the amount of cereal bran in their diets (43). These patients had adult onset diabetes, a milder form of the disease than that which appears in children. Nevertheless, the fact that a high intake of bran relieves these patients of the need for daily insulin injections is a welcome form of therapy. Practically all evidence indicates that dietary fiber decreases the maximum blood glucose level following a dose of glucose as in the typical glucose tolerance test. The lower level of blood glucose presumably occurs because its absorption from the intestinal tract is slowed. The mechanism whereby dietary fiber decreases the rate of absorption of glucose from the intestinal tract is not completely understood. One possibility is that the large molecules of dietary fiber physically impede the movement of the glucose molecules to the absorptive surface of the intestine. Another theory suggests that dietary fibers increase the viscosity of the intestinal contents and thereby slow the movement of glucose molecules (44).

Cardiovascular Diseases.   For sometime it was believed that dietary fiber protected the individual from various heart disturbances, especially ischemic heart disease. That presumably occurred as a result of the action of dietary fiber in lowering serum cholesterol levels. This effect results, according to reports, from the increased rate of passage of food through the intestinal tract and consequent incomplete absorption of dietary cholesterol and some bile acids. Since some of the bile acids are secreted into the lower intestinal tract to be reabsorbed at a lower site and since cholesterol is the precursor of bile acids, this would lower blood cholesterol levels. However, some investigators now suggest that it may be other differences in dietary constituents or lifestyle that are responsible for the low blood cholesterol levels among primitive people whose diet is high in fiber (45). Inclusion of large amounts of high fiber foods in the diet may decrease the intake of cholesterol-containing foods and those high in saturated fats. The intake of these two dietary components is probably closely related to the level and nature of blood cholesterol (45).

The implications of the role that dietary fiber may play in the maintenance of human health should increase the demand for cereal products. It may well have an influence on the milling industry, especially if the demand for cereal products that are less highly milled becomes a reality.

Wheat: Production, Trade, and Uses

Wheat is cultivated in most countries on all continents. World wheat production, consumption, and net exports (imports) are summarized in Table 1.

The top five wheat producing countries are the former Soviet Union, the People's Republic of China, the United States, India, and Canada. Of these five countries, only the United States and Canada grow more wheat than they use

Table 1. Wheat: World Production, Consumption, and Net Exports, 1985/1986 (million metric tons)[a,b]

Country	Production	Consumption	Net Exports
Principal Exporters			
United States	64.68	30.21	32.53
Canada	23.50	5.50	17.50
Australia	17.00	3.10	15.20
EC-10	70.07	53.07	14.30
Argentina	11.50	4.85	6.70
Turkey	13.00	13.70	−0.15
Principal Importers			
USSR (former)	83.00	100.00	−19.00
China	87.00	94.00	−7.00
Eastern Europe	38.98	38.97	−0.21
other Western Europe	9.57	9.68	−0.29
Brazil	2.20	6.40	−4.40
Mexico	4.40	4.50	−0.30
Other Latin America	0.05	2.72	−2.69
Japan	0.79	6.35	−5.50
India	45.00	43.00	1.40
South Korea	0.02	2.37	−2.30
Indonesia	0	1.37	−1.15
other Asia	17.56	25.21	−8.14
Egypt	1.85	8.45	−6.70
Morocco	1.83	4.14	−2.30
other northern Africa–Mideast	11.83	25.47	−14.23
other Africa	3.58	9.48	−5.85
Residual	2.43	10.34	−3.92
Total world	*509.84*	*502.88*	

[a]*Source*: Grain and Feed Division, USDA.
[b]Trade on July–June years.

and export to other countries. The other three nations are large wheat importers. Some of the top wheat customers of the United States have been China, India, the former Soviet Union, Japan, and Brazil.

A nation of one billion people, China is traditionally regarded as a rice-eating nation. But China grows almost as much wheat as the United States and buys and uses more wheat than any other country in the world. Each person in China on the average consumes 180 lb of wheat every year, mostly in the form of noodles. The average American eats only about 116 lb of wheat flour per year in all types of wheat-based products. Some nations have much higher per capita consumption, up to 300 lb of wheat per year per person (46–48).

Wheat is grown in most of the 50 states of the United States. The kind of wheat grown and quantity vary widely from one region to another. In total over 200 varieties are grown annually. Winter wheats are planted in the fall. After the grasslike seedlings emerge from the ground, they lie dormant during the winter. They come up again in the spring, ripen, and are harvested in early summer. Spring wheats are planted in the spring and harvested in late summer. Spring wheats grow best in the northern areas of the United States where the summers are not too hot for the young plants. Winter wheats grow best in areas of the country where the winters are not too harsh for the young plants.

The many varieties of winter and spring wheat are grouped into five official classes. The class a variety fits into is determined by the hardness, the color of its kernels, and its planting time. Each class of wheat has its own relatively uniform characteristics, including those related to milling, baking, or other food uses (46–48). Protein range and flour use of principal wheat classes are listed in Figure 1 (49).

Hard red winter (HRW) is an important bread wheat that accounts for more than 40% of the United States' wheat crop and wheat exports. This fall-seeded wheat is produced in the Great Plains, which extend from the Mississippi River west to the Rocky Mountains, and from the Dakotas and Montana south to Texas. Significant quantities are also produced in California. HRW has moderately high protein content, usually averaging 11–12%, and good milling and baking characteristics.

Hard red spring (HRS), another important bread wheat, has the highest protein content, usually 13–14%, in addition to good milling and baking characteristics. This spring-seeded wheat is primarily grown in the north central United States—North Dakota, South Dakota, Minnesota, and Montana. HRS constitutes about 15% of U.S. wheat exports. Subclasses based on the dark, hard, and vitreous (DHV) content, include Dark Northern Spring, Northern Spring, and Red Spring.

White wheat (WW) is a preferred wheat for noodles, flat breads, and bakery products other than loaf bread. WW, which includes both fall and spring-seeded varieties, is grown mainly in the Pacific Northwest. This low-protein wheat, usually about 10%, comprises about 15% of U.S. wheat exports, destined

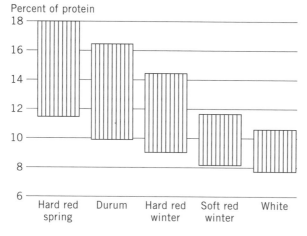

Fig. 1. Protein range and flour uses of principal wheat classes. Flour uses are listed according to the approximate level of protein required for specified wheat products. Durum is not traded on the basis of protein content. From Ref. 49.

primarily for East Asia and the Middle East. Subclasses include hard white, soft white, western white, and white club.

Soft red winter (SRW), which is grown in the eastern third of the United States, is a high-yielding wheat, but relatively low in protein, usually about 10%. SRW best provides flour for cakes, pastries, quick breads, crackers, and snack foods. This fall-seeded wheat constitutes about one-quarter of U.S. wheat exports.

Durum, the hardest of all U.S. wheats, provides semolina for spaghetti, macaroni, and other pasta products. This spring-seeded wheat is grown primarily in the same northern areas as hard red spring, but small winter sown quantities are also grown in Arizona and California. Durum represents about 5% percent of total U.S. wheat exports. Subclasses are hard amber durum, amber durum, and durum.

Grades and grade requirements for all U.S. wheat classes are given in Table 2. Values of the total domestic consumption of wheat products as foods in the United States are listed in ref. 50.

Milling of Wheat

Conventional, Modern Milling.   Before wheat can be used in the production of most foods, it must undergo several mechanical and chemical changes. The first change involves milling of wheat into flour (46,51). The steps involved in wheat-to-flour production are wheat selection, blending, cleaning, conditioning, milling, and maturing. The endosperm, which forms about 83% of the kernel, is the source of white flour and contains 70–75% of the kernel's protein. The bran, forming about 14% of the kernel, is included in whole wheat flour, but is more often removed and used in animal or poultry feed. Because the cellulosic material of the bran cannot be digested and tends to accelerate the passage of food through the human digestive tract, the total nutritive contribution of whole wheat flour is less than that found in enriched white flour products. The germ, forming about 3% of the kernel, is the embryo or sprouting tissue of the seed. It is usually separated out because it contains oil, which limits the keeping quality of flours. Although the germ is available as human food, it is usually added to animal or poultry feed.

The mill flow diagram begins with a separator, where the wheat first passes through a vibrating screen that removes straw and other coarse materials, and then over a second screen through which drop small foreign materials like seeds. An aspirator lifts off lighter impurities in the wheat. After the aspirator, wheat moves into a disk separator, consisting of disks revolving on a horizontal axis. The disk surfaces are indented to catch individual grains of wheat but reject larger or smaller material. The blades push the wheat from one end of the machine to the other. The revolving disks discharge the wheat into a hopper or into the continuing stream. The wheat then moves into a scourer—a machine in which beaters attached to a central shaft throw the wheat against a surrounding drum. Scourers may be either horizontal or upright, with or without brushes, and adjusted for mild, medium, or hard scouring. Air currents carry off the dust and loosened particles of bran coating.

grades shown in Figure 2 (52). The remaining percentage of the wheat berry is classified as millfeed. The flour can be classified in several ways. Straight flour is all the flour produced, with various streams of flour mixed into one. Flour emerges at a number of points in the milling process, with the purified middlings yielding the extrashort or patent or bread flour. In hard wheat mills, as much as 75–80% may be run together as first clear or split into fancy clear and second clear.

In a soft wheat mill, 40–60% of the fancy patent may be taken off separately, leaving about 55% of the remaining flour to be classified as fancy clear. The chart (Fig. 2) is a generalization rather than an exact description of the yield of and single mill from a particular kind of wheat. It shows how the various streams of flour may be classified, starting with fancy patent, through short, medium, and long patent flours, leaving less and less to be classified as clear flour. The extrashort or fancy patent is the finest, with grades dropping down the scale to clears (53,54).

Toward the end of the millstream, the finished flour flows through a device that releases a bleaching–maturing agent in measured amounts. For bakery customers, the finished flour flows into hoppers for bulk storage, since most bakers add their own form of the enrichment formula to dough—a combination of thiamine, niacin, riboflavin, and iron. For packing as family flour, the enrichment ingredients are added in another mixing machine as the flour flows to the packing room. If the flour is self-rising, a leavening agent and salt are also added.

In milling of durum wheat for the macaroni trade, special equipment is required, especially additional purifiers to separate the bran from the semolina, a coarse granulation of the endosperm (55). By federal definition, semolina is

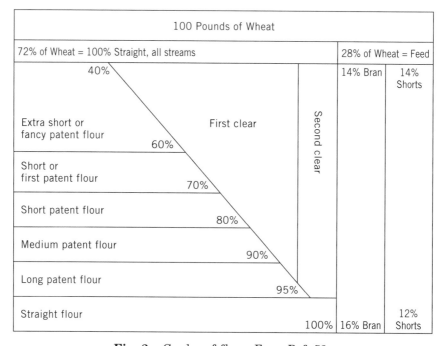

Fig. 2.   Grades of flour. From Ref. 52.

prepared by grinding and bolting durum wheat, separating the bran and germ, to produce a granular product with no more than 3% flour. Durum millers also make granulars, or a coarse product with greater amounts of flour; or they grind the wheat into flour for special use in macaroni products, particularly in noodles.

Other Milling Developments. In the early 1960s it was found that wheat could be ground to a very fine flour by impact milling and that the product could be further separated into products varying widely in protein. Finished flour from a conventional roller mill is further reduced in particle size in a high speed grinder, an impact or pin mill. Disintegration of the flour particles takes place as they strike one another, the surface of the rotors, and the pins. The flour is fractured in granular form rather than pressed and broken as in a roller system. The reground flour contains a mixture of relatively coarse endosperm chunks, intermediate fragments, and fines.

The reground flour is channeled to a classifier, where swirling air funnels the larger particles down and away while the smaller fines are lifted up and separated. Repeating the process 20–30% of the flour is separated into a low-protein product suitable for cakes and pastries; 5–15% makes up the fine fraction, containing 15–22% protein.

The high-protein flour may be used to fortify or blend with other flours. Recombining some of the high-protein fraction with the coarse portions permits a miller to tailor a flour of protein value to a buyer's specifications.

Fine grinding and air classification make possible the production of some cake flour from hard wheat and some bread flour or high-protein fractions from soft wheat. Application of the process theoretically frees the miller from dependence on different wheats, either hard or soft, that change each crop year. The problem is how to market the larger volume of low-protein or starch fractions at prices adequate to justify the installation and operation of the special equipment (46).

Bleaching and Maturing Agents. Flour is often bleached, both for use by bakers and for the family flour trade. Bleaching improves the color of the flour, and some bleaching agents mature the flour and condition the gluten, improving baking quality. Flour fresh from the mill may be aged in storage or treated with a bleaching–maturing agent to improve baking quality. In aged, unbleached flour or in treated flour, the proteins are slightly oxidized by oxygen from air. The oxidation of flour makes gluten stronger or more elastic and produces better baking results. Cookie, pie crust, and cracker flours usually perform better when no bleach is used.

Dough handling properties can be modified beneficially, or adversely, by the addition of minute amounts of reducing agents (such as cysteine). In addition, the performance of a flour can be improved significantly by the addition of certain oxidizing agents. A list of oxidizing and reducing agents used to improve the baking quality of wheat flour is given in Table 3.

Flour Types

Hard-wheat flours are usually higher in protein than are soft-wheat flours. They may be milled from either winter or spring wheat varieties. Those with highest

Table 3. Flour Treatment Agents[a,b]

	Bromate	Iodate	Azodicarbonamide	Chlorine dioxide
			Compound	
structure	$KBrO_3$	KIO3	$$H_2N-\overset{\overset{O}{\|\|}}{C}-N{=}N-\overset{\overset{O}{\|\|}}{C}-NH_2$$	ClO_2
form	powder	powder	powder	gas
introduced	1916	1916	1962 patent	1948
regulations				
flour	50 ppm max U.S. whole wheat, 75 ppm max	not allowed	45 ppm max	sufficient for bleaching
bread	U.S.—75 ppm max (see below) Can.—100 ppm max (see below)	45 ppm max (see below)	U.S. and Canada, 45 ppm max	
labeling	potassium bromate or bromated flour	potassium iodate	bleached flour	bleached flour
used in	bread flour and dough	bread dough	bread flour and dough	bread flour
action	oxidant— improver	oxidant— improver, mix reducer	oxidant— improver, maturing	oxidant— maturing, bleaching
reaction rate	slow	fast	fast	fast
reaction product	bromides	iodides	biurea	chlorides
usage rates, ppm				
flour, normal	15	0	5	15
total, range	10–75	10–20	2–20	
overtreatment effects	open grain holes keyholing incorrect volume	low volume weak, sticky doughs	dry, bucky dough; low volume; poor grain; crust checking	poor volume; poor grain
testing				
flour, spot	AACC 48-02		no	no
flour quantitative	AACC 48-42		yes—difficult	no
bread	very difficult	difficult	no	no
Pennwalt products	Bromolux No—D-Lay		Maturox Dependox	Dyox

[a] Courtesy: Pennwalt, Buffalo, NY.

[b] These oxidating–reducing agents are used to improve the baking quality of wheat flour. They are added at the flour mill, the bakery, or both. Having the proper oxidation treatment is critical: it ranks second only to protein quality and quantity in determining flour quality.

Government regulations on bakery use are as follows: United States—potassium bromate, calcium bromate, potassium iodate, calcium iodate, and calcium peroxide, or any combination of these, cannot

Table 3. (*Continued*)

	Compound		
Chlorine	Benzoyl peroxide	L-Ascorbic acid	L-Cysteine
Cl_2	benzoyl peroxide structure: two benzene rings joined by $-C(=O)-O-O-C(=O)-$	CH_2OH, $HCOH$, lactone ring with O, $=O$, OH, OH	$HS-CH_2-CH{<}^{COOH}_{NH_2}$
gas	powder	powder	powder
1912	1921	1935	1962 patent
sufficient for bleaching	U.S.—sufficient for bleaching; Canada—150 ppm max	200 ppm max	U.S.—not allowed; Canada—90 ppm max
		U.S.—no limit; Canada—200 ppm max	U.S. and Canada—90 ppm max
bleached flour	bleached flour	ascorbic acid added as a dough conditioner	L-cysteine
cake flour	bread and cake flour	bread flour and dough	flour (Canada only), miscellaneous doughs
oxidant— maturing, bleaching	oxidant—bleaching	reductant–oxidant —mix reducer, improver	reductant—mix reducer, relaxer
fast	1–3 days	medium	fast
chlorides	benzoic acid		
120	50	70	30
25–500	30–100	50–100	15–70
low volume, dense crumb	none	none	weak, sticky doughs; poor grain and texture
no	AACC 48-06B	yes—good	yes—fair
pH used	flour color	yes	no
no	benzoic acid—hard	no	no
flour	Novadelox	No—D-Lay	No—D-Lay
chlorination systems	Zerolux	AA-25	Assist

exceed 75 ppm. This includes any bromate added to the flour. Canada—potassium bromate, calcium peroxide, ammonium persulfate and potassium persulfate, or any combination of these, cannot exceed 100 ppm. This includes any bromate added to the flour. Calcium iodate, potassium iodate or any combination of iodates cannot exceed 45 ppm.

protein content, characterized by their capacity to develop the strongest gluten, are used in commercial bread production where doughs must withstand the rigors of machine handling. Other hard-wheat flours with more mellow gluten, easier to develop in kneading by hand, are packed as family flour, all-purpose flour, and self-rising flour. The protein in hard spring wheat flour runs from 11 to 16%; in hard winter wheat flour from 10–14% (47,48).

Soft-wheat flours are sold for general family use, as biscuit or cake flours, and for the commercial production of crackers, pretzels, cakes, cookies, and pastry. The protein in soft wheat flour runs from 7 to 10%. There are differences in appearance, texture, and absorption capacity between hard- and soft-wheat flour subjected to the same milling procedures. Hard-wheat flour falls into separate particles if shaken in the hand whereas, soft-wheat flour tends to clump and hold its shape if pressed together. Hard-wheat flour feels slightly coarse and granular when rubbed between the fingers; soft-wheat flour feels soft and smooth. Hard-wheat flour absorbs more liquid than does soft-wheat flour. Consequently, many recipes recommend a variable measure of either flour or liquid to achieve a desired consistency.

Whole wheat flour, according to FDA specifications, is a coarse-textured flour ground from the entire wheat kernel. It contains the bran, germ, and endosperm. The presence of bran reduces gluten development. Baked products made from whole wheat flour tend to be heavier and denser than those made from white flour. Whole wheat flour is rich in B-complex vitamins, vitamin E, fat, protein, and contains more trace minerals and dietary fiber than does white flour. In most recipes, whole wheat flour can be mixed half and half with white flour for a satisfactory product. Graham flour is synonymous with whole wheat flour, named after a physician, Dr. Sylvester Graham, who advocated the use of whole wheat flour in the first half of the nineteenth century (47,48).

According to FDA standards, wheat flour must be milled from cleaned wheat, essentially free of bran and germ, and be ground to such a degree that the product will pass through a sieve having 65 openings per lineal inch (or 4,225 openings per square inch). The moisture content must not exceed 15%.

All-purpose flours are designed for home baking of a wide range of products—yeast breads, quick breads, cakes, cookies, and pastries. Such flours may be made from low-protein hard wheat, from soft or intermediate wheats, or from blends of hard and soft wheat designed to achieve mellow gluten a homemaker can manipulate, and tenderness of structure in the final baked product. All-purpose flour, sometimes called general-purpose or family flour, is most commonly used for home baking. Most modern recipes for home baking are designed for use with all-purpose flour.

Self-rising flours are all-purpose flours to which leavening agents and salt have been added. The leavening agents used are sodium bicarbonate and an acid-reacting substance, monocalcium phosphate, sodium acid pyrophosphate, sodium aluminum phosphate, or a combination of these acids. The sodium bicarbonate and the acid ingredient react in the presence of liquid to release carbon dioxide. The leavening agent and salt are added in amounts sufficient to produce not less than 0.5% of their combined weight of carbon dioxide. Their combined weight must not exceed 4.5 parts per hundred parts of flour. Phosphated flour is all-purpose flour to which the acid-reacting ingredient, monocalcium phosphate, has been added in a quantity of not more than 0.75% of the weight of the finished phosphated flour. It assists in stabilizing gluten and helps nourish yeast. Bromated flour contains potassium bromate in a quantity not exceeding 50 ppm. The bromate has an oxidizing effect that improves the baking qualities of the flour. Such flour must be labeled as bromated.

Cake flours are milled from low-protein soft wheat especially suitable for baking cakes and pastries or from low-protein fractions derived in the milling process. Cake flours are usually not enriched, but are bleached. The sale of cake flours declined during the 1950s as packaged cake mixes became more popular. Instantized, instant blending, or quick-mixing flour is a granular or more dispersible type of product for home use. It is free-pouring like salt, and dust-free compared to regular flour. It eliminates the need for sifting, since it does not pack down in the package and since it pours right through a screen or sieve. Granular flour instantly disperses in cold liquid rather than balling or lumping as does regular flour. The granular flour is produced by special processes of grinding and bolting, or from regular flour subjected to a controlled amount of finely dispersed moisture that causes the flour to clump or agglomerate. It is then dried to a normal moisture level (47,48). Gluten flour is milled to have a high wheat gluten and a low starch content and is used primarily by bakers for dietetic breads, or mixing with other flours of a lower protein content.

Semolina is the coarsely ground endosperm of durum wheat. High in protein, it is used by U.S. and Italian manufacturers for high-quality pasta products such as macaroni and spaghetti. In Africa and Latin America it is also used for a dish called couscous. Durum flour, a by-product in the production of semolina, is used to make commercial American noodles. Farina is the coarsely ground endosperm of hard wheats. It is the prime ingredient in many American breakfast cereals. It is also used by manufacturers for inexpensive pasta.

Additional basic wheat products are wheat berry (kernel), bulgar, cracked wheat, wheat germ, bran, and commercial cereals. See ref. 56 for more information on wheat science technology.

Rice: Production and Consumption

Rice is grown in more than 100 countries and on every continent except Antarctica. In the world economy rice is an extremely important food, second only to wheat in total world production, and its yield per hectare exceeds that of wheat (57). Rice is the main staple food for more than half of the world's population and it accounts for one-third to one-half of the daily caloric intake in many Asian countries. It is also the major source of protein for the masses of Asian people. In many African and South American countries rice is rapidly becoming the staple food for much of the population.

World rice production has been increasing at approximately 7 million t/yr since 1950. Land use for rice production increased from 103 million ha in 1950 to 146 million in 1978. Since 1978 the acreage has remained fairly constant. The increase in production since 1978 has been the result of higher yields per ha (increasing from 2.58 in 1978 to 3.16 million t/ha in 1988).

The major rice-exporting countries are Thailand, United States, Pakistan, EC-12, China, Burma, and Australia.

It should also be understood that in the rice industry a rice mill may refer to a machine for removing bran or it may include a series of processing operations that, in general, consists of cleaning, shelling, milling, sizing, packaging, and several auxiliary operations. The objective of these rice mill operations is to produce a white, whole-grain product that is essentially free of bran and foreign matter and that contains a minimum of broken kernels.

In some rice growing areas rice milling is accomplished by very primitive methods such as pounding the rough rice in a wooden mortar and pestle followed by winnowing. At the other extreme are very modern methods where milling is accomplished in large, highly automated plants. Thus there is no typical rice mill. However, the modern processing of rice consists of essentially these steps: (1) cleaning the incoming rough rice, (2) shelling the rough rice, (3) milling to remove the bran from the brown rice, (4) grading the milled rice by length into whole grain and different sizes of brokens, (5) mixing milled whole grain and brokens to meet specifications of buyers, and (6) packaging.

Parboiling. A hydrothermic process, parboiling will greatly improve the milling quality of rice such that head yields will approach total yields (ie, zero breakage). Comprehensive quantitative studies on the effects of parboiling on the breakage of rice have been reported and it has been noted that kernel defects such as cracks, chalkiness, and incomplete grain filling are completely healed during the parboiling process (63,64). When properly dried, the rice kernels are resistant to mechanical breakage. The milling quality of parboiled rough rice is determined largely by drying conditions following parboiling rather than the previous history or condition of the rice. Consequently, for rough rice that is to be parboiled the optimum harvesting, drying, and storage conditions should be selected on some basis other than that of preserving the milling quality. For the same reasons, parboiling is an excellent means for salvaging rice whose milling quality has been inadvertently damaged by improper handling or processing. For more information on rice processing and uses, see ref. 65.

Corn: Production and Economics

Every continent, except Antarctica, grows corn; 40% of the present world crop is produced in the United States. In the 1987—1988 crop year, 12 states (Iowa, Ill., Nebr., Minn., Ind., Ohio, Wis., Mo., S. Dak., Mich., Kans., and Tex. in order of production) produced 157.5 million metric tons (6.2 billion bushels) that was 88% of the United States and 36% of the world's crop (66). Yield is influenced by many factors, including climate, pest control, planting density, and fertilization. Yield in the United States has increased from about 1.5 metric tons/hectare in the 1930s to about 7.5 metric tons/hectare. In 1985, a test plot produced 23.2 metric tons/hectare and yields approaching 40 metric tons/hectare are considered possible; corn is the most productive of the principal food crops.

A crop of corn is always maturing somewhere in the world. It grows from north latitude 58° to south latitude 40° and from below sea level to altitudes of 4,000 meters. It is adapted to areas with fewer than 10 in. of rainfall and regions having more than 400 in. Early varieties, that have been adapted to cold climates, mature in as little as 60 days. Late varieties grown in the tropics need nearly a year to reach maturity. Corn can grow to as little as 60 cm in height or as tall as 6.5 meters. Ears of corn can be as small as your thumb for some popcorn varieties or as large as 60 cm as those grown in the Jala Valley of Mexico (67).

A century ago, all corn was harvested by hand and stored on the cob in drying cribs. Today, one machine that picks and shells in the field can do the work of 50 workers. After harvesting, kernels are dried to less than 15% moisture content to maintain grain quality and prevent long term storage spoilage. Drying must be done carefully to allow the use in wet milling (68,69).

As with most major agricultural commodities, corn economics are tied to governmental policy. These policies, designed to achieve domestic objectives, generally force local prices above free market levels. A worldwide network of grain exchanges provides instantaneous response to perceived changes in supply, demand, and harvest estimates throughout the growing season. Seasonally produced commodities like corn generally are least expensive at harvest and rise in price to compensate for storage charges.

Corn Use in Feed

Although corn was originally grown as food, the single largest use today is as feed for farm animals. In the United States, 80% of the domestic corn usage is for this purpose. Swine consume ca one-third; beef cattle eat one-fourth; and dairy cattle and poultry each account for about one-fifth of the feed corn. Corn is an important feed ingredient because it supplies the energy component and a large portion of the protein input to the animal's diet. It has been a dominant force in feed manufacture, consistent in quality and composition, and never presenting its customers with a shortage of supply.

Also, the co-products of the various industries that use corn to produce beverage alcohol, starch, corn sweeteners, corn oil, and dry milled products provide the concentrated protein and vitamin content of corn, making them valuable feed ingredients as well. Corn gluten feed is a source of additional protein in feeds compounded for beef cattle and dairy herds. The fiber content is readily converted by these ruminants. The xanthophylls in corn gluten meal provide the coloration in chicken and eggs so desired by many United States consumers. The distilling industry provides distillers dried grain while dry millers produce high-fiber, high-calorie hominy feed.

Wet Milling

The United States wet millers buy ca 15% of the corn used in the U.S. (worldwide, wet millers consume about 10% of the corn used). There are two dozen corn wet mills in the United States ranging in capacity from 600–10,000 metric tons/day. Shelled corn is shipped to wet millers by truck, rail, or barge. After cleaning to remove coarse material, ie, cobs, and fines (broken corn, dust, etc), the corn is steeped in a sulfurous acid solution to soften the corn and render the starch granules separable from the protein matrix that envelopes them. About 7% of the kernel's dry substance is leached out during this step, forming protein-rich steep-water, a valuable feed ingredient and fermentation adjunct.

The softened kernels are coarsely ground (first grind) to release the germs. Because of their high oil content, the germs are lighter than the starch, protein, and fiber fractions and can be separated in hydrocyclones. The germs are washed free of remaining starch, dried, and the valuable corn oil is removed by expelling, solvent extraction, or a combination of both. The spent germs are a valuable feed ingredient.

The starch-protein-fiber slurry is subjected to an intense milling to release additional starch from the fiber. The fiber is then wet-screened from the starch-protein slurry, washed free of starch, and dried to form the major component of corn gluten feed—a valuable feed ingredient. The best fiber can be additionally purified to become corn bran, a dietary fiber ingredient that has been shown to lower serum cholesterol and triglycerides (70). The starch-protein slurry is separated into its component parts, again taking advantage of the density difference between the heavier starch and the lighter protein (gluten) particles. Separation is usually done with combinations of disk-nozzle centrifuges and banks of hydrocyclones.

The protein fraction is filtered and dried to become high (60%) protein content corn gluten meal. The starch slurry can be dewatered and dried to produce regular corn starch. Dry starch can be sold as is or heat treated in the presence of acid catalysts to produce dextrins. Or, it is chemically modified before dewatering and drying to produce modified starches used in food and industrial applications. Lastly, it can be hydrolyzed to produce corn sweeteners.

Starch.   Corn starch is a principal ingredient in many food products, providing texture and consistency, as well as energy. More than half of the corn starch sold is used in industrial applications, primarily in paper, textile weaving, adhesives, and coatings. Starch is a polymer consisting of α-linked anhydroglucopyranose units. Two forms exist: amylose is an essentially linear molecule in which the anhydroglucopyranose units are linked almost exclusively via α-1,4 bonds. Amylopectin is a much larger, branched molecule (the mol wt is ca 1,000 times greater than amylose) (71). α-1,4 linkages predominate but there is a significant number of α-1,6 linkages that result in the branched structure. Although the ratio of amylose to amylopectin is quite consistent in normal corn varieties, it varies considerably when starch is obtained from either waxy or high amylose varieties.

When heated in the presence of water to 62–72°C, normal starch granules swell, forming high viscosity pastes or gels. This process is called gelatinization. Starch from normal corns form characteristic firm, opaque gels because of the amylose fraction. The linear molecules align on cooling after gelatinization in a process called retrogradation, forming a thick, rubbery mass. The bushy amylopectin molecules in waxy starch cannot align to form such a mass, resulting in softer, translucent salve-like gels. High amylose starches are difficult to fully gelatinize and provide little viscosity unless cooked above the boiling point of water. These vastly different characteristics are further enhanced by modification of the native granules, resulting in starches with a wide range of properties for industrial and food applications (see also STARCH).

Corn Starch-Based Sweeteners.   Acid or enzyme catalysts can be used to break the linkages between the anhydroglucopyranose units in the starch molecule with the addition of a molecule of water at the break site. This process, called hydrolysis, produces a variety of corn-based sweeteners. The first sweetener from starch (arrowroot, not corn) was produced in Japan in the ninth century. By the nineteenth century, starch sugars were being produced in Europe and the United States. Because glucose is not as sweet as sucrose, products made from corn syrups are not cloyingly sweet, allowing delicate flavors to reach the palate. However, enzymatic isomerization of glucose to fructose produces high fructose corn syrups (HFCS) that are as sweet as sucrose syrups, thus allowing corn sweeteners to replace sugar in liquid applications (such as soft drinks).

To produce sweeteners from corn starch, the starch is gelatinized in the presence of a catalyst under conditions that promote hydrolysis. Acid is usually used to make slightly converted or low dextrose equivalent (DE) syrups. Enzymatic conversion, using thermostable alpha-amylases (for liquefaction), beta-amylases (for maltose production), and glucoamylase (for high glucose content) is widely practiced to produce a full range of saccharide compositions. After the desired degree of hydrolysis is achieved, insolubles are separated by centrifugation or filtration (or both). Soluble impurities are removed using activated carbon or ion exchange (either singly or in combination) before evaporation of the purified syrup to the desired solids concentration. Pure glucose is obtained by crystallization from highly converted starch syrups. Highly converted glucose syrups can also be enzymatically isomerized to 42% (dry basis) fructose content, significantly increasing their sweetness. Further increases in fructose content are possible using chromatographic enrichment (72). Pure fructose is produced by crystallization from syrups enriched to fructose contents above 90% (dry basis).

Corn Oil. The crude corn oil recovered from the germs consists of a mixture of triglycerides, free fatty acids, phospholipids, sterols, tocopherols, waxes, and pigments (73). Refining removes the substances that detract from the quality, resulting in a nearly pure (99%) triglyceride stream. The first refining step is degumming; 1–3% water is added and the dense, hydrated gums are removed by centrifugation. The degummed oil is then refined. Treatment with dilute (12–13 wt %) NaOH forms water soluble soaps of the free fatty acids, allowing centrifugal separation. An alternative physical refining process steam strips the volatile components, primarily free fatty acids.

Pigments are removed by sorption on bleaching clay that is then separated from the oil by filtration. The oil is then winterized by cooling to ca 4°C, precipitating the waxes that are then filtered from the oil. Winterization is not required if the oil is to be hydrogenated. Deodorization, a steam stripping process similar to physical refining, removes volatile impurities, resulting in an oil with lighter color and improved oxidative stability.

Corn oil's flavor, color, stability, retained clarity at refrigerator temperatures, polyunsaturated fatty acid composition, and vitamin E content make it a premium vegetable oil. The major uses are frying or salad applications (50%) and margarine formulations (35%).

Drying Milling of Corn

The United States dry millers buy ca 3% of the corn used in the U.S. There are 71 dry corn mills in 22 states in the United States (74). They tend to be smaller than wet mills; while the largest U.S. dry mill has a capacity of 1,750 metric tons/day, most grind fewer than 250 metric tons/day. About one-fifth of the corn ground is white corn, the remainder being yellow dent. Dry millers use three processes; tempering-degerming, stone-grinding or nondegerming, and alkali-cooking.

Food Uses for Corn

Corn is used directly and ingredients produced from corn are widely used for food in Asia, Africa, North and South America, and parts of the former USSR. Much of the corn consumed in the United States is in the form of ready-to-eat cereal.

Corn as corn flakes, sweet corn, corn as various types of flour and meal, popcorn, other snacks foods such as chips, and corn juice as sweeteners, corn used in fermentation for beer and in the production of alcohol, and corncobs and stalks used as carriers for various chemicals and medications, as fiber sources, and for the improvement of soil condition by plowing under stalks, are some of the uses for this versatile crop. See ref. 75 for more information on corn.

BIBLIOGRAPHY

"Cereals" in *ECT* 1st ed., Vol. 3, pp. 591–634, by W. F. Geddes, University of Minnesota, and F. L. Dunlap, Wallace & Tiernan Co., Inc.; "Wheat" in *ECT* 2nd ed., Vol. 22, pp. 253–307, by Y. Pomeranz, U.S. Department of Agriculture, and M. M. MacMasters, Kansas State University; "Wheat and Other Cereal Grains", *ECT* 3rd ed., Vol. 24, pp. 522–549, by Olaf Mickelsen, Consultant.

1. W. R. Aykroyd and J. Doughty, *Wheat in Human Nutrition*, FAO Nutritional Studies, No. 23, Food and Agricultural Organization, Rome, 1970, p. 1.
2. J. Storck and W. D. Teague, *A History of Milling Flour for Man's Bread*, University of Minnesota Press, Minneapolis, Minn., 1952, p. 27.
3. W. J. Darby, P. Ghalioungiu, and L. Grivetti, *Food: The Gift of Osiris*, Vol. 2, Academic Press, New York, 1977, pp. 492, 493.
4. Ref. 3, p. 460.
5. H. A. Wallace and W. L. Brown, *Corn and Its Early Fathers*, Michigan State University Press, East Lansing, Mich., 1956, p. 35.
6. Ref. 5, p. 33.
7. Ref. 2, p. 37.
8. D. K. Mecham in Y. Pomeranz, ed., *Wheat, Chemistry and Technology*, American Association of Cereal Chemists, Inc., St. Paul, Minn., 1971, p. 395.
9. L. Zeleny in ref. 8, p. 34.
10. S. Davidson, R. Passmore, J. F. Brock, and A. S. Truswell, *Human Nutrition and Dietetics*, 7th ed., Churchill Livingstone, Edinburgh, 1979, p. 43.
11. R. R. Williams, *Toward the Conquest of Beriberi*, Harvard University Press, Cambridge, Mass., 1961, p. 38.
12. Ref. 11, p. 41.
13. K. Y. Guggenheim, *Nutrition and Nutritional Diseases*, D. C. Heath & Co., Lexington, Mass., 1981, p. 172.
14. C. A. Elvehjem, R. J. Madden, F. M. Strong, and D. W. Wooley, *J. Am. Chem. Soc.* **59**, 1767 (1937).
15. E. V. McCollum, *A History of Nutrition*, Houghton Mifflin Co., Boston, Mass., 1957, p. 306.
16. J. B. Mason, N. Gibson, and E. Kodicek, *Br. J. Nutr.* **30**, 297 (1973).
17. A. S. Prasad in A. S. Prasad, ed., *Trace Elements in Human Health and Disease*, Vol. 1, Academic Press, New York, 1976, pp. 1–20.

18. D. Oberleas and A. S. Prasad in ref. 17, pp. 155–162.

19. A. R. P. Walker, F. W. Fox, and J. T. Irving, *Biochem. J.* **42**, 452 (1948).

20. H. H. Mitchell, *Nutr. Rev.* **3**, 130 (1945).

21. A. O'Boyle, T. Watkins, and O. Mickelsen, *Fed. Proc. Fed. Am. Soc. Exp. Biol.* **41**, 394A (1982).

22. S. Bolourchi, J. S. Feurig, and O. Mickelsen, *Am. J. Clin. Nutr.* **21**, 836 (1968).

23. T. Addis, E. Barrett, L. J. Poo, and D. W. Yuen, *J. Clin. Invest.* **26**, 869 (1947).

24. D. D. Makdani, M. Ahmad, B. H. Selleck, D. R. Rovner, J. S. Feurig, and O. Mickelsen, to be published in *Am. J. Clin. Nutr.*

25. R. P. Heaney, *Clin. Obstet. Gynecol.* **19**, 791 (1976).

26. E. T. Jensen and E. Østergaard, *Am. J. Obstet. Gynecol.* **67**, 1094 (1954); D. C. Smith, *N. Eng. J. Med.* **293**, 1164 (1975); T. M. Mack and co-workers, *N. Eng. J. Med.* **294**, 1262 (1976).

27. E. C. Reifenstein and F. Albright, *J. Clin. Invest.* **26**, 24 (1947).

28. T. V. Sanchez, O. Mickelsen, A. G. Marsh, S. M. Garn, and G. H. Mayor, *Proceedings of the Fourth International Conference on Bone Measurement*, National Institute of Health Publication #80-1938, Public Health Services, U.S. Department of Health and Human Services, Washington, D.C., May 1980.

29. O. Mickelsen and D. D. Makdani, *Proceedings of the National Conference on Wheat Utilization Research*, ARS-NC-40, Agricultural Research Services, U.S. Department of Agriculture, Washington, D.C., 1975.

30. O. Mickelsen, D. D. Makdani, R. H. Cotton, S. T. Titcomb, J. C. Colmey, and R. Gatty, *Am. J. Clin. Nutr.* **32**, 1703 (1979).

31. *Pediatrics* **23**, 400 (1981).

32. J. H. Shaw, *J. Am. Med. Assoc.* **166**, 633 (1958).

33. B. E. Gustafsson and co-workers, *Acta Odontol. Scand.* **11**, 232 (1954).

34. B. H. Landis in *Products of the Corn Refining Industry in Food, Seminar Proceedings*, Corn Refiners Association, Washington, D.C., May 9, 1978, pp. 47–54.

35. J. S. Trier in M. H. Sleisenger and J. S. Fortran, eds., *Gastrointestinal Disease, Pathophysiology, Diagnosis, Management*, 2nd ed., Vol. II, W. B. Saunders Co., Philadelphia, Pa., 1978, pp. 1029–1053.

36. W. K. Dicke, H. A. Weijers, and J. H. van de Kamer, *Acta Paediat.* **42**, 34 (1953).

37. U.S. Pat. 1,367,530 (1921), J. C. Baker (to Wallace and Tiernan Co.).

38. *Fed. Regist.* **13**, 6969 (Nov. 27, 1948).

39. H. E. Magee, *Mon. Bull. Minist. Health Public Health Lab. Ser.*, 205 (Sept. 1950).

40. J. C. Drummond and A. Wilbraham, *The Englishman's Food*, Jonathan Cape, London, 1939, pp. 298–299.

41. D. P. Burkitt in D. P. Burkitt and H. C. Trowell, eds., *Refined Carbohydrate Foods and Disease, Some Implications of Dietary Fibre*, Academic Press, London, 1975, pp. 3–20.

42. D. P. Burkitt and H. C. Trowell in ref. 41.

43. T. G. Kiehm, J. W. Anderson, and K. Ward, *Am. J. Clin. Nutr.* **29**, 895 (1976).

44. D. J. A. Jenkins, A. R. Leeds, M. A. Gassull, B. Cochet, and K. G. M. M. Alberti, *Ann. Intern. Med.* **86**, 20 (1977).

45. M. Stasse-Wolthuis, *World Rev. Nutr. Diet.* **36**, 130 (1981).

46. *From Wheat to Flour*, Wheat Flour Institute, Chicago, Ill., 1965.

47. *From Flour to Bread*, Wheat Flour Institute, Chicago, Ill., 1971.

48. *Nature's Best Wheat Foods*, Wheat Foods Council, Washington, D.C., 1984.

49. W. G. Heid, *U.S. Wheat Industry*, Agricultural Economics Report No. 432, U.S. Department of Agricultural Economics, Statistics and Cooperative Service, Washington, D.C., 1979.

50. National Food Review, *The Wheat Grower* **8**(9), 28 (1985).

History

The important role wines and their production have had in ancient times as well as in more modern periods of history, particularly in relation to science and technology, deserves wider appreciation.

Winemaking extends back to unrecorded times, perhaps 8000 BC. It was certainly well developed by 3000 BC in ancient Egypt, Mesopotamia, and the other areas considered the cradle of Western civilization. At the beginning of recorded history, wines were described, their production portrayed, and their properties critically examined. By 2500 BC, the Egyptians had evolved hieroglyphics describing various types of wine. As part of the funerary goods in King Tutankhamon's 1339 BC burial (7), included were wine amphorae stamped with the region of growth, the estate, the vintage, and the winemaker's name. One was 35 years old.

The progenitors of the present wine grapes still grow wild in the Near East. In the wild state, grapevines grow in trees and are propagated by seed-scattering birds. A sugar content higher in the ripe grape than in all the other common fruits adapted grapes especially to attract birds. It also adapted them to ferment into wine. Broken open, grapes and their associated yeast ferment spontaneously to wine. One can imagine a small amount of wine forming briefly in a depression in a rock under wild vines. A passing person would likely find it pleasant and be entranced by the euphoria it produced. The impetus to reproduce this phenomenon must have been very great, in fact and soon proved successful.

A case can be made that winemaking was a cause, not just the result, of the early transition from nomadic to specialized, sedentary agricultural civilization. Cereals can be grown in many places. The plants have a relatively short lifetime, but the ripe grain can be gleaned over a considerable period and is readily transported. Grapes, however, are perennials taking a number of years to develop good crops of fruit from new plantings. Vines need to be pruned and tended to fruit well and must be guarded against depredations by other people as well as by birds and animals, particularly during the short and intense ripening stage. The fruit and wine are perishable.

Compared to grain, wine is less easily portable. Although beer can be made from grain throughout the year, there is the involved but necessary process to hydrolyze the grain's starch before beer can even be made. The grape's sugar, in contrast, is directly fermentable to wine. Both wine and beer were very early beverages, but of the two wine deserves the greater consideration in assessing the origins of agriculture and associated civilization. It is the easiest to make, and was probably the first alcoholic beverage where grapes grew.

In addition to its ancient origins, part of the traditional mysticism about wine relates to its euphoric effect. Certainly this would have seemed magical in earliest times. It contributed to involvement of wine in religion, in rituals and in celebration. This fact today is still reflected in the special ritualistic place accorded wines.

Grape cultivation is very ancient. One knows this from the steps necessary to convert the wild to the domesticated vine as well as from early writings and drawings. The earliest Egyptian tomb paintings show all vines productive with sizable crops. Wild grapevines grow as separate male and female plants, only the latter bearing fruit. Spontaneous mutants are hermaphroditic males, but they usually bear only a small amount of fruit. There were many varieties of grapes described, especially by Roman times. A long period of human intervention would have been required to convert the dioecious wild European vine into a large number of distinguishable *Vitis vinifera* varieties all self-fertile and bearing good yields. The term "variety of grape" signifies a cultivar (cultivated variety) representing a vegetatively propagated selection with specific characteristics (including, perhaps, a distinguishable varietal flavor in its wines) such as Cabernet Sauvignon or Chardonnay (Table 8) analogous to Red Delicious or Granny Smith apples.

Seeds cannot be used to propagate a new vine, if the characteristics of the mother vine are to be maintained. Rather, a clonal vine must be grown by rooting a cutting from a cane of the mother vine. The grapevine is naturally adapted to such propagation. Canes touching the ground easily take root and can be transplanted. A new vine takes at least three years to produce an appreciable amount of fruit from small rootings. Furthermore, if managed by pruning, the vine produces more consistent crops of better quality fruit. Earliest depictions and writings, especially from Rome and Egypt, show a well-developed agriculture, utilizing specialized tools for pruning and otherwise managing vineyards. These and other considerations (8) make wine an important but neglected contributor to the development and maintenance of civilization.

Writings from Ancient Greece and Rome frequently mention wine as part of everyday life (9). The Old and New Testaments contain many references to wine. The universality of knowledge about wine was used in them to make scriptural ideas clear to the common people. For example, Jesus' first miracle involved turning water into wine when the supply of wine ran out at a wedding feast. The account not only shows the importance of wine in the celebration, but also it tells that the guests were surprised the later wine was better than the first. Good wine was recognized but uncommon, and lesser or spoiled wine was served after the guests were less critical.

During the Middle (Dark) Ages, wine and winemaking along with other sacraments and knowledge were husbanded in cloisters and enclaves which maintained the pockets of sophistication that enabled the later flowering of the Renaissance (9,10). Wine became further associated with art, letters, religion, and culture, and it remains so.

The development of modern microbiology resulted from initial studies focusing on wines and their yeasts. Being relatively large compared to bacteria, yeasts were early observed microscopically. Their causative role in fermentation was shown by Pasteur, insofar as sterilization by heating killed them and prevented fermentation as long as they were not reintroduced. The investigation into the mildest conditions needed to achieve this prevention so as to minimize flavor changes, the procedure now called pasteurization, was originally developed for wine, not for milk. Pasteur's work on the so-called diseases of wine, ie, acetic or undesirable malolactic fermentation, etc, led directly to clarification by him and others of the microbiological nature of anthrax and other infectious diseases of humans and other animals.

Biochemistry resulted from the early elucidation of the pathway of enzymatic conversion of glucose to ethanol by yeasts and its relation to carbohydrate metabolism in animals. The word enzyme means "in yeast," and the earlier word ferment has an obvious connection. Partly because of the importance of wine and related products and partly because yeasts are relatively easily studied, yeasts and fermentation were important in early scientific development and still figure widely in studies of biochemical mechanisms, genetic control, cell characteristics, etc. Fermentation yeast was the first eukaryote to have its genome elucidated.

Wine studies were crucial in the development of organic chemistry as well as microbiology and biochemistry. Polarized light was known to be rotated by products from living systems, but not by apparently identical synthesized compounds. The molecular reasons for this were unknown. Tartaric acid [87-69-4] is the major acid of grapes, but is present in very few other plant sources. It occurs in the L(+) form. Racemic acid [133-37-9] was the optically inactive form produced by racemization of the natural acid in alkali, raceme being the term for the kind of stem a grape cluster has, derived from a Latin word for grape clusters. Pasteur found that when the racemic derivative was allowed to crystallize slowly, two mirror-image crystals were produced. Separated by hand, the two forms rotated polarized light in opposite directions. The crystal structure mirrored the asymmetry of the molecules of tartaric acid. From such studies the whole field of stereochemistry was derived. Pasteur became famous for these studies in 1847 at the age of 26. It is said that is was a good thing Pasteur worked under less than ideal conditions, because, if his laboratory temperature had exceeded about 20°C, separate crystals would not have formed.

More recently, studies of wine and beer have initiated techniques of statistically valid sensory analysis. Scientific studies involving wine continue in these areas, building on past discoveries. Natural phenols as desirable dietary components and monitors of storage and aging reactions are currently active fields. Viticultural research, as well as enological, continues to improve grapes and the wines made from them (11).

Classification of Types and Styles of Wine

In classifying wines, many parameters might be used: fruit (species, variety, or condition), composition (color, alcohol, sugar, acid, or carbon dioxide), fermentation procedure (carbonic maceration, alcoholic, aerobic yeast film, malolactic), regulations (taxes, etc), geography, climate, weather (vintage). The last three are considered important in determining prices and reputations for individual wines, but are not very helpful in wine group classification. All of the above, and legally permitted variations, as well as different producer and advertising descriptions contribute to an almost infinite number of kinds of wines.

An effort to categorize all the significant classes of wine in the general form of a dichotomous key follows. A number of all-or-none alternatives are useful here, eg, does or does not the wine have added flavors, added alcohol, red color, significant sweetness, appreciable carbon dioxide, oxidized character, or recognizable flavor from the grape variety. It must be borne in mind that there is considerable leeway among styles within each type of wine. Not every wine fits neatly into such a classification. Regulations, taxation levels, geographical specifics, and tradition impose some limits on wines available in or from a specific locale.

Most wines with <14% alcohol are classed as table wines because they are usually consumed with meals. Note that as used here, premium wines are included. In the countries of the European Union (EU), table wine means only ordinary or everyday wine. Sparkling wines are included in this group because producing the sparkle and retaining it during consumption of a bottle by few people necessitates a modest alcohol level. The "generous" group of wines have had distilled grape spirits added in order to reach, usually, about 18% ethanol. Such dessert and aperitif wines are intended to be consumed after or before a meal. Sweet and otherwise strongly flavored wines have gravitated here because they are more stable and thus can be consumed over several sittings. Other pertinent notes follow the list.

I. "Natural" wines, <14% alcohol. Their nature and keeping qualities have traditionally depended heavily on complete fermentation and protection from air.
   A. Wines without obvious carbon dioxide
      1. Wines containing anthocyanin (red) pigments
         a) Pink wines; (can be made from most red grapes, by limiting extraction).
            (1) Dry: rosé types including Tavel and varietals such as Gamay and Grenache
            (2) Sweet: sweet rosés, "white" Zinfandel, etc
         b) Full red color
            (1) Dry (<0.5% sugar)
               (a) Having usually distinguishable varietal aromas: Barbera, Barolo, Beaujolais, Bordeaux (Médoc, St. Émilion, etc), Burgundy, Cabernet Sauvignon, Châteauneuf-du-Pape, Chelois, Chianti, Concord, Gamay, Hermitage, Petite Syrah, Pinot noir, Pinotage, Rioja, Ruby Cabernet, Syrah (Shiraz), Zinfandel, etc
               (b) Without normally distinguishable varietal aromas: California (etc) dry red table wine (burgundy, claret, chianti), Carignane, Charbono, Cinsaut, Malvoisie, Mourastel, Valdepenas.
            (2) Sweet. California (etc) red table and sweet red table, proprietary types; generally without specific fruit or varietal aroma unless Concord-type or fruit wines (blackberry, raspberry, strawberry, etc)
      2. White wines not containing anthocyanin (red) pigments
         a) With (usually) distinguishable varietal aromas
            (1) Not containing obvious sugar: Catawba, Chablis, Chardonnay, Chenin blanc, Delaware, Gewürztraminer, Graves, Moselle, Muscadet, Pinot blanc, Pinot gris, Rhine (Rheingau, etc.), Sauvignon blanc, Sémillon, Seyval blanc, Viognier, White Riesling, etc
            (2) Containing considerable sugar: "Late Harvest" types of California (etc), German and similar (Auslese, Beerenauslese, Trockenbeerenauslese), Sauternes, Hungarian Tokay, sweet

Catawba, light sweet muscat, sweet Sauvignon blanc, sweet Sémillon, and various proprietary wines

    b) Without (usually) distinguishable varietal aromas

        (1) Not containing obvious sugar: California (etc) dry white table wines (chablis, rhine, etc), Burger, Colombard, Ugni blanc (Trebbiano), etc

        (2) Containing considerable sugar: California (etc) sweet white table wines ("chateau" types, etc), various proprietarily labeled types

  3. Specialty products (partial list)

    a) Flavored table wines: may be white, amber, tawny, or red. Flavors must be from a natural source such as a spice or fruit juices (no single other fruit can be imitated). Formula must be approved by U.S. Bureau of Alcohol, Tobacco and Firearms (BATF): proprietary brands. See also category III. B.

    b) Nouveau-styles: red, prepared by carbonic maceration of the grapes, giving a special flavor; rushed to market: Beaujolais nouveau, California (etc) nouveau

    c) Coolers, refreshment wines: made from white or red wine according to a proprietary formula. Low alcohol, often because of blending with various fruit juices: California Cooler brand and others.

 B. Wines with obvious carbon dioxide

  1. From fermentation of added sugar; usual gauge pressure about 2.026–6.078×10^5 Pa (2–6 atm)

    a) Containing anthocyanins and related (red) pigments

        (1) Pink: sparkling rosé, pink champagnes

        (2) Red: sparkling burgundy, champagne rouge, Cold Duck

    b) Not containing red pigments

        (1) With muscat flavor: moscato spumante, sparkling muscat

        (2) Without muscat flavor: Champagne, California (etc) champagne (bottle or bulk refermented), Cava (Spain), Sekt (Germany), espumante, shampanski, spumante, sparkling Catawba, etc

            (*a*) Below about 1.5% sugar: brut style

            (*b*) Above 1.5% sugar: (increasing) sec, demi-sec, doux

  2. Wines with excess carbon dioxide, not from refermentation of added sugar; Usual gauge pressure about 0.2026–2.026×10^5 Pa (0.2–2.0 atm)

    a) Gassiness from fermentation of residual grape sugar: occasional wines from several countries sometimes called pearl or perle wines, California moscato amabile type

    b) Gassiness from malolactic fermentation: vinhos verdes wines (from Portugal, white and red)

  3. Carbonated wines: relatively rare; may be white or red

II. Wines with 14–17% alcohol

 A. Miscellaneous, white and red, usually sweet types, mainly with proprietary names

 B. Special types

    1. Blending: to increase the alcohol (or other component) or to impart a special character to other wines or even other products, eg, whiskey

    2. Sacramental, kosher, ecclesiastical, etc: usually sweet, eg, *vino santo*, with dessert-type names produces for special markets

    3. Refermented with aerobic yeast: light (white) flor (Fino) sherries, Spanish manzanilla types, California (etc) submerged culture dry sherry. See also III.A.2.*b*) (1) (*b*) i)

III. "Generous" or Fortified Wines, 17–21% alcohol. Their nature and keeping qualities depend heavily on the addition of distilled wine spirits.

  A. Without added food flavoring materials

    1. Containing anthocyanins and related (red) pigments

      a) With a muscat flavor: certain pink and "black" muscatels from Aleatico or Muscat Hamburg, Mavrodaphne, etc

      b) Without a muscat flavor: ruby, tawny, vintage, port-type wines

    2. Not containing red pigments

      a) With a muscat flavor: Australian and California (etc) muscatel, Malaga (also raisiny), Muscat blanc (Muscat Frontignan, Muscat Canelli), Samos, Setubal, Sitges, etc

      b) Without a muscat flavor

        (1) With a special (usually oxidized or heated) odor, the result of treatment or aging

          (*a*) Without film yeast involvement

            i) Raisin-like flavor: usually an overlaid flavor, see 2a)

            ii) Baked odor: Madeira (usually sweet), California (etc) baked sherry (dry, medium, or sweet)

            iii) Cooked down must odor (caramelized): Marsala

            iv) Aged, oxidized, or rancio odor: Oloroso sherries, Banyuls (possibly tawny color from some red grapes), Tarragona (Priorato), etc

          (*b*) With film (aerobic) yeast involvement

            i) Dry (<1% sugar): Spanish (flor or fino, amontillado, etc) sherries, various cocktail sherry-types from Australia, California (etc, including aerobic submerged cultured), Cyprus, South Africa, and the former USSR, Chateau Chalon (France). See also II.B.3.

            ii) Sweet: California medium or sweet flor sherry, some sweet Spanish types

        (2) Without a special odor resulting from treatment or aging

          (*a*) With amber color: Angelica (California), white port (Portugal), etc

          (*b*) Without amber color: California white port

  B. With added (natural) food flavors or plant extracts

    1. With a red color

      a) Proprietary types: includes wines containing quinine and similar additives (Byrrh, Campari, Dubonnet, etc)

      b) Medicinal or homemade types: iron-containing wines, etc

    2. Without a red color

      a) Nearly dry: dry or French-type vermouth

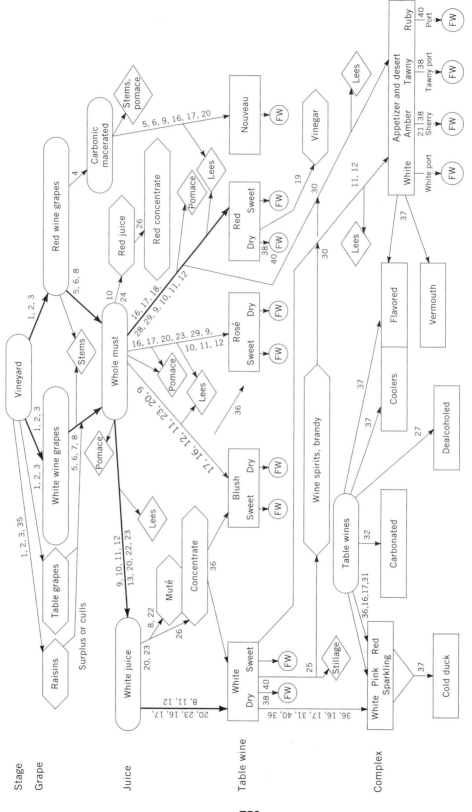

Stage

Grape

Juice

Table wine

Complex

750

Fig. 1. An amplified outline scheme of the making of various wines, alternative products, by-products, and associated wastes (23). Ovals = raw materials, sources; rectangles = wines; hexagon = alternative products (decreasing wine yield); diamond = wastes. To avoid some complexities, eg, all the wine vinegar and all carbonic maceration are indicated as red. This is usual, but not necessarily true. Similarly, malolactic fermentation is desired in some white wines. FW = finished wine and always involves clarification and stabilization, as in 8, 11, 12, 13, 14, 15, 33, 34, followed by 39, 41, 42. It may or may not include maturation (38) or bottle age (40), as indicated for usual styles. Stillage and lees may be treated to recover potassium bitartrate as a by-product. Pomace may also yield red pigment, seed oil, seed tannin, and wine spirits as by-products. Sweet wines are the result of either arresting fermentation at an incomplete stage (by fortification, refrigeration, or other means of yeast inactivation) or addition of juice or concentrate.

Operations or actions:

1. analyze
2. select
3. harvest
4. carbonic maceration
5. destem
6. crush
7. pectinase treat
8. SO_2 addition
9. drain
10. press
11. settle
12. rack
13. centrifuge
14. filter
15. microfilter (microbially stabilize)
16. inoculate (yeast, lactic bacteria)
17. alcoholic fermentation
18. malolactic fermentation
19. acetification
20. protect from air
21. aerate
22. refrigerate
23. temperature control (generally low)
24. heat (pasteurize)
25. distill
26. concentrate (usually vacuum distill)
27. alcohol removal (reverse osmosis, vacuum distill)
28. cap management (punch down, pump over, irrigate)
29. pomace contact
30. fortify (wine spirits addition)
31. champagnization, CO_2 retention
32. carbonation
33. fine, protein stabilize
34. chemically stabilize (refrigerate, ion exchange)
35. dehydrate
36. sweeten
37. blend
38. bulk maturation (barrel, tank)
39. bottle
40. bottle age
41. case
42. market

751

Table 7. California Wine Shipments to All Markets[a]

Year	Total, 10^6 L	Relative amount[b]	Table, %	Sparkling, %	Dessert, %	Vermouth, %	SNW=>14,[b] %	SNW=<14,[b] %	Coolers, %
1985	1453.8	93	65.6	6.8	4.3	0.9	2.9	3.1	15.3
1986	1571.6	100	61.2	6.4	3.7	0.8	2.8	2.7	20.6
1987	1559.3	99	63.0	6.0	3.9	0.8	3.1	2.3	20.0
1988	1505.7	96	67.0	6.2	3.8	0.8	3.2	2.0	16.8
1989	1441.6	92	68.2	6.0	3.4	0.9	3.3	2.9	15.3
1990	1410.1	90	69.2	5.9	3.4	0.8	2.5	2.8	15.4
1991	1324.6	84	71.6	6.2	3.0	0.7	2.7	4.1	11.6
1992	1360.0	87	75.0	6.1	2.9	0.8	2.4	4.5	8.3
1993	1288.1	82	78.6	6.4	2.9	0.8	2.0	4.3	5.0
1994	1298.0	83	81.1	5.8	2.9	0.7	1.6	4.2	3.6

Wine type or parameter

[a]Ref. 6.
[b]1986 = 100.
[c]SNW = Special Natural Wines, flavored with natural flavorings; >< 14% alcohol.

California, which are essentially the same as the U.S. as a whole, are given in Table 8. For the representative years shown, the fraction of the total crush attributable to wine-grape varieties rose greatly, as the fraction attributable to raisin and table-grape varieties fell. These latter varieties, except for the small proportion of muscats, are lacking in distinctive flavor and are made into white dessert (and appetizer) wines, generic white table wines, distilled for beverage and fortifying brandy, and used for juice and concentrate production. The great decrease in winery use of these varieties occurred despite a large increase in consumption of white table wines. This decrease reflects both the greatly decreased interest in appetizer/dessert wines (less base wine and less fortifying spirits needed) and the shift to varietally named premium wines from the less expensive generic wines (red or white table wine, California burgundy, etc). In California, 1993 was the first year total varietal wine shipments exceeded those of generically named wines.

In much of the period shown, white wine varieties gained in share of the total, whereas red wine varieties did not. Trends in production in response to changing consumer interests are also seen in the usage of individual wine-grape varieties as a percentage of all wine varieties crushed (Table 8). Palomino, desirable for sherry, dropped 6-fold. Similarly Carignane and Grenache, often used in port-type wines, dropped among the reds. Also the portion contributed by miscellaneous white and red wine varieties fell as attention shifted to the

Table 8. Grapes Processed in California Wineries by Type and Variety[a]

Parameter	Year				
	1994	1990	1985	1980	1975
varietal type, %					
raisin	9.4	10.3	19.6	26.8	33.6
table grape	4.8	6.6	7.6	6.8	9.4
wine varieties,	85.8	83.0	72.8	66.4	57.0
red wine varieties,	36.9	31.2	30.3	42.8	38.7
white wine varieties,	48.9	51.8	42.5	23.6	18.3
Total, 10^6 t	*2.303*	*2.337*	*2.577*	*2.627*	*1.968*
wine varieties, %, red					
Barbera	3.7	4.3	6.1	8.6	6.9
Cabernet Sauvignon	7.9	4.4	3.5	3.3	3.1
Carignane	3.3	3.9	5.8	9.6	15.8
Grenache	5.4	5.3	6.1	8.8	10.0
Pinot noir	1.5	1.5	1.5	1.3	1.1
Ruby Cabernet	1.9	2.5	3.7	7.5	6.3
Zinfandel	10.3	8.7	6.0	6.0	5.8
other red	9.0	7.0	8.9	19.4	18.9
wine varieties, %, white					
Chardonnay	14.1	7.9	4.0	1.6	0.8
Chenin blanc	9.6	12.9	14.4	8.6	6.4
Colombard	24.8	32.1	28.3	14.8	14.1
Palomino	0.4	0.4	0.7	1.7	2.6
Sauvignon blanc	3.2	2.8	2.7	0.9	0.4
other white	4.9	6.2	8.3	7.9	7.8

[a]Ref. 6 and compiled.

better varieties for table wine. Among the white, Colombard and Chenin blanc peaked early and then gave way a bit for Chardonnay and Sauvignon blanc. Among the red wine varieties, Cabernet Sauvignon, Zinfandel, and Pinot noir have increased at the expense of others less noted for premium wines.

Note most of these varieties make desirably distinctive wines, and the few that usually do not, eg, Carignane and Colombard, have special characteristics such as high yield and relative quality under difficult circumstances. With most vinifera varieties the flavor is subtle and more intensity is often sought. With varieties and hybrids from other species, such as Concord from *Vitis labrusca* or *labruscana*, the flavor may be too strong and soon satiating with a meal.

Varietal labeling is an important quality factor in the United States, and indirectly elsewhere because only certain specific varieties are planted in each prestigious foreign area. U.S. law currently requires that 75% of the wine must come from the *V. vinifera* variety named on the label. Concord-type varieties only require 51%, owing to their intense, distinctive flavor. If more than one variety is named, the relative amounts must total 100%.

Vineyard site is important to wine quality and character and interacts with variety. The general climate must not be too cold, too hot, or too humid. A mild, dry climate that still induces a dormant season, like the Mediterranean area and California, is desirable. A relatively constant weather pattern year-to-year is also sought. The nearer to the limits of cold tolerance, for example, that the climate comes, the more likely are disastrous vintages. The modifying influence of close bodies of water, sun-facing slopes, or frost-resisting air drainage can make one vineyard more desirable than another nearby.

Among satisfactory vineyards, the warmer site or vintage ripens the fruit earlier with more sugar, less acid, and less flavonoids including anthocyanins. If the situation is cool, the fruit may be held on the vine as long as possible to get adequate sugar with as much decrease in acid as possible. In a warm vineyard or vintage, sugar comes up early and fruit should be picked before acidity has declined unduly.

Proper management of the vines and vineyard are important for yield of high quality wine grapes. Pruning of the dormant vine controls the next year's fruit set because cluster primordia for next year's crop are formed during the current year. Pruning canes too long will allow more fruit to start than the photosynthetic capacity of the vine can ripen. Such overcropped fruit develops sufficient sugar for good wine either late or never and has other deficiencies, such as greater loss of acid in hot vineyards owing to delayed harvesting.

There are many details of good vineyard management as there are for any crop, but grapes are relatively tolerant, being perennials with deep roots and not high in water or fertility requirements. A few pests are special, notably grape phylloxera, a root louse native to America, but spread nearly worldwide. It devastates *Vitis vinifera* and requires that the vine be grafted to a resistant rootstock derived from certain native vines of other *Vitis* species.

Having made the most of the available vineyard, chosen the variety and rootstock, and properly managed the vineyard, the winegrower must watch the ripening fruit and determine the optimum stage for harvest. Varieties differ in ripening date and one variety in the same vineyard may ripen at considerably different times in different years. The proper stage of grape ripeness is different for different wines. Sparkling wines need lower sugar and more tartness; dessert wines need more sugar but need to avoid raisining. Acid drops as sugar accumulates, more so in warmer places. At a minimum, the decision to harvest is based on sugar content, acidity, and close observation of the fruit's manner of changing in other parameters, including taste.

Depending upon a number of variables, sugar in the range of about 20–25 Brix is chosen, with sparkling wines at the low end and dessert wines at the high end of this range. Brix is calibrated using solutions of sucrose by weight in water, ie, 1 Brix = 1 g sucrose/100 g of solution. It is measured by hydrometry or refractometry of a juice sample and includes dissolved solids other than sugar, although sugars are on the order of 90% of the value. Higher than about 25 Brix indicates shriveling or raisining (desirable for a few wines); lower than about 20 Brix indicates underripe or overcropped grapes.

Acid content calculated as tartaric acid is about 6–7 g/L for best flavor and stability. It is higher for tart low Brix musts and less important for sweet high Brix musts. High acid levels coincide with a higher level of the second acid of grapes, malic acid.

Must Processing. Ideally the grapes are picked quickly at the chosen stage and transported to the winery so as to not cause unprotected damage to them. Mechanical harvesting is suitable, given adequate prevention of oxidation, premature fermentation, etc. As soon as possible after harvest, the grapes are ordinarily passed through destemmer–crusher machinery to remove the stems and break each berry open. If white table wine is to be made, rapid juice separation is recommended, whereas for red, fermentation of the whole pomace (pulp, skins, seeds) is usual. The method and extent of juice (or later wine) recovery from the crushed grape is important. Free-running, readily separated fluid gives the lighter products; heavier pressing leads to heavier and eventually coarser wines. There are several types and capacities of equipment for destemming, crushing, pumping, pressing, juice clarification, etc (23). Much of it is specially designed for winery use, with diverse types and capacities for special circumstances. General principles common to other food-processing equipment are considered with special care to avoid generating fine dispersal of solids, unwanted incorporation of air, or contribution of contaminants (including iron and copper) to the wines.

Because all the red color is in the epidermis of most wine grapes, proper conditions can give a white or pale pink wine from red grapes. This is the source of the recently popular "white" Zinfandel wine. The must, ie, the mixture to be fermented, is often immediately clarified for white wine by settling or centrifugation. Pectinase enzymes may be added to produce water-clear juice and better juice yield. The must comprises the whole crushed mass (pomace, ie, seeds and skins, plus juice) of red grapes, if the wine is to be red. Frequently, but not invariably, a small amount, 60 mg/L or so, of sulfur dioxide in one form or another is added to minimize any effects of oxidation and inhibit any undesirable microorganisms. If it is added, it should be as or before the grapes are crushed, because an additional effect is to inhibit polyphenol oxidase and browning.

Fermentation. Today (ca 1997) it is almost universal to inoculate the must with a selected yeast strain. Yeasts are chosen for conducting predictable, prompt, and complete fermentations under the conditions applicable for the

ferment if not prevented from doing so. In bottle, gassy and cloudy wines result. Sulfur dioxide is not effective against wine yeasts, but helps inhibit malolactic bacteria. Sorbic acid added at about 150 mg/L can control yeasts, at least long enough for wines marketed quickly. Properly used, dimethyldicarbonate (DMDC, at <200 mg/L) can kill yeasts without undesirable residue. Heat can kill the microorganisms, but flavor change is then a problem. The most satisfactory procedure is to sterilize the wine by very close filtration, but this option requires a high level of expertise to execute. All equipment must be sterilized so that the wine does not encounter any source of viable cells downstream from the filter, until it is finally enclosed in a sterile bottle. Steam, sterilant solutions such as very high ethanol concentrations, etc may be used.

Table wines of higher sweetness can be made by extension of the same treatments, but Sauternes (French), Trockenbeerenauslesen (German), and specialty wines elsewhere are made from shriveled grapes. Such grapes have been colonized by the mold *Botrytis cinerea*, which makes the skin porous. In dry weather the grapes then lose water and the must becomes concentrated. Not only are special, concentrated flavors produced, but also the sugar becomes so high the yeasts cannot ferment it all and the wine remains sweet. The mold metabolizes some of the acid, so that is not concentrated equivalently to the sugar. Raisin flavor is to be avoided in such wines, as it is in the few wines made from partially shade-dried grapes, but it is featured in a few wines.

A third method of making sweet wines is by arresting the fermentation through use of alcohol so that the level reached prevents further fermentation and leaves the wine sweet. Most of the traditional sweet wines of the world, port and other types, are made by such fortification. Apparently owing to osmotic effect, both high sugar and high alcohol inhibit fermentation and more of one can substitute for less of the other. If a nearly sugar-free fermentation is coaxed along by small additions of sugar, alcohol can reach about 18% by volume. Fortifications are made to this minimum level, as a rule. In the United States, the alcohol used must come from the type of fruit making up the wine, ie, in the context being discussed, grape-distilled spirits of high purity, but this is not always the case elsewhere.

Port-type red dessert wines require skin contact time to extract the anthocyanins, but the fermentation must be short to retain the sugar level near the 6–10% level desired. The winemaker cannot always achieve desired composition in individual lots. In order to reach the desired standard, it is necessary to make new lots to enable blending to that standard. The right volume of a redder, less sweet wine will need to be made to bring to standard a lot with low color and more sugar, for example, while keeping the alcohol also within the desired limits.

Oxidized Wines.   Oxidized color and flavor are defects in most wines, but certain ones capitalize upon oxidation. Sherries fall in this group and include wines that are at least moderately high in alcohol and may or may not be sweet. Access to air is required during processing or maturation and may take the form of chemical autoxidation (as in oloroso sherries) or aerobic microbiological metabolism. Aerobically, yeasts produce aldehydes from ethanol and other alcohols in the production of flor sherry types (fino, Manzanilla, etc). This, accompanied by more or less aging in wooden barrels, yeast autolysis, and involved blending results in a complex flavor. Some fino types are relatively light (about 15%) in alcohol, but most of these wines are fortified. Flor sherries are first fermented dry and to about 15% alcohol. This level is high enough to inhibit acetic acid bacteria (also aerobic) and yet allow aerobic growth of yeast, either as a surface film on the wine or submerged in aerated wine.

Other wines falling in this class are amber and usually sweet (Madeira-types, Malaga, Marsala, etc). Their oxidation also results from exposure to air and often long maturation, but may also involve heating, cooking must, or raisining to produce the particular type. Considering the wide variety of interesting oxidized and sweet wines made, it is unfortunate that interest in them has fallen to such a low level currently. The first year post-Prohibition in California that the shipment of fortified dessert wines did not exceed that of the table wines was 1969, and it has fallen steadily since then.

Post-Fermentation Processing. When the wine has been fermented and matured to the desired stage, it is clarified and stabilized so that it will remain clear and not be undesirably changed when bottled and marketed. Processing is held to the essential minimum to avoid flavor change and loss. Microbiological stabilization has already been mentioned, and growth of microbes in bottles is invariably undesirable in modern wines. At the end of fermentation and at each transfer thereafter wines are "racked," ie, the relatively clear portion is transferred without disturbing the yeasts and the other sediments in the lees. Much of the early clarification is accomplished by this method.

Chemical stabilization is considered necessary in the United States to prevent formation of crystals of potassium bitartrate in the bottled wine. This is commonly done by cooling the wine to barely above its freezing point, holding it to allow crystallization, and filtering it cold. Contact with preformed seed crystals and tests to verify that stability has been reached are helpful. Ion-exchange can be used, depending upon the regulations of the country, but if hydrogen ions are exchanged, the acidity is raised. If sodium ions are exchanged for the excess potassium, the nutritional advantage of wine as a low sodium, high potassium food can be lost. Sometimes acidity is adjusted downward by the use of such agents as potassium bicarbonate or calcium carbonate. This procedure also affects stability, and calcium tartrate is relatively insoluble, but slow to precipitate from wine at very low levels. Sometimes in Europe tartrate crystals are tolerated or even welcomed as evidence of genuine wine.

In addition to microbes and tartrate crystals, precipitates and hazes can be formed in wines by proteins, phenolic substances, pectins and related carbohydrates, and combinations thereof. Finer and finer particles are removed by coarse or rough through-polishing and sterilizing filtration. A wide range of types and capacities of filters have been used in the wine industry, often incorporating a precoat or mixed feed, using diatomaceous earth as a filter acid. Finer particles often quickly plug and stop filters, so that a sequence of decreasing porosity filters is often needed.

Incipient hazes may not be removed at all by simple filtration. An array of fining procedures have been developed to achieve stable clarity in such cases. Fining agents are substances that are or become insoluble in wines, and, as they precipitate, adsorb or coprecipitation incipient sources of cloudiness. Properly used, the fining agents themselves are not retained in the wines and their effect is subtractive rather than additive.

Minerals, particularly Bentonite, are used to remove proteins that tend to cause haze in white wines. The natural tannin of red wines usually removes unstable proteins from them. Excess tannin and related phenols can be removed and haze from them prevented by addition of proteins or adsorbents such as polyvinylpyrrolidone. Addition of protein such as gelatin along with tannic acid can even be used to remove other proteins from white wines. Egg whites or albumen are often used to fine red wines. Casein can be used for either process, because it becomes insoluble in acidic solutions like wines.

Because each wine is likely to differ, the treatment is chosen after experimental fining tests to determine the best agent and the minimum satisfactory level. Sediment (lees) volume is held to the minimum level possible in order to avoid extra loss of wine. Usually a single well-chosen treatment is sufficient, and some wines are not fined at all. Proprietary fining agents are usually mixtures and include other agents, such as carbohydrate polymers from seaweed. These and other specialty chemicals sold for wine treatment are not large in tonnage, but are costly because of their special preparation for wine and food use.

Waste Disposal.   Table-wine wineries and other large-volume food processors generate proportionally little waste. It can create a nuisance if allowed to develop off-odors, insect attraction, or oxygen demand in runoff, but generally it is innocuous. Wastes consist of stems, pomace, lees, and stillage from the production of beverage or fortifying brandy. In many instances, potential problems are avoided by frequent removal to and scattering in the vineyards. In others, stillage and wash water particularly, a chain of sealed-bottom ponds with mechanical aeration purify the water. Use in animal feeds of pomace and yeast lees is sometimes practiced. By-product recovery, whether of tartrates, distillate from fermented pomace, red pigment for food coloring, grape-seed oil, or antioxidant flavonoids and tannins, is occasionally warranted.

Storage (Maturation, Aging) and Blending.   One of the prime requisites of an interesting premium wine is complexity of flavor. Maturation (bulk storage and associated final processing), aging (properly speaking, the storage of packaged wine ready for the consumer), and blending are the principal ways of achieving that complexity. Ideally, the varietal and desirable vinous characters of the wine remain prominent and recognizable, but are supplemented by these grace notes.

Maturation is conducted in closed, full containers to prevent oxidation and aerobic growth of microorganisms. Free air contact with low alcohol wine soon leads to vinegar. Except for those sherry types already mentioned, wines are exposed to air minimally and temporarily. During transfers incident to bulk storage and processing, some air exposure is almost inevitable, more in total the longer the wine is held. In the cases of white and pink table wines, it is ordinarily as near zero as possible, and stainless steel or other impermeable containers, inert gas headspace, etc are employed. Red wines withstand and even benefit from small but repeated exposures to air.

In oaken barrels, slow evaporation of water and alcohol through the staves is ordinarily compensated for by refilling and topping each barrel by addition of the required amount of the same wine every week or so. This special process and the transfers necessary in the course of normal processing inevitably slowly permit a little oxidation, but wet wood and cork essentially do not pass oxygen

(23). Oak barrels, unless exhausted by previous extraction, contribute extractives and flavor to wines. Such flavors, unless they overwhelm the other desirable characteristics of the wine in question, contribute desirably to complexity and quality.

Maturation regimes vary from as little change as possible in many white and pink wines (stainless steel tanks, cool storage, minimum time) to considerable modification in red table and a few white table wines. Fermentation and storage in fairly new 200-L barrels for about 6 mo is not uncommon for Chardonnay and white Burgundy wines. Many robust red table wines such as those from Cabernet Sauvignon grapes are often stored similarly, after fermentation and initial clarification, for up to about 3 yr in such barrels.

When they have suitably matured and been fully processed and blended, wines are bottled. Ideally, all the changes which should result during the bulk stage have been completed and the bottling itself should preserve all the desirable characteristics. Freshly bottled, the wine must be aseptic if not sterile, with minimum headspace preferably filled with nitrogen or other inert gas, and well sealed. Changes which are too slow to have been completed continue after bottling, eg, formation of ethyl tartrate, and new reactions dependent on the lowered oxidation–reduction potential can occur. Development in bottle, ie, aging proper as opposed to bulk maturation, is limited by economics to the essential few weeks or months at the winery. It may be continued in merchant and customer cellars for much longer. Special bouquets develop over several years and are the source of endless admiration and debate by the cognoscenti. The specific chemistry has not been completely clarified, but after about four years in bottle a "sun-dried linens" bouquet can be noted in many white table wines and, usually longer, a "fruity-cedar-lacquer" one in certain old red table wines. Both can be very attractive even to the uninitiated, despite some inevitable loss in grape aroma.

Conditions for proper storage of bottled wine include a fairly low and constant (about 13°C) temperature, restricted light and agitation, and bottles stored with corks wetted by the contents. Long aging of red wines often leads to some phenol-derived precipitate which can be avoided by decanting before service. Not every wine improves to the same degree, and complete bouquet development is likely to take 4–8 years for white wines and longer for red.

Regulations

Wine, along with other alcoholic beverages, was nationally prohibited in the United States by Constitutional Amendment during the period 1919 to 1933. Wine did get a unique dispensation to allow production at home for family use of 200 gallons (757 L) per year. This appears to have been in recognition of traditions of wine consumption on the part of many recent immigrants to the United States during that period. Also, there was a very small amount of commercial production for sacramental or medicinal use, but winemaking outside the home was essentially defunct. That it has been so quickly reborn and now thrives is a great tribute to the resiliency of American winemakers and their willingness to innovate.

When federal prohibition was repealed, states and localities were allowed to retain local prohibition. Today most of the states have similar laws to the federal ones, except for the imposition of additional taxes. A state can be more restrictive, but not more lenient, because then federal law would supervene. Although universal prohibition has long since been abolished, alcoholic beverages remain subject to very stringent controls and a high level of taxation, compared to other food industries. Their production is described as a permissive industry, ie, unless specifically permitted, all practices are prohibited. Some attitudes in the United States are still to an extent, holdovers from the Prohibition era; for example, a U.S. citizen can marry or become a soldier at age 18, but cannot legally consume alcohol until age 21. In the European Union, in contrast, 16 is generally the minimum legal age for drinking alcoholic beverages.

A special agency now called the Bureau of Alcohol, Tobacco, and Firearms (BATF) within the U.S. Treasury Department was empowered to regulate the alcoholic beverage industries. Although less adversarial, but strictly enforced even today (ca 1997), the regulations and their application remain voluminous and detailed. They specify not only label compliance and matters relating to taxation that are of direct interest to consumers, but contain all the details of permitted processes for and additions to wines.

The regulations have several underlying purposes: to hold composition within specified boundaries, to allow good commercial winemaking practices, to assure identity and authenticity of the product, to prevent unfair or fraudulent practices for the benefit of both the consumer and the conscientious producer, to ensure healthfulness, and to produce tax revenue. BATF regulations are not identical to those in other countries, but the United States has generally allowed import of wines made legally in other countries as long as they are properly labeled and healthfulness has not been compromised. The reverse is often not the case.

The U.S. regulations are, in effect, being continually considered for change, but the requirements for holding public hearings, the likelihood of predictable opposition, etc, make changes slow. The most recent sweeping change in the law occurred in 1990, when the entire wine code was revised and updated (25). Amendments have been (26,27) and will continue to be made, as industry practices and the prevailing legal climate evolve and change over time.

Designation was allowed for specific vineyard areas other than states and counties in 1978. These appellations have now grown to about 150 nationwide and include legally defined "Napa Valley," "Sonoma Valley," subdivisions thereof, etc, designations. A wine may be labeled as the product of a politically defined area (state, county), if 75% of the grapes from which it was made grew there. If they did not, it is labeled "American." If a specific approved appellation is designated, eg, Napa Valley, the minimum requirement is 85% from that area. If a specific vineyard estate is named, 95% of the wine must be from it. If the wine is labeled with a vintage date, 95% of the wine must have been fermented that year. Opportunity for blending of wines having these labels is severely restricted, but diversity among commercial wines is promoted.

Codification to protect regional names for wines was perfected in France and, by treaty, extended for their wines internationally under *appelation d'origine contrôlée* nomenclature (AOC, or just AC). In general these regulations delimit the region protected, limit the varieties of grapes that can be used, restrict the wine production per hectare, and require approved enological practices. Several other countries, notably Italy, Germany, and South Africa, have adopted somewhat similar regulations. It should be clearly noted that these regulations are aimed at guaranteeing type and authenticity, not necessarily quality, although there is a panel that disapproves a few wines most years which are outside the permitted range for the region.

Under these systems for assuring authenticity in regional names for wine, the larger region indicated is assumed to produce the normal minimum standard wine, and smaller subdivisions within the regions customarily attract more avid followers and command higher prices. To give a very simple example, Bordeaux includes any wine produced in the delimited district of Bordeaux. Delimited subdistricts of Bordeaux like the Médoc are further divided into specific properties, called there châteaux. The proprietor of a specific château zealously guards its reputation. If the wine of a given vintage from that property does not come up to the expected quality, it can be labeled and sold as Médoc, and if it is of an even more questionable quality, as Bordeaux. Owing to vagaries of vintage and the changing economic straits of owners, not to mention differences of style preference, quality fluctuates even if authenticity does not.

Every aspect of winemaking and marketing is regulated; to fully enumerate these regulations is well beyond the scope of this article. In the United States, winemakers must meet all requirements even before beginning production and must follow in detail the regulations of the BATF, particularly as listed in Title 27 of the *U.S. Code of Federal Regulations* (28). A permit must be obtained and the winery bonded before the making of wine for sale may be begun. The winery must be posted, equipped, and secured. Federal Inspectors must be allowed to make unannounced inspections. Of course, applicable state and local regulations must also be followed. For regulations in force within or those applying to the exporting of wine to European countries, the Office International de la Vigne et du Vin in Paris or the laws of the particular country in question should be consulted.

Compositional aspects that are regulated include a label statement within $\pm 1.5\%$ for table wine (for dessert wines, $\pm 1.0\%$) of the wine's alcohol content. For tax identity reasons, alcohol of $7-14\%$ is required for table wines and $17-21\%$ for dessert and appetizer wines. Federal excise tax rates are \$0.28/L for table wines, \$0.41/L for wines $14-21\%$ alcohol, and \$0.83/L for those (rare) $>21-24\%$. Coolers made with wine become taxable at 0.5% and are taxed as table wine above 7% alcohol. Sparkling wine is taxed at \$0.90/L and carbonated at \$0.87/L. The borderline to incur these taxes is CO_2 above 3.92 g/L.

BATF permits no more acetic acid than 1.4 g/L in red table and 1.2 g/L in white and dessert wines, California and the European Union slightly less. California requires a minimum fixed acidity as tartaric of 4.0 g/L for red table, 3.0 g/L for white table, and 2.5 g/L for dessert wines. California also requires a minimum extract in dry wines of 18 g/L for red and 17 g/L for white, but other states generally do not specify a minimum. In the United States, maximum total sulfur dioxide is 350 mg/L. Far less is usually used today. European maxima are lower for dry wines and higher for sweet table wines.

Regulations specify a considerable list of additives and treatments which may be permitted under controlled limits and conditions. It is important to note that no wine receives more than a few of these treatments, and many have none.

Most fibers develop so that their strongest, lengthwise direction is parallel to the vertical direction of the stem. A substance called lignin is formed to reinforce and bond the fibers to one another. As the tree grows, these layers of fibers develop outward and upward to form the stem of the tree. These natural, renewable structures (fibers and wood) are the basis of virtually all major uses of forest products.

As with all living things, trees have a finite lifespan. Only a few species will flourish for more than 100 years, the majority passing into old age at 30 to 80 years. Because wood, wood products, and paper make up such an important and integral part of our lives, it is equally important to have well-conceived forest management plans that provide for growth management, timely removals, and speedy regeneration. Only in this way can adequate supplies for future generations be assured. Fortunately, the United States is endowed with some of the best lands and climate for forest production in the world. Properly managed, these lands could supply the future needs not only of the United States but a significant portion of the needs of other parts of the world as well. In addition, these lands can provide significant employment and resultant contributions to the economy.

HISTORICAL DEVELOPMENT

Wood is one of the oldest construction materials in human use and continues to be an extremely valuable material to this day. Wood is lightweight, strong, stable, easily worked and fastened, a good insulator, warm to the touch, and pleasing in appearance, among many other attributes.

Logs and rough beams were probably the first uses of wood, employed in large construction projects as soon as simple tools became available. Prior to the industrial revolution, sawn lumber was available only in limited quantities. Paper, also in small quantities, was made from cotton rags, or papyrus, the inner bark of some species of trees, or rice straw in China and Japan. The advent of large machinery allowed production of lumber as a commodity. The manufacture of large quantities of wood veneer followed quickly. The discovery of chemical pulping of wood to separate the fibers and lignin coincided with the availability of large machinery to make continuous lengths of paper. Thus, the wood products and paper industries began, without which a modern way of life would have been impossible to achieve.

The vast natural resources of timber in the United States, through the foresight of early foresters and political leaders, led to the availability of large quantities of lumber at reasonable cost. This abundance has made the United States the best-housed and most comfortable of all the areas of the world. However, in the conversion of timber to lumber or veneer, only about one-half of the volume of the log actually becomes lumber or veneer. The remainder is in the form of bark, slabs, edgings, sawdust, planer shavings, veneer round-up waste and clippings, veneer cores, and plywood trim waste. For some years, these residues were largely unused except for small amounts diverted to pulp and paper mills. Then, in the early 1900s, various developers and entrepreneurs began discovering new and useful products which could be made from these residues.

Today (ca 1997), virtually the entire woody stem of the tree is converted to a family of wood and wood-based products. These developments have literally doubled, and perhaps with recycling, more than doubled the yield of usable products from a given volume of wood, contributing even more greatly to the comfort of ordinary people and to the economy. The products that have resulted are treated herein in chronological order of development as much as possible. Inasmuch as most of these products are bonded together with adhesives, the major types of wood-bonding adhesives are also treated briefly.

Adhesives for Wood

The following list includes most of the adhesives which can be or have been used in wood and wood composite bonding applications.

Natural adhesives

> starch
>
> soy flour
>
> blood
>
> casein
>
> hide/bone

Synthetic adhesives

> urea–formaldehyde
>
> melamine–formaldehyde
>
> phenol–formaldehyde
>
> resorcinol–formaldehyde
>
> isocyanate/urethanes
>
> poly(vinyl acetate) (PVA)
>
> contact adhesives
>
> epoxy resins
>
> hot-melt adhesives

In the natural adhesive group, only limited quantities are still used today, largely in conjunction with a synthetic adhesive where the purpose of the mixture is to improve some characteristic of the adhesive. In early plywood hot-pressing applications, blood and starch or soy flour were used. Casein, a milk-based product, was a long-time favorite adhesive for cold-pressed plywood and lumber laminates. Soy flour glues were used in the early years of cold-pressed softwood plywood. Hide and bone glues constituted the original version of the hot-melt glue and were widely used as assembly glues in furniture construction. A disadvantage of the natural glues was that they were suitable only for interior uses, where exposure to liquid water would not be encountered. Casein and blood-soya glues were the most water-resistant, but neither could be classed as truly suitable for exterior exposure.

Synthetic adhesives have resulted from developments in the field of organic chemistry. Virtually all are obtained from petroleum, natural gas, or coal by-products.

Formaldehyde, HCHO, is a primary and necessary constituent of the first five synthetic adhesives in the listing. It is a simple organic chemical first identified during the latter half of the 1800s. Its irritating and toxic odor and preservative properties were known from the time of its early development. It is a ubiquitous chemical, formed naturally in small quantities by every process of incomplete combustion as well as in normal biologic processes. The human body has a natural formaldehyde level of about 3 μg/g, ie, 3 parts per million (ppm) in the blood at all times.

During the late 1970s, concerns were raised about levels of airborne formaldehyde in buildings resulting primarily from construction using composite panels bonded with urea–formaldehyde resins and combined with energy-efficient building practices which reduced air losses.

In an effort to address these concerns, regulations and guidelines were established. The U.S. Department of Housing and Urban Development (HUD) regulation 24CFR 3280.208, *Manufactured Home Construction and Safety Standard*, promulgated in 1985 (1), required emissions of 0.2 μL/L (ppm) or less from hardwood plywood and 0.3 μL/L (ppm) or less from particleboards, when both products are made with urea–formaldehyde adhesives and ultimately used in manufactured homes. The industries implemented voluntary standards, and formaldehyde emission levels were significantly reduced. In 1993, the National Particleboard Association voluntarily reduced emissions to 0.2 μL/L (ppm) or less for particleboards used in flooring applications. Airborne formaldehyde levels in most studies of conventional homes have shown average levels of less than 0.1 μL/L (ppm), a level which also is generally accepted by the U.S. Environmental Protection Agency as a level of concern. It should be noted that the average person cannot detect formaldehyde in air below levels of about 0.5 μL/L (ppm), although there have been reports of hypersensitive individuals. Another factor which should ease concern about formaldehyde is that most emissions occur as a result of formaldehyde trapped in the product during manufacture. As the formaldehyde is slowly released over time the natural emission rates decrease, resulting over time in decreasing levels in the surrounding air. Today there should be little concern for irritation problems from formaldehyde in most current construction and there should be no concern about cancer caused by breathing formaldehyde in normal air.

Phenol–formaldehyde (PF) was the first of the synthetic adhesives developed. By combining phenol with formaldehyde, which has exceptional cross-linking abilities with many chemicals and materials, and a small amount of sodium hydroxide, a resin was obtained. The first resins solidified as they cooled, and it was discovered that if it was ground to a powder with a small amount of additional formaldehyde and the application of more heat, the mixture would liquify and then convert to a permanently hard material. Upon combination of the powdered resin mixture with a filler material such as wood flour, the result then being placed in a mold and pressed under heat and pressure, a hard, durable, black plastic material was found to result. For many years these resulting products were called Bakelite, the trade name of the inventor. Bakelite products are still produced today, but this use accounts for only a small portion of the PF resins used.

Subsequently it was discovered that liquid PF resins could be produced and used as binders for other products. The primary uses for one class of PF resins is in foundry applications where the resin is employed in binding the sands used in the molds. The resin burns away during the molding process, allowing removal and recycling of the sand. Today the major use of PF resins is in bonded wood products, ie, plywood, hardboard, hardboard siding, oriented strandboard, and waferboard. PF resins produce strong, durable, and waterproof bonds; primary characteristics required in any product which may be exposed to the weather or to liquid water for significant time periods. PF resins generally produce a dark-colored glueline, require hot-pressing to cure, and are much more expensive than the popular urea-based adhesives used for interior exposures.

Urea–formaldehyde (UF) was also developed as an early synthetic adhesive by combining the two chemicals under carefully controlled heating (cooking) conditions. Small amounts of other chemicals are added to obtain other desired characteristics in the resin adhesives. Urea–formaldehyde resins are inexpensive, fast-bonding, produce a colorless glueline, are water-resistant, and tolerate a fairly wide range of bonding conditions. These resins are an almost ideal interior-grade adhesive, with the single exception of formaldehyde emissions. However, developments in improved resin chemistry have reduced potential emissions to unnoticeable levels, with a possible exception occurring in the case of a few hypersensitive individuals. Urea–formaldehyde resins cure at acid conditions which means that they can cure at room temperature with the simple addition of a small amount of acid catalyst, or more quickly under heat and pressure using the natural acidity of the wood as a catalyst.

Both melamine–formaldehyde (MF) and resorcinol–formaldehyde (RF) followed the earlier developments of phenol–, and urea–formaldehyde. Melamine has a more complex structure than urea and is also more expensive. Melamine-base resins require heat to cure, produce colorless gluelines, and are much more water-resistant than urea resins but still are not quite waterproof. Because of melamine's similarity to urea, it is often used in fairly small amounts with urea to produce melamine–urea–formaldehyde (MUF) resins. Thus, the improved characteristics of melamine can be combined with the economy of urea to provide an improved adhesive at a moderate increase in cost. The improvement is roughly proportional to the amount of melamine used; the range of addition may be from 5 to 35%, with 5–10% most common.

Resorcinol is to phenol as melamine is to urea. Resorcinol–formaldehyde (RF) is very expensive, produces dark and waterproof gluelines, but will cure at room temperature. As with melamine and urea, resorcinol is often combined with phenol to produce phenol–resorcinol–formaldehyde (PRF) adhesives, thus producing an excellent adhesive with some of the economy of phenol. These adhesives are the mainstay of the laminated timber industry which generally requires a room-temperature cure with durable, waterproof gluelines.

Another class of adhesives has been gaining more widespread use in the wood laminates and composites industries. These adhesives are isocyanates, the common name of a chemical group of diphenylmethane-p,p-diisocyanate polymers. They are excellent adhesives, fast-curing, and will produce durable, waterproof bonds if used in sufficient quantity. Considerations which have slowed wider usage of isocyanates are high cost (about 6× urea resins and 3× phenol resins); adhesive qualities which bonds to press plates as easily as to wood, thus requiring a totally reliable press release additive system; and finally, their toxic characteristics require a high level of manufacturing safeguards. The liquid adhesive should not contact the skin, and whereas only very minor amounts of odorless emissions occur during application and curing, about 4% of exposed persons are sensitive to these fumes and may develop a life-threatening asthmatic condition on continued exposure. However, their fast curing rate offsets some of the higher cost as compared to phenolics, and the other primary concerns can be handled by special care in monitoring, handling, and dust/ventilation controls. Provided they are used properly and safely, isocyanates have been found to be excellent bonding agents.

Four other groups of synthetic adhesives find uses in secondary processing, ie, overlaying, assembly gluing, etc, and in furniture and cabinet manufacture. Poly(vinyl acetate) (PVA) adhesives are widely used in application of veneers and other overlays to panel substrates and in some unit-assembly operations. PVA adhesives are an emulsion of polyvinyl acetate in water and cure by loss of water. The PVA adhesives are somewhat expensive, but are extremely versatile, cure at room temperature, and provide a colorless glueline. The curing reaction time may be decreased by addition of mild heat, or by using a catalyst in the case of special types of PVA resins. Table 1 lists the chemical formulas of all of the preceding chemicals widely used in wood-bonding applications.

Another widely used overlay adhesive is the contact type. These specialized adhesives, in the same group as rubber cement, may be of the solvent-base or water-base types. They are often used to bond overlays such as wood veneer, vinyl (poly(vinyl chloride)) films, or high pressure laminates such as countertop overlays.

Epoxy resins are also used in special applications, such as an overlaying procedure requiring a durable, heat-resistant bond of a difficult-to-bond overlay on a wood-base panel substrate. Metal sheets used as overlays, for example, often require an epoxy adhesive.

Finally, a large variety of hot-melt or thermoplastic adhesives have been developed in recent years. These are solid at normal temperatures, but melt and flow if heated and resolidify when cooled. A wide range of melting points and bond strengths are available, depending on the requirements of the application. These adhesives are widely used in furniture and cabinet construction and have largely replaced the hide/bone glues formerly used in these applications.

Assuming correct adhesive application and bonding conditions, it is generally agreed that the bonding mechanisms are the result of forces of molecular attraction between the adhesive and the wood surfaces. Any normal structural wood adhesive should be capable of producing bonds which are stronger than the shear strength of the wood. Treatments of the manufacture and uses of the various composite and laminated products given herein provide more specific details on the applications and uses of adhesives.

Table 1. Chemical Structures of Constituents of Principal Wood Adhesives

Name	Chemical formula	Chemical structure
formaldehyde	(HCHO)	$\begin{matrix} H \\ H \end{matrix}\!\!>\!C{=}O$
urea	(H_2NCONH_2)	$\begin{matrix} H_2N \\ H_2N \end{matrix}\!\!>\!C{=}O$
phenol	(C_6H_5OH)	OH on benzene ring
melamine	$(C_3H_6N_6)$	triazine ring with three NH_2 groups
resorcinol	$(1,3\text{-}(HO)_2C_6H_4)$	benzene ring with two OH groups (1,3)
isocyanate	$(NCO)_2CH_2(C_6H_4)_2$	$OCN\text{—}\!\!\bigcirc\!\!\text{—}CH_2\text{—}\!\!\bigcirc\!\!\text{—}NCO$
poly(vinyl acetate)	$(CH_3CO_2CH{=}CH_2)_n$	$CH_2{=}\overset{H}{C}OOCH_3$

MANUFACTURING PROCESSES

SAFETY

An extremely important safety issue with respect to all wood product manufacturing processes is personal worker safety. All of the processes use much moving machinery, usually including many saws or knives. Workers must continually remember the inherent dangers these machines involve as well as other possible dangerous situations which could result from malfunctions or other errors. In addition, most processes are more or less dusty and noisy. Most employers require use of safety glasses and many require hearing protection, safety shoes, and hardhats as well as other kinds of protection needed for specific jobs.

Plywood

Plywood is a panel made from wood veneers (thin slices or sheets) bonded to one another. Generally each ply is oriented at right angles to the adjacent ply, and the two face plies should have the grain direction parallel to each other. Thus most plywood will have an uneven number of plies, such as 3, 5, 7, or more.

SOFTWOOD PLYWOOD

Softwood is generally considered to be that coming from a coniferous tree, ie, an evergreen tree having needle-like or scale-like leaves. There are exceptions to the evergreen rule, however. In addition, many hardwoods also may now be used in softwood plywood as core veneers.

Softwood Plywood Processing and Products. The first softwood plywood was made in 1905 using Douglas-fir veneer, soya flour adhesive, and cold-press system. Figure 3 illustrates the traditional softwood plywood process. Veneer logs are kept wet until cut to peeler block lengths, about 2500 mm (100 in.). The peeler block may be stored in cold or hot water, or placed in large, steam-heated rooms until it is removed to be used for veneer. The block is placed in a charger, which quickly loads and positions the block into the lathe after the previous block is finished. The charger electronically scans each block and determines a point at each end of the log which will be centered in the lathe and will allow the optimum yield of veneer from the block.

After the block is chucked in the lathe, the lathe turns the block against the knife and peels the veneer in a continuous sheet as the knife moves toward the center of the block. When the knife cannot advance further without moving into the metal chucks, the lathe is stopped, the core of the block is dropped, the lathe is recharged, and the cycle repeated.

As the veneer leaves the lathe, it moves into a series of trays, long storage conveyors which provide short-term storage while the veneer moves through the clipper. The clipper is a high speed knife which chops the veneer into sheets which will maximize the yield of usable veneer. Unusable veneer is chopped into chips for use in paper or composites. The veneer sheets move along another long conveyer table where they are pulled and sorted into stacks by grade and width. The stacked veneer is moved to the dryer area where the sheets are placed end-to-end and moved through the long conveyor dryer. Hot air heated by steam, oil, or gas is blown onto the surfaces of the veneer to hasten the drying process. As the veneer emerges from the dryer at a desired moisture content (mc) of 3–5%, a moisture sensor marks those veneers which are not dry enough. These will be redried. The dry veneers are again sorted by grade and width and moved to the splicing and patching area. Here the narrow veneers are spliced together into full-size sheets. Some stacks of narrow veneers are cut in half, from 2550 mm (ca 8 ft.) to 1250 mm (ca 4 ft.) to make veneers for cross-plies. Some veneers are patched to make higher quality sheets for face veneers. The patching machines punch out defects such as large knots and refill the hole with a similarly sized piece of clear veneer. Then the stacks of veneer are moved to the gluing and lay-up area. Plywood manufacture is a labor-intensive operation. No other wood-base composite manufacturing operation requires as much hand labor, which is excellent from the viewpoint of providing jobs, but contributes significantly to the final cost of the end-product.

The adhesive used in virtually all softwood plywood has a phenol–formaldehyde (PF) base to provide an exterior-grade, durable, waterproof bond. Thus, most grades of plywood can be used in structural applications. A very small percentage of softwood plywood is made using interior-grade adhesive systems, and this material is used in interior cabinetry, furniture, and shelving.

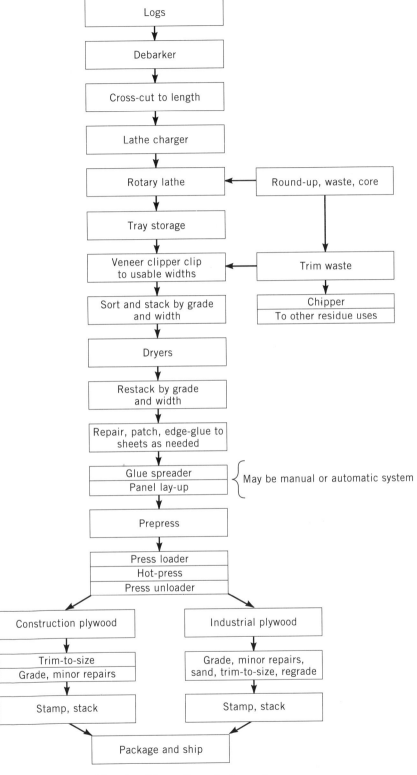

Fig. 3. The softwood plywood process.

workmanship, scarf and finger-jointed panels, dimensional tolerances, moisture content, and packaging and loading

Testing—includes test specimen preparation, bond durability tests, and structural performance tests. It should be noted that formaldehyde emission tests of phenolic bonded products such as structural plywood are not required because emissions are normally about 0.02–.03 μl/L (ppm), well below the previously noted safe level of 0.10 μL/L (ppm).

Grademarking and certification—includes certification procedures performed by a qualified inspection and testing agency, and definitions of panel marking requirements

Insulation Board and Structural Fiberboards

Production, Processing and Shipment. *Insulation Board.* The panel products known as insulation board were the earliest commodity products made from fibers or particles in the composite panel area. These are fiber-base products with a density less than 500 kg/m^3. Early U.S. patents were obtained in 1915 and production began soon thereafter. The initial production used wood fiber as a raw material, but later products were made of recycled paper, bagasse (sugar cane residue), and straw. Schematics of the two major processes still in use are shown in Figure 4.

In the process using wood residues, the raw material (usually wood chips) is passed through a chip washer to remove extraneous materials such as dirt, sand, stones, or metal. The washed chips move to a refiner where the chips pass between rotating steel disks with molded patterns in the faces of the disks. The disk faces are maintained at a small, controlled distance, 0.3–0.5 mm, which results in conversion of the chips into fiber and small fiber bundles as they pass between the disks. The fiber drops into a stock chest, where it is mixed with a large quantity of water. Large paddle mixers are used to keep the fiber in a uniform suspension. This fiber slurry is then moved to a smaller stock chest, where additives are metered into the slurry. Additives may be wax emulsion, starch or PF resin as binder, and alum. Alum is used to change the acidity (pH) of the slurry, which causes the PF resin to precipitate from solution and deposit onto the fiber. One process uses a linseed oil emulsion as a binder. Only small amounts of binder are required in these fiberboards because most bonding is the result of hydrogen bonding, similar to that used in paper-making.

The fiber slurry is pumped to the mat-forming box of a wet-process forming machine. This is a modification of the Fourdrinier machine, widely used in the paper industry. A wide endless wire screen belt, usually about 2540 mm (100 in.) wide, moves under the flow of fiber and water. The water drains through the fiber and through the belt to be recirculated back to the stock chest. The fibers are deposited on the belt at a rate consistent with the speed of the belt to produce a finished panel of the desired thickness and density. The belt then passes over a number of suction boxes, in which a partial vacuum is maintained. The air drawn through the fiber mat draws more water from the mat through the screen and through a series of holes in the top of the suction boxes, further consolidating the mat as well as removing excess moisture. It should be noted that this process

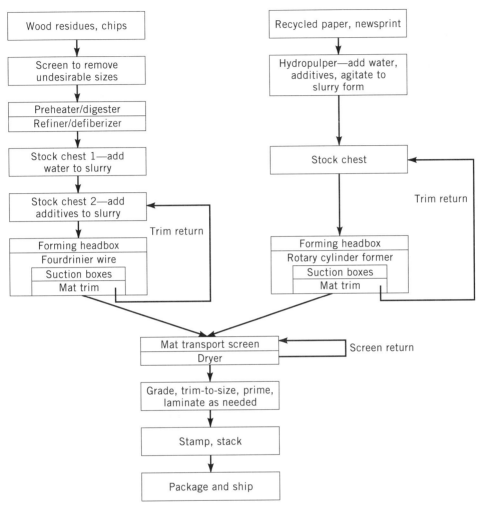

Fig. 4. Insulation board processes.

uses large volumes of water. Whereas most of the water is recirculated, current regulations concerning clean water have hampered manufacturers, ensuring that in all probability no more of these plants will be built.

The formed mat is carried along on the wire screen and is trimmed to the desired width and length by high pressure water jets. The mats then pass onto another endless wire screen which carries the mats into a long heated drying chamber. The mat may contain 200–300% moisture based on the dry weight of the mat as it enters the dryer. Fans circulate hot air within the dryer, a procedure which removes the majority of the water from the fiber. Also, bonding occurs at points within the mat where individual fibers or fiber bundles touch one another. Some of the bonding is a result of hydrogen bonding at fiber-to-fiber interfaces and some bonding is a result of the presence of additives at these points. As the dry (5–7% mc) mat emerges from the dryer, it is already a useful product, having its own inherent properties and characteristics.

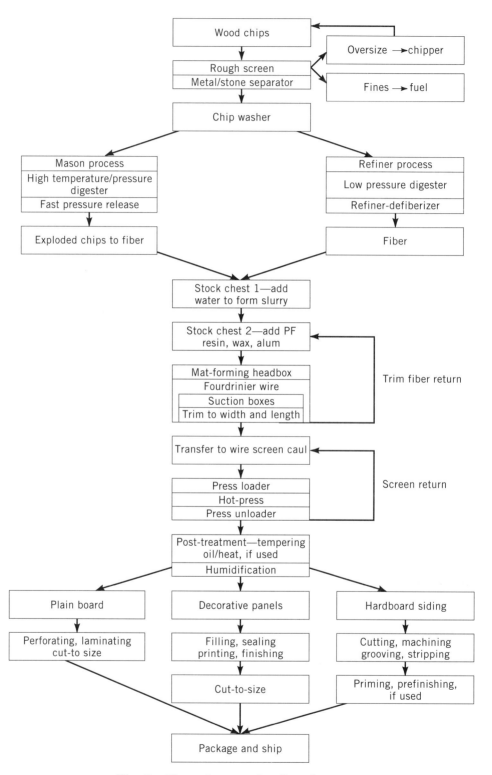

Fig. 5. The wet-process hardboard process.

softening and transferring of lignin under moist heat conditions and then drying on adjacent fiber surfaces as the mat dries in the press. The other bonding process is similar to that occurring in paper manufacturer, where adjacent, reactive hydroxyl, −OH, units bond to one another and other reactive sites during the drying process. To aid in removal of water and steam during pressing, it is a common practice to use a breathe cycle. This is done by quickly reducing the press pressure to the point where only a minimal pressure is exerted on the board. Steam escapes with a loud squeal as it passes between the board surface and the press platen surface. After a few seconds, the press is quickly returned to high pressure and remains at proper thickness position for the remainder of the press cycle. It is critical to insert the breathe cycle at the optimum point in the press cycle in order to remove as much moisture as possible in a short time period and then return the press to position before the resin binder cures. Then, when the pressed mats are dry and the press is opened, strong, stable, and durable panels are removed from the press without the requirement of added binder. The original Masonite process, used for many years, followed this procedure. It should be noted that both PF resin and wax were added in small amounts in later years to provide improvements in strength, stability, and durability. The Masonite process is an example of a wet–wet hardboard process; that is, the mat is formed wet and pressed wet.

Wet-Process Hardboard From Refined Fiber. This hardboard is another form of wet–wet hardboard. The principal difference between this process and the Masonite process lies in the method of preparing fiber and the additives used in the boards. Fiber is prepared by processing wood chips through a pressurized refiner. In this machine, chips are metered into a chamber under steam pressure at 520–830 kPa (75–120 psi). As the chips move slowly through the chamber, they are heated and softened by the steam. They then move from the pressurized cylinder (digestor) into the refiner, where they pass between serrated discs rotating at high speed and maintained in close proximity. One or both discs may rotate, depending on the design of the refiner, although most modern refiners use only one drive motor and rotating disc. The rubbing action between the discs separates the chips into fibers, which are actually bundles of fibers mixed with individual fibers. The fibers move out of the refiner into a stock chest, where they are mixed with water into a fiber slurry.

The slurry is pumped into another stock chest, where wax in emulsion form, usually about 0.5–1.0% wax-to-fiber weight, and 1–3% PF resin are added. PF resin is also added on the basis of resin solids-to-dry fiber. Then a small amount of alum is added, which changes the pH (acidity) of the slurry, causing the resin to precipitate from solution and deposit on the fibers. Resin is required in greater quantity than in the Masonite process because only light bonding occurs between fibers prepared in a refiner. The fiber slurry is then pumped to the headbox of a Fourdrinier mat former, and from this point the process is similar to the Masonite process.

Dry-Process Hardboard. Dry-process hardboard is produced by a dry-dry system where dry fiber is formed into mats, which are then pressed in a dry condition. A flow diagram of this process is shown in Figure 6. In this process, wood chips, sawdust, or other residues are refined to fiber in pressurized refiners. Wax and PF resin may be added in the refiner or immediately outside

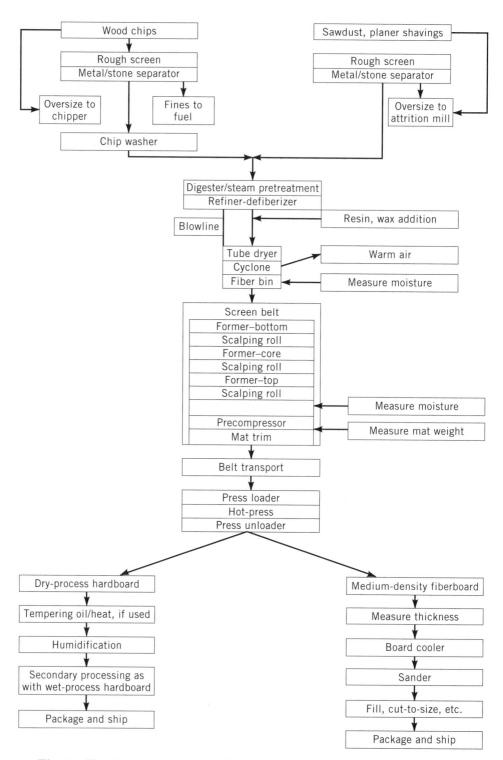

Fig. 6. The dry-process hardboard (and medium-density fiberboard) process.

of the refiner, in the fiber-ejection tube or "blowline." It is also noted that a small amount of dry-process hardboard is made with UF resin binders. UF resins, because of their inherent faster curing at lower temperatures, can be added only at the blowline or in a blender located after the dryer.

As the fiber exits the refiner, it moves at high velocity and turbulence into and through the blowline. Blowlines may be 15–30 m (ca 50–100 ft.) in length and 100–150 mm (ca 4–6 in.) in diameter. During passage through the blowline, a sharp decrease in steam pressure occurs. This results in flash-off of moisture and a decrease in temperature of the fiber. Fiber is generally at only 30–40% mc as it exits the blowline. The blowline is an excellent point for addition of wax and resin to the fiber because of the extreme turbulence and mixing which can occur in the blowline. Generally, because little natural bonding occurs in this process, dry-process hardboard requires more wax and resin than wet-process hardboard. Usually 0.5–1.5% wax and 5–7% PF are used (9–11% UF, if used).

From the blowline, the fiber is blown into a tube dryer. Tube dryers are 760–1520 mm (30–60 in.) in diameter and up to 100 m or more in length. Air heated by steam heat exchangers or direct flame is blown through the dryers and moves the fibers in suspension as they dry and pass through the dryer. Temperatures are in the 120–150°C range at the inlet to the dryer and about 70°C at the dryer exit. Fiber moisture should be in the 6–12% range at this point if blowline blending is used. It must be lower if machine blending is to be used later because of water added with the adhesive resin. The resin will not cure in the dryer because of both the short residence time and the cooling effect of evaporating water in the dryer. Where there is moisture in the fiber, this drying and evaporating effect prevents the resin from becoming hot enough to cure.

From the tube dryer, fiber drops into a cyclone which separates the fiber from the moist, warm air. The fiber falls from the bottom of the cyclone into a large bin and in then moved to the mat forming line. If machine-blending of fiber with additives is used, it will occur at this point, immediately before the forming line.

Mat forming is usually accomplished by 2–5 similar machines in sequence, each depositing a portion of the fiber. A plastic or wire screen moves under each forming head, which results in a cloud of fiber falling onto the screen. Under the screen are suction boxes which aid in pulling the loose fiber toward the screen and providing a small amount of mat densification. Immediately past each forming head is a rotating toothed roll called a scalping roll. The scalping roll is adjusted to remove the uneven top of the mat so that the mat will be flat and uniform as it reaches the next forming head. Moisture control is extremely critical, so there will also be moisture-sensing devices situated along the former. These devices provide a measurement of the mat moisture content in order to ensure that moisture is in the proper range at this point and to alert operators of any need for adjustment of moisture. As the mat emerges from the last forming machine and scalper, it should contain enough fiber per unit area to produce the desired thickness and density of the final product. The mat may be 10–20 times thicker than the final board at this point and quite loose in composition. In order to make the mat more compact and to improve the handling capabilities of the mat as it moves toward the press, the mat now moves through a precompressor. This machine is a pair of opposing wide belts which gradually compress the mat to a minimum thickness as it moves between the belts. A thinner mat, 3–5 times the final board thickness, emerges from the precompressor. The mat is then trimmed to standard width and length and from this point is transported on wide belt sections. Usually, there is a device which continually measures the weight of fiber per unit area at this point to ensure that the desired mat weight and uniform cross-panel density are being formed. Mats which are off-weight or uneven in weight will be rejected at this point and the fiber returned to the system for recycling into other mats.

The mats are moved along the line to the press loader. When the loader is filled and the press opens to remove the load of freshly pressed boards, the loader pushes the new boards into the unloader and deposits the load of mats on the press platens. The press closes as quickly as possible to the desired panel thickness. More pressure, as much as 4.8–6.9 MPa (700–1000 psi) is required to press high density dry-process hardboard, because the dry fiber exhibits much more resistance to compression and densification than wet fiber. Press temperatures are also higher, in the range of 220–246°C. No screens are used in the dry-process, but the moisture in the mats requires a breathe cycle during pressing to avoid blowing the boards apart at the end of the cycle. Because no screens are used, the products are called smooth-two-sides (S-2-S), in contrast to the wet-process boards, which have a screen pattern embossed into the back side and are known as smooth-one-side (S-1-S).

Wet–Dry Hardboard. Wet–dry hardboards are a special class of boards in which the fiber is processed through the mat-forming stage in the same manner as in the case of wet-process board. However, an emulsion of a drying oil such as linseed oil is used as the binder and is applied in the stock chest preceding the former. Because this oil requires a long drying and cure time, the wet mats are passed into and through a conveyor dryer, as in the insulation board process. The mats are often sprayed on one surface with the emulsion immediately before the dryer. As the mats emerge from the dryer, every other mat is turned over so that the two oil-sprayed faces are adjacent. This assembly of two mats proceeds to the press loader and press, where it is pressed into one board. The drying oil completes the final cure and bonding action in the press. This process requires extremely high press temperatures, usually about 246°C, and great care must be taken to avoid press fires.

Dry–Wet Hardboard. Dry–wet hardboards are the other possible manufacturing alternative. Fiber is processed as in the dry process up through the mat-forming stage. PF binders are used in this system. Then, as the mats are ready to enter the press loader, a large quantity of water is sprayed onto the surface of the mat. This water saturates the top surface of the mat and is sufficient to raise the total moisture content of the mat to 20–40% mc, depending on the thickness and density of the product. Thinner, more dense products require more water. The press operates at about 200°C, and as the press closes on the mats, the water is heated to steam. The steam quickly heats and softens the mat, requiring less pressure to compress the mat and produces a hard, dense top surface much like a wet-process surface. However, because there is much less total water in the mat in this system compared to the wet-process system, press cycles are significantly shorter, which produces more product per unit of press surface per unit of time.

Hardboard Siding.   Hardboard siding is, as the name implies, a hardboard intended for use as an exterior siding material for buildings. These products have been made using all of the previously outlined hardboard processes, using PF or drying oil binders. Two products, one of which is no longer made, used a powdered thermoplastic binder as the durable adhesive. All of these products were and are made to perform adequately as siding materials for the life of the buildings on which they are installed. A normal life cycle, assuming proper installation and finishing, is at least 25 yr.

Several additional facts are worthy of mention in regard to hardboard siding. The majority of the product is made in the medium-density range, between 640–880 kg/m^3; thicknesses are normally 11.1–12.7 mm. In addition to the normal smooth panels, there are many embossed surface patterns available which simulate rough sawn lumber siding as well as stucco, stone, brick, and shingle patterns. Embossed panels are made by attaching patterned steel plates to the press platens. A mirror-image of this pattern is then embossed into the top surface of the siding as it is being pressed. Almost all siding is preprimed and possibly prefinished before delivery to market. A variation of the siding product is also available as a roofing material which resembles a shingle or shake roof when installed.

Regardless of which of the various types of production processes hardboard panels follow, processing and shipment are generally quite similar for each kind of hardboard panel. Depending on the projected end-use of the panels, some will be tempered or heat-treated immediately after pressing. These treatments provide an added measure of resistance to water, or retard the rate at which the products will absorb water during exposure. The treatment is especially useful in the production of siding or roofing products. Tempering can be a two-stage operation, where the panels are passed through a bath of hot drying oils and then are placed in large racks and passed through a heated chamber. The chamber is heated to about 135°C, and products require 3–4 h to pass through the chamber. Some products may not receive the tempering oil dip, and will only receive the baking or heat treatment. Almost all hardboard products are humidified after pressing and heat treatment, if they are used. Humidification is necessary because the products emerge from the press and heat-treatment, if used, in an almost oven-dry condition and the moisture content of the panels needs to be increased to a level close to that expected in actual use, or about 6–10% mc. Humidification helps to avoid many swelling problems, ie, warping or buckling, which do occur if the panels are installed dry and then allowed to find their own equilibrium in use. The products are generally placed in large racks and passed through a large chamber maintained at high humidity and high temperature conditions.

After humidification, the products are trimmed to size and stacked. The stacks are then moved to the next processing step and many of the secondary treatments of hardboard will take place at the panel production site. These latter may include the following:

For basic hardboard

packaging and shipment of raw panels

touch-sanding to close thickness tolerances

punching to produce perforated board (pegboard)

sealer coating

laminating to produce thick panels

decorative finishes such as wood grain or other decorative designs similar to those used in the hardwood plywood industry

cut-to-size for specific customer orders.

For hardboard siding

sanding or planing to thickness

cutting or machining decorative groove patterns

cutting into strips for lap siding

cutting into shingle/shake patterns

sealing/priming with exterior-grade primer

prefinishing with exterior-grade topcoat

Upon completion of the various treatments the products may receive, the panels or strips are stacked, wrapped, labelled, and shipped.

Secondary Treatments and Uses. Because hardboard products are utilized in a myriad of different ways, the variety of secondary treatments used by customers are practically unlimited. Hardboards are used in furniture, cabinets, paneling, doors, toys, and a host of other uses. Post-treatments may include cutting-to-size, finishing treatments with roll-applied patterns, melamine overlays, printed paper overlays, paints, and even some extremely durable and water-resistant coatings used in tub and shower linings or other uses where water contact is frequent and extreme.

A separate mention is merited for a special molded hardboard product. These are made by a process in which either a fiber mat or hardboard panel is placed between two shaped platens and press-molded to a three-dimensional configuration. The most common resulting shape is a doorskin which resembles a wood panel door. The doorskins are bonded to wood frames to make an excellent, attractive, and relatively inexpensive door. This fiber/panel molding process is also used to make a wide variety of molded interior linings used in automobile manufacture.

Economic Aspects. In 1994, there were 16 operating hardboard and hardboard siding mills in the United States. Production was $1.535 \times m^3$ (2) in standard hardboard products. These figures do not include the significant quantities of door skin products made, for which production quantities are not tabulated. Production of hardboards has been relatively stable in recent years, considering them as a group. There have been a few new mill closings and a few mill start-ups. In addition, imports of hardboard have also become more common in recent years.

Specifications, Standards, Quality Control, and Health and Safety Factors. The hardboard industry is represented by the American Hardboard Association (AHA). Specifications and standards are contained in several ANSI standards (8–11). These standards define the various hardboard product categories as well as specific product qualities required for each group.

There are five classes of basic hardboard for which ANSI A135.4 (8) summarizes required performance levels of water resistance in terms of water absorption and thickness swelling, modulus of rupture (breaking strength), and tensile strength, both parallel and perpendicular to the surface. These test procedures are found in ASTM D1037-93 (12). For prefinished hardboard paneling, ANSI A135.5 (9) specifies qualifications which are primarily related to the surface finish. These are abrasion resistance, adhesion of finish, fade resistance, gloss, heat resistance, humidity resistance, scrape adhesion, and stain resistance. The AHA ANSI standard for medium-density fiberboard for interior use, ANSI A208.2 (10), has been superseded by the NPA standard for medium-density fiberboard, also ANSI A208.2 (13). This latter standard represents the small amount of hardboard made with UF resin binders which are specifically limited to interior uses, most of which falls into the medium-density group; this standard is discussed in detail herein. The standard for hardboard siding, ANSI A135.6 (11), defines the principal types of siding and the inspection and normal quality control test procedures related to the product. These tests are weatherability of both unprimed and primed surfaces, water resistance, linear expansion, nail-holding tests, bending strength, hardness, and impact resistance.

In addition to the previously noted safety factors associated with these processes, there are additional needs for dust control and ventilation for dissipation of various vapors from pressing, tempering/heat treatment, and machining and finishing operations.

Particleboard

Production, Processing, and Shipment. Particleboards are composites made from particles or small pieces of wood or other lignocellulosic residues, in contrast to the fibers used in the various types of hardboard, and the former are bonded together with an adhesive under heat and pressure. There has been a long debate over the origin of particleboards. At least 5 patents between 1880–1930 describe a form of particleboard, long before an economical adhesive was available. When it was discovered that the polymer of urea–formaldehyde (UF) could be used as an adhesive, it quickly became the binder of choice. Particleboards were first made commercially in Europe in the late 1930s and the industry expanded significantly in Germany during World War II. The industrial technology came to the United States after 1945 and has grown steadily in both size and product quality. The primary benefit of the product is that it provides a necessary and useful outlet for the majority of the previously unused wood residues such as sawdust and planer shavings, as well as some chips from sawlogs and veneer logs. Small amounts of urban wood waste and low value logs are also used.

Particleboards are made in thicknesses of 3.2–44.4 mm (1/8–1 3/4 in.), with the bulk of production going into the 12.7–19.1-mm (1/2–3/4-in.) range. Particleboards are made across a wide range of densities, 415–1000 kg/m^3, with a generally inverse relationship between thickness and density prevailing. Figure 7 shows a general flow diagram of the particleboard process.

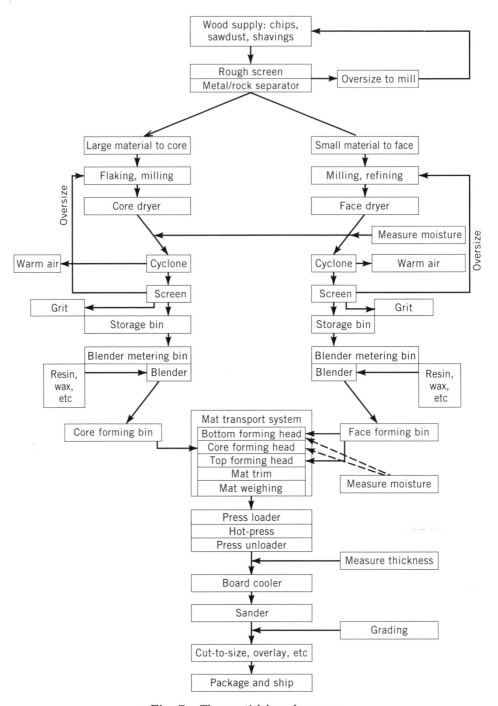

Fig. 7. The particleboard process.

To begin the process, the particulate raw materials are rough-screened to remove oversize materials. The oversize materials may be used as fuel or processed through a grinder and returned to the screen. Also, there should be a unit in this area to remove rocks and trash metal. The material then moves to the milling/drying/screening area (material preparation area). It should be noted that from this point to the mat-formers, materials usually flow in two streams, one of smaller particles for the surfaces of the board and one of larger particles for the core or middle of the board. Milling generally precedes drying because a better mix of particle sizes is achieved if material is milled when still wet or moist. However, dry milling requires less power and some manufacturers prefer this sequence, even though more fines (small particles) are generated. If dry planer shavings are part of the mix, they will necessarily be milled dry. Another advantage of drying preceding milling is that it allows for separation of grit (dirt, sand, etc) following drying, which reduces wear on grinding equipment and also reduces the chance of fires occurring during milling. Grit removal is not easily done on wet or moist particles. Particulate milling is done by hammermills, impact mills, refiners with special plates, or knife-ring flakers. The knife-ring flaker is a machine designed to make flakes (long, thin, flat particles) from larger residues, such as chips. The other forms of mills use attrition or impact to break up the particles, and a much smaller mix of particles is generally obtained. In a few cases where small logs are used as raw material, these are first debarked and then flaked in large drum or disk flakers. The bark is usually used as fuel.

Drying is almost always done in large, rotating drum dryers. Heat can be provided from direct flame oil, gas, or wood dust burners, or from heat exchangers using these same sources of heat. A modern mill is typically designed to generate as much heat as possible from otherwise unusable wood and bark sources. The wood particles enter the dryers at the hottest (heat inlet) end and proceed through the dryers in a tumbling fashion, carried along by the flow of hot air. Temperatures will be about 540–760°C at the inlet and 200–430°C at the outlet. Dry moisture content of surface material will be at 4–6% mc and core materials at 2–4% mc. After the dryer, particles are separated from the moist, warm air in a series of cyclones. The dryer air then passes through a pollution-control system to remove dust and volatiles, so that the only significant emissions from the dryers are warm air and steam.

A measurement of particle moisture content will normally be taken at the exit of the dryer. This allows the process operators to make such adjustments as may be needed to maintain moisture within the desired range. Various instruments are used, none of which are entirely satisfactory, and periodic hand samples are used in some mills. Considering the importance of moisture sensing and control at the dryers, it is unfortunate that a truly efficient, consistent, and accurate sensing system is not yet available to the industry. The primary reasons for the difficulty of measuring moisture at the dryer exit are the extreme and adverse conditions of heat, dust, and moisture present at this location.

At this point, the dry particles and flakes may pass through a system designed to remove grit. To avoid excessive wear on saws and milling machinery downline and especially for the sake of safeguarding customers' equipment, it is necessary to have a grit (silica) level less than 0.05%, and preferably less than 0.03%. In these days, when more emphasis is being placed on recycling of urban waste woods, grit and metal removal equipment is a necessity for those mills reusing this resource. If the grit is primarily very fine material, it can be removed by a small screen with only small wood losses, and these can be used for fuel. If larger grit is present, more sophisticated equipment is required.

The dry material then passes to the screening area where separations are made, based primarily on suitability as surface (fine fractions) or core (coarse fractions) materials. Oversize materials also is removed and returned to the milling area. After screening, the materials are stored in large bins or silos which provide several hours of running inventory so that the manufacturing process can continue if a breakdown of short duration occurs in the milling/drying area. The storage capacity should not be too large, as there is a danger that the raw materials will adsorb moisture and become too moist to allow the process to continue efficiently.

From the storage units, the raw materials are metered at desired and uniform flow rates to blending area. Blenders are normally large tubes employing a rapidly rotating central shaft with mixing paddles attached. Additives such as UF resin, wax, scavenger, and catalyst or buffer are metered into the blender and are mixed by the paddles as the material proceeds through the blender with the desired objective of uniformly mixing the proper amount of additives on all surfaces of the particles. Typical additives used in relative amounts as a percent of dry wood weight are given in Table 2.

A small amount of water may also be added to improve blending or to assure proper furnish (blended materials) mc. Actual amounts of resin and wax used in a specific operation are based on amounts needed to produce the desired strength and water-resistance properties. Wax provides, in addition to water resistance, a small amount of lubricity, which aids in moving the materials through the process. The scavenger, normally urea, is added to react with excess formaldehyde from the resin and prevent excessive formaldehyde emissions from the product. In some cases, resin suppliers have been able to formulate resins with extremely low emission potential and thus avoid the additional step of adding an external scavenger. The amounts of catalyst (cure accelerator) and buffer (cure retardant) used must be balanced to meet the requirements of the system. Catalyst is used to hasten the cure of the core adhesive, along with the possible use of small amounts in the surface. Because catalyzed resins are sensitive to tempera-

Table 2. Additives Used in Production of Particleboard as a Percentage of Dry Wood Weight

Component	Face, wt %	Core, wt %
UF resin[a]	7–11	5–9
wax[b]	0.25–0.75	0.0–0.5
scavenger (urea)[c]	0.4–0.8	0.2–0.5
catalyst (ammonium sulfate)[d]	0.0–0.2	0.0–2.0
buffer (ammonium hydroxide)[e]	0.0–0.3	0.0–0.3

[a]Resin solids based on dry weight of wood.
[b]Wax solids based on dry weight of wood.
[c]Urea solids based on dry weight of wood.
[d]Catalyst based on liquid-to-liquid weight with resin.
[e]Buffer based on liquid-to-liquid weight with resin.

ture, amounts are often changed on a summer-to-winter basis. Careful control is required to balance the need for faster press times and more production, with possible precure of the resin occurring before pressing. Resin durability problems may also be associated with too much catalyst remaining in the board after pressing. Resin suppliers also have proprietary catalysts and buffers which can be added during resin manufacture and in some cases can tailor a resin to the specific needs of the mill so that external catalysts and buffers are unnecessary. Moisture content from the blenders after wax and resin addition should be 9–11% for surfaces and 6–8% for core.

The blended materials, now called furnish, move from the blenders to the formers. A typical mill will have 3–4 formers, one for each surface (top and bottom layers) and one or two for core (middle layer). A mat-carrying plate or caul will pass under the former. This caul will be preweighed and the weight entered on the forming line computer. The caul may be a sheet of steel or aluminum, a wire screen, or a plastic sheet and will generally be the size of the hot-press. On some of the newer systems, the mat-forming and transport system may be a continuous wire screen. The speed of the caul system is such that enough mats may be formed in the time required to press a load of boards, the objective being to maintain the formers in a constant and uniform state of operation. As the caul enters the first former, a layer of fine surface particles is spread uniformly on the caul. The layer will be 20–25% of the panel weight and, depending on the type of former used, may be distributed in such a way that there is a gradation of finer particles on the bottom surface, ranging to coarser particles at the top of this layer. A desirable characteristic of particleboards is a smooth surface, and finer particles on each surface help to achieve this goal. A moisture sensor is often located between the first and second formers.

The caul now moves under the core former(s), where a layer of coarser particles is distributed. This layer represents 50–60% of the board weight. A moisture sensor is often located after the core former(s). A weight sensor may be under the line to sense the additional weight on the caul at this point. Finally, the caul moves under the top surface former and a final layer of fines is distributed atop the core layer. This layer will also be 20–25% of the total mat weight, generally equivalent to the amount formed as the bottom surface. As the caul moves out from the last core former, there is another weight sensor. Given the initial caul weight plus the two weights during forming, the system can easily be adjusted to maintain uniform mat weights by ensuring that the proper amounts are applied by the face and core systems.

After the mat is formed, it may move through a mat precompressor, if this is a feature of the line. Then the mat is trimmed to width and length. If the line does not have the in-line weight sensors noted previously, the trimmed mat will be weighed in total at this point. Mats which are off-weight for the target density and thickness will be dumped and the furnish recycled through the system. Normally, when the line is in uniform operation, very few mats will be off-weight. Mats then proceed to the press loader and into the press. Blended furnish should optimally be formed into mats and be ready to press within 15–20 min after blending. Unless special precautions are taken and special resin formulations, including buffers, are used, material which does not reach the press within 30 min should be dumped and then recycled through the process. It should be apparent how even a short (10–15-min) interruption in operation (downtime) can easily extend to 45–60 min or longer, owing to complications incidental to the interruption.

The pressing operation in a composites facility is a critical one from many standpoints. The amount of high quality board made determines the productivity and profitability of the operation. Thus, it is important to use a press cycle which is as short as possible while producing high quality board and maintaining a consistently uniform operation. But a short press cycle does not ensure high productivity if a number of short interruptions in production occur on each shift. The operators must learn what the optimum speed of operation is in their facility, with the objective of using the press as the slowest point or "bottleneck" in the process. In this way, the press is the only unit in the system which is operating at maximum speed, and thus allows the other equipment to operate at lower and generally more uniform and efficient rates.

Hot presses vary widely in size and number of openings (panels pressed during one cycle). A small size would be 10 openings of 1220 × 2440 mm (4 × 8 ft.). The largest size is 24 openings of 2440 × 7315 mm (8 × 24 ft.). A few single-opening presses may still be found. A recent development is the continuous press, which has two opposing heated steel belts which transport, press, and cure the adhesives as the mat is moved through the press. This is an expensive but highly efficient method of pressing.

The press may operate over a wide range of temperatures, 132–190°C a range of 154–177°C being most common. Higher temperatures allow faster production, but can create other problems if the facility is not specifically designed for these temperatures. Low temperatures are generally only used for thin, high density products which will blister (develop small bubbles) at higher temperatures. A typical press cycle for a 19 mm (3/4 in.) board is as follows, beginning with the press in the open position:

Step	Operation	Time, s
1.	unloading and loading the press	15–30
2.	closing the press to mat contact with platens	5–15
3.	closing the press to final board thickness	30–60
4.	maintaining thickness as adhesive cures	180–210
5.	decompression	10–20
6.	opening the press	10–20

Typically, steps 3–5 require 10–12 s/mm of thickness (16–19 s/1/16 in.). It should be apparent that reducing time in steps 1, 2, and 6 to minimums is well worth the effort. The combined time in these steps is known as dead time, and mills which have dead times of 60–70 s are at a distinct disadvantage compared to those with dead times of 30–40 s. Another type of press worthy of mention is the steam-through or steam press. In the case of this press, thus far limited to single-opening presses, live steam is forced into the mat through a pattern of numerous small holes in the platen surface as the press is closing. The steam heat almost instantly raises the temperature in the entire board cross-section to a point at which the adhesive cures and thus decreases pressing times dramatically. The process works equally well with particles or fiber, but not as well as with large flakes or strands. One possible disadvantage is that density gradients through the thickness are very uniform and thus the products are not well suited to applications requiring high surface densities. However, for thick, low density products such as door cores, the steam press process has no competitors in terms of overall efficiency.

Desired mat moisture contents of 9–11% for faces and 6–8% for cores are generally accepted. UF resins cure best in a moisture regime of 8–9% mc. An 11% mc surface and 6% mc core are an excellent combination because the high moisture in the surface heats quickly, allowing the surface to compress optimally as well as converting the moisture to steam, which penetrates and heats the core quickly. Too little moisture, as in the case of a dry surface, slows the rate of heat penetration and extends the press cycle. Too much moisture (a more common condition) often requires a significant extension of pressing steps 4 and especially 5 to allow moisture to escape before opening the press. If this is not done, the boards will delaminate (blow) when the press is opened, which may result in the loss of the entire pressload. A severe blow may also cause downtime as the damaged boards are removed from the system. Thus it should be emphasized that success in moisture control is the most important factor in determining the ultimate efficiency and profitability of the operation.

As the panels leave the press area, they may pass through two sensing units. The first is a blow detector, which can locate delaminations that may or may not be seen by visual inspection. Boards with blow areas are marked for removal and then are usually ground up, remilled, and recycled through the process. The second unit is an automatic thickness sensor which, by means of several sensing heads across the board, measures and averages the thickness across and along the board. These thickness measurements and the mat weights taken before the press inform the operators if the product is in the proper thickness and density range and whether or not adjustments need to be made. This is the first product quality monitoring step and it is critical in achieving maximum production of on-grade panel materials.

Panels then move into a cooling device, normally a wheel or rack, where they are held individually and air is circulated between them to remove the majority of heat remaining in the boards after pressing. It is desirable to reduce the average board surface temperature to about 55°C. This temperature is sufficient to complete the cure of adhesive in the core of the board. The heat also helps to redistribute moisture uniformly within the boards, because the board surfaces are drier than the core when the boards come out of the press. Warm boards are normally stacked for several hours to a day to allow for resin cure and moisture equalization.

From these stacks the boards are sanded to final thickness. Most modern sanders can sand to a panel average target thickness ±0.2 mm (±0.008 in.). Sanders are multihead machines, with sanding grits proceeding from coarse to medium to fine. The total thickness removal (sandoff) needed to reach a uniform, smooth sanded surface can be from only 0.75 mm (0.030 in.) on boards from a continuous press to more than 2.5 mm (0.100 in.) on an older multiopening press. Average sandoff on panels from multiopening presses is 1.8–2.1 mm (0.070–.085 in.). The sanderdust may be recycled into the product, but is more

pressures are needed, and presses are not quite so robust compared to the dry-process hardboard presses. Also, some early MDF presses used primary steam heating and secondary heating systems based on radio-frequency (RF) heating. The use of RF heating allowed the center of the board to heat almost as quickly as the surface, and this occurred during closure of the press. Thus, the heated core, would compress almost as easily as the faces, which resulted in a quite uniform density profile through the thickness of the board. This then produced the appearance of a uniform, smooth edge when cut or machined.

The other major benefit of RF heating was in reduced presstimes. A typical steam-heated MDF press was operated at about 163°C. Presstimes, not including deadtime, for 19-mm (3/4 in.) board would be about 7 min. With RF, this time could be reduced to about 5 min. It will be noted that these presstimes, even with the use of RF, are longer than those required for particleboards and this, in addition to the more costly base fiber and the higher resin requirements, explains much of the manufacturing cost differential between MDF and particleboard.

After pressing, the MDF process basically duplicates the particleboard process with the steps of cooling, sanding, trimming, cut-to-size, stacking, strapping, and shipping.

Secondary Treatments and Uses.   MDF competes with particleboard in virtually every application, especially in furniture and cabinetry. However, MDF is not used as floor underlayment or decking except on special request. The popularity of MDF is the result of the smooth surfaces and smooth-cut or machined edges. For applications such as direct grain printing, laminating with thin overlay papers or foils, application of smooth, gloss finishes, or uses requiring smooth, highly-machined edges, MDF is the product of choice.

The exterior form of MDF is used in special applications requiring durability and resistance to water or weather exposure. Highway signs would be an example of this use of exterior MDF. It is an extremely expensive product and thus is used only for special applications requiring its special properties. Another example of use would be where a customer would be willing to pay the additional cost to use a composite which has the exceptional qualities of MDF, but also has virtually no formaldehyde emissions.

Economic Aspects.   In 1994, there were 14 producing MDF mills in the United States. These mills produced 2.240 million m^3 of product (2). The market for MDF in the United States is fairly well saturated at this time and for this reason the industry is expanding only slowly. However, as noted herein, the world market is still growing rapidly and the manufacturers are building to satisfy this market. Currently, some U.S. manufacturers are also exporting MDF into these markets.

Specifications, Standards, Quality Control, and Health and Safety Aspects.   The MDF industry is represented in the areas of specifications and standards by the NPA. Specifications and standards are found in *Medium-Density Fiberboard (MDF)*, ANSI A208.2-1994 (13). There are four classes of MDF, three classes divided by density and one for exterior-bonded MDF. Primary properties outlined in the standard and used in QC tests are bending strength and stiffness, bond strength perpendicular to the surface, screwholding properties, and formaldehyde emissions. Strength tests are described in ASTM D1037-93 (12) and formaldehyde emissions in ASTM E1333-90 (4).

The health and safety issues outlined herein for particleboard also apply to MDF. A special note should be made of the fact that, because the MDF raw material is of dry fiber base, there exists in MDF a large component of very small, broken, dust-like wood fibers. These contribute to the dust concerns in the manufacturing areas, requiring excellent dust-control systems, good housekeeping, and personal protection.

Waferboards and Oriented Strand Boards

Production, Processing, and Shipment. The waferboard and oriented strand board (OSB) industries are based upon the use of special forms of wood flakes generated from small logs. A flake is a long, flat section of wood which may be 25–100 mm (1–4 in.) in length and in the grain (longitudinal) directions of the wood. Thickness may be 0.25–1.00 mm (0.010–0.040 in.), and width is usually variable. Normally, a good flake has a length-to-thickness ratio of at least 100.

Wafers as used in the industry were large, flat flakes of about 0.6–1.0 mm (0.025–0.040 in.) in thickness, 38–50 mm (1 1/2–2 in.) in length, and 13–50 mm (1/2–2 in.) in width. Strands are long, narrow flakes of about 0.6–1.0 mm (0.025–0.040 in.) in thickness, 75–100 mm (3–4 in.) in length, and 6–25 mm (1/4–1 in.) in width. Variation will occur around these dimensions, but these will apply in the case of the majority of the waferboard and OSB products.

Flakes and flakeboards have long been a source of attention and interest to the industry, because excellent strength and stability can be achieved with boards made from flakes. However, the requirement of solid wood as a starting material and uneven swelling properties of the boards prevented them from becoming an economical or desirable product for use in many particleboard applications. Nevertheless, at least seven 1950s era particleboard plants operated on large flakes as part or all of their raw material. Decorative paneling was probably the most successful product of this production, but eventually all went out of production because they could not compete in the standard particleboard markets. In 1955, a plant was built in Canada utilizing the technology of an earlier U.S. mill: large flakes or wafers bonded with powdered PF resin adhesive. The market was low cost, exterior sheathing panels called Waferboard for barns, sheds, fences, etc. Time spent in exposure proved that the panels were entirely satisfactory for these uses; in fact, many of these structures are still in use today. Over the succeeding years, a few more plants were built, mostly in Canada.

Then, in the early 1980s the concept of OSB was realized in the construction and operation of large-size mills. OSB is a panel product made from wood strands and somewhat like plywood in that the strands on the two faces are oriented in the long direction of the panel and the core strands are oriented in the cross-panel direction. The use of orientation yields panels having excellent directional properties, much like plywood, and thus an excellent and economical structural sheathing material is created.

The manufacture of waferboard and OSB has many of the same process steps as particleboard, but adapted to the special needs of producing an exterior quality panel with large wafers or strands. This discussion focuses on OSB, because waferboard has been almost entirely replaced by OSB and most of the early waferboard mills have now been converted to production of OSB. The OSB process is outlined in Figure 8.

The process begins with small logs of almost any species. Aspen was the preferred species for years but now the industry has spread well beyond the range of aspen growth and any species which can be converted to strands has been used. High density hardwoods such as oak, hickory, and hard maple are

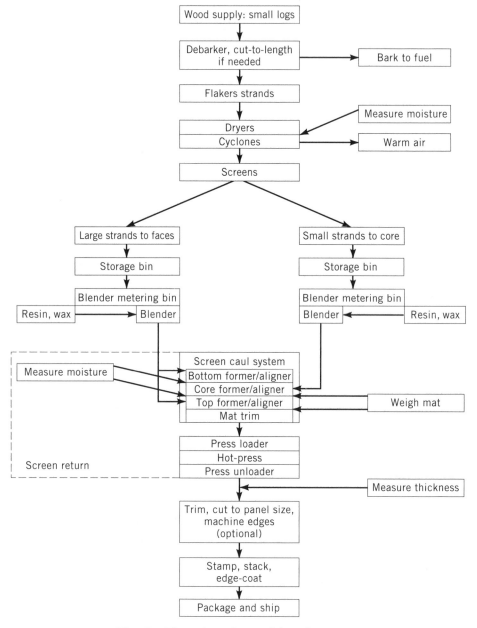

Fig. 8. The oriented strand board process.

avoided where possible, because of the difficulties in flaking. The logs to be processed are debarked and the bark used for fuel. The debarked logs are placed in soaking tanks which will be heated in winter to thaw and heat the frozen logs. As the logs reach the end of the soaking tanks, they are removed and prepared for the flakers. Flakers are of two types, the first using huge rotating disks containing a number of long knives extending outward from the axis of the disk. These knives protrude from the face of the disk by the desired flake thickness. There are also small, sharp spur knives protruding from the disk at intervals equal to the desired flake length. Log sections are fed into the disk and each knife removes a slice from the log. The spur knives cut the slice into many separate lengths, and these pass through an opening in the disk. As the slices pass through the disk, they encounter special metal shapes which bend, buckle, and break the slices into narrow strands which fall into a conveyor below and behind the disk. Another type of flaker is a revolving ring flaker. The ring is about 600 mm (24 in.) in depth, with many flaker knives and spur knives protruding inward on the inner periphery of the ring. Logs or log sections are pressed against the inner periphery as the ring rotates. Slices of wood are removed and are again broken into strands as they pass through the ring and fall into a transport conveyor. Recent developments have made it possible to flake whole logs one section at a time in both of these flaker types.

The strands move through large drum dryers which reduce the moisture content to about 2 or 6%, the difference being whether liquid or dry resin is to be used. Because a desired moisture level into the press is about 6–7% mc, a liquid resin adds water to the system and requires a lower flake moisture than a dry resin.

From the dryer, the strands are screened to remove fines and small particles, which would detract from board quality and economy. These fines are burned for fuel or possibly sold to a nearby particleboard mill as raw material. The larger strands are used as surface material and the smaller strands as core material.

The strands are metered from the bins into the blenders, which are large rotating drums which tumble the flakes as they move from the infeed to the exit end. Resin adhesive and wax are applied; if dry, powdered PF is applied, this is accomplished by a simple metering system which meters the powder at a rate proportional to strand flowrate and at a level of 2–2.5% by weight of dry wood. Other liquid additives are normally applied by centrifugal disk atomizers rotating at high speed and generating a mist of additives which attach to the surfaces of the tumbling strands. Wax is applied at a 1–1.5% rate to provide water repellancy in the finished product. Liquid PF adhesives are applied to rates of about 3.5–4.0%. Another binder being used in some locations is isocyanate resin. It may be used in the entire board or it may only be used in the core of the board with PF-bonded surfaces. Usage rates are 1.5–2.0%. Isocyanates have several desirable features, such as excellent bonding abilities, high moisture tolerance, faster curing than PF, and lower adhesive usage than PF. Less desirable attributes are press sticking potential, high cost, and workplace health and safety concerns. Press sticking can be avoided by use of isocyanates in the core only where faster cure rate also helps to offset the added cost.

Strands move from the blenders to the formers. Mats are formed on wire-screen cauls. The major benefits of screen cauls are long working life and as an aid to removal of moisture in steam formed during and immediately after the press cycle. This dramatically reduces the tendency of these large strand products to blow after pressing. An unintentional benefit was the embossed screen imprint in the board, which when placed screen-side-up on roof sheathing installations, provides a slip-resistant walking surface for workers. As the screen-cauls pass under the formers, the lower surface strands are placed on the screen-caul such that they are generally aligned with the long direction of the screen. This is done by dropping the strands between rows of closely spaced, rotating metal disks located just above the screen caul. The core strands are then laid down so that their lengthwise dimension is generally perpendicular to the surface strands. Finally, the top layer of strands is applied to the mat, also in the long direction of the screen and mat.

Mats then move to the press loader and into the press. Presses in OSB plants are usually very large, with panels of 2440 × 7320 mm (8 × 24 ft) a common size. Newer plants produce panels of 3660 × 7320 mm (12 × 24 ft). Presses will produce 8–16 panels of these sizes in one pressload. Press temperatures are 204–218°C for pressing PF-bonded panels, but temperatures as low as 190°C can be used when isocyanate adhesives are used. Press times, not including deadtimes, with PF-bonded panels are about 14–15 s/mm of thickness. With isocyanate adhesives, presstimes are reduced to about 12 s/mm of thickness. The presses are equipped with elaborate position-control systems so that all panels will be close to desired thickness as they exit the press. Panel densities will range from 575–675 kg/m^3, depending on wood species used and end-product use requirements. The objectives are to reduce panels to minimum densities consistent with good quality and end-use requirements. This extends the resource, as well as easing handling in the field.

After the press, the panels are immediately cut into ordered sizes, the most common of which is 1200 × 2440 mm (4 × 8 ft.). Panels are then printed with appropriate use information, stacked, and usually edge coated with a moisture-proof barrier coating prior to banding and shipping. The edge coating provides protection from moisture damage on panel edges during shipping and field installation of the panels.

Secondary Treatment and Uses. The vast majority of OSB panels are used "as is," without further processing or treatment. Primary uses are as wall and roof sheathing, floor decking, and other construction panel uses in home and commercial construction. OSB products are effectively filling in for the decline in plywood production. Small amounts of OSB are used in furniture, primarily as frame stock, and in other uses in which plywood might be used.

The major secondary use of OSB has been the production and marketing of an OSB exterior wall-paneling or siding. This was done by applying a PF-resin impregnated paper on the top of the strand mat immediately before pressing. The paper overlay was then bonded to the mat during pressing and embossed with a rough-sawn wood grain pattern in a manner similar to that used for hardboard siding. Panels were then sawn into strips, coated with a paint primer, and sold as lap siding. Another competitive product was made with a layer of adhesive-coated fiber bonded to the top of the strand mat in the press. This product was also embossed in the press and then primed and sold as lap siding strips.

Economic Aspects. There were 31 OSB plants in the United States in 1994 and these produced 6.625×10^6 m^3 of OSB products (2). This industry is growing rapidly in both Canada and the United States. In fact, many of the composite mills currently under construction are designed to produce OSB or similar products based on strands. Outside of North America, where building practices are not yet extensively utilizing the distinct advantages of the stud wall and plywood/OSB sheathing, there are only a few operating OSB plants. There are also small export markets for OSB products in Europe and the Far East.

Specifications, Standards, Quality Control and Health and Safety Aspects. The OSB industry is represented by the APA—The Engineered Wood Association in areas of specifications and standards. These are outlined in APA *Performance Standard for Wood-Based Structural-Use Panels*, PS2-92 (15). The standard defines specifications, qualifications, and tests for panels of any type which are to be used in structural-use applications. Thus, structural plywood, OSB, and all other similar products are considered to be covered by this standard. The major areas discussed in the standard are product classification, general requirements, and product evaluation and qualification. Twenty test methods are outlined and can be placed in the following general groups: bending and load testing, fastener holding, moisture and dimensional stability, mold and bacteria tests, probe tests, and cyclic durability tests.

Health and safety factors in the OSB industry are similar to those in other composite mills. Worker safety cannot be stressed too highly, a main component of which is to develop an awareness in the workers to be prepared for danger at all times. Many accidents are the result of a moment of careless or unthinking activity on the part of the person injured or another nearby person. An area of special concern in some OSB mils, those using isocyanate adhesives, is awareness of the toxic nature of this adhesive in the uncured state and the requirements of personal care, housekeeping, and ventilation and air handling in the process areas from blending through pressing.

Structural Composite Lumber

Production, Processing, and Shipment. Structural composite lumber is a group of composite or veneer products which can be used in place of structural lumber in many applications. These products and markets have been developed to fit specific needs of the construction industry where solid lumber of desired grades is no longer available or at least, very difficult to obtain. An important factor in the acceptance of these solid lumber substitutes is their uniformity. While composites in general may not be as strong as clear, straight-grain solid wood, they may be significantly stronger than knotty or angled-grain lower grades. Also, the variation in properties is much less than in run-of-the-mill solid lumber. Thus, in the design of wood structures for which a composite with known and predictable properties can be selected, as compared to a batch of

7. *Cellulosic Fiberboard*, ANSI/AHA A194.1, American National Standards Institute, New York, 1985.

8. *Basic Hardboard*, ANSI/AHA A135.4-1982, American National Standards Institute, New York, 1982.

9. *Prefinished Hardboard Paneling*, ANSI/AHA A 135.5-1988, American National Standards Institute, New York, 1988.

10. *Medium-Density Fiberboard for Interior Use*, ANSI A208.2-1980, American National Standards Institute, New York, 1980.

11. *Hardboard Siding*, ANSI/AHA A 135.6-1990, American National Standards Institute, New York, 1990.

12. *Standard Methods of Evaluating the Properties of Wood-Base Fiber and Particle Panel Materials*, ASTM D1037-93, American Society for Testing and Materials, Philadelphia, Pa.,

13. *Medium-Density Fiberboard (MDF)*, ANSI A208.2-1994, American National Standards Institute, New York, 1994.

14. *Particleboard*, ANSI A208.1-1993, American National Standards Institute, New York, 1993.

15. ASTM 1037-93, American Society for Testing and Materials, Philadelphia, Pa.,

16. *Performance Standard for Wood-Based Structural-Use Panels*, PS2-92, APA—The Engineered Wood Association, Tacoma, Wash., 1992.

17. *Standard Specification for Evaluation of Structural Lumber Composite Products*, ASTM D5456-93, American Society for Testing and Materials, Philadelphia, Pa., 1993.

General References

For more information, see publications of the following associations.

American Hardboard Association (AHA), 887-B Wilmette Road, Palatine, IL 60067, Tel. (708) 934-8800.

American National Standards Institute, 1430 Broadway, New York, NY 10018.

APA—The Engineered Wood Association, 7011 So. 19th St., Tacoma, WA 98411, Tel. (206) 565-6600.

The American Society for Testing and Materials (ASTM), 1916 Race St., Philadelphia, PA 19103-1187.

National Particleboard Association, 18928 Premiere Court, Gaithersburg, MD 20879, Tel. (301) 670–0604.

The Hardwood Plywood and Veneer Association (HPVA), 1825 Michael Faraday Drive, P.O. Box 2789, Reston, VA 22090, (703) 435-2900.

WILLIAM F. LEHMANN
Consultant

WOOD PULP. See PULP.

X

XANTHENE DYES

Xanthene dyes are those containing the xanthylium [261-23-4] (**1a**) or dibenzo-γ-pryan nucleus [92-83-1] (xanthene) (**1b**) as the chromophore with amino or hydroxy groups meta to the oxygen as the usual auxochromes. They are

(**1a**)   (**1b**)

important because of their brilliant hues, and shades between greenish yellows to dark violets and blues are obtainable, the most important being reds and pinks. They are generally very strong, with much higher oscillator strengths (and hence higher absorbances) than, for example, an anthraquinone dye of the same shade (see DYES, ANTHRAQUINONE). As a consequence of their rigid chromophoric nucleus, xanthenes are often fluorescent, which also adds to their strength and brightness, but, as is often the case with fluorescent dyes, they have lower lightfastness compared to other chromophores. Their use then is concentrated on those areas in which lightfastness is relatively unimportant compared to economy (eg, paper dyes) or where lightfastness can be achieved by modification by metallization. They are used for the direct dyeing of wool and silk and mordant dyeing of cotton. Paper, leather, woods, food, drugs, and cosmetics are dyed with xanthene dyes (see DYES, APPLICATION AND EVALUATION–APPLICATION). Brilliant insoluble lakes are used in paints and varnishes. In addition, several new applications for xanthene dyes have emerged, for example in ink-jet printers, as markers in biological and medical research, and even as insecticides.

Diphenylmethane Derivatives

Pyronines. Pyronines are diphenylmethane derivatives synthesized by the condensation of m-dialkylaminophenols with formaldehyde, followed by oxidation of the xanthene derivative (**12**) to the corresponding xanthydrol (**13**) which in the presence of acid forms the dye (**14**). If R is methyl, the dye produced is

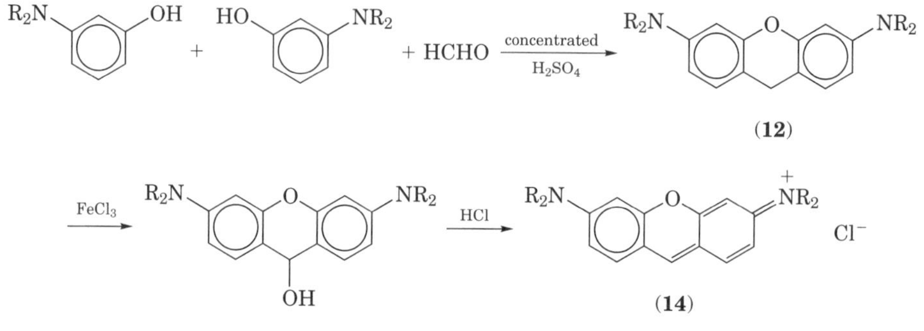

(**12**)

(**13**)

(**14**)

pyronine G (CI 45005); if R is ethyl, pyronine B (CI 45010) is obtained. If pyronine G is oxidized with potassium permanganate, two methyl groups are eliminated to give Acridine Red 3B (CI 45000) (**15**). Pyronine B and Acridine Red 3B

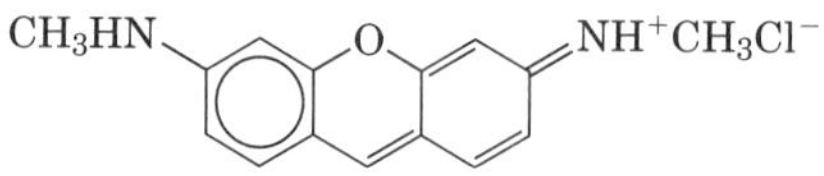

(**15**)

are used as biological stains and particularly in wet staining for the direct microscopic observations of living cells (23).

Succineins. Succineins are carboxyethyl-substituted pyronines made by substituting succinic anhydride for formaldehyde in the basic synthesis, as, for example, in Basic Red 11 [72968-14-0] (CI 45050) (Rhodamine S) (**16**).

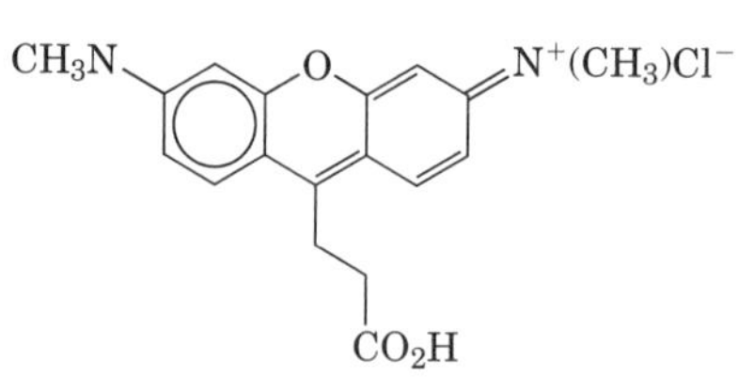

(**16**)

Triphenylmethane Derivatives

Amino-Derivatives. *Rhodamines.* Rhodamines are commercially the most important aminoxanthenes. If phthalic anhydride is used in place of formaldehyde in the above condensation reaction with m-dialkylaminophenol, a triphenylmethane analogue, 9-phenylxanthene, is produced. Historically, these have been called rhodamines. Rhodamine B (Basic Violet 10, CI 45170) (**17**) is usually manufactured by the condensation of two moles of m-diethylaminophenol with phthalic anhydride (24). An alternative route is the reaction of diethylamine with fluorescein dichloride [*630-88-6*] (3,6-dichlorofluoran) (**18**) under pressure.

The free base of compound (**17**) is Rhodamine B base [*509-34-2*] (Solvent Red 49; CI 45170:1). The phosphotungstomolybdic acid salt of (**17**) is Pigment Violet 1 [*1326-03-0*] (CI 45170:2). Pigment Red 173 [*12227-77-9*] (CI 45170:3) is the corresponding aluminum salt.

Esterification of the carboxyl group also yields commercially useful dyes. If Rhodamine B is esterified with ethyl chloride or ethanol at 160–170°C under pressure, Basic Violet 11 [*2390-63-8*] (**19**) (CI 45175) forms. Another commercially important esterified aminoxanthene is Rhodamine 6G (Basic Red 1; CI 45160) [*989-38-8*] (**20**). This is manufactured by condensing 3-ethylamino-p-cresol with phthalic anhydride, then esterifying the product with ethanol and a mineral acid. The phosphotungstomolybdic acid salt of Rhodamine 6G is Pigment Red 81 [*12224-98-5*] (CI 45160:1). The copper ferrocyanide complex of structure (**20**) is Pigment Red 169 [*122237-63-7*] (CI 45160:2). Highly concentrated stable liquid forms of rhodamines can be prepared on a commercial scale by the reaction of the rhodamine base with dialkyl sulfate and a saturated aliphatic glycol at 100–160°C (25,26). These solutions are particularly suited for dyeing paper.

The rhodamines described thus far are basic rhodamines. They are used primarily for the dyeing of paper and the preparation of lakes for use as pigments. They are also used in the dyeing of silk and wool where brilliant shades with fluorescent effects are required, but where lightfastness is unimportant. Many new uses for rhodamine dyes have been reported. For example, when vacuum-sublimed onto a video disk, Rhodamine B loses its color to form a clear stable film which becomes permanently colored on exposure to uv light (27). This can be used in optical recording for computer storage or video recording (28). Fluorescent coloration of rigid or nonplasticized poly(vinyl chloride) (PVC) by the addition of selected rhodamines to the PVC resin before sheet or film formation has been reported (29). The addition of rhodamines to liquefied bis(hydroxyalkyl)aromatic dicarboxylic acid esters prior to their condensation to form polyesters to produce tinted polymer sheets is also used (30). Rhodamine 6G can be used in ink-jet printing where fluorescence under uv conditions is a desired property (31). Selected xanthenes, including fluorescein, Rhodamine B, and Rhodamine 6G, have been used as laser dyes (32).

Recently, several rhodamines have been used as magentas in ink-jet printing. These have been chosen primarily for their brilliant magenta shades, but with the disadvantage of very poor light and waterfastness on ordinary copy papers. On specially coated color ink-jet papers the waterfastness improves, but the lightfastness is still much lower than comparable azo dyes, and, on this media, the difference in brightness between azo and xanthene chromophores is also much smaller. Attempts to improve the waterfastness of xanthene dyes for ink-jet have been made as in (**21**) (33,34). Improvements in lightfastness have been claimed by introducing branching and rings in the alkyl chains on the nitrogens and by using a polymeric counterion, $(An^-)_n$, eg, (**22**) (35).

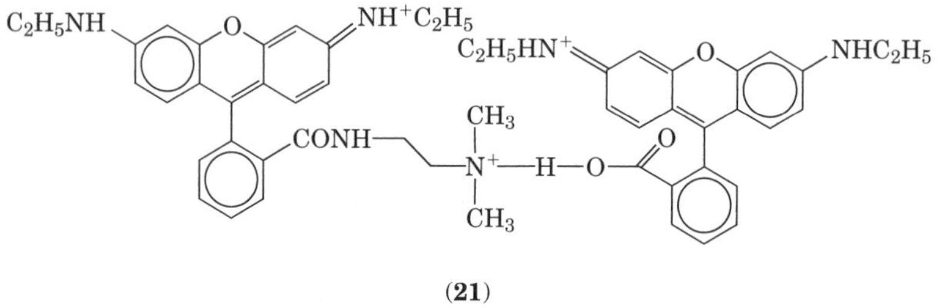

(**21**)

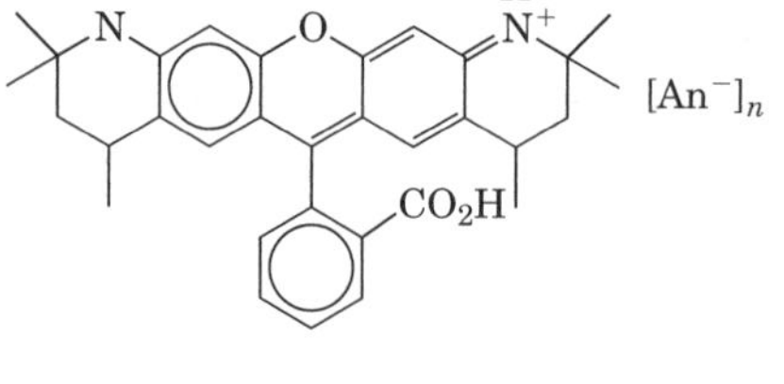

(**22**)

Acid rhodamines are made by the introduction of the sulfonic acid group to the aminoxanthene base. The preferred route is the reaction fluorescein (**2**) with phosphorous pentachloride to give 3,6-dichlorofluoran (fluorescein dichloride) (**23**), which is then condensed with a primary aromatic amine in the presence of zinc chloride and quicklime. This product is then sulfonated. For example, if compound (**23**) (fluorescein dichloride) is condensed with aniline and the product is sulfonated, Acid Violet 30 (CI 45186) (**24**) is produced.

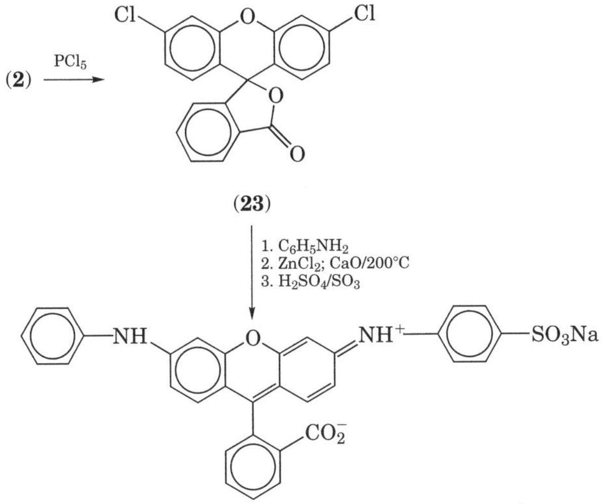

(**23**)

(**24**)

These acid rhodamines are usually used for silk and wool because they have level dyeing properties and show good fastness to alkali; however, they have poor lightfastness. An improved process for manufacturing 3,6-diaminosubstituted xanthenes is reaction of the inner salts of 3,6-dihalo-9(2-sulfophenyl)xanthene-9-ols with a primary or secondary amine in stoichiometric amounts in the presence of an inorganic acid-binding agent or acid-binding tertiary aliphatic or tertiary nitrogen-containing heterocyclic amine (36).

Highly substituted acid rhodamines have been reported for fiber-reactive dye applications. For example, if 3,6-dichloro-9-(2-sulfophenyl)xanthene is first condensed with 1,4-phenylenediamine-3-sulfonic acid and then N-methyltaurine and then acylated with cyanuric chloride, compound (**25**) is produced. This sulfurein derivative dyes cellulosic fibers blue; shades have good washfastness and improved lightfastness (37). Another route to fiber-reactive xanthenes is exemplified by the condensation of 4-nitro-2-sulfobenzaldehyde and 3-N-ethylamino-4-methylphenol. The product is reduced and then reacts with an acylating agent formed by the reaction of 1-amino-3,5-disulfonic acid and cyanuric chloride to produce compound (**26**), which gives brilliant red shades with good washfastness and moderate fastness to light (38). In general then, it appears that the primary fading mechanisms for rhodamines is N-dealkylation, and substitution of the N-alkyl groups with N-aryl groups, ring fusions, or sterically hindered alkyls, which are less easily removed causes an increase in lightfastness.

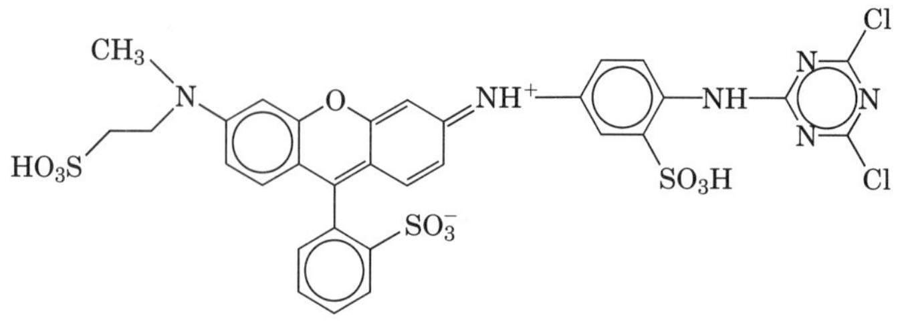

(**25**)

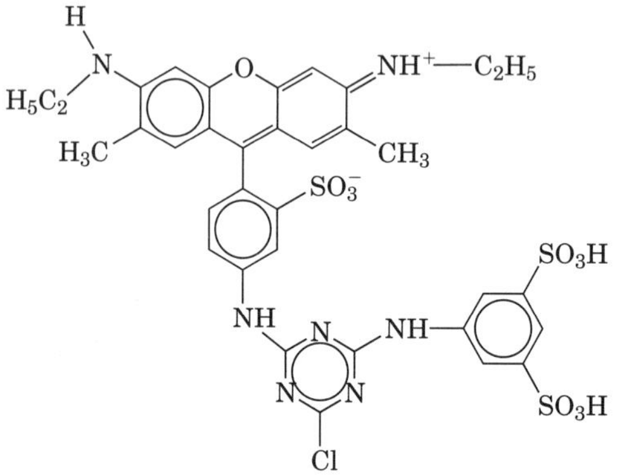

(**26**)

Unsymmetrical rhodamines can be prepared by the condensations of one mole of a *m*-aminophenol with phthalic anhydride to give an *o*-benzoyl benzoic acid (**27**) which is then further condensed with a different *m*-aminophenol to give the required product, Rhodamine 3GO (**28**) (2). A general route to asymmetrical acid xanthenes has been patented (39).

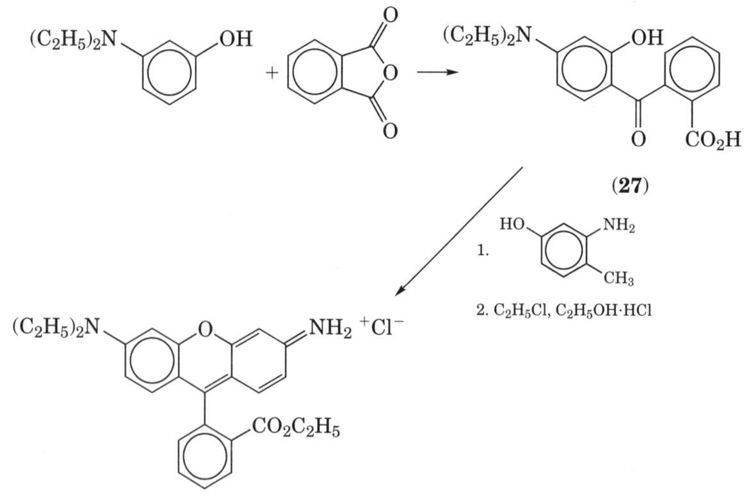

(**27**)

(**28**)

Reactive xanthene dyes with β-hydroxyethylsulfonyl groups, as exemplified by structure (**29**), provide brilliant shades and excellent washfastness on cotton (40). The sulfurein derivative (**29**) is synthesized by condensing 3-aminophenol-β-hydroxyethylsulfone with 3,6-dichloroxanthene-9-phenyl-2'-sulfonate at 90°C in N-methylpyrrolidone or dimethylformamide, then condensing with N-methyl-taurine, and finally esterifying with chlorosulfonic acid (41). Condensation of 3,6-dihaloxanthene-9-phenyl-2'-sulfonic acid with an aromatic amine yields compound (**30**); a second condensation with an appropriately substituted aliphatic amine, usually with subsequent sulfonation, yields the acid rhodamine (**31**).

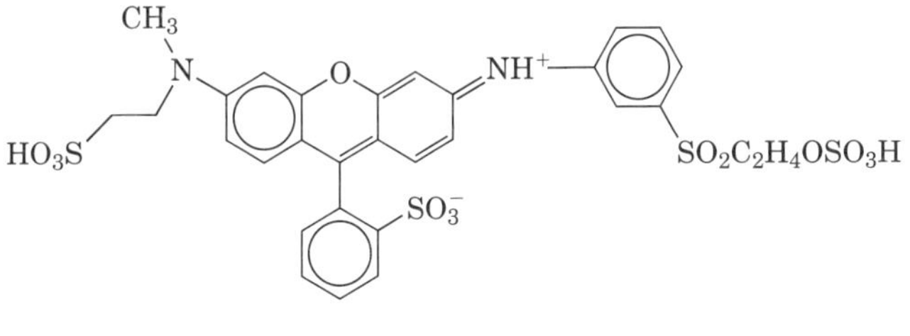

(**29**)

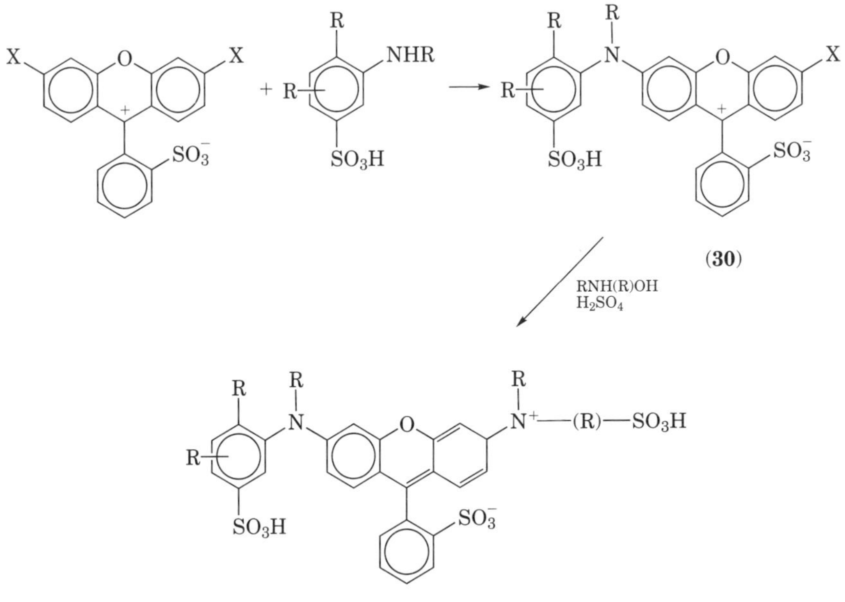

(**30**)

(**31**)

Rosamines. Rosamines are 9-phenylxanthene derivatives prepared from substituted benzaldehydes instead of phthalic anhydride. Condensing benzaldehyde-2,4-disulfonic acid with m-diethylaminophenol, dehydrating the product with sulfuric acid, and oxidizing with ferric chloride yields Sulforhodamine B [3520-42-1] (Acid Red 52; CI 45100) (**32**) which is the most important

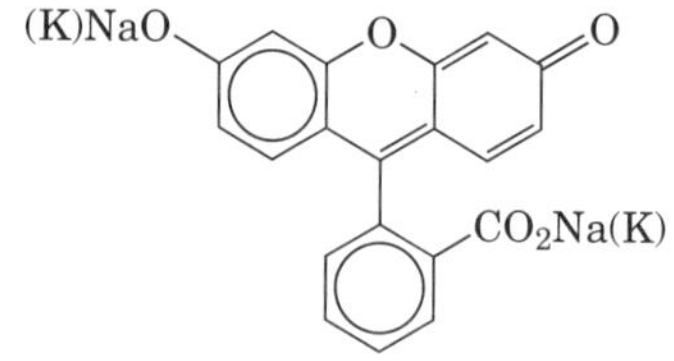

(37)

A modern use of uranine is in the manufacture of fluorescent laminates, eg, sheets, glass, and plastic films, that are transparent to electromagnetic waves and visible light rays (45). Such material might be used in windows, viewing partitions, and optical lenses.

The principal use of fluorescein is as an intermediate for more highly substituted hydroxyxanthenes. When fluorescein is brominated in ethanolic solution and converted to the sodium salt with sodium chlorate, eosine [17372-87-1] (Acid Red 87; CI 45380) (38) forms. This has been shown to be the 2',4',5',7'-tetrabromo analogue. It is used for dyeing silk red with a brilliant yellow fluorescence, and for coloring inks, dyeing paper, and coloring cosmetics. Another use is as an indicator in the analytical determination of polymeric biguanide concentrations in aqueous solutions (see ANALYTICAL METHODS), which is important in controlling the growth of bacteria and algae in swimming pools (46) (see WATER, TREATMENT OF SWIMMING POOLS, SPAS AND HOT TUBS). The lead salt of eosine is Pigment Red 90 [1326-05-2] (CI 45380:1), the free acid is Solvent Red 43 [15086-94-9] (CI 45380:2), and the aluminum salt is Pigment Red 90:1 [17372-87-1] (CI 45380:3).

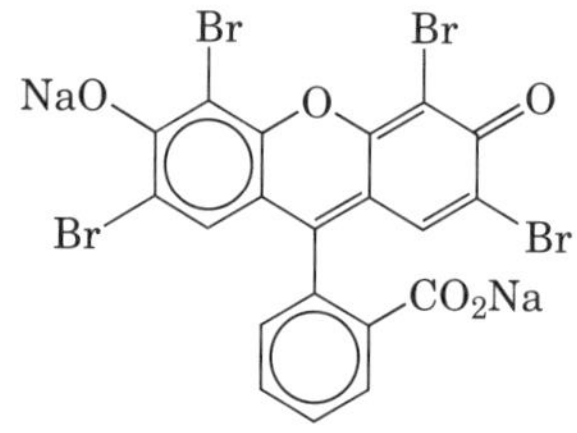

(38)

When fluorescein is reacted with iodine and potassium iodate in an ethanolic solution and converted to the sodium salt, the tetraiodo analogue erythrosine [16423-68-0] (Acid Red 51, CI 45430) (39) forms. This is used as a food coloring, a sensitizer in photographic plates, and in microscopical stains (see COLORANTS FOR FOOD, DRUGS, AND COSMETICS; PHOTOGRAPHY). Nitrated fluoresceins are used in dye applications. For example, dibrominations of fluorescein in aqueous sodium hydroxide followed by treatment with mixed sulfuric–nitric acid yields the dibromo–dinitro analogue saffrosine [548-24-3] (Acid Red 91; CI 45400) (40). Saffrosine is used to make fade-resistant electrophotographic sheet material by incorporating such nitro-substituted xanthenes in the photoconductive layer (47).

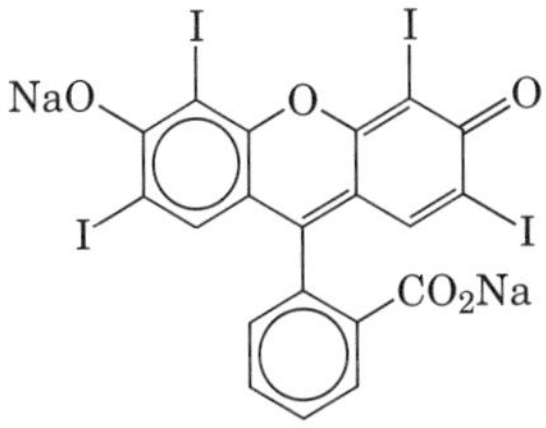

(39)

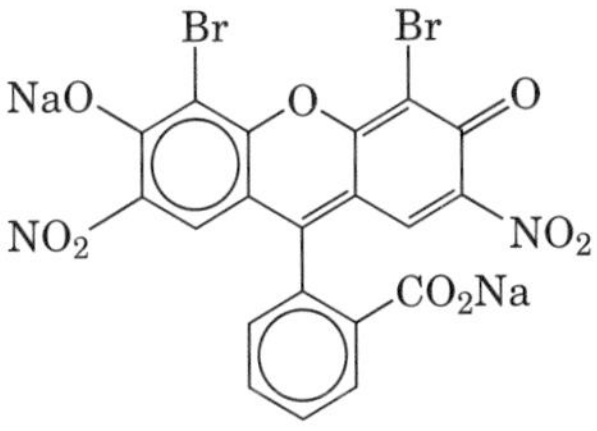

(40)

Treatment of this same 4',5'-dibromofluorescein intermediate with mercuric acetate and conversion to the disodium salt yields the hydroxymercuric analogue merbromin or mercurochrome [129-16-8] (41). It was once a widely used antiseptic, especially for skin disinfection, and was even administered internally. However, it has been replaced by more effective antibacterial agents.

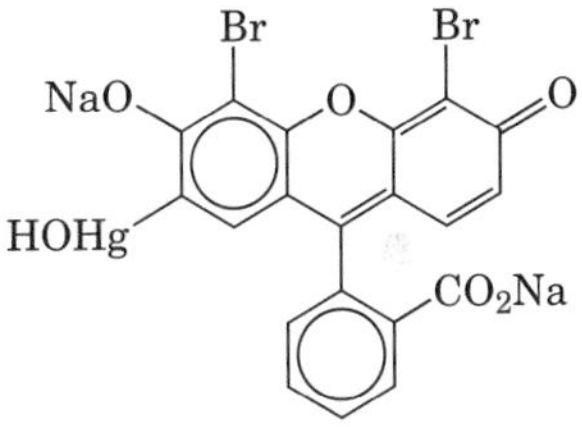

(41)

Another group of halogenated fluorescein dyes is prepared by condensing chloro derivatives of phthalic anhydride with resorcinol, followed by bromination or iodination. Thus Phloxine B [18472-87-2] (Acid Red 92, CI 45410) (42) is prepared by condensing tetrachlorophthalic anhydride with resorcinol followed by tetrabromination. Phloxine B undergoes ethylation to yield the yellowish red acid dye Cyanosine B [6441-80-1] (43).

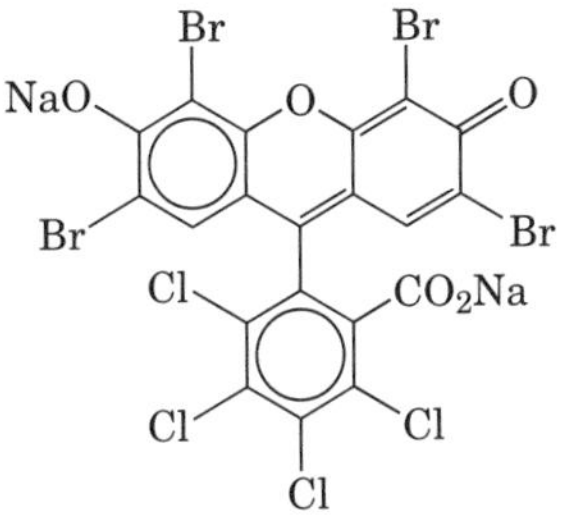

(42)

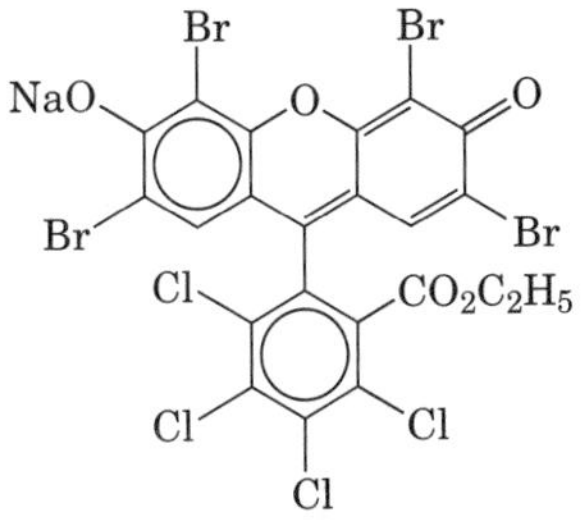

(43)

Another important polyhalogenated fluorescein dye is Rose Bengal [632-68-8] (Acid Red 94, CI 45440) (44). This is synthesized by condensation of resorcinol with tetrachlorophthalic anhydride, tetraiodination, and conversion to the potassium salt. A new use for Rose Bengal and other tetrabromo- or tetraiodofluoresceins involves their chemiluminescent reaction with ozone. Measuring the intensity of emitted light makes possible a quantitative determination of ozone concentration in the atmosphere (48) (see CHEMILUMINESCENCE). Another recent use is as a nonsilver halide photographic system, particularly for use in making direct prints for microfilm enlargements (49).

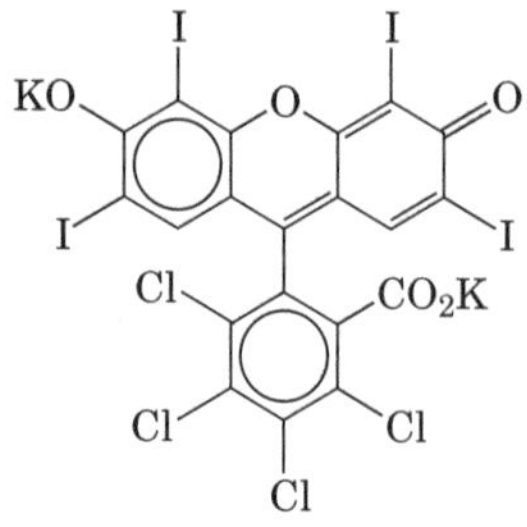

(**44**)

Aminohydroxy Derivatives

Aminohydroxy-substituted xanthenes are of little commercial importance. They are synthesized by condensing one mole of *m*-dialkylaminophenol with phthalic anhydride, and then condensing that product with an appropriately substituted phenol. For example, Mordant Red 77 [*6528-43-4*] (CI 45300) (**45**) is prepared by condensing *m*-dimethylaminophenol with phthalic anhydride, and then condensing the product with 2,4-dihydroxybenzenesulfonic acid.

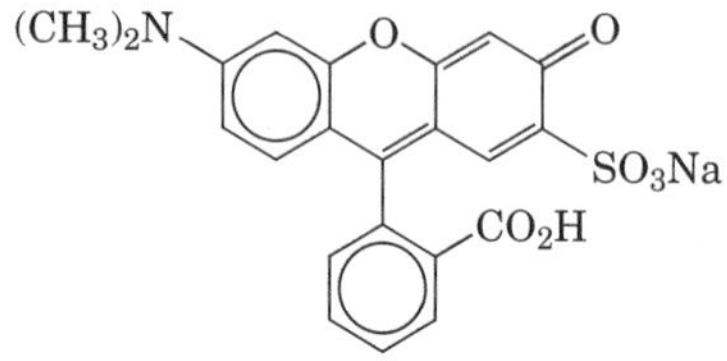

(**45**)

Miscellaneous Derivatives

Two additional xanthene analogues are termed fluorescent brighteners (see BRIGHTENERS, FLUORESCENT). Condensation of two moles of *p*-cresol with one mole of phthalic anhydride yields 2′,7′-dimethyl fluoran which is cyclized with 24% oleum and then reduced with zinc dust and ammonia under pressure to form Fluorescent Brightener 74 (CI 45550) (**46**). If the reduction is carried out with zinc dust and caustic soda in the presence of pyridine, followed by acetylation with acetic anhydride, Fluorescent Brightener 155 (CI 45555) (**47**) is obtained. These are used in the formulation of solid dielectric compositions for application in high voltage cables to prevent conductive treeing (50). The naphthalene analogue of fluorescein can be made by melting 1,6-dihydroxynaphthalene and phthalic anhydride (51) and has been sold as Scheckfarbe AS, for security applications.

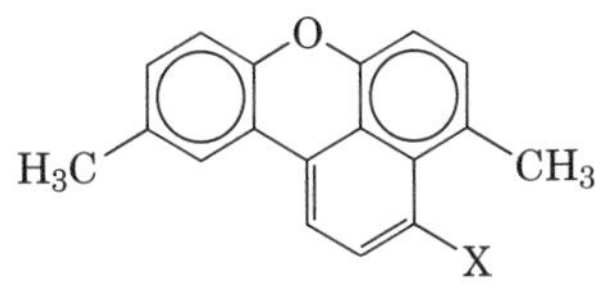

X = H (**46**)
X = O$_2$CCH$_3$ (**47**)

Another series of ring-closed xanthenes starting with benzoxanthene- and benzothioxanthenedicarboxylic acid hydrazides provides shades of bright yellow to red on cellulose acetate, polyamide and polyesters with excellent sublimation fastness and unusually good lightfastness. Because of their high fluorescence, they can be used for preparing daylight fluorescent pigments. For example, condensation of 8-methoxybenzo[d,e]xanthene-3,4-dicarboxylic acid hydrazide with acetylacetone in the presence of toluene sulfonic acid followed by cyclization of the hydrazone in N-methylimidizolidinone yields a mixture of isomers (**48**) and (**49**) (**52**).

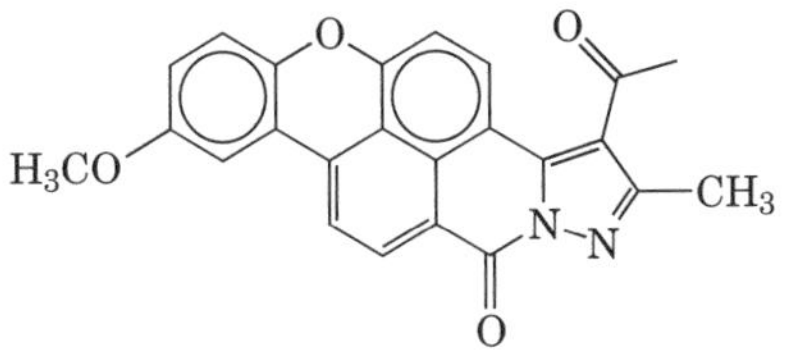

(**48**)

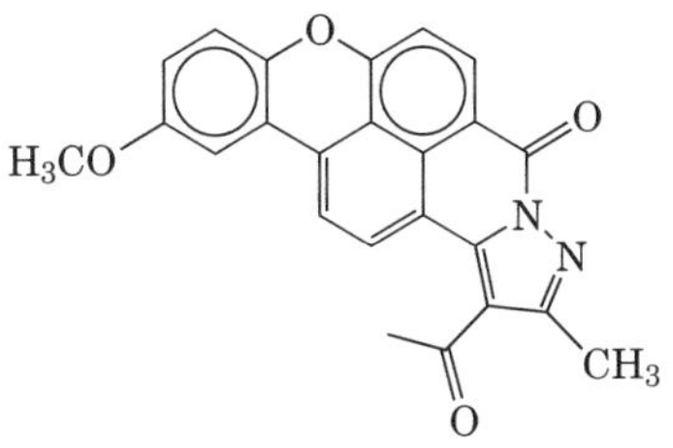

(**49**)

A series of water-soluble fiber-reactive xanthene dyes has been prepared from the reaction of benzoxanthenedicarboxylic acid anhydride disulfonic acid with, for example, 3-aminophenyl-β-hydroxyethyl sulfone to yield dyes, eg, compound (**50**), with high brilliance and good fastness properties for dyeing of or printing on leather, wool, silk, or cellulosic fibers (**53**).

Economic Aspects

Since 1973, the U.S. International Trade Commission has reported the manufacture and sales of dyes by application class only. During the latter half of the 1970s and all through the 1980s, annual dye production in the United States, including xanthene dyes, changed very little. Statistics for the production of basic dyes include those products listed as cationic dyes, eg, cyanines, for dyeing polyacrylonitrile fibers and the classical triarylmethane dyes, eg, malachite green, for coloring paper and other applications. Furthermore, statistics for xanthene dyes are also hidden in the production figures for acid, solvent, mordant, and food dyes and organic pigments. Between 1975 and 1984, the production of basic dyes in the United States varied from (11–17 million lbs). However, from 1985–1990, production of basic dyes varied from (11–12.5 million lbs) with an increase in sales value from $56 to $73 million per annum. The production figures for xanthenes in 1980 are reproduced in Table 1.

Table 1. Economic Aspects of Xanthene Dyes

CI name	CAS Registry Number	CI Number	Common name	U.S. Imports (U.S. Production), t					U.S. suppliers
				1976	1977	1978	1979	1980	
Acid Yellow 73	[2321-07-5]	45350	fluorescein	0.45	0.7	1.6	1.7	2.0	American Cyanimid Co.; Leeben Color, Div. of Tricon Colors, Inc.; International Dyestuffs Corp.; Hilton-Davis Chemical Group of Sterling Drug, Inc.
Solvent Yellow 94	[518-47-8]	45350:1	uranine						
Acid Red 52	[3520-42-1]	45100	Sulforhoda-mine B	2.7	5.4	3.6	9.1	7.3	Atlantic Chemical Corp.; Carolina Color and Chemical Corp.; American Hoechst Corp.; Organic Chemical Corp.; Sandoz Colours and Chemicals, Inc.
Acid Red 87	[17372-87-1]	45380	eosine	0.45		0.7	3.2	2.7	Hilton-Davis Chemical Group of Sterling Drug, Inc. H. Kohnstann & Co.
Pigment Red 90	[1326-05-2]	45380:1							Hilton-Davis Chemical Group of Sterling Drug, Inc.; Sun Chemical Corp.
Solvent Red 43	[15086-94-9]	45380:2							
Acid Red 92	[18472-87-2]	45410	Phloxine B	3.4	1.1	0.2	0.2	0.2	Hilton-Davis Chemical Group of Sterling Drug, Inc.
Solvent Red 48	[13473-26-2]	45410:1							
Pigment Red 174	[15876-58-1]	45410:2							
Acid Red 94	[632-68-8]	45440	Rose Bengal	6.8	0.9	1.4	1.4	0.1	
Acid Red 289	[12220-28-9]						2.7	4.5	American Hoechst Corp.
Acid Violet 9	[6252-76-2]	45190		1.4	0.2	2.7	2.3	0.9	International Dyestuffs Corp.
Solvent Violet 10	[66225-66-9]								

Name	CI Number	CAS Number	Colorant	1	2	3	4	5	Manufacturers
Basic Red 1	45160	[989-38-8]	Rhodamine 6G	34	41	113	141	140	Atlantic Chemical Corp.; C. Lever Co.; International Dyestuffs Corp.; BASF Wyandotte Corp.; Dye Specialties, Inc.; C. Lever Co.; Sun Chemical Corp.; Mobay Chemical Corp.; CIBA-GEIGY Corp.; BASF Wyandotte Corp.
Pigment Red 81	45160:1	[12224-98-5]		(263)	(233)	(254)	(227)	(186)	
Basic Violet 10	45170	[81-88-9]	Rhodamine B	91	82	102	154	75	Atlantic Chemical Corp.; American Cyanimid Co.; Buffalo Color Corp.; Sun Chemical Corp.; Leeben Color, Div. of Tricon Colors, Inc.; International Dyestuffs Corp.; BASF Wyandotte Corp.; Dye Specialties, Inc.; Mobay Chemical Corp.; American Cyanimid Co.; C. Lever Co.; Buffalo Color Corp.; International Dyestuffs Corp.; Dye Specialties, Inc.; CIBA-GEIGY-Corp.; Sun Chemical Corp.
Solvent Red 49	45170:1	[509-34-2]	Rhodamine B Base	6.8	11	21	2.7	14	
Pigment Violet 1	45170:2	[1326-03-0]			(64)	(62)	(120)	(104)	
Basic Violet 11	45175	[2390-63-8]	Fanal Red 6BM (IG)	29	14	22	27	16	
Solvent Green 4	45550	[81-37-8]	Fluorescent Brightener 74	0.05	0.2	0.1			
Mordant Red 27	45180	[6539-22-4]	Chromoxane Brilliant Red				4.5	4.5	American Hoechst Corp.
Food Red 14	45430	[16423-68-0]	erthrosine						Warner Jenkinson Co.; Hilton-Davis Chemical Group of Sterling Drug, Inc.; Crompton and Knowles Corp.; Leeben Color, Division of Tricon Colors, Inc.

The rhodamines are economically the most important amino-substituted xanthene dyes. The total sales of Rhodamine B in the United States in 1980 were over 10^7. The total domestic market for fluorescein and uranine was estimated to be over $\$0.5 \times 10^6$/yr.

Health and Safety Factors, Toxicology

Xanthene dyes have not exhibited health or safety properties warranting special precautions; however, standard chemical labeling instructions are required. Toxicological properties of important dyes are listed in Table 2 (12).

Table 2. Toxicological Properties of Selected Xanthene Dyes[a]

Compound	Structure	Property	Value, mg/kg
xanthene	(**1b**)	LD$_{50}$ (mouse), subcutaneous	690
fluorescein	(**5**)	LD$_{Lo}$ (rat), intraperitoneal	600
		LD$_{Lo}$ (mouse)[b]	600
		LD$_{Lo}$ (rabbit), intravenous	300
		LD$_{Lo}$ (guinea pig)[b]	400
eosine	(**30**)	LD$_{Lo}$ (rat), intraperitoneal	500
		LD$_{Lo}$ (rat), subcutaneous	1,500
		TD$_{Lo}$ (rat), subcutaneous	1,300
		LD$_{50}$ (mouse), intravenous	550
		LD$_{Lo}$ (rabbit), intravenous	300
erythrosine	(**31**)	LD$_{50}$ (rat), intraperitoneal	300
		LD$_{Lo}$ (rat), intravenous	200
		LD$_{Lo}$ (rabbit), intravenous	200
		LD$_{Lo}$ (mouse), oral	2,500
		LD$_{50}$ (mouse), intravenous	370
Phloxine B	(**34**)	LD$_{50}$ (mouse), intravenous	310
		TD$_{Lo}$ (mouse), oral	39,600
		TD$_{Lo}$ (mouse)[b]	66,000
		TD$_{Lo}$ (rat)[b]	63,000
merbromin	(**33**)	LD$_{Lo}$ (mouse), subcutaneous	20
		LD$_{Lo}$ (rabbit), intravenous	15
		LD$_{Lo}$ (mouse), intravenous	50
uranine	(**29**)	LD$_{50}$ (rat), intraperitoneal	1,700
		TD$_{Lo}$ (rat), subcutaneous	19
		LD$_{50}$ (mouse), intraperitoneal	1,800
		LD$_{50}$ (mouse), oral	4,700
		LD$_{50}$ (rat), oral	6,700
		LD$_{Lo}$ (guinea pig), intraperitoneal	1,000
Rhodamine B	(**15**)	LD$_{Lo}$ (rat), oral	500
		TD$_{Lo}$ (rat), subcutaneous	360
		LD$_{50}$ (rat), intravenous	89,500
		LD$_{Lo}$ (mouse), intraperitoneal	128
Rhodamine 6G	(**18**)	TD$_{Lo}$ (rat), subcutaneous	100
		LD$_{Lo}$ (mouse), intraperitoneal	2

[a]Ref. 12.   [b]Administration method unknown.

BIBLIOGRAPHY

"Xanthene Dyes" in *ECT* 1st ed., Vol. 15, pp. 136–149 by W. G Huey and S. K. Morse, General Aniline and Film Corp.; in *ECT* 2nd ed., Vol. 22, pp. 430–437 by F. F. Cesark, American Cyanamid Co.; in *ECT* 3rd ed., Vol. 24, pp. 662–677 by R. E. Farris, Sandoz Colors and Chemicals.

1. A. Baeyer, *Chem. Ber.* **4**, 558, 662 (1871).
2. H. A. Lubs, ed., *The Chemistry of Synthetic Dyes and Pigments*, American Chemical Society Monograph Series, Reinhold Publishing Corp., New York, 1955.
3. O. Valdes-Aguilera and D. C. Neckers, *Acc. Chem. Res.* **22**(5), 171–177 (1989).
4. O. Valdes-Aguilera and D. C. Neckers, *Adv. Photoechem.* **18**, 315–394 (1993).
5. A. K. Chibisov and G. V. Zakharova, *Usp. Nauchn. Fotogr.* **25**, 114–136 (1989).
6. I. Carmichael and G. L. Hug, *J. Phys. Chem. Ref. Data* **15**(1), 1–250 (1986).
7. Y. Y. Marchant, *ACS Symp. Ser.* **339**, 168–175 (1987).
8. D. C. Neckers, *J. Photochem. Photobiol. A*, **471**(1), 1–29 (1989).
9. A. F. Lopez, A. T. Lopez, L. E. Gil, and A. I. Lopez, *Appl. Fluoresc. Technol.* **2**(5), 8–12 (1990).
10. J. R. Heitz, *ACS Symp. Ser.* **616**, 1–16 (1995).
11. J. R. Heitz, in J. R. Coats, ed., *Insecticide Mode Action*, Academic Press, New York, 1982, pp. 429–457.
12. D. V. Sweet, *Registry of Toxic Effects of Chemical Substances*, 1986 ed., U.S. Dept. of Health and Human Services, National Institute for Occupational Safety and Health, Cincinnati, Ohio.
13. X. Zhang and T. Shen, *Hauxue Tongbao*, (6), 8–14 (1995).
14. S. Speiser and F. L. Chisena, *Spec. Pub. R. Chem. Soc.* **69**, 211–216 (1989).
15. Jpn. Pat. 6227850 (Dec. 3, 1987), N. Nakayama, T. Isoda, Y. Watanabe, M. Aoki, and co-workers (to Ricoh Co.).
16. Brit. Pat. 2283744 (May 17, 1995), R. P. Haugland, M. N. Malekzadeh, and Y. Zhang (to Molecular Probes Inc.).
17. X. He, X. Gu, G. Zhao, and G. Dai *Wuli Huaxue Xuebao*, **11**(6), 504–508 (1995).
18. Brit. Pat. 2277097 (Oct. 19, 1994), M. A. Kuhn and R. P. Haughland (to Molecular Probes Inc.).
19. M. Torneiro and W. C. Still, *J. Am. Chem. Soc.* **117**(21), 5887–5888 (1995).
20. E. Reisfeld, *J. Phys. IV*, **4** (1994).
21. U.S. Pat. 5,111,472 (May 5, 1992), P. R. Hammond and J. F. Feeman (to U.S. Dept. of Energy).
22. U.S. Pat. 4,945,176 (July 31, 1990), P. R. Hammond and J. F. Feeman (to U.S. Dept. of Energy).
23. U.S. Pat. 3,961,039 (June 1, 1976), R. Sternheimer.
24. *BIOS Report 959*, British Intelligence Objectives Subcommittees, U.K., 1946, pp. 8 and 15.
25. U.S. Pat. 3,849,065 (Nov. 19, 1974), K. Schmeidl (to BASF).
26. U.S. Pat. 3,767,358 (Oct. 23, 1973), H. I. Stryker (to DuPont).
27. U.S. Pat. 3,690,889 (Sept. 12, 1972), S. E. Harrison and R. Drake (to RCA Corp.).
28. U.S. Pat. 3,767,408 (Oct. 23, 1973), S. E. Harrison and J. E. Goldmacher (to RCA Corporation).
29. U.S. Pat. 3,796 (Mar. 12, 1974), R. T. Hickcox (to Herculea, Inc.).
30. U.S. Pat. 3,644,270 (Feb. 22, 1972), G. D. V. Aliaveedam (to DuPont).
31. Brit. Pat. 1,494,768 (Dec. 14, 1977), D. M. Zabiak and K. S. Hwang (to A. B. Dick Co.).
32. U.S. Pat. 3,541,470 (Nov. 17, 1970), J. R. Lankard and P. P. Sorokin (to IBM Corp.).
33. U.S. Pat. 4935059 (June 23, 1988), U. Mayer and A. Oberlinner (to BASF).
34. U.S. Pat. 4647675 (July 12, 1984), U. Mayer and A. Oberlinner (to BASF).

35. U.S. Pat. 5410053 (May 11, 1992), B. Albert, W. Denziger, E. Kahn, and C. Kraeh (to BASF).
36. Brit. Pat. 1,503,380 (Mar. 8, 1978) (to Hoechst AG).
37. U.S. Pat. 3,956,300 (May 11, 1976), P. W. Austin, A. T. Costello, and A. Crabtree (to ICI Ltd.).
38. Brit. Pat. 1,377,695 (Dec. 18, 1974), P. W. Austin and A. T. Costello (to ICI Ltd.).
39. Brit. Pat. 1,586,820 (Mar. 25, 1981) (to Hoechst AG).
40. Brit. Pat. 1,471,452 (Apr. 27, 1977) (to Hoechst AG); Brit. Pat. 1,471,453 (Apr. 27, 1977) (to Hoechst AG).
41. U.S. Pat. 3,772,335 (Nov. 13, 1973), F. Meininger and F. Kohlhass (to Hoechst AG).
42. U.S. Pat. 3,883,529 (May 13, 1975), P. W. Austin (to ICI Ltd.).
43. U.S. Pat. 3,988,492 (Oct. 26, 1976), S. M. Spatz (to Mead Corp.).
44. U.S. Pat. 3,929,825 (Dec. 30, 1975), S. M. Spatz (to Mead Corp.).
45. Brit. Pat. 1,429,597 (Mar. 24, 1978), G. O. Okikiolu (to Okikiolu Scientific and Industrial Organization).
46. Brit. Pat. 1,533,255 (Nov. 22, 1978), G. G. Barraclough, D. Myles, D. A. Reilly, and P. Tomlinson (to ICI Ltd.).
47. U.S. Pat. 3,951,655 (Apr. 20, 1976), B. Schoustra and H. Roncken (to Oce-van der Grinten NV, Netherlands).
48. U.S. Pat. 3,975,159 (Aug. 17, 1976), S. Van Heusden (to U.S. Philips Corp.).
49. U.S. Pat. 3,615,566 (Oct. 26, 1971), I. D. Robinson (to Arthur D. Little, Inc.).
50. U.S. Pat. 4,216,101 (Aug. 5, 1980), H. J. Davis (to Canada Wire and Cable Ltd.).
51. K. Ventkataraman, ed., *The Chemistry of Synthetic Dyes*, Vol. 2, Academic Press, Inc., New York, 1952.
52. U.S. Pat. 3,853,884 (Dec. 10, 1974), H. Troster (to Hoechst AG).
53. U.S. Pat. 3,888,862 (June 10, 1975), F. Meininger and F. Kohlhass (to Hoechst AG).

General References

D. R. Waring and G. Hallas, ed., *The Chemistry and Applications of Dyes*, Plenum Publishing Corp., New York, 1990.

P. Gregory, *High-Technology Applications of Organic Colorants*, Plenum Publishing Corp., New York, 1990.

G. Booth, *The Manufacture of Organic Colorants and Intermediates*, The Society of Dyers and Colourists, Bradford, U.K., 1988.

P. F. Gordon and P. Gregory, *Organic Chemistry in Colour*, Springer Verlag, Berlin, 1983.

G. Hallas, in J. Shore, *Colorants and Auxiliaries*, Vol. 1, Society of Dyers and Colourists, 1990, pp. 279–294.

E. N. Abrahart, *Dyes and their Intermediates*, 2nd ed., Chemical Publishing, New York, 1977.

R. L. M. Allen, *Color Chemistry*, Appleton-Century-Crofts, New York, 1971.

K. Ventkataraman, ed., *The Chemistry of Synthetic Dyes*, Vol. 2, Academic Press Inc., New York, 1952.

Color Index, 3rd ed., The Society of Dyers and Colourists, Bradford, U.K. and the American Associations of Textile Chemists and Colorists, N.C., Vols. 1–6, 1971.

P. Rys and H. Zollinger, *Fundamentals of the Chemistry and Applications of Dyes*, Wiley-Interscience, New York, 1972.

H. A. Lubs, ed., *The Chemistry of Synthetic Dyes and Pigments*, American Chemical Society Monograph Series, Reinhold Publishing Corp., New York, 1955.

K. Venkataraman, ed., *The Analytical Chemistry of Synthetic Dyes*, Wiley-Interscience, New York, 1977.

J. Fabian and H. Hartmann, *Light Absorption of Organic Colorants*, Springer Verlag, Berlin, 1980.

T. E. Furia, ed., *Handbook of Food Additives*, 2nd ed., CRC Press, Cleveland, Ohio, 1972.

D. M. Marmion, *Handbook of U.S. Colorants for Food, Drugs, and Cosmetics*, John Wiley & Sons, Inc., New York, 1979.

E. Gurr, *Synthetic Dyes in Biology, Medicine and Chemistry*, Academic Press, London, 1971.

R. Raue, *Rev. Prog. Coloration*, **14**, 187 (1984).

PAUL WIGHT
Zeneca Specialties

XYLENES AND ETHYLBENZENE

Xylenes and ethylbenzene [*100-41-4*] (EB) are C_8 aromatic isomers having the molecular formula C_8H_{10}. The xylenes consist of three isomers: *o*-xylene [*95-47-6*] (OX), *m*-xylene [*108-38-3*] (MX), and *p*-xylene [*106-42-3*] (PX). These differ in the positions of the two methyl groups on the benzene ring. The molecular structures are shown below.

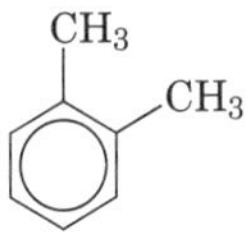

     o-Xylene (OX) 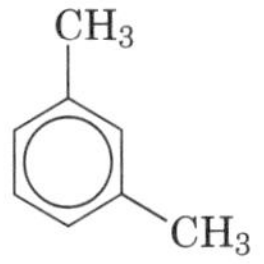   *m*-Xylene (MX)    *p*-Xylene (PX) 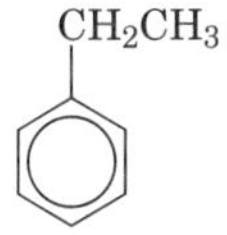   Ethylbenzene (EB)

Sources and Uses

The term mixed xylenes describes a mixture containing the three xylene isomers and usually EB. Commercial sources of mixed xylenes include catalytic reformate, pyrolysis gasoline, toluene disproportionation product, and coke-oven light oil. Ethylbenzene is present in all of these sources except toluene disproportionation product. Catalytic reformate is the product obtained from catalytic reforming processes. In catalytic reforming, a low octane naphtha cut (typically a straight run or hydrocracked naphtha) is converted into high octane aromatics, including, benzene, toluene, and mixed xylenes (see BTX PROCESSING). Aromatics are separated from the reformate using a solvent such as diethylene glycol or sulfolane and then stripped from the solvent. Distillation is then used to separate the BTX into its components. The amount of xylenes contained in the catalytic reformate depends on the fraction and type of crude oil, the reformer operating conditions, and the catalyst used. The amount of xylenes produced can vary

PX is not separated via distillation because its boiling point is too close to that of MX. Instead, the differences in freezing points and adsorption characteristics are exploited commercially, as described in detail herein.

Since xylenes are important components of gasoline, their combustion and octane characteristics are of interest. The critical compression ratios are 14.2, 1.36, and 9.6 for PX, MX, and OX, respectively. The research octane numbers are 116.4, 117.5, 107.4, and 113 for PX, MX, OX, and EB, respectively (7). The motor octane numbers are 110.0, 111.5, 100, and 105 for PX, MX, OX, and EB respectively.

Chemical Properties

Chemical reactions that the xylenes participate in include (1) migration of the methyl groups, (2) reaction of the methyl groups, (3) reaction of the aromatic ring, and (4) complex formation.

Migration of the Methyl Groups.   Reactions that involve migration of the methyl groups include isomerization, disproportionation, and dealkylation. The interconversion of the three xylene isomers via isomerization is catalyzed by acids. The acids can be liquids or solids. One example of an acidic liquid-phase system is hydrogen flouride–boron trifluoride (8). At low boron trifluoride concentrations, the xylenes isomerize to near equilibrium levels. At high boron trifluoride concentrations, a complex containing MX–hydrogen fluoride–boron trifluoride is formed which can be decomposed to produce high purity MX. Two other acidic liquid-phase systems that can isomerize the xylenes are hydrogen bromide in toluene and aluminum bromide in toluene (9). Examples of solid acids include aluminum-based materials and zeolites.

The mechanism of these reactions involves the rapid and reversible addition of a proton to the aromatic ring, followed by 1,2-intramolecular methyl shifts (10):

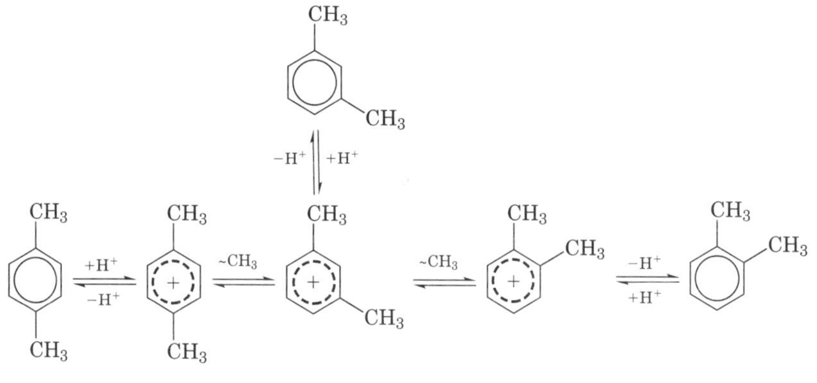

As shown in Figure 1, the equilibrium concentration is affected slightly by temperature (11). The actual concentration is affected by the reaction rate and the initial concentration of each isomer. Deviations beyond equilibrium can be achieved when zeolites are used, owing to shape selectivity (see MOLECULAR SIEVES). The thermal isomerization of the three xylenes has been studied at

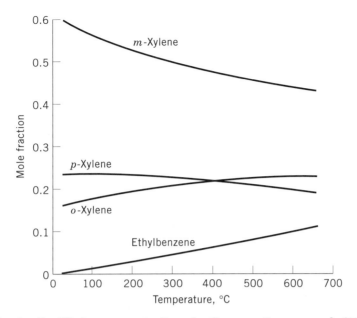

Fig. 1. Equilibrium concentrations for C_8 aromatic compounds (11).

1000°C (12). Side reactions predominated, and only a small percentage of xylenes was interconverted.

Transalkylation is also catalyzed by acids, but requires more severe conditions than isomerization. As shown below, the methyl migration is intermolecular and ultimately produces a mixture of aromatic compounds ranging from benzene to hexamethylbenzene. The overall equilibrium constants for all possible methylbenzenes have been determined experimetally and calculated theoretically (Fig. 2 and Table 3).

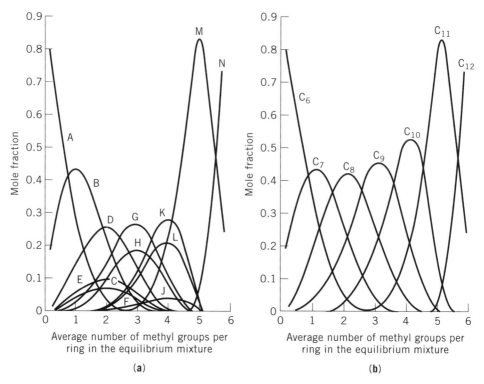

Fig. 2. Equilibrium concentrations of methyl benzenes in ideal gas state at 25°C (13). A = benzene; B = toluene; C = OX; D = MX; E = PX; F = hemimellitene; G = pseudocumene; H = mesitylene; J = prehnitene; K = isodurene; L = durene; M = pentamethylbenzene; N = hexamethylbenzene.

Table 3. Overall Equilibrium Constants of Methylbenzenes[a]

	$K_{eq}^{b,c}$			
Hydrocarbon	Reference 15	Reference 16	Reference 17	HCJ[d] and inductive model[a]
benzene		0.09	2×10^{-4}	$1-2 \times 10^{-4}$
toluene	~0.01	0.63	0.25	0.11−0.13
OX	2	1.1		1.8
MX	20	26	300	90−110
PX	1	1	1	1
pseudocumene	40	63		110−140
hemimellitene	40	69		200−310
durene	120	140		510−810
prehnitene	170	400		960−1700
mesitylene	2800	1.3×10^4	2×10^5	$2-3.6 \times 10^4$
isodurene	5600	1.6×10^4		$0.6-1 \times 10^5$
pentamethylbenzene	8700	2.9×10^4		$1.1-2.5 \times 10^5$
hexamethylbenzene	8.0×10^4	9.7×10^4	1×10^7	$0.7-1.5 \times 10^6$

[a]Ref. 14.
[b]Compared to PX.
[c]Column headings indicate sources of various test results.
[d]HCJ = hyperconjugative.

Reactions of the Methyl Groups. These reactions include oxidation, poly-condensation, and ammoxidation. PX can be oxidized to both terephthalic acid and dimethyl terephthalate, which are then condensed with ethylene glycol to form polyesters. Oxidation of OX yields phthalic anhydride, which is used in the production of esters. These are used as plasticizers for synthetic polymers. MX is oxidized to isophthalic acid, which is also converted to esters and eventually used in plasticizers and resins (see PHTHALIC ACIDS AND OTHER BENZENEPOLY-CARBOXYLIC ACIDS).

In a study of the slow combustion of the three xylenes, it was observed that OX is much more reactive towards oxygen than MX and PX (18). Under identical conditions, OX was approximately ten times as reactive as its isomers. It was proposed that the initial steps in the mechanisms of formation of each isomer are the same.

In the oxidation of PX and MX, formaldehyde is a degenerate branching inter-mediate, whereas phthalan is formed from OX:

phthalan

PX forms *p*-xylylene when heated above 1200°C. The structure of *p*-xylylene is represented by a *p*-quinoid structure or as a *p*-benzenoid biradical. Condensation yields poly(*p*-xylylene) (19–22) (see XYLENE POLYMERS).

p-quinoid structure   *p*-benzenoid biradical   poly(*p*-xylylene)

The methyl groups on xylenes can undergo ammoxidation, reaction with ammonia and oxygen (23).

$$2\ RCH_3 + 2\ NH_3 + 3\ O_2 \longrightarrow 2\ RCN + 6\ H_2O$$

The Showa Denka Co. practices this reaction with a PX–MX mixture (24), whereas Mitsubishi Gas Chemical Company uses high purity MX first to form the dicyanide (25). In both processes, hydrogenation to the diamine follows. *m*-Xylenediamine is reacted with phosgene to give *m*-xylene diisocyanate, which is used in urethane resins (26–28).

Reactions of the Aromatic Ring.   The reactions of the aromatic ring of the C_8 aromatic isomers are generally electrophilic substitution reactions. All of

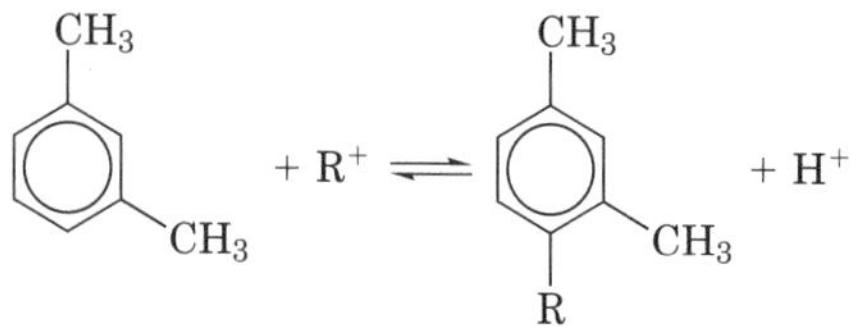

the classical electrophilic substitution reactions are possible (see FRIEDEL-CRAFTS REACTIONS), but in most instances they are of little practical significance. The relative nuclear chlorination rates of polymethylbenzenes have been studied (29,30). The higher the degree of substitution, the higher is the rate of chlorination.

As in most electrophilic reactions, the ability to stabilize the positive charge generated by the initial addition strongly affects the relative rates. MX reacts faster than OX and PX because both methyl groups work in conjunction to stabilize the charge on the next-but-one carbon. Sulfonation was, at one time, used to separate MX from the other C_8 aromatic isomers. MX reacts most rapidly to form the sulfonic acid which remains in the aqueous phase. The sulfonation reaction is reversible, and MX can be regenerated.

Hydrogenation of the aromatic ring to form naphthenic compounds has been proposed as a route to facilitate the separation of the C_8 aromatic isomers (31). The spread in boiling points of the naphthenic compounds is 12°C vs a spread of 8°C for the aromatic compounds. However, the cycloparaffinic products obtained from OX and EB boil only 3°C apart, impeding the separation.

Complex Formation.   All four C_8 aromatic isomers have a strong tendency to form several different types of complexes. Complexes with electrophilic agents are utilized in xylene separation. The formation of the HF–BF$_3$–MX complex is the basis of the Mitsubishi Gas–Chemical Company (MGCC) commercial process for MX recovery, discussed herein. Equimolar complexes of MX and HBr (mp -77°C) and EB and HBr (mp -103°C) have been reported (32,33). Similarly, HCl complexes undergo rapid formation and decomposition at -80°C (34).

Werner complexes can be used to form clathrates with the C_8 aromatic isomers (35–42). The aromatic compounds are released upon heating. Since the uptake and release characteristics of the four C_8 aromatic isomers are each different, this method has been suggested as a means of separating the isomers.

Inclusion compounds of the C_8 aromatic compounds with tris(o-phenyl-enedioxy)cyclotriphosphazene have been used to separate the individual isomers (43–47). The Schardinger dextrins, such as alpha-cyclodextrin, beta-dextrin, and gamma-dextrin are used for clathration; alpha-dextrin is particularly useful for recovering PX from a C_8 aromatic mixture (48,49). Pyromellitic dianhydride (50) and beryllium oxybenzoate (51) also form complexes, and procedures for separations were developed.

A 1:1 complex melting at 24.8°C is formed between PX and carbon tetra-chloride (52). The other C_8 aromatic compounds do not form these complexes. Carbon tetrabromide and chloral (CCl_3CHO) form addition compounds with PX.

Manufacture of Xylenes

The initial manufacture of mixed xylenes and the subsequent production of high purity PX and OX consists of a series of stages in which (1) the mixed xylenes are initially produced; (2) PX and/or OX are separated from the mixed xylenes stream; and (3) the PX- (and perhaps OX-) depleted xylene stream is isomerized back to an equilibrium mixture of xylenes and then recycled back to the separation step. These steps are discussed below.

Mixed Xylenes Production Via Reforming. Again, two principal methods for producing xylenes are catalytic reforming and toluene disproportionation. A general schematic for the production of PX and OX (along with benzene and toluene) via catalytic reforming is shown in Figure 3 (see BTX PROCESSING). In

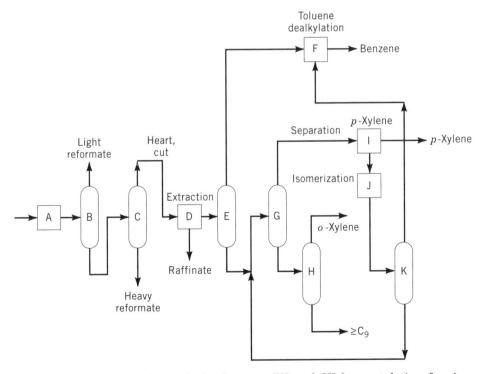

Fig. 3. General scheme for producing benzene, PX, and OX from catalytic reforming.

this example, a light fraction (ie, 65–175°C) from a straight run petroleum fraction or from an isocracker is fed to a catalytic reformer, unit A. This is followed by heart-cutting and extraction in units B, C, and D. The mixed xylenes stream must then be processed further to produce high purity PX and/or OX. As discussed herein, high purity OX can be produced via distillation. However, because of the close boiling points of PX and MX, using distillation to produce high purity PX is impractical. Instead, other separation methods such as crystallization and adsorption are used. These separation processes are discussed herein.

Xylenes Production Via Toluene Transalkylation and Disproportionation. The toluene that is produced from processes such as catalytic reforming can be converted into xylenes via transalkylation and disproportionation. Toluene disproportionation is defined as the reaction of 2 mol of toluene to produce 1 mol of xylene and 1 mol of benzene. Toluene transalkylation is defined as the reaction of toluene with C_9 or higher aromatics to produce xylenes:

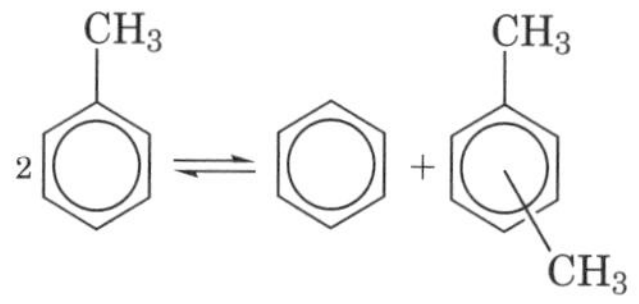

Disproportionation

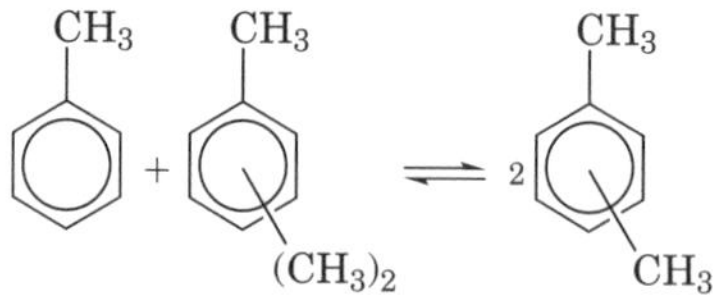

Transalkylation

Other species that are also present in the feed, such as ethylbenzene and methylethylbenzenes will also undergo transalkylation reactions. These reactions tend to approach an equilibrium that depends on the operating conditions.

There are several commercial processes that produce xylenes via disproportionation or transalkylation. These include: UOP's Tatoray and PX-Plus, ARCO's Xylenes Plus, and Mobil's MTDP and STDP.

The Tatoray process was originally developed by Toray and is currently licensed by UOP (53–57). A schematic of the process is shown in Figure 4. In this process, toluene or a mixture of toluene and C_{9+} aromatics are reacted to form primarily xylenes and benzene. An equilibrium distribution of xylenes is produced. As shown in Table 4, the ratio of xylenes to benzene can be adjusted by altering the feed ratio to toluene to C_9 aromatics. Trimethylbenzenes are the preferred C_9 aromatic compound.

Catalyst improvements have been made since the process was first developed. The current catalyst is designated TA-4 (57). It has a high per pass conversion and good stability. Yields to xylenes are reported to be over 97%. The pelleted catalyst is used in a fixed bed reactor in the presence of hydrogen. Typical operating conditions are: 350–530°C, 1–5 MPa (10–50 atm), and H_2/hydrocarbon

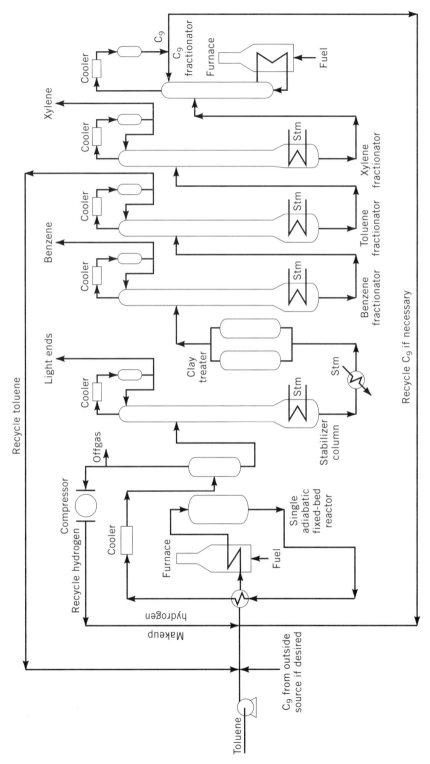

Fig. 4. UOP Tatoray process for xylenes production.

process known as the (T2PX) process in 1984 (70). It uses a proprietary catalyst to react toluene at 42–48% conversion with selectivities to benzene of 42 wt % and to xylenes of 46 wt %. The xylenes produced are at equilibrium. Typical commercial operating conditions of 390–495°C, H_2 partial pressure of 4.1 MPa, H_2/hydrocarbon molar ratio of 4:1, and LHSV of 1–2/h. Fina's first commercial implementation occurred in 1985 at their Port Arthur refinery.

Separation Processes for PX.   There are essentially two methods that are currently used commercially to separate and produce high purity PX: (*1*) crystallization and (*2*) adsorption. A third method, a hybrid crystallization/adsorption process, has been successfully field-demonstrated and the first commercial unit is expected in the near future.

Crystallization.   Low temperature fractional crystallization was the first and for many years the only commercial technique for separating PX from mixed xylenes. As shown in Table 2, PX has a much higher freezing point than the other xylene isomers. Thus, upon cooling, a pure solid phase of PX crystallizes first. Eventually, upon further cooling, a temperature is reached where solid crystals of another isomer also form. This is called the eutectic point. PX crystals usually form at about −4°C and the PX–MX eutectic is reached at about −68°C. In commercial practice, PX crystallization is carried out at a temperature just above the eutectic point. At all temperatures above the eutectic point, PX is still soluble in the remaining C_8 aromatics liquid solution, called mother liquor. This limits the efficiency of crystallization processes to a per pass PX recovery of about 60–65%.

The solid PX crystals are typically separated from the mother liquor by filtration or centrifugation. Good solid/liquid separation is important for obtaining high purity PX. One key to good separation is crystal size. The larger the crystal, the better the separation. Crystal size is affected by the degree of supersaturation and nucleation, which in turn is affected by a number of parameters, including temperature, agitation, and the presence of crystal growth sites.

PX crystals are typically produced in two or more stages of crystallization, separated by centrifuges. Commercial crystallizers use either direct contact or indirect refrigeration. The latter has the disadvantage that the walls of the cooled surface tend to foul, which reduces heat transfer. The first crystallizer stage is usually at the lowest temperature. The cake from this stage has a purity of about 80–90%. The impurity arises from the mother liquor which wets the crystal surface or is occluded in the crystal cake. The efficiency of the solid–liquid separation depends on the temperature and the loading of the centrifuges. As temperature falls, the viscosity and density of the mother liquor rise sharply. Thus, it becomes more difficult for the centrifuges to achieve effective separation.

In the second crystallizer stage, the crystals are usually reslurried with a higher purity PX stream from a later stage of purification. A second stage of centrifugation is sufficient in most cases to give PX purity >99%.

Currently, about 40% of the PX produced worldwide uses crystallization technology. A number of crystallization processes have been commercialized over the years. The more common ones are those developed by Chevron, Krupp, Amoco, ARCO (Lyondell), and Phillips. Some of the features of these processes are discussed herein.

The Chevron process (71) is shown in Figure 5. It consists of two crystallizers in series operated at difficult pressures. Direct contact cooling is used. This is accomplished by injecting liquid CO_2 with the feed to the crystallizer. As the slurry rises, part of the CO_2 vaporizes, causing the temperature to drop below the saturation temperature, and crystallization occurs. Because cooling is gradual, the degree of supersaturation is low and thus crystal growth occurs on the existing crystals. This leads advantageously to the formation of relatively large crystals, rather than many small ones. The crystals and slurry move down from the crystallizer body. Most of the slurry is recycled, but some is withdrawn and sent to the second crystallizer, which is operated under vacuum. The operation of the second crystallizer is similar to the first, except that typically it is not necessary to inject additional CO_2. The crystals are separated from the mother liquor in two stages. The first stage uses screen bowl centrifuges, and the second uses pusher centrifuges. The Chevron process offers the advantage that large crystals are obtained in a relatively short residence time, which permits good solid–liquid separation in the centrifuges. Plants using this process have been licensed and built in the United States, Germany, Mexico, Japan, and in the United Kingdom.

The Krupp process for PX crystallization uses scraped chillers for crystallization (72). Rotary drum filters and centrifuges are used for phase separation. The first-stage solids are reslurried in a PX-rich filtrate from the final PX

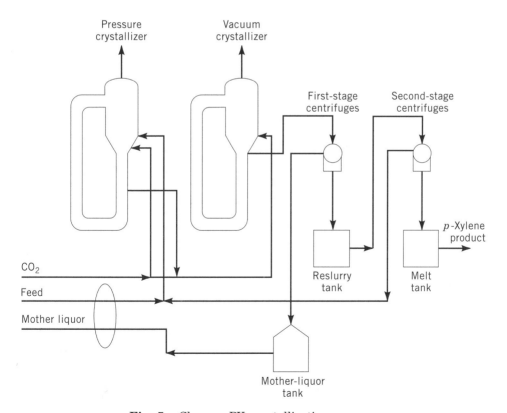

Fig. 5. Chevron PX crystallization process.

product centrifuge. Another feature is that the second-stage mother liquor is further chilled and centrifuged before being passed back to the first-stage feed. The resultant solids are added to the first-stage crystals.

The Amoco PX crystallization process (73,74) is a two-stage process that operates with indirect cooling. A schematic of this process is shown in Figure 6. Ethylene is used as the coolant in the first stage and propane is used in the second stage. In the first-stage crystallizer, the temperature is brought down in stages to near the PX–MX eutectic. The first stage cake is melted and sent to a second-stage crystallizer, which is designed like the first, but uses propane refrigerant instead of ethylene. The crystallizers are fitted with scrapers mounted on a central shaft, which provides agitation and maintains a good heat-exchange surface. The residence time in each of the two crystallizers is about 3 h, in order to encourage crystal growth.

The ARCO process is similar to the Amoco process in that it is a two-stage crystallization process that uses external ethylene refrigerant in the first stage and propane in the second stage (75,76). The first stage consists of two crystallizers which operate at −40 to −50°C and −55 to −70°C, respectively. The first-stage slurry flows to a continuous centrifuge where crystals of 85–90% purity are removed from the filtrate. The crystals are remelted and then sent to a single second-stage crystallizer operated at −20 to 0°C, depending on the feedstock and desired purity. The crystal slurry flows from the crystallizer to a second-stage continuous centrifuge. The crystals are washed with toluene in the centrifuge and melted. The liquid is then fed to a PX–toluene splitter where high purity PX is taken as a bottoms product. Eight plants have been built worldwide.

The Maruzen process uses ethylene gas as a direct refrigerant in a two-stage process (77,78). The first stage slurry is centrifuged, partially remelted, and fed to the second centrifuge.

The Phillips process is a two-stage crystallization process that uses a pulsed column in the second stage to purify the crystals (79,80). In the pulsed column, countercurrent contact of the high purity PX liquid with cold crystals results in displacement of impurities. In the first stage, a rotary filter is used. In both stages, scraped surface chillers are used. This process was commercialized in 1957, but no plants in operation as of 1996 use this technology.

Relatively recently, several companies have commercialized static crystallization processes based on progressive freezing. In progressive freezing, crystals are allowed to form on a cooled surface until a certain proportion of the original batch is frozen. The remaining melt is drained away from the solid layer. In some versions, the frozen layer is also gently warmed to sweat out impurities caught in the crystal structure. Finally, the PX is recovered by heating the surface to completely melt off the frozen layer. In systems developed by BEFS and Sulzer Chemtech, the crystal layer is grown on vertical cooled plates. In the BEFS PROKEM PROABD process, feed containing about 80–99.5% PX is further purified to 99.9+% in a single-stage melt static crystallizer (81). After crystallization, the crystalline mass is purified by a partial melting of the crystals to wash out adhering impurities. This process can be used as a finishing step to further improve the purity of PX product from adsorption, crystallization, or MSTDP units. An 8000-t/yr plant has been in operation for three years in

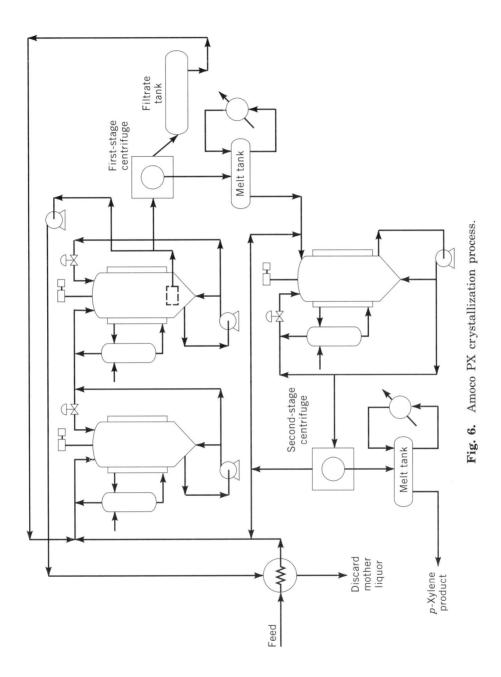

Fig. 6. Amoco PX crystallization process.

847

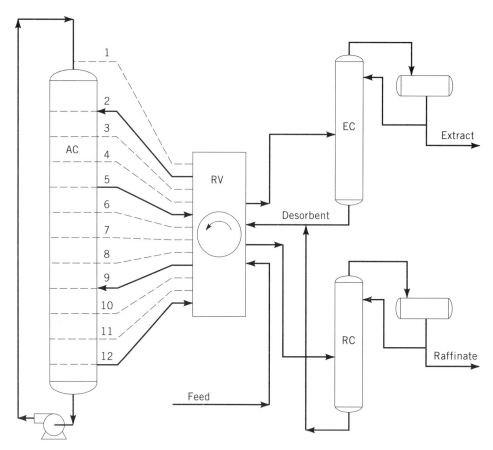

Fig. 8. UOP Parex simulated moving bed for adsorptive separation. AC = adsorbent chamber; RV = rotary valve; EC = extract column; RC = raffinate column. Lines: 2-desorbent; 5-extract; 9-feed; 12-raffinte. All other ports are closed at this time.

The Aromax process was developed in the early 1970s by Toray Industries, Inc. in Japan (95–98). The adsorption column consists of a horizontal series of independent chambers containing fixed beds of adsorbent. Instead of a rotary valve, a sequence of specially designed on–off valves under computer control is used to move inlet and withdrawal ports around the bed. Adsorption is carried out in the liquid phase at 140°C, 785–980 kPA, and 5–13 L/h. PX yields per pass is reported to exceed 90% with a typical purity of 99.5%. The first Aromax unit was installed at Toray's Kawasaki plant in March 1973. In 1994, IFP introduced the Eluxyl adsorption process (59,99). The proprietary adsorbent used is designated SPX 3000. Individual on–off valves controlled by a microprocessor are used. Raman spectroscopy to used to measure concentration profiles in the column. A 10,000 t/yr demonstration plant was started and successfully operated at Chevron's Pascagoula plant from 1995–96. IFP has licensed two hybrid units.

Asahi Chemical Industry Co. Ltd. was working to develop an adsorption process in the late 1970s and early 1980s that was to produce high purity EB as well as PX (100–103). In 1981 they reported that pilot plants results were being confirmed in larger equipment. However, this process does not appear to have been commercialized.

Hybrid Crystallization/Adsorption Process. In 1994, IFP and Chevron announced the development of a hybrid process that reportedly combines the best features of adsorption and crystallization (59,99). In this option of the Eluxyl process, the adsorbent bed is used to initially produce PX of 90–95% purity. The PX product from the adsorption section is then further purified in a small single-stage crystallizer and the filtrate is recycled back to the adsorption section. It is reported that ultra-high (99.9+%) purity PX can be produced easily and economically with this scheme for both retrofits of existing crystallization units as well as grass-roots units. A demonstration plant was built at Chevron's Pascagoula refinery in 1994.

MX Separation Process. The Mitsubishi Gas–Chemical Company (MGCC) has commercialized a process for separating and producing high purity MX (104–113). In addition to producing MX, this process greatly simplifies the separation of the remaining C_8 aromatic isomers. This process is based on the formation of a complex between MX and HF–BF_3. MX is the most basic xylene and its complex with HF–BF_3 is the most stable. The relative basicities of MX, OX, PX, and EB are 100, 2, 1, and 0.14, respectively.

MX of >99% purity can be obtained with the MGCC process with <1% MX left in the raffinate by phase separation of hydrocarbon layer from the complex-HF layer. The latter undergoes thermal decomposition, which liberates the components of the complex.

A schematic of the MGCC process is shown in Figure 9. The mixed C_8 aromatic feed is sent to an extractor (unit A) where it is in contact with HF–BF_3 and hexane. The MX–HF–BF_3 complex is sent to the decomposer (unit B) or the isomerization section (unit D). In the decomposer, BF_3 is stripped and taken overhead from a condensor–separator (unit C), whereas HF in hexane is recycled from the bottom of C. Recovered MX is sent to column E for further purification. The remaining C_8 aromatic compounds and hexane are sent to raffinate column F where residual BF_3 and HF are separated, as well as hexane for recycle. Higher boiling materials are rejected in column H, and EB and OX are recovered in columns I and J. The overhead from J is fed to unit K for PX separation. The raffinate or mother liquor is then recycled for isomerization.

The MGCC process is used in Japan, the United States, and Spain.

Xylene Isomerization. After separation of the preferred xylenes, ie, PX or OX, using the adsorption or crystallization processes discussed herein, the remaining raffinate stream, which tends to be rich in MX, is typically fed to a xylenes isomerization unit in order to further produce the preferred xylenes. Isomerization units are fixed-bed catalytic processes that are used to produce a close-to-equilibrium mixture of the xylenes. To prevent the buildup of EB in the recycle loop, the catalysts are also designed to convert EB to either xylenes, benzene and lights, or benzene and diethylbenzene.

Historically, the isomerization catalysts have included amorphous silica–aluminas, zeolites, and metal-loaded oxides. All of the catalysts contain acidity, which isomerizes the xylenes and if strong enough can also crack the EB and xylenes to benzene and toluene. Dual functional catalysts additionally contain a metal that is capable of converting EB to xylenes.

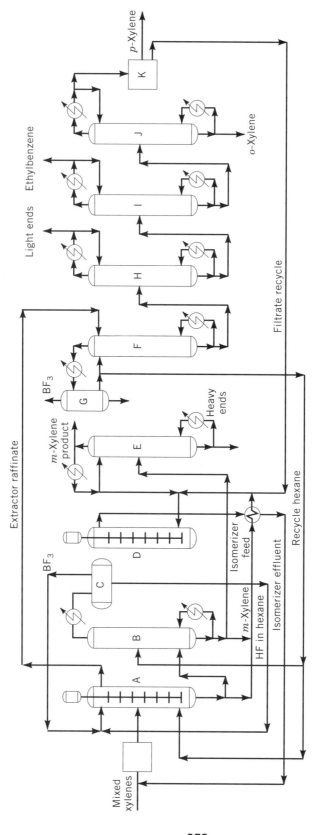

Fig. 9. Xylenes separation via Mitsubishi Gas–Chemical Co. HF-BF₃ extraction–isomerization process (107). A = extractor; B = decomposer; C = separator; D = isomerization reactor; E = heavy ends tower; F = raffinate tower; G = separator; H = light ends fractionator; I = ethylbenzene fractionator; J = OX fractionator; K = PX crystallizer.

The three major commercial licensors of xylenes isomerization processes are Engelhard, UOP, and Mobil. Several other companies have developed and used their own catalysts. These companies include Mitsubishi Gas–Chemical, Toray, ICI, Amoco, and Shell. All of these processes are discussed herein.

Dual Function Catalytic Processes. Dual-function catalytic processes use an acidic oxide support, such as alumina, loaded with a metal such as Pt to isomerize the xylenes as well as convert EB to xylenes. These catalysts promote carbonium ion-type reactions as well as hydrogenation–dehydrogenation. In the mechanism for the conversion of EB to xylenes shown, EB is converted to xylenes

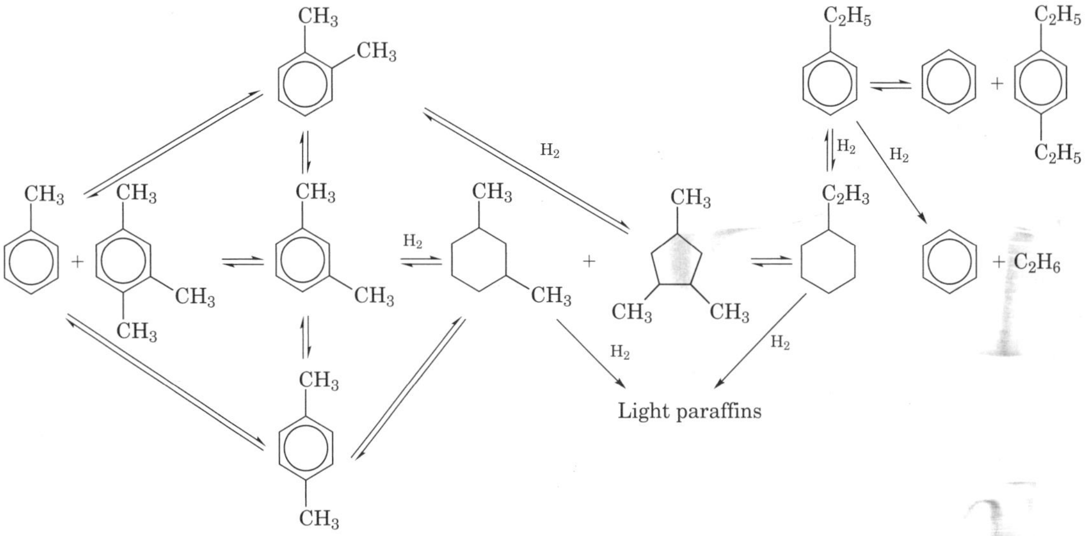

by hydrogenation of the aromatic ring and formation of naphthenic intermediates such as ethylcyclohexane, dimethylcyclohexanes, and trimethylcyclohexanes. To provide an effective pathway between EB and xylenes, a certain concentration of the naphthenic intermediates must be maintained. This is accomplished by proper selection of operating conditions and separation and recycle of the naphthenic intermediates. Side reactions include disproportionation and hydrodealkylation of aromatic compounds, and hydrodealkylation of naphthenes. Effective catalysts minimize the loss of xylenes and EB via these side reactions.

Commercial processes which use a dual-functional catalyst are Octafining, Isomar, and Isolene.

The Octafining process (114–116) was developed and commercialized by Atlantic Richfield and Engelhard in the early 1960s. The first-generation catalyst was prepared by mixing equal amounts of a silica–alumina cracking catalyst with Pt on alumina. The Pt content of the mixture was about 0.5 wt %. The EB approach to equilibrium was 88%, with an 80% selectivity to xylenes. Reaction conditions consist of temperature of 425–480°C; pressure of 1.14–2.51 MPa; H_2/hydrocarbon ratio $< 10:1$ (preferably 4–6:1); and LHSV = 0.6–1.6/h. An equilibrium mixture of xylenes is produced. To maintain conversion, the reaction temperature is gradually increased as the catalyst deactivates up to a maximum temperature of about 480°C. The catalyst is then regenerated. The first Octafining unit was placed onstream in 1960. As of 1995, 30 units had been licensed in 17 countries (99).

Recently, a new second-generation catalyst has been commercialized. This new catalyst, O-750, is formulated by loading Pt on alumina and combining the mixture with H−mordenite such that the final Pt content is <0.5 wt % (101). This catalyst, along with several other process improvements, forms the basis of the Octafining II process. Compared to Octafining I, yields and per pass conversions have been improved. Catalyst cycle lengths typically exceed 3 yr. The catalyst is capable of cracking C_9 nonaromatics, thereby increasing xylene purity. IFP has recently acquired the rights to license the Octafining II process worldwide (99).

UOP's Isomar process (56,117−119) was originally developed to use dual-functional catalysts. The first-generation catalyst contained Pt and halogen on alumina. Operating conditions using this catalyst were 399°C; 1.25 MPa; 2 LHSV; and H_2/hydrocarbon ratio of 6:1. A C_8 naphthene concentration of 2−9% was maintained in the process loop. In the mid-1980s UOP introduced an improved catalyst, I-9. Unlike the original catalyst, this catalyst does not require chlorine addition. Process conditions were modified to 388°C and 1.68 MPa.

In 1993, UOP commercialized an improved Pt-based catalyst, I-210. This catalyst is based on a molecular sieve, but not an aluminosilicate zeolite. UOP claims that yields are about 10% better than those for I-9 catalyst. EB to xylenes conversion is about 22−25% with a C_8 aromatics per pass loss of about 1.2−1.5%. As discussed below, UOP's Isomar process can also use zeolite catalysts which convert EB to benzene rather than to xylenes. UOP has licensed over 40 Isomar units.

The Isolene II process was commercialized in 1971 by Toray Industries (120−122). The catalyst is Pt on an acidic support. Operating conditions are reported to be 250−500°C and 1−3 MPa. The first Isolene II plant was built at Toray's Kawasaki complex.

Zeolite and Molecular Sieve Based Process.   Mobil has commercialized several xylene isomerization processes that are based on ZSM-5. Amoco has developed a process based on a medium-pore borosilicate molecular sieve.

Mobil's Low Pressure Isomerization Process (MLPI) was developed in the late 1970s (123,124). Two unique features of this process are that it is operated at low pressures and no hydrogen is used. In this process, EB is converted to benzene and diethylbenzene via disproportionation. The patent believed to be the basis for the MLPI process (123) discusses the use of H-ZSM-5 zeolite with an alumina binder. The reaction conditions described are start-of-run temperatures of 290−380°C, a pressure of 273 kPa; and WHSV of 5−8.5/h. The EB conversion is about 25−40% depending on reaction conditions, with xylene losses of 2.5−4%. The PX approach to equilibrium is about 99−101%. The first commercial unit was licensed in 1978. A total of four commercial plants have been built.

A second Mobil process is the Mobil's Vapor Phase Isomerization Process (MVPI) (125,126). This process was introduced in 1973. Based on information in the patent literature (125), the catalyst used in this process is believed to be composed of NiHZSM-5 with an alumina binder. The primary mechanism of EB conversion is the disproportionation of two molecules of EB to one molecule of benzene and one molecule of diethylbenzene. EB conversion is about 25−40%, with xylene losses of 2.5–4%. PX is produced at concentration levels of 102–104% of equilibrium. Temperatures are in the range of 315–370°C, pressure is generally 1480 kPa, the H_2/hydrocarbon molar ratio is about 6:1, and WHSV is dependent on temperature, but is in the range of 2–50, although normally it is 5–10.

Mobil's High Temperature Isomerization (MHTI) process, which was introduced in 1981, uses Pt on an acidic ZSM-5 zeolite catalyst to isomerize the xylenes and hydrodealkylate EB to benzene and ethane (126). This process is particularly suited for unextracted feeds containing C_{8+} aliphatics, because this catalyst is capable of cracking them to light paraffins. Reaction occurs in the vapor phase to produce a PX concentration slightly higher than equilibrium, ie, 102–104% of equilibrium. EB conversion is about 40–65%, with xylene losses of about 2%. Reaction conditions are temperature of 427–460°C, pressure of 1480–1825 kPa, WHSV of 10–12, and a H_2/hydrocarbon molar ratio of 1.5–2:1. Compared to the MVPI process, the MHTI process has lower xylene losses and lower formation of heavy aromatics.

In the early 1990s, Mobil commercialized their High Activity Isomerization (MHAI) process (124,127,128). A patent corresponding to this process (127) describes the use of a two catalyst system in which the catalyst beds are separated from each other but can be located in the same reactor. The first catalyst consists of a noble metal (preferably Pt) on ZSM-5 zeolite having a crystal size of at least 1 micron with an alpha value >100. The primary function of this catalyst is to hydrodealkylate the EB to benzene and ethane. The second catalyst consists of a noble metal, preferably Pt, on a ZSM-5 zeolite having a crystal size <1 micron and an alpha value <100. The primary function of this catalyst is to isomerize the xylenes. Preferred operating conditions are reported to be 400–480°C; 450–2860 kPa, WHSV of 3–50; and a H_2/hydrocarbon molar ratio of 1–5:1. For a nonextracted feed containing 15% EB, typical EB conversions are 65%, with xylene losses of 1.8%. Compared to MHTI, for a given EB conversion MHAI xylene losses are lower and nonaromatics conversion higher. As of 1995, 9 MHAI units had been commercialized (127).

Amoco has developed a series of xylene isomerization catalysts called AMSAC catalysts (129). These catalysts are based on a medium-pore borosilicate molecular sieve, designated AMS-1B. They are reported to retard xylenes destruction by suppressing the transfer of methyl groups. Amoco claims that the required intermediate is too bulky to form within the pores of AMS-1B. EB is reacted via transfer of ethyl groups in a reaction scheme that involves dealkylation, formation of ethylene, and realkylation. Optionally, a hydrogenating metal such as Pt is added to inhibit realkylation by intercepting the ethylene and converting it to ethane, which is essentially unreactive. This has the advantage of further limiting xylene losses by preventing the alkylation of xylenes with C_2 groups to ethyl methyl benzenes. Amoco claims to have used its AMSAC catalysts in its PX units since 1980.

Amorphous Silica–Alumina Based Processes. Amorphous silica–alumina catalysts had been used for many years for xylene isomerization. Examples are the Chevron (130), Maruzen (131), and ICI (132–135). The primary advantage of these processes was their simplicity. No hydrogen was required and the only side reaction of significance was disproportionation. However, in the absence of H_2,

prices generally declined as availability increased and crude oil prices declined, except in 1989 when derivatives demand was strong and xylenes supply was tight.

Analytical Methods

Typical ASTM methods for analyzing the C_8 aromatic isomers are listed in Table 9.

   p-Xylene purities can be determined by both freezing point and chromatographic methods. One advantage of freezing point methods is that they can detect the presence of water and carbon dioxide impurities, which chromatographic methods do not. Chromatographic methods exhibit another disadvantage, ie, for the analysis of very high purity PX streams, in that the PX peak tends to tail over the MX peak. Conditions for one chromatographic method for separating C_8 aromatic isomers are given in Table 10. This procedure readily separates benzene, toluene, and the C_8 aromatic isomers. Di-n-propyl tetrachlorophthalate is also a useful column coating because it reverses the positions of PX and MX. Thus, in the case of this column, MX elutes ahead of PX and provides better analysis for streams containing high purity PX.

   Bentone-34 has commonly been used in packed columns (138–139). The retention indices of many benzene homologues on squalane have been determined (140). Gas chromatography of C_6–C_{11} aromatic compounds using a Ucon B550X-coated capillary column is discussed in Reference 141. A variety of other separation media have also been used, including phthalic acids (142), liquid crystals (143), and Werner complexes (144). Gel permeation chromatography of alkylbenzenes and the separation of the C_8 aromatics treated with zeolites are described in References 145–148.

Table 9. Analytical Methods for C_8 Aromatic Isomers

Assay	ASTM method
purity, mol %	D1016
distillation range, °C	D850
specific gravity, 15.5°C/15.5°C	D891
color Saybolt	D156
flash point, °C	D56

Table 10. Chromatographic Separation of C_8 Aromatic Isomers

Parameter	Value or measure
column[a]	30.5 m × 0.5 mm dia
sample size, microliter	0.2
detector	flame ionization
temperature, °C	70
pressure, kPa (psig)	156 (8)
helium flow, mL/min	8

[a]Stainless steel column coated with m-bis(m-phenoxyphenoxy)benzene.

Health and Safety Factors

The xylene isomers are flammable liquids and should be stored in approved closed containers with appropriate labels and away from heat and open flames. Limits for transportation by air are 5 L on passenger planes and 60 L on cargo planes.

Flash points and autoignition temperatures are given in Table 11. The vapor can travel along the ground to an ignition source. In the event of fire, foam, carbon dioxide, and dry chemical are preferred extinguishers. The lower and upper explosion limits are 1% and 7%.

The xylenes are mildly toxic. They are mild skin irritants, and skin protection and the cannister-type masks are recommended. The oral LD_{50} value for rats is 4300 ppm. The STEL for humans is 150 ppm. Xylenes show only mild toxicity to fish, and the threshold limit for crop damage is 800–2400 ppm. Biodegradation with activated seed is slow, and sewage digestion is impaired by 0.1% concentrations. In the event of a spill, oil-skimming equipment, adsorbent foam, and charcoal may be used for cleanup.

Table 11. Flash Points and Autoignition Temperatures of the C_8 Aromatic Compounds

Compound	Flash point, °C	Autoignition temperature, °C
PX	27	530
MX	27	530
OX	32	460
EB	15	432

Uses

The majority of xylenes, which are mostly produced by catalytic reforming or petroleum fractions, are used in motor gasoline (see GASOLINE AND OTHER MOTOR FUELS). The majority of the xylenes that are recovered for petrochemicals use are used to produce PX and OX. PX is the most important commercial isomer. Almost all of the PX is converted to terephthalic acid and dimethylterephthalate, and then to poly(ethylene terephthalate) for ultimate use in fibers, films, and resins.

Almost all of the OX that is recovered is used to produce phthalic anhydride. Phthalic anhydride is a basic building block for plasticizers used in flexible PVC resins, for polyester resins used in glass-reinforced plastics, and for alkyd resins used for surface coatings. OX is also used to manufacture phthalonitrile, which is converted to copper phthalocyanine, a pigment.

Most of the MX that is contained in recovered mixed xylenes is isomerized to PX and OX. MX is also used to produce isophthalic acid and isophthalonitrile. Isophthalic acid is the base of unsaturated polyester resins which have good corrosion resistance, and greater strength and higher modulus than resins derived from phthalic anhydride. Isophthalonitrile is the starting material for the fungicide tetrachloroisophthalonitrile.

Some of the mixed xylenes that are produced are used as solvents in the paints and coatings industry (see SOLVENTS, INDUSTRIAL). However, this use has declined, particularly in the United States as environmental efforts to reduce hydrocarbon emissions into the air have increased.

The EB present in recovered mixed xylenes is largely converted to xylenes or benzene. The EB used to make styrene is predominately manufactured by the alkylation of benzene with ethylene.

BIBLIOGRAPHY

"Xylenes and Ethylbenzene" in *ECT* 1st ed., Vol. 15, pp. 186–194, by M. Lapeyrouse, Esso Research and Engineering Co.; in *ECT* 2nd ed., Vol. 22, pp. 467–507, by H. E. Cier, Esso Research and Engineering Co.; in *ECT* 3rd ed., Vol. 24, pp. 709–744, by D. L. Ransley, Chevron Research Co.

1. *Chemical Economics Handbook*, SRI International, Menlo Park, Calif., 1993.
2. C. L. Yaws, *Chem. Eng.*, 113 (July 21, 1975).
3. C. L. Yaws, *Chem. Eng.*, 73 (Sept. 29, 1975).
4. J. Chao and co-workers, *Hydrocarbon Process.* **59** (8), 117 (1980).
5. J. Chao, *Hydrocarbon Process.* 295 (Nov. 1979).
6. D. R. Stull and co-workers, *The Chemical Thermodynamics of Organic Compounds*, John Wiley & Sons, Inc., New York, 1968.
7. W. L. Nelson, *Oil Gas J.*, 68 (Mar. 19, 1970).
8. D. A. McCauley and A. P. Lien, *J. Am. Chem. Soc.* **74**, 6246 (1950).
9. H. C. Brown and H. Jungk, *J. Am. Chem. Soc.* **77**, 5579 (1955).
10. R. H. Allen and L. D. Yats, *J. Am. Chem. Soc.* **81**, 5289 (1959).
11. *Selected Values of Physical and Thermodynamic Properties of Hydrocarbons and Related Compounds*, API Research Project 44, Carnegie Press, 1953.
12. W. D. Crow and C. Wentrup, *Tetrahedron Lett.* **27**, 3111 (1968).
13. C. J. Egan, *J. Chem. Eng. Data* **5**(3), 298 (1960).
14. S. Ehrenson, *J. Am. Chem. Soc.* **84**, 2681 (1962).
15. A. McCaulay and A. P. Lien, *J. Am. Chem. Soc.* **73**, 2013 (1951).
16. M. Kilpatrick and F. E. Luborsky, *J. Am. Chem. Soc.* **75**, 577 (1953).
17. E. L. Mackor, A. Hofstra, and J. H. van der Waals, *Trans. Faraday Soc.*, 186 (1958).
18. J. Barnard and B. M. Sankey, *Combust. Flame* **12**(4), 345 (1968).
19. L. Errede and M. Swarc, *Q. Rev.* **12**, 301 (1958).
20. L. Errede, R. Gregorian, and J. Hoyt, *J. Am. Chem. Soc.* **82**, 5218 (1960).
21. L. Errede and N. Kroll, *J. Polym. Sci.* **60**, 33 (1962).
22. U.S. Pat. 3,412,167 (Nov. 19, 1968), J. W. Lewis (to Union Carbide Co.).
23. D. J. Hadley, *Chem. Ind.*, 238 (Feb. 25, 1961).
24. *Hydrocarbon Process.*, 252 (Nov. 1969).
25. *Hydrocarbon Process.*, 238 (Nov. 1981).
26. E. Kamegaya, *Chem. Econ. Eng.* **2**(2), 30 (1970).
27. S. Ariki and A. Ohira, *Chem. Econ. Eng.* **5**(7), 39 (1973).
28. N. E. Ockerbloom, *Hydrocarbon Process.*, 101 (Feb. 1972).
29. P. B. de la Mare and J. Robertson, *J. Chem. Soc.*, 279 (1943).
30. F. Condon, *J. Am. Chem. Soc.* **70**, 1963 (1948).
31. U.S. Pat. 2,282,231 (May 5, 1942), W. J. Mattox (to UOP).
32. O. Maas and J. Russell, *J. Am. Chem. Soc.* **40**, 1561 (1918).
33. O. Maas, E. H. Boomer, and D. M. Morrison, *J. Am. Chem. Soc.* **45**, 1433 (1923).
34. H. C. Brown and H. W. Pearsall, *J. Am. Chem. Soc.* **74**, 191 (1952).
35. W. D. Schaeffer and W. S. Dorsey, *Advances in Petroleum Chemistry and Refining*, Vol. 6, Wiley-Interscience, New York, 1962.
36. W. D. Schaeffer and co-workers, *J. Am. Chem. Soc.* **79**, 5870 (1957).
37. U.S. Pat. 2,798,103 (July 2, 1957) and U.S. Pat. 2,951,104 (Aug. 30, 1960), W. D. Schaeffer and J. D. Wordie (to Union Oil Co.).

38. M. J. Minton and co-workers, *J. Phys. Chem.* **71**(11), 3618 (1967).
39. J. Menyhart, *Magy. Kem. Lapja* **22**(1), 18 (1967).
40. F. Casellato, *Erdoel Kohle* **22**, 2 (1969).
41. E. Ger. Pat. 56,236 (June 6, 1967), W. Jugel.
42. Brit. Pat. 1,151,606 (May 14, 1969), J. Lumsden (to Imperial Smelting Co.).
43. H. R. Allcock and L. A. Siegel, *J. Am. Chem. Soc.* **86**, 5140 (1964).
44. U.S. Pat. 3,472,762 (Oct. 14, 1969), A. Goldrup and M. T. Westaway (to British Petroleum).
45. U.S. Pat. 3,484,500 (Dec. 16, 1969), A. Goldrup, A. B. Morrison, and M. T. Westaway (to British Petroleum).
46. Brit. Pat. 1,183,524 (Mar. 11, 1970) (to British Petroleum Co.).
47. U.S. Pat. 3,504,047 (Mar. 31, 1970), J. N. Haresnape (to British Petroleum Co.).
48. U.S. Pat. 3,456,028 (July 15, 1969), C. G. Gerhold and D. B. Broughton (to UOP).
49. U.S. Pat. 3,465,055 (Sept. 2, 1969), W. K. T. Gleim, R. C. Wackler, and R. C. Ranquist (to UOP).
50. U.S. Pat. 2,900,428 (Aug. 18, 1959), C. D. Heaton and W. E. Toland (to Chevron Research Co.).
51. U.S.S.R. Pat. 176,277 (Nov. 2, 1965), V. S. Bogdanov and co-workers (to Scientific Research Institute for Petrochemical Industries).
52. C. J. Egan and R. V. Luthy, *Ind. Eng. Chem.* **47**, 250 (1955).
53. T. Iwamura, S. Otani, and M. Sato, *Bull. Jpn. Pet. Inst.* **13**(1), 116 (1971).
54. S. Otani, *Chem. Eng.*, 118 (July 20, 1970).
55. *Hydrocarbon Process.*, 133 (Nov. 1977).
56. UOP Petrochemical Technology Conference, 1992, Houston, Tex.
57. J. J. Jeanneret, in R. A. Meyers, ed., *Handbook of Petroleum Refining Processes*, 2nd ed., McGraw-Hill Book Co., Inc., New York, 1997, p. 2.55.
58. J. A. Verdol, *Oil Gas J.*, 63 (June 9, 1969).
59. HRI/IFP Petrochemical Seminar, May 8–9, 1995, Houston, Tex.
60. P. Grandio and F. H. Schneider, *Oil Gas J.*, 62 (Nov. 29, 1971).
61. P. Grandio and co-workers, *Hydrocarbon Process.* **51**, 85 (Aug. 1972).
62. S. Han and co-workers, *Oil Gas J.*, 83 (Aug. 21, 1989).
63. Brochure on MTDP-3, Mobil Oil Corp., 1996.
64. F. Gorra, L. Breckeridge, R. Sailor, *Oil Gas J.* **90**(41), 60 (Oct. 12, 1992).
65. R. Sailor, presentation at Chiyoda Forum, Saudi Arabia, 1993.
66. Brochure on MSTDP, Mobil Oil Corp., 1996.
67. D. Rothman, *Chem. Week*, 18 (Aug. 30/Sept. 6, 1995).
68. U.S. Pat. 5,243,117 (Sept. 7, 1993), C. D. Chang and D. S. Shihabi (to Mobil Oil Corp.).
69. Technical data, Mobil Technology Sales and Licensing, Mobil Oil Co., May, 1996.
70. K. Menard, *Oil Gas J.*, 46 (Mar. 16, 1987).
71. U.S. Pat. 3,467,724 (Sept. 16, 1969), S. A. Laurich (to Chevron Research Co.).
72. *Hydrocarbon Process.*, 253 (Nov. 1979).
73. U.S. Pat. 3,177,265 (Apr. 6, 1965), G. C. Lammens (to Standard Oil Co., Indiana).
74. Belg. Pat. 617,795 (Sept. 14, 1962), (to Standard Oil Co., Indiana).
75. R. J. Desiderio and co-workers, *Hydrocarbon Process.*, **53**(8), 81 (1974).
76. *Hydrocarbon Process.*, 239 (1981).
77. Y. Hatanaka and T. Nakamura, *Oil Gas J.* **70**(47), 60 (1972).
78. *Hydrocarbon Process.*, 254 (Nov. 1979).
79. D. L. McKay and H. W. Goard, *Chem. Eng. Prog.* **61**(11), 99 (1966).
80. D. L. McKay and co-workers, *Chem. Eng. Prog.* **62**(11), 104 (1966).
81. *Hydrocarbon Process.*, 132 (Mar. 1995).
82. G. Parkinson, *Chem. Eng.* **103**(9), 23 (Sept. 1996).

83. N. P. Wynn, *Chem. Eng. Prog.* **88**(3), 52 (1992).

84. D. B. Broughton and co-workers, *The Separation of p-Xylene from C₈ Hydrocarbon Mixtures by Parex Process*, AIChE, Puerto Rico, May 1970.

85. D. B. Broughton and co-workers, *Chem. Eng. Prog.* **66** (9), 70 (1970).

86. D. P. Thornton, Jr., *Hydrocarbon Process.*, 151, (Nov. 1970).

87. D. B. Broughton, *Chem. Eng. Prog.* **73**(10), 49 (1977).

88. *Hydrocarbon Process.*, 256 (Nov. 1979).

89. U.S. Pat. 3,663,638 (May 16, 1972), R. W. Neuzil (to UOP).

90. U.S. Pat. 3,665,046 (May 23, 1972), A. J. DeRosset (to UOP).

91. U.S. Pat. 3,686,342 (Aug. 22, 1972), R. W. Neuzil (to UOP).

92. U.S. Pat. 3,636,180 (Jan. 18, 1972), D. B. Broughton (to UOP).

93. U.S. Pat. 3,696,107 (Oct. 3, 1972), R. W. Neuzil (to UOP).

94. Ref. 57, p. 2.45.

95. S. Otani and co-workers, *Chem. Econ. Eng. Rev.*, 56 (June 1971).

96. S. Otani, *Chem. Eng.*, 106 (Sept. 17, 1973).

97. S. Otani and co-workers, *Jpn. Pet. Inst. Bull.* **16**(1), 60 (1974).

98. U.S. Pat. 3,761,533 (Sept. 25, 1973), S. Otani and co-workers (to Toray Industries, Inc.).

99. G. Ash and co-workers, *Rev. de IFP* **49**(5), 541 (Sept.–Oct. 1994).

100. M. Seko, *Oil Gas J.*, 81 (July, 1979).

101. M. Seko, T. Miyake, and K. Inada, *IEC Prod. Res. Dev.* **18** (4), 263 (1979).

102. M. Seko, T. Miyake, and K. Inada, *Hydrocarbon Process.*, 133 (Jan. 19, 1980).

103. M. Seko, H. Takeuchi, and T. Inada, *IEC Prod. Res. Dev.* **21**, 656 (1982).

104. Y. Igarashi and T. Ueno, *A New Xylene Separation Process*, ACS Meeting, Atlantic City, N.J., 1968.

105. *Hydrocarbon Process.*, 254 (Nov. 1969).

106. Y. Igarashi, *Jpn. Chem. Q.* **IV**(4), 27 (1969).

107. T. Ueno, *Bull. Jpn. Pet. Inst.* **12** (May 1970).

108. G. R. Herrin and E. H. Martel, *Pet. Petrochem. Int.* **12**(7), 74 (1972).

109. J. C. Davis, *Chem. Eng.*, 77 (Aug. 9, 1971).

110. T. Ueno and T. Nakana, *Eighth World Pet. Congr. Proc.* **4**, 187 (1971).

111. S. Ariki and A. Ohira, *Chem. Econ. Eng. Rev.* **5**(7), 30 (1973).

112. J. J. H. Masseling, *Chem. Tech.*, 714 (Nov. 1976).

113. *Hydrocarbon Process.*, 255 (Nov. 1979).

114. H. F. Uhlig and W. C. Pfefferle, *Am. Chem. Soc. Div. Pet. Chem. D*, 154 (Sept. 1969).

115. P. M. Pitts, Jr., J. E. Connor, Jr., and L. N. Leum, *Ind. Eng. Chem.* **47**(4), 770 (1955).

116. U.S. Pat. 2,976,332 (Mar. 21, 1961), L. N. Leum and J. E. Connor, Jr. (to Atlantic Refining Co.).

117. C. V. Berger, *Hydrocarbon Process.*, 173 (Sept. 1973).

118. *Hydrocarbon Process.*, 215 (Mar. 1993).

119. Ref. 57, p. 2.37.

120. S. Otani and co-workers, *Chem. Econ. Eng. Rev.*, 56 (June 1971).

121. S. Otani, *Chem. Eng.*, 106 (Sept. 17, 1973).

122. Y. Tsuchiya and co-workers, *Jpn. Pet. Inst. Bull.* **16**(1), 60 (1974).

123. U.S. Pat. 4,101,596 (July 18, 1978), K. M. Mitchell and J. J. Wise (to Mobil Oil Corp.).

124. Brochure on MHAI and MLPI, Mobil Oil Corp., 1996.

125. U.S. Pat. 3,856,872 (Dec. 24, 1974), R. A. Morrison (to Mobil Oil Corp.).

126. N. Y. Chen, W. E. Garwood, F. G. Dwyer, *Shape Selective Catalysis in Industrial Applications*, Marcel Dekker, Inc., New York, 1989.

127. U.S. Pat. 4,899,011, (Feb. 6, 1990), Chu and co-workers (to Mobil Oil Corp.).

128. Y. Y. Huang and co-workers, **30d**, 1991 AIChE Spring National Meeting, Houston, Apr. 7–11, 1991.

129. Amelse and co-workers, **33a**, *1988 AIChE Summer National Meeting*, Denver, Aug. 21–24, 1988.
130. H. C. Ries, *Process Economics Program 25A*, Stanford Research Institute, Menlo Park, California, July 1970, pp. 41–53.
131. *Hydrocarbon Process.*, 240 (Nov. 1981).
132. *Hydrocarbon Process.*, 109 (Aug. 1969).
133. *Hydrocarbon Process.*, 253 (Nov. 1969).
134. U.S. Pat. 3,793,384 (Feb. 19, 1974), J. G. Chenoweth and co-workers (to Imperial Chemical Industries).
135. U.S. Pat. 3,860,668 (Jan. 14, 1975), J. K. January and A. Marchant (to Imperial Chemical Industries).
136. *World Petrochemicals*, SRI International, Menlo Park, Calif., 1996.
137. *Chemical Economics Handbook*, SRI International, Menlo Park, Calif., 1996.
138. J. M. Vergnaud, *J. Chromatogr.* **27**, 54 (1967).
139. C. F. Raley and J. W. Kaufman, *Anal. Chem.* **40**(8), 1371 (1968).
140. M. Dimov and D. Papzova, *J. Chromatogr.* **137**, 265 (1977).
141. F. Baumann and S. M. Csicsery, *J. Chromatogr.* **26**, 262 (1967).
142. M. F. Burke and L. B. Rogers, *J. Chromatogr. Sci.* **6**(1), 75 (1968).
143. A. B. Richmond, *J. Chromatogr. Sci.* **9**(9), 571 (1971).
144. A. C. Bhattacharyya and A. Bhattacherjee, *J. Chromatogr.* **41**(3–4), 446 (1969).
145. M. Popl, J. Coupek, and S. Pokorny, *J. Chromatogr.* **104**, 135 (1975).
146. U.S. Pat. 3,653,184 (Apr. 4, 1972), B. M. Drinkard, P. T. Allen, and E. H. Unger (to Mobil Oil Corp.).
147. U.S. Pat. 3,656,278 (Apr. 18, 1972), B. M. Drinkard, P. T. Allen, and E. H. Unger (to Mobil Oil Corp.).
148. U.S. Pat. 3,724,170 (Apr. 3, 1973), P. T. Allen, B. M. Drinkard, and E. H. Unger (to Mobil Oil Corp.).

WILLIAM J. CANNELLA
Chevron Research & Technology Co.

XYLYLENE POLYMERS

In a process capable of producing pinhole-free coatings of outstanding conformality and thickness uniformity through the unique chemistry of *p*-xylylene (PX) [*502-86-3*] (**1**), a substrate is simply exposed to a controlled atmosphere of pure gaseous monomer. The coating process is best described as a vapor deposition polymerization (VDP). The monomer molecule is thermally stable, but kinetically very reactive toward polymerization with other molecules of its kind. Although it is stable as a rarified gas, upon condensation it polymerizes spontaneously to produce a coating of high molecular weight, linear poly(*p*-xylylene) (PPX) [*25722-33-2*] (**2**). This article emphasizes recent VDP developments. There have been several reviews of the subject (1), which offer a more thorough treatment of early developments in the field.

xylylene) (PPX) (**2**). This of course requires the cleavage of the original dimer into two fragments.

Experiments with monoethyl and monocarbomethoxy di-*p*-xylylene (**4**) gave similar results. These experiments do not, however, shed any light on whether the rupture of the methylene–methylene bonds in the dimer upon pyrolysis is simultaneous or sequential.

Only one exception to the clean production of two monomer molecules from the pyrolysis of dimer has been noted. When α-hydroxydi-*p*-xylylene (**9**) is subjected to the Gorham process, no polymer is formed, and the 16-carbon aldehyde (**10**) is the principal product in its stead, isolated in greater than 90% yield. This transformation indicates that, at least in this case, the cleavage of dimer proceeds in stepwise fashion rather than by a concerted process in which both methylene–methylene bonds are broken at the same time. This is consistent with the predictions of Woodward and Hoffmann from orbital symmetry considerations for such [6 + 6] cycloreversion reactions in the ground state (**5**).

Monomer Properties. Despite difficulties involved in studying it owing to its great reactivity, a great deal is known about the structure of the parylene process monomer PX. The eight-carbon framework is planar (6). The molecule is diamagnetic, ie, all electron spins are paired in the ground state (spectroscopically, a singlet). Although many have ascribed its reactivity to its so-called biradical nature, the true biradical (triplet) form (**11**) of the molecule, an electronically excited state, is substantially more energetic, estimated at ca 50 kJ/mol (12 kcal/mol), and therefore cannot contribute to the monomer at equilibrium to any appreciable extent, even at pyrolysis temperatures. The PX molecule is instead a conjugated tetra olefin whose particular arrangement gives it extreme reactivity at its end carbons.

$$H_2\dot{C}-\langle\bigcirc\rangle-\dot{C}H_2$$

(**11**)

This extreme reactivity of PX has precluded many experimental approaches that otherwise would have been useful in studying it. Most of the present structural knowledge has been gleaned from spectroscopic studies and molecular orbital calculations. A noteworthy exception is an electron diffraction study (7) in which an electron beam was directed at a stream of gaseous PX, generated much as it is in the parylene process, issuing from a nozzle in a specially constructed apparatus. The results of the study are shown in Figure 2. Although the study was unable to resolve the lengths of the two different C=C and C—H bonds, it clearly distinguished between the C—C and C=C bond lengths. Thus p-xylene is experimentally demonstrated to have an olefinic geometry rather than that of an aromatic biradical.

By trapping PX at liquid nitrogen temperature and transferring it to THF at −80°C, the ^{1}H nmr spectrum could be observed (9). It consists of two sharp peaks of equal area at chemical shifts of 5.10 and 6.49 ppm downfield from tetramethylsilane (TMS). The fact that any sharp peaks are observed at all attests to the absence of any significant concentration of unpaired electron spins, such as those that would be contributed by the biradical (**11**). Furthermore, the chemical shift of the ring protons, 6.49 ppm, is well upfield from the typical aromatic range and more characteristic of an olefinic proton. Thus the olefin structure (**1**) for PX is also supported by nmr.

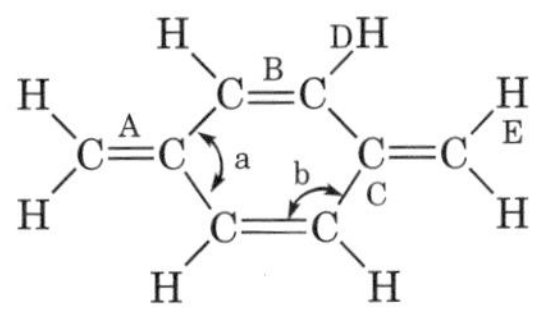

Fig. 2. Structure of PX monomer molecule from electron diffraction (8). Bond lengths: C=C, A = B + 0.1381 ± 0.008; C—C, C = 0.1451 ± 0.0007 nm; C—H, D = E = 0.1116 ± 0.0035 nm. Bond angles: a = 122.2 ± 3.7°, b = 118.9 ± 1.9°.

A particularly useful property of the PX monomer is its enthalpy of formation. Conventional means of obtaining this value, such as through its heat of combustion, are, of course, excluded by its reactivity. An experimental attempt was made to obtain this measure of chemical reactivity with the help of ion cyclotron resonance; a value of 209 ± 17 kJ/mol (50 ± 4 kcal/mol) was obtained (10). Unfortunately, the technique suffers from lack of resolution in addition to experimental imprecision. It is perhaps better to rely on molecular orbital calculations for the formation enthalpy. Using a semiempirical molecular orbital technique, which is tuned to give good values for heat of formation on experimentally accessible compounds, the heat of formation of *p*-xylylene has been computed to be 234.8 kJ/mol (56.1 kcal/mol) (11).

Successful *p*-Xylylene VDP Monomers. Within the limits mentioned above, it is frequently possible, and often desirable, to modify the *p*-xylylene monomer by attaching to it certain substituents. Limitations on such modifications lie in three areas: reactivity, performance in the coater, and cost.

Reactivity. Although the reactivity which enables the gas–solid polymerization to proceed is a characteristic of the eight-carbon *p*-xylylene tetraolefin system, it is possible to subdue that reactivity. For example, by attaching electron-withdrawing substituents to the alpha positions and thereby further delocalizing the π-electrons of the highly reactive *p*-xylylene nucleus, it is in several instances possible to prepare *p*-xylylenes that are so stable that they can be isolated and handled as normal organic compounds (Fig. 3). These sorts of substitutions must of course be avoided if the goal is to make polymer.

It is also possible to interfere with the polymerization by attaching at the alpha positions either too many groups, or groups which are too bulky. Four chlorine atoms (12) or four methyl groups (13) seem to be sufficient to hinder the

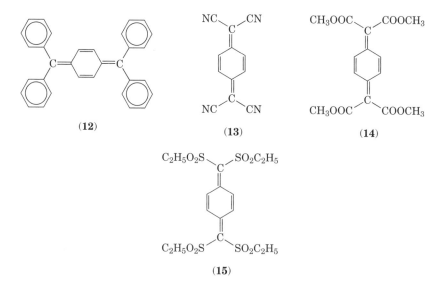

Fig. 3. Isolatable *p*-xylylene derivatives: (**12**), Thiele's hydrocarbon - 1904 [*26392-12-1*]; (**13**), tetracyanoquinodimethane [*1518-16-7*] (TCNQ); (**14**), tetrakis(methoxycarbonyl)-quinodimethan [*65649-20-9*]; (**15**), tetrakis(ethylsulfonyl)quinodimethan [*84928-90-5*].

production of polymer. These crowded *p*-xylylene monomers can be polymerized, but not through a VDP process.

Thus, except for electron-withdrawing or bulky substituents, at least from the standpoint of reactivity towards polymerization, modification by most other substituents is possible.

Performance in Coater. The modified monomer should perform well in commercial deposition equipment. Performance considerations include the growth rate of the coating, the uniformity of thickness of the coating over the chamber volume, and the efficiency with which the dimer is converted to useful coatings on the substrates.

An important further constraint is the fact that economic considerations in the construction of deposition equipment normally lead to a preference for an ambient temperature deposition chamber. Control of deposition temperature is possible, but it adds both equipment expense and operational complexity.

The vapor pressure of a parylene monomer is a prime factor in determining how rapidly a coating grows when exposed to an atmosphere of monomer at a given pressure. Vapor pressure is reduced as molecular weight increases, thereby increasing the monomer's tendency to condense and, along with it, increasing the VDP growth rate. The presence of polar functionality in the molecule further depresses vapor pressure. But too low a vapor pressure makes it difficult to transport gaseous monomer from point to point in the deposition chamber. Hence, some optimum value of monomer volatility is expected.

The widely used Parylene C owes its popularity principally to the room temperature volatility of its monomer. The Parylene C monomer, chloro-*p*-xylylene, has become the de facto performance standard. By comparison, the Parylene N monomer, *p*-xylylene itself, is too volatile and would perform better in a sub-ambient temperature deposition system. The Parylene D monomer, dichloro-*p*-xylylene [85586-88-5] is too heavy, and causes distribution problems in larger deposition systems.

Cost. It is necessary to produce the feedstock from which the monomer is generated, viz, the dimer, at a cost which can be supported by the commercial application, and yet allow it to be economically competitive with all other alternative ways to achieve the same end result. This factor often, but not always, seriously limits the amount of effort that can be put into dimer synthesis and purification.

Other, Related Processes

VDP processes using means other than the pyrolytic cleavage of DPX (Gorham process) to generate the reactive monomer are also known, although none are practiced commercially at the time of this writing (ca 1997).

Photopolymerization and Plasma Polymerization. The use of ultraviolet light alone (14) as well as the use of electrically excited plasmas or glow discharges to generate monomers capable of undergoing VDP have been explored. The products of these two processes, called plasma polymers, continue to receive considerable scientific attention. Interest in these approaches is enhanced by the fact that the feedstock material from which the monomer capable of VDP is generated is often inexpensive and readily available. In spite of these widespread scientific efforts, however, commercial use of the technologies is quite limited.

When *p*-xylene is used as the monomer feed in a plasma polymer process, PX may play an important role in the formation of the plasma polymer. The plasma polymer from *p*-xylene closely resembles the Gorham process polymer in the infrared, although its spectrum contains evidence for minor amounts of non-linear, branched, and cross-linked chains as well. Furthermore, its solubility and low softening temperature suggest a material of very low molecular weight (15).

VDP Polyimides. Polyimide films have also been prepared by a kind of VDP (16). The poly(amic acid) layer is first formed by the coevaporation and condensation of two monomers, followed by copolymerization on the substrate. The imidization is carried out in a separate baking step (see POLYIMIDES).

***o*-Xylylene/BCB.** Thermosetting resins based on benzocyclobutene (BCB)

chemistry have been reported (17). In these condensed phase cures, the *ortho*-xylylene isomer is the key reactive intermediate. From the behavior of this energetically similar ortho isomer, the value of the para configuration's rendering any ring closure reaction, analogous to cyclobutene formation from the ortho isomer, geometrically forbidden can be appreciated.

Dimer

In contrast to the extreme reactivity of the monomeric PX (**1**) generated from it, the dimer DPX (**3**) feedstock for the parylene process is an exceptionally stable compound. Because of their chemical inertness, dimers in general do not exhibit shelf-life limitations. Although a variety of substituted dimers are known in the literature, at present only three are commercially available: DPXN, DPXC, and DPXD, which give rise to Parylene N, Parylene C, and Parylene D, respectively.

The unsubstituted C-16 hydrocarbon, [2.2]paracyclophane (**3**), is DPXN. Both DPXC and DPXD are prepared from DPXN by aromatic chlorination and differ only in the extent of chlorination; DPXC has an average of one chlorine atom per aromatic ring and DPXD has an average of two.

Manufacture. For the commercial production of DPXN (di-*p*-xylylene) (**3**), two principal synthetic routes have been used: the direct pyrolysis of *p*-xylene (**4**, X = Y = H) and the 1,6-Hofmann elimination of ammonium (HNR_3^+) from a quaternary ammonium hydroxide (**4**, X = H, Y = NR_3^+). Most of the routes to DPX share a common strategy: PX is generated at a controlled rate in a dilute medium, so that its conversion to dimer is favored over the conversion to polymer. The polymer by-product is of no value because it can neither be recycled nor processed into a commercially useful form. Its formation is minimized by careful attention to process engineering. The chemistry of the direct pyrolysis route is shown in equation 1:

$$H_3C-\langle\bigcirc\rangle-CH_3 \xrightarrow[\text{steam}]{950^\circ C} CH_2=\langle\bigcirc\rangle=CH_2 \xrightarrow[\text{quench}]{p\text{-xylene}} \text{[structure 3]} \tag{1}$$

(**4**, X = Y = H)   (**1**)

(**3**)

First, p-xylene is dehydrogenated pyrolytically in the presence of steam at about 950°C to give p-xylylene (PX), which in turn forms di-p-xylylene (DPX) when quenched in liquid xylene. The xylene is recycled to the pyrolysis vessel. Yields and conversion efficiency are satisfactory. However, several engineering challenges need to be overcome, including the choice of a suitable diluent; establishing optimal residence time, vapor velocity, and operating pressure during pyrolysis; and the design and construction of novel equipment to withstand the highly corrosive reaction environment.

The Hofmann elimination route, of which many versions exist, can be carried out at much lower temperatures in conventional equipment. The PX is generated by a 1,6-Hofmann elimination of amine from a quaternary ammonium hydroxide in the presence of a base. This route gives yields of 17–19%. Undesired polymeric products can be as high as 80% of the product. In the presence of a polymerization inhibitor, such as phenothiazine, DPXN yields can be increased to 50%.

In the 1,6-elimination of p-trimethylsilylmethylbenzyltrimethylammonium iodide with tetrabutylammonium fluoride, yields as high as 56% have been reported (18). The starting materials are not readily accessible, however, and are costly.

The yield can be raised to 28% if the Hofmann elimination is conducted in the presence of a water-soluble copper or iron compound (19). Further improvements up to 50% were reported when the elimination was carried out in the presence of ketone compounds (20). Further beneficial effects have been found with certain cosolvents, with reported yields of greater than 70% (8).

DPXC and DPXD. The economic pressure to control dimer costs has had an important effect on what is in use today (ca 1997). Attaching substituents to the ring positions of a [2.2]paracyclophane does not proceed with isomeric exclusivity. Indeed, isomeric purity in the dimer is not an essential requirement for the obtaining of isomeric purity, eg, monosubstituted monomer, in the pyrolysis. Any mixture of the four possible heteronuclearly disubstituted dichloro[2.2]paracyclophanes, will, after all, if pyrolyzed produce the same monomer molecule, chloro-p-xylene [10366-09-3] (**16**) (Fig. 4).

Although DPXC and DPXD prepared by the chlorination of DPXN are relatively complex mixtures, after pyrolytic cleavage the resulting mixture of monomers is considerably simpler. Thus DPXC, when pyrolyzed, gives predominantly monochloro PX, which is accompanied by small but significant amounts of PX and dichloro PX. The resulting polymer, Parylene C, consequently has an average of about one chlorine atom per repeat unit. However, it contains significant amounts of unchlorinated, as well as dichlorinated, repeat units.

DPXC and DPXD are prepared from DPXN by chlorinating to different extents. The conditions are controlled to favor aromatic ring chlorination to the

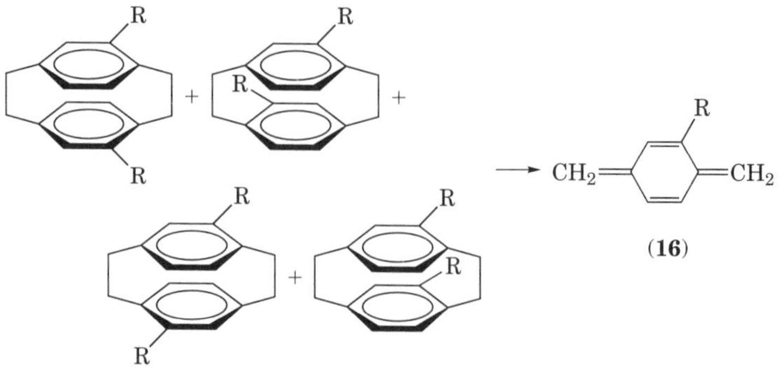

(16)

Fig. 4. Isomeric dichloro[2.2]paracyclophanes produce the same xylylene.

exclusion of the free-radical chlorination of the ethylene bridges. However, the chlorination products are complex mixtures of the homologues DPXN, monochloro DPX, dichloro DPX, trichloro DPX, and tetrachloro DPX, and even higher homologues, as well as the several possible isomers of each.

New synthetic routes for the preparation of homologously pure dichloro DPX and tetrachloro DPX have been reported through the 1,6-Hofmann elimination of chlorinated *p*-methylbenzyltrimethylammonium hydroxide. In the case of dichloro DPX, yields of 30% were reported (21). In the presence of ketone compounds, yields were increased to 50% (20).

Purification.   Unsubstituted di-*p*-xylylene (DPXN) is readily purified by recrystallization from xylene. It is a colorless, highly crystalline solid. The principal impurity is polymer, which fortunately is insoluble in the recrystallization solvent and easily removed by hot filtration.

In purifying DPXC and DPXD, care must be taken not to disturb the homologue composition, so that product uniformity is maintained. For example, a recrystallization of DPXC from ethanol would give a higher melting, more crystalline dimer material, at the expense of a decrease in yield owing to the removal of otherwise useful isomers, but the polymer made from it would not be identical to the historical Parylene C, as defined by its preparation from the chlorination mixture. The real purification issues are the removal of insoluble residues and any components that contain aliphatic side-chain chlorine. Although ring-substituted chlorine is stable, side-chain chlorine can give rise to hydrogen chloride gas under the conditions of the parylene process, or subsequent to it, which in certain applications could initiate substrate corrosion. Fortunately, the aliphatic chlorine problem can be minimized by proper attention to process detail.

Properties.   The DPXs are all crystalline solids; melting points and densities are given in Table 1. Their solubility in aromatic hydrocarbons is limited. At 140°C, the solubility of DPXN in xylene is only about 10%. DPXC is more readily soluble in chlorinated solvents, eg, in methylene chloride at 25°C its solubility is 10%. In contrast, the corresponding figure for DPXN is 1.5%.

The structure of DPXN was determined in 1953 from x-ray diffraction studies (22). There is considerable strain energy in the buckled aromatic rings and distorted bond angles. The strain has been experimentally quantified at

Table 1. Properties of Parylene Dimers

Dimer	Melting point, °C	Density, g/cm^3
DPXN	284^a	1.22
DPXC	140–160^b	1.30
DPXD	170–195^b	1.45

^{a}Decomposes.
^{b}Mixture of homologues and their isomers.

130 kJ/mol (31 kcal/mol) by careful determination of the formation enthalpy through heat of combustion measurements (23). The release of this strain energy is doubtless the principal reason for success in the particularly convenient preparation of monomer in the parylene process.

Polymer

The linear polymer of PX, poly(p-xylylene) (PPX) (**2**), is formed as a VDP coating in the parylene process. The energetics of the polymerization set it apart from all other known polymerizations and enable it to proceed as a vapor deposition polymerization.

Thermodynamic Considerations. On the basis of the value for the enthalpy of formation of p-xylylene, $\Delta H_f^0(\text{PX})$, the enthalpy of polymerization, $\Delta H_{\text{polym}}^0 = \Delta H_f^0(\text{PPX}) - \Delta H_f^0(\text{PX})$, can be estimated. No experimental combustion data are available for high molecular weight poly(p-xylylene) as it is formed in the parylene process, $\Delta H_f^0(\text{PPX})$.

For crystalline [2.2]paracyclophane [(1), DPXN], a ΔH_f^0 of +154.4 kJ/mol (+36.9 kcal/mol) is reported (23). The hypothetical transformation of crystalline DPXN into polymer is accompanied by the release of 129.7 kJ/mol (31.0 kcal/mol) of paracyclophane strain energy per mole of paracyclophane, and 12.6 kJ/mol (3.0 kcal/mol) per polymer repeat unit as a result of the bibenzyl hyperconjugative stabilization, which is permitted in the polymer but excluded by geometry in the dimer. Thus the standard enthalpy of formation for the hypothetical 100% crystalline poly(p-xylylene) is estimated to be −0.3 kJ/mol (−0.05 kcal/mol), assuming that the energies associated with crystallinity are the same in both cases. Although it might be acceptable to assume that such energies per repeat unit are similar in the crystalline polymer and crystalline dimer, Parylene N, as produced by the parylene process, is typically only about 57% crystalline. Using a value of 14.1 kJ/mol (3.37 kcal/mol) for the heat of fusion for poly(p-xylylene) (24), the standard formation enthalpy for Parylene N, as it is typically deposited in the parylene process, $\Delta H_f^0(\text{Parylene N})$, is +5.7 kJ/mol (+1.4 kcal/mol).

In estimating the enthalpy of polymerization, the physical state of both starting monomer and polymer must be specified. Changes in state are accompanied by ethalpy changes. Therefore, they also affect the level of the polymerization enthalpy. The ΔH_f^0 for p-xylylene previously mentioned is applicable to the monomer as an ideal gas. To make comparisons with other polymerization processes, most of which start with condensed monomer, a heat of vaporization for p-xylylene is needed. It is assumed herein that it is the same as that for p-xylene,

42.4 kJ/mol (10.1 kcal/mol). Thus the ΔH_f^0 of the liquid monomer p-xylylene is 192.3 kJ/mol (46.0 kcal/mol).

The enthalpy of polymerization of unannealed (57% crystalline) Parylene N, as it is deposited, starting with liquid monomer, $\Delta H_{\text{polym(lu)}}^0$, is -186.6 kJ/mol (-44.6 kcal/mol). This is an exceptionally high value compared with those of other addition polymers, which generally fall in the -60 to -100 kJ/mol (-14.3 to -23.8 kcal/mol) range. It quantifies the vigor of the polymerization. Because the source of polymerization enthalpy is within the p-xylylene system, substituents affect it only to a minor extent. All parylenes are expected to have a similar molar enthalpy of polymerization. An experimental value for the heat of polymerization of Parylene C has appeared. Using the gas evolution from the liquid nitrogen cold trap to measure thermal input from the polymer, and taking advantage of a peculiarity of Parylene C at $-196°C$ to polymerize abruptly, perhaps owing to the arrival of a free radical, a $\Delta H_{\text{polym}}^0 = -152 \pm 8$ kJ/mol (-36.4 ± 2.0 kcal/mol) at $-196°C$ was reported (25). The correction from $-196°C$ to room temperature is estimated at -17 kJ/mol, bringing this experimental value for Parylene C closer to the calculated value for Parylene N. It is assumed that S_{polym} is 0 at 0 K (3rd law), 125 J/(mol·K) [30 cal/(mol·K)] at 298 K, and proportional to T in between, a crude assumption, but appropriate to the current level of knowledge. Thus experiment and calculation are in harmony in quantifying the exceptional exothermicity of parylene polymerization.

The thermodynamic ceiling temperature (26) T_c for a polymerization is computed by dividing the $\Delta H_{\text{polym}}^0$ by the standard entropy of polymerization, $\Delta S_{\text{polym}}^0$. The T_c is the temperature at which monomer and polymer are in equilibrium in their standard states at 25°C (298.15 K) and 101.3 kPa (1 atm). (In the case of p-xylylene, such a state is, of course, purely hypothetical.) The T_c quantifies the binding forces between monomer units in a polymer and measures the tendency of the polymer to revert back to monomer. In other systems, the T_c indicates a temperature above which the polymer is unstable with respect to its monomer, but in the case of parylene it serves rather as a means of comparing the relative stability of the polymer with respect to its reversion to monomer. For computing the T_c, however, the standard entropies of polymerization are required.

The standard polymerization entropies can be estimated from the following. The standard entropy S^0 for PX as an ideal gas is computed by a group-contribution method (27) to be 310.6 J/(mol·K) [74.24 cal/(mol·K)]. The entropy of vaporization for PX is assumed to be the same as that of p-xylene, 104.7 J/(mol·K) [25.03 cal/(mol·K)] (28). Therefore, the S^0 for liquid PX is 205.9 J/(mol·K) [49.21 cal/(mol·K)]. Noting that the experimental specific heat C_p of PPX follows that of polystyrene over the range of 160 to 340 K (29), it can be assumed that the proportionality continues down to 0 K and that the factor 135/116 at 298 K can be applied to the known S^0 for polystyrene [$S = 128.5$ J/(mol·K) or 30.70 cal/(mol·K)] (30). It follows that the S^0 for as-deposited 57% crystalline Parylene N is 149.5 J/(mol·K) [35.73 cal/(mol·K)]. Therefore, $\Delta S_{\text{polym}(g)}^0 = -161.1$ J/(mol·K) [-38.50 cal/(mol·K)] and $\Delta S_{\text{polym}(l)}^0 = -56.4$ J/(mol·K) [-13.48 cal/(mol·K)].

The results of the above polymerization thermodynamics calculations for parylene are compared to similar data for typical addition polymers in Table 2.

Table 2. Entropies, Enthalpies, and Ceiling Temperatures for the Polymerization of Various Monomers at 25°C (298.15 K) and 101.3 kPa (1 atm)[a]

Monomer	Liquid			Gas		
	$-\Delta H^0$, kJ/mol[b]	$-\Delta S^0$, J/(mol·K)[b]	T_c, °C	$-\Delta H^0$, kJ/mol[b]	$-\Delta S^0$, J/(mol·K)[b]	T_c, °C
ethylene	108.4	173.6	351			
propylene	81.6	116.3	429			
isoprene	74.9	101.3	467	101.3	187.0	268
styrene	69.9	104.6	395	113.4	212.1	262
methyl methacrylate	55.2	117.2	198			
α-methylstyrene	35.1	103.8	65			
p-xylylene	186.6	56.4	3035	229.1	161.1	1149

[a]Ref. 26.
[b]To convert J to cal, divide by 4.184.

The T_c quantifies the stability of the polymer only with respect to reversion to monomer. When PPX is thermally degraded (ca 500°C), a mixture of degradation products including hydrogen gas, p-xylene, toluene, and p-methylstyrene is observed (31), suggesting that the path taken in thermal degradation requires the cleavage of bonds other than those formed in the polymerization, very likely starting with the methylene C—H bond. Complete replacement of the methylene hydrogens in PPX with fluorine gives a polymer with substantially better stability at elevated temperatures (32).

The enthalpy liberated on the VDP of parylene is real and in an adiabatic situation causes a rise in temperature of the coated substrate. For Parylene C, 229.1 kJ/mol (54.7 cal/mol) corresponds to 1654 J/g (395 cal/g) whereas its specific heat at 25°C is only 1.00 J/(g·K) [0.239 cal/(g·K)] (33). In most practical situations, however, the mass of parylene deposited is dwarfed by the substrate mass, and the heat of polymerization is dissipated within the coated substrate over the time required to deposit the coating with minimal actual temperature rise.

Polymerization Mechanism. The physical processes of condensation and diffusion must be considered along with the p-xylylene polymerization chemistry for a proper understanding of what happens microscopically during vapor deposition polymerization (34). These processes point to an important distinction between VDP and vacuum metallization, ie, that in the latter, adsorption is followed by a surface reorganization of the existing deposited material, and diffusion of incoming species through the bulk is nonexistant. In most parylene depositions, a coating forms from gaseous monomer under steady-state conditions.

Gaseous monomer is transported to the location within the coating where it is to be consumed to produce polymer by an initial condensation, followed by diffusion. The net flux of monomer molecules through the growth interface, ie, the outer boundary of the coating, between the gaseous and condensed phases, needed to sustain growth at a given rate can be readily calculated [for Parylene C, 10 μm/h requires 1.55×10^{15}/(cm²·s)]. Comparing a net flux so obtained with the flux of molecules that according to the kinetic theory of gases are striking the growth surface [$Z = PN_0/\sqrt{2\pi MRT}$] for the conditions typical of parylene deposition, a large difference (two or three orders of magnitude) is observed. For Parylene C monomer at a pressure of 1.3 Pa (10 μm Hg) and 25°C, $Z = 6.7 \times 10^{17}/(\text{cm}^2 \cdot \text{s})$. For each molecule that eventually enters the coating, some hundred or thousand molecules strike the growth interface. Those that condense and do not react must, of course, evaporate. The term "sticking coefficient" has sometimes been borrowed from vacuum metallization to describe this ratio of incident molecules to consumed molecules. However, the VDP situation is not adequately described by hard spheres bouncing off a growth interface. Every incident molecule spends at least some time in the polymeric coating phase beyond the growth interface before it is lost again to the gas phase.

Because most of the condensing molecules evaporate, condensation equilibrium at the growth surface can be assumed, to a good approximation. The concentration of monomer dissolved in the coating near the growth interface is, therefore, governed by Henry's law, and monomer concentration in polymer solution increases proportionately to the partial pressure of monomer in the gas phase. Furthermore, as the temperature is lowered, or as higher molecular-weight monomers of lower volatility are selected, monomer concentration at the growth interface increases. In most practical situations, these Henry's law effects dominate in determining growth rates for VDP coatings by regulating monomer concentration within the coating. For each monomer, there exists a threshold condensation temperature, T_{tc}, above which the rate of growth of coating is, for all practical purposes, zero (Table 3), but this phenomenon is governed by the competition between initiation and propagation chemistries, discussed herein.

Once it is in "solution" in the coating, the monomer moves about in random directions by diffusion until it evaporates or is consumed by chemical reaction. The polymer molecules that have already grown to higher molecular weight cannot relocate appreciably owing to entanglement with their neighbors. The rate of diffusion of monomer through the polymer bulk is adequate for the participation of diffusive transport in the mechanism of VDP (ca 10^{-10} cm^2/s at room temperature). This can be confirmed in swelling-rate experiments with solvents having similar physical properties, such as p-xylene.

The monomer is consumed by two chemical reactions: initiation, in which new polymer molecules are generated, and propagation, in which existing polymer molecules are extended to higher molecular weight. In steady-state VDP,

**Table 3. Threshold Condensation Temperatures
T_{tc} for Substituted p-Xylene Monomers**

Monomer	T_{tc}, °C
p-xylene	30
2-methyl-p-xylylene	60
2-ethyl-p-xylylene	90
2-chloro-p-xylylene	90
2-acetyl-p-xylylene	130
2-cyano-p-xylylene	130
2-bromo-p-xylylene	130
dichloro-p-xylylene	130

both reactions proceed continuously inside polymeric coating, in the reaction zone just behind the growth interface.

The first step of the initiation reaction is the coupling of two monomer molecules to form the dimer diradical (**5**). The formation of this diradical is energetically uphill, ie, the energy of two benzyl radicals is greater than that of two starting p-xylylene systems. The rate of destruction greatly exceeds the rate of formation. Only a trace concentration of the dimer diradical species exists at equilibrium. Further reaction of the dimer diradical with monomer gives more stable diradicals. In these subsequent transformations, a p-xylylene is converted into a benzene with a net stabilizing effect. At some stage of oligomerization, the resulting n-mer diradical becomes more stable than the n p-xylylene molecules from which it was constructed. At this point, the new polymer molecule is formed. Thus the overall order of the initiation reaction, the reaction in which new polymer molecules are generated, is some $n > 2$. Initiation chemistry requires no species other than monomer, another unusual aspect of the polymerization chemistry of p-xylylene.

The order n of the initiation reaction has an important influence on the manner in which the VDP occurs. Because monomer molecules, even in solution at low concentration, are closer together in the condensed phase than they are in the gaseous phase, the rate of initiation is greater in the condensed phase than in the gaseous phase. The higher the order n, the more the condensed phase is favored. The order n, according to the mathematical model (34) of p-xylylene VDP, at the same time governs the effect of monomer pressure on growth rate at a given deposition temperature. The model predicts that growth rate should vary with the pressure raised to the power $((n + 3)/4$. Thus, if $n = 3$, the growth rate should be proportional to $p^{1.5}$. In an early attempt to determine the pressure dependence of parylene growth rate γ, an expression of $\gamma = k \cdot p^2$ was reported (35). A pressure exponent of 2 would be interpreted as an initiation order of $n = 5$. Although such a high order would favorably deemphasize "snow," consideration of the energetics of oligomeric p-xylylene diradicals would seem to place the order nearer to 3. Perhaps the early investigators did not anticipate a nonintegral order for pressure dependence. A more recent report (36) places n at 3 for Parylenes N and C, and 4 for Parylene D. Thus, with $n \geq 3$, the parylenes are more likely to form a continuous coating than a dust or a snow, the physical form of the product of a gas-phase polymerization. To the extent that snow is included in the formation of a coating (ie, dual-phase polymerization), haze develops.

In the propagation reaction, the monomer molecule reacts with an existing free-radical polymer chain end to make the chain one repeat unit longer. The polymer chains have two active ends, and they grow from both ends at the same time. Under normal coating conditions, the consumption of monomer by propagation must be much higher than its consumption by initiation to obtain high molecular weight polymer. In fact, the number-average molecular weight is determined by the proportion of monomer consumed by the two reactions, and is diminished by increases in deposition temperature or monomer partial pressure.

The concentration of monomer within the coating decreases approximately exponentially with distance from the growth interface. With this decrease in monomer concentration, the rates of initiation and propagation reactions also decrease. Moving back into the polymer from the growth interface, through the reaction zone where polymer is being manufactured, a region in which the polymer formation is essentially complete is gradually entered. Because initiation is of higher order in monomer concentration, it tends to occur closer to the growth interface than does propagation. Under conditions prevailing during a typical deposition, the characteristic depth of the reaction zone is a few hundred nanometers, and the maximum concentration of monomer, ie, the concentration at the growth interface, is of the order of a few tenths percent by weight. Thus the parylene polymerization takes place just behind the growth interface in a medium that is best described as a slightly swollen, solid polymer.

During the vapor deposition process, the polymer chain ends remain truly alive, ceasing to grow only when they are so far from the growth interface that fresh monomer can no longer reach them. No specific termination chemistry is needed, although subsequent to the deposition, reaction with atmospheric oxygen, as well as other chemical conversions that alter the nature of the free-radical chain ends, is clearly supported experimentally.

Polymer Properties.   The single most important feature of the parylenes, that feature which dominates the decision for their use in any specific situation, is the vapor deposition polymerization (VDP) process by which they are applied. VDP provides the room temperature coating process and produces the films of uniform thickness, having excellent thickness control, conformality, and purity. The engineering properties of commercial parylenes once they have been formed are given in Table 4. As crystalline polymers, the parylenes retain useful physical integrity up to temperatures approaching their crystalline melting points. However, their glass-transition temperatures, T_g, the temperature spans over

Table 4. Typical Engineering Properties of Commercial Parylenes

Property	Parylene N	Parylene C	Parylene D	ASTM method
	General			
density, g/cm^3	1.110	1.289	1.418	D1505
refractive index, n_D^{23}	1.661	1.639	1.669	
	Mechanical			
tensile modulus, GPa^a	2.4	3.2	2.8	D882
tensile strength, MPa^b	45	70	75	D882
yield strength, MPa^b	42	55	60	D882
elongation to break, %	30	200	10	D882
yield elongation, %	2.5	2.9	3	D882
Rockwell hardness	R85	R80		D785
coefficient of friction				D1894
static	0.25	0.29	0.35	
dynamic	0.25	0.29	0.31	
	Thermal			
melting point, °C	420	290	380	
linear coefficient of expansion at 25°C × 10^5, K^{-1}	6.9	3.5		
heat capacity at 25°C, J/(g·K)^c	1.3^d	1.0^e		
thermal conductivity at 25°C, W/(m·K)	0.12	0.082		